Brief Contents

Contents

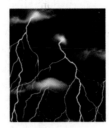
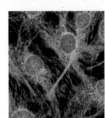

BIOLOGY

Seventh Edition

Peter H. Raven

Director, Missouri Botanical Gardens;
Engelmann Professor of Botany
Washington University

George B. Johnson

Professor Emeritus of Biology
Washington University

Jonathan B. Losos

Professor of Biology
Washington University

Susan R. Singer

Professor of Biology
Carleton College

Illustration Authors
William C. Ober, M.D.
and
Claire W. Garrison, R.N.

 **Higher Education**

Boston   Burr Ridge, IL   Dubuque, IA   Madison, WI   New York   San Francisco   St. Louis
Bangkok   Bogotá   Caracas   Kuala Lumpur   Lisbon   London   Madrid   Mexico City
Milan   Montreal   New Delhi   Santiago   Seoul   Singapore   Sydney   Taipei   Toronto

Higher Education

BIOLOGY, SEVENTH EDITION

Published by McGraw-Hill, a business unit of The McGraw-Hill Companies, Inc., 1221 Avenue of the Americas, New York, NY 10020. Copyright © 2005, 2002, 1999, 1996 by The McGraw-Hill Companies, Inc. All rights reserved. No part of this publication may be reproduced or distributed in any form or by any means, or stored in a database or retrieval system, without the prior written consent of The McGraw-Hill Companies, Inc., including, but not limited to, in any network or other electronic storage or transmission, or broadcast for distance learning.

Some ancillaries, including electronic and print components, may not be available to customers outside the United States.

❂ This book is printed on recycled, acid-free paper containing 10% postconsumer waste.

International   1 2 3 4 5 6 7 8 9 0 VNH/VNH 0 9 8 7 6 5 4 3
Domestic   1 2 3 4 5 6 7 8 9 0 VNH/VNH 0 9 8 7 6 5 4 3

ISBN 0–07–243731–6
ISBN 0–07–111182–4 (ISE)

Publisher: *Martin J. Lange*
Senior sponsoring editor: *Patrick E. Reidy*
Developmental editor: *Anne L. Winch*
Marketing manager: *Tami Petsche*
Lead project manager: *Peggy J. Selle*
Production supervisor: *Kara Kudronowicz*
Senior media project manager: *Jodi K. Banowetz*
Senior media technology producer: *John J. Theobald*
Senior coordinator of freelance design: *Michelle D. Whitaker*
Cover/interior designer: *Christopher Reese*
Senior photo research coordinator: *Lori Hancock*
Photo research: *Meyers Photo-Art*
Supplement producer: *Brenda A. Ernzen*
Compositor: *Carlisle Communications, Ltd.*
Typeface: *10/12 Janson Text Roman*
Printer: *Von Hoffmann Corporation*

Cover images: DNA: © *Doug Struthers/Getty Images*; Pollen: *S. Lowry, Univ. Ulster/Getty Images*; Beetle: © *Davies + Starr/Getty Images*; Leopard: © *Ryan McVay/Getty Images*; Man's Profile: © *Suza Scalora/Getty Images*; Leaf: *Christopher Reese*

The credits section for this book begins on page C-1 and is considered an extension of the copyright page.

3605780523

Library of Congress Cataloging-in-Publication Data

Biology / Peter H. Raven . . . [et al.]. 7th ed.
   p. cm.
  Rev. ed. of: Biology / Peter H. Raven and George B. Johnson. 6th ed., © 2002.
  ISBN 0–07–243731–6 (hard : alk. paper)
  1. Biology. I. Raven, Peter H. II. Raven, Peter H. Biology.

QH308.2.R38   2005
570—dc22   2003016998
                                    CIP

INTERNATIONAL EDITION ISBN 0–07–111182–4
Copyright © 2005. Exclusive rights by The McGraw-Hill Companies, Inc., for manufacture and export. This book cannot be re-exported from the country to which it is sold by McGraw-Hill. The International Edition is not available in North America.

www.mhhe.com

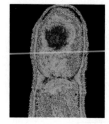

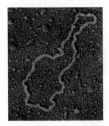

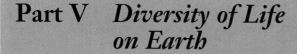

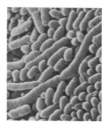

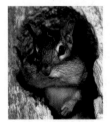

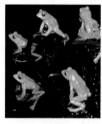

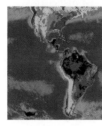

About the Authors

Dr. Peter Raven is director of the Missouri Botanical Garden and Engelmann Professor of Botany at Washington University. A distinguished scientist, Dr. Raven is a member of the National Academy of Sciences, the National Research Council, and is a MacArthur and a Guggenheim fellow. He has received numerous honors and awards for his botanical research and work in tropical conservation, including the National Medal of Science. In addition to coauthoring this text, Raven has authored twenty other books and several hundred scientific articles.

George Johnson is professor emeritus of biology at Washington University in St. Louis, where he has taught genetics and general biology to undergraduates for 30 years. Also professor of genetics at Washington University School of Medicine, he is a student of population genetics and evolution. He has authored more than fifty scientific publications, and several high school and college texts, including *The Living World*, a very successful non-majors college biology text. He has pioneered the development of interactive CD-ROM and web-based investigations for biology teaching.

Jonathan Losos is a professor in the Department of Biology at Washington University and is also chair of the undergraduate Environmental Studies program. An evolutionary biologist, Losos's research has focused on studying patterns of adaptive radiation and evolutionary diversification in lizards. The recipient of several awards including the prestigious Theodosius Dobzhanksy and David Starr Jordan Prizes for outstanding young evolutionary biologists, Losos has published more than eighty scientific articles. He is currently the editor of the *American Naturalist*, a leading journal integrating the fields of evolutionary biology, behavior and ecology.

Susan Singer is professor of biology at Carleton College in Northfield, Minnesota, where she has taught introductory biology, plant biology, plant development, and developmental genetics for 18 years. Her research interests are focused on the development and evolution of flowering plants. Singer has authored numerous scientific publications on plant development, contributed chapters to developmental biology texts, and been actively involved with the education efforts of several professional societies. She serves on the NRC Committee on Undergraduate Science Education.

Preface

We first began work on this text in 1982, over twenty years ago. We set out to write a text that explained biology the way we taught it in the classroom—as the product of evolution. Most texts in 1982 relegated evolution to a few chapters in the diversity section. But evolution pervades biology, and is just as evident in the bacterial character of the mitochondria within your cells, in the biochemical similarities of photosynthesis and glycolysis, in the evolution of genes that control development—everywhere you look in biology, you see Darwin staring back at you. This evolutionary approach has proven popular among our nation's biology faculty, and most texts to greater or lesser degrees now adopt it.

Our text has changed a lot over twenty years, reflecting great changes in biology. The book has become more molecular, as biology has. In particular, a lot more is said about cell biology and development, areas where biology has made enormous strides. But our text remains fundamentally an evolutionary explanation of biology. In this edition, for example, while there is a new chapter on genomes, there is also a new "evo-devo" chapter that examines how genomes and developmental control mechanisms have evolved. This is just one example of our efforts to integrate biological questions and approaches at multiple levels of organization throughout the text, as we strive to guide students toward a connected understanding of biology.

This new seventh edition marks perhaps the greatest change in our text: the addition of two new biologists to the author team, Jonathan Losos, also at Washington University, and Susan Singer of Carleton College. Both made major contributions to the previous edition—Jonathan to the chapter on evolution and ecology, and Susan to the chapters on plant biology—and we are delighted to welcome them as full-fledged authors. In this edition their responsibilities have broadened to include the revolution that is ongoing in our understanding of systematics and evolution at the DNA level, matters that affect many chapters of this book.

Text development today involves an even greater number of people, as instructors from across the country are continually invited to share their knowledge and experience with us through reviews and focus groups. All of the feedback we have received has shaped this edition, resulting in new chapters, reorganization of the table of contents, and expanded coverage in key areas. This edition also incorporated the expertise of three consultants: Randy Di Domenico, University of Colorado—Boulder; Kenneth Mason, Purdue University; and Randall Phillis, University of Massachusetts—Amherst, who provided detailed suggestions for improving the clarity, flow and accuracy of large portions of the text.

How We Have Responded to You

Perhaps more than any other text on the market, this text has continued to evolve as a result of feedback from instructors teaching majors' biology. Overwhelmingly, they have told us that up-to-date content, a clear writing style, quality illustrations and dynamic presentation materials are the most important factors they consider when evaluating textbooks. We have let those values guide our revision of the text, as McGraw-Hill Education worked with those same instructors to create supplements that will help them in the classroom.

Up-to-date Treatment

The core of any majors' biology course is the exploration of cells and genetics, always covered in the first half of any majors' text. This book has been particularly aggressive in keeping its treatment of cell biology and genetics comprehensive and up-to-date. It was the first to present a chapter on cell communication, for example (other books soon followed). We are continuing in that tradition by incorporating such cutting edge topics as the structure of ATP synthetase, small RNAs and RNA editing. This edition also includes a new chapter 17 that explores what we can learn about genomes, covering topics ranging from human health issues and concerns about privacy, agricultural applications, and the potential of genomics in minimizing bioterrorism.

We did not contain this revision to a few select chapters in the first half of the text, however, and it's possible to point to many areas where treatment of recent breakthroughs has been integrated. By concluding our evolution section with a comparative approach to genomes and evolution of development, we were able to connect new breakthroughs in these areas and provide a springboard into the diversity section. Major changes in our understanding of phylogenetic relationships among land plants, protists, and fungi along with other major groups, are reflected in the extensive revision of the diversity chapters. Rapid advances in our knowledge of plant defense responses led to a new chapter on this topic.

Writing Style

Students of biology are responsible for an ever-growing volume of information, and that amount of detail is reflected in today's textbooks, which are increasingly becoming encyclopedic references as opposed to teaching texts.

But students are more likely to succeed with a text that they enjoy reading, that gives them a sense of the wonderment that inspired their own instructors to study biology. For this reason, we have endeavored to strike a balance; an inviting and accessible writing style with the level of authority and rigor expected of a majors' level text.

To further aid the student, every page or two-page spread in this book functions as a semi-independent learning module, organized under its own heading at the top of the left-hand page, with its own summary at the bottom of the right-hand page. This modular presentation makes the conceptual organization of the chapter clear, greatly enhancing student learning.

Illustrations

This book is set apart form others in that its artists, William Ober, M.D. and Claire Garrison, R.N., are part of the author team. Their respective backgrounds as a practicing physician and pediatric registered nurse, and their experience creating art for highly successful anatomy and physiology, zoology and marine biology textbooks, bring an invaluable contribution to the text. The close collaboration between text and illustration authors results in dynamic, accurate figures that aid student understanding and instructor presentation.

- **Combination Figures** These pieces combine a photo or micrograph with a line drawing, to make the connection between conceptual figures and what the student may encounter in lab (Figure 5.10, page 88).
- **Biochemistry Pathway Icons** Found in the discussion of metabolism, these icons help students follow complex processes by highlighting the step currently under discussion (Figure 9.15, page 174).
- **Phylogeny Guideposts** This icon is presented as each group is introduced, to remind students of relationships among diverse organisms (TA 32.1, page 636).
- **Process Boxes** These figures include step-by-step descriptions to walk the student through a compact summary of important concepts (Figure 6.18, page 121)

We have also been fortunate in that this collaboration has allowed us to carefully integrate explanatory text into the figures. The end results are uncluttered, easy to follow illustrations that guide a student through a concept. They also benefit the instructor, as figures without distracting captions can be used for presentation while still allowing instructors to tell their own story.

What Sets this Book Apart

Those who have not used or reviewed previous editions will want to know how this book differs from others.

Evolutionary Focus

The treatment of evolution in this book differs from others in a simple but very important way: Evolution is the organizing principle guiding the teaching of each chapter. Instead of leaping from chemistry directly to cell structure as in other books, this book uses the chemistry of the first chapters to examine the origin of life and the evolution of cells; the cell chapters that follow can then be seen in a broader evolutionary context. Similarly, the treatment of animal anatomy and physiology in other books is largely limited to structure and function—this is the organ and this is how it functions. This book examines each animal body system in terms of how it has evolved. Every section of this book, whether it is genetics or plant biology, presents biology from an evolutionary perspective.

Chemistry in a Biological Context

In talking to students over 30 years of teaching freshman biology, a consistent student complaint has been that the introductory biology course begins with a heavy jolt of chemistry. In other books, only after as many as 100 pages of chemistry do students encounter any biology. This is very off-putting for many students, and gets the course off to a rocky start. This book, by contrast, integrates the chemistry of the first section with biological themes. The treatment of macromolecules in Chapter 3 starts with proteins, which can be easily understood without detailed knowledge of carbohydrates. This arrangement has the distinct advantage of starting the student off with material of obvious relevance to biology.

A Modern Approach

Some of the most obvious differences between this book and others can be seen in the second half of the book, that part devoted to coverage of evolution, diversity, plant biology, anatomy and physiology, and ecology.

Evolution. Our approach to evolutionary biology is unique in two respects. First, we strongly emphasize the role of experiments in studying evolution. Although much of evolutionary biology concerns the study of what happened in the past, that does not mean that experimental approaches are impossible. We emphasize the role that experiments play in studying evolutionary phenomena. More generally, like any detective story, we point out how various approaches must be integrated to fully understand evolutionary diversity.

Second, our book devotes an entire chapter to the evidentiary basis for evolutionary biology. Unlike other aspects of biology, or science in general, the factual basis of evolutionary biology is disputed by some segments of society. Thus, we feel that it is important to clearly present the diversity and depth of evolution that leads almost all biologists to conclude that evolution has occurred. We feel that it is essential for all college biology students—regardless of their own opinions—to understand the scientific basis for this view.

Diversity. Our text has been organized so the diversity section is framed by a discussion of the revolution in taxonomy and phylogenetics (Chapter 25). Complete with vivid examples of dramatic changes in our understanding of relationships among organisms, this chapter can be used alone as an abbreviated approach to diversity or as a foundation for a more comprehensive evolutionary investigation of the diversity of life in the chapters that follow.

The book also differs from other majors' biology textbooks in that its coverage of diversity is more extensive. Consider for example the invertebrates. Other books devote as few as 30 pages to the invertebrates, presenting only the briefest of sketches of what used to be the core of traditional biology courses. This book devotes more than twice as many pages to the invertebrates, followed by a more comprehensive chapter on the vertebrates than is found in other books. Why is this more extensive treatment of diversity important? Even in courses that don't cover diversity in detail, it is important that students be able to uncover for themselves the relations among animal groups.

Plant Biology. The plant biology chapters have undergone extensive revision, and are now organized to lead the student through the plant life cycle. In addition, we have carefully integrated both developmental and genetic perspectives, a fusion not found in other texts. For example, Chapter 36, Vegetative Plant Development, explores root formation in the context of the *monopterous* mutant of *Arabidopsis* that fails to make a root. The shift from its developmental role to its functional role as an auxin receptor begins to move students toward a physiological understanding of plant function.

Anatomy and Physiology. Most books devote nearly the same amount of space to anatomy and physiology, about 250 pages. The differences lie primarily in approach, this book having a more evolutionary focus than others, and in its emphasis on fundamentals.

Ecology and Behavior. We take an integrative view to understanding how the environment functions and how organisms interact with it. This section is broken into different chapters, such as behavioral ecology, population ecology, and community ecology, but the topics are carefully integrated. Moreover, we apply this information extensively in Chapters 56 and 57 (The Biosphere and Conservation Biology) to address the environmental issues facing our planet. We believe it is of the utmost importance that students understand the scientific bases to current problems so they can evaluate efforts to solve them.

Changes to the Seventh Edition

The seventh edition of *Biology* is the result of extensive analysis of the text and evaluation of input from biology instructors who conscientiously reviewed chapters during various stages of this revision. We have utilized the constructive comments provided by these professionals in our continuing efforts to enhance the strengths of the text. Listed first are general changes that have been made to the entire text, which is then followed by specific changes for each part.

End-of-chapter Pedagogy

The end-of-chapter student review has been greatly expanded, offering students a full-page chapter review and three assessment tools: Self Test, multiple choice questions; Test Your Visual Understanding, questions based on a figure from the chapter; and Apply Your Knowledge, critical thinking questions. The assessment doesn't end there, however. These tools are carried over to the web where the student can take an interactive version of the test and receive instructional feedback.

Inquiry Questions

In this edition we have developed Inquiry Questions, which follow the legend in figures presenting graphed data. These questions require the student to think about the information contained in the graph in even greater depth, increasing their understanding of, and facility with, the material.

Answers to both end of chapter questions and Inquiry Questions are found on the Online Learning Center at www.mhhe.com/raven7.

Volumes

We recognize that instructors don't always use the entire text, so we now offer *Biology* in the following volumes:

Volume 1 Chapters 1–20 Chemistry, Cell and Genetics
Volume 2 Chapters 35–51 Plant Biology and Animal Biology
Volume 3 Chapters 21–34, 52–57 Evolution, Diversity and Ecology

Content Changes by Part
Part I The Origin of Living Things

Part I was revised with the intention of creating a more solid foundation of key concepts in biology, which students can then build on in later chapters. Discussions are now clearer and better supported with illustrations.

New Topics and Revised Treatments

Chapter 1	Properties of life, hierarchical organization *Revised;* Additional topics in evolution *New*
Chapter 2	The Nature of Molecules *Entire chapter revised*
Chapter 3	Figures on chaperons and protein denaturing *New;* Protein folding, lipids *Revised*
Chapter 4	Figures on endosymbiosis, Domains/Kingdoms, phylogenetic tree of life *New* Bacteria and archaebacteria, microfossils *Revised*

Part II Biology of the Cell

Randall Phillis assisted in the revision of Part II by directing the authors to concepts that needed additional detail, and by providing suggestions for improving the accuracy and parsimony of the narrative. Concepts that were covered too briefly in previous editions are now supported with more extensive discussion and new illustrations.

New Topics and Revised Treatments

Chapter 6	Membrane microdomains *New* Osmosis; coupled transport *Revised*
Chapter 7	Signal amplification, expression of cellular identification *Revised*
Chapter 8	Redox reactions, ATP functioning *Revised*
Chapter 9	ATP synthetase *New;* Electron transport; reducing power; chemiosmosis *Revised*
Chapter 10	The Calvin cycle *Revised*
Chapter 11	Cell cycle control *New;* Chromosome structure *Revised*

Part III Genetic and Molecular Biology

With the help of Kenneth Mason, Part III was carefully updated to incorporate the most current research. Chapter 13, Patterns of Inheritance, was rewritten for better organization and clearer presentation. Two new chapters provide expanded discussion in fields where our knowledge has grown exponentially.

Chapter 17, "Genomes" integrates plant and animal genomics, functional genomics and proteomics in a chapter that is inquiry driven rather than a discussion of techniques.

Chapter 20, "Cancer Biology and Cell Technology" explores two areas where recent advances in cell and molecular biology have the potential of revolutionizing medicine. The first is cancer, where research into the molecular events leading to cancer is beginning to suggest effective therapies. The second is cell technology, including cloning, embryonic stem cells, and the exciting and controversial proposal of therapeutic cloning.

New Topics and Revised Treatments

Chapter 12	Meiotic prophase *Revised*
Chapter 13	Patterns of inheritance *Entire chapter revised*
Chapter 14	Eukaryotic DNA replication *Revised*
Chapter 15	Eukaryotic transcription *New*
Chapter 16	The tools of genetic engineering *New*
Chapter 18	Small RNAs, iRNA, RNA editing *New* Transcriptional control in prokaryotes *Revised*
Chapter 19	Vertebrate embryonic axis formation, evolution of homeotic genes *New* Cell movement; cell induction; embryonic determination; pattern formation *Revised*

Part IV Evolution

The Evolution section has been revised to bring more experimental data and analysis into the discussions. Because presentation of the experimental data used to derive conclusions and concepts is key to understanding how the concepts arose from the research, you will see that graphs and charts have become more plentiful in these chapters. The evolution of many groups is reassessed in light of new molecular data.

Chapter 24, "Evolution of Genomes and Developmental Mechanisms" is a new comparative genomics chapter that addresses our emerging understanding of the evolution of development, and helps to provide a conceptual framework for the diversity chapters that follow. The chapter was developed in conjunction with Chapter 17, Genomes, to first provide students with an understanding of what we can learn about genomes, and then having learned about evolution, delve into a deeper discussion of how development has evolved to yield novel phenotypes.

New Topics and Revised Treatments

Chapter 21	Measuring fitness, components of fitness, role of selection in maintaining variation *New* Hardy-Weinberg *Revised*
Chapter 22	Darwin's finches, industrial melanism *New* Evidence from developmental biology for evolution *Revised*
Chapter 23	Plant speciation by chromosomal change, the future of evolution *New*

Part V Diversity of Life on Earth

The fungi chapter has been moved to the phylogenetically appropriate place in the diversity section following plant diversity. Where appropriate, chapters in the diversity unit have been updated to reflect phylogenetic changes. The thoroughly revised and rewritten **Chapter 25, "Systematics and the Phylogenetic Revolution"** addresses the current tension between taxonomy and systematics. The chapter can be used alone, to teach the basic concepts of diversity, or can be used as a starting place for a more in-depth study of this area of biology.

New Topics and Revised Treatments

Chapter 25 Phylogenetics and classification *Entire chapter revised*

Chapter 26 Virus genomes *New;* Viral diseases and HIV *Revised*

Chapter 27 Term "eubacteria" replaced with "bacteria," figures of cell structure and clades *New*

Chapter 28 Phylogenetic approach *New;* Protist disease in South, Central and North America, relationships between algae and land plants *Revised*

Chapter 29 Fossil evidence of ancient angiosperm *Archaefructacea*, evolution of triploid endosperm *New*
Monophyletic relationships between ferns and horsetails *Revised*

Chapter 30 Fungi *Entire chapter revised*

Chapter 31 Protostomes and deuterostomes *New and expanded*
Classification *Revised*

Chapter 32 Protostome phylogeny, rotifers and cycliophora *Revised*

Chapter 33 Mollusks, annelids, arthropods, and echinoderms combined into one chapter *New*

Chapter 34 Characteristics and phylogeny illustration, primate evolution *New*

Part VI Plant Form and Function

The plant biology chapters have been revised so that the traditional discussion of evolutionary influences on plant form and function are brought into a developmental context. Evolution is still presented as the underlying explanation for the character of vascular tissues, seeds, flowers, and fruits, however the developmental processes that produce these organs are now given more emphasis. Two previously combined topics, transport and nutrition, have been split into separate chapters allowing for more in-depth discussion of both topics.

Chapter 39, "Plant Defense Responses" is a new chapter that provides a thorough discussion of secondary compounds and their roles in both plant defense and human applications. Wound responses and R gene mediated responses are explored in depth with an emphasis on signaling pathways.

New Topics and Revised Treatments

Chapter 35 Updated photographs, discussion of genetic regulation of trichomes *Revised*

Chapter 36 Discussion of signal transduction in germination, comparison of roles of *Hox* genes in plant and animal development *Revised*

Chapter 37 Water relations problems, mRNA transport in phloem *New*

Chapter 38 Newly expanded chapter on plant nutrition. Effects of global change on photosynthesis and balance of plant nutrients, phytoremediation *New*
Nutritional symbioses *Revised and expanded*

Chapter 40 Signal transduction mediated by light including phot1 *New*
Light responses *Revised*

Chapter 41 Plant reproduction *Entire chapter revised and reorganized*

Part VII Animal Form and Function

With the assistance of Randy DiDomenico, many discussions were rewritten for better organization and clarity. Previous chapters on circulation and respiration were combined into one chapter, as were chapters on body organization and locomotion.

New Topics and Revised Treatments

Chapter 42 Combined organization of the animal body and locomotion into one chapter, coordination of organ systems *New*

Chapter 43 Neural, hormonal and accessory organ regulation *New*
Small intestine discussion reorganized to group all functions together *Revised*

Chapter 44 Maximizing rate of gas diffusion *New*
Integration of circulation and respiration chapter *Revised*

Chapter 45 Graded potentials *New;* Membrane and action potentials, synapses and drug addiction *Revised*

Chapter 46 Sensory transduction *Revised*

Chapter 47 The Endocrine System *Entire chapter revised*

Chapter 48 Immunoglobulins, illustrated table *New*
AIDS *Revised*

Chapter 49 Discussion of ammonia, urea and uric acid reorganized, nephron *Revised*

Chapter 50 Sex determination, reptiles and birds *Revised*
Human intercourse *Omitted*

Chapter 51 Combined discussion of chick and mammalian extraembryonic membranes, neurulation *Revised*

Part VIII Ecology and Behavior

The ecology and behavior chapters were moved to follow diversity and physiology, where these chapters are more often taught. There is now an even greater emphasis on experimental data and analysis.

New Topics and Revised Treatments

Chapter 52 Animal cognition *New;* Integration of animal behavior and behavioral ecology chapters *Revised*

Chapter 53 Introduction to ecology, integration of autoecology and population ecology *New* Population regulation and limitation, human population growth *Revised*

Chapter 54 Introduction, definition of community *New* Parasitism, succession, disturbance *Revised*

Chapter 55 Geochemical cycles, energy flow, species richness *Revised*

Chapter 56 Differences between aquatic and terrestrial ecosystems *New* Integration of biosphere and future of the biosphere chapters, global climate change, El Niño *Revised*

Chapter 57 Chapter organization, biodiversity hotspots, amphibian extinctions, invasive species *New* Economic benefits of biodiversity *Revised*

Overview of Changes to BIOLOGY, Seventh Edition

All Cell & Genetics Chapters Extensively Revised

In addition to discussing important advances, many sections have been reworked for improved clarity.

Chapter 17, "Genomes"

This new chapter describes how researchers sequence entire genomes, and how the information is being used.

Chapter 20, "Cancer Biology and Cell Technology"

This new chapter updates progress in understanding cancer, and introduces many new advances in cloning and stem cell technology.

Evolution & Diversity Sections Extensively Revised

New RNA and genomic information is leading to a reassessment of traditional evolutionary phylogenies.

Chapter 24, "Evolution of Genomes and Developmental Mechanisms"

This new chapter explores the wealth of new information on genome sequences, and introduces the new and exciting field of "evo/devo", the evolution of development.

Treatment of Plant Biology Expanded

A total of seven plant chapters provide extensive plant biology coverage with a molecular development point of view. Chapters have been organized to lead the student through the life cycle of a plant.

Chapter 39, "Plant Defense Responses"

This new chapter captures the excitement of this area of biology, which has seen rapid advances and recent breakthroughs in understanding plant defense responses.

Ecology Chapters Updated and Expanded

Up-to-date examples have been integrated into all chapters; note the use of case histories in Chapter 57.

Physiology Chapters Reworked

Discussions of processes like nervous conduction reworked for increased clarity, and related subjects like circulation and respiration treated together.

End-of-Chapter Assessment

Two pages are now devoted to student review and assessment. A full-page chapter summary is followed by multiple choice, illustration-based and application questions.

Illustrations

Many new illustrations clarify difficult concepts; others illustrate tables to aid understanding. Wherever data are presented in graphs, the figure is accompanied by an Inquiry Question to test the student's understanding.

Teaching and Learning Supplements

McGraw-Hill offers various tools and technology products to support *Biology*. Students can order supplemental study materials by contacting their local bookstore or by calling 800-262-4729. Instructors can obtain teaching aids by calling the Customer Service Department at 800-338-3987, visiting our website at www.mhhe.com/biology, or contacting their local McGraw-Hill sales representative.

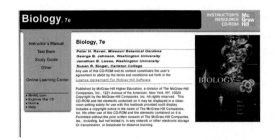

For the Instructor:

Digital Content Manager CD-ROM

This multimedia collection of visual resources allows instructors to utilize artwork from the text in multiple formats to create customized classroom presentations, visually based tests and quizzes, dynamic course website content, or attractive printed support material. The digital assets on this cross-platform CD-ROM include:

Art Library   Color-enhanced, digital files of all illustrations in the book, plus the same art saved in unlabeled and gray scale versions, can be readily incorporated into lecture presentations, exams, or custom-made classroom materials. Upsized labels make the images appropriate for use in large lecture halls.

TextEdit Art Library   Every line art piece is placed into a PowerPoint presentation that allows the user to revise, move, or delete labels as desired for creation of customized presentations or for testing purposes.

Active Art Library   Active Art consists of art files that have been converted to a format that allows the artwork to be edited inside of PowerPoint. Each piece can be broken down to its core elements, grouped or ungrouped, and edited to create customized illustrations.

Animations Library   Full color presentations involving key process figures in the book have been brought to life via animation. These animations offer flexibility for instructors and were designed to be used in lecture. Instructors can pause, rewind, fast forward, and turn audio off/on to create dynamic lecture presentations.

PowerPoint Lecture Outlines   These ready-made presentations combine art and lecture notes for each of the 57 chapters of the book. The presentations can be used as they are, or can be customized to reflect your preferred lecture topics and organization.

PowerPoint Outlines   The art, photos, and tables for each chapter are inserted into blank PowerPoint presentations to which you can add your own notes.

Photo Library   Like the Art Library, digital files of all photographs from the book are available.

Table Library   Every table that appears in the book is provided in electronic form.

Video Library   Contains digitized video clips that can be inserted into a PowerPoint lecture.

Additional Photo Library   Over 700 photos, not found in *Biology*, are available for use in creating lecture presentations.

Instructor's Testing and Resource CD-ROM

The cross-platform CD-ROM contains the Instructor's Manual and Test Item File, both available in both Word and PDF formats. The manual contains chapter synopses, objectives, key terms, outlines, instructional strategies and sources for additional visual resources. The Test Bank offers questions that can be used for homework assignments or the preparation of exams. The computerized test bank utilizes Brownstone Diploma testing software, which allows the user to quickly create customized exams. This user-friendly program allows instructors to search questions by topic, format, or difficulty level; edit existing questions or add new ones; and scramble questions and answer keys for multiple versions of the same test.

Transparencies

A set of 1300 transparency overheads includes every piece of line art and table in the text. The images are printed with better visibility and contrast than ever before, and labels are large and bold for clear projection.

Online Learning Center
www.mhhe.com/raven7

Instructor resources at this site include access to online laboratories, course-specific current articles, real-time news feeds, course updates and research links.

Course Delivery Systems

With help from our partners, WebCT, Blackboard, Top-Class, eCollege, and other course management systems, instructors can take complete control over their course content. These course cartridges also provide online testing and powerful student tracking features. The *Biology* Online Learning Center is available within all of these platforms.

For the Student:

Online Learning Center
www.mhhe.com/raven7

The site includes quizzes for each chapter, interactive activities, and answers to questions from the text. Turn to the inside cover of the text to learn more about the exciting features provided for students through the enhanced *Biology* Online Learning Center.

Student Study Guide

This student resource contains activities and questions to help reinforce chapter concepts. The guide provides students with tips and strategies for mastering the chapter content, concept outlines, concept maps, key terms and sample quizzes.

Acknowledgements

Our goal for *Biology* has always been to present the science in an interesting and engaging way while maintaining a comprehensive and authoritative text. This is a lofty goal considering the mountain of information and research we must go through just to update the text from one edition to the next. This seventh edition would not have been possible without the contributions of many. We are indebted to our colleagues across the country and around the globe that provided numerous suggestions on how to improve on the sixth edition. We wish particularly to thank Kenneth Mason of Purdue University, Randy DiDomenico of the University of Colorado, Boulder, and Randall Phyllis of the University of Massachusetts, Amherst for very detailed advice on how to improve large sections of the text.

As any author knows, a textbook is made not only by an author team aided by their colleagues, but also by a publishing team, a group of people that guide the raw book written by the authors through a yearlong process of reviewing, editing, fine-tuning and production. This edition was particularly fortunate in its book team, led by Patrick Reidy, sponsoring editor; Anne Winch, developmental editor; Tami Petsche, marketing manager; Peggy Selle, project manager; Michelle Whitaker, designer; Megan Jackman and Elizabeth Sievers, off-site editors; Kennie Harris, copy editor, and many more people behind the scenes.

The illustrations are critically important to a biology text, and ours continue to be superbly conceived and rendered by Bill Ober and Claire Garrison.

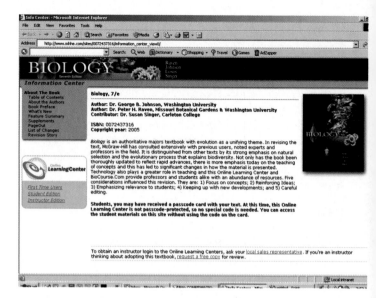

As always, we have had the support of spouses and children, who have seen less of us than they might have liked because of the pressures of getting this revision completed. They have adapted to the many hours this book draws us away from them and, even more than us, look forward to its completion.

As with every edition, acknowledgements would not be complete without thanking the generations of students who have used the many editions of this text. They have taught us as least as much as we have taught them, and, thanks to e-mail, are an increasing part of our lives.

Finally, we need to thank our reviewers. Every text owes a great deal to those instructors across the country who review it. Serving as sensitive antennae for errors and omissions, and as sounding boards for new approaches, reviewers are among the most valuable tools at an author's disposal. Many improvements in this edition are the direct result of their suggestions. Every one of them has our heartfelt thanks.

Reviewers of the Seventh Edition

Heather Addy   *University of Calgary*
Lawrence A. Alice   *Western Kentucky University*
Terry C. Allison   *The University of Texas–Pan American*
Loran C. Anderson   *Florida State University*
Mohammad Ashraf   *City Colleges of Chicago*
Ellen Baker   *Santa Monica College*
R. Neal Band   *Michigan State University*
Dale L. Barnard   *Utah State University*
Diane C. Bassham   *Iowa State University*
Wayne M. Becker   *University of Wisconsin–Madison*
Robert L. Beckmann   *North Carolina State University*
Gerald Bergtrom   *University of Wisconsin–Milwaukee*
Cheryl Briggs   *University of California–Berkeley*
Trey Broadhurst   *Montgomery College*
Arthur L. Buikema, Jr.   *Virginia Tech*
Ann B. Burgess   *University of Wisconsin–Madison*
Carol A. Burkart   *Mountain Empire Community College*
D. Brent Burt   *Stephen F. Austin State University*
David Byres   *Florida Community College at Jacksonville*
Les Chappell   *University of Aberdeen, UK*

Jung Choi   *Georgia Institute of Technology*
Don Cipollini   *Wright State University*
Richard J. Cogdell   *University of Glasgow*
Jerry Cook   *Sam Houston State University*
David T. Corey   *Midlands Technical College*
George Cornwall   *University of Colorado–Boulder/Metropolitan State College of Denver*
Francie Smith Cuffney   *Meredith College*
Paul V. Cupp, Jr.   *Eastern Kentucky University*
James A. Danoff-Burg   *Columbia University*
Sandra L. Davis   *University of Louisiana at Monroe*
Mark D. Decker   *University of Minnesota*
Mary B. Dettman   *Seminole Community College*
John Dickerman   *Northern Illinois University*
Cathy Donald-Whitney   *Collin County Community College*
Thomas W. Dreschel   *Brevard Community College, Kennedy Space Center*
Carolyn S. Dunn   *University of North Carolina at Wilmington*
Roland R. Dute   *Auburn University*
Frederick B. Essig   *University of South Florida*
Bruce E. Felgenhauer   *University of Louisiana at Lafayette*
James Franzen   *University of Pittsburgh*
Andrea Gargas   *University of Wisconsin–Madison*
John V. Gartner, Jr.   *St. Petersburg College*
John R. Geiser   *Western Michigan University*
Florence K. Gleason   *University of Minnesota*
John S. Graham   *Bowling Green State University*
John S. Greenwood   *University of Guelph*
Peggy J. Guthrie   *University of Central Oklahoma*
Adrian Hailey   *University of Bristol*
Dana Brown Haine   *Central Piedmont Community College*
Robert O. Hall   *University of Wyoming*
Robert W. Hamilton   *Loyola University Chicago*
David S. Hibbett   *Clark University*
Leland N. Holland, Jr.   *Pasco-Hernando Community College*
Eva A. Horne   *Kansas State University*
Jeffrey Jack   *University of Louisville*
Lee F. Johnson   *The Ohio State University*
Gregory A. Jones   *Santa Fe Community College*
Walter S. Judd   *University of Florida*
Richard R. Jurin   *University of Northern Colorado*
Thomas C. Kane   *University of Cincinnati*
Ronald Keiper   *Valencia Community College*
John J. Kelly   *Loyola University Chicago*
Cheryl A. Kerfeld   *University of California, Los Angeles*
David J. Kittlesen   *University of Virginia*
William Kroll   *Loyola University Chicago*
Harry D. Kurtz, Jr.   *Clemson University*
Roberta Lammers-Campbell   *Loyola University Chicago*
Peter Lavrentyev   *The University of Akron*
Michael Lawson   *Missouri Southern State College*
Roger M. Lloyd   *Florida Community College at Jacksonville*
David Magrane   *Morehead State University*
Richard Malkin   *University of California–Berkeley*
Terry C. Maxwell   *Angelo State University*
Michael McLeod   *Belmont Abbey College*
Frank J. Messina   *Utah State University*
Sandra Millward   *University of Cincinnati*
Jacalyn S. Newman   *University of Pittsburgh*
Janice Moore   *Colorado State University*
Deborah A. Neher   *University of Toledo*

Erik T. Nilsen   *Virginia Tech*
T. Mark Olsen   *University of Notre Dame*
John C. Osterman   *University of Nebraska–Lincoln*
Daniel M. Pavuk   *Bowling Green State University*
Andrew J. Pease   *Villa Julie College*
Rhoda E. Perozzi   *Virginia Commonwealth University*
Carolyn Peters   *Spoon River College*
Susan Phillips   *Brevard Community College*
Eric R. Pianka   *University of Texas at Austin*
Aleksandar Popadic   *Wayne State University*
Angela R. Porta   *Kean University*
Calvin A. Porter   *Xavier University of Louisiana*
Elena Pravosudova   *Sierra College*
Linda R. Richardson   *Blinn College*
Laurel Roberts   *University of Pittsburgh*
Charles L. Rutherford   *Virginia Tech University*
Erik P. Scully   *Towson University*
Wendy E. Sera   *Seton Hall University*
Alison M. Shakarian   *Salve Regina University*
Neil F. Shay   *University of Notre Dame*
Shree R. Singh   *Alabama State University*
David A. Smith   *Lock Haven University of Pennsylvania*
Willie Smith   *Brevard Community College*
Nancy G. Solomon   *Miami University*
Alan J. Spindler   *Brevard Community College*
Ann Springer   *Hillsborough Community College*
Amy C. Sprinkle   *Jefferson Community College Southwest*
Bruce Stallsmith   *University of Alabama in Huntsville*
John D. Story   *North West Arkansas Community College*
Robert Sullivan   *Marist College*
Marshall D. Sundberg   *Emporia State University*
Pamela S. Thomas   *University of Central Florida*
Patrick A. Thorpe   *Grand Valley State University*
Rani Vajravelu   *University of Central Florida*
Carol M. F. Wake   *South Dakota State University*
Jane Waterman   *University of Central Florida*
Cindy Martinez Wedig   *University of Texas–Pan American*
Olivia Masih White   *University of North Texas*
Lance R. Williams   *The Ohio State University*
Michael Zimmerman   *University of Wisconsin–Oshkosh*

General Biology Symposium

Each year McGraw-Hill holds a General Biology Symposium, which is attended by instructors from across the country. These events are an opportunity for editors from McGraw-Hill to gather information about the needs and challenges of instructors teaching the major's biology course, however it also offers professors a forum for exchanging ideas and experiences with colleagues they might not have otherwise met. The feedback we have received has been invaluable, and has contributed to the success of *Biology* and its supplements.

2003

Marc Ammerlaan   *University of Michigan–Ann Arbor*
Scott Chandler   *University of California, Los Angeles*
Bill Collins   *SUNY at Stony Brook*
Elizabeth Connor   *University of Massachusetts–Amherst*
Steve Connor   *University of South Florida*

Robert Fulginiti   *Xavier University*
Florence Gleason   *University of Minnesota*
Carla Haas   *Penn State University*
David Julian   *University of Florida*
Steve Kelso   *University of Illinois at Chicago*
Bob Locy   *Auburn University*
Kenneth Mason   *Purdue University*
Nancy Solomon   *Miami University–Oxford*
Bill Stein   *SUNY at Binghamton*
Sally Swain   *Middle Tennessee State University*
Linda Waters   *University of Central Florida*

2002

Richard J. Cyr   *Penn State University*
Randy DiDomenico   *University of Colorado–Boulder*
Doug Gaffin   *University of Oklahoma*
Marielle Hoefnagels   *University of Oklahoma*
Jan Jenner   *Science Writer*
Cheryl A. Kerfeld   *University of California, Los Angeles*
Kenneth Mason   *Purdue University*
Michael Meighan   *University of California–Berkeley*
Jane Phillips   *University of Minnesota*
Randall W. Phillis   *University of Massachusetts–Amherst*
Joelle Presson   *University of Maryland*
Leslie Winemiller   *Texas A & M University*
Denise Woodward   *Penn State University*

2001

Mark Ammerlaan   *University of Michigan–Ann Arbor*
Doug Gaffin   *University of Oklahoma*
Jon C. Glase   *Cornell University*
Richard Hallik   *University of Arizona*
Marielle Hoefnagels   *University of Oklahoma*
Fernan Jaramillo   *Carleton College*
Randall Johnson   *University of California, San Diego*
Kenneth Mason   *Purdue University*
Sally Frost-Mason   *Purdue University*
Jorge Moreno   *University of Colorado–Boulder*
Tom Owens   *Cornell University*
Deanna Raineri   *University of Illinois at Urbana-Champaign*
Jon Ruehle   *University of Central Arkansas*
Steven A. Wasserman   *University of California, San Diego*

Reviewers of the Sixth Edition

Michael Adams   *Pasco-Hernando Community College*
Sylvester Allred   *Northern Arizona University*
Lon Alterman   *Clarke College*
Elena Amesbury   *University of Florida*
William Anyonge   *University of California–Los Angeles*
Amir Assadirad   *Delta College*
Gary I. Baird   *Brigham Young University*
Ellen Baker   *Santa Monica College*
Stephen W. Banks   *Louisiana State University–Shreveport*
Ruth Beattie   *University of Kentucky*
Samuel N. Beshers   *University of Illinois*
Christine Konicki Bieszczad   *Saint Joseph College*
John Birdsell   *University of Arizona*
Brenda C. Blackwelder   *Central Piedmont Community College*

Sandra Bobrick   *Community College of Allegheny County
   Allegheny Campus*
Randall Breitwisch   *University of Dayton*
Mark Browning   *Purdue University*
Roger Buckanan   *Arkansas State University*
Theodore Burk   *Creighton University*
John S. Campbell   *Northwest College*
John R. Capeheart   *University of Houston–Downtown*
Michael S. Capp   *Carlow College*
Jeff Carmichael   *University of North Dakota*
George P. Chamuris   *Bloomsburg University*
Susan Cockayne   *Brigham Young University*
William Cohen   *University of Kentucky*
W. Wade Cooper   *Shelton State Community College*
Lisa M. Coussens   *University of California–San Francisco,
   Cancer Research Institute*
Wilson Crone   *Hudson Valley Community College*
Paul V. Cupp Jr.   *Eastern Kentucky University*
Richard Cyr   *The Pennsylvania State University*
Grayson Davis   *Trinity University*
Mark A. DeCrosta   *University of Tampa*
David L. Denlinger   *Ohio State University*
C. Lynn Dorn   *Valencia Community College*
Charles D. Drewes   *Iowa State University*
Sondra Dubowsky   *Allen County Community College*
Peter I. Ekechukwu   *Horry-Georgetown Technical College*
Dennis Emery   *Iowa State University*
Frederick B. Essig   *University of South Florida*
Bruce Evans   *Huntington College*
Deborah Fahey   *Wheaton College*
Linda E. Fisher   *University of Michigan–Dearborn*
Rob Fitch   *Wenatchee Valley College*
Robert Fogel   *University of Michigan*
James Franzen   *University of Pittsburgh–Pittsburgh Campus*
William Friedman   *University of Colorado*
Lawrence Fritz   *Northern Arizona University*
Bernard Frye   *University of Texas at Arlington*
Robert J. Full   *University of California–Berkeley*
Warren Gallin   *University of Alberta*
Darrell Galloway   *The Ohio State University*
Ted Gish   *St. Mary's College*
Donald Glassman   *Des Moines Area Community College*
Jim Glenn   *Red Deer College*
Jim R. Goetze   *Laredo Community College*
Jack M. Goldberg   *University of California–Davis*
Elizabeth Godrick   *Boston University*
Dalton Gossett   *Louisiana State University–Shreveport*
John Griffis   *Joliet Junior College*
Kathryn Gronlund   *New Mexico State University–Carlsbad*
Elizabeth L. Gross   *The Ohio State University*
Patricia A. Grove   *College of Mount St. Vincent*
Randolph Hampton   *University of California–San Diego*
Sehoya E. Harris   *The Pennsylvania State University*
Carla Ann Hass   *The Pennsylvania State University*
Chris Haynes   *Shelton State Community College*
Albert A. Herrera   *University of Southern California*
Pamela Higgins   *Allentown College of St. Francis DeSales*
Richard Hill   *Michigan State University*
Phyllis Hirsch   *East Los Angeles College*
Victoria Hittinger   *Rhode Island College*
Nan Ho   *Las Positas College*

Leland N. Holland, Jr.   *Pasco-Hernando Community College–West Campus*
Elisabeth A. Hooper   *Truman State University*
Terry L. Hufford   *The George Washington University*
Allen Hunt   *Elizabethtown Community College*
Sobrasua E. M. Ibin   *Morris Brown College*
Louis Irwin   *University of Texas at El Paso*
Laurie E. Iten   *Purdue University*
Jeffrey Jack   *College of Arts & Sciences*
James B. Jensen   *Brigham Young University*
Judy Jernstedt   *University of California - Davis*
George P. Johnson   *Arkansas Tech University*
Kenneth V. Kardong   *Washington State University*
Cheryl Kerfeld   *University of California–Los Angeles*
Joanne M. Kilpatrick   *Auburn University at Montgomery*
Peter King   *Francis Marion University*
Edward C. Kisailus   *Canisius College*
Robert M. Kitchin   *University of Wyoming*
Will Kleinelp   *Middlesex County College*
Kenton Ko   *Queen's University*
Ross E. Koning   *Eastern Connecticut State University*
Karen L. Koster   *University of South Dakota*
V.A. Langman   *Louisiana State University–Shreveport*
Simon Lawrance   *Otterbein College*
Jeffrey N. Lee   *Essex County College*
Laura G. Leff   *Kent State University*
Mary E. Lehman   *Longwood College*
Niles Lehman   *University at Albany   SUNY*
Michael Lema   *Midlands Technical College*
Charles Kingsley Levy   *Boston University*
Leslie Lichtenstein   *Massasoit Community College*
Harvey Liftin   *Broward Community College*
Richard Londraville   *University of Akron*
Sonja L. Maki   *Clemson University*
Bradford D. Martin   *La Sierra University*
Barbara Maynard   *Colorado State University*
Deanna McCullough   *University of Houston Downtown*
L. R. McEdward   *University of Florida*
Michael Ray Meighan   *University of California–Berkeley*
John Merrill   *Michigan State University*
Harry A. Meyer   *McNeese State University*
Dennis J. Minchella   *Purdue University*
Jonathan D. Monroe   *James Madison University*
David L. Moore   *Utica College of Syracuse University*
Tony E. Morris   *Fairmont State College*
Roger N. Morrissette   *Framingham State College*
Richard Mortensen   *Albion College*
William H. Nelson   *Morgan State University*
Peter H. Niewiarowski   *University of Akron*
Colleen J. Nolan   *St. Mary's University*
John C. Osterman   *University of Nebraska–Lincoln*
Thomas G. Owens   *Cornell University*
Bruce Parker   *Utah Valley State University*
Dustin Penn   *University of Utah*
Stacia Pieffer-Schneider   *Marquette University*
Carl S. Pike   *Franklin and Marshall College*
Nancy A. Perigo   *Willamette University*
Greg Phillips   *Blinn College–Brenham Campus*
Jon Pigage   *University of Colorado at Colorado Springs*
Barbara Pleasants   *Iowa State University*
John Pleasants   *Iowa State University*

Peggy Pollack   *Northern Arizona University*
Mitch Price   *The Pennsylvania State University*
Margene Ranieri   *Bob Jones University*
Arthur Raske   *Northland Baptist Bible College*
Keith Redetzke   *University of Texas at El Paso*
Peter J. Rizzo   *Texas A&M University*
Ellison Robinson   *Midlands Technical College*
Lyndell P. Robinson   *Lincoln Land Community College*
Angel M. Rodriguez   *Broward Community College*
June R. P. Ross   *Western Washington University*
Patricia Rugaber   *Coastal Georgia Community College*
Connie Rye   *Bevill State Community College*
Nancy K. Sanders   *Truman State University*
Robert B. Sanders   *University of Kansas–Main Campus*
Lisa M. Sardinia   *Pacific University*
Brian W. Schwartz   *Columbus State University*
Bruce S. Serlin   *DePauw University*
Mark A. Sheridan   *North Dakota State University*
Janet Anne Sherman   *Penn College of Technology*
Louis Sherman   *Purdue University*
Jim Shinkle   *Trinity University*
Richard Shippee   *Vincennes University*
Brian Shmaefsky   *Kingwood College*
Michele Shuster   *University of Pittsburgh*
Robert C. Sizemore   *Alcorn State University*
Mark Smith   *Victor Valley College*
Nancy Solomon   *Miami University*
Norm Stacey   *University of Alberta*
Ruth Stutts-Moseley   *Bishop State Community College*
Kathy Sympson   *Florida Keys Community College*
Stan Szarek   *Arizona State University*
Robert H. Tamarin   *University of Massachusetts Lowell*
Michael Tenneson   *Evangel University*
Sharon Thoma   *Edgewood College*
Joanne Kivela Tillotson   *Purchase College   State University of New York*
Maurice Thomas   *Palm Beach Atlantic College*
Thomas Tomasi   *Southwest Missouri State University*
Leslie Towill   *Arizona State University*
Akif Uzman   *University of Houston–Downtown*
Thomas J. Volk   *University of Wisconsin–La Crosse*
Keith D. Waddington   *University of Miami*
D. Alexander Wait   *Southwest Missouri State University*
Timothy S. Wakefield   *Auburn University*
Charles Walcott   *Cornell University*
Eileen Walsh   *Westchester Community College*
Frederick Wasserman   *Boston University*
Steven A. Wasserman   *University of California–San Diego*
Robert F. Weaver   *University of Kansas*
Andrew N. Webber   *Arizona State University*
Harold J. Webster   *Penn State DuBois*
Mark Wheelis   *University of California–Davis*
Lynn D. Wike   *University of South Carolina at Aiken*
William Williams   *Saint Mary's College of Maryland*
Mary L. Wilson   *Gordon College*
Kevin Winterling   *Emory & Henry College*
E. William Wischusen   *Louisiana State University and Agricultural and Mechanical College*
Kenneth Wunch   *Tulane University*
Mark L. Wygoda   *McNeese State University*
Roger Young   *Drury College*

Instructive Art Program

The core of every biology textbook is its art program, and the text and illustration authors of *Biology* have worked together to create a dynamic program of full-color illustrations and photographs that support and further clarify the text explanations. Brilliantly rendered and meticulously reviewed for accuracy and consistency, the carefully conceived illustrations and accompanying photos provide concrete, visual reinforcement of the topics discussed throughout the text.

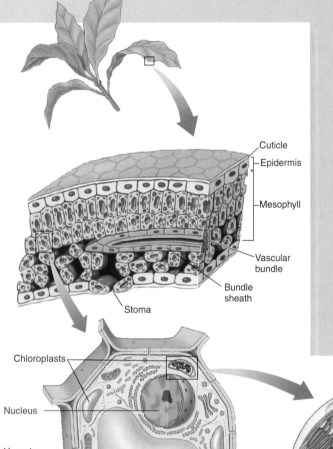

Cuticle
Epidermis
Mesophyll
Vascular bundle
Bundle sheath
Stoma

Multi-Level Perspective

Illustrations depicting complex structures or processes combine macroscopic and microscopic views to help you see the relationship between increasingly detailed images.

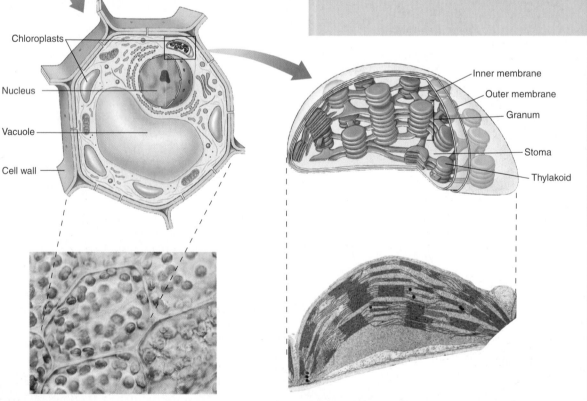

Chloroplasts
Nucleus
Vacuole
Cell wall

Inner membrane
Outer membrane
Granum
Stoma
Thylakoid

Light micrographs, as well as scanning and transmission electron micrographs, are used in conjunction with illustrations to present a true picture of what you would encounter in lab. A micron bar is added whenever the magnification is known.

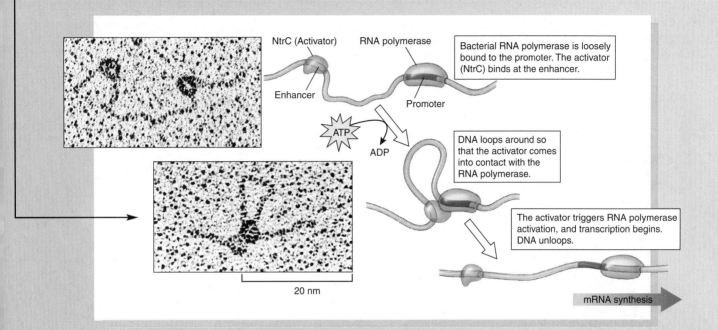

NtrC (Activator)   RNA polymerase

Enhancer   Promoter

Bacterial RNA polymerase is loosely bound to the promoter. The activator (NtrC) binds at the enhancer.

ATP   ADP

DNA loops around so that the activator comes into contact with the RNA polymerase.

The activator triggers RNA polymerase activation, and transcription begins. DNA unloops.

20 nm

mRNA synthesis

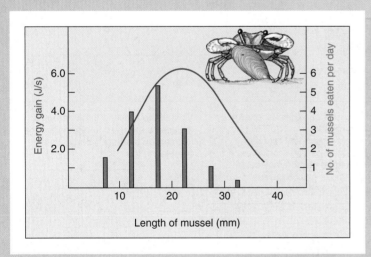

Length of mussel (mm)

FIGURE 52.26
Optimal diet. The shore crab selects a diet of energetically profitable prey. The curve describes the net energy gain (equal to energy gained minus energy expended) derived from feeding on different sizes of mussels. The bar graph shows the numbers of mussels of each size in the diet. Shore crabs tend to feed on those mussels that provide the most energy.
What factors might be responsible for the slight difference in peak prey length relative to the length optimal for maximum energy gain?

Instructive Art Program

Explanatory text boxes describe the action depicted in each step. The discrete, carefully placed boxes guide you through the process, without cluttering the image.

Instructors benefit from this style as well, as the images can be used for presentation without the distraction of extraneous captions.

THE CALVIN CYCLE

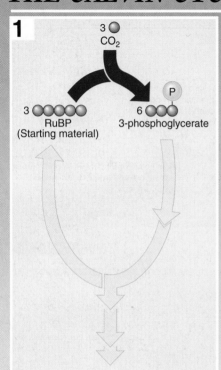

1

The Calvin cycle begins when a carbon atom from a CO_2 molecule is added to a five-carbon molecule (the starting material). The resulting six-carbon molecule is unstable and immediately splits into three-carbon molecules.

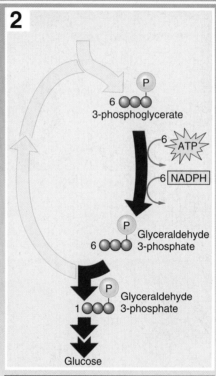

2

Then, through a series of reactions, energy from ATP and hydrogens from NADPH (the products of the light-dependent reactions) are added to the three-carbon molecules. The now-reduced three-carbon molecules either combine to make glucose or are used to make other molecules.

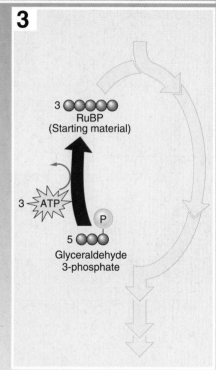

3

Most of the reduced three-carbon molecules are used to regenerate the five-carbon starting material, thus completing the cycle.

FIGURE 10.17
How the Calvin cycle works.

Process Boxes

Process Boxes break down complex processes into a series of small steps, allowing you to track the key occurrences and learn them as you go.

Phylogeny Guideposts

Phylogeny Guideposts are used in the diversity chapters to help you track relationships among diverse organisms. As each group is introduced, the appropriate branch on the phylogenetic tree is highlighted.

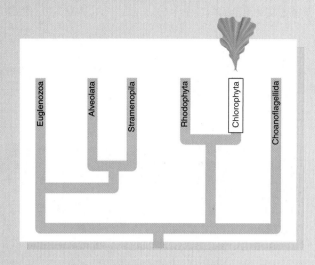

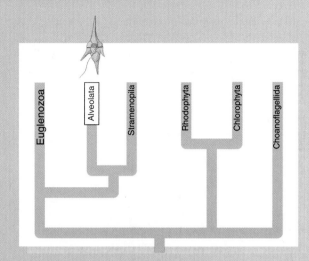

Biochemistry Pathway Icons

These icons are paired with more detailed illustrations to assist you in keeping the big picture in mind when learning complex metabolic processes. The icon highlights which step the main illustration represents, and where that step occurs in the complete process.

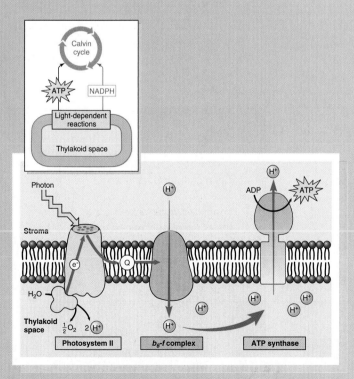

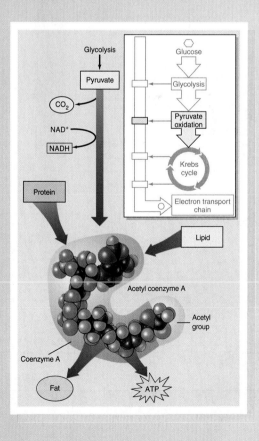

The Learning System

This text is designed to help you learn in a systematic fashion. Simple facts are the building blocks for developing explanations of more complex concepts. The text discussion is presented within a supporting framework of learning aids that help organize studying, reinforce learning, and promote problem-solving skills.

Numbered Headings

The numbered headings employed in the modules form the backbone of the Concept Outline. This consistency makes it easier to identify the key concepts for each chapter, and to then manage the supporting details for each concept.

11

How Cells Divide

Concept Outline

11.1 Prokaryotes divide far more simply than do eukaryotes.

Cell Division in Prokaryotes. Prokaryotic cells divide by splitting in two.

11.2 The chromosomes of eukaryotes are highly ordered structures.

Discovery of Chromosomes. All eukaryotic cells contain chromosomes, but different organisms possess differing numbers of chromosomes.
The Structure of Eukaryotic Chromosomes. Proteins play an important role in packaging DNA in chromosomes.

11.3 Mitosis is a key phase of the cell cycle.

The Cell Cycle. The cell cycle consists of three growth phases, a nuclear division phase, and a cytoplasmic division stage.
Interphase: Preparing for Mitosis. In interphase, the cell grows, replicates its DNA, and prepares for cell division.
Mitosis. In prophase, the chromosomes condense, and microtubules attach sister chromosomes to opposite poles of the cell. In metaphase, the chromosomes align along the center of the cell. In anaphase, the chromosomes separate; in telophase, the spindle dissipates and the nuclear envelope re-forms.
Cytokinesis. In cytokinesis, the cytoplasm separates into two roughly equal halves.

11.4 The cell cycle is carefully controlled.

General Strategies of Cell Cycle Control. At three points in the cell cycle, feedback from the cell determines whether the cycle will continue.
Molecular Mechanisms of Cell Cycle Control. Special proteins regulate the checkpoints of the cell cycle.
Cancer and the Control of Cell Proliferation. Cancer results from damage to genes encoding proteins that regulate the cell division cycle.

11.1 Prokaryotes divide far more simply than do eukaryotes.

Cell Division in Prokaryotes

The end result of cell division in both prokaryotic and eukaryotic cells is two daughter cells, each with the same genetic information as the original cell. The differences between these two basic cell types lead to large differences in how this process occurs. Despite these differences, the essentials of the process are the same: duplication and segregation of genetic information into daughter cells, and division of cellular contents. We will begin by looking at the simpler process, which occurs in prokaryotes: division by **binary fission.**

Most prokaryotes have a genome made up of a single, circular DNA molecule. Despite its apparent simplicity, the DNA molecule of the bacterium *Escherichia coli* is actually on the order of 500 times longer than the cell itself! Thus, this "simple" structure is actually exquisitely packaged to fit into the cell. Although not found in a nucleus, the DNA is in a compacted form called a *nucleoid* that is distinct from the cytoplasm around it.

For many years, it was believed that the *E. coli* DNA molecule was passively segregated by attachment to the membrane and growth of the membrane as the cell elongates. More recently, a more complex picture is emerging that involves both active partitioning of the DNA and formation of a septum that divides the elongated cell in half. Although the details differ, species as different as *E. coli* and *Bacillus subtilis* both exhibit active partitioning of the newly replicated DNA molecules during the division process. This requires both specific sites on the chromosomes and a number of proteins actively involved in the process.

Binary fission begins with the replication of the prokaryotic DNA at a specific site—the origin of replication (see chapter 15)—and proceeds bidirectionally around the circular DNA to a specific site of termination (figure 11.2). Growth of the cell results in elongation, and the newly replicated DNA molecules are actively partitioned to one-fourth and three-quarters of the cell length. This process requires sequences near the origin of replication and results in these sequences being attached to the membrane. The cell itself is partitioned by the growth of new membrane and cell material called a septum (see figure 11.2). This process of septation is complex and under control of the cell as well.

The site of septation is usually the midpoint of the cell and begins with the formation of a ring composed of the molecule FtsZ (figure 11.3). This then results in the accumulation of a number of other proteins, including ones embedded in the membrane. The exact mechanism of septation is not known, but this structure grows inward radially until the cells pinch off into new cells.

FIGURE 11.2
Binary fission. Prior to cell division, the prokaryotic DNA molecule replicates. The replication of the double-stranded, circular DNA molecule *(blue)* that constitutes the genome of a prokaryote begins at a specific site, called the origin of replication. The replication enzymes move out in both directions from that site and make copies *(red)* of each strand in the DNA duplex. The enzymes continue until they meet at another specific site, the terminus of replication. After the DNA is replicated, the cell elongates, and the DNA is partitioned in the cell. Septation then begins, in which new cell membrane material begins to grow and form a septum at approximately the midpoint of the cell. A protein molecule called FtsZ facilitates this process. When the septum is complete, the cell pinches in two, and two daughter cells are formed, each containing a prokaryotic DNA molecule.

208   Part II   Biology of the Cell

FIGURE 11.1
Cell division in prokaryotes. It's hard to imagine fecal coliform bacteria as being beautiful, but here is *Escherichia coli*, inhabitant of the large intestine and the biotechnology lab, spectacularly caught in the act of fission.

All species of organisms—bacteria, alligators, the weeds in a lawn—grow and reproduce. From the smallest creature to the largest, all species produce offspring like themselves and pass on the hereditary information that makes them what they are. In this chapter, we examine how cells divide and reproduce (figure 11.1). The mechanism of cell reproduction and its biological consequences have changed significantly during the evolution of life on earth. The process is complex in eukaryotes, involving both the replication of chromosomes and their separation into daughter cells. Much of what we are learning about the causes of cancer relates to how cells control this process, and in particular their propensity to divide, a mechanism that in broad outline remains the same in all eukaryotes.

207

Concept Outline

Each chapter begins with an outline that gives you an overview of the content contained within that chapter. Reviewing the concept outline before reading the chapter will help focus your attention on the major concepts you should take away from the chapter.

Modular Format

Each page or two-page spread in *Biology* is organized as an independent module, with its own numbered heading at the top of the left-hand page, and a highlighted summary at the bottom of the right-hand page. This system organizes the information in the chapter within a clear conceptual framework, which in turn helps you learn and retain the material.

Section Summaries

Each module ends with a summary intended to reinforce the key concepts from that section. Reviewing the summary after reading the section will indicate whether you learned the main ideas presented in the module.

Vocabulary Boxes

These boxes are found throughout the text, in chapters that require you to learn many new terms. This saves you time when studying, by placing the definitions you need in one location. It is also a handy study tool, as it reinforces the key terms for that chapter.

FIGURE 11.3
The FtsZ protein. In these dividing *E. coli* bacteria, the FtsZ protein is fluorescent, and its location during binary fission can be seen. The protein assembles into a ring at approximately the midpoint of the cell, where it facilitates septation and cell division. Bacteria in which the *ftsZ* gene is mutated cannot divide.

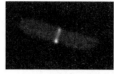 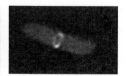

The FtsZ molecule is interesting for a number of reasons. It is highly conserved evolutionarily, having been identified in most prokaryotes, including archaebacteria. It shows some small similarity to eukaryotic tubulin and can form filaments and rings. Recent 3-D crystals show similarity to tubulin as well. It is interesting to speculate that the elaborate spindle found in eukaryotic division may be related to this simple prokaryotic precursor (figure 11.4).

The evolution of eukaryotic cells led to much more complex genomes composed of multiple linear chromosomes housed in a membrane-bounded nucleus. These chromosomes contain even more DNA, and thus pose packaging problems that are solved by DNA being complexed with protein and packaged into functionally distinct chromosomes. This creates more challenges both for the replication of the genome and for its accurate segregation during cell division. The process that evolved to accomplish this segregation of chromosomes is called mitosis.

> Prokaryotes divide by binary fission. Fission begins in the middle of the cell. An active partitioning process ensures that one genome will end up in each daughter cell.

FIGURE 11.4
A comparison of protein assemblies during cell division among different organisms. The prokaryotic protein FtsZ has a structure that is similar to that of the eukaryotic protein tubulin. Tubulin is the protein component of microtubules, which are fibers that play an important role in eukaryotic cell division.

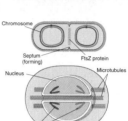
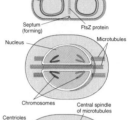
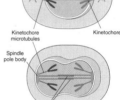
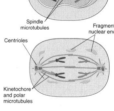

Prokaryotes
No nucleus; single circular chromosome. After DNA is replicated, it is partitioned in the cell. After cell elongation, FtsZ protein assembles into a ring and facilitates septation and cell division.

Some protists
Nucleus present and nuclear envelope remains intact during cell division. Chromosomes linear. Fibers called microtubules, composed of the protein tubulin, pass through tunnels in the nuclear membrane and set up an axis for separation of replicated chromosomes, and cell division.

Other protists
A spindle of microtubules forms between two pairs of centrioles at opposite ends of the cell. The spindle passes through one tunnel in the intact nuclear envelope. Kinetochore microtubules form between kinetochores on the chromosomes and the spindle poles and pull the chromosomes to each pole.

Yeasts
Nuclear envelope remains intact; spindle microtubules form inside the nucleus between spindle pole bodies. A single kinetochore microtubule attaches to each chromosome and pulls each to a pole.

Animals
Spindle microtubules begin to form between centrioles outside of nucleus. As these centrioles move to the poles, the nuclear envelope breaks down, and kinetochore microtubules attach kinetochores of chromosomes to spindle poles. Polar microtubules extend toward the center of the cell and overlap.

Chapter 11 How Cells Divide **209**

A Vocabulary of Cell Division

binary fission Reproduction of a cell by division into two equal or nearly equal parts. Prokaryotes divide by binary fission.

centromere A constricted region of a chromosome about 220 nucleotides in length, composed of highly repeated DNA sequences. During mitosis, the centromere joins the two sister chromatids and is the site to which the kinetochores are attached.

chromatid One of the two copies of a replicated chromosome, joined by a single centromere to the other strand.

chromatin The complex of DNA and proteins of which eukaryotic chromosomes are composed.

chromosome The structure within cells that contains the genes. In eukaryotes, it consists of a single linear DNA molecule associated with proteins. The DNA replicates during S phase, and the replicas separate during M phase.

cytokinesis Division of the cytoplasm of a cell after nuclear division.

euchromatin The portion of a chromosome that is extended except during cell division, and from which RNA is transcribed.

heterochromatin The portion of a chromosome that remains permanently condensed and, therefore, is not transcribed into RNA. Most centromere regions are heterochromatic.

homologues Homologous chromosomes; in diploid cells, one of a pair of chromosomes that carry equivalent genes.

kinetochore A disk of protein bound to the centromere and attached to microtubules during mitosis, linking each chromatid to the spindle apparatus.

microtubule A hollow cylinder, about 25 nanometers in diameter, composed of subunits of the protein tubulin. Microtubules lengthen by the addition of tubulin subunits to their end(s) and shorten by the removal of subunits.

mitosis Nuclear division in which replicated chromosomes separate to form two genetically identical daughter nuclei. When accompanied by cytokinesis, it produces two identical daughter cells.

nucleosome The basic packaging unit of eukaryotic chromosomes, in which the DNA molecule is wound around a cluster of histone proteins. Chromatin is composed of long strings of nucleosomes that resemble beads on a string.

214 Part II Biology of the Cell

The Learning System

Concept Review

An expanded version of the Concept Outline, the Concept Review details each numbered section head followed by its supporting ideas. Each supporting idea is page referenced to allow you to focus your time on areas where you need additional study.

Concept Review

For interactive testing, visit the Online Learning Center with PowerWeb at www.mhhe.com/Raven7

11.1 Prokaryotes divide far more simply than do eukaryotes.

Cell Division in Prokaryotes
- Most prokaryotes have a genome made up of a single, circular DNA molecule, and replicate via binary fusion. (p. 208)
- Binary fusion begins with DNA replication, which starts at the origin site and proceeds bidirectionally around the circular DNA to a specific site of termination. (p. 208)
- The evolution of eukaryotic cells led to much more complex genomes and, thus, new and different ways to replicate and segregate the genome during cell division. (p. 209)

11.2 The chromosomes of eukaryotes are highly ordered structures.

Discovery of Chromosomes
- Chromosomes were first discovered in 1882 by Walther Fleming. (p. 210)
- The number of chromosomes varies from one species to another. Humans have 23 nearly identical pairs for a total of 46 chromosomes. (p. 210)

The Structure of Eukaryotic Chromosomes
- The DNA is a very long, double-stranded fiber extending unbroken through the entire length of the chromosome. A typical human chromosome contains about 140 million nucleotides. (p. 211)
- Every 200 nucleotides, the DNA duplex is coiled around a core of eight histone proteins, forming a nucleosome. (p. 211)
- The particular array of chromosomes an individual possesses is its karyotype. (p. 212)
- The number of different chromosomes a species contains is known as its haploid (*n*) number, and is considered one complete set of chromosomes. (p. 212)
- Humans are diploid, with homologues coming from both [...]

Interphase: Preparing for Mitosis
- The cell grows throughout interphase. The G_1 and G_2 phases are periods of protein synthesis and organelle production, while the S phase is when DNA replication occurs. (p. 214)

Mitosis
- Chromatin condensation continues into prophase. The spindle apparatus is assembled, and sister chromatids are linked to opposite poles of the cell by microtubules. The nuclear envelope breaks down. (p. 215)
- During metaphase, chromosomes align in the center of the cell along the metaphase plate. (p. 215)
- Anaphase begins when centromeres divide, freeing the two sister chromatids from each other. Sister chromatids are pulled to opposite poles as the attached microtubules shorten. (pp. 216–217)
- In telophase, the spindle apparatus disassembles, and the nuclear membrane begins to re-form. (p. 217)

Cytokinesis
- Cytokinesis is the phase of the cell cycle when the cell actually divides. Cytokinesis generally involves the cleavage of the cell into roughly equal halves, forming two daughter cells. (p. 218)

11.4 The cell cycle is carefully controlled.

General Strategies of Cell Cycle Control
- A cell uses three main checkpoints to both assess the internal state of the cell and integrate external signals. The G_1/S checkpoint is the primary point at which the cell decides to divide; the G_2/M checkpoint represents a commitment to mitosis; and the spindle checkpoint ensures that all chromosomes are attached to the spindle in preparation for anaphase. (p. 219)

Molecular Mechanisms of Cell Cycle Control
- Two groups of proteins, cyclins and Cdk's, interact and regulate the cell cycle. (p. 220)
- Cells also receive protein signals (growth factors) that affect cell division. (p. 222)

Cancer and the Control of Cell Proliferation
- Cancer is failure of cell division control. (p. 223)
- It is believed that a malfunction in the *p53* gene may allow cells to go through repeated cell division without being stopped at the appropriate checkpoints. (p. 223)
- Proto-oncogenes are normal cellular genes that become oncogenes when mutated. Proto-oncogenes can encode growth factors, protein relay switches, and kinase enzyme. (p. 224)
- Tumor-suppressor genes can also lead to cancer when they are mutated. (p. 224)

Test Your Understanding

For interactive testing, visit the Online Learning Center with PowerWeb at www.mhhe.com/Raven7

Self Test

1. Bacterial cells divide by
   a. mitosis.
   b. replication.
   c. cytokinesis.
   d. binary fission.
2. Most eukaryotic organisms have _____ chromosomes in their cells.
   a. 1–5
   b. 10–50
   c. 100–500
   d. over 1000
3. Replicate copies of each chromosome are called _____ and are joined at the _____.
   a. homologues/centromere
   b. sister chromatids/kinetochore
   c. sister chromatids/centromere
   d. homologues/kinetochore
4. During which phase of the cell cycle is DNA synthesized?
   a. G_1
   b. G_2
   c. S
   d. M
5. Chromosomes are visible under a light microscope
   a. during mitosis.
   b. during interphase.
   c. when they are attached to their sister chromatids.
   d. All of these are correct.
6. During mitosis, the sister chromatids are separated and pulled to opposite poles during which stage?
   a. interphase
   b. metaphase
   c. anaphase
   d. telophase
7. Cytokinesis is
   a. the same process in plant and animal cells.
   b. the separation of cytoplasm and the formation of two cells.
   c. the final stage of mitosis.
   d. the movement of kinetochores.
8. The eukaryotic cell cycle is controlled at several points; which of these statements is *not* true?
   a. Cell growth is assessed at the G_1/S checkpoint.
   b. DNA replication is assessed at the G_2/M checkpoint.
   c. Environmental conditions are assessed at the G_0 checkpoint.
   d. The chromosomes are assessed at the spindle checkpoint.
9. What proteins are used to control cell growth specifically in *multicellular* eukaryotic organisms?
   a. Cdk
   b. MPF
   c. cyclins
   d. growth factors
10. What causes cancer in cells?
    a. damage to genes
    b. chemical damage to cell membranes
    c. UV damage to transport proteins
    d. All of these cause cancer in cells.

Test Your Visual Understanding

a   b

c   d

e

1. Match the mitotic and cell cycle phases with the appropriate figure.
   anaphase
   interphase
   metaphase
   prophase
   telophase

Apply Your Knowledge

1. An ancient plant called horsetail contains 216 chromosomes. How many homologous pairs of chromosomes does it contain? How many chromosomes are present in its cells during metaphase?
2. Colchicine is a poison that binds to tubulin and prevents its assembly into microtubules; cytochalasins are compounds that bind to the ends of actin filaments and prevent their elongation. What effects would these two substances have on cell division in animal cells?
3. If you could construct an artificial chromosome, what elements would you introduce into it, at a minimum, so that it could function normally in mitosis?

226   Part II   Biology of the Cell

Testing Yourself

Each chapter concludes with a set of questions designed to test your knowledge of the content, including multiple choice questions, illustration-based questions, and application questions. Answers to these questions are found on the *Biology* Online Learning Center at: *www.mhhe.com/raven7*. At the site you can take an interactive version of the end-of-chapter quiz that provides you with hints and instructional feedback.

BIOLOGY

1

The Science of Biology

Concept Outline

1.1 Biology is the science of life.

Organization of Living Things. Biology is the science that studies living organisms and how they interact with one another and with their environment. There are several characteristics that define life.

1.2 Scientists form generalizations from observations.

The Nature of Science. Science employs both deductive reasoning and inductive reasoning.

How Science Is Done. Scientists construct hypotheses from systematically collected objective data. They then perform experiments designed to disprove the hypotheses.

1.3 Darwin's theory of evolution illustrates how science works.

Charles Darwin. On a round-the-world voyage, Darwin made observations that eventually led him to formulate the hypothesis of evolution by natural selection.

Darwin's Evidence. The fossil and geographic patterns of life he observed convinced Darwin that a process of evolution had occurred.

Inventing the Hypothesis of Natural Selection. The Malthus idea that populations cannot grow unchecked led Darwin, and another naturalist named Wallace, to propose the hypothesis of natural selection.

Evolution After Darwin: More Evidence. In the century since Darwin, a mass of scientific advances and discoveries have supported his theory of evolution, which is now accepted by practically all practicing biologists.

1.4 Four themes unify biology as a science.

Core Themes Unite Biology. Living things all exhibit cellular organization, a mechanism for heredity (DNA), adaptation to produce unique features as the result of evolution, and the conservation of key features during evolution.

FIGURE 1.1
A replica of the *Beagle*, off the southern coast of South America. The famous English naturalist Charles Darwin set forth on H.M.S. *Beagle* in 1831, at the age of 22.

You are about to embark on a journey—a journey of discovery about the nature of life. Nearly 180 years ago, a young English naturalist named Charles Darwin set sail on a similar journey on board H.M.S. *Beagle*; figure 1.1 shows a replica of the *Beagle*. What Darwin learned on his five-year voyage led directly to his development of the theory of evolution by natural selection, a theory that has become the core of the science of biology. Darwin's voyage seems a fitting place to begin our exploration of biology, the scientific study of living organisms and how they have evolved. Before we begin, however, let's take a moment to think about what biology is and why it's important.

Part I *The Origin of Living Things*

1.1 Biology is the science of life.

Organization of Living Things

In its broadest sense, biology is the study of living things—*the science of life*. Living things come in an astounding variety of shapes and forms, and biologists study life in many different ways. They live with gorillas, collect fossils, and listen to whales. They read the messages encoded in the long molecules of heredity and count how many times a hummingbird's wings beat each second.

Properties of Life

What makes something "alive"? Anyone could deduce that a galloping horse is alive and a car is not, but why? We cannot say, "If it moves, it's alive," because a car can move, and gelatin can wiggle in a bowl. They certainly are not alive. What characteristics *do* define life? All living organisms share a family of basic characteristics:

1. **Cellular organization.** All organisms consist of one or more cells. Often too tiny to see, cells carry out the basic activities of living. Each cell is bounded by a membrane that separates it from its surroundings.
2. **Order.** All living things are highly ordered. Your body is composed of many different kinds of cells, each containing many complex molecular structures.
3. **Sensitivity.** All organisms respond to stimuli. Plants grow toward a source of light, and your pupils dilate when you walk into a dark room.
4. **Growth, development, and reproduction.** All organisms are capable of growing and reproducing, and they all possess hereditary molecules that are passed to their offspring, ensuring that the offspring are of the same species.
5. **Energy utilization.** All organisms take in energy and use it to perform many kinds of work. Every muscle in your body is powered with energy you obtain from the food you eat.
6. **Evolutionary adaptation.** All organisms interact with other organisms and the environment in ways that influence survival, and as a consequence, organisms evolve adaptations to their environments.
7. **Homeostasis.** All organisms maintain relatively constant internal conditions, different from their environment, a process called homeostasis.

Hierarchical Organization

The organization of the biological world is hierarchical—that is, each level builds on the level below it.

The Cellular Level. At the cellular level (figure 1.2), atoms, the fundamental elements of matter, are joined together into clusters called **molecules.** Complex biological molecules are assembled into tiny structures called **organelles** within membrane-bounded units we call **cells.**

The cell is the basic unit of life. Many organisms are composed of single cells. Bacteria are single cells, for example. All animals and plants, as well as most fungi and algae, are multicellular—composed of more than one cell.

The Organismal Level. Cells are organized into three levels of organization. The most basic level is that of **tissues,** which are groups of similar cells that act as a functional unit. Tissues, in turn, are grouped into **organs,** which are body structures composed of several different tissues grouped together in a structural and functional unit. Your brain is an organ composed of nerve cells and a variety of connective tissues that form protective coverings and contribute blood. At the third level of organization, organs are grouped into **organ systems.** The nervous system, for example, consists of sensory organs, the brain and spinal cord, and neurons that convey signals to and from them.

The Populational Level. Individual organisms are organized into several hierarchical levels within the living world. The most basic of these is the **population,** which is a group of organisms of the same species living in the same place. All the populations of a particular kind of organism together form a **species,** its members similar in appearance and able to interbreed. At a higher level of biological organization, a **biological community** consists of all the populations of different species living together in one place.

At the highest tier of biological organization, a biological community and the physical habitat within which it lives together constitute an ecological system, or **ecosystem.** For example, the soil and water of a mountain ecosystem interact with the biological community of a mountain meadow in many important ways.

Emergent Properties

At each higher level in the living hierarchy, novel properties emerge. These **emergent properties** result from the way in which components interact, and often cannot be guessed just by looking at the parts themselves. Examining the cells gives little clue of what the animal is like. You have the same array of cell types as a giraffe. It is because the living world exhibits many emergent properties that it is difficult to define "life."

All living things share certain key characteristics including: cellular organization, sensitivity, growth, development and reproduction, adaptation, and homeostasis.

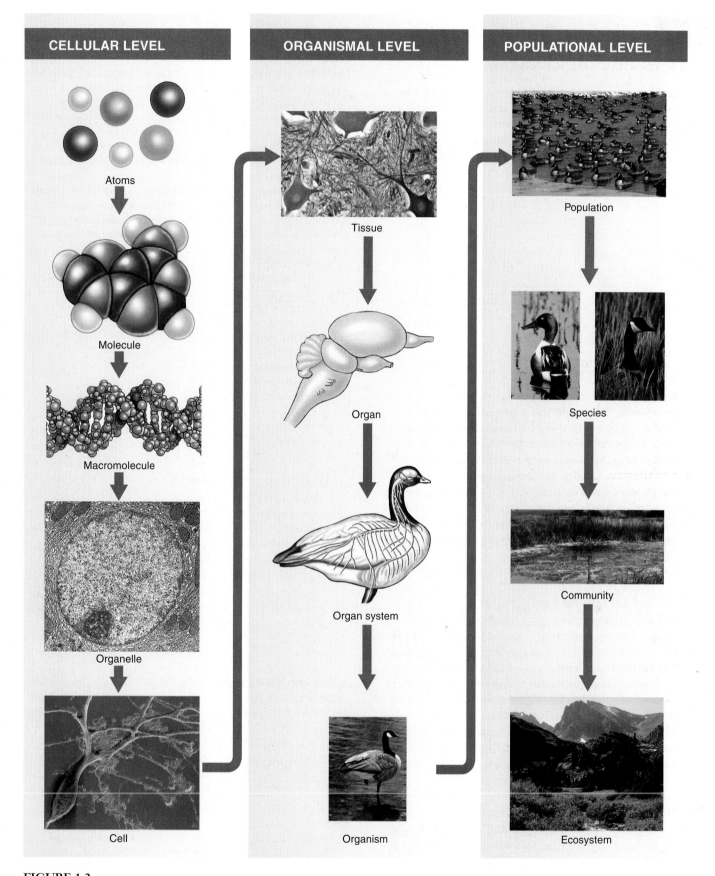

FIGURE 1.2
Hierarchical organization of living things.   Life is highly organized—from small and simple to large and complex, within cells, within multicellular organisms, and among populations of organisms.

1.2 Scientists form generalizations from observations.

The Nature of Science

Biology is a fascinating and important subject because it dramatically affects our daily lives and our futures. Many biologists are working on problems that critically affect our lives, such as the world's rapidly expanding population and diseases like cancer and AIDS. The knowledge these biologists gain will be fundamental to our ability to manage the world's resources in a suitable manner, to prevent or cure diseases, and to improve the quality of our lives and those of our children and grandchildren.

Biology is one of the most successful of the "natural sciences," those devoted to explaining what our world is like. To understand biology, you must first understand the nature of science. Because the basic tool a scientist uses is thought, to understand the nature of science, it is useful to focus for a moment on how scientists think. They reason in two ways: deductively and inductively.

Deductive Reasoning

Deductive reasoning applies general principles to predict specific results. The logic flows from the general to the specific. Over 2200 years ago, the Greek Eratosthenes used Euclidean geometry and deductive reasoning to accurately estimate the circumference of the earth (figure 1.3). This sort of analysis of specific cases using general principles is an example of deductive reasoning. It is the reasoning of mathematics and philosophy and is used to test the validity of general ideas in all branches of knowledge. A biologist uses deductive reasoning to infer the species of a specimen from its characteristics.

Inductive Reasoning

In **inductive reasoning,** the logic flows in the opposite direction, from the specific to the general. Inductive reasoning uses specific observations to construct general scientific principles. If cats possess hair, and dogs possess hair, and every other mammal you observe has hair, then you may infer that perhaps *all* mammals have hair. Inductive reasoning leads to generalizations that can then be tested.

Webster's Dictionary defines science as systematized knowledge derived from observation and experiment carried on to determine the principles underlying what is being studied. In other words, a scientist determines principles from observations, discovering general principles by carefully examining specific cases. Inductive reasoning first became important to science in the 1600s in Europe, when Francis Bacon, Isaac Newton, and others began to use the results of particular experiments to infer general principles about how the world operates. If you release an apple from

FIGURE 1.3

Deductive reasoning: How Eratosthenes estimated the circumference of the earth using deductive reasoning. **1.** On a day when sunlight shone straight down a deep well at Syene in Egypt, Eratosthenes measured the length of the shadow cast by a tall obelisk in the city of Alexandria, about 800 kilometers away. **2.** The shadow's length and the obelisk's height formed two sides of a triangle. Using the recently developed principles of Euclidean geometry, Eratosthenes calculated the angle, a, to be 7° and 12′, exactly $\frac{1}{50}$ of a circle (360°). **3.** If angle $a = \frac{1}{50}$ of a circle, then the distance between the obelisk (in Alexandria) and the well (in Syene) must equal $\frac{1}{50}$ of the circumference of the earth. **4.** Eratosthenes had heard that it was a 50-day camel trip from Alexandria to Syene. Assuming that a camel travels about 18.5 kilometers per day, he estimated the distance between obelisk and well as 925 kilometers (using different units of measure, of course). **5.** Eratosthenes thus deduced the circumference of the earth to be 50 × 925 = 46,250 kilometers. Modern measurements put the distance from the well to the obelisk at just over 800 kilometers. Employing a distance of 800 kilometers, Eratosthenes's value would have been 50 × 800 = 40,000 kilometers. The actual circumference is 40,075 kilometers.

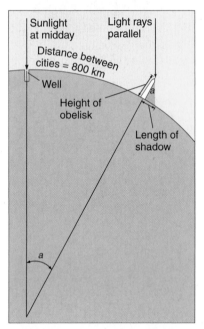

your hand, what happens? The apple falls to the ground. From a host of simple, specific observations like this, Newton inferred a general principle: All objects fall toward the center of the earth. What Newton did was construct a mental model of how the world works, a family of general principles consistent with what he could see and learn. Scientists do the same today. They use specific observations to build general models, and then test the models to see how well they work.

Science is a way of viewing the world that focuses on objective information, putting that information to work to build understanding.

How Science Is Done

How do scientists establish which general principles are true from among the many that might be true? They do this by systematically testing alternative proposals. If these proposals prove inconsistent with experimental observations, they are rejected as untrue. Figure 1.4 illustrates the process. After making careful observations concerning a particular area of science, scientists construct a **hypothesis,** which is a suggested explanation that accounts for those observations. A hypothesis is a proposition that might be true. Those hypotheses that have not yet been disproved are retained. They are useful because they fit the known facts, but they are always subject to future rejection if, in the light of new information, they are found to be incorrect.

Testing Hypotheses

We call the test of a hypothesis an **experiment.** Suppose that a room appears dark to you. To understand why it appears dark, you propose several hypotheses. The first might be, "There is no light in the room because the light switch is turned off." An alternative hypothesis might be, "There is no light in the room because the lightbulb is burned out." And yet another alternative hypothesis might be, "I am going blind." To evaluate these hypotheses, you would conduct an experiment designed to eliminate one or more of the hypotheses. For example, you might test your hypotheses by reversing the position of the light switch. If you do so and the light does not come on, you have disproved the first hypothesis. Something other than the setting of the light switch must be the reason for the darkness. Note that a test such as this does not prove that any of the other hypotheses are true; it merely demonstrates that one of them is not. A successful experiment is one in which one or more of the alternative hypotheses is demonstrated to be inconsistent with the results and is thus rejected.

As you proceed through this text, you will encounter many hypotheses that have withstood the test of experiment. Many will continue to do so; others will be revised as new observations are made by biologists. Biology, like all science, is in a constant state of change, with new ideas appearing and replacing old ones.

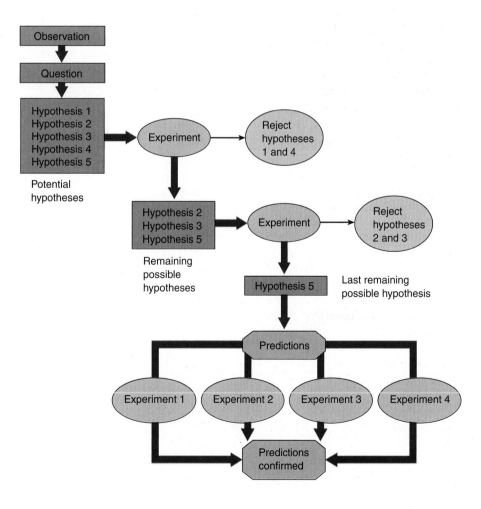

FIGURE 1.4

How science is done. This diagram illustrates how scientific investigations proceed. First, scientists make observations that raise a particular question. They develop a number of potential explanations (hypotheses) to answer the question. Next, they carry out experiments in an attempt to eliminate one or more of these hypotheses. Then, predictions are made based on the remaining hypotheses, and further experiments are carried out to test these predictions. As a result of this process, the least unlikely hypothesis is selected.

Establishing Controls

Often we are interested in learning about processes that are influenced by many factors, or **variables.** To evaluate alternative hypotheses about one variable, all other variables must be kept constant. This is done by carrying out two experiments in parallel: In the first experiment, one variable is altered in a specific way to test a particular hypothesis; in the second experiment, called the **control experiment,** that variable is left unaltered. In all other respects the two experiments are identical, so any difference in the outcomes of the two experiments must result from the influence of the variable that was changed. Much of the challenge of experimental science lies in designing control experiments that isolate a particular variable from other factors that might influence a process.

Using Predictions

A successful scientific hypothesis needs to be not only valid but useful—it needs to tell you something you want to know. A hypothesis is most useful when it makes predictions, because those predictions provide a way to test the validity of the hypothesis. If an experiment produces results inconsistent with the predictions, the hypothesis must be rejected. On the other hand, if the predictions are supported by experimental testing, the hypothesis is supported. The more experimentally supported predictions a hypothesis makes, the more valid the hypothesis is. For example, Einstein's hypothesis of relativity was at first provisionally accepted because no one could devise an experiment that invalidated it. The hypothesis made a clear prediction: that the sun would bend the path of light passing by it. When this prediction was tested in a total eclipse, the light from background stars was indeed bent. Because this result was unknown when the hypothesis was being formulated, it provided strong support for the hypothesis, which was then accepted with more confidence.

Developing Theories

Scientists use the word **theory** in two main ways. A "theory" is a proposed explanation for some natural phenomenon, often based on some general principle. Thus we speak of the principle first proposed by Newton as the "theory of gravity." Such theories often bring together concepts that were previously thought to be unrelated, and offer unified explanations of different phenomena. Newton's theory of gravity provided a single explanation for objects falling to the ground and the orbits of planets around the sun. "Theory" is also used to mean the body of interconnected concepts, supported by scientific reasoning and experimental evidence, that explains the facts in some area of study. Such a theory provides an indispensable framework for organizing a body of knowledge. For example, quantum theory in physics brings together a set of ideas about the nature of the universe, explains experimental facts, and serves as a guide to further questions and experiments.

To a scientist, theories are the solid ground of science, that of which we are most certain. In contrast, to the general public, "theory" implies just the opposite—a *lack* of knowledge, or a guess. Not surprisingly, this difference often results in confusion. In this text, theory will always be used in its scientific sense, in reference to an accepted general principle or body of knowledge.

To suggest, as many critics outside of science do, that evolution is "just a theory" is misleading. The hypothesis that evolution has occurred is an accepted scientific fact; it is supported by overwhelming evidence. Modern evolutionary theory is a complex body of ideas whose importance spreads far beyond explaining evolution; its ramifications permeate all areas of biology, and it provides the conceptual framework that unifies biology as a science.

Research and the Scientific Method

It used to be fashionable to speak of the "scientific method" as consisting of an orderly sequence of logical, either/or steps. Each step would reject one of two mutually incompatible alternatives, as if trial-and-error testing would inevitably lead a researcher through the maze of uncertainty that always impedes scientific progress. If this were indeed so, a computer would make a good scientist. But science is not done this way. As the British philosopher Karl Popper has pointed out, successful scientists without exception design their experiments with a pretty fair idea of how the results are going to come out. They have what Popper calls an "imaginative preconception" of what the truth might be. A hypothesis that a successful scientist tests is not just any hypothesis; rather, it is an educated guess or a hunch, in which the scientist integrates all that he or she knows and allows his or her imagination full play, in an attempt to get a sense of what *might* be true (see Box: How Biologists Do Their Work). It is because insight and imagination play such a large role in scientific progress that some scientists are so much better at science than others, just as Beethoven and Mozart stand out among most other composers.

Some scientists perform what is called *basic research,* which is intended to extend the boundaries of what we know. These individuals typically work at universities, and their research is usually financially supported by their institutions and by external sources, such as the government, industry, and private foundations. Basic research is as diverse as its name implies. Some basic scientists attempt to find out how certain cells take up specific chemicals, while others count the number of dents in tiger teeth. The information generated by basic research contributes to the growing body of scientific knowledge, and it provides the scientific foundation utilized by *applied research.* Scientists who conduct applied research are often employed in some kind of industry. Their work may

How Biologists Do Their Work

The Consent

Late in November, on a single night
Not even near to freezing, the ginkgo trees
That stand along the walk drop all their leaves
In one consent, and neither to rain nor to wind
But as though to time alone: the golden and green
Leaves litter the lawn today, that yesterday
Had spread aloft their fluttering fans of light.
What signal from the stars? What senses took it in?
What in those wooden motives so decided
To strike their leaves, to down their leaves,
Rebellion or surrender? And if this
Can happen thus, what race shall be exempt?
What use to learn the lessons taught by time,
If a star at any time may tell us: Now.

                              Howard Nemerov

What is bothering the poet Howard Nemerov is that life is influenced by forces he cannot control or even identify. It is the job of biologists to solve puzzles such as the one he poses, to identify and try to understand those things that influence life.

Nemerov asks why ginkgo trees (figure 1.A) drop all their leaves at once. To find an answer to questions such as this, biologists and other scientists pose *possible* answers and then try to determine which answers are false. Tests of alternative possibilities are

FIGURE 1.A
A ginkgo tree.

called experiments. To learn why the ginkgo trees drop all their leaves simultaneously, a scientist would first formulate several possible answers, called hypotheses:

Hypothesis 1: Ginkgo trees possess an internal clock that times the release of leaves to match the season. On the day Nemerov describes, this clock sends a "drop" signal (perhaps a chemical) to all the leaves at the same time.

Hypothesis 2: The individual leaves of ginkgo trees are each able to sense day length, and when the days get short enough in the fall, each leaf responds independently by falling.

Hypothesis 3: A strong wind arose the night before Nemerov made his observation, blowing all the leaves off the ginkgo trees.

Next, the scientist attempts to eliminate one or more of the hypotheses by conducting an experiment. In this case, one might cover some of the leaves so that they cannot use light to sense day length. If hypothesis 2 is true, then the covered leaves should not fall when the others do, because they are not receiving the same information. Suppose, however, that despite the covering of some of the leaves, all the leaves still fall together. This result would eliminate hypothesis 2 as a possibility. Either of the other hypotheses, and many others, remain possible.

This simple experiment with ginkgoes points out the essence of scientific progress: Science does not prove that certain explanations are true; rather, it proves that others are not. Hypotheses that are inconsistent with experimental results are rejected, while hypotheses that are not proven false by an experiment are provisionally accepted. However, hypotheses may be rejected in the future when more information becomes available, if they are inconsistent with the new information. Just as you can find the correct path through a maze by trying and eliminating false paths, scientists work to find the correct explanations of natural phenomena by eliminating false possibilities.

involve the manufacture of food additives, the creation of new drugs, or the testing of environmental quality.

After developing a hypothesis and performing a series of experiments, a scientist writes a paper carefully describing the experiment and its results. He or she then submits the paper for publication in a scientific journal, but before it is published, it must be reviewed and accepted by other scientists who are familiar with that particular field of research. This process of careful evaluation, called *peer review*, lies at the heart of modern science, fostering careful work, precise description, and thoughtful analysis. When an important discovery is announced in a paper, other scientists attempt to reproduce the result, providing a check on accuracy and honesty. Nonreproducible results are not taken seriously for long.

The explosive growth in scientific research during the second half of the twentieth century is reflected in the enormous number of scientific journals now in existence. Although some, such as *Science* and *Nature*, are devoted to a wide range of scientific disciplines, most are extremely specialized: *Cell Motility and the Cytoskeleton, Glycoconjugate Journal, Mutation Research,* and *Synapse* are just a few examples.

> **The scientific process involves rejecting hypotheses that are inconsistent with experimental results or observations. Hypotheses that are consistent with available data are conditionally accepted. The formulation of a hypothesis often involves creative insight.**

1.3 Darwin's theory of evolution illustrates how science works.

Charles Darwin

Darwin's theory of evolution explains and describes how organisms on earth have changed over time and acquired a diversity of new forms. This famous theory provides a good example of how a scientist develops a hypothesis and how a scientific theory grows and wins acceptance.

Charles Robert Darwin (1809–1882; figure 1.5) was an English naturalist who, after 30 years of study and observation, wrote one of the most famous and influential books of all time. This book, *On the Origin of Species by Means of Natural Selection, or The Preservation of Favoured Races in the Struggle for Life*, created a sensation when it was published, and the ideas Darwin expressed in it have played a central role in the development of human thought ever since.

In Darwin's time, most people believed that the various kinds of organisms and their individual structures resulted from direct actions of the Creator (and to this day many people still believe this). Species were thought to be specially created and unchangeable, or immutable, over the course of time. In contrast to these views, a number of earlier philosophers had presented the view that living things must have changed during the history of life on earth. Darwin proposed a concept he called natural selection as a coherent, logical explanation for this process, and he brought his ideas to wide public attention.

Darwin's book, as its title indicates, presented a conclusion that differed sharply from conventional wisdom. Although his theory did not directly challenge the existence of a Divine Creator, Darwin argued that the operation of *natural* laws produced change over time, or **evolution.** These views put Darwin at odds with most people of his day, who believed in a literal interpretation of the Bible and thus accepted the idea of a fixed and constant world, largely unchanged since it was created by God.

The story of Darwin and his theory begins in 1831, when he was 22 years old. On the recommendation of one of his professors at Cambridge University, he was selected to serve as naturalist on a five-year navigational mapping expedition around the coasts of South America (figure 1.6), aboard H.M.S. *Beagle* (figure 1.7). During this long voyage, Darwin had the chance to study a wide variety of plants and animals on continents and islands

FIGURE 1.5
Charles Darwin. This newly rediscovered photograph taken in 1881, the year before Darwin died, appears to be the last ever taken of the great biologist.

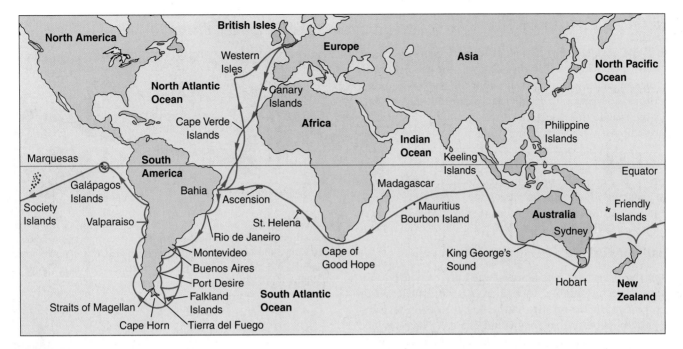

FIGURE 1.6
The five-year voyage of H.M.S. *Beagle*. Most of the time was spent exploring the coasts and coastal islands of South America, such as the Galápagos Islands. Darwin's studies of the animals of the Galápagos Islands played a key role in his eventual development of the concept of evolution by means of natural selection.

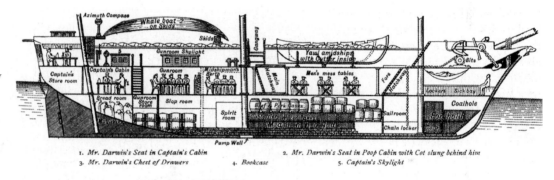

FIGURE 1.7
Cross section of the *Beagle*. A 10-gun brig of 242 tons, only 90 feet in length, the *Beagle* had a crew of 74 people! After he first saw the ship, Darwin wrote to his college professor Henslow: "The absolute want of room is an evil that nothing can surmount."

and in distant seas. He was able to explore the biological richness of the tropical forests, examine the extraordinary fossils of huge extinct mammals in Patagonia at the southern tip of South America, and observe the remarkable series of related but distinct forms of life on the Galápagos Islands, off the west coast of South America. Such an opportunity clearly played an important role in the development of his thoughts about the nature of life on earth.

When Darwin returned from the voyage at the age of 27, he began a long period of study and contemplation. During the next 10 years, he published important books on several different subjects, including the formation of oceanic islands from coral reefs and the geology of South America. He also devoted eight years of study to barnacles, a group of small marine animals with shells that inhabit rocks and pilings, eventually writing a four-volume work on their classification and natural history. In 1842, Darwin and his family moved out of London to a country home at Down, in the county of Kent. In these pleasant surroundings, Darwin lived, studied, and wrote for the next 40 years.

Darwin was the first to propose natural selection as an explanation for the mechanism of evolution that produced the diversity of life on earth. His hypothesis grew from his observations on a five-year voyage around the world.

Darwin's Evidence

One of the obstacles that had blocked the acceptance of any theory of evolution in Darwin's day was the incorrect notion, widely believed at that time, that the earth was only a few thousand years old. Evidence discovered during Darwin's time made this assertion seem less and less likely. The great geologist Charles Lyell (1797–1875), whose *Principles of Geology* (1830) Darwin read eagerly as he sailed on the *Beagle*, outlined for the first time the story of an ancient world of plants and animals in flux. In this world, species were constantly becoming extinct while others were emerging. It was this world that Darwin sought to explain.

What Darwin Saw

When the *Beagle* set sail, Darwin was fully convinced that species were immutable. Indeed, it was not until two or three years after his return that he began to consider seriously the possibility that they could change. Nevertheless, during his five years on the ship, Darwin observed a number of phenomena that were of central importance to him in reaching his ultimate conclusion. For example, in the rich fossil beds of southern South America, he observed fossils of extinct armadillos similar to the armadillos that still lived in the same area (figure 1.8). Why would similar living and fossil organisms be in the same area unless the earlier form had given rise to the other?

Repeatedly, Darwin saw that the characteristics of similar species varied somewhat from place to place. These geographical patterns suggested to him that organismal lineages change gradually as species migrate from one area to another. On the Galápagos Islands, 900 kilometers (540 miles) off the coast of Ecuador, Darwin encountered a variety of different finches on the various islands. The 14 species, although related, differed slightly in appearance, particularly in their beaks (figure 1.9). Darwin felt it most reasonable to assume all these birds had descended from a common ancestor blown by winds from the South Ameri-

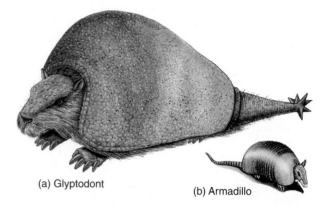

FIGURE 1.8
Fossil evidence of evolution. The now-extinct glyptodont (*a*) was a 2000-kilogram South American armadillo, much larger than the modern armadillo (*b*), which weighs an average of about 4.5 kilograms. (Drawings are not to scale.)

can mainland several million years ago. Eating different foods on different islands, the species had changed during their descent—"descent with modification," or evolution. These finches are discussed in more detail on pages 454 and 483.

In a more general sense, Darwin was struck by the fact that the plants and animals on these relatively young volcanic islands resembled those on the nearby coast of South America. If each one of these plants and animals had been created independently and simply placed on the Galápagos Islands, why didn't they resemble the plants and animals of islands with similar climates, such as those off the coast of Africa, for example? Why did they resemble those of the adjacent South American coast instead?

The fossils and patterns of life that Darwin observed on the voyage of the *Beagle* eventually convinced him that evolution had taken place.

Large ground finch (seeds)

Cactus ground finch
(cactus fruits and flowers)

Vegetarian finch (buds)

Woodpecker finch (insects)

FIGURE 1.9
Four Galápagos finches and what they eat. On the Galápagos Islands, Darwin observed 14 different species of finches differing mainly in their beaks and feeding habits. These four finches eat very different food items, and Darwin surmised that the different shapes of their bills represented evolutionary adaptations that improved their ability to eat the foods available in their specific habitats.

Inventing the Hypothesis of Natural Selection

It is one thing to observe the results of evolution, but quite another to understand how it happens. Darwin's great achievement lies in his formulation of the hypothesis that evolution occurs because of natural selection.

Darwin and Malthus

Of key importance to the development of Darwin's insight was his study of Thomas Malthus's *Essay on the Principle of Population* (1798). In his book, Malthus pointed out that populations of plants and animals (including human beings) tend to increase geometrically, while humans are able to increase their food supply only arithmetically. A *geometric progression* is one in which the elements increase by a constant *factor*; for example, in the progression 2, 6, 18, 54, . . . , each number is three times the preceding one. An *arithmetic progression*, in contrast, is one in which the elements increase by a constant *difference*; in the progression 2, 4, 6, 8, . . . , each number is two greater than the preceding one (figure 1.10).

Because populations increase geometrically, virtually any kind of animal or plant, if it could reproduce unchecked, would cover the entire surface of the world within a surprisingly short time. Instead, populations of species remain fairly constant year after year, because death limits population numbers. Malthus's conclusion provided the key ingredient that was necessary for Darwin to develop the hypothesis that evolution occurs by natural selection.

Sparked by Malthus's ideas, Darwin saw that although every organism has the potential to produce more offspring than can survive, only a limited number actually do survive and produce further offspring. Combining this observation with what he had seen on the voyage of the *Beagle*, as well as with his own experiences in breeding domestic animals, Darwin made an important association (figure 1.11): Those individuals that possess superior physical, behavioral, or other attributes are more likely to survive than those that are not so well endowed. By surviving, they gain the opportunity to pass on their favorable characteristics to their offspring. As the frequency of these characteristics increases in the population, the nature of the population as a whole will gradually change. Darwin called this process selection. The driving force he identified has often been referred to as survival of the fittest.

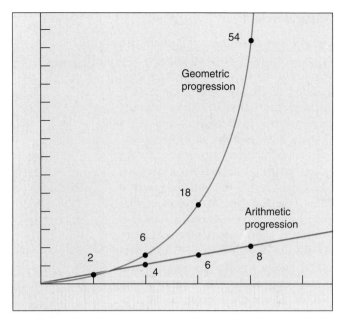

FIGURE 1.10
Geometric and arithmetic progressions. A geometric progression increases by a constant factor (for example, × 2 or × 3 or × 4), while an arithmetic progression increases by a constant difference (for example, units of 1 or 2 or 3). Malthus contended that the human growth curve was geometric, but the human food production curve was only arithmetic.
What is the effect of reducing the constant factor by which the geometric progression increases? Might this effect be achieved with humans? How?

"Can we doubt . . . that individuals having any advantage, however slight, over others, would have the best chance of surviving and procreating their kind? On the other hand, we may feel sure that any variation in the least degree injurious would be rigidly destroyed. This preservation of favorable variations, I call Natural Selection."

FIGURE 1.11
An excerpt from Charles Darwin's *On the Origin of Species*.

Natural Selection

Darwin was thoroughly familiar with variation in domesticated animals and began *On the Origin of Species* with a detailed discussion of pigeon breeding. He knew that breeders selected certain varieties of pigeons and other animals, such as dogs, to produce certain characteristics, a process Darwin called **artificial selection.** Once this had been done, the animals would breed true for the characteristics that had been selected. Darwin had also observed that the differences purposely developed between domesticated races or breeds were often greater than those that separated wild species. Domestic pigeon breeds, for example, show much greater variety than all of the hundreds of wild species of pigeons found throughout the world. Such relationships suggested to Darwin that evolutionary change could occur in nature too. Surely if pigeon breeders could foster such variation by artificial selection, nature could do the same, playing the breeder's role in selecting the next generation—a process Darwin called **natural selection.**

Darwin's theory incorporates the hypothesis of evolution, the process of natural selection, and the mass of new evidence for both evolution and natural selection that he had compiled. Thus, Darwin's theory provides a simple and direct explanation of biological diversity, or why animals are different in different places. Because habitats differ in their requirements and opportunities, the organisms with characteristics favored locally by natural selection will tend to vary in different places.

Darwin Drafts His Argument

Darwin drafted the overall argument for evolution by natural selection in a preliminary manuscript in 1842. After showing the manuscript to a few of his closest scientific friends, however, Darwin put it in a drawer, and for 16 years turned to other research. No one knows for sure why Darwin did not publish his initial manuscript—it is very thorough and outlines his ideas in detail. Some historians have suggested that Darwin was shy of igniting public criticism of his evolutionary ideas because there could have been little doubt in his mind that his hypothesis of evolution by natural selection would spark controversy. Others have proposed that Darwin was simply re-

FIGURE 1.12
Darwin greets his monkey ancestor.
In his time, Darwin was often portrayed unsympathetically, as in this drawing from an 1874 publication.

fining his hypothesis all those years, although there is little evidence that he altered his initial manuscript in all that time.

Wallace Has the Same Idea

The stimulus that finally brought Darwin's hypothesis into print was an essay he received in 1858. A young English naturalist named Alfred Russel Wallace (1823–1913) sent the essay to Darwin from Malaysia; it concisely set forth the hypothesis of evolution by means of natural selection, a hypothesis Wallace had developed independently of Darwin. Like Darwin, Wallace had been greatly influenced by Malthus's 1798 essay. Colleagues of Wallace, knowing of Darwin's work, encouraged him to communicate with Darwin. After receiving Wallace's essay, Darwin arranged for a joint presentation of their ideas at a seminar in London. Darwin then completed his own book, expanding the 1842 manuscript he had written so long ago, and submitted it for publication.

Publication of Darwin's Hypothesis

Darwin's book appeared in November 1859 and caused an immediate sensation. Many people were deeply disturbed by the suggestion that human beings were descended from the same ancestor as apes (figure 1.12). Darwin did not actually discuss this idea in his book, but it followed directly from the principles he outlined. In a subsequent book, *The Descent of Man*, Darwin presented the argument directly, building a powerful case that humans and living apes have common ancestors. Although people had long accepted that humans closely resembled apes in many characteristics, the possibility that a direct evolutionary relationship might exist was unacceptable to many. Darwin's arguments for the theory of evolution by natural selection were so compelling, however, that his views were almost completely accepted within the intellectual community of Great Britain after the 1860s.

The fact that populations do not really expand geometrically implies that nature acts to limit population numbers. The traits of organisms that survive to produce more offspring will be more common in future generations—a process Darwin called natural selection.

Evolution After Darwin: More Evidence

More than a century has elapsed since Darwin's death in 1882. During this period, the evidence supporting his theory has grown progressively stronger. Also, many significant advances in our understanding of how evolution works have occurred. Although these advances have not altered the basic structure of Darwin's theory, they have taught us a great deal more about the mechanisms by which evolution occurs. We will briefly explore some of this evidence here; in chapter 22, we will return to the theory of evolution and examine the evidence in more detail.

The Fossil Record

Darwin predicted that the fossil record would yield intermediate links between the great groups of organisms—for example, between fishes and the amphibians thought to have arisen from them, and between reptiles and birds. We now know the fossil record to a degree that was unthinkable in the nineteenth century. Recent discoveries of microscopic fossils have extended the known history of life on earth back to about 2.5 billion years ago. The discovery of other fossils has supported Darwin's predictions and has shed light on how organisms have, over this enormous time span, evolved from the simple to the complex. For vertebrate animals especially, the fossil record is rich and exhibits a graded series of changes in form, with the evolutionary parade visible for all to see.

The Age of the Earth

In Darwin's day, some physicists argued that the earth was only a few thousand years old. This bothered Darwin, because the evolution of all living things from some single original ancestor would have required a great deal more time. Using evidence obtained by studying the rates of radioactive decay, we now know that the physicists of Darwin's time were wrong, very wrong: The earth was formed about 4.5 billion years ago.

The Mechanism of Heredity

Darwin received some of his sharpest criticism in the area of heredity. At that time, no one had any concept of genes or of how heredity works, so it was not possible for Darwin to explain completely how evolution occurs. Theories of heredity in Darwin's day seemed to rule out the possibility of genetic variation in nature, a critical requirement of Darwin's theory. Genetics was established as a science only at the start of the twentieth century, 40 years after the publication of Darwin's *On the Origin of Species*. When scientists began to understand the laws of inheritance (discussed in chapter 13), the heredity problem with Darwin's theory vanished. Genetics accounts in a neat and orderly way for the production of new variations in organisms.

Comparative Anatomy

Comparative studies of animals have provided strong evidence for Darwin's theory. In many different types of vertebrates, for example, the same bones are present, indicating their evolutionary past. Thus, the forelimbs shown in figure 1.13 are all constructed from the same basic array of bones, modified in one way in the wing of a bat, in another way in the fin of a porpoise, and in yet another way in the leg of a horse. The bones are said to be **homologous** in the different vertebrates; that is, they have the same evolutionary origin, but they now differ in structure and function. This contrasts with **analogous** structures, such as the wings of birds and butterflies, which have similar structure and function but different evolutionary origins.

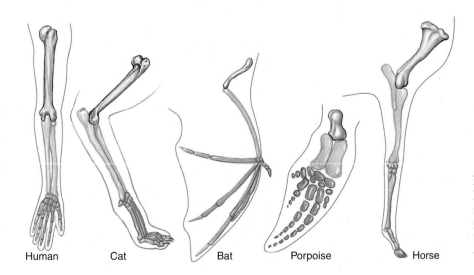

Human   Cat   Bat   Porpoise   Horse

FIGURE 1.13
Homology among vertebrate limbs. The forelimbs of these five vertebrates show the ways in which the relative proportions of the forelimb bones have changed in relation to the particular way of life of each organism.

Molecular Evidence

Evolutionary patterns are also revealed at the molecular level. By comparing the genomes (that is, the sequences of all the genes) of different groups of animals or plants, we can specify the degree of relationship among the groups more precisely than by any other means. A series of evolutionary changes over time should involve a continual accumulation of genetic changes in the DNA. Organisms that are more distantly related should have accumulated a greater number of these evolutionary differences, while two species that are more closely related will share a greater proportion of unchanged DNA. Thus gorillas, which the fossil record indicates diverged from humans between 6 and 8 million years ago, differ from the human genome in 1.6% of its DNA, while chimpanzees, which diverged about 5 million years ago, differ by only 1.2%. This same difference can be seen clearly in the protein hemoglobin (figure 1.14). The macaques, which like humans are primates, have fewer differences from humans in the 146 amino acid hemoglobin beta chain than do more distantly related mammals, such as dogs. Nonmammalian vertebrates, such as birds and frogs, differ even more.

Molecular Clocks. The consistent pattern emerging from a growing mountain of data (discussed in detail in chapter 24) is one of progressive change over time, with more distantly related species showing more differences in their DNA than closely related ones, just as Darwin's theory predicts. For example, the longer the time since two organisms diverged, the greater the number of differences in the nucleotide sequence of the cytochrome *c* gene (figure 1.15). This gene plays a key role in the oxidative metabolism of all terrestrial vertebrates, and appears to have accumulated changes at a constant rate, a phenomenon sometimes referred to as a **molecular clock.** All proteins for which data are available appear to accumulate changes over time, although different proteins evolve at different rates.

Phylogenetic Trees. The sequences of some genes, such as the ones specifying the hemoglobin proteins, have been determined in many organisms, and the entire time course of their evolution can be laid out with confidence by tracing the origins of particular nucleotide changes in the gene sequence. The pattern of descent obtained is called a **phylogenetic tree.** It represents the evolutionary history of the gene, its "family tree." Molecular phylogenetic trees agree well with those derived from the fossil record, which is strong direct evidence of evolution. The pattern of accumulating DNA changes represents, in a real sense, the footprints of evolutionary history.

Since Darwin's time, new discoveries in the fossil record, genetics, anatomy, and molecular biology all strongly support Darwin's theory.

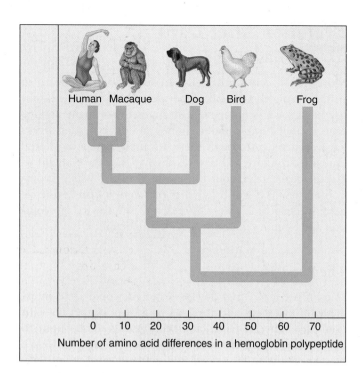

FIGURE 1.14
Molecules reflect evolutionary patterns. Vertebrates that are more distantly related to humans have a greater number of amino acid differences in this vertebrate hemoglobin polypeptide.
Where do you imagine a snake might fall on the graph? Why?

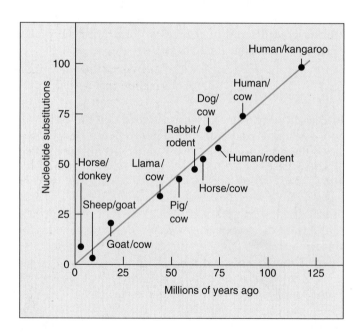

FIGURE 1.15
The molecular clock of cytochrome *c*. When the time since each pair of organisms diverged in the fossil record is plotted against the number of nucleotide differences in the cytochrome *c* gene, the result is a straight line, suggesting this gene is evolving at a constant rate.
Do you think dog differs from cow more than might be expected based on other vertebrates? What might explain this?

1.4 Four themes unify biology as a science.

Core Themes Unite Biology

Organization of Life: The Cell Theory

As was stated at the beginning of this chapter, all organisms are composed of cells, life's basic units (figure 1.16). Cells were discovered by Robert Hooke in England in 1665. Hooke was using one of the first microscopes, one that magnified 30 times. Looking through a thin slice of cork, he observed many tiny chambers, which reminded him of monks' cells in a monastery. Not long after that, the Dutch scientist Anton van Leeuwenhoek used microscopes capable of magnifying 300 times, and discovered an amazing world of single-celled life in a drop of pond water. However, it took almost two centuries before biologists fully understood the significance of cells. In 1839, the German biologists Matthias Schleiden and Theodor Schwann, summarizing a large number of observations by themselves and others, concluded that all living organisms consist of cells. Their conclusion forms the basis of what has come to be known as the **cell theory.** Later, biologists added the idea that all cells come from other cells. The cell theory, one of the basic ideas in biology, is the foundation for understanding the reproduction and growth of all organisms.

Continuity of Life: The Molecular Basis of Inheritance

Even the simplest cell is incredibly complex—more intricate than a computer. The information that specifies what a cell is like—its detailed plan—is encoded in a long, cable-like molecule called **DNA (deoxyribonucleic acid).** Each DNA molecule is formed from two long chains of building blocks, called nucleotides, wound around each other (figure 1.17). The two chains face each other, like two lines of people holding hands. The chains contain information in the same way this sentence does—as a sequence of letters. There are four different nucleotides in DNA, and the sequence in which they occur encodes this information. Specific sequences of several hundred to many thousand nucleotides make up a **gene,** a discrete unit of information. A gene might encode a particular protein or a different kind of unique molecule called RNA, or a gene might act to regulate other genes. The proteins and RNA molecules that are produced determine what the cell will be like.

The continuity of life from one generation to the next—heredity—depends upon the faithful copying of a cell's DNA into daughter cells. The entire set of DNA instructions that specifies a cell is called its **genome.** The sequence of the human genome, 3 billion nucleotides long, was decoded in rough draft form in 2001, a triumph of scientific investigation.

51 μm

FIGURE 1.16
Life in a drop of pond water. All organisms are composed of cells. Some organisms, including these protists, are single-celled, while others, such as plants, worms, and mushrooms, consist of many cells.

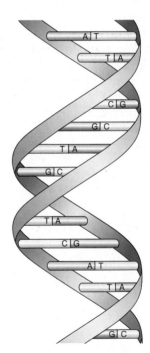

FIGURE 1.17
Genes are made of DNA. Winding around each other like the rails of a spiral staircase, the two strands of DNA make a double helix. Because of their size and shape, the nucleotide represented by the letter A can only pair with the nucleotide represented by the letter T, and likewise, the letter G and the letter C. This means that whatever the sequence is on one strand, the other strand will be its mirror image. From each strand, the other can be easily assembled.

Diversity of Life: Evolutionary Change

The unity of life that we see in the retention of certain key characteristics among many related life-forms contrasts with the incredible diversity of living things that have evolved to fill the varied environments of earth. Biologists divide life's great diversity into three great groups, called domains: Bacteria, Archaea, and Eukarya. The domains Bacteria and Archaea are composed of prokaryotes (single-celled organisms with little internal structure), while domain Eukarya is made up of eukaryotes, organisms composed of a complexly organized cell or multiple complex cells. However, Archaea seem more closely related to Eukarya than to Bacteria. Within the Eukarya are four main groups called kingdoms (figure 1.18). Kingdom Protista consists of all the unicellular eukaryotes except yeasts, as well as the multicellular algae. Because of the great diversity among the protists, many biologists feel kingdom Protista should be split into several kingdoms. Kingdom Plantae consists of organisms that have cell walls of cellulose and obtain energy by photosynthesis. Organisms in the kingdom Fungi have cell walls of chitin and obtain energy by secreting digestive enzymes onto organisms and then absorbing the products they release. Kingdom Animalia contains organisms that lack cell walls and obtain energy by first ingesting other organisms and then digesting them internally.

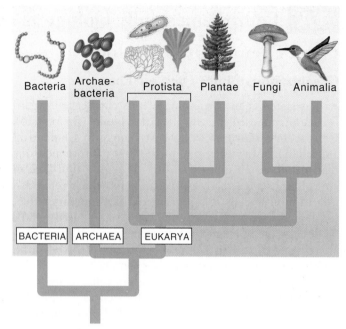

FIGURE 1.18
The diversity of life. Biologists categorize all living things into three overarching groups called domains: Bacteria, Archaea, and Eukarya. Domain Eukarya is composed of four kingdoms: Protista, Plantae, Fungi, and Animalia.

Unity of Life: Evolutionary Conservation

Biologists believe that all living things have descended from some simple cellular creature that arose about 2.5 billion years ago. Some of the characteristics of that earliest organism have been preserved in all things alive today. The storage of hereditary information in DNA, for example, is common to all living things. Also, all eukaryotes possess a nucleus that contains chromosomes, and flagellae throughout the animal kingdom possess the same 9 + 2 arrangement of microtubules. The retention of these conserved characteristics in a long line of descent usually reflects a fundamental role in the biology of the organism, one not easily changed once adopted. A good example is provided by the homeodomain proteins, proteins that play a critical role in early development in eukaryotes. Conserved characteristics can be seen in approximately 1850 homeodomain proteins, distributed among three different kingdoms of organisms (figure 1.19)! The homeodomain proteins are powerful developmental tools that evolved early, and for which no better alternative has arisen.

Cellular organisms store hereditary information in DNA. Sometimes DNA alterations occur, which when preserved result in evolutionary change. Today's biological diversity is the product of a long evolutionary journey.

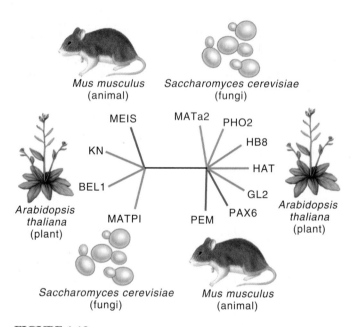

FIGURE 1.19
Tree of homeodomain proteins. Homeodomain proteins are found in fungi (*brown*), plants (*green*), and animals (*blue*). Based on their sequence similarities, these 11 different homeodomain proteins (uppercase letters at the ends of branches) fall into two groups, with representatives from each kingdom in each group. That means, for example, the mouse homeodomain protein PAX6 is more closely related to fungal and flowering plant proteins, such as PHO2 and GL2, than it is to the mouse protein MEIS.

Concept Review

For interactive testing, visit the Online Learning Center with PowerWeb at www.mhhe.com/Raven7

1.1 Biology is the science of life.

Organization of Living Things

- All living organisms share a collection of characteristics: cellular organization, order, sensitivity, growth, development and reproduction, energy utilization, evolutionary adaptation, and homeostasis. (p. 2)
- The biological world is hierarchical from the cellular level to the population level, with emergent properties entering at each higher level. (p. 2)

1.2 Scientists form generalizations from observations.

The Nature of Science

- Deductive reasoning applies general principles to predict specific results, while inductive reasoning uses specific observations to construct general principles. (p. 4)

How Science Is Done

- Scientists make observations and then construct a suggested explanation (hypothesis) to account for those observations. (p. 5)
- A successful experiment rejects one or more alternative hypotheses. (p. 5)
- Often scientists will conduct two experiments in parallel. In one experiment, all the variables are kept constant except one, while in the other experiment, the control experiment, that variable is left unaltered. (p. 6)
- Predictions provide a way to test the validity of a hypothesis. (p. 6)
- A theory is a proposed explanation for some natural phenomenon, but the term is also used to refer to a body of interconnected concepts supported by scientific reasoning and experimental evidence. (p. 6)
- After developing a hypothesis and performing a series of experiments, a paper is written describing the experiment and the results; the paper is then submitted for peer review. (p. 7)

1.3 Darwin's theory of evolution illustrates how science works.

Charles Darwin

- In Darwin's time, most people believed species to be specially created and unchangeable. (p. 8)

- Darwin proposed the concept of natural selection to account for his view that living things must have changed during their history on earth. (p. 8)
- Darwin served as naturalist on a five-year navigational mapping expedition around the coasts of South America. (p. 8)

Darwin's Evidence

- Repeatedly, Darwin found that characteristics of similar species varied from place to place, suggesting that organismal lineages change gradually as species migrate, and that plants and animals on young volcanic islands resembled those on nearby coasts of South America. (p. 10)

Inventing the Hypothesis of Natural Selection

- Darwin was influenced by Thomas Malthus and his idea that plant and animal populations increase geometrically. (p. 11)
- Darwin reasoned that because only a limited number of offspring can survive in each generation, those possessing superior physical or behavioral attributes are more likely to survive. (p. 11)
- Darwin knew of extensive examples of artificial selection in domesticated animals, and he reasoned that such evolutionary change could also occur in nature. (p. 12)

Evolution After Darwin: More Evidence

- The increasingly complete fossil record, the determination that the earth is 4.5 billion years old, the discovery of heredity mechanisms, comparative anatomy studies differentiating homologous and analogous structures, and molecular evidence such as the discovery of molecular clocks all support Darwin's ideas. (pp. 13–14)

1.4 Four themes unify biology as a science.

Core Themes Unite Biology

- The four themes uniting the field of biology are the cell theory, the molecular basis of inheritance, evolutionary change, and evolutionary conservation. (pp. 15–16)

Self Test

1. Which of the following is *not* a property of life?
   a. responding to stimuli
   b. dividing in two
   c. regulating internal conditions
   d. All of these are characteristics that define life.
2. The process of inductive reasoning
   a. involves the observation of specific occurrences to construct a general principle.
   b. involves taking a general principle and applying it to a specific situation.
   c. is not used very often in the study of biology.
   d. is all of the above.
3. A goal of a scientist is to formulate a hypothesis
   a. that will never be proven false.
   b. that is essentially a theory explaining an observation.
   c. that is just a wild guess.
   d. that will be tested by experimentation.
4. An experiment testing a hypothesis will
   a. always support the hypothesis.
   b. always disprove the hypothesis.
   c. include a variable and a control.
   d. not be successful if the hypothesis is rejected.
5. Darwin's proposal that evolution occurs through natural selection caused controversy because
   a. it challenged the existence of a Divine Creator.
   b. it challenged the views of earlier philosophers.
   c. it challenged a literal interpretation of the Bible.
   d. the explanation wasn't based on any observations.
6. Darwin was convinced that evolution had occurred based on his observations that
   a. armadillos on the different islands off the coast of South America had slightly different physical characteristics.
   b. tortoises found on the different Galápagos Islands had different and identifiable shells.
   c. bird fossils showed modifications from modern-day birds found in the same areas.
   d. All of these observations were made by Darwin.
7. A key piece of information for Darwin's hypothesis that evolution occurs through natural selection was
   a. the fossil evidence.
   b. his work breeding pigeons through artificial selection.
   c. Malthus's proposal that death, due to limited food supply, restricts population size.
   d. geographic distribution of similar animals with slight variations in physical characteristics.
8. Which of the following "new" areas of scientific study could Darwin have used to strengthen his hypothesis?
   a. the age of the earth
   b. the mechanism of inheritance.
   c. the expanded fossil record
   d. the geographic distribution of animal species
9. What are homologous anatomical structures?
   a. structures that look different but have similar evolutionary origins
   b. structures that have similar functions but different evolutionary origins
   c. a bat's wing and a butterfly's wing
   d. a bat's wing and a human's leg
10. The "themes" of biology include all of the following except
    a. chemistry.   c. genetics.
    b. cell biology.   d. evolution.

Test Your Visual Understanding

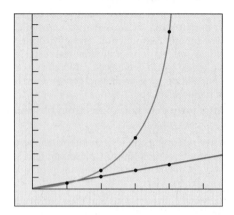

1. Which line on the graph indicates the growth of a population if resources were unlimited? Name three factors that limit population growth.

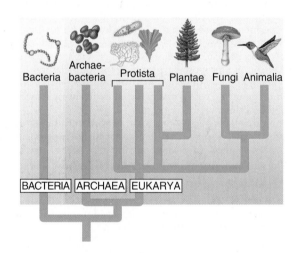

2. By following the general path of evolution, would you say that fungi are more closely related to Archaebacteria or Bacteria? Why?

Apply Your Knowledge

1. When breeding hunting dogs, the parents are selected for their keen sense of smell, which allows them to find the game (i.e., birds) faster. Breeders predict that this trait will be passed on to the dogs' offspring, producing a family of good hunting dogs. This is artificial selection. Explain how this same trait could be selected through natural selection in a population of wild dogs.
2. Assume that four people are stranded on a deserted island and they reproduce once a year. How many people will be on the island after ten years using:
   a. a geometric progression with a factor of 2.
   b. an arithmetic progression with a factor of 2.

2

The Nature of Molecules

Concept Outline

2.1 Atoms are nature's building material.

    Atoms. All substances are composed of tiny particles called atoms, each a positively charged nucleus around which negative electrons orbit.

    Electrons Determine the Chemical Behavior of Atoms. The closer an electron's orbit is to the nucleus, the lower is its energy level.

2.2 The atoms of living things are among the smallest.

    Kinds of Atoms. Of the 92 naturally occurring elements, only 11 occur in organisms in significant amounts.

2.3 Chemical bonds hold molecules together.

    Ionic Bonds Form Crystals. Atoms are linked together into molecules, joined by chemical bonds that result from forces such as the attraction of opposite charges or the sharing of electrons.

    Covalent Bonds Build Stable Molecules. Chemical bonds formed by the sharing of electrons can be very strong, and require much energy to break.

2.4 Water is the cradle of life.

    Chemistry of Water. Water forms weak chemical associations that are responsible for much of the organization of living chemistry.

    Water Atoms Act Like Tiny Magnets. Because electrons are shared unequally by the hydrogen and oxygen atoms of water, a partial charge separation occurs. Each water atom acquires a positive and negative pole and is said to be "polar."

    Water Clings to Polar Molecules. Because the opposite partial charges of polar molecules attract one another, water tends to cling to itself and to other polar molecules while excluding nonpolar molecules.

    Water Ionizes. Because its covalent bonds occasionally break, water contains a low concentration of hydrogen (H^+) and hydroxide (OH^-) ions, the fragments of broken water molecules.

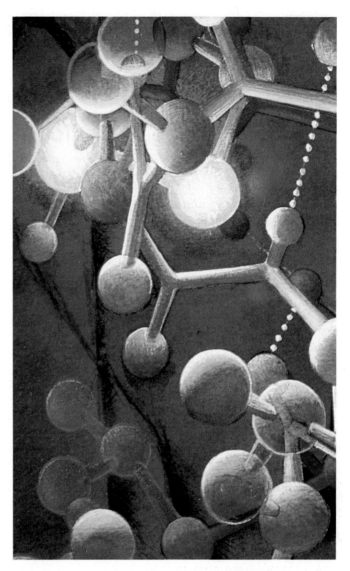

FIGURE 2.1
Cells are made of molecules. Specific, often simple combinations of atoms yield an astonishing diversity of molecules within the cell, each with unique functional characteristics.

About 14 billion years ago, an enormous explosion likely marked the beginning of the universe. With this explosion began a process of star building and planetary formation that eventually led to the formation of earth about 4.5 billion years ago. Starting about 2.5 billion years ago, life began on earth and started to diversify. When viewed from the perspective of 14 billion years, life within our solar system is a recent development, but to understand the origin of life, we need to consider events that took place much earlier. The same processes that led to the evolution of life were responsible for the evolution of molecules (figure 2.1). Thus, our study of life on earth begins with physics and chemistry. As chemical machines ourselves, we must understand chemistry to begin to understand our origins.

2.1 Atoms are nature's building material.

Atoms

Any substance in the universe that has mass and occupies space is defined as **matter**. All matter is composed of extremely small particles called **atoms**. Because of their size, atoms are difficult to study. Not until early in the last century did scientists carry out the first experiments suggesting what an atom is like.

The Structure of Atoms

Objects as small as atoms can be "seen" only indirectly, by using very complex technology such as tunneling microscopy. We now know a great deal about the complexities of atomic structure, but the simple view put forth in 1913 by the Danish physicist Niels Bohr provides a good starting point. Bohr proposed that every atom possesses an orbiting cloud of tiny subatomic particles called **electrons** whizzing around a core like the planets of a miniature solar system. At the center of each atom is a small, very dense nucleus formed of two other kinds of subatomic particles, **protons** and **neutrons** (figure 2.2).

Within the nucleus, the cluster of protons and neutrons is held together by a force that works only over short subatomic distances. Each proton carries a positive (+) charge, and each electron carries a negative (−) charge. Typically, an atom has one electron for each proton. The number of protons an atom has is called its **atomic number**. This number indirectly determines the chemical character of the atom, because it dictates the number of electrons orbiting the nucleus that are available for chemical activity. Neutrons, as their name implies, possess no charge.

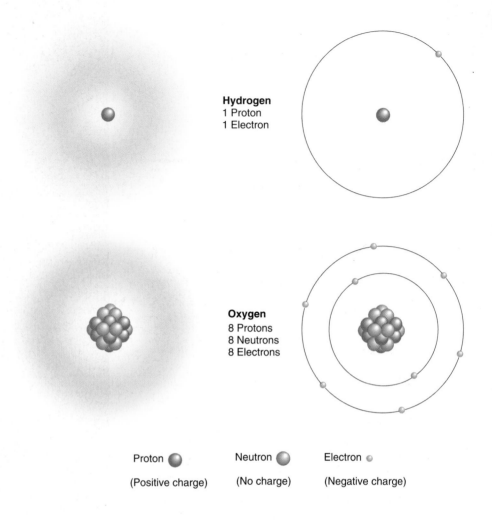

FIGURE 2.2
Basic structure of atoms. All atoms have a nucleus consisting of protons and neutrons, except hydrogen, the smallest atom, which has only one proton and no neutrons in its nucleus. Oxygen, for example, has eight protons and eight neutrons in its nucleus. In the simple "Bohr model" of atoms pictured here, electrons spin around the nucleus at a relatively far distance.

Atomic Mass

The terms mass and weight are often used interchangeably, but they have slightly different meanings. *Mass* refers to the amount of a substance, while *weight* refers to the force gravity exerts on a substance. Hence, an object has the same mass whether it is on the earth or the moon, but its weight will be greater on the earth because the earth's gravitational force is greater than the moon's. The **atomic mass** of an atom is equal to the sum of the masses of its protons and neutrons. Atoms that occur naturally on earth contain from 1 to 92 protons and up to 146 neutrons.

The mass of atoms and subatomic particles is measured in units called *daltons*. To give you an idea of just how small these units are, note that it takes 602 million million billion (6.02×10^{23}) daltons to make 1 gram. A proton weighs approximately 1 dalton (actually 1.009 daltons), as does a neutron (1.007 daltons). In contrast, electrons weigh only $\frac{1}{1840}$ of a dalton, so their contribution to the overall mass of an atom is negligible.

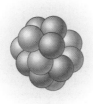

FIGURE 2.3
The three most abundant isotopes of carbon. Isotopes of a particular element have different numbers of neutrons.

Carbon-12
6 Protons
6 Neutrons
6 Electrons

Carbon-13
6 Protons
7 Neutrons
6 Electrons

Carbon-14
6 Protons
8 Neutrons
6 Electrons

Isotopes

Atoms with the same atomic number (that is, the same number of protons) have the same chemical properties and are said to belong to the same **element.** Formally speaking, an element is any substance that cannot be broken down to any other substance by ordinary chemical means. However, while all atoms of an element have the same number of protons, they may not all have the same number of neutrons. Atoms of an element that possess different numbers of neutrons are called **isotopes** of that element. Most elements in nature exist as mixtures of different isotopes. Carbon (C), for example, has three isotopes, all containing six protons (figure 2.3). Over 99% of the carbon found in nature exists as an isotope with six neutrons. Because its total mass is 12 daltons (6 from protons plus 6 from neutrons), this isotope is referred to as carbon-12, and symbolized ^{12}C. Most of the rest of the naturally occurring carbon is carbon-13, an isotope with seven neutrons. The rarest carbon isotope is carbon-14, with eight neutrons. Unlike the other two isotopes, carbon-14 is unstable: Its nucleus tends to break up into elements with lower atomic numbers. This nuclear breakup, which emits a significant amount of energy, is called radioactive decay, and isotopes that decay in this fashion are **radioactive isotopes.**

Some radioactive isotopes are more unstable than others and therefore decay more readily. For any given isotope, however, the propensity to decay is constant. The decay time is usually expressed as the **half-life,** the time it takes for one-half of the atoms in a sample to decay. Carbon-14, for example, has a half-life of about 5600 years. A sample of carbon containing 1 gram of carbon-14 today would contain 0.5 gram of carbon-14 after 5600 years, 0.25 gram 11,200 years from now, 0.125 gram 16,800 years from now, and so on. By determining the ratios of the different isotopes of carbon and other elements in biological samples and in rocks, scientists are able to accurately determine when these materials formed.

While there are many useful applications of radioactivity, there are also harmful side effects that must be considered in any planned use of radioactive substances. Radioactive substances emit energetic subatomic particles that have the potential to severely damage living cells, producing mutations in their genes and, at high doses, cell death. Consequently, exposure to radiation is now very carefully controlled and regulated. Scientists who work with radioactivity (basic researchers as well as applied scientists such as X-ray technologists) wear radiation-sensitive badges to monitor the total amount of radioactivity to which they are exposed. Each month the badges are collected and scrutinized. Thus, employees whose work places them in danger of excessive radioactive exposure are equipped with an "early warning system."

Electrons

The positive charges in the nucleus of an atom are counterbalanced by negatively charged electrons located in regions called orbitals at varying distances around the nucleus. Thus, atoms with the same number of protons and electrons are electrically neutral, having no net charge, and are called **neutral atoms.**

Electrons are maintained in their orbits by their attraction to the positively charged nucleus. Sometimes other forces overcome this attraction and an atom loses one or more electrons. In other cases, atoms gain additional electrons. Atoms in which the number of electrons does not equal the number of protons are known as **ions,** and they carry a net electrical charge. An atom that has more protons than electrons has a net positive charge and is called a **cation.** For example, an atom of sodium (Na) that has lost one electron becomes a sodium ion (Na^+), with a charge of +1. An atom that has fewer protons than electrons carries a net negative charge and is called an **anion.** A chlorine atom (Cl) that has gained one electron becomes a chloride ion (Cl^-), with a charge of −1.

An atom consists of a nucleus of protons and neutrons surrounded by a cloud of electrons. The number of its electrons largely determines the chemical properties of an atom. Atoms that have the same number of protons but different numbers of neutrons are called isotopes. Isotopes of an atom differ in atomic mass but have similar chemical properties.

Electrons Determine the Chemical Behavior of Atoms

The key to the chemical behavior of an atom lies in the number and arrangement of its electrons in their orbits. It is convenient to visualize individual electrons as following discrete circular orbits around a central nucleus, as in the Bohr model of the atom. However, such a simple picture is not realistic. It is not possible to precisely locate the position of any individual electron at any given time. In fact, a particular electron can be anywhere at a given instant, from close to the nucleus to infinitely far away from it.

Nevertheless, a particular electron is more likely to be located in some positions than in others. The area around a nucleus where an electron is most likely to be found is called the **orbital** of that electron. Some electron orbitals near the nucleus are spherical (*s* orbitals), while others are dumbbell-shaped (*p* orbitals) (figure 2.4). Still other orbitals, more distant from the nucleus, may have different shapes. Regardless of its shape, no orbital may contain more than two electrons.

Almost all of the volume of an atom is empty space, because the electrons are quite far from the nucleus relative to its size. If the nucleus of an atom were the size of a golf ball, the orbit of the nearest electron would be more than a kilometer away. Consequently, the nuclei of two atoms never come close enough in nature to interact with each other. It is for this reason that an atom's electrons, not its protons or neutrons, determine its chemical behavior. This also explains why the isotopes of an element, all of which have the same arrangement of electrons, behave the same way chemically.

Energy Within the Atom

All atoms possess *energy*, defined as the ability to do work. Because electrons are attracted to the positively charged nucleus, it takes work to keep them in orbit, just as it takes work to hold a grapefruit in your hand against the pull of gravity. The grapefruit is said to possess *potential energy* because of its position; if you were to release it, the grapefruit would fall, and its energy would be reduced. Conversely, if you were to move the grapefruit to the top of a building, you would increase its potential energy. Similarly, electrons have potential energy of position. To oppose the attraction of the nucleus and move the electron to a more distant orbital requires an input of energy and results in an electron with greater potential energy. This is how chlorophyll captures energy from light during photosynthesis (see chapter 10)—the light excites electrons in the chlorophyll. Moving an electron closer to the nucleus has the opposite effect: Energy is released, usually as heat, and the electron ends up with less potential energy (figure 2.5).

A given atom can possess only certain discrete amounts of energy. Like the potential energy of a grapefruit in your hand, the potential energy contributed by the position of an electron in an atom can have only certain values. Every

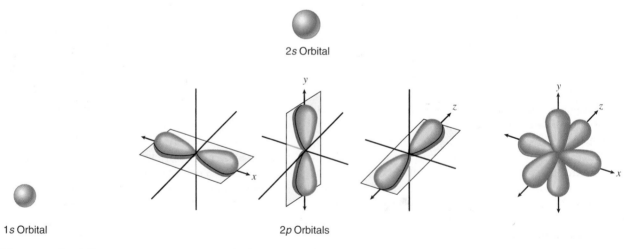

2s Orbital

1s Orbital

2p Orbitals

Orbital for energy level K:
one spherical orbital (1s)

Orbitals for energy level L:
one spherical orbital (2s) and
three dumbbell-shaped orbitals (2p)

**Composite of
all p orbitals**

FIGURE 2.4

Electron orbitals. The lowest energy level or electron shell—the one nearest the nucleus—is level *K*. It is occupied by a single *s* orbital, referred to as 1*s*. The next highest energy level, *L*, is occupied by four orbitals: one *s* orbital (referred to as the 2*s* orbital) and three *p* orbitals (each referred to as a 2*p* orbital). The four *L*-level orbitals compactly fill the space around the nucleus, like two pyramids set base-to-base. Thus, the lowest energy level, *K*, is populated by two electrons, while the next highest energy level, *L*, is populated by a total of eight electrons.

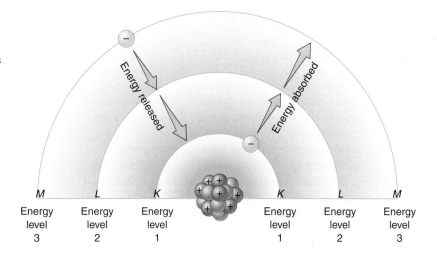

FIGURE 2.5
Atomic energy levels. When an electron absorbs energy, it moves to higher energy levels, farther from the nucleus. When an electron releases energy, it falls to lower energy levels, closer to the nucleus.

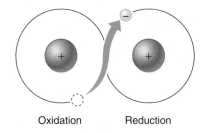

Oxidation   Reduction

FIGURE 2.6
Oxidation and reduction. Oxidation is the loss of an electron; reduction is the gain of an electron. In either reaction, the electron keeps its energy of position.

atom exhibits a ladder of potential energy values, a discrete set of orbits at particular distances from the nucleus.

During some chemical reactions, electrons are transferred from one atom to another. In such reactions, the loss of an electron is called **oxidation,** and the gain of an electron is called **reduction** (figure 2.6). It is important to realize that when an electron is transferred in this way, it keeps its energy of position. In organisms, chemical energy is stored in high-energy electrons that are transferred from one atom to another in reactions involving oxidation and reduction.

Because the amount of energy an electron possesses is related to its distance from the nucleus, electrons that are the same distance from the nucleus have the same energy, even if they occupy different orbitals. Such electrons are said to occupy the same **energy level.** In a schematic diagram of an atom (figure 2.7), the nucleus is represented as a small circle, and the electron energy levels are drawn as concentric rings, with the energy level increasing with distance from the nucleus. Be careful not to confuse energy levels, which are drawn as rings to indicate an electron's *energy*, with orbitals, which have a variety of three-dimensional shapes and indicate an electron's most likely *location*.

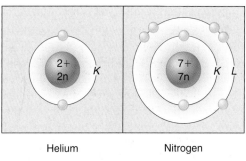

Helium   Nitrogen

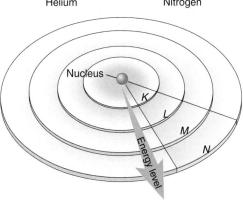

FIGURE 2.7
Electron energy levels for helium and nitrogen. *Top:* Gold balls represent electrons. *Bottom:* Each concentric circle represents a different distance from the nucleus and so a different electron energy level.

Electrons are localized about a nucleus in regions called orbitals. No orbital can contain more than two electrons, but many orbitals may be the same distance from the nucleus and, thus, contain electrons of the same energy.

2.2 The atoms of living things are among the smallest.

Kinds of Atoms

There are 92 naturally occurring elements, each with a different number of protons and a different arrangement of electrons. When the nineteenth-century Russian chemist Dmitri Mendeleev arranged the known elements in a table according to their atomic number, he discovered one of the great generalizations in all of science: The elements in his table exhibited a pattern of chemical properties that repeated itself in groups of eight elements. This periodically repeating pattern lent the table its name: the periodic table of elements (figure 2.8).

The Periodic Table

The eight-element periodicity that Mendeleev found is based on the interactions of the electrons in the outer energy levels of the different elements. These electrons are called **valence electrons,** and their interactions are the basis for the differing chemical properties of the elements. For most of the atoms important to life, an outer energy level can contain no more than eight electrons; the chemical behavior of an element reflects how many of the eight positions are filled. Elements possessing all eight electrons in their outer energy level (two for helium) are **inert,** or nonreactive; they include helium (He), neon (Ne), argon (Ar), krypton (Kr), xenon (Xe), and radon (Rn). In sharp contrast, elements with seven electrons (one fewer than the maximum number of eight) in their outer energy level, such as fluorine (F), chlorine (Cl), and bromine (Br), are highly reactive. They tend to gain the extra electron needed to fill the energy level. Elements with only one electron in their outer energy level, such as lithium (Li), sodium (Na), and potassium (K), are also very reactive; they tend to lose the single electron in their outer level.

Mendeleev's periodic table thus leads to a useful generalization, the **octet rule** (Latin *octo,* "eight"), or *rule of eight:* Atoms tend to establish completely full outer energy levels. Most chemical behavior of biological interest can be predicted quite accurately from this simple rule, combined with the tendency of atoms to balance positive and negative charges.

Only 11 elements are found in significant amounts in living organisms.

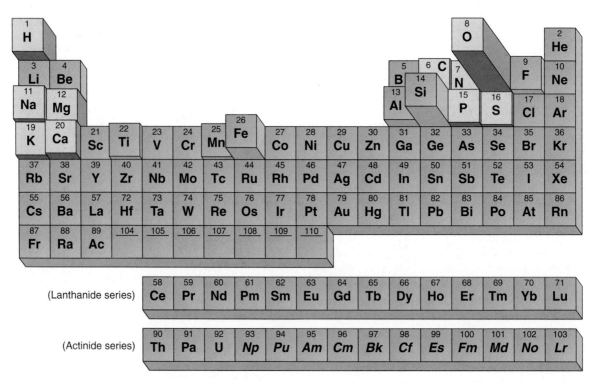

FIGURE 2.8
Periodic table of the elements.   In this representation, the frequency of elements that occur in the earth's crust is indicated by the height of the block. Of the 92 naturally occurring elements on earth, only 11 are found in organisms in more than trace amounts (0.01% or higher). These elements, shaded in *blue,* have atomic numbers less than 21 and, thus, have low atomic masses. Four elements—nitrogen, oxygen, carbon, and hydrogen (NOCH for short)—constitute 96.3% of the weight of your body.

Chemical bonds hold molecules together.

Ionic Bonds Form Crystals

A group of atoms held together by energy in a stable association is called a **molecule.** When a molecule contains atoms of more than one element, it is called a **compound.** The atoms in a molecule are joined by **chemical bonds;** these bonds can result when atoms with opposite charges attract (ionic bonds), when two atoms share one or more pairs of electrons (covalent bonds), or when atoms interact in other ways. We will start by examining **ionic bonds,** which form when atoms with opposite electrical charges (ions) attract.

A Closer Look at Table Salt

Common table salt, the molecule sodium chloride (NaCl), is a lattice of ions in which the atoms are held together by ionic bonds (figure 2.9). Sodium has 11 electrons: 2 in the inner energy level, 8 in the next level, and 1 in the outer (valence) level. The valence electron is unpaired (free) and has a strong tendency to join with another electron. A stable configuration can be achieved if the valence electron is lost to another atom that also has an unpaired electron. The loss of this electron results in the formation of a positively charged sodium ion, Na^+.

The chlorine atom has 17 electrons: 2 in the inner energy level, 8 in the next level, and 7 in the outer level. Hence, one of the orbitals in the outer energy level has an unpaired electron. The addition of another electron to the outer level fills that level and causes a negatively charged chloride ion, Cl^-, to form.

When placed together, metallic sodium and gaseous chlorine react swiftly and explosively, as the sodium atoms donate electrons to chlorine, forming Na^+ and Cl^- ions. Because opposite charges attract, the Na^+ and Cl^- remain associated in an **ionic compound,** NaCl, which is electrically neutral. However, the electrical attractive force holding NaCl together is not directed specifically between particular Na^+ and Cl^- ions, and no discrete sodium chloride molecules form. Instead, the force exists between any one ion and all neighboring ions of the opposite charge, and the ions aggregate in a crystal matrix with a precise geometry. Such aggregations are what we know as salt crystals. If a salt such as NaCl is placed in water, the electrical attraction of the water molecules, for reasons we will point out later in this chapter, disrupts the forces holding the ions in their crystal matrix, causing the salt to dissolve into a roughly equal mixture of free Na^+ and Cl^- ions.

> An ionic bond is an attraction between ions of opposite charge in an ionic compound. Such bonds are not formed between particular ions in the compound; rather, they exist between an ion and all of the oppositely charged ions in its immediate vicinity.

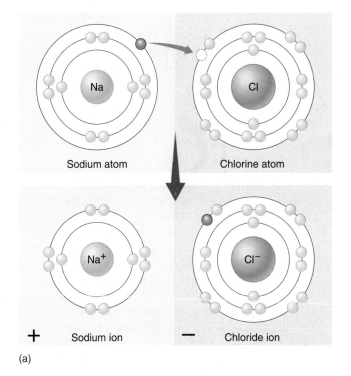

(a)

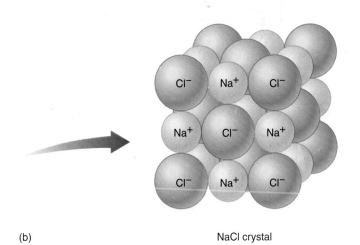

(b)   NaCl crystal

FIGURE 2.9
The formation of ionic bonds by sodium chloride. (*a*) When a sodium atom donates an electron to a chlorine atom, the sodium atom becomes a positively charged sodium ion, and the chlorine atom becomes a negatively charged chloride ion. (*b*) Sodium chloride (NaCl) forms a highly regular lattice of alternating sodium ions and chloride ions.

Covalent Bonds Build Stable Molecules

Covalent bonds form when two atoms share one or more pairs of valence electrons. Consider hydrogen (H) as an example. Each hydrogen atom has an unpaired electron and an unfilled outer energy level; for these reasons, the hydrogen atom is unstable. However, when two hydrogen atoms are close to each other, each atom's electron is attracted to both nuclei. In effect, the nuclei are able to share their electrons. The result is a diatomic (two-atom) molecule of hydrogen gas (figure 2.10*a*).

The molecule formed by the two hydrogen atoms is stable for three reasons:

1. **It has no net charge.** The diatomic molecule formed as a result of this sharing of electrons is not charged because it still contains two protons and two electrons.
2. **The octet rule is satisfied.** Each of the two hydrogen atoms can be considered to have two orbiting electrons in its outer energy level. This satisfies the octet rule, because each shared electron orbits both nuclei and is included in the outer energy level of *both* atoms.
3. **It has no free electrons.** The bond between the two atoms also pairs the two free electrons.

Unlike ionic bonds, covalent bonds are formed between two specific atoms, giving rise to true, discrete molecules.

The Strength of Covalent Bonds

The strength of a covalent bond depends on the number of shared electrons. Thus **double bonds,** which satisfy the octet rule by allowing two atoms to share two pairs of electrons, are stronger than **single bonds,** in which only one electron pair is shared. This means more chemical energy is required to break a double bond than a single bond. The strongest covalent bonds are **triple bonds,** such as those that link the two nitrogen atoms of nitrogen gas molecules. Covalent bonds are represented in chemical formulations as lines connecting atomic symbols, where each line between two bonded atoms represents the sharing of one pair of electrons. The **structural formulas** of hydrogen gas and oxygen gas are H—H and O=O, respectively, while their **molecular formulas** are H_2 and O_2.

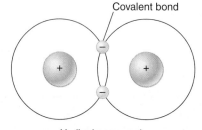

Covalent bond

H_2 (hydrogen gas)

(a)

(b)

FIGURE 2.10
Hydrogen gas. (*a*) Hydrogen gas is a diatomic molecule, meaning that it is composed of two hydrogen atoms, each sharing its electron with the other. (*b*) The flash of fire that consumed the *Hindenburg* occurred when the hydrogen gas used to inflate the dirigible combined explosively with oxygen gas in the air to form water.

Molecules with Several Covalent Bonds

Molecules often consist of more than two atoms. One reason that larger molecules may be formed is that a given atom is able to share electrons with more than one other atom. An atom that requires two, three, or four additional electrons to fill its outer energy level completely may acquire them by sharing its electrons with two or more other atoms.

For example, the carbon atom (C) contains six electrons, four of which are in its outer energy level. To satisfy the octet rule, a carbon atom must gain access to four additional electrons; that is, it must form four covalent bonds. Because four covalent bonds may form in many ways, carbon atoms are found in many different kinds of molecules.

Chemical Reactions

The formation and breaking of chemical bonds, the essence of chemistry, is called a **chemical reaction.** All chemical reactions involve the shifting of atoms from one molecule or ionic compound to another, without any change in the number or identity of the atoms. For convenience, we refer to the original molecules before the reaction starts as **reactants,** and the molecules resulting from the chemical reaction as **products.** For example:

$$A—B + C—D \rightarrow A—C + B + D$$
$$\text{reactants} \qquad \text{products}$$

The extent to which chemical reactions occur is influenced by several important factors:

1. **Temperature.** Heating the reactants increases the rate of a reaction (as long as the temperature isn't so high as to destroy the molecules).
2. **Concentration of reactants and products.** Reactions proceed more quickly when more reactants are available. An accumulation of products typically speeds reactions in the reverse direction.
3. **Catalysts.** A catalyst is a substance that increases the rate of a reaction. It doesn't alter the reaction's equilibrium between reactants and products, but it does shorten the time needed to reach equilibrium, often dramatically. In organisms, proteins called enzymes catalyze almost every chemical reaction.

A covalent bond is a stable chemical bond formed when two atoms share one or more pairs of electrons.

(a)

(b)

(c)

FIGURE 2.11
Water takes many forms. (*a*) When water cools below 0° C, it forms beautiful crystals, familiar to us as snow and ice. (*b*) Ice turns to liquid when the temperature is above 0°C; in icy water, this orca must find holes in the covering of ice in order to surface and breathe. (*c*) Liquid water becomes steam when the temperature rises above 100°C, as seen in this hot spring at Yellowstone National Park.

Of all the molecules that are common on earth, only **water** exists as a liquid at the relatively low temperatures that prevail on the earth's surface, three-fourths of which is covered by liquid water (figure 2.11). When life was originating, water provided a medium in which other molecules could move around and interact without being held in place by strong covalent or ionic bonds. Life evolved in water for 2 billion years before spreading to land. And even today, life is inextricably tied to water. About two-thirds of any organism's body is composed of water, and no organism can grow or reproduce in any but a water-rich environment. It is no accident that tropical rain forests are bursting with life, while dry deserts appear almost lifeless except when water becomes temporarily plentiful, such as after a rainstorm.

Chemistry of Water

Water has a simple atomic structure. It consists of an oxygen atom bound to two hydrogen atoms by two single covalent bonds (figure 2.12). The resulting molecule is stable: It satisfies the octet rule, has no unpaired electrons, and carries no net electrical charge.

The single most outstanding chemical property of water is its ability to form weak chemical associations with only 5 to 10% of the strength of covalent bonds. This property, which derives directly from the structure of water, is responsible for much of the organization of living chemistry.

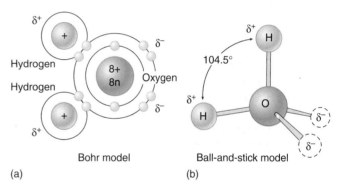
(a) Bohr model (b) Ball-and-stick model

FIGURE 2.12
Water has a simple molecular structure. (*a*) Each water molecule is composed of one oxygen atom and two hydrogen atoms. The oxygen atom shares one electron with each hydrogen atom. (*b*) The greater electronegativity of the oxygen atom makes the water molecule polar: Water carries two partial negative charges (δ^-) near the oxygen atom and two partial positive charges (δ^+), one on each hydrogen atom.

The chemistry of life is water chemistry. The way life first evolved was determined in large part by the chemical properties of the liquid water in which that evolution occurred.

Water Atoms Act Like Tiny Magnets

In a water molecule, both the oxygen and the hydrogen atoms attract the electrons they share in the covalent bonds; this attraction is called **electronegativity.** However, the oxygen atom is more electronegative than the hydrogen atoms, so it attracts the electrons more strongly than do the hydrogen atoms. As a result, the shared electrons in a water molecule are far more likely to be found near the oxygen nucleus than near the hydrogen nuclei. This stronger attraction for electrons gives the oxygen atom two partial negative δ^- charges, as though the electron cloud were denser near the oxygen atom than around the hydrogen atoms. The Greek letter delta (δ) signifies a partial charge, much weaker than the full unit charge of an ion. Because the water molecule as a whole is electrically neutral, each hydrogen atom carries a partial positive charge (δ^+).

What would you expect the shape of a water molecule to be? Each of water's two covalent bonds has a partial charge at each end, δ^- at the oxygen end and δ^+ at the hydrogen end. The most stable arrangement of these charges is a *tetrahedron,* in which the two negative and two positive charges are approximately equidistant from one another. The oxygen atom lies at the center of the tetrahedron, the hydrogen atoms occupy two of the apexes, and the partial negative charges occupy the other two apexes (figure 2.12*b*). This results in a bond angle of 104.5° between the two covalent oxygen-hydrogen bonds. (In a regular tetrahedron, the bond angles would be 109.5°; in water, the partial negative charges occupy more space than the hydrogen atoms, and therefore, they compress the oxygen-hydrogen bond angle slightly.)

The water molecule, thus, has distinct "ends," each with a partial charge. A rough analogy might be the two poles of

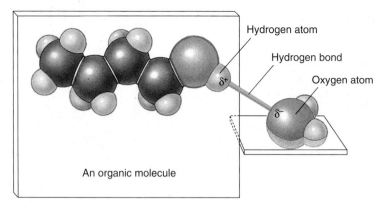

FIGURE 2.13
Structure of a hydrogen bond. The polar end of this organic molecule is interacting with a water molecule. This interaction bridges a hydrogen atom and is called a hydrogen bond.

a magnet. (These partial charges are much less than the unit charges of ions, however.) Molecules that exhibit charge separation are called **polar molecules** because of these partially charged poles, and water is one of the most polar molecules known. *The polarity of water underlies its chemistry and the chemistry of life.*

Polar molecules interact with one another, as the δ^- of one molecule is attracted to the δ^+ of another. Many of these interactions involve bridging hydrogen atoms; those that do are called **hydrogen bonds** (figure 2.13). Each hydrogen bond is individually very weak and transient, lasting on average only $\frac{1}{100,000,000,000}$ second (10^{-11} sec). However, the cumulative effects of large numbers of these bonds can be enormous. Water forms an abundance of hydrogen bonds, which are responsible for many of its important physical properties (table 2.1).

The water molecule is very polar, with ends that exhibit partial positive and negative charges. Opposite charges attract, forming weak linkages called hydrogen bonds.

Table 2.1 The Properties of Water

Property	Explanation	Example of Benefit to Life
Cohesion	Hydrogen bonds hold water molecules together	Leaves pull water upward from the roots; seeds swell and germinate
High specific heat	Hydrogen bonds absorb heat when they break, and release heat when they form, minimizing temperature changes	Water stabilizes the temperature of organisms and the environment
High heat of vaporization	Many hydrogen bonds must be broken for water to evaporate	Evaporation of water cools body surfaces
Lower density of ice	Water molecules in an ice crystal are spaced relatively far apart because of hydrogen bonding	Because ice is less dense than water, lakes do not freeze solid
Solubility	Polar water molecules are attracted to ions and polar compounds, making them soluble	Many kinds of molecules can move freely in cells, permitting a diverse array of chemical reactions

Water Clings to Polar Molecules

The polarity of water causes it to be attracted to other polar molecules. When the other molecules are also water, the attraction is referred to as **cohesion.** When the other molecules are of a different substance, the attraction is called **adhesion.** Because water is cohesive, it is a liquid, not a gas, at moderate temperatures.

The cohesion of liquid water is also responsible for its **surface tension.** Small insects can walk on water (figure 2.14) because at the air-water interface all of the hydrogen bonds in water face downward, causing the molecules of the water surface to cling together. Water is adhesive to any substance with which it can form hydrogen bonds. That is why substances containing polar molecules get "wet" when they are immersed in water, while those that are composed of nonpolar molecules (such as oils) do not.

The attraction of water to substances such as glass having surface electrical charges is responsible for capillary action: If a glass tube with a narrow diameter is lowered into a beaker of water, water will rise in the tube above the level of the water in the beaker, because the adhesion of water to the glass surface, drawing it upward, is stronger than the force of gravity, drawing it down. The narrower the tube, the greater the electrostatic forces between the water and the glass, and the higher the water rises (figure 2.15).

Water Stores Heat

Water moderates temperature through two properties: its high specific heat and its high heat of vaporization. The temperature of any substance is a measure of how rapidly its individual molecules are moving. Because of the many hydrogen bonds that water molecules form with one another, a large input of thermal energy is required to break these bonds before the individual water molecules can begin moving about more freely and thus have a higher temperature. Therefore, water is said to have a high **specific heat,** which is defined as the amount of heat that must be absorbed or lost by 1 gram of a substance to change its temperature by 1 degree Celsius (°C). Specific heat measures the extent to which a substance resists changing its temperature when it absorbs or loses heat. Because polar substances tend to form hydrogen bonds, and energy is needed to break these bonds, the more polar a substance is, the higher is its specific heat. The specific heat of water (1 calorie/gram/°C) is twice that of most carbon compounds and nine times that of iron. Only ammonia, which is more polar than water and forms very strong hydrogen bonds, has a higher specific heat than water (1.23 calories/gram/°C). Still, only 20% of the hydrogen bonds are broken as water heats from 0° to 100°C.

Because of its high specific heat, water heats up more slowly than almost any other compound and holds its temperature longer when heat is no longer applied. This characteristic enables organisms, which have a high water con-

FIGURE 2.14
Cohesion. Some insects, such as this water strider, literally walk on water. In this photograph you can see how the insect's feet dimple the water as its weight bears down on the surface. Because the surface tension of the water is greater than the force that one foot brings to bear, the strider glides atop the surface of the water rather than sinking.

FIGURE 2.15
Capillary action. Capillary action causes the water within a narrow tube to rise above the surrounding water; the adhesion of the water to the glass surface, which draws water upward, is stronger than the force of gravity, which tends to draw it down. The narrower the tube, the greater the surface area available for adhesion for a given volume of water, and the higher the water rises in the tube.

tent, to maintain a relatively constant internal temperature. The heat generated by the chemical reactions inside cells would destroy the cells if not for the high specific heat of the water within them.

A considerable amount of heat energy (586 calories) is required to change 1 gram of liquid water into a gas. Hence, water also has a high **heat of vaporization.** Because the transition of water from a liquid to a gas requires the input of energy to break its many hydrogen bonds, the evaporation of water from a surface causes cooling of that surface. Many organisms dispose of excess body heat by evaporative cooling; for example, humans and many other vertebrates sweat.

At low temperatures, water molecules are locked into a crystal-like lattice of hydrogen bonds, forming the solid we call ice (figure 2.16). Interestingly, ice is less dense than liquid water because the hydrogen bonds in ice space the water molecules relatively far apart. This unusual feature enables icebergs to float. Were it otherwise, ice would cover nearly all bodies of water, with only the shallow surface melting annually.

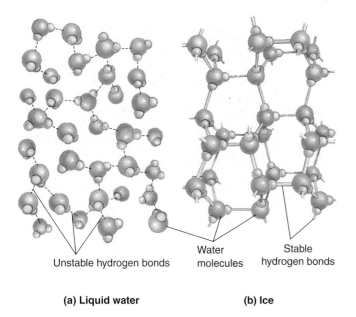

(a) Liquid water   (b) Ice

FIGURE 2.16
The role of hydrogen bonds in an ice crystal. (*a*) In liquid water, hydrogen bonds are not stable and constantly break and reform. (*b*) When water cools below 0°C, the hydrogen bonds are more stable, and a regular crystalline structure forms in which the four partial charges of one water molecule interact with the opposite charges of other water molecules. Because water forms a particularly open crystal latticework, ice is less dense than liquid water and floats. If it did not, inland bodies of water far from the earth's equator might never fully thaw.

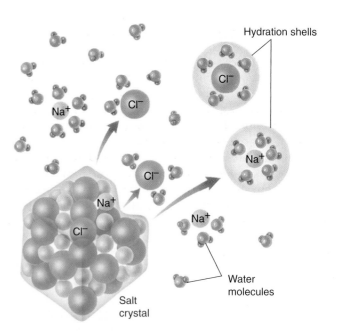

FIGURE 2.17
Why salt dissolves in water. When a crystal of table salt dissolves in water, individual Na^+ and Cl^- ions break away from the salt lattice and become surrounded by water molecules. Water molecules orient around Cl^- ions so that their partial positive poles face toward the negative Cl^- ion; water molecules surrounding Na^+ ions orient in the opposite way, with their partial negative poles facing the positive Na^+ ion. Surrounded by hydration shells, Na^+ and Cl^- ions never reenter the salt lattice.

Water Is a Powerful Solvent

Water is an effective solvent because of its ability to form hydrogen bonds. Water molecules gather closely around any substance that bears an electrical charge, whether that substance carries a full charge (ion) or a charge separation (polar molecule). For example, sucrose (table sugar) is composed of molecules that contain slightly polar hydroxyl (OH) groups. A sugar crystal dissolves rapidly in water because water molecules can form hydrogen bonds with individual hydroxyl groups of the sucrose molecules. Therefore, sucrose is said to be *soluble* in water. Sugar water is a solution of sugar (the solute) dissolved in water (the solvent). Every time a sucrose molecule dissociates or breaks away from the crystal, water molecules surround it in a cloud, forming a **hydration shell** and preventing it from associating with other sucrose molecules. Hydration shells also form around ions such as Na^+ and Cl^- (figure 2.17).

Water Organizes Nonpolar Molecules

Water molecules always tend to form the maximum possible number of hydrogen bonds. When nonpolar molecules such as oils, which do not form hydrogen bonds, are placed in water, the water molecules act to exclude them. The nonpolar molecules are forced into association with one another, thus minimizing their disruption of the hydrogen bonding of water. In effect, they shrink from contact with water, and for this reason they are referred to as **hydrophobic** (Greek *hydros*, "water," and *phobos*, "fearing"). In contrast, polar molecules, which readily form hydrogen bonds with water, are said to be **hydrophilic** ("water-loving").

The tendency of nonpolar molecules to aggregate in water is known as **hydrophobic exclusion.** By forcing the hydrophobic portions of molecules together, water causes these molecules to assume particular shapes. Different molecular shapes have evolved by alteration of the location and strength of nonpolar regions. As you will see, much of the evolution of life reflects changes in molecular shape that can be induced in just this way.

Because polar water molecules cling to one another, it takes considerable energy to separate them. Water also clings to other polar molecules, causing them to be soluble in a water solution, but water tends to exclude nonpolar molecules.

Water Ionizes

The covalent bonds within a water molecule sometimes break spontaneously. In pure water at 25°C, only 1 out of every 550 million water molecules undergoes this process. When it happens, the proton (hydrogen atom nucleus) dissociates from the molecule. Because the dissociated proton lacks the negatively charged electron it was sharing in the covalent bond with oxygen, its own positive charge is no longer counterbalanced, and it becomes a positively charged **hydrogen ion, H⁺**. The rest of the dissociated water molecule, which has retained the shared electron from the covalent bond, is negatively charged and forms a **hydroxide ion, OH⁻**. This process of spontaneous ion formation is called **ionization**:

$$H_2O \rightarrow OH^- + H^+$$
water hydroxide ion hydrogen ion (proton)

At 25°C, a liter of water contains $\frac{1}{10,000,000}$ (or 10^{-7}) mole of H⁺ ions. (A **mole** is defined as the weight in grams that corresponds to the summed atomic masses of all of the atoms in a molecule. In the case of H⁺, the atomic mass is 1, and a mole of H⁺ ions would weigh 1 gram. One mole of any substance always contains 6.02×10^{23} molecules of the substance.) Therefore, the **molar concentration** of hydrogen ions (represented as [H⁺]) in pure water is 10^{-7} mole/liter. Actually, the hydrogen ion usually associates with another water molecule to form a hydronium (H_3O^+) ion.

pH

A more convenient way to express the hydrogen ion concentration of a solution is to use the **pH scale** (figure 2.18). This scale defines pH as the negative logarithm of the hydrogen ion concentration in the solution:

$$pH = -\log [H^+]$$

Because the logarithm of the hydrogen ion concentration is simply the exponent of the molar concentration of H⁺, the pH equals the exponent times −1. Thus, pure water, with an [H⁺] of 10^{-7} mole/liter, has a pH of 7. Recall that for every H⁺ ion formed when water dissociates, an OH⁻ ion is also formed, meaning that the dissociation of water produces H⁺ and OH⁻ in equal amounts. Therefore, a pH value of 7 indicates neutrality—a balance between H⁺ and OH⁻—on the pH scale.

Note that, because the pH scale is *logarithmic*, a difference of 1 on the scale represents a tenfold change in hydrogen ion concentration. This means that a solution with a pH of 4 has 10 times the concentration of H⁺ of a solution with a pH of 5.

Acids. Any substance that dissociates in water to increase the concentration of H⁺ ions is called an acid. Acidic solu-

H⁺ Ion Concentration	pH Value	Examples of Solutions
10^0	0	Hydrochloric acid
10^{-1}	1	
10^{-2}	2	Stomach acid / Lemon juice
10^{-3}	3	Vinegar, cola, beer
10^{-4}	4	Tomatoes
10^{-5}	5	Black coffee / Normal rainwater
10^{-6}	6	Urine / Saliva
10^{-7}	7	Pure water / Blood
10^{-8}	8	Seawater
10^{-9}	9	Baking soda
10^{-10}	10	Great Salt Lake
10^{-11}	11	Household ammonia
10^{-12}	12	Household bleach
10^{-13}	13	Oven cleaner
10^{-14}	14	Sodium hydroxide

FIGURE 2.18
The pH scale. The pH value of a solution indicates its concentration of hydrogen ions. Solutions with a pH less than 7 are acidic, while those with a pH greater than 7 are basic. The scale is logarithmic, so that a pH change of 1 means a tenfold change in the concentration of hydrogen ions. Thus, lemon juice is 100 times more acidic than tomato juice, and seawater is 10 times more basic than pure water, which has a pH of 7.

tions have pH values below 7. The stronger an acid is, the more H⁺ ions it produces and the lower its pH. For example, hydrochloric acid (HCl), which is abundant in your stomach, ionizes completely in water. This means a dilution of 10^{-1} mole per liter of HCl will dissociate to form 10^{-1} mole per liter of H⁺ ions, giving the solution a pH of 1. The pH of champagne, which bubbles because of the carbonic acid dissolved in it, is about 2.

Bases. A substance that combines with H⁺ ions when dissolved in water is called a base. By combining with H⁺ ions, a base lowers the H⁺ ion concentration in the solution. Therefore, basic (or alkaline) solutions have pH values above 7. Very strong bases, such as sodium hydroxide (NaOH), have pH values of 12 or more.

Buffers

The pH inside almost all living cells, and in the fluid surrounding cells in multicellular organisms, is fairly close to 7. Most of the biological catalysts (enzymes) in living systems are extremely sensitive to pH; often even a small change in pH will alter their shape, thereby disrupting their activities and rendering them useless. For this reason, it is important that a cell maintain a constant pH level.

Yet the chemical reactions of life constantly produce acids and bases within cells. Furthermore, many animals eat substances that are acidic or basic; cola, for example, is a moderately strong (although dilute) acidic solution. Despite such variations in the concentrations of H^+ and OH^-, the pH of an organism is kept at a relatively constant level by buffers (figure 2.19).

A **buffer** is a substance that acts as a reservoir for hydrogen ions, donating them to the solution when their concentration falls and taking them from the solution when their concentration rises. What sort of substance will act in this way? Within organisms, most buffers consist of pairs of substances, one an acid and the other a base. The key buffer in human blood is an acid-base pair consisting of carbonic acid (acid) and bicarbonate (base). These two substances interact in a pair of reversible reactions. First, carbon dioxide (CO_2) and H_2O join to form carbonic acid (H_2CO_3), which in a second reaction dissociates to yield bicarbonate ion (HCO_3^-) and H^+ (figure 2.20). If some acid or other substance adds H^+ ions to the blood, the HCO_3^- ions act as a base and remove the excess H^+ ions by forming H_2CO_3. Similarly, if a basic substance removes H^+ ions from the blood, H_2CO_3 dissociates, releasing more H^+ ions into the blood. The forward and reverse reactions that interconvert H_2CO_3 and HCO_3^- thus stabilize the blood's pH.

The reaction of carbon dioxide and water to form carbonic acid is important because it permits carbon, essential to life, to enter water from the air. The earth's oceans are rich in carbon because of the reaction of carbon dioxide with water.

In a condition called blood acidosis, human blood, which normally has a pH of about 7.4, drops 0.2 to

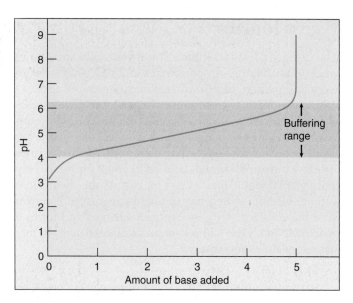

FIGURE 2.19
Buffers minimize changes in pH. Adding a base to a solution neutralizes some of the acid present, and so raises the pH. Thus, as the curve moves to the right, reflecting more and more base, it also rises to higher pH values. A buffer makes the curve rise or fall very slowly over a portion of the pH scale, called the "buffering range" of that buffer.

For this buffer, adding base raises pH more rapidly below pH 4 than above it; what might account for this behavior?

0.4 points on the pH scale. This condition is fatal if not treated immediately. The reverse condition, blood alkalosis, involves an increase in blood pH of a similar magnitude and is just as serious.

The pH of a solution is the negative logarithm of the H^+ ion concentration in the solution. Thus, low pH values indicate high H^+ concentrations (acidic solutions), and high pH values indicate low H^+ concentrations (basic solutions). Even small changes in pH can be harmful to life.

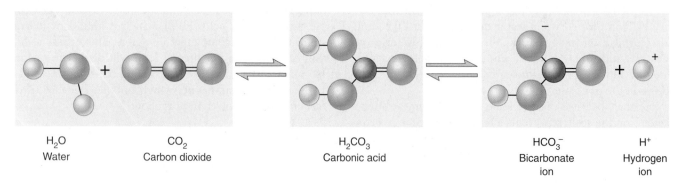

FIGURE 2.20
Buffer formation. Carbon dioxide and water combine chemically to form carbonic acid (H_2CO_3). The acid then dissociates in water, freeing H^+ ions. This reaction makes carbonated beverages acidic, and produced the carbon-rich early oceans where many scientists believe life began.

Concept Review

For interactive testing, visit the Online Learning Center with PowerWeb at www.mhhe.com/Raven7

2.1 Atoms are nature's building material.

Atoms

- All substances are composed of matter, and all matter is composed of atoms. (p. 20)
- Negatively charged electrons circle the nucleus of the atom, while the nucleus is composed of positively charged protons and neutral neutrons. (p. 20)
- An atom's atomic number refers to its number of protons, while its atomic mass refers to the number of protons and neutrons. (p. 20)
- Isotopes are atoms of an element that possess different numbers of neutrons. (p. 21)
- Radioactive isotopes contain nuclei that spontaneously break up into elements with lower atomic numbers. The decay rate is expressed in terms of a half-life. (p. 21)

Electrons Determine the Chemical Behavior of Atoms

- Electrons circle the nucleus of an atom in orbitals. Because the orbitals are so large and are mostly composed of empty space, the nuclei of two atoms never come close enough in nature to interact with each other. (p. 22)
- Because electrons are attracted to the positively charged nucleus, work is necessary so that they stay in their orbits, and thus they have potential energy of position. Moving an electron to a more distant orbital requires energy, while moving an electron closer to the nucleus releases energy. (p. 22)
- Oxidation refers to the loss of an electron; reduction refers to the gain of an electron. (p. 23)

2.2 The atoms of living things are among the smallest.

Kinds of Atoms

- There are 92 naturally occurring elements, arranged in a periodic table based on the interactions of their valence electrons. (p. 24)

2.3 Chemical bonds hold molecules together.

Ionic Bonds Form Crystals

- A molecule is a group of atoms held together by energy in a stable association and joined by chemical bonds. (Compounds are composed of atoms of more than one element.) (p. 25)
- Ionic bonds are formed as attractions between ions of opposite charge, such as those in sodium chloride. (p. 25)

Covalent Bonds Build Stable Molecules

- Covalent bonds form when two atoms share one or more pairs of valence electrons and give rise to true, discrete molecules. (p. 26)

- Covalent bonds are relatively strong, and the strength increases with the number of shared electron pairs. (p. 26)
- A chemical reaction is formed during the formation or breaking of chemical bonds. A reaction may be influenced by several factors, including temperature, concentration of reactants and products, and the presence of catalysts. (p. 26)

2.4 Water is the cradle of life.

Chemistry of Water

- The most outstanding chemical property of water is its ability to form weak chemical associations. (p. 27)

Water Atoms Act Like Tiny Magnets

- A water molecule exhibits electronegativity as both the oxygen and the hydrogen atoms attract the electrons they share in covalent bonds. However, the oxygen atom is more electronegative than the hydrogen atoms. (p. 28)
- Water molecules are polar and exhibit distinct ends with partial charges. (p. 28)
- Hydrogen bonds are formed as opposite charges of bridging hydrogen atoms are attracted, and although each hydrogen bond is relatively weak, the cumulative effect of large numbers of them can be very strong. (p. 28)

Water Clings to Polar Molecules

- Cohesion refers to the attraction of water molecules to other water molecules; adhesion refers to the attraction of other molecules to water molecules. (p. 29)
- The cohesion of water is responsible for its surface tension. (p. 29)
- Water moderates temperatures through its high specific heat and its high heat of vaporization. (p. 29)
- Water is an effective solvent because of its ability to form hydrogen bonds. (p. 30)

Water Ionizes

- A solution's pH is defined as the negative logarithm of its H^+ ion concentration. (p. 31)
- Acids are solutions with high H^+ concentrations (pH < 7), while bases are solutions with low H^+ concentrations (pH > 7). (p. 31)
- Buffers are hydrogen ion reservoirs that either accept or donate H^+ as needed. (p. 32)

Chapter 2 The Nature of Molecules   **33**

Self Test

1. An atom with a neutral charge must contain
   a. the same number of protons and neutrons.
   b. the same number of protons and electrons.
   c. more neutrons because they are neutral.
   d. the same number of neutrons and electrons.
2. Electrons determine the chemical behavior of an atom because
   a. they interact with other atoms.
   b. they determine the charge of the atom (positive, negative, or neutral).
   c. they can be exchanged between atoms.
   d. All of these are correct.
3. The elements within the periodic table are organized by
   a. the number of protons.
   b. the number of neutrons.
   c. the mass of protons and neutrons.
   d. the mass of electrons.
4. Which of the following statements is *not* true?
   a. Molecules held together by ionic bonds are called ionic compounds.
   b. In NaCl, both sodium and chloride have completely filled outer energy levels of 8 electrons.
   c. A sodium atom is able to form an ionic bond with chloride because sodium gives up an electron and chloride gains an electron.
   d. Ionic bonds can form between any two atoms.
5. Oxygen has 6 electrons in its outer energy level; therefore,
   a. it has a completely filled outer energy level.
   b. it can form one double covalent bond or two single covalent bonds.
   c. it does not react with any other atom.
   d. it has a positive charge.
6. The atomic structure of water satisfies the octet rule by
   a. filling the hydrogen atoms' outer energy levels with 8 electrons each.
   b. having electrons shared between the two hydrogen atoms.
   c. having oxygen form covalent bonds with two hydrogen atoms.
   d. having each hydrogen atom give up an electron to the outer energy level of the oxygen atom.
7. The partial charge separation of H_2O results from
   a. the electrons' greater attraction to the oxygen atom.
   b. oxygen's higher electronegativity.
   c. a denser electron cloud near the oxygen atom.
   d. All of these are correct.
8. The attraction of water molecules to other water molecules is
   a. cohesion.
   b. adhesion.
   c. capillary action.
   d. surface tension.
9. What two properties of water help it to moderate changes in temperature?
   a. formation of hydration shells and high specific heat
   b. high heat of vaporization and hydrophobic exclusion
   c. high specific heat and high heat of vaporization
   d. formation of hydration shells and hydrophobic exclusion
10. A substance with a high concentration of hydrogen ions is
    a. called a base.
    b. capable of acting as a buffer.
    c. called an acid.
    d. said to have a high pH.

Test Your Visual Understanding

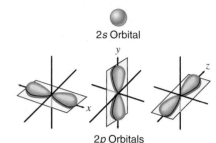

2s Orbital

1s Orbital

2p Orbitals

Orbital for energy level *K:* one spherical orbital (1*s*)

Orbitals for energy level *L:* one spherical orbital (2*s*) and three dumbbell-shaped orbitals (2*p*)

1. This figure shows two energy levels (*K*, which has an *s* orbital labeled 1*s*, and *L*, which has an *s* orbital labeled 2*s* and three *p* orbitals labeled 2*p*). Knowing that the lower level fills with electrons first, followed by the 2*s* orbital and then the *p* orbitals, indicate the number and placement of electrons for the following elements: carbon (C), hydrogen (H), fluorine (F), and neon (Ne).

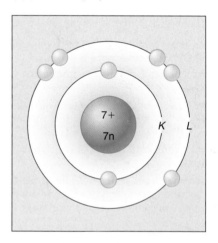

2. This atom has 7 protons and 7 neutrons. What is its atomic number? What is its atomic mass? Can you predict the number of covalent bonds it will form and explain why? What is this element?

Apply Your Knowledge

1. The half-life of radium-226 is 1620 years. If a sample of material contains 16 milligrams of radium-226, how much will it contain in 1620 years? How much will it contain in 3240 years? How long will it take for the sample to contain 1 milligram of radium-226?
2. Scientists find a fossil of a marine animal in the middle of a desert. They use carbon-14 dating to determine how old the fossil is, which would indicate when an ocean covered the desert. There was a 10% (0.10) ratio of ^{14}C compared to ^{12}C in the sample (N_f/N_o) with a ^{14}C half-life ($t_{\frac{1}{2}}$) of 5730 years. Using the equation: $t = [\ln(N_f/N_o)/(-0.693)] \times t_{\frac{1}{2}}$, find the age of the fossil.

3

The Chemical Building Blocks of Life

Concept Outline

3.1 Molecules are the building blocks of life.

Carbon: The Framework of Biological Molecules.
Because individual carbon atoms can form multiple
covalent bonds, biological molecules can be quite complex.

3.2 Proteins perform the chemistry of the cell.

The Many Functions of Proteins. Proteins can be
catalysts, transporters, supporters, and regulators.
Amino Acids: The Building Blocks of Proteins. Proteins
are long chains of various combinations of amino acids.
Protein Structure. A protein's shape is determined by its
amino acid sequence.
How Proteins Fold into Their Functional Shapes.
Chaperone proteins assist in protein folding.
How Proteins Unfold. When conditions such as pH or
temperature fluctuate, proteins may denature or unfold.

3.3 Nucleic acids store and transfer genetic information.

Information Molecules. Nucleic acids store information
in cells. RNA is a single-chain polymer of nucleotides,
while DNA consists of a double helix of nucleotides.

3.4 Lipids make membranes and store energy.

Phospholipids Form Membranes. The aggregation of
phospholipids in water forms biological membranes.
Fats and Other Kinds of Lipids. Organisms utilize a wide
variety of water-insoluble molecules.
Fats as Energy-Storage Molecules. Fats are very efficient
energy-storage molecules because of their high
proportion of C—H bonds.

**3.5 Carbohydrates store energy and provide building
materials.**

Sugars. Sugars are simple carbohydrates, often consisting
of six-carbon rings. The rings can be linked together to
form sugar polymers, or polysaccharides.
Transport and Storage Carbohydrates. Sugars can be
transported as disaccharides and can be stored using the
complex polysaccharides starch and glycogen.
Structural Carbohydrates. Structural carbohydrates, such
as cellulose, are chains of sugars linked in a way that
enzymes cannot easily attack.

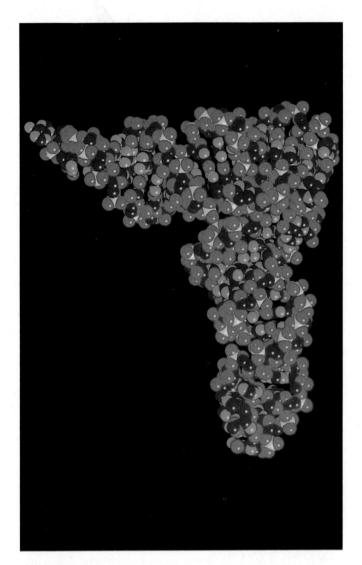

FIGURE 3.1
A macromolecule: tRNA. Some of the macromolecules that
are critical to living things play unique roles as molecular tools.
Many of these are proteins that catalyze chemical reactions, but
increasingly we are coming to appreciate the diverse roles of
nucleic acids, such as this transfer RNA, or tRNA, molecule.
tRNA is a type of RNA, or ribonucleic acid, that assists in the
assembly of proteins.

Molecules are extremely small compared to the familiar objects we see around us. Imagine: There are
more molecules in a cup of water than there are stars in the
sky. But some molecules are much larger than water molecules; they consist of thousands of atoms, forming hundreds of molecules that are linked together into long
chains. These enormous assemblies, almost always synthesized by living things, are macromolecules. As we shall see,
there are four types of biological macromolecules, and they
are the basic chemical building blocks from which all organisms are assembled (figure 3.1).

3.1 Molecules are the building blocks of life.

Carbon: The Framework of Biological Molecules

In chapter 2, we discussed how atoms combine to form molecules. In this chapter, we will focus on biological molecules, those chemical compounds that contain carbon. The frameworks of biological molecules consist predominantly of carbon atoms bonded to other carbon atoms or to atoms of oxygen, nitrogen, sulfur, or hydrogen. Because carbon atoms possess four valence electrons and so can form four covalent bonds, molecules containing carbon can form straight chains, branches, or even rings.

Biological molecules consisting only of carbon and hydrogen are called **hydrocarbons.** Covalent bonds between carbon and hydrogen are energy-rich. Propane gas, for example, is a hydrocarbon consisting of a chain of three carbon atoms, with eight hydrogen atoms bound to it:

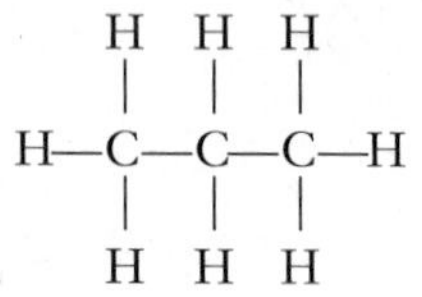

```
      H   H   H
      |   |   |
  H — C — C — C — H
      |   |   |
      H   H   H
```

Because carbon-hydrogen covalent bonds store considerable energy, hydrocarbons make good fuels. Gasoline, for example, is rich in hydrocarbons.

Functional Groups

Carbon and hydrogen atoms both have very similar electronegativities, so electrons in C—C and C—H bonds are evenly distributed, and there are no significant differences in charge over the molecular surface. For this reason, hydrocarbons are nonpolar. Most biological molecules that are produced by cells, however, also contain other atoms. Because these other atoms often have different electronegativities, molecules containing them exhibit regions of positive or negative charge, and so are polar. These molecules can be thought of as a C—H core to which specific groups of atoms called **functional groups** are attached. For example, a hydrogen atom bonded to an oxygen atom (—OH) is a functional group called a *hydroxyl group.*

Functional groups have definite chemical properties that they retain no matter where they occur. The hydroxyl group, for example, is polar, because its oxygen atom, being very electronegative, draws electrons toward itself (as described in chapter 2). Figure 3.2 illustrates the hydroxyl group and other biologically important functional groups.

Building Biological Macromolecules

Some biological molecules in organisms are small and simple, containing only one or a few functional groups. Others are large, complex assemblies called **macromolecules.** Biological macromolecules are traditionally grouped into four major categories: proteins, nucleic acids, lipids, and carbohydrates (table 3.1). In many cases, these macromolecules are **polymers.** A polymer is a long molecule built by linking together a large number of small, similar chemical subunits, like railroad cars coupled to form a train. For example, complex carbohydrates such as starch are polymers of simple ring-shaped sugars, proteins are polymers of amino acids, and nucleic acids (DNA and RNA) are polymers of nucleotides (see table 3.2).

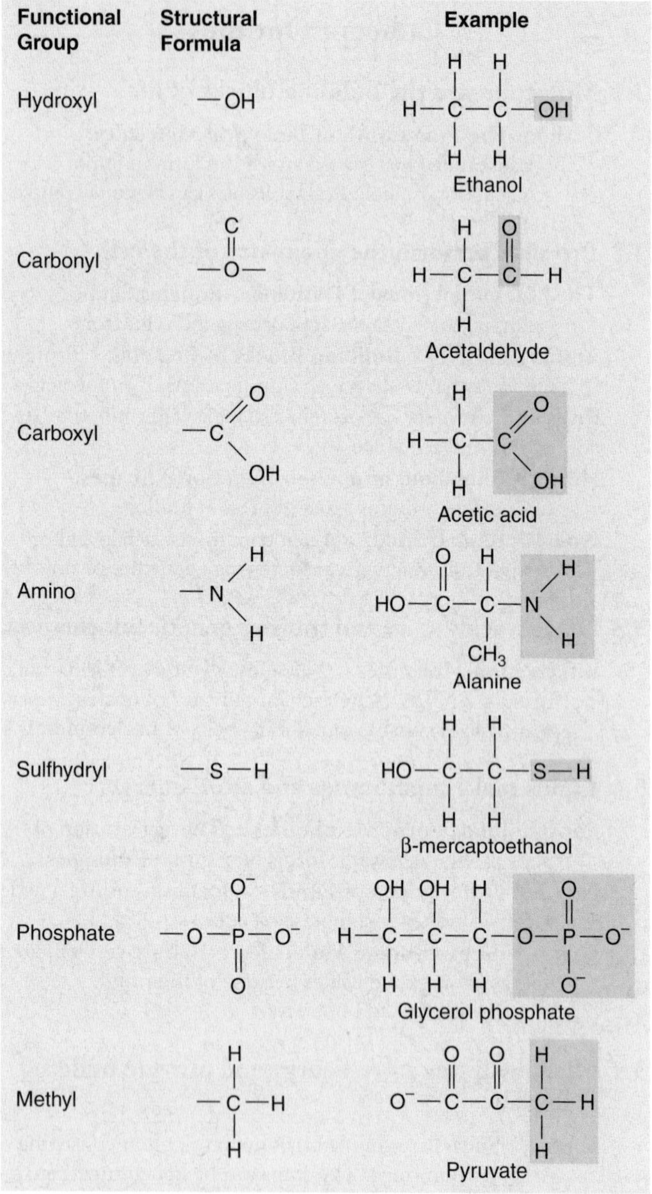

FIGURE 3.2
The primary functional chemical groups.   These groups tend to act as units during chemical reactions and confer specific chemical properties on the molecules that possess them. Amino groups, for example, make a molecule more basic, while carboxyl groups make a molecule more acidic.

Table 3.1 Macromolecules

Macromolecule		Subunit	Function	Example
PROTEINS	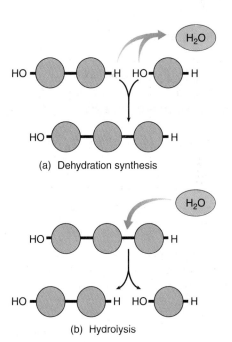			
Functional		Amino acids	Catalysis; transport	Hemoglobin
Structural		Amino acids	Support	Hair; silk
NUCLEIC ACIDS				
DNA		Nucleotides	Encodes genes	Chromosomes
RNA		Nucleotides	Needed for gene expression	Messenger RNA
LIPIDS				
Fats		Glycerol and three fatty acids	Energy storage	Butter; corn oil; soap
Phospholipids		Glycerol, two fatty acids, phosphate, and polar R groups	Cell membranes	Lecithin
Prostaglandins		Five-carbon rings with two nonpolar tails	Chemical messengers	Prostaglandin E (PGE)
Steroids		Four fused carbon rings	Membranes; hormones	Cholesterol; estrogen
Terpenes		Long carbon chains	Pigments; structural	Carotene; rubber
CARBOHYDRATES				
Starch, glycogen		Glucose	Energy storage	Potatoes
Cellulose		Glucose	Cell walls	Paper; strings of celery
Chitin		Modified glucose	Structural support	Crab shells

Although the four categories of macromolecules contain different kinds of subunits, they are all assembled in the same fundamental way: To form a covalent bond between two subunit molecules, an —OH group is removed from one subunit and a hydrogen atom (H) is removed from the other (figure 3.3a). This condensation reaction is called **dehydration synthesis,** because the removal of the —OH group and H during the synthesis of a new molecule in effect constitutes the removal of a molecule of water (H_2O). For every subunit that is added to a macromolecule, one water molecule is removed. These and other biochemical reactions require that the reacting substances be held close together and that the correct chemical bonds be stressed and broken. This process of positioning and stressing, termed *catalysis,* is carried out in cells by a special class of proteins known as enzymes.

Cells disassemble macromolecules into their constituent subunits by performing reactions that are essentially the reverse of dehydration—a molecule of water is added instead of removed (figure 3.3b). In this process, which is called **hydrolysis** (Greek *hydro,* "water," + *lyse,* "break"), a hydrogen atom is attached to one subunit and a hydroxyl group to the other, breaking a specific covalent bond in the macromolecule.

Polymers are large molecules consisting of long chains of similar subunits joined by dehydration reactions.

(a) Dehydration synthesis

(b) Hydrolysis

FIGURE 3.3

Making and breaking macromolecules. (*a*) Biological macromolecules are polymers formed by linking subunits together. The covalent bond between the subunits is formed by dehydration synthesis, a process that creates a water molecule for every bond formed. (*b*) Breaking the bond between subunits requires the returning of a water molecule, a process called hydrolysis.

Table 3.2 Polymer Macromolecules

Monomer	Polymer	Cellular structure
Amino acid	Polypeptide	Intermediate filament
Alanine		
Nucleotide	DNA strand	Chromosome
Fatty acid	Fat molecule	Adipose cells with fat droplets
Monosaccharide	Starch	Starch grains in a chloroplast

Starch grains

3.2 Proteins perform the chemistry of the cell.

The Many Functions of Proteins

We will begin our discussion of biological macromolecules with proteins. The proteins within living organisms are immensely diverse in structure and function. Here we break down their functions into seven categories (table 3.3).

1. **Enzyme catalysis.** We have already encountered one class of proteins, enzymes, which are biological catalysts that facilitate specific chemical reactions. Because of this property, the appearance of enzymes was one of the most important events in the evolution of life. Enzymes are globular proteins, with a three-dimensional shape that fits snugly around the chemicals they work on, facilitating chemical reactions by stressing particular chemical bonds.

2. **Defense.** Other globular proteins use their shapes to "recognize" foreign microbes and cancer cells. These cell surface receptors form the core of the body's hormone and immune systems.

3. **Transport.** A variety of globular proteins transport specific small molecules and ions. The transport protein hemoglobin, for example, transports oxygen in the blood, and myoglobin, a similar protein, transports oxygen in muscle. Iron is transported in blood by the protein transferrin.

Table 3.3   The Many Functions of Proteins

Function	Class of Protein	Examples	Examples of Use
Enzyme catalysis	Enzymes	Hydrolytic enzymes	Cleave polysaccharides
		Proteases	Break down proteins
		Polymerases	Synthesize nucleic acids
		Kinases	Phosphorylate sugars and proteins
Defense	Immunoglobulins	Antibodies	Mark foreign proteins for elimination
	Toxins	Snake venom	Blocks nerve function
	Cell surface antigens	MHC proteins	"Self" recognition
Transport	Circulating transporters	Hemoglobin	Carries O_2 and CO_2 in blood
		Myoglobin	Carries O_2 and CO_2 in muscle
		Cytochromes	Electron transport
	Membrane transporters	Sodium-potassium pump	Excitable membranes
		Proton pump	Chemiosmosis
		Glucose transporter	Transports sugar into cells
Support	Fibers	Collagen	Forms cartilage
		Keratin	Forms hair, nails
		Fibrin	Forms blood clots
Motion	Muscle	Actin	Contraction of muscle fibers
		Myosin	Contraction of muscle fibers
Regulation	Osmotic proteins	Serum albumin	Maintains osmotic concentration of blood
	Gene regulators	*lac* repressor	Regulates transcription
	Hormones	Insulin	Controls blood glucose levels
		Vasopressin	Increases water retention by kidneys
		Oxytocin	Regulates uterine contractions and milk production
Storage	Ion binding	Ferritin	Stores iron, especially in spleen
		Casein	Stores ions in milk
		Calmodulin	Binds calcium ions

FIGURE 3.4
Support. A single functional class of proteins may contain very different members. (*a*) keratin: a peacock feather; (*b*) fibrin: scanning electron micrograph of a blood clot (3000×); (*c*) collagen: strings of a tennis racket from gut tissue; (*d*) silk: a spider's web; (*e*) keratin: human hair.

4. **Support.** Protein fibers play structural roles (figure 3.4). These fibers include keratin in hair, fibrin in blood clots, and collagen, which forms the matrix of skin, ligaments, tendons, and bones, and is the most abundant protein in a vertebrate body.

5. **Motion.** Muscles contract through the sliding motion of two kinds of protein filaments: actin and myosin. Contractile proteins also play key roles in the cell's cytoskeleton and in moving materials in cells.

6. **Regulation.** Small proteins called hormones serve as intercellular messengers in animals. Proteins also play many regulatory roles within the cell, turning on and shutting off genes during development, for example. In addition, proteins also receive information, acting as cell surface receptors.

7. **Storage.** Calcium and iron are stored by binding as ions to specific storage proteins.

Proteins carry out a diverse array of functions, including enzyme catalysis, defense, transport, support, motion, regulation, and storage.

Amino Acids: The Building Blocks of Proteins

Although proteins are complex and versatile molecules, they are all polymers of only 20 different kinds of amino acids, in a specific order. Many scientists believe amino acids were among the first molecules formed on the early earth. It seems highly likely that the oceans that existed early in the history of the earth contained a wide variety of amino acids.

Amino Acid Structure

An **amino acid** molecule contains an amino group (—NH_2), a carboxyl group (—COOH), and a hydrogen atom, all bonded to a central carbon atom:

$$
\begin{array}{c}
R \\
| \\
H_2N-C-COOH \\
| \\
H
\end{array}
$$

Each amino acid has unique chemical properties determined by the nature of the side group (indicated by R) covalently bonded to the central carbon atom. For example, when the side group is —CH_2OH, the amino acid (serine) is polar, but when the side group is —CH_3, the amino acid (alanine) is nonpolar. The 20 common amino acids are grouped into five chemical classes, based on their side groups:

1. Nonpolar amino acids, such as leucine, often have R groups that contain —CH_2 or —CH_3.
2. Polar uncharged amino acids, such as threonine, have R groups that contain oxygen (or only —H).
3. Charged amino acids, such as glutamic acid, have R groups that contain acids or bases.
4. Aromatic amino acids, such as phenylalanine, have R groups that contain an organic (carbon) ring with alternating single and double bonds.
5. Special-function amino acids have unique individual properties; methionine is often the first amino acid in a chain of amino acids, proline causes kinks in chains, and cysteine links chains together.

Each amino acid affects the shape of a protein differently, depending on the chemical nature of its side group. For example, portions of a protein chain with numerous nonpolar amino acids tend to fold into the interior of the protein by hydrophobic exclusion.

Proteins Are Polymers of Amino Acids

In addition to its R group, each amino acid, when ionized, has a pronated amino ($NH_3{}^+$) group at one end and a negative carboxyl (COO^-) group at the other end. The amino and carboxyl groups on a pair of amino acids can undergo a

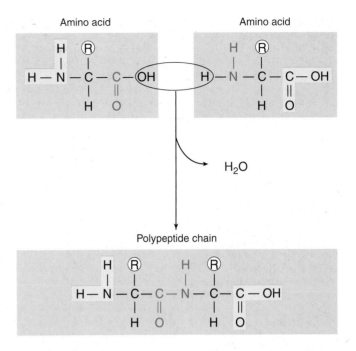

FIGURE 3.5
The peptide bond. A peptide bond forms when the —NH_2 end of one amino acid joins to the —COOH end of another. Because of the partial double-bond nature of peptide bonds, the resulting peptide chain cannot rotate freely around these bonds.

condensation reaction, losing a molecule of water and forming a covalent bond. A covalent bond that links two amino acids is called a **peptide bond** (figure 3.5). The two amino acids linked by such a bond are not free to rotate around the N—C linkage because the peptide bond has a partial double-bond character, unlike the N—C and C—C bonds to the central carbon of the amino acid. The stiffness of the peptide bond is one factor that determines the structural character of the coils and other regular shapes formed by chains of amino acids.

A protein is composed of one or more long chains, or **polypeptides,** composed of amino acids linked by peptide bonds. It was not until the pioneering work of Frederick Sanger in the early 1950s that it became clear that each kind of protein had a specific amino acid sequence. Sanger succeeded in determining the amino acid sequence of insulin and in so doing demonstrated clearly that this protein had a defined sequence, the same for all insulin molecules in the solution. Although many different amino acids occur in nature, only 20 commonly occur in proteins. Figure 3.6 illustrates these 20 "common" amino acids and their side groups.

A protein is a polymer containing a combination of up to 20 different kinds of amino acids. The amino acids fall into five chemical classes, each with different properties. These properties determine the nature of the resulting protein.

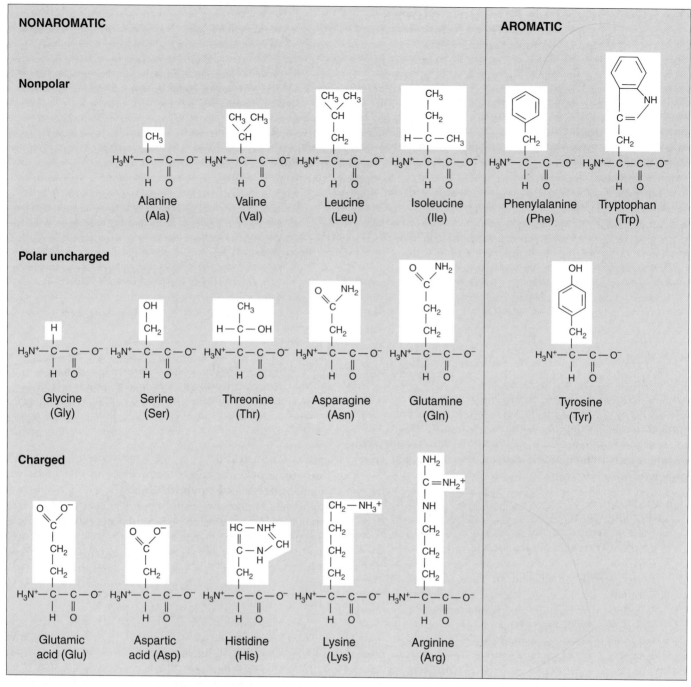

NONAROMATIC

Nonpolar

Alanine (Ala) Valine (Val) Leucine (Leu) Isoleucine (Ile)

AROMATIC

Phenylalanine (Phe) Tryptophan (Trp)

Polar uncharged

Glycine (Gly) Serine (Ser) Threonine (Thr) Asparagine (Asn) Glutamine (Gln)

Tyrosine (Tyr)

Charged

Glutamic acid (Glu) Aspartic acid (Asp) Histidine (His) Lysine (Lys) Arginine (Arg)

SPECIAL FUNCTION

Proline (Pro) Methionine (Met) Cysteine (Cys)

FIGURE 3.6

The 20 common amino acids. Each amino acid has the same chemical backbone, but differs in the side, or R, group it possesses. Six of the amino acids are nonpolar because they have —CH₂ or —CH₃ in their R groups, which are not electronegative. Two of the six are bulkier because they contain ring structures, which classifies them also as aromatic. Another six are polar because they have oxygen or just hydrogen in their R groups; these amino acids, which are uncharged, differ from one another in how polar they are. Five other polar amino acids have a terminal acid or base in their R group, and so are capable of ionizing to a charged form. The remaining three special-function amino acids have chemical properties that allow them to help form links between protein chains or kinks in proteins.

Protein Structure

The shape of a protein is very important because it determines the protein's function. Proteins consist of long amino acid chains folded into complex shapes. What do we know about the shape of these proteins? One way to study the shape of something as small as a protein is to look at it with very short wavelength energy—in other words, with X rays. X-ray diffraction is a painstaking procedure that allows the investigator to build up a three-dimensional picture of the position of each atom. The first protein to be analyzed in this way was myoglobin, soon followed by the related protein hemoglobin. As more and more proteins were studied, a general principle became evident: In every protein studied, essentially all the internal amino acids are nonpolar ones—amino acids such as leucine, valine, and phenylalanine. Water's tendency to hydrophobically exclude nonpolar molecules literally shoves the nonpolar portions of the amino acid chain into the protein's interior (figure 3.7). This positions the nonpolar amino acids in close contact with one another, leaving little empty space inside. Polar and charged amino acids are restricted to the surface of the protein except for the few that play key functional roles.

Levels of Protein Structure

The structure of proteins has traditionally been discussed in terms of four levels of structure: *primary*, *secondary*, *tertiary*, and *quaternary*. Because of progress in our knowledge of protein structure, two additional levels of structure are increasingly distinguished by molecular biologists: *motifs* and *domains* (figure 3.8), two elements that play important roles in coming chapters.

Primary Structure. The specific amino acid sequence of a protein is its primary structure. This sequence is determined by the nucleotide sequence of the gene that encodes the protein. Because the R groups that distinguish the various amino acids play no role in the peptide backbone of proteins, a protein can consist of any sequence of amino acids. Thus, because any of 20 different amino acids might appear at any position, a protein containing 100 amino acids could form any of 20^{100} different amino acid sequences (that's the same as 10^{130}, or 1 followed by 130 zeros—more than the number of atoms known in the universe). This is an important property of proteins because it permits great diversity.

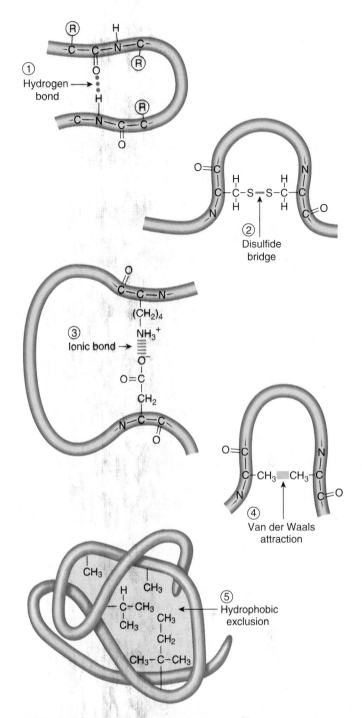

FIGURE 3.7

Interactions that contribute to a protein's shape. Aside from the bonds that link the amino acids in a protein together, several other weaker forces and interactions determine how a protein will fold. (*1*) Hydrogen bonds can form between the different amino acids. (*2*) Fairly strong disulfide bridges can form between two cysteine side chains. (*3*) Ionic bonds can form. (*4*) Van der Waals attractions occur—that is, weak attractions between atoms due to oppositely polarized electron clouds. (*5*) Polar portions of the protein tend to gather on the outside of the protein and interact with water, while the hydrophobic portions of the protein, including nonpolar amino acid chains, are shoved toward the interior of the protein.

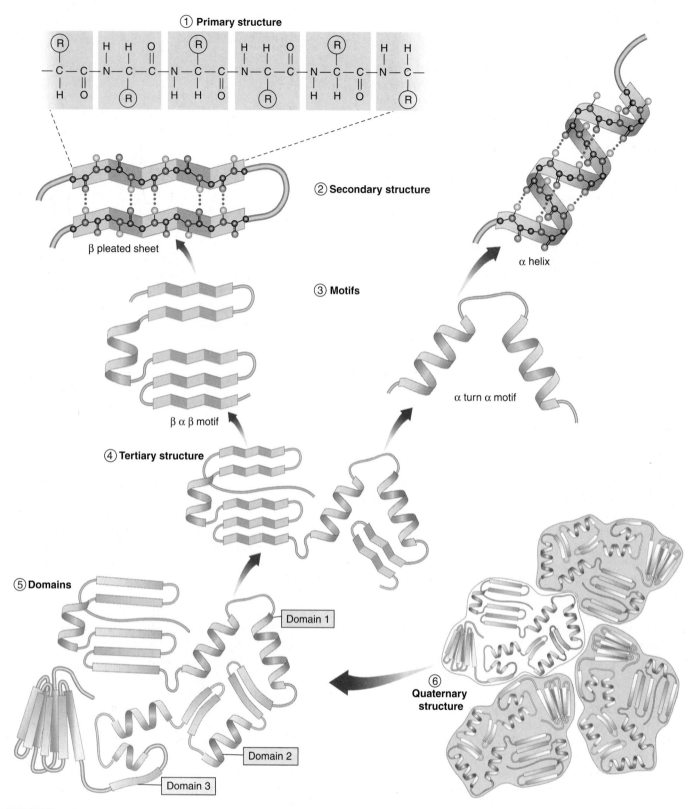

FIGURE 3.8

Six levels of protein structure. The amino acid sequence of a protein is called its (*1*) primary structure. Hydrogen bonds form between nearby amino acids, producing (*2*) foldbacks called beta (β) pleated sheets and coils called alpha (α) helices. These foldbacks and coils constitute the protein's secondary structure. The elements of secondary structure can combine, fold, or crease to form (*3*) motifs. A globular protein folds up on itself further to assume a three-dimensional (*4*) tertiary structure, several of which can be linked together in a protein (*5*) domain. Many proteins aggregate with other polypeptide chains in clusters; this clustering is called the (*6*) quaternary structure of the protein.

Secondary Structure. The amino acid side groups are not the only portions of proteins that form hydrogen bonds. The peptide groups of the main chain also form quite good hydrogen bonds—so good that their interactions with water might be expected to offset the tendency of nonpolar side groups to be forced into the protein interior. Inspection of the protein structures determined by X-ray diffraction reveals why they don't—the polar groups of the main chain form hydrogen bonds with each other! Two patterns of hydrogen bonding occur. In one, hydrogen bonds form along a single chain, linking one amino acid to another farther down the chain. This tends to pull the chain into a coil called an alpha (α) helix. In the other pattern, hydrogen bonds occur across two chains, linking the amino acids in one chain to those in the other. Often, many parallel chains are linked, forming a pleated, sheet-like structure called a beta (β) pleated sheet. The folding of the amino acid chain by hydrogen bonding into these characteristic coils and pleats is called a protein's secondary structure.

Motifs. The elements of secondary structure can combine in proteins in characteristic ways called motifs, or sometimes "supersecondary structure." One very common motif is the $\beta \alpha \beta$ motif, which creates a fold or crease; the so-called "Rossmann fold" at the core of nucleotide binding sites in a wide variety of proteins is a $\beta \alpha \beta \alpha \beta$ motif. A second motif that occurs in many proteins is the β barrel, a β sheet folded around to form a tube. A third type of motif, the α turn α motif, is important because many proteins use it to bind the DNA double helix.

Tertiary Structure. The final folded shape of a globular protein, which positions the various motifs and folds nonpolar side groups into the interior, is called a protein's tertiary structure. A protein is initially driven into its tertiary structure by hydrophobic exclusion from water. Ionic bonds between oppositely charged side groups bring regions into close proximity, and disulfide bonds (covalent links between two cysteine side groups) lock particular regions together. The final folding of a protein is determined by its primary structure—the chemical nature of its side groups (see figure 3.7). Many small proteins can be fully unfolded ("denatured") and will spontaneously refold into their characteristic shape.

The stability of a protein, once it has folded into its 3-D shape, is strongly influenced by how well its interior fits together. When two nonpolar chains in the interior are in very close proximity, they experience a form of molecular attraction called van der Waal's forces. Individually quite weak, these forces can add up to a strong attraction when many of them come into play, like the combined strength of hundreds of hooks and loops on a strip of Velcro. They are effective forces only over short distances, however; no "holes" or cavities exist in the interior of proteins. That is why there are so many different nonpolar amino acids (ala-nine, valine, leucine, isoleucine). Each has a different-sized R group, allowing very precise fitting of nonpolar chains within the protein interior. Now you can understand why a mutation that converts one nonpolar amino acid within the protein interior (alanine) into another (leucine) very often disrupts the protein's stability; leucine is a lot bigger than alanine and disrupts the precise way the chains fit together within the protein interior. A change in even a single amino acid can have profound effects on protein shape and can result in lost or altered function of the protein.

Domains. Many proteins in your body are encoded within your genes in functional sections called exons (exons will be discussed in detail in chapter 15). Each exon-encoded section of a protein, typically 100 to 200 amino acids long, folds into a structurally independent functional unit called a **domain.** As the polypeptide chain folds, the domains fold into their proper shape, each more-or-less independent of the others. This can be demonstrated experimentally by artificially producing the fragment of a polypeptide that forms the domain in the intact protein, and showing that the fragment folds to form the same structure as it does in the intact protein.

A single polypeptide chain connects the domains of a protein, like a rope tied into several adjacent knots. Often the domains of a protein have quite separate functions— one domain of an enzyme might bind a cofactor, for example, and another the enzyme's substrate.

Quaternary Structure. When two or more polypeptide chains associate to form a functional protein, the individual chains are referred to as subunits of the protein. The subunits need not be the same. Hemoglobin, for example, is a protein composed of two α chain subunits and two β chain subunits. A protein's subunit arrangement is called its quaternary structure. In proteins composed of subunits, the interfaces where the subunits contact one another are often nonpolar, and play a key role in transmitting information between the subunits about individual subunit activities.

A change in the identity of one of these amino acids can have profound effects. Sickle cell hemoglobin is a mutation that alters the identity of a single amino acid at the corner of the β subunit from polar glutamate to nonpolar valine. Putting a nonpolar amino acid on the surface creates a "sticky patch" that causes one hemoglobin molecule to stick to another, forming long, nonfunctional chains and leading to the cell sickling characteristic of the hereditary disorder sickle cell anemia.

Protein structure can be viewed at six levels: (1) the amino acid sequence, or primary structure; (2) coils and sheets, called secondary structure; (3) folds or creases, called motifs; (4) the three-dimensional shape, called tertiary structure; (5) functional units, called domains; and (6) individual polypeptide subunits associated in a quaternary structure.

How Proteins Fold into Their Functional Shapes

How does a protein fold into a specific shape? Nonpolar amino acids play a key role. Until recently, investigators thought that newly made proteins fold spontaneously as hydrophobic interactions with water shove nonpolar amino acids into the protein interior. We now know this view is too simple. Protein chains can fold in so many different ways that trial and error would simply take too long. In addition, as the open chain folds its way toward its final form, nonpolar "sticky" interior portions are exposed during intermediate stages. If these intermediate forms are placed in a test tube in the same protein environment that occurs in a cell, they stick to other unwanted protein partners, forming a gluey mess.

Chaperone Proteins

How do cells avoid this? A vital clue came in studies of unusual mutations that prevented viruses from replicating in bacterial cells—it turned out that the virus proteins could not fold properly! Further study revealed that normal cells contain special proteins called **chaperone proteins,** which help new proteins fold correctly. When the bacterial gene encoding its chaperone protein is disabled by mutation, the bacteria die, clogged with lumps of incorrectly folded proteins. Fully 30% of the bacterial proteins fail to fold to the right shape.

Molecular biologists have now identified more than 17 kinds of proteins that act as molecular chaperones. Many are heat shock proteins, produced in greatly elevated amounts if a cell is exposed to elevated temperature; high temperatures cause proteins to unfold, and heat shock chaperone proteins help the cell's proteins refold (figure 3.9).

There is considerable controversy about how chaperone proteins work. It was first thought that they provided a protected environment within which folding could take place unhindered by other proteins, but it now seems more likely that chaperone proteins rescue proteins that are caught in a wrongly folded state, giving them another chance to fold correctly. When investigators "fed" a deliberately misfolded protein, malate dehydrogenase, to chaperone proteins, the protein was rescued, refolding to its active shape.

Protein Folding and Disease

There are tantalizing suggestions that chaperone protein deficiencies may play a role in certain diseases by failing to facilitate the intricate folding of key proteins. Cystic fibrosis is a hereditary disorder in which a mutation disables a protein that plays a vital part in moving ions across cell membranes. In at least some cases, the vital membrane protein appears to have the correct amino acid sequence, but fails to fold to its final form. It has also been speculated that chaperone deficiency may be a cause of the protein clumping in brain cells that produces the amyloid plaques characteristic of Alzheimer disease.

Chaperone proteins help newly produced proteins fold properly.

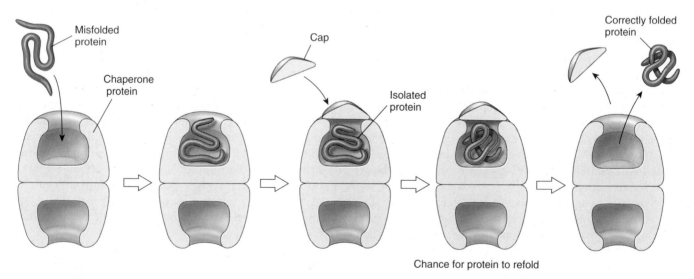

FIGURE 3.9
How one type of chaperone protein works.   This barrel-shaped chaperone protein is a heat shock protein, produced in elevated amounts at high temperatures. An incorrectly folded protein enters one chamber of the barrel, and a cap seals the chamber and confines the protein. The isolated protein is now prevented from aggregating with other misfolded proteins, and it has a chance to refold properly. After a short time period, the protein is ejected, folded or unfolded, and the cycle can repeat itself.

How Proteins Unfold

If a protein's environment is altered, the protein may change its shape or even unfold. This process is called **denaturation** (figure 3.10). Proteins can be denatured when the pH, temperature, or ionic concentration of the surrounding solution is changed. When proteins are denatured, they are usually rendered biologically inactive. This is particularly significant in the case of enzymes. Because practically every chemical reaction in a living organism is catalyzed by a specific enzyme, it is vital that a cell's enzymes remain functional. That is the rationale behind traditional methods of salt-curing and pickling: Prior to the ready availability of refrigerators and freezers, the only practical way to keep microorganisms from growing in food was to keep the food in a solution containing a high concentration of salt or vinegar, which denatured the enzymes of microorganisms and kept them from growing on the food.

Most enzymes function within a very narrow range of physical parameters. Blood-borne enzymes that course through a human body at a pH of about 7.4 would rapidly become denatured in the highly acidic environment of the stomach. On the other hand, the protein-degrading enzymes that function at a pH of 2 or less in the stomach would be denatured in the basic pH of the blood. Similarly, organisms that live near oceanic hydrothermal vents have enzymes that work well at the temperature of this extreme environment (over 100°C). They cannot survive in cooler waters, because their enzymes do not function properly at lower temperatures. Any given organism usually has a tolerance range of pH, temperature, and salt concentration. Within that range, its enzymes maintain the proper shape to carry out their biological functions.

When a protein's normal environment is reestablished after denaturation, a small protein may spontaneously refold into its natural shape, driven by the interactions between its nonpolar amino acids and water (figure 3.11). Larger proteins can rarely refold spontaneously because of the complex nature of their final shape. It is important to distinguish denaturation from **dissociation**. The four subunits of hemoglobin may dissociate into four individual molecules (two α-globin and two β-globin) without denaturation of the folded globin proteins, and will readily reassume their four-subunit quaternary structure.

Every globular protein has a narrow range of conditions in which it folds properly; outside that range, proteins tend to unfold.

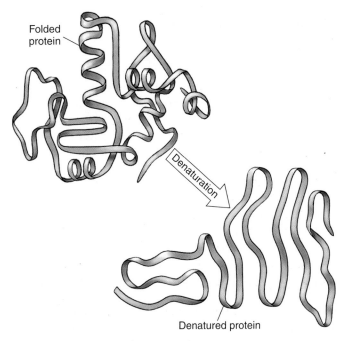

FIGURE 3.10
Protein denaturation. Changes in a protein's environment, such as variations in temperature or pH, can cause a protein to unfold and lose its shape in a process called denaturation. In this denatured state, proteins are biologically inactive.

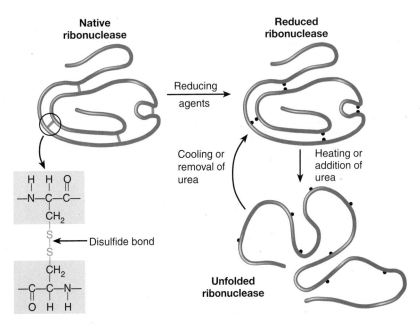

FIGURE 3.11
Primary structure determines tertiary structure. When the protein ribonuclease is treated with reducing agents to break the covalent disulfide bonds that cross-link its chains, and then placed in urea or heated, the protein denatures (unfolds) and loses its enzymatic activity. Upon cooling or removal of urea, it refolds and regains its enzymatic activity. This demonstrates that no information except the amino acid sequence of the protein is required for proper folding: The primary structure of the protein determines its tertiary structure.

3.3 Nucleic acids store and transfer genetic information.

Information Molecules

The biochemical activity of a cell depends on production of a large number of proteins, each with a specific sequence. The ability to produce the correct proteins is passed between generations of organisms, even though the protein molecules themselves are not.

Nucleic acids are the information storage devices of cells, just as disks store the information that computers use, blueprints store the information that builders use, and road maps store the information that tourists use. There are two varieties of nucleic acids: **deoxyribonucleic acid (DNA;** figure 3.12) and **ribonucleic acid (RNA).** The way DNA encodes the information used to assemble proteins is similar to the way the letters on a page encode information (see chapter 14). Unique among macromolecules, nucleic acids are able to serve as templates to produce precise copies of themselves, so that the information that specifies what an organism is can be copied and passed down to its descendants. For this reason, DNA is often referred to as the hereditary material. Cells use the alternative form of nucleic acid, RNA, to read the cell's DNA-encoded information and direct the synthesis of proteins. RNA is similar to DNA in structure and is made as a transcribed copy of portions of the DNA. This transcript passes out into the rest of the cell, where it serves as a blueprint specifying a protein's amino acid sequence. This process will be described in detail in chapter 15.

"Seeing" DNA

DNA molecules cannot be seen with an optical microscope, which is incapable of resolving anything smaller than 1000 atoms across. An electron microscope can image structures as small as a few dozen atoms across, but still cannot resolve the individual atoms of a DNA strand. This limitation was finally overcome in the mid-1980s with the introduction of the scanning-tunneling microscope (figure 3.13).

How do these microscopes work? Imagine you are in a dark room with a chair. To determine the shape of the chair, you could shine a flashlight on it, so that the light bounces off the chair and forms an image on your eye. That's what optical and electron microscopes do; in the latter, the "flashlight" emits a beam of electrons instead of light. You could, however, also reach out and feel the chair's surface with your hand. In effect, you would be putting a probe (your hand) near the chair and measuring how far away the surface is. In a scanning-tunneling microscope, computers advance a probe over the surface of a molecule in steps smaller than the diameter of an atom.

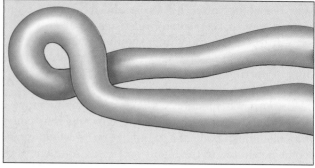

FIGURE 3.12
The first photograph of a DNA molecule. This micrograph, with sketch below, shows a section of DNA magnified a million times. The molecule is so slender that it would take 50,000 of them to equal the diameter of a human hair.

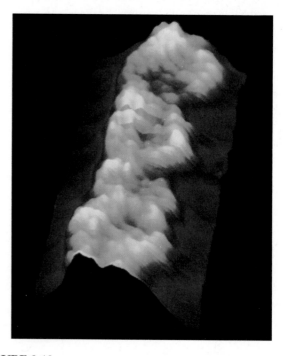

FIGURE 3.13
A scanning-tunneling micrograph of DNA (false color; 2,000,000×). The micrograph shows approximately three turns of the DNA double helix (see figure 3.16).

The Structure of Nucleic Acids

Nucleic acids are long polymers of repeating subunits called **nucleotides.** Each nucleotide consists of three components: a five-carbon sugar (ribose in RNA and deoxyribose in DNA); a phosphate (—PO₄) group; and an organic nitrogenous (nitrogen-containing) base (figure 3.14). When a nucleic acid polymer forms, the phosphate group of one nucleotide binds to the hydroxyl group of another, releasing water and forming a phosphodiester bond. A **nucleic acid,** then, is simply a chain of five-carbon sugars linked together by phosphodiester bonds with an organic base protruding from each sugar (figure 3.15).

Two types of organic bases occur in nucleotides. The first type, *purines,* are large, double-ring molecules found in both DNA and RNA; they are adenine (A) and guanine (G). The second type, *pyrimidines,* are smaller, single-ring molecules; they include cytosine (C, in both DNA and RNA), thymine (T, in DNA only), and uracil (U, in RNA only).

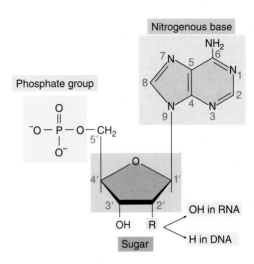

FIGURE 3.14
Structure of a nucleotide. The nucleotide subunits of DNA and RNA are made up of three elements: a five-carbon sugar, an organic nitrogenous base (adenine is shown here), and a phosphate group.

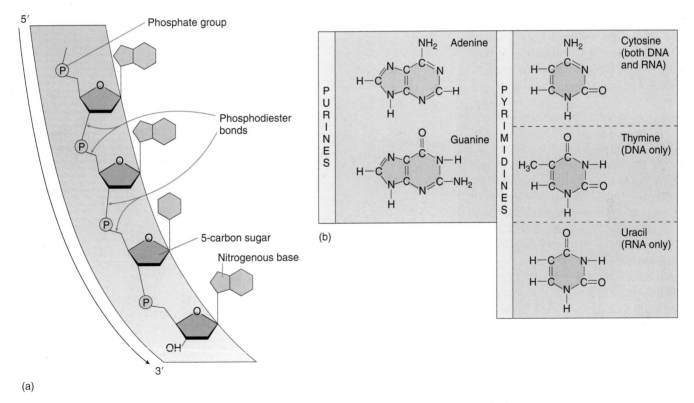

FIGURE 3.15
The structure of a nucleic acid and the organic nitrogen-containing bases. (*a*) In a nucleic acid, nucleotides are linked to one another via phosphodiester bonds, with organic bases protruding from the chain. (*b*) The organic nitrogenous bases can be either purines or pyrimidines. In DNA, thymine replaces the uracil found in RNA.

DNA

Organisms encode the information specifying the amino acid sequences of their proteins as sequences of nucleotides in the DNA. This method of encoding information is very similar to that by which the sequences of letters encode information in a sentence. While a sentence written in English consists of a combination of the 26 different letters of the alphabet in a specific order, the code of a DNA molecule consists of different combinations of the four types of nucleotides in specific sequences, such as CGCTTACG. The information encoded in DNA is used in the everyday metabolism of the organism and is passed on to the organism's descendants.

DNA molecules in organisms exist not as single chains folded into complex shapes, like proteins, but rather as double chains. Two DNA polymers wind around each other like the outside and inside rails of a spiral staircase. Such a winding shape is called a helix, and a helix composed of two chains winding about one another, as in DNA, is called a **double helix.** Each step of DNA's helical staircase is a base-pair, consisting of a base in one chain attracted by hydrogen bonds to a base opposite it on the other chain. These hydrogen bonds hold the two chains together as a duplex (figure 3.16). The base-pairing rules are rigid: Adenine can pair only with thymine (in DNA) or with uracil (in RNA), and cytosine can pair only with guanine. The bases that participate in base-pairing are said to be **complementary** to each other. Additional details of the structure of DNA and how it interacts with RNA in the production of proteins are presented in chapters 14 and 15.

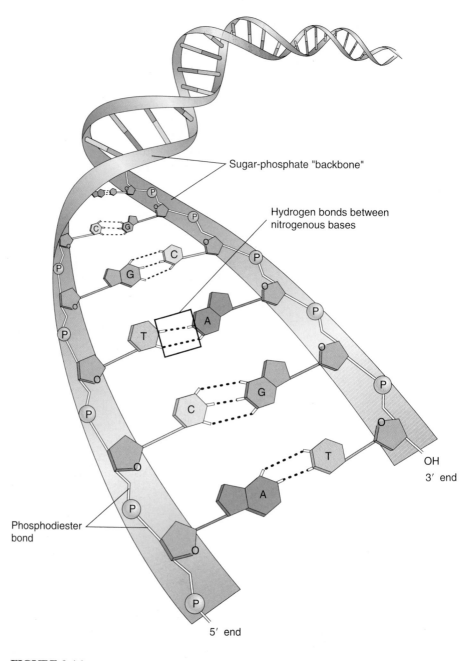

Sugar-phosphate "backbone"

Hydrogen bonds between nitrogenous bases

Phosphodiester bond

OH
3′ end

5′ end

FIGURE 3.16
The structure of DNA. Hydrogen bond formation (dashed lines) between the organic bases, called base-pairing, causes the two chains of a DNA duplex to bind to each other and form a double helix.

RNA

RNA is similar to DNA, but with two major chemical differences. First, RNA molecules contain ribose sugars in which the number 2 carbon is bonded to a hydroxyl group. In DNA, this hydroxyl group is replaced by a hydrogen atom. Second, RNA molecules utilize uracil in place of thymine. Uracil has the same structure as thymine, except that one of its carbons lacks a methyl (—CH$_3$) group.

Transcribing the DNA message into a chemically different molecule such as RNA allows the cell to tell which is the original information storage molecule and which is the

DNA

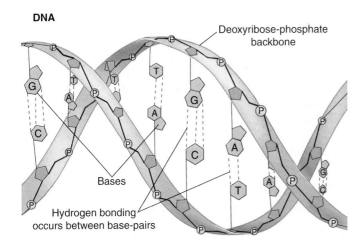

- Deoxyribose-phosphate backbone
- Bases
- Hydrogen bonding occurs between base-pairs

RNA

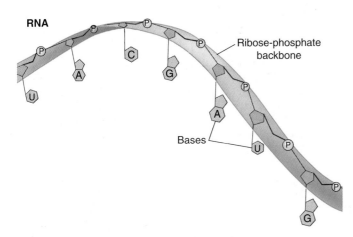

- Ribose-phosphate backbone
- Bases

FIGURE 3.17
DNA versus RNA. DNA forms a double helix, uses deoxyribose as the sugar in its sugar-phosphate backbone, and utilizes thymine among its nitrogenous bases. RNA, on the other hand, is usually single-stranded, uses ribose as the sugar in its sugar-phosphate backbone, and utilizes uracil in place of thymine.

transcript. DNA molecules are always double-stranded (except for a few single-stranded DNA viruses that will be discussed in chapter 26), while the RNA molecules transcribed from DNA are typically single-stranded (figure 3.17). RNA cannot form double helices as DNA does, because of the steric hindrance of the 2′ —OH group. Using two different molecules, one single-stranded and the other double-stranded, separates the role of DNA in storing hereditary information from the role of RNA in using this information to specify protein structure.

Which Came First, DNA or RNA?

The information necessary for the synthesis of proteins is stored in the cell's double-stranded DNA base sequences. The cell uses this information by first making an RNA

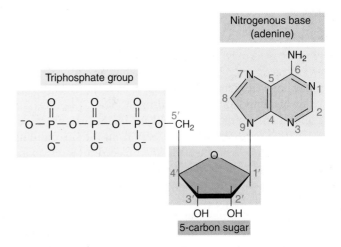

FIGURE 3.18
ATP. Adenosine triphosphate (ATP) contains adenine, a five-carbon sugar, and three phosphate groups. This molecule serves to transfer energy rather than store genetic information.

transcript of it: RNA nucleotides pair with complementary DNA nucleotides. By storing the information in DNA while using a complementary RNA sequence to actually direct protein synthesis, the cell does not expose the information-encoding DNA chain to the dangers of single-strand cleavage every time the information is used. Therefore, DNA is thought to have evolved from RNA as a means of preserving the genetic information, protecting it from the ongoing wear and tear associated with cellular activity. This genetic system has come down to us from the very beginnings of life.

The cell uses the single-stranded, short-lived RNA transcript to direct the synthesis of a protein with a specific sequence of amino acids. Thus, the information flows from DNA to RNA to protein, a process that has been termed the "central dogma" of molecular biology.

ATP

In addition to serving as subunits of DNA and RNA, nucleotide bases play other critical roles in the life of a cell. For example, adenine is a key component of the molecule *adenosine triphosphate* (ATP; figure 3.18), the energy currency of the cell. It also occurs in the molecules *nicotinamide adenine dinucleotide* (NAD^+) and *flavin adenine dinucleotide* (FAD), which carry electrons whose energy is used to make ATP.

A nucleic acid is a long chain of five-carbon sugars with an organic base protruding from each sugar. DNA is a double-stranded helix that stores hereditary information as a specific sequence of nucleotide bases. RNA is a single-stranded molecule that transcribes this information to direct protein synthesis.

3.4 Lipids make membranes and store energy.

Lipids are a loosely defined group of molecules with one main characteristic: They are insoluble in water. The most familiar lipids are fats and oils. Lipids have a very high proportion of nonpolar carbon-hydrogen (C—H) bonds, and so long-chain lipids cannot fold up like a protein to sequester their nonpolar portions away from the surrounding aqueous environment. Instead, when placed in water, many lipid molecules spontaneously cluster together and expose what polar (hydrophilic) groups they have to the surrounding water while sequestering the nonpolar (hydrophobic) parts of the molecules together within the cluster. This spontaneous assembly of lipids is of paramount importance to cells, as it underlies the structure of cellular membranes.

Phospholipids Form Membranes

Complex lipid molecules called **phospholipids** are among the most important molecules of the cell, because they form the core of all biological membranes. An individual phospholipid is a composite molecule, made up of three kinds of subunits:

1. *Glycerol*, a three-carbon alcohol, with each carbon bearing a hydroxyl group. Glycerol forms the backbone of the phospholipid molecule.

2. *Fatty acids*, long chains of —CH_2 groups (hydrocarbon chains) ending in a carboxyl (—COOH) group. Two fatty acids are attached to the glycerol backbone in a phospholipid molecule.

3. *Phosphate group*, attached to one end of the glycerol. The charged phosphate group usually has a charged organic molecule linked to it, such as choline, ethanolamine, or the amino acid serine.

The phospholipid molecule can be thought of as having a polar "head" at one end (the phosphate group) and two long, very nonpolar "tails" at the other (figure 3.19). In water, the nonpolar tails of nearby lipid molecules aggregate away from the water, forming spherical *micelles*, with the tails inward. Phospholipids form two layers with tails pointed toward each other—a lipid bilayer (figure 3.20). Lipid bilayers are the basic framework of biological membranes, discussed in detail in chapter 6.

Because the C—H bonds in lipids are very nonpolar, they are not water-soluble, and aggregate together in water. This kind of aggregation by phospholipids forms biological membranes.

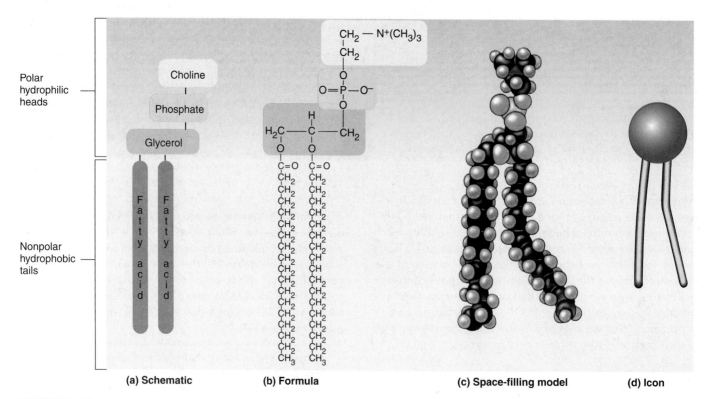

(a) Schematic   (b) Formula   (c) Space-filling model   (d) Icon

FIGURE 3.19
Phospholipids. The phospholipid phosphatidylcholine is shown as (*a*) a schematic, (*b*) a formula, (*c*) a space-filling model, and (*d*) an icon.

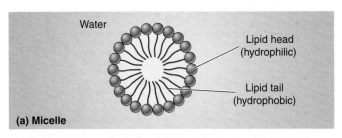

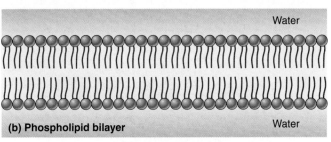

FIGURE 3.20
Lipids spontaneously form micelles or lipid bilayers in water. In an aqueous environment, lipid molecules will orient so that their polar (hydrophilic) heads are in the polar medium, water, and their nonpolar (hydrophobic) tails are held away from the water. (*a*) Droplets called micelles can form, or (*b*) phospholipid molecules can arrange themselves into two layers; in both structures, the hydrophilic heads extend outward and the hydrophobic tails inward.

Fats and Other Kinds of Lipids

Fats are another kind of lipid, but unlike phospholipids, fat molecules do not have a polar end. **Fats** consist of a glycerol molecule to which is attached three fatty acids, one to each carbon of the glycerol backbone. Because it contains three fatty acids, a fat molecule is called a **triglyceride,** or more properly, a **triacylglycerol** (see figure 3.22). The three fatty acids of a triglyceride need not be identical, and often they differ markedly from one another. Organisms store the energy of certain molecules for long periods in the many C—H bonds of fats.

Because triglyceride molecules lack a polar end, they are not soluble in water. Placed in water, they spontaneously clump together, forming fat globules that are very large relative to the size of the individual molecules. Because fats are insoluble, they can be deposited at specific locations within an organism.

Storage fats such as animal fat are one kind of lipid. Oils such as olive oil, corn oil, and coconut oil are also lipids, as are waxes such as beeswax and earwax. The hydrocarbon chains of fatty acids vary in length; the most common are even-numbered chains of 14 to 20 carbons. If all of the internal carbon atoms in the fatty acid chains are bonded to at least two hydrogen atoms, the fatty acid is said to be

saturated, because it contains the maximum possible number of hydrogen atoms (see figure 3.22). If a fatty acid has double bonds between one or more pairs of successive carbon atoms, the fatty acid is said to be **unsaturated.** If a given fatty acid has more than one double bond, it is said to be **polyunsaturated.** Fats made from polyunsaturated fatty acids have low melting points because their fatty acid chains bend at the double bonds, preventing the fat molecules from aligning closely with one another. Consequently, a polyunsaturated fat such as corn oil is usually liquid at room temperature and is called an oil. In contrast, most saturated fats such as those in butter are solid at room temperature.

Organisms contain many other kinds of lipids besides fats (figure 3.21). *Terpenes* are long-chain lipids that are components of many biologically important pigments, such as chlorophyll and the visual pigment retinal. Rubber is also a terpene. *Steroids*, another type of lipid found in membranes, are composed of four carbon rings. Most animal cell membranes contain the steroid cholesterol. Other steroids, such as testosterone and estrogen, function in multicellular organisms as hormones. *Prostaglandins* are a group of about 20 lipids that are modified fatty acids, with two nonpolar "tails" attached to a five-carbon ring. Prostaglandins act as local chemical messengers in many vertebrate tissues.

Cells contain a variety of different lipids, in addition to membrane phospholipids, that play many important roles in cell metabolism.

FIGURE 3.21
Other kinds of lipids. (*a*) Terpenes are found in biological pigments, such as chlorophyll and retinal, and (*b*) steroids play important roles in membranes and in chemical signaling.

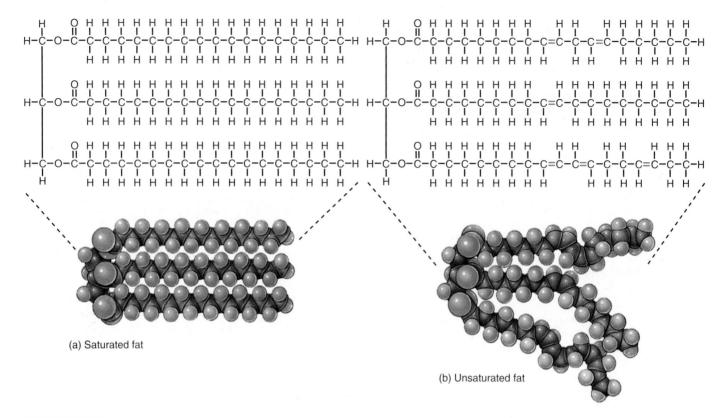

FIGURE 3.22
Saturated and unsaturated fats. (*a*) A saturated triacylglycerol contains three saturated fatty acids, those with no double bonds and, thus, a maximum number of hydrogen atoms bonded to the carbon chain. Many animal triacylglycerols (fats) are saturated. Because their fatty acid chains can fit closely together, these triacylglycerols form immobile arrays called hard fat. (*b*) An unsaturated triacylglycerol contains three unsaturated fatty acids, those with one or more double bonds and, thus, fewer than the maximum number of hydrogen atoms bonded to the carbon chain. Plant fats are typically unsaturated. The many kinks the double bonds introduce into the fatty acid chains prevent the triacylglycerols from closely aligning and instead produce oils, which are liquid at room temperature.

Fats as Energy-Storage Molecules

Most fats contain over 40 carbon atoms. The ratio of energy-storing C—H bonds to carbon atoms in fats is more than twice that of carbohydrates (see section 3.5), making fats much more efficient molecules for storing chemical energy. On the average, fats yield about 9 kilocalories (kcal) of chemical energy per gram, as compared with somewhat less than 4 kcal per gram for carbohydrates.

All fats produced by animals are saturated (except some fish oils), while most plant fats are unsaturated (figure 3.22). The exceptions are the tropical oils (palm oil and coconut oil), which are saturated despite their fluidity at room temperature. It is possible to convert an oil into a solid fat by adding hydrogen. Peanut butter sold in stores is usually artificially hydrogenated to make the peanut fats solidify, preventing them from separating out as oils while the jar sits on the store shelf. However, artificially hydrogenating unsaturated fats seems to eliminate the health advantage they have over saturated fats, by making both

equally rich in C—H bonds. Therefore, it now appears that margarine made from hydrogenated corn oil is no better for your health than butter.

When an organism consumes excess carbohydrate, it is converted into starch, glycogen, or fats and reserved for future use. The reason that many humans gain weight as they grow older is that the amount of energy they need decreases with age, while their intake of food does not. Thus, an increasing proportion of the carbohydrate they ingest is available to be converted into fat.

A diet rich in fats is one of several factors that are thought to contribute to heart disease, particularly atherosclerosis, a condition in which deposits of fatty tissue called plaque adhere to the lining of blood vessels, blocking the flow of blood. Fragments of plaque, breaking off from a deposit, are a major cause of strokes.

Fats are efficient energy-storage molecules because of their high concentration of C—H bonds.

3.5 Carbohydrates store energy and provide building materials.

Sugars

The **carbohydrates** are a loosely defined group of molecules that contain carbon, hydrogen, and oxygen in the molar ratio 1:2:1. Their empirical formula (which lists the atoms in the molecule with subscripts to indicate how many there are of each) is $(CH_2O)_n$, where n is the number of carbon atoms. Because they contain many carbon-hydrogen (C—H) bonds, which release energy when they are oxidized, carbohydrates are well suited for energy storage. Among the most important energy-storage molecules are sugars, which exist in several different forms.

Monosaccharides

The simplest of the carbohydrates are the simple sugars, or **monosaccharides** (Greek *mono*, "single," + Latin *saccharum*, "sugar"). Simple sugars may contain as few as three carbon atoms, but those that play the central role in energy storage have six (figure 3.23). The empirical formula of six-carbon sugars is:

$$C_6H_{12}O_6, \text{ or } (CH_2O)_6$$

Six-carbon sugars can exist in a straight-chain form, but in an aqueous environment they almost always form rings. The most important of these for energy storage is *glucose* (figure 3.24), a six-carbon sugar that has seven energy-storing C—H bonds.

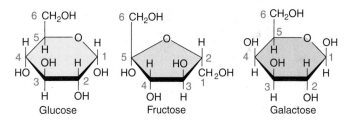

FIGURE 3.23
Monosaccharides. Monosaccharides, or simple sugars, can contain as few as three carbon atoms and are often used as building blocks to form larger molecules. The five-carbon sugars ribose and deoxyribose are components of nucleic acids (see figure 3.17). The six-carbon sugar glucose is a component of large energy-storage molecules. The green numbers refer to the carbon atoms; monosaccharides are conventionally numbered from the more oxidized end.

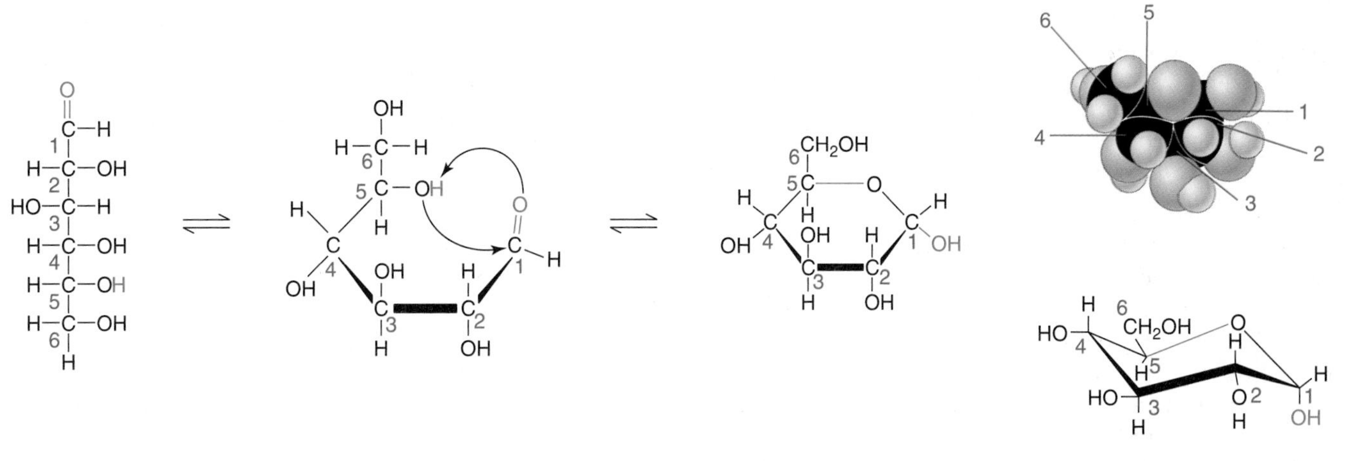

FIGURE 3.24
Structure of the glucose molecule. Glucose is a linear, six-carbon molecule that forms a ring shape in solution. The structure of the ring can be represented in many ways; the ones shown here are the most common, with the carbons conventionally numbered (in *green*) so that the forms can be compared easily. The bold, darker lines represent portions of the molecule that are projecting out of the page toward you—remember, these are three-dimensional molecules!

Disaccharides

Many familiar sugars, such as sucrose, are "double sugars," two monosaccharides joined by a covalent bond (figure 3.25). Called **disaccharides,** they often play a role in the transport of sugars, as we will discuss shortly.

Polysaccharides

Polysaccharides are macromolecules made up of monosaccharide subunits. Starch is a polysaccharide used by plants to store energy. It consists entirely of glucose molecules, linked one after another in long chains. Cellulose is a polysaccharide that serves as a structural building material in plants. It too consists entirely of glucose molecules linked together into chains, and special enzymes are required to break the links.

Sugar Isomers

Glucose is not the only sugar with the formula $C_6H_{12}O_6$. Other common six-carbon sugars such as fructose and galactose also have this empirical formula (figure 3.26). These sugars are **isomers,** or alternative forms, of glucose or other six-carbon monosaccharides. Even though isomers have the same empirical formula, their atoms are arranged in different ways; that is, their three-dimensional structures are different. These structural differences often account for substantial functional differences between the isomers. Glucose and fructose, for example, are *structural isomers.* In fructose, the double-bonded oxygen is attached to an internal carbon rather than to a terminal one. Your taste buds can tell the difference, as fructose tastes much sweeter than glucose, despite the fact that both sugars have the same chemical composition. This structural difference also has an important chemical consequence: The two sugars form different polymers.

Unlike fructose, galactose has the same bond structure as glucose; the only difference between galactose and glucose is the orientation of one hydroxyl group. Because the hydroxyl group positions are mirror images of each other, galactose and glucose are called *stereoisomers.*

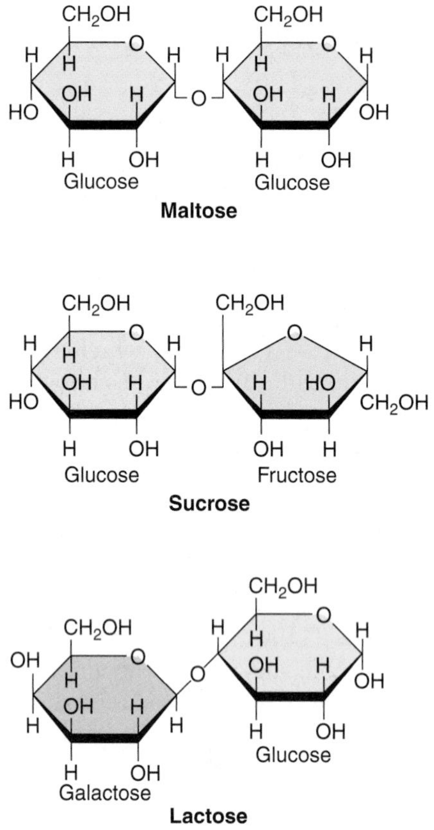

FIGURE 3.25
Disaccharides. Sugars such as maltose, sucrose, and lactose are disaccharides, composed of two monosaccharides linked by a covalent bond.

Sugars are among the most important energy-storage molecules in organisms, containing many energy-storing C—H bonds. The structural differences among sugar isomers can confer substantial functional differences upon the molecules.

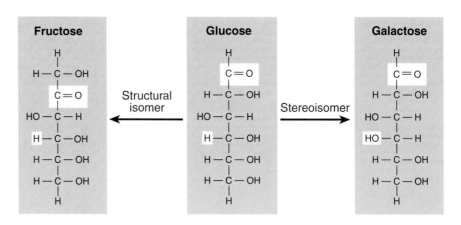

FIGURE 3.26
Isomers and stereoisomers. Glucose, fructose, and galactose are isomers with the empirical formula $C_6H_{12}O_6$. A structural isomer of glucose, such as fructose, has identical chemical groups bonded to different carbon atoms, while a stereoisomer of glucose, such as galactose, has identical chemical groups bonded to the same carbon atoms but in different orientations.

Transport and Storage Carbohydrates

Transport Disaccharides

Most organisms transport sugars within their bodies. In humans, the glucose that circulates in the blood does so as a simple monosaccharide. In plants and many other organisms, however, glucose is converted into a transport form before it is moved from place to place within the organism. In such a form, it is less readily metabolized (used for energy) during transport. Transport forms of sugars are commonly made by linking two monosaccharides together to form a disaccharide (Greek *di*, "two"). Disaccharides serve as effective reservoirs of glucose because the normal glucose-utilizing enzymes of the organism cannot break the bond linking the two monosaccharide subunits. Enzymes that can do so are typically present only in the tissue where the glucose is to be used.

Transport forms differ depending on which monosaccharides link to form the disaccharide. Glucose forms transport disaccharides with itself and many other monosaccharides, including fructose and galactose. When glucose forms a disaccharide with its structural isomer, fructose, the resulting disaccharide is *sucrose*, or table sugar (figure 3.27*a*). Sucrose is the form in which most plants transport glucose and the sugar that most humans (and other animals) eat. Sugarcane and sugar beets are rich in sucrose.

When glucose is linked to its stereoisomer, galactose, the resulting disaccharide is *lactose*, or milk sugar. Many mammals supply energy to their young in the form of lactose. Adults have greatly reduced levels of lactase, the enzyme required to cleave lactose into its two monosaccharide components, and thus cannot metabolize lactose

efficiently. Most of the energy that is channeled into lactose production is therefore reserved for their offspring.

Storage Polysaccharides

Organisms store the metabolic energy contained in monosaccharides by converting them into disaccharides, such as *maltose* (figure 3.27*b*), which are then linked together into insoluble forms that are deposited in specific storage areas in their bodies. These insoluble polysaccharides are long polymers of monosaccharides formed by dehydration synthesis. Plant polysaccharides formed from glucose are called **starches.**

The starch with the simplest structure is *amylose*, which is composed of many hundreds of glucose molecules linked together in long, unbranched chains. Each linkage occurs between the number 1 carbon of one glucose molecule and the number 4 carbon of another, so that amylose is, in effect, a longer form of maltose. The long chains of amylose tend to coil up in water, a property that renders amylose insoluble. Potato starch is about 20% amylose.

Most plant starch, including the remaining 80% of potato starch, is a somewhat more complicated variant of amylose called *amylopectin*. Pectins are branched polysaccharides with short, linear amylose branches consisting of 20 to 30 glucose subunits.

The animal version of starch is **glycogen.** Like amylopectin, glycogen is an insoluble polysaccharide containing branched amylose chains. In glycogen, the average chain length is much greater, and there are more branches than in plant starch.

Starches are glucose polymers. Most starches are branched, rendering the polymer insoluble.

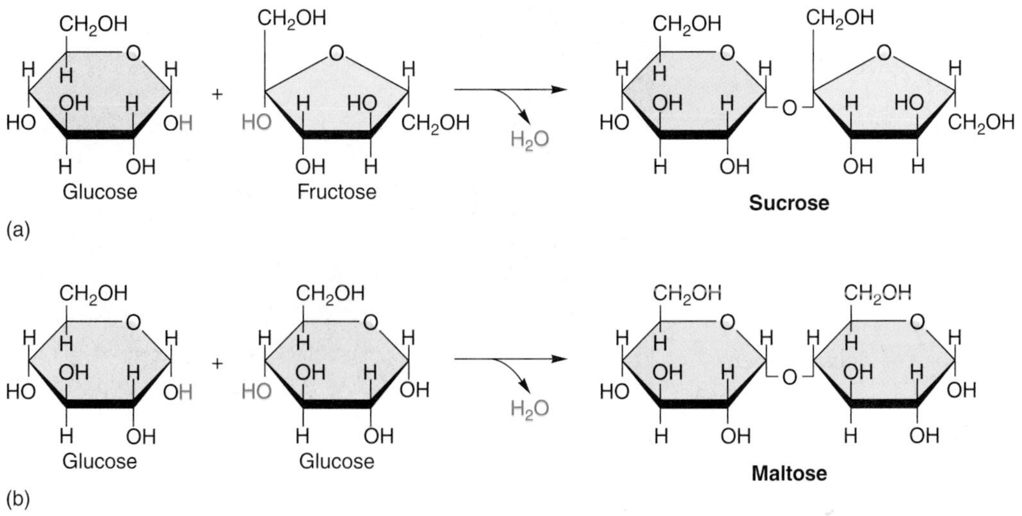

(a)

(b)

FIGURE 3.27
How disaccharides form.
Some disaccharides are used to transport glucose from one part of an organism's body to another; one example is sucrose (*a*), which is found in sugarcane. Other disaccharides, such as maltose in grain (*b*), are used for storage.

Structural Carbohydrates

Cellulose

While some chains of sugars store energy, others serve as structural material for cells. For two glucose molecules to link together, the glucose subunits must be the same form. Glucose can form a ring in two ways, with the hydroxyl group attached to the carbon where the ring closes, being locked into place either below or above the plane of the ring. If below, it is called the **alpha form (α)**, and if above, the **beta form (β)**. All of the glucose subunits of the starch chain are alpha-glucose (figure 3.28). When a chain of glucose molecules consists of all beta-glucose subunits, a polysaccharide with very different properties results. This structural polysaccharide is *cellulose*, the chief component of plant cell walls (figure 3.28). Cellulose is chemically similar to amylose, with one important difference: the starch-degrading enzymes that occur in most organisms cannot break the bond between two beta-glucose sugars. This is not because the bond is stronger, but rather because its cleavage requires an enzyme most organisms lack. Because cellulose cannot be broken down readily, it works well as a biological structural material and occurs widely in this role in plants. Those few animals able to break down cellulose find it a rich source of energy. Certain vertebrates, such as cows, can digest cellulose by means of bacteria and protists harbored in their digestive tracts that provide the necessary enzymes.

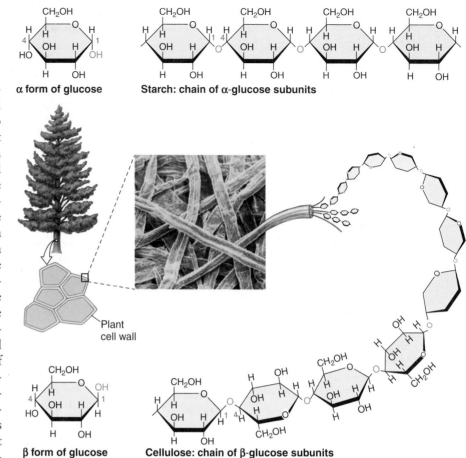

FIGURE 3.28
Polymers of glucose.   While starch chains consist of alpha-glucose subunits, cellulose chains consist of beta-glucose subunits. Cellulose fibers can be very strong and are quite resistant to metabolic breakdown, which is one reason wood is such a good building material.

Chitin

The structural building material in arthropods and many fungi is called chitin. *Chitin* is a modified form of cellulose with a nitrogen group added to the glucose units. When cross-linked by proteins, it forms a tough, resistant surface material that serves as the hard exoskeleton of arthropods such as insects and crustaceans (figure 3.29; see chapter 33). Few organisms are able to digest chitin and use it as a major source of nutrition, but most possess a chitinase enzyme, probably to protect against fungi.

> **Structural carbohydrates are chains of sugars that are not easily digested. They include cellulose in plants and chitin in arthropods and fungi.**

FIGURE 3.29
Chitin.   Chitin, which might be considered a modified form of cellulose, is the principal structural element in the external skeletons of many invertebrates, such as this lobster.

Concept Review

For interactive testing, visit the Online Learning Center with PowerWeb at www.mhhe.com/Raven7

3.1 Molecules are the building blocks of life.

Carbon: The Framework of Biological Molecules

- Biological molecules consist predominantly of carbon atoms bonded to other carbon atoms or to atoms of oxygen, nitrogen, sulfur, or hydrogen. (p. 36)
- Hydrocarbons are molecules consisting only of carbon and hydrogen; thus, they store considerable energy. (p. 36)
- Functional groups have definite chemical properties that are retained no matter where they occur, and tend to act as units during chemical reactions, conferring specific chemical properties to molecules that possess them. (p. 36)
- Polymers are composed of long chains of similar subunits joined by dehydration synthesis. (pp. 36–37)

3.2 Proteins perform the chemistry of the cell.

The Many Functions of Proteins

- The major functions of proteins within living organisms include enzyme catalysis, defense, transport, support, motion, regulation, and storage. (pp. 39–40)

Amino Acids: The Building Blocks of Proteins

- Proteins are polymers of 20 different kinds of amino acids, arranged in specific orders. (p. 41)
- The 20 common amino acids are grouped into five chemical classes: nonpolar, polar uncharged, charged, aromatic, and special function. (p. 41)
- A peptide bond is a covalent bond that links two amino acids together. (p. 41)

Protein Structure

- Proteins consist of long amino acid chains folded into complex shapes, and the shape of a protein determines its function. (p. 43)
- The specific amino acid sequence of a protein forms its primary structure. (p. 43)
- The folding of the amino acid chain by hydrogen bonding into coils and pleats forms a protein's secondary structure. (p. 45)
- The elements of secondary structure can combine in proteins in characteristic ways (motifs). (p. 45)
- The final folded shape of a globular protein forms a protein's tertiary structure. (p. 45)
- Each exon-encoded section of a protein folds into a structurally independent functional unit (domain). (p. 45)
- Quaternary structure refers to individual polypeptide subunits associated to form functional proteins. (p. 45)

How Proteins Fold into Their Functional Shapes

- Chaperone proteins help new proteins fold correctly. (p. 46)

How Proteins Unfold

- Denaturation occurs when a protein changes shape, or unfolds, and can be caused by a change in environmental pH, temperature, or ionic concentration. (p. 47)
- Denaturation usually renders a protein biologically inactive. (p. 47)

3.3 Nucleic acids store and transfer genetic information.

Information Molecules

- Deoxyribonucleic acid (DNA) is often referred to as heredity material, and encodes the information used to assemble proteins. (p. 48)
- Cells use ribonucleic acid (RNA) to read the DNA-encoded information and direct protein synthesis. (p. 48)
- Nucleic acids are long polymers of repeating subunits (nucleotides) that consist of a five-carbon sugar, a phosphate, and a nitrogen-containing base. (p. 49)
- DNA exists in a double-stranded helix, while RNA is a single-stranded molecule. (pp. 50–51)

3.4 Lipids make membranes and store energy.

Phospholipids Form Membranes

- A phospholipid is a composite molecule made up of glycerol, fatty acids, and a phosphate group. It contains a polar head and two nonpolar tails, forming the core of all biological membranes. (p. 52)

Fats and Other Kinds of Lipids

- A triglyceride is a fat, consisting of a glycerol molecule and three fatty acids. Because triglyceride molecules lack a polar end, they are not soluble in water. (p. 53)
- Other kinds of lipids include terpenes, steroids, and prostaglandins. (p. 53)

Fats as Energy-Storage Molecules

- Fats have a high concentration of C—H bonds, and are thus efficient energy-storage molecules. (p. 54)

3.5 Carbohydrates store energy and provide building materials.

Sugars

- Carbohydrates are a loosely defined group of molecules containing carbon, hydrogen, and oxygen in 1:2:1 ratio. (p. 55)
- Structural differences between sugar isomers confer functional differences between the molecules. (p. 56)

Transport and Storage Carbohydrates

- Starches are glucose polymers found in plants, while glycogen serves as the animal version of starch. (p. 57)

Structural Carbohydrates

- Structural carbohydrates such as cellulose (in plants) and chitin (in arthropods) are chains of sugars that resist digestion because most organisms lack the necessary enzymes. (p. 58)

Self Test

1. Proteins, nucleic acids, lipids, and carbohydrates all have certain characteristics in common. Which of the following is *not* a common characteristic?
   a. They are organic, which means they are all living substances.
   b. They all contain the element carbon.
   c. They contain simpler units that are linked together, making larger molecules.
   d. They all contain functional groups.

2. A peptide bond forms by
   a. a condensation reaction.
   b. dehydration synthesis.
   c. the formation of a covalent bond.
   d. all of these.

3. Protein motifs are considered a type of
   a. primary structure.
   b. secondary structure.
   c. tertiary structure.
   d. quaternary structure.

4. The substitution of one amino acid for another
   a. will change the primary structure of the polypeptide.
   b. can change the secondary structure of the polypeptide.
   c. can change the tertiary structure of the polypeptide.
   d. All of these are correct.

5. Chaperone proteins function by
   a. providing a protective environment in which proteins can fold properly.
   b. degrading proteins that have folded improperly.
   c. rescuing proteins that folded incorrectly and allowing them to refold into the proper configuration.
   d. providing a template for how the proteins should fold.

6. Which of the following lists the purine nucleotides?
   a. adenine and cytosine
   b. guanine and thymine
   c. cytosine and thymine
   d. adenine and guanine

7. The two strands of a DNA molecule are held together through base-pairing. Which of the following best describes the base-pairing in DNA?
   a. Adenine forms two hydrogen bonds with thymine.
   b. Adenine forms two hydrogen bonds with uracil.
   c. Cytosine forms two hydrogen bonds with guanine.
   d. Cytosine forms two hydrogen bonds with thymine.

8. A characteristic common to all lipids is
   a. that they contain long chains of C—H bonds.
   b. that they are insoluble in water.
   c. that they have a glycerol backbone.
   d. All of these are characteristics of all lipids.

9. Carbohydrates have many functions in the cell. Which of the following is an *incorrect* match of the carbohydrate with its function?
   a. sugar transport in plants—disaccharides
   b. energy storage in plants—starches
   c. energy storage in plants—lactose
   d. sugar transport in humans—glucose

10. Which of the following carbohydrates is associated with plants?
    a. glycogen
    b. amylopectin
    c. chitin
    d. levo-glucose

Test Your Visual Understanding

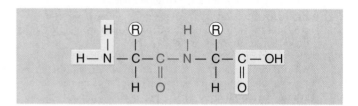

1. The amino acids of a polypeptide affect the shape of the protein. Assume this is a section of a longer polypeptide chain. Predict where each of the following amino acid pairs would be found in the protein—facing toward the outside or folded toward the interior—and explain why:
   a. Both animo acids are valine.
   b. One amino acid is aspartic acid, and the other is serine.
   c. Both amino acids are glycine.
   d. One amino acid is alanine, and the other is isoleucine.

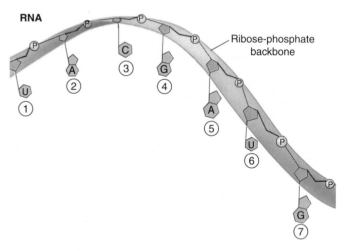

2. Describe the DNA template that produced this molecule of RNA by indicating the DNA bases that are complementary to the numbered RNA bases.

Apply Your Knowledge

1. How many molecules of water are used up in the breakdown of a polypeptide 15 amino acids in length?
2. Why do you suppose the monosaccharide glucose is circulated in the blood of humans rather than a disaccharide, such as sucrose, which is the transport sugar found in plants?

4

The Origin and Early History of Life

Concept Outline

4.1 All living things share key characteristics.

What Is Life? All known organisms share certain general properties, and to a large degree these properties define what we mean by life.

4.2 There is considerable disagreement about the origin of life.

Hypotheses About the Origin of Life. There are both religious and scientific views about the origin of life. This text treats only the latter—the scientifically testable.

Scientists Disagree About Where Life Started. The atmosphere of the early earth was rich in hydrogen, providing a ready supply of energetic electrons with which to build organic molecules.

The Miller-Urey Experiment. Experiments by Miller and Urey and others have attempted to duplicate the conditions of the early earth and produce many of the key molecules of living organisms.

4.3 There are many hypotheses about the origin of cells.

Ideas About the Origin of Cells. The first cells are thought to have arisen spontaneously, but there is little agreement as to the mechanism.

4.4 Cells became progressively more complex as they evolved.

The Earliest Cells. The earliest fossils are of prokaryotes too small to see with the unaided eye.

The First Eukaryotic Cells. Fossils of the first eukaryotic cells do not appear in rocks until 1.5 billion years ago, over 1 billion years after prokaryotes appear. Multicellular life is restricted to the four eukaryotic kingdoms of life.

4.5 Scientists are beginning to take the possibility of extraterrestrial life seriously.

Has Life Evolved Elsewhere? It seems probable that life has evolved on other planets besides our own. The possible presence of life in the warm waters beneath the surface of Europa, a moon of Jupiter, is a source of current speculation.

FIGURE 4.1
The origin of life. The fortuitous mix of physical events and chemical elements at the right place and time created the first living cells on earth.

A great many scientists have intriguing ideas that explain how life may have originated on earth, but we know very little for sure. New hypotheses are being proposed constantly, and old ones reevaluated. By the time this text is published, some of the ideas presented here about the origin of life will surely be obsolete. Thus, the contesting ideas are presented in this chapter in an open-ended format, attempting to make clear that there is as yet no single answer to the question of how life originated on earth. Although recent photographs taken by the Hubble Space Telescope have revived controversy about the age of the universe, it seems clear that the earth itself was formed about 4.5 billion years ago. The oldest clear evidence of life—microfossils in ancient rock—are 2.5 billion years old. The origin of life seems to have been sparked by just the right combination of physical events and chemical processes (figure 4.1).

4.1 All living things share key characteristics.

The earth formed as a hot mass of molten rock about 4.5 billion years ago. As the earth cooled, much of the water vapor present in its atmosphere condensed into liquid water that accumulated on the surface in chemically rich oceans. One scenario for the origin of life is that it originated in this dilute, hot, smelly soup of ammonia, formaldehyde, formic acid, cyanide, methane, hydrogen sulfide, and organic hydrocarbons. Whether at the oceans' edge, in hydrothermal deep-sea vents, or elsewhere, the general consensus among researchers is that life arose spontaneously from these early waters. While the way in which this happened remains a puzzle, we cannot escape a certain curiosity about the earliest steps that eventually led to the origin of all living things on earth, including ourselves. How did organisms evolve from the complex molecules that swirled in the early oceans?

What Is Life?

Before we can address the question "What *is* life?", we must first consider what qualifies something as "living." This is not a simple concept because of the loose manner in which the term "alive" is used. Imagine a situation in which two astronauts encounter a large, amorphous blob on the surface of a planet. How would they determine whether it is alive?

Movement. One of the first things the astronauts might do is observe the blob to see if it moves. Most animals move about (figure 4.2), but movement from one place to another in itself is not diagnostic of life. Most plants and even some animals do not move about, while numerous nonliving objects, such as clouds, do move. The criterion of movement is thus neither *necessary* (possessed by all life) nor *sufficient* (possessed only by life).

Sensitivity. The astronauts might prod the blob to see if it responds. All living things respond to stimuli (figure 4.3). Plants grow toward light, and animals retreat from fire. Not all stimuli produce responses, however. Imagine kicking a redwood tree or singing to a hibernating bear. This criterion, although superior to the first, is still inadequate to define life.

Death. The astronauts might attempt to kill the blob. All living things die, while inanimate objects do not. Death is not easily distinguished from disorder, however; a car that breaks down has not died because it was never alive. Death is simply the loss of life, so this is a circular definition at best. Unless one can detect life, death is a meaningless concept, and hence a very inadequate criterion for defining life.

Complexity. Finally, the astronauts might cut up the blob, to see if it is complexly organized. All living things are complex. Even the simplest bacteria contain

FIGURE 4.2
Movement. Animals have evolved mechanisms that allow them to move about in their environment. While some animals, like this giraffe, move on land, others move through water or air.

FIGURE 4.3
Sensitivity. This father lion is responding to a stimulus: He has just been bitten on the rump by his cub. As far as we know, all organisms respond to stimuli, although not always to the same ones or in the same way. Had the cub bitten a tree instead of its father, the response would not have been as dramatic.

a bewildering array of molecules, organized into many complex structures. However, a computer is also complex, but not alive. Complexity is a *necessary* criterion of life, but it is not *sufficient* in itself to identify living things because many complex things are not alive.

To determine whether the blob is alive, the astronauts would have to learn more about it. Probably the best thing they could do would be to examine it more carefully and determine whether it resembles the organisms we are familiar with, and if so, how.

Fundamental Properties of Life

As we discussed in chapter 1, all known organisms share certain general properties. To a large degree, these properties define what we mean by life. The following fundamental properties are shared by all organisms on earth.

Cellular organization.   All organisms consist of one or more *cells*—complex, organized assemblages of molecules enclosed within membranes (figure 4.4).

Sensitivity.   All organisms respond to stimuli—though not always to the same stimuli in the same ways.

Growth.   All living things assimilate energy and use it to maintain order and grow, a process called **metabolism.** Plants, algae, and some bacteria use sunlight to create covalent carbon–carbon bonds from CO_2 and H_2O through photosynthesis. This transfer of the energy in covalent bonds is essential to all life on earth.

Development.   Both unicellular and multicellular organisms undergo systematic, gene-directed changes as they grow and mature.

Reproduction.   All living things reproduce, passing on individuals from one generation to the next.

Regulation.   All organisms have regulatory mechanisms that coordinate internal processes.

Homeostasis.   All living things maintain relatively constant internal conditions, different from their environment.

The Key Role of Heredity

Are these properties adequate to define life? Is a membrane-enclosed entity that grows and reproduces alive? Not necessarily. Soap bubbles and protein microspheres spontaneously form hollow bubbles that enclose a small volume of air or water. These spheres can enclose energy-processing molecules, and they may also grow and subdivide. Despite these features, they are certainly not alive. Therefore, the criteria just listed, although necessary for life, are not sufficient to define life. One ingredient is missing—a mechanism for the preservation of improvement.

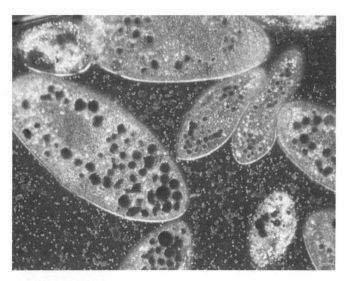

FIGURE 4.4
Cellular organization (150×).   These *Paramecia*, which are complex, single-celled organisms called protists, have just ingested several yeast cells. The yeasts, stained red in this photograph, are enclosed within membrane-bounded sacs called digestive vacuoles. A variety of other organelles are also visible.

Heredity.   All organisms on earth possess a *genetic system* that is based on the replication of a long, complex molecule called DNA. This mechanism allows for adaptation and evolution over time, and is a distinguishing characteristic of living organisms.

To understand the role of heredity in our definition of life, let us return for a moment to protein microspheres. When we examine an individual microsphere, we see it at that precise moment in time but learn nothing of its predecessors. It is likewise impossible to guess what future droplets will be like. The droplets are the passive prisoners of a changing environment, and in this sense they are not alive. The essence of being alive is the ability to encompass change and to reproduce the results of change permanently. Heredity, therefore, provides the basis for the great division between the living and the nonliving. Viruses, while possessing nucleic acids, do not possess a functional genetic system, as they cannot use or reproduce their genes outside of a living cell. A genetic system is the sufficient condition of life. Some changes are preserved because they increase the chances of survival in a hostile world, while others are lost. Not only did life evolve, but evolution is the very essence of life.

All living things on earth are characterized by heredity and possess a handful of other characteristics that serve to define the term life.

4.2 There is considerable disagreement about the origin of life.

Hypotheses About the Origin of Life

The question of how life originated is not easy to answer because it is impossible to go back in time and observe life's beginnings; nor are there any witnesses. Testimony exists in the rocks of the earth, but it is not easily read, and often it is silent on issues crying out for answers. There are, in principle, at least three possibilities: special creation, extraterrestrial origin, and spontaneous origin.

Special Creation

Life-forms may have been put on earth by supernatural or divine forces. The hypothesis that a divine God created life is at the core of most major religions. The oldest hypothesis about life's origins, it is also the most widely accepted. Far more Americans, for example, believe that God created life on earth than believe in the other two hypotheses. Many take an extreme position, accepting the biblical account of life's creation as factually correct. This viewpoint forms the basis for the very unscientific "scientific creationism" viewpoint discussed in chapter 22.

Extraterrestrial Origin

Life may not have originated on earth at all; instead, life may have infected earth from some other planet. This hypothesis, called **panspermia,** proposes that meteors or cosmic dust may have carried significant amounts of complex organic molecules to earth, kicking off the evolution of life. Hundreds of thousands of meteorites and comets are known to have slammed into the early earth, and recent findings suggest that at least some may have carried organic materials. Nor is life on other planets ruled out. For example, the discovery of liquid water under the surface of Jupiter's ice-shrouded moon Europa and suggestions of fossils in rocks from Mars lend some credence to this idea. The hypothesis that an early source of carbonaceous material is extraterrestrial is testable, although it has not yet been proven. Indeed, NASA is plan-

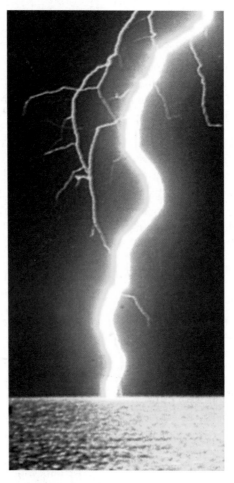

FIGURE 4.5
Lightning. Before life evolved, the simple molecules in the earth's atmosphere combined to form more complex molecules. The energy that drove these chemical reactions may have come from lightning and forms of geothermal energy.

ning to land on Europa, drill through the surface, and send a probe down to see if there is life.

Spontaneous Origin

Most scientists tentatively accept the hypothesis of spontaneous origin—that life evolved from inanimate matter as associations among molecules became more and more complex. In this view, the force leading to life was selection. As changes in molecules increased their stability and caused them to persist longer, these molecules could initiate more and more complex associations, culminating in the evolution of cells.

The Scientific Viewpoint

In this book, we focus on the second and third possibilities, attempting to understand whether natural forces could have led to the origin of life and, if so, how the process might have occurred. This is not to say that the first possibility is definitely incorrect. Any one of the three possibilities might be true. Nor do the second and third possibilities preclude religion—for example, a divine agency might have acted via evolution. However, we are limiting the scope of our inquiry to scientific matters, and only the second and third possibilities permit testable hypotheses to be constructed—that is, explanations that can be tested and potentially disproved.

In our search for understanding, we must look back to early times. Fossils of simple living things have been found in rocks 2.5 billion years old. They tell us that life originated during the early years of the history of our planet. As we attempt to determine how this process took place, we will first focus on how organic molecules may have originated (figure 4.5), and then we will consider how those molecules might have become organized into living cells.

Panspermia and spontaneous origin are the only testable hypotheses of life's origin currently available.

Scientists Disagree About Where Life Started

While most researchers agree that life first appeared as the primitive earth cooled and its rocky crust formed, there is little agreement as to just where this occurred.

Did Life Originate in a Reducing Atmosphere?

The more we learn about the earth's early history, the more likely it seems that earth's first organisms emerged and lived at very high temperatures. Rubble from the forming solar system slammed into the early earth beginning about 4.6 billion years ago, keeping the surface molten hot. As the bombardment slowed down, temperatures dropped. By about 3.8 billion years ago, ocean temperatures are thought to have dropped to a hot 49° to 88°C (120° to 190°F). Between 3.8 and 2.5 billion years ago, life first appeared, promptly after the earth was inhabitable. Thus, as intolerable as early earth's infernal temperatures seem to us today, they gave birth to life.

Very few geochemists agree on the exact composition of the early atmosphere. One popular view is that it contained principally carbon dioxide (CO_2) and nitrogen gas (N_2), along with significant amounts of water vapor (H_2O). It is possible that the early atmosphere also contained hydrogen gas (H_2) and compounds in which hydrogen atoms were bonded to the other light elements (sulfur, nitrogen, and carbon), producing hydrogen sulfide (H_2S), ammonia (NH_3), and methane (CH_4).

We refer to such an atmosphere as a *reducing atmosphere* because of the ample availability of hydrogen atoms and their electrons. In a reducing atmosphere, it would not take as much energy as it would today to form the carbon-rich molecules from which life evolved.

The key to this reducing-atmosphere hypothesis is the assumption that there was very little oxygen around. In an atmosphere with oxygen, amino acids and sugars react spontaneously with the oxygen to form carbon dioxide and water. Therefore, the building blocks of life, the amino acids, would not last long, and the spontaneous formation of complex macromolecules could not occur. Our atmosphere changed once organisms began to carry out photosynthesis, harnessing the energy in sunlight to split water molecules and form complex carbon molecules, giving off gaseous oxygen molecules in the process. The earth's atmosphere is now approximately 21% oxygen.

Critics of the reducing-atmosphere hypothesis point out that no carbonates have been found in rocks dating back to the early earth. This suggests that at that time CO_2 was locked up in the atmosphere, and if that was the case, the prebiotic atmosphere would not have been reducing.

Another problem for the reducing-atmosphere hypothesis is that because a prebiotic reducing atmosphere would have been oxygen free, there would have been no ozone. Without the protective ozone layer, any organic compounds that might have formed would have been broken down quickly by ultraviolet radiation.

Where on Earth Did Life Originate?

Many ideas have been advanced about where on earth life first appeared.

At the ocean's edge. Some scientists have suggested that life arose in bubbles that are constantly forming at the ocean's edge.

Under frozen oceans. One hypothesis proposes that life originated under a frozen ocean, not unlike the one that covers Jupiter's moon Europa today. All evidence suggests, however, that the early earth was quite warm and frozen oceans therefore quite unlikely.

Deep in the earth's crust. Another hypothesis is that life originated deep in the earth's crust. In 1988 Gunter Wachtershauser proposed that life might have formed as a by-product of volcanic activity, with iron and nickel sulfide minerals acting as chemical catalysts to recombine gases spewing from eruptions into the building blocks of life. In later work, he and co-workers were able to use this unusual chemistry to build precursors for amino acids (although they did not actually succeed in making amino acids), and to link amino acids together to form peptides. Critics of this hypothesis point out that the concentrations of chemicals used in their experiments greatly exceed those thought to have been present in nature 3–4 billion years ago.

Within clay. Other researchers have proposed the unusual hypothesis that life is the result of silicate surface chemistry. The surfaces of clays have positive charges to attract organic molecules and exclude water, providing a potential catalytic surface on which life's early chemistry might have occurred. While interesting conceptually, there is little evidence that this sort of process could actually occur.

At deep-sea vents. Becoming more popular is the hypothesis that life originated at deep-sea hydrothermal vents, with the necessary prebiotic molecules being synthesized on metal sulfides in the vents. The positive charge of the sulfides would have acted as a magnet for negatively charged biological molecules. In part, the current popularity of this hypothesis comes from the new science of genomics, which suggests that the ancestors of today's prokaryotes are most closely related to the archaebacteria that live on the deep-sea vents.

No one is sure whether life originated in a reducing atmosphere, under frozen ocean, deep in the earth's crust, within clay, or at deep-sea vents. Perhaps one of these hypotheses will be proven correct or perhaps the correct theory has not yet been proposed.

When life first appeared on earth, the environment was very hot. Most researchers assume that the organic chemicals that were the building blocks of life arose spontaneously at that time. How is a matter of considerable disagreement.

The Miller-Urey Experiment

An early attempt to see what kinds of organic molecules might have been produced on the early earth was carried out in 1953 by Stanley L. Miller and Harold C. Urey. In what has become a classic experiment, they attempted to reproduce the conditions in the earth's primitive oceans under a reducing atmosphere. Even if this assumption proves incorrect—the jury is still out on this—their experiment is critically important because it ushered in the whole new field of prebiotic chemistry.

To carry out their experiment, Miller and Urey (1) assembled a reducing atmosphere rich in hydrogen and excluding gaseous oxygen; (2) placed this atmosphere over liquid water; (3) maintained this mixture at a temperature somewhat below 100°C; and (4) simulated lightning by bombarding it with energy in the form of sparks (figure 4.6).

They found that within a week, 15% of the carbon originally present as methane gas (CH_4) had converted into other simple carbon compounds. Among these compounds were formaldehyde (CH_2O) and hydrogen cyanide (HCN; figure 4.7). These compounds then combined to form simple molecules, such as formic acid (HCOOH) and urea (NH_2CONH_2), and more complex molecules containing carbon–carbon bonds, including the amino acids glycine and alanine.

In similar experiments performed later by other scientists, more than 30 different carbon compounds were identified, including the amino acids glycine, alanine, glutamic acid, valine, proline, and aspartic acid. As we saw in chapter 3, amino acids are the basic building blocks of proteins, and proteins are one of the major kinds of molecules of which organisms are composed. Other biologically important molecules were also formed in these experiments. For example, hydrogen cyanide contributed to the production of a complex ring-shaped molecule called adenine—one of the bases found in DNA and RNA. Thus, the key molecules of life could have formed in the atmosphere of the early earth.

The Path of Chemical Evolution

A raging debate among biologists who study the origin of life concerns which organic molecules came first, the nucleic acid RNA or proteins. Scientists are divided into three camps, those that focus on RNA, those that support protein, and those that believe in a combination of the two. All three arguments have their strong points. Like the hypotheses that try to account for where life originated, these competing hypotheses are diverse and speculative.

An RNA World. The "RNA world" group feels that without a hereditary molecule, other molecules could not have formed consistently. This argument earned support when Thomas Cech at the University of Colorado discovered ribozymes, RNA molecules that can behave as enzymes, catalyzing their own assembly. Recent work has

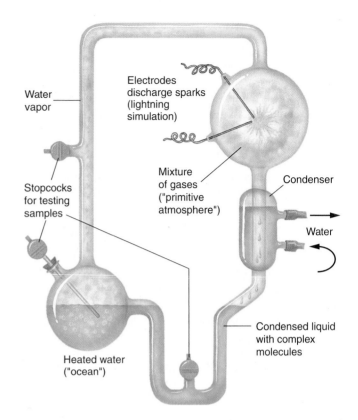

FIGURE 4.6
The Miller-Urey experiment. The apparatus consisted of a closed tube connecting two chambers. The upper chamber contained a mixture of gases thought to resemble the primitive earth's atmosphere. Electrodes discharged sparks through this mixture, simulating lightning. Condensers then cooled the gases, causing water droplets to form, which passed into the second heated chamber, the "ocean." Any complex molecules formed in the atmosphere chamber would be dissolved in these droplets and carried to the ocean chamber, from which samples were withdrawn for analysis.

shown that the RNA contained in ribosomes (discussed in chapter 5) catalyzes the chemical reaction that links amino acids to form proteins. Therefore, the RNA in ribosomes also functions as an enzyme. If RNA has the ability to pass on inherited information and the capacity to act like an enzyme, were proteins really needed?

A Protein World. The "protein-first" group argues that without enzymes (which are proteins), nothing could replicate at all, heritable or not. The protein-first proponents argue that nucleotides, the individual units of nucleic acids such as RNA, are too complex to have formed spontaneously, and certainly too complex to form spontaneously again and again. While there is no doubt that simple proteins are easier to synthesize from abiotic components than nucleotides, both can form in the laboratory under the right conditions. Deciding which came first is a chicken-and-egg paradox. In an effort to shed light on this problem, Julius Rebek and a number of

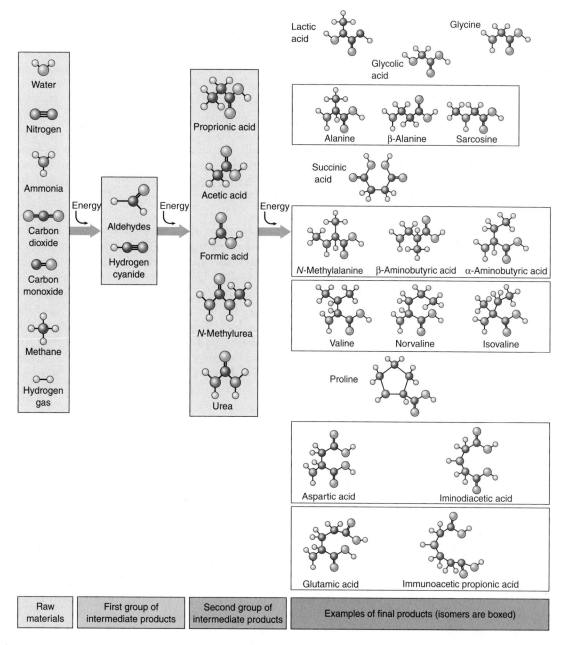

FIGURE 4.7
Results of the Miller-Urey experiment. Seven simple molecules, all gases, were included in the original mixture. Note that oxygen was not among them; instead, the atmosphere is thought to have been rich in hydrogen. At each stage of the experiment, more complex molecules were formed: first aldehydes, then simple acids, then more complex acids. Among the final products, the molecules that are structural isomers of one another are grouped together in boxes. In most cases, only one isomer of a compound is found in living systems today, although many may have been produced in the Miller-Urey experiment.

other chemists have created synthetic nucleotide-like molecules in the laboratory that are able to replicate. Moving even further, Rebek and his colleagues have created synthetic molecules that could replicate and "make mistakes." This simulates mutation, a necessary ingredient for the process of evolution.

A Peptide–Nucleic Acid World. Another important and popular theory about the first organic molecules assumes key roles for both protein and nucleic acids. Because RNA is so complex and unstable, this theory assumes there must have been a pre-RNA world where the protein-nucleic acid (PNA) was the basis for life. PNA is stable and simple enough to have formed spontaneously, and is also a self-replicator.

> Molecules that are the building blocks of living organisms have been shown to form spontaneously under conditions designed to simulate those of the primitive earth.

There are many hypotheses about the origin of cells.

Ideas About the Origin of Cells

The evolution of cells required early organic molecules to assemble into a functional, interdependent unit. Cells, discussed in chapter 5, are essentially little bags of fluid. What the fluid contains depends on the individual cell, but every cell's contents differ from the environment outside the cell. Therefore, an early cell may have floated along in a dilute "primordial soup," but its interior would have had a higher concentration of specific organic molecules.

The Importance of Bubbles

How did these "bags of fluid" evolve from simple organic molecules? As you can imagine, this question is a matter for debate. Scientists favoring an "ocean's edge" scenario for the origin of life have proposed that bubbles may have played a key role in this evolutionary step. A bubble is a hollow, spherical structure. Certain molecules, particularly those with hydrophobic regions, spontaneously form bubbles in water. The structure of the bubble shields the hydrophobic regions of the molecules from contact with water. If you have ever watched the ocean surge upon the shore, you may have noticed the foamy froth created by the agitated water. The edges of the primitive oceans were more than likely very frothy places bombarded by ultraviolet and other ionizing radiation, and exposed to an atmosphere that may have contained methane and other simple organic molecules.

Oparin's Bubble Hypothesis. The first bubble hypothesis is attributed to Alexander Oparin, a Russian chemist with extraordinary insight. In the mid-1930s, Oparin suggested that the present-day atmosphere was incompatible with the creation of life. He proposed that life must have arisen from nonliving matter under a set of very different environmental circumstances some time in the distant history of the earth. His was the theory of **primary abiogenesis** (primary because all living cells are now known to come from previously living cells, except in that first case). At the same time, J. B. S. Haldane, a British geneticist, was independently espousing the same views.

Oparin decided that in order for cells to evolve, they must have had some means of developing chemical complexity, separating their contents from their environment by means of a cell membrane, and concentrating materials within themselves. He termed these early, chemical-concentrating, bubblelike structures **protobionts**.

Oparin's hypothesis was published in English in 1938, and for awhile most scientists ignored them. However, Harold Urey, an astronomer at the University of Chicago, was quite taken with Oparin's ideas. He convinced one of his graduate students, Stanley Miller, to follow Oparin's rationale and see if he could "create" life. The Urey-Miller experiment has proven to be one of the most significant experiments in the history of science. As a result, Oparin's ideas became better known and more widely accepted.

A Host of Bubble Hypotheses. Different versions of "bubble hypotheses" have been championed by numerous scientists since Oparin. The bubbles they propose go by a variety of names; they may be called *microspheres, protocells, protobionts, micelles, liposomes,* or *coacervates,* depending on the composition of the bubbles (lipid or protein) and how they form. In all cases, the bubbles are hollow spheres, and they exhibit a variety of cell-like properties. For example, the lipid bubbles called **coacervates** form an outer boundary with two layers that resembles a biological membrane. They grow by accumulating more subunit lipid molecules from the surrounding medium, and they can form budlike projections and divide by pinching in two, like bacteria. They also can contain amino acids and use them to facilitate various acid-base reactions, including the decomposition of glucose. Although they are not alive, they obviously have many of the characteristics of cells.

A Bubble Scenario. It is not difficult to imagine that a process of chemical evolution involving bubbles or microdrops preceded the origin of life (figure 4.8). The early oceans must have contained untold numbers of these microdrops, billions in a spoonful, each one forming spontaneously, persisting for a while, and then dispersing. Some would, by chance, have contained amino acids with side groups able to catalyze growth-promoting reactions. Those microdrops would have survived longer than ones that lacked those amino acids, because the persistence of both protein microspheres and lipid coacervates is greatly increased when they carry out metabolic reactions such as glucose degradation and when they are actively growing.

Over millions of years, then, the complex bubbles that were better able to incorporate molecules and energy from the lifeless oceans of the early earth would have tended to persist longer than the others. Also favored would have been the microdrops that could use these molecules to expand in size, growing large enough to divide into "daughter"

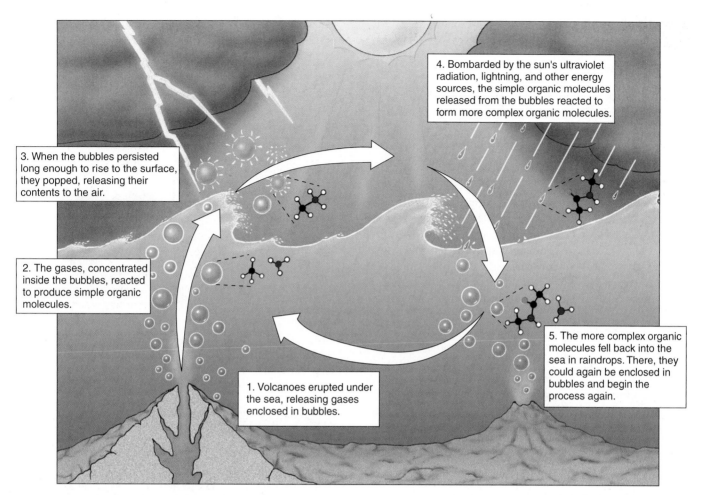

FIGURE 4.8

A current bubble hypothesis. In 1986, geophysicist Louis Lerman proposed that the chemical processes leading to the evolution of life took place within bubbles on the ocean's surface.

microdrops with features similar to those of their "parent" microdrop. The daughter microdrops have the same favorable combination of characteristics as their parent, and would have grown and divided, too. When a way to facilitate the reliable transfer of new ability from parent to offspring developed, heredity—and life—began.

Current Thinking

Whether the early bubbles that gave rise to cells were lipid or protein remains an unresolved argument. While it is true that lipid microspheres (coacervates) will form readily in water, there appears to be no mechanism for their heritable replication. On the other hand, we *can* imagine a heritable mechanism for protein microspheres. Although protein microspheres do not form readily in water, Sidney Fox and his colleagues at the University of Miami have shown that they can form under dry conditions.

The discovery that RNA can act as an enzyme to assemble new RNA molecules on an RNA template has raised the interesting possibility that neither coacervates nor protein microspheres were the first step in the evolution of life. Perhaps the first components were RNA molecules, and the initial steps on the evolutionary journey led to increasingly complex and stable RNA molecules. Later, stability might have improved further when a lipid (or possibly protein) microsphere surrounded the RNA. At present, those studying this problem have not arrived at a consensus about whether RNA evolved before or after a bubblelike structure that likely preceded cells.

Eventually, DNA took the place of RNA as the replicator in the cell and the storage molecule for genetic information. DNA, because it is a double helix, stores information in a more stable fashion than RNA, which is single-stranded.

> Little is known about how the first cells originated. Current hypotheses involve chemical evolution within bubbles, but there is no general agreement about their composition or about how the process occurred.

4.4 Cells became progressively more complex as they evolved.

The Earliest Cells

What do we know about the earliest life-forms? The fossils found in ancient rocks show an obvious progression from simple to complex organisms.

Microfossils

Paleontologists have found uncontestable microfossils in rocks as old as 2.5 billion years (figure 4.9). **Microfossils** are fossilized forms of microscopic life. What do we know about these early microscopic life-forms? Many microfossils are small (1 to 2 micrometers in diameter) and single-celled, lack external appendages, and have little evidence of internal structure. Thus, the organisms of microfossils often resemble present-day bacteria. We call organisms with this simple body plan **prokaryotes,** from the Greek words meaning "before," and "kernel" or "nucleus." The name reflects their lack of a **nucleus,** a spherical organelle characteristic of the more complex cells of **eukaryotes,** to be discussed later in this section.

The record of microfossils older than 2.5 billion years is sparse and controversial. The oldest of them, microscopic squiggles in 3.5-billion-year-old rock from western Australia, proved not to be ancient filamentous cyanobacteria as had been claimed, but lifeless mineral artifacts. The half-dozen other groups of microfossils from this time period are only poorly characterized. Life may have been present earlier than 2.5 billion years ago, but the fossil record provides us with scant evidence.

Judging from the fossil record, eukaryotes did not appear until about 1.5 billion years ago. Therefore, for at least 1 billion years, prokaryotes were the only organisms that existed.

Ancient Prokaryotes: Archaebacteria

Most organisms living today are adapted to the relatively mild conditions of present-day earth. However, if we look in unusual environments, we encounter organisms that are quite remarkable, differing in form and metabolism from other living things. Sheltered from evolutionary alteration in unchanging habitats that resemble earth's early environment, these living relics are the surviving representatives of the first ages of life on earth. In places such as the oxygen-free depths of the Black Sea or the boiling waters of hot springs and deep-sea vents, we can find prokaryotes living at very high temperatures without oxygen.

These unusual prokaryotes are called **archaebacteria,** from the Greek word for "ancient ones." Among the first to be studied in detail have been the **methanogens,** or methane-producing archaebacteria, which are among the most primitive prokaryotes that exist today. These organisms are typically simple in form and able to grow only in an oxygen-free environment; in fact, oxygen poisons them.

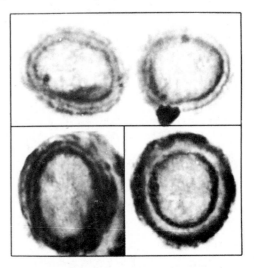

FIGURE 4.9
Cross sections of fossil prokaryotes. These microfossils from the Bitter Springs formation of Australia are of ancient cyanobacteria, far too small to be seen with the unaided eye. In these electron micrographs, the cell walls are clearly evident.

FIGURE 4.10
Methanogens. These archaebacteria (*Methanosarcina barkeri*) are called methanogens because their metabolic activities produce methane gas.

For this reason, they are said to grow "without air," or **anaerobically** (Greek *an,* "without," + *aer,* "air," + *bios,* "life"). Methanogens convert CO_2 and H_2 into methane gas (CH_4) (figure 4.10). Although primitive, they resemble all other prokaryotes in having DNA, a lipid cell membrane, an exterior cell wall, and a metabolism based on an energy-carrying molecule called ATP.

Unusual Cell Structures

When the details of cell wall and membrane structure of the methanogens were examined, they proved to be different from those of all other prokaryotes. Archaebacteria are characterized by a conspicuous lack of a protein cross-linked carbohydrate material called **peptidoglycan** in their cell walls, a key compound in the cell walls of most modern prokaryotes. Archaebacteria also have unusual lipids in their cell membranes that are not found in any other group of organisms as well as major differences in some of their fundamental biochemical processes of metabolism.

Earth's First Organisms?

Other archaebacteria are some of those that live in very salty environments, such as the Dead Sea (**extreme halophiles**—"salt lovers"), or very hot environments, such as hydrothermal volcanic vents under the ocean (**extreme thermophiles**—"heat lovers"). Thermophiles have been found living comfortably in boiling water. Indeed, many kinds of thermophilic archaebacteria thrive at temperatures of 110°C (230°F). Because these thermophiles live at high temperatures similar to those that may have existed in the earth's early oceans, microbiologists speculate that thermophilic archaebacteria may be relics of earth's first organisms.

Just how different are extreme thermophiles from other organisms? A methane-producing type of archaebacteria called *Methanococcus* isolated from deep-sea vents provides a startling picture. These prokaryotes thrive at temperatures of 88°C (185°F) and crushing pressures 245 times greater than at sea level. In 1996 molecular biologists announced that they had succeeded in determining the full nucleotide sequence of *Methanococcus*. This was possible because archaebacterial DNA is relatively small—it has only 1700 genes, coded in a DNA molecule only 1,739,933 nucleotides long (a human cell has 2000 times more). The thermophile nucleotide sequence proved to be astonishingly different from the DNA sequence of any other organism ever studied; fully two-thirds of its genes are unlike any ever known to science before. Clearly, these archaebacteria separated from other life on earth a long time ago. Preliminary comparisons to the gene sequences of other prokaryotes suggest that archaebacteria split from a second larger group of prokaryotes over 2 billion years ago, soon after life began.

Bacteria

The second major group of prokaryotes, the **bacteria,** have very strong cell walls and a simpler gene architecture. Most prokaryotes living today are bacteria. Included in this group are bacteria that have evolved the ability to capture the energy of light and transform it into the en-

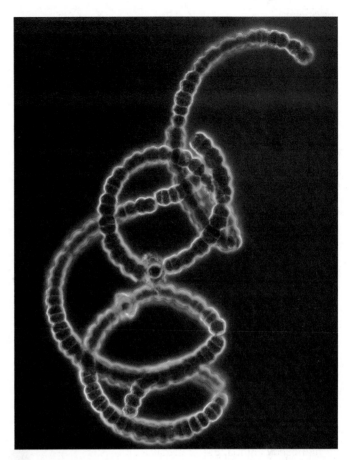

FIGURE 4.11
Living cyanobacteria. Although not multicellular, these bacteria often aggregate into chains such as those seen here.

ergy of chemical bonds within cells. These organisms are *photosynthetic*, as are plants and algae.

One type of photosynthetic bacteria that has been important in the history of life on earth are the cyanobacteria, sometimes called "blue-green algae" (figure 4.11). They have the same kind of chlorophyll pigment that is most abundant in plants and algae, as well as other pigments that are blue or red. Cyanobacteria produce oxygen as a result of their photosynthetic activities, and when they appeared at least 2 billion years ago, they played a decisive role in increasing the concentration of free oxygen in the earth's atmosphere from below 1% to the current level of 21%. As the concentration of oxygen increased, so did the amount of ozone in the upper layers of the atmosphere. The thickening ozone layer afforded protection from most of the ultraviolet radiation from the sun, radiation that is highly destructive to proteins and nucleic acids. Certain cyanobacteria are also responsible for the accumulation of massive limestone deposits.

All prokaryotes now living are members of either archaebacteria or bacteria.

The First Eukaryotic Cells

All fossils more than 1.5 billion years old are generally similar to one another structurally. They are small, simple cells; most measure 0.5 to 2 micrometers in diameter, and none are more than 6 micrometers in diameter. These simple cells eventually evolved into larger, more complex forms—the first eukaryotic cells.

Indirect chemical traces hint that eukaryotes may go as far back as 2.7 billion years, but no fossils as yet support such an early appearance. In rocks about 1.5 billion years old, we begin to see the first microfossils that are noticeably different in appearance from the earlier, simpler forms (figure 4.12). These cells are much larger than those of prokaryotes and have internal membranes and thicker walls. Evidence indicates that cells more than 10 micrometers in diameter rapidly increased in abundance. Some fossilized cells 1.4 billion years old are as much as 60 micrometers in diameter; others, 1.5 billion years old, contain what appear to be small, membrane-bounded structures.

These early fossils mark a major event in the evolution of life: a new kind of organism had appeared. These new cells are called **eukaryotes,** from the Greek words for "true" and "nucleus," because they possess an internal structure called a nucleus. All organisms other than prokaryotes are eukaryotes.

Origin of the Nucleus and ER

Many prokaryotes have infoldings of their outer membranes extending into the cytoplasm and serving as passageways to the surface. The network of internal membranes in eukaryotes called endoplasmic reticulum (ER)

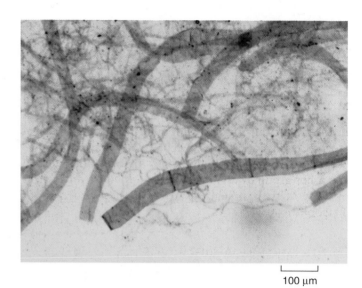

100 µm

FIGURE 4.12
Microfossil of a primitive eukaryote. This multicellular alga is between 900 million and 1 billion years old.

and the nuclear envelope, an extension of the ER network that isolates and protects the nucleus, are thought to have evolved from such infoldings (figure 4.13).

Origin of Mitochondria and Chloroplasts

Bacteria that live within other cells and perform specific functions for their host cells are called *endosymbiotic bacteria.* Their widespread presence in nature led Lynn Margulis in the early 1970s to champion the theory of **endosymbiosis** (Latin, *endo,* "inside," + Greek, *syn,* "together with," + *bios,* "life"), which means living together in close association.

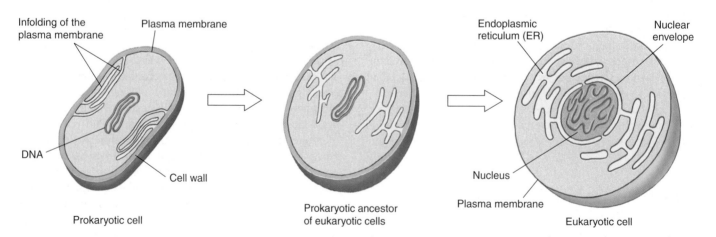

FIGURE 4.13
Origin of the nucleus and endoplasmic reticulum. Many prokaryotes today have infoldings of the plasma membrane (see also figure 27.6). The eukaryotic internal membrane system, called the endoplasmic reticulum (ER), and the nuclear envelope may have evolved from such infoldings of the plasma membrane encasing prokaryotic cells that gave rise to eukaryotic cells.

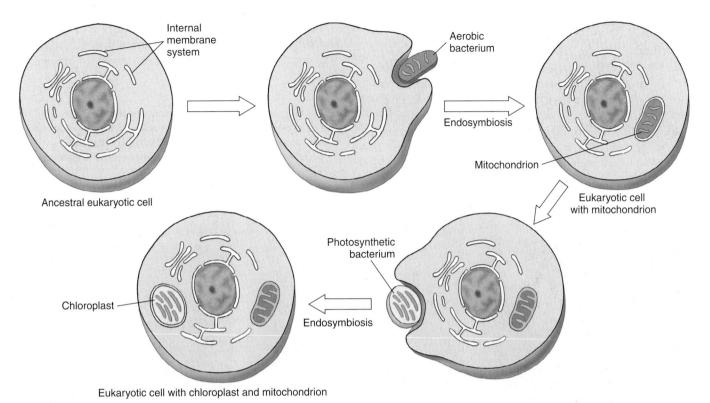

FIGURE 4.14
The theory of endosymbiosis. Scientists propose that ancestral eukaryotic cells, which already had an internal system of membranes, engulfed aerobic bacteria, which then became mitochondria in the eukaryotic cell. Chloroplasts may also have originated this way, with eukaryotic cells engulfing photosynthetic bacteria.

This theory, now widely accepted, suggests that a critical stage in the evolution of eukaryotic cells involved endosymbiotic relationships with prokaryotic organisms. According to this theory, energy-producing bacteria may have come to reside within larger bacteria, eventually evolving into what we now know as mitochondria. Similarly, photosynthetic bacteria may have come to live within other larger bacteria, leading to the evolution of chloroplasts, the photosynthetic organelles of plants and algae (figure 4.14). Bacteria with flagella, long whiplike cellular appendages used for propulsion, may have become symbiotically involved with nonflagellated bacteria to produce larger, motile cells. The fact that we now witness so many symbiotic relationships lends general support to this theory. Even stronger support comes from the observation that present-day organelles such as mitochondria, chloroplasts, and centrioles contain their own DNA, which is remarkably similar to the DNA of bacteria in size and character.

Sexual Reproduction

Most eukaryotic cells also possess the ability to reproduce sexually, something prokaryotes cannot do effectively. **Sexual reproduction** is the process of producing offspring by fertilization, the union of two cells that each have one copy of each chromosome. The great advantage of sexual reproduction is that it allows for frequent genetic recombination, which generates the variation that is the raw material for evolution. Not all eukaryotes reproduce sexually, but most have the capacity to do so. The evolution of meiosis and sexual reproduction (discussed in chapter 12) led to the tremendous explosion of diversity among the eukaryotes.

Multicellularity

Diversity was also promoted by the development of **multicellularity.** Some single eukaryotic cells began living in association with others, in colonies. Eventually, individual members of the colony began to assume different duties, and the colony began to take on the characteristics of a single individual. Multicellularity has arisen many times among the eukaryotes. Practically every organism big enough to be seen with the unaided eye is multicellular, including all animals and plants. The great advantage of multicellularity is that it fosters specialization; some cells devote all of their energies to one task, other cells to another. Few innovations have had as great an impact on the history of life as the specialization made possible by multicellularity.

The Diversity of Life

Confronted with the great diversity of life on earth today, biologists have attempted to categorize similar organisms in order to better understand them, giving rise to the science of taxonomy. In later chapters, we will discuss taxonomy and classification in detail, but for now we can generalize that all living things fall into one of three domains (not to be confused with protein domains described in chapter 3), which include six kingdoms (figure 4.15).

As more is learned about living things, particularly from the newer evidence that DNA studies provide, scientists will continue to reevaluate the relationships among the kingdoms of life (figure 4.16).

For at least the first 1 billion years of life on earth, all organisms were prokaryotes. About 1.5 billion years ago, the first eukaryotes appeared. Biologists place living organisms into six general categories called kingdoms.

Domain Bacteria

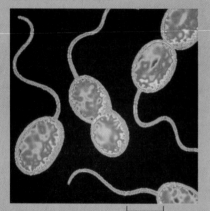

3.8 μm

Kingdom Bacteria. Prokaryotic organisms with a peptidoglycan cell wall, including cyanobacteria, soil bacteria, nitrogen-fixing bacteria, and pathogenic (disease-causing) bacteria.

Domain Archaea

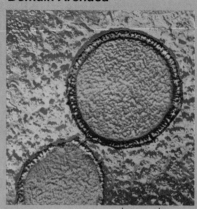

1.7 μm

Kingdom Archaebacteria. Prokaryotes that lack a peptidoglycan cell wall, including the methanogens and extreme halophiles and thermophiles.

Domain Eukarya

Kingdom Protista. Eukaryotic, primarily unicellular (although some algae are multicellular), photosynthetic or heterotrophic organisms, such as amoebas and paramecia.

Kingdom Fungi. Eukaryotic, mostly multicellular (although yeasts are unicellular), heterotrophic, usually nonmotile organisms, with cell walls of chitin, such as mushrooms.

Kingdom Plantae. Eukaryotic, multicellular, nonmotile, usually terrestrial, photosynthetic organisms, such as trees, grasses, and mosses.

Kingdom Animalia. Eukaryotic, multicellular, motile, heterotrophic organisms, such as sponges, spiders, newts, penguins, and humans.

FIGURE 4.15
The three domains of life.

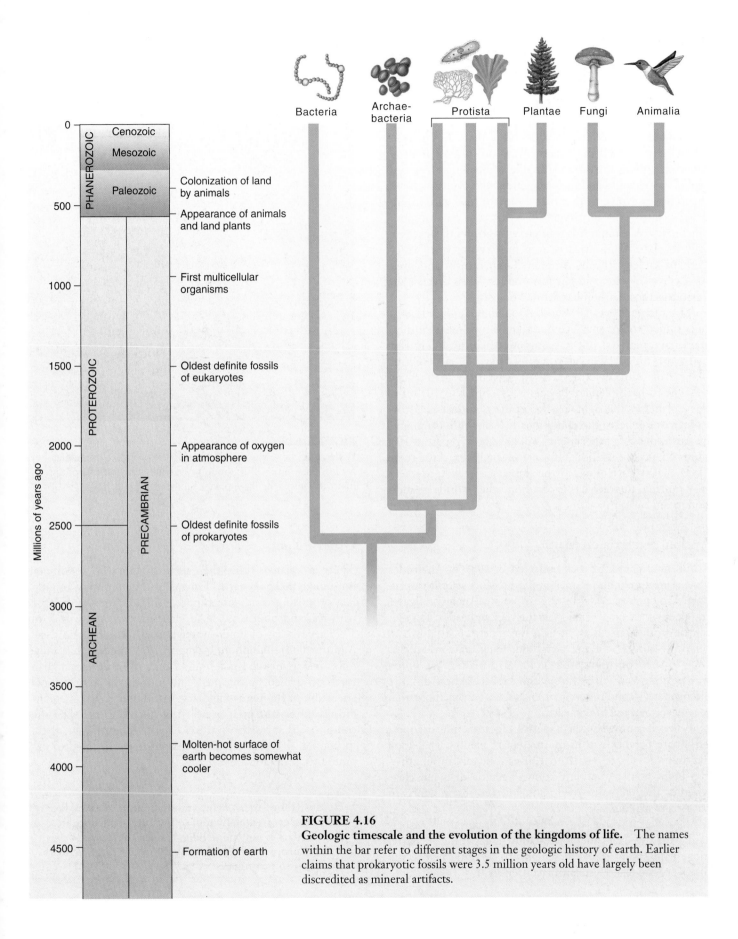

FIGURE 4.16
Geologic timescale and the evolution of the kingdoms of life. The names within the bar refer to different stages in the geologic history of earth. Earlier claims that prokaryotic fossils were 3.5 million years old have largely been discredited as mineral artifacts.

4.5 Scientists are beginning to take the possibility of extraterrestrial life seriously.

Has Life Evolved Elsewhere?

We should not overlook the possibility that life processes might have evolved in different ways on other planets. A functional genetic system, capable of accumulating and replicating changes and thus adapting and evolving, could theoretically evolve from molecules other than carbon, hydrogen, nitrogen, and oxygen in a different environment. Silicon, like carbon, needs four electrons to fill its outer energy level, and ammonia is even more polar than water. Perhaps under radically different temperatures and pressures, these elements might form molecules as diverse and flexible as those carbon has formed on earth.

The universe has 10^{20} (100,000,000,000,000,000,000) stars similar to our sun. We don't know how many of these stars have planets, but it seems increasingly likely that many do. Since 1996, astronomers have been detecting planets orbiting distant stars. At least 10% of stars are thought to have planetary systems. If only 1 in 10,000 of these planets is the right size and at the right distance from its star to duplicate the conditions in which life originated on earth, the "life experiment" will have been repeated 10^{15} times (that is, a million billion times). It does not seem likely that we are alone in the universe. In fact, it seems very possible that life has evolved on other worlds in addition to our own.

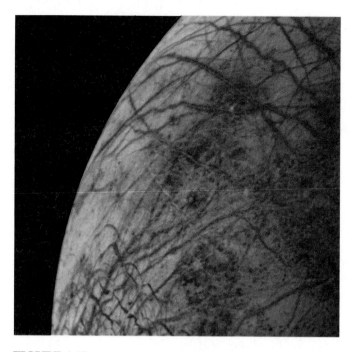

FIGURE 4.17
Is there life elsewhere? Currently the most likely candidate for life elsewhere within the solar system is Europa, one of the many moons of the large planet Jupiter.

Ancient Bacteria on Mars?

A dull gray chunk of rock collected in 1984 in Antarctica ignited an uproar about ancient life on Mars after it was reported to contain evidence of possible life. Analysis of gases trapped within small pockets of the rock indicate it is a meteorite from Mars. It is, in fact, the oldest rock known to science—fully 4.5 billion years old. Back then, when this rock formed on Mars, that cold, arid planet was much warmer, flowed with water, and had a carbon dioxide atmosphere—conditions not too different from those that may have spawned life on earth.

When examined with powerful electron microscopes, carbonate patches within the meteorite exhibit what look like microfossils, some 20 to 100 nanometers in length—100 times smaller than any known bacteria. It is not clear whether they actually are fossils, but the resemblance to bacteria is striking.

Viewed as a whole, the evidence of bacterial life associated with the Mars meteorite is not compelling. Clearly, more painstaking research remains to be done before the discovery can claim a scientific consensus.

Other Planets

There are planets other than ancient Mars with conditions not unlike those on earth. Europa, a large moon of Jupiter, is a promising candidate (figure 4.17). Europa is covered with ice, and photos taken in close orbit in the winter of 1998 reveal seas of liquid water beneath a thin skin of ice. Additional satellite photos taken in 1999 suggest that a few miles under the ice lies a liquid ocean of water larger than earth's, warmed by the push and pull of the gravitational attraction of Jupiter's many large satellite moons. The conditions on Europa now are far less hostile to life than the conditions that existed in the oceans of the primitive earth. In coming decades, satellite missions are scheduled to explore this ocean for life.

Because there are so many stars similar to our sun, life may have evolved many times. Although evidence for life on Mars is not compelling, the seas of Jupiter's moon Europa offer a promising candidate that scientists are eager to investigate.

4.1 All living things share key characteristics.

What Is Life?
- All organisms on the earth share several fundamental properties of life. The foremost of these properties is heredity; others include cellular organization, sensitivity, growth, development, reproduction, regulation, and homeostasis. (p. 63)

4.2 There is considerable disagreement about the origin of life.

Hypotheses About the Origin of Life
- In principle, the three possible explanations for the origin of life are special creation, extraterrestrial origin, and spontaneous origin. The scientific viewpoint can only address the second and third possibilities because they permit testable hypotheses. (p. 64)

Scientists Disagree About Where Life Started
- Many scientists believe the earth's early atmosphere was a reducing atmosphere of carbon dioxide and nitrogen gas, with very little oxygen. (p. 65)
- Other locations where life may have originated include at the ocean's edge, under frozen seas, deep in the earth's crust, within clay, and at deep-sea vents. (p. 65)

The Miller-Urey Experiment
- Miller and Urey's classic experiment attempted to reproduce a model of earth's possible reducing atmosphere and produce complex organic molecules. (p. 66)
- Subsequent chemical evolution on the earth is argued among many groups, including the RNA-first group, the protein-first group, and a peptide-nucleic acid group. (pp. 66–67)

4.3 There are many hypotheses about the origin of cells.

Ideas About the Origin of Cells
- Many different versions of bubble hypotheses have been championed by numerous scientists. (p. 68)
- Current hypotheses encompass chemical evolution within bubbles, but whether early bubbles were lipid or protein remains unsolved. (p. 69)

4.4 Cells became progressively more complex as they evolved.

The Earliest Cells
- Documented microfossils have been found that are as old as 2.5 billion years. (p. 70)
- Currently, prokaryotes are either archaebacteria or bacteria. (pp. 70–71)

The First Eukaryotic Cells
- Eukaryotes seemed to appear about 1.5 billion years ago. They seem to be much larger than prokaryotes and contain internal membranes and thicker cell walls. (p. 72)
- Endosymbiotic theory suggests that energy-producing bacteria may have come to reside within larger bacteria, eventually evolving into mitochondria. (pp. 72–73)
- Sexual reproduction allows for frequent genetic recombination and serves as the raw material for evolution. (p. 73)
- Multicellularity has arisen many times among the eukaryotes, and has promoted the development of diversity. (p. 73)

4.5 Scientists are beginning to take the possibility of extraterrestrial life seriously.

Has Life Evolved Elsewhere?
- Because at least 10% of stars are thought to have planetary systems, it is theoretically possible that life may have evolved many times. (p. 76)
- Currently, the Jupiter moon Europa is the most promising candidate to hold life because satellite images have revealed liquid oceans under the ice. (p. 76)

Self Test

1. Which of the following is necessary and sufficient for life?
   a. complexity
   b. heredity
   c. growth
   d. all of these
2. Which of the following most accurately describes the origins of life on earth?
   a. Special creation: Life-forms were put on earth by supernatural or divine forces.
   b. Extraterrestrial origin: The earth may have been infected by life from another planet.
   c. Spontaneous origin: Life evolved from inanimate matter to increasing levels of complexity.
   d. None of these are accurate.
3. Which of the following proposals assumes that the earth had a reducing atmosphere low in oxygen?
   a. life evolved deep in the earth's crust
   b. life evolved under frozen oceans
   c. life evolved at the ocean's edge
   d. life evolved at deep-sea vents
4. Scientists have created synthetic nucleotide-like molecules in the laboratory that are able to replicate. This seems to support which hypothesis of chemical evolution?
   a. an RNA-first hypothesis
   b. a protein-first hypothesis
   c. a peptide–nucleic acid hypothesis
   d. All of the hypotheses are supported by these results.
5. A key link needed between bubbles and cells is
   a. that they needed a hereditary molecule.
   b. that they needed to grow and bud off new bubbles.
   c. that they needed to incorporate enzymes.
   d. that they needed to form spontaneously.
6. Prokaryotes were the only life-form on the earth
   a. for about 1.5 billion years.
   b. for about 1 billion years.
   c. for about 2.5 billion years.
   d. for more than 2.5 billion years.
7. The first organisms may have been a type of archaebacteria similar to present-day organisms called
   a. extreme halophiles.
   b. prokaryotes.
   c. extreme thermophiles.
   d. eubacteria.
8. The infolding of outer membranes seen in prokaryotes is believed to have given rise to which of the following?
   a. mitochondrion
   b. chloroplast
   c. endoplasmic reticulum
   d. all of these structures
9. All organisms fall into one of _____ domains, which include _____ kingdoms.
   a. six/three
   b. six/six
   c. three/three
   d. three/six
10. Which of the following statements is false?
    a. Life may have evolved on Mars.
    b. In order for life to evolve anywhere, carbon is required.
    c. A large ocean exists under the icy surface of Europa.
    d. It is possible that life evolved on some other planet.

Test Your Visual Understanding

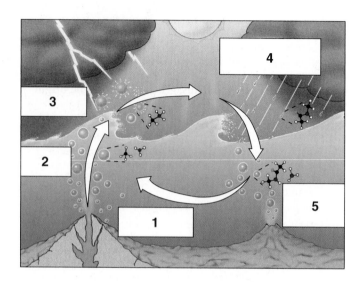

1. Match the following descriptions with the appropriate numbered step in the figure that explains a bubble hypothesis.
   a. Bombarded by the sun's ultraviolet radiation, lightning, and other energy sources, the simple organic molecules released from the bubbles reacted to form more complex organic molecules.
   b. Volcanoes erupted under the sea, releasing gases enclosed in bubbles.
   c. The more complex organic molecules fell back into the sea in raindrops. There, they could be enclosed in bubbles and begin the process again.
   d. When the bubbles persisted long enough to rise to the surface, they popped, releasing their contents to the air.
   e. The gases, concentrated inside the bubbles, reacted to produce simple organic molecules.

Apply Your Knowledge

1. In Fred Hoyle's science fiction novel *The Black Cloud*, the earth is approached by a large interstellar cloud of gas that orients itself around the sun. Scientists soon discover that the cloud is feeding on the sun, absorbing the sun's energy through the excitation of electrons in the outer energy levels of cloud molecules, in a process similar to the photosynthesis that occurs on earth. Different portions of the cloud are isolated from each other by associations of ions created by this excitation. Electron currents pass between these portions, much as they do on the surface of the human brain, endowing the cloud with self-awareness, memory, and the ability to think. Using electrical discharges, the cloud is able to communicate with humans and describe its history. It tells scientists that it originated as a small extrusion from an ancestral cloud, and that since then, it has grown by absorbing molecules and energy from stars such as our sun, on which it has been feeding. Soon the cloud moves off in search of other stars. Is it alive?

5

Cell Structure

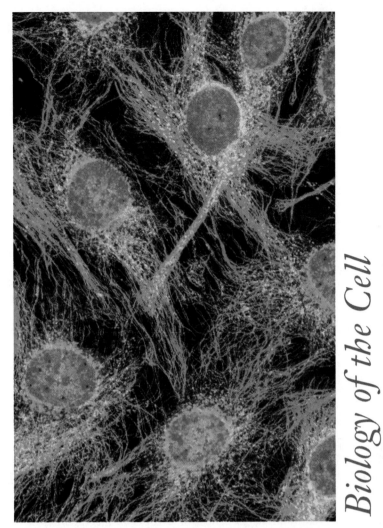

FIGURE 5.1
Fluorescence image of a cell. This amazing image of
mammalian fibroblast cells that are found in connective tissue was
obtained using a technique called fluorescence microscopy. The
use of a variety of fluorescent stains makes it possible to
distinguish different components of the cell through a
microscope. In this image, the nuclei and DNA are stained blue,
and cytoskeletal proteins are stained red.

Biology of the Cell

Part II

All organisms are composed of cells. The gossamer
wing of a butterfly is a thin sheet of cells, and so is the
glistening outer layer of your eyes. The hamburger or
tomato you eat is composed of cells, and its contents soon
become part of your cells. Some organisms consist of a single cell too small to see with the unaided eye, while others,
such as humans, are composed of many cells (figure 5.1).
Cells are so much a part of life as we know it that we cannot imagine an organism that is not cellular in nature. In
this chapter, we take a close look at the internal structure of
cells. In chapters 6–11, we will focus on cells in action—
how they communicate with their environment, grow, and
reproduce.

5.1 All organisms are composed of cells.

Characteristics of Cells

What is a typical cell like, and what would we find inside it? The general plan of cellular organization varies in the cells of different organisms, but despite these modifications, all cells resemble each other in certain fundamental ways. Before we begin our detailed examination of cell structure, let's first summarize three major features all cells have in common: a nucleoid or nucleus, cytoplasm, and a plasma membrane.

The Central Portion of the Cell Contains the Genetic Material

Every cell contains DNA, the hereditary molecule. In **prokaryotes,** most of the genetic material lies in a single circular molecule of DNA. It typically resides near the center of the cell in an area called the **nucleoid,** but this area is not segregated from the rest of the cell's interior by membranes. By contrast, the DNA of **eukaryotes** is contained in the **nucleus** (figure 5.2), which is surrounded by a double membrane structure called the **nuclear envelope.** In both types of organisms, the DNA contains the genes that code for the proteins synthesized by the cell.

The Cytoplasm Comprises the Rest of the Cell's Interior

A semifluid matrix called the **cytoplasm** fills the interior of the cell, exclusive of the nucleus (nucleoid in prokaryotes) lying within it. The cytoplasm contains the chemical wealth of the cell: the sugars, amino acids, and proteins the cell uses to carry out its everyday activities. In eukaryotic cells, the cytoplasm also contains specialized membrane-bounded compartments called **organelles.**

The Plasma Membrane Surrounds the Cell

The **plasma membrane** encloses a cell and separates its contents from its surroundings. The plasma membrane is a phospholipid bilayer about 5 to 10 nanometers (5 to 10 billionths of a meter) thick, with proteins embedded in it. Viewed in cross section with the electron microscope, such membranes appear as two dark lines separated by a lighter area. This distinctive appearance arises from the tail-to-tail packing of the phospholipid molecules that make up the membrane (figure 5.2). The proteins of a membrane may have large hydrophobic domains, which associate with and become embedded in the phospholipid bilayer.

The proteins of the plasma membrane are in large part responsible for a cell's ability to interact with its environment. *Transport proteins* help molecules and ions move

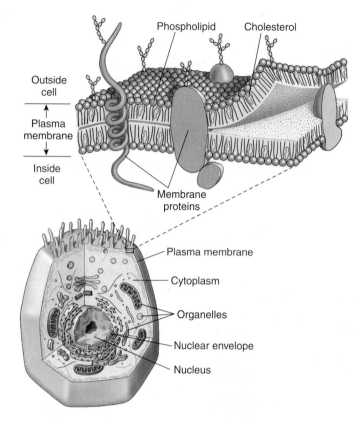

FIGURE 5.2

A generalized eukaryotic cell. In eukaryotes, a nucleus houses the DNA and is located near the center of the cell. Cytoplasm fills the rest of the cell's interior, which also contains specialized organelles. The plasma membrane encases the cell and consists of a phospholipid bilayer embedded with proteins.

across the plasma membrane, either from the environment to the interior of the cell or vice versa. *Receptor proteins* induce changes within the cell when they come in contact with specific molecules in the environment, such as hormones, or with molecules on the surface of neighboring cells. These molecules can function as *markers* that identify the cell as a particular type. This interaction between cell surface molecules is especially important in multicellular organisms, whose cells must be able to recognize each other as they form tissues.

We'll examine the structure and function of cell membranes more thoroughly in chapter 6.

The Cell Theory

A general characteristic of cells is their microscopic size. While there are a few exceptions—the marine alga *Acetabularia* can be up to 5 centimeters long—a typical eukaryotic cell is 10 to 100 micrometers (10 to

FIGURE 5.3
Surface area-to-volume ratio. As a cell gets larger, its volume increases at a faster rate than its surface area. If the cell radius increases by 10 times, the surface area increases by 100 times, but the volume increases by 1000 times. A cell's surface area must be large enough to meet the needs of its volume.

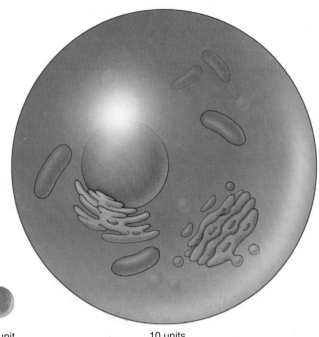

Cell radius (r)	1 unit	10 units
Surface area ($4\pi r^2$)	12.57 units^2	1257 units^2
Volume ($\frac{4}{3}\pi r^3$)	4.189 units^3	4189 units^3

100 millionths of a meter) in diameter; most prokaryotic cells are only 1 to 10 micrometers in diameter.

Because cells are so small, no one observed them until microscopes were invented in the mid-seventeenth century. Robert Hooke first described cells in 1665, when he used a microscope he had built to examine a thin slice of cork, a nonliving tissue found in the bark of certain trees. Hooke observed a honeycomb of tiny, empty (because the cells were dead) compartments. He called the compartments in the cork *cellulae* (Latin, "small rooms"), and the term has come down to us as *cells*. The first living cells were observed a few years later by the Dutch naturalist Antonie van Leeuwenhoek, who called the tiny organisms that he observed "animalcules," meaning little animals. For another century and a half, however, biologists failed to recognize the importance of cells. In 1838, botanist Matthias Schleiden stated that all plants "are aggregates of fully individualized, independent, separate beings, namely the cells themselves." In 1839, Theodor Schwann reported that all animal tissues also consist of individual cells.

These were the earliest statements of the **cell theory**, which in its modern form, includes the following three principles:

1. All organisms are composed of one or more cells, and the life processes of metabolism and heredity occur within these cells.
2. Cells are the smallest living things, the basic units of organization of all organisms.
3. Cells arise only by division of a previously existing cell. Although life likely evolved spontaneously in the environment of the early earth, biologists have

concluded that no additional cells are originating spontaneously at present. Rather, life on earth represents a continuous line of descent from those early cells.

Why Aren't Cells Larger?

Most cells are not large for practical reasons. Proteins and organelles are being synthesized, and materials are continually entering and leaving the cell. All of these processes involve the diffusion of substances at some point, and the larger a cell is, the longer it takes for substances to diffuse from the plasma membrane to the center of the cell. For this reason, an organism made up of many relatively small cells has an advantage over one composed of fewer, larger cells.

The advantage of small cell size is readily visualized in terms of the **surface area-to-volume ratio.** As a cell's size increases, its volume increases much more rapidly than its surface area. For a spherical cell, the increase in surface area is equal to the square of the increase in radius, while the increase in volume is equal to the cube of the increase in radius. Thus, if two cells differ by a factor of 10 in radius, the larger cell will have 10^2, or 100 times, the surface area, but 10^3, or 1000 times, the volume of the smaller cell (figure 5.3). A cell's surface provides its only opportunity for interaction with the environment, since all substances enter and exit a cell via the plasma membrane. This membrane plays a key role in controlling cell function, and because small cells have more surface area per unit of volume than large ones, the control is more effective when cells are relatively small.

Although most cells are small, some cells are quite large and have apparently overcome the surface area-to-volume problem by one or more adaptive mechanisms. For example, some cells, such as muscle cells, have more than one nucleus, allowing genetic information to be spread around a large cell. Some other large cells, such as your own neurons, are long and skinny so that any given point in the cytoplasm is close to the plasma membrane, and thus diffusion between the inside and outside of the cell can still be rapid.

A cell is a membrane-bounded unit that contains the DNA hereditary machinery and cytoplasm. All organisms are cells or aggregates of cells. Multicellular organisms usually consist of many small cells rather than a few large ones because small cells allow more rapid movement of molecules between the center of the cell and the environment.

Visualizing Cells

How many cells are big enough to see with the unaided eye? Other than egg cells, not many (figure 5.4). Most are less than 50 micrometers in diameter, far smaller than the period at the end of this sentence.

The Resolution Problem

How do we study cells if they are too small to see? The key is to understand why we can't see them. The reason we can't see such small objects is the limited resolution of the human eye. **Resolution** is defined as the minimum distance two points can be apart and still be distinguished as two separated points. When two objects are closer together than about 100 micrometers, the light reflected from each strikes the same "detector" cell at the rear of the eye. Only when the objects are farther than 100 micrometers apart will the light from each strike different cells, allowing your eye to resolve them as two objects rather than one.

Microscopes

One way to increase resolution is to increase magnification so that small objects appear larger. Robert Hooke and Antonie van Leeuwenhoek used glass lenses to magnify small cells and cause them to appear larger than the 100-micrometer limit imposed by the human eye. The glass lens adds additional focusing power. Because the glass lens makes the object appear closer, the image on the back of the eye is bigger than it would be without the lens.

Modern light microscopes use two magnifying lenses (and a variety of correcting lenses) to achieve very high magnification and clarity (table 5.1). The first lens focuses the image of the object on the second lens, which magnifies it again and focuses it on the back of the eye. Microscopes that magnify in stages using several lenses are called **compound microscopes.** They can resolve structures that are separated by more than 200 nanometers (nm).

Increasing Resolution

Light microscopes, even compound ones, are not powerful enough to resolve many of the structures within cells. For example, a membrane is only 5 nanometers thick. Why not just add another magnifying stage to the microscope and so increase its resolving power? Because when two objects are closer than a few hundred nanometers, the light beams reflecting from the two images start to overlap. The only way two light beams can get closer together and still be resolved is if their wavelengths are shorter.

One way to avoid overlap is by using a beam of electrons rather than a beam of light. Electrons have a much shorter wavelength, and a microscope employing electron beams has 1000 times the resolving power of a light microscope. **Transmission electron microscopes,** so called because the electrons used to visualize the specimens are transmitted through the material, are capable of resolving objects only 0.2 nanometer apart—just twice the diameter of a hydrogen atom!

A second kind of electron microscope, the **scanning electron microscope,** beams the electrons onto the surface of the specimen. The electrons reflected back from the surface of the specimen, together with other electrons that the specimen itself emits as a result of the bombardment, are amplified and transmitted to a screen, where the image can be viewed and photographed.

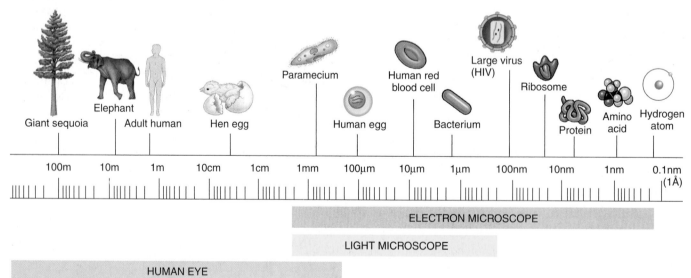

FIGURE 5.4
The size of cells and their contents. Most cells are microscopic in size, although vertebrate eggs are typically large enough to be seen with the unaided eye. Prokaryotic cells are generally 1 to 10 micrometers (μm) across.

Table 5.1 Types of Microscopes

Light Microscopes

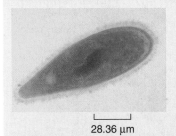

28.36 μm

Bright-field microscope: Light is simply transmitted through a specimen in culture, giving little contrast. Staining specimens improves contrast but requires that cells be fixed (not alive), which can cause distortion or alteration of components.

Fluorescence microscope: A set of filters transmits only light that is emitted by fluorescently stained molecules or tissues.

67.74 μm

Dark-field microscope: Light is directed at an angle toward the specimen; a condenser lens transmits only light reflected off the specimen. The field is dark, and the specimen is light against this dark background.

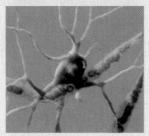

Confocal microscope: Light from a laser is focused to a point and scanned across the specimen in two directions. Clear images of one plane of the specimen are produced, while other planes of the specimen are excluded and do not blur the image. Fluorescent dyes and false coloring enhances the image.

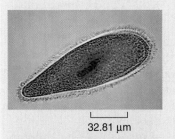

32.81 μm

Phase-contrast microscope: Components of the microscope bring light waves out of phase, which produces differences in contrast and brightness when the light waves recombine.

Electron Microscopes

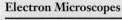

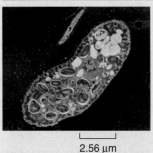

2.56 μm

Transmission electron microscope: A beam of electrons is passed through the specimen. Electrons that pass through are used to form an image. Areas of the specimen that scatter electrons appear dark. False coloring enhances the image.

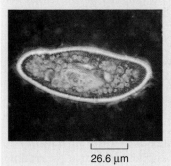

26.6 μm

Differential-interference-contrast microscope: Out-of-phase light waves to produce differences in contrast are combined with two beams of light traveling close together, which create even more contrast, especially at the edges of structures.

6.76 μm

Scanning electron microscope: An electron beam is scanned across the surface of the specimen, and electrons are knocked off the surface. Thus, the surface topography of the specimen determines the contrast and the content of the image. False coloring enhances the image.

Scanning electron microscopy yields striking three-dimensional images and has improved our understanding of many biological and physical phenomena (see table 5.1).

Visualizing Cell Structure by Staining Specific Molecules

A powerful tool for analyzing cell structure has been the use of stains that bind to specific molecular targets. Staining has been used in the analysis of tissue samples, or histology, for many years and has been improved dramatically with the use of antibiotics that bind to very specific molecular structures. This process, called *immunocytochemistry*, uses antibodies generated in animals such as rabbits or mice. When these animals are injected with specific pro-

teins, they produce antibodies that specifically bind to the injected protein, which can be purified from their blood. These purified antibodies can then be chemically bonded to enzymes, stains, or fluorescent molecules that glow when exposed to specific wavelengths of light. When cells are washed in a solution containing the antibodies, they bind to cellular structures that contain the target molecule and can be seen with light microscopy. This approach has been used extensively in the analysis of cell structure and function.

Most cells and their components are so small they can only be viewed using microscopes. Different kinds of microscopes and different staining procedures are used, depending on what kind of image is desired.

5.2 Eukaryotic cells are more structurally complex than prokaryotic cells.

Prokaryotes: Simple Cells with Complex Physiology

Prokaryotes are the simplest organisms. Prokaryotic cells are small, consisting of cytoplasm surrounded by a plasma membrane and encased within a rigid cell wall, with no distinct interior compartments (figure 5.5). A prokaryotic cell is like a one-room cabin in which eating, sleeping, and watching TV all occur. Prokaryotes are very important in the economy of living organisms. They harvest light in photosynthesis, break down dead organisms and recycle their components, cause disease, and are involved in many important industrial processes. As we mentioned in chapter 4, there are two main groups of prokaryotes: archaebacteria and bacteria; prokaryotes are the subject of chapter 27.

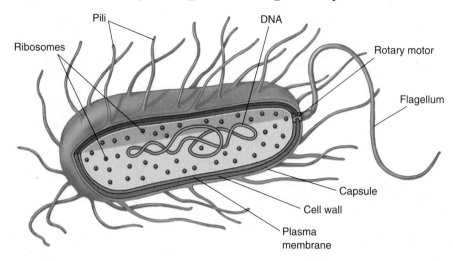

FIGURE 5.5
Structure of a prokaryotic cell. Generalized cell organization of a prokaryote. Some prokaryotes have hairlike growths on the outside of the cell called pili.

Strong Cell Walls

Most prokaryotic cells are encased by a strong **cell wall.** The cell wall of bacteria is composed of *peptidoglycan*, which consists of a carbohydrate matrix (polymers of sugars) that is cross-linked by short polypeptide units. The cell walls of archaebacteria have varying chemical compositions, but they all lack peptidoglycan. Cell walls protect the cell, maintain its shape, and prevent excessive uptake of water. Plants, fungi, and most protists also have cell walls of a different chemical structure that lack peptidoglycan, which we will discuss in later chapters. With a few exceptions such as the TB- and leprosy-causing bacteria, all bacteria may be classified into two types based on differences in their cell walls detected by the Gram staining procedure. The name refers to the Danish microbiologist Hans Christian Gram, who developed the procedure to detect the presence of certain disease-causing bacteria. **Gram-positive** bacteria have a thick, single-layered cell wall that retains a violet dye from the Gram stain procedure, causing the stained cells to appear purple under a microscope. More complex cell walls have evolved in other groups of bacteria. In them, the wall is multilayered and does not retain the purple dye after Gram staining; such bacteria exhibit the background red dye and are characterized as **gram-negative.**

The susceptibility of bacteria to antibiotics often depends on the structure of their cell walls. The drugs penicillin and vancomycin, for example, interfere with the ability of bacteria to cross-link the peptide units that hold the carbohydrate chains of the wall together. Like removing all the nails from a wooden house, this destroys the integrity of the matrix, which can no longer prevent water from rushing in, swelling the cell to bursting.

Long chains of sugars called polysaccharides cover the cell walls of many bacteria. They enable a bacterium to adhere to teeth, skin, food—or to practically any surface that will support their growth. Many disease-causing bacteria secrete a jellylike protective capsule of polysaccharide around the cell.

Rotating Flagella

Some prokaryotes use a **flagellum** (plural, *flagella*) to move. Flagella are long, threadlike structures protruding from the surface of a cell that are used in locomotion. Prokaryotic flagella are protein fibers that extend out from a bacterial cell. There may be one or more per cell, or none, depending on the species. Bacteria can swim at speeds of up to 20 cell diameters per second by rotating their flagella like screws (figure 5.6). The rotary motor uses the energy stored in a gradient of protons across the plasma membrane to power the movement of the flagellum. Interestingly, the same principle, in which a proton gradient powers the rotation of a molecule, is used in eukaryotic mitochondria and chloroplasts by an enzyme that synthesizes ATP.

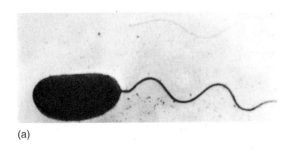

(a)

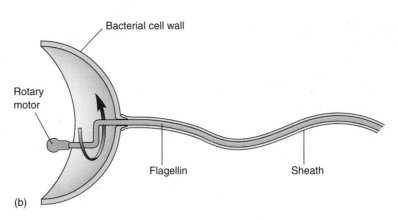

Bacterial cell wall

Rotary
motor

Flagellin

Sheath

(b)

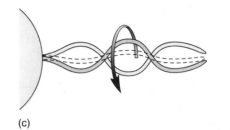

(c)

FIGURE 5.6

Prokaryotes swim by rotating their flagella. (*a*) The photograph shows *Vibrio cholerae*, the microbe that causes the serious disease cholera. The unsheathed core visible at the top of the photograph is composed of a single crystal of the protein flagellin. (*b*) In intact flagella, the flagellin core is surrounded by a flexible sheath. (*c*) Imagine that you were standing inside the *Vibrio* cell, turning the flagellum like a turbine. You would create a spiral wave that travels down the flagellum, just as if you were turning a wire within a flexible tube. This movement propels the cell forward when it swims.

Simple Interior Organization

If you were to look at an electron micrograph of a prokaryotic cell, you would be struck by the cell's simple organization. There are few, if any, internal compartments, and while prokaryotic cells contain complex structures like ribosomes, most have no membrane-bounded organelles, the kinds so characteristic of eukaryotic cells. Nor do prokaryotes have a true nucleus. The entire cytoplasm of a prokaryotic cell is one unit with no internal support structure. Consequently, the strength of the cell comes primarily from its rigid cell wall (see figure 5.5).

The plasma membrane of a prokaryotic cell carries out some of the functions organelles perform in eukaryotic cells. For example, before a prokaryotic cell divides, the prokaryotic hereditary material, a simple circle of DNA, replicates. The two DNA molecules that result from the replication attach to the plasma membrane at different points, ensuring that each daughter cell will contain one of the identical units of DNA. Moreover, some photosynthetic bacteria, such as cyanobacteria and *Prochloron* (figure 5.7), have an extensively folded plasma membrane, with the folds extending into the cell's interior. These membrane folds contain the bacterial pigments connected with photosynthesis.

Because a prokaryotic cell contains no membrane-bounded organelles, the DNA, enzymes, and other cytoplasmic constituents have access to all parts of the cell. Reactions are not compartmentalized as they are in eukaryotic cells, and the whole prokaryote operates as a single unit.

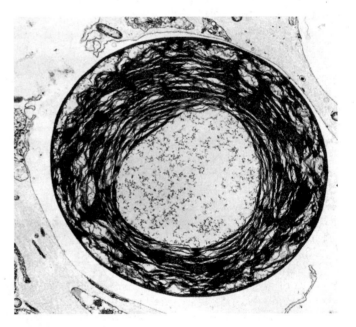

FIGURE 5.7

Electron micrograph of a photosynthetic bacterial cell. Extensive folded photosynthetic membranes are visible in this *Prochloron* cell (14,500×).

Prokaryotes are small cells that lack complex interior organization. They are encased by an exterior wall composed of carbohydrates cross-linked by short polypeptides, and some are propelled by rotating flagella.

Eukaryotes: Cells with Complex Interiors

Eukaryotic cells (figures 5.8 and 5.9) are far more complex than prokaryotic cells. The hallmark of the eukaryotic cell is compartmentalization, achieved by an extensive **endomembrane system** that weaves through the cell interior and by numerous **organelles,** membrane-bounded structures that close off compartments within which multiple biochemical processes can proceed simultaneously and independently. Plant cells often have a large, membrane-bounded sac called a **central vacuole,** which stores proteins, pigments, and waste materials. Both plant and animal cells contain **vesicles,** smaller sacs that store and transport a variety of materials. Inside the nucleus, the DNA is

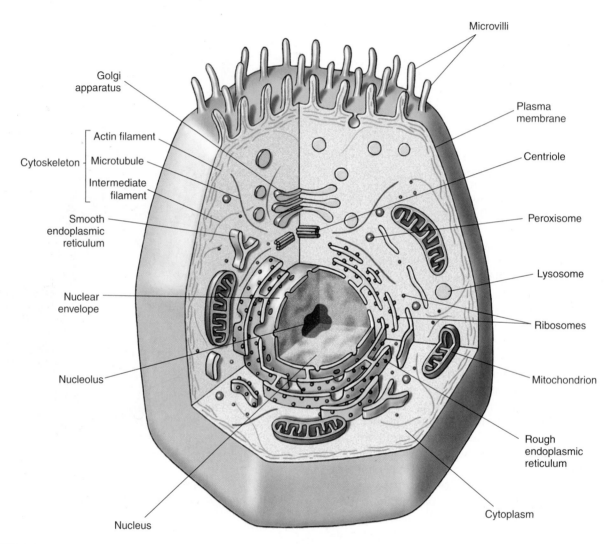

FIGURE 5.8
Structure of an animal cell. In this generalized diagram of an animal cell, the plasma membrane encases the cell, which contains the cytoskeleton and various cell organelles and interior structures suspended in a semifluid matrix called the cytoplasm. Some kinds of animal cells possess fingerlike projections called microvilli. Other types of eukaryotic cells—for example, many protist cells—may possess flagella, which aid in movement, or cilia, which can have many different functions.

wound tightly around proteins and packaged into compact units called **chromosomes.** All eukaryotic cells are supported by an internal protein scaffold, the **cytoskeleton.** While the cells of animals and some protists lack cell walls, the cells of fungi, plants, and many protists have strong **cell walls** composed of cellulose or chitin fibers embedded in a matrix of other polysaccharides and proteins. In the remainder of this chapter, we will examine the internal components of eukaryotic cells in more detail.

Eukaryotic cells contain membrane-bounded organelles that carry out specialized functions.

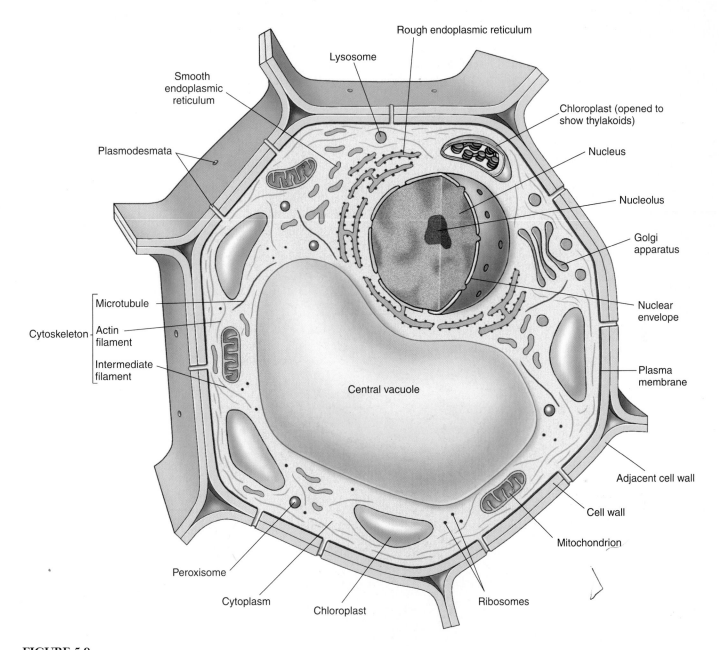

FIGURE 5.9
Structure of a plant cell.   Most mature plant cells contain large central vacuoles, which occupy a major portion of the internal volume of the cell, and organelles called chloroplasts, within which photosynthesis takes place. The cells of plants, fungi, and some protists have cell walls, although the composition of the walls varies among the groups. Plant cells have cytoplasmic connections through openings in the cell wall called plasmodesmata. Flagella occur in sperm of a few plant species, but are otherwise absent in plant and fungal cells. Centrioles are also absent in plant and fungal cells.

5.3 Take a tour of a eukaryotic cell.

The Nucleus: Information Center for the Cell

The largest and most easily seen organelle within a eukaryotic cell is the **nucleus** (Latin, "kernel" or "nut"), first described by the English botanist Robert Brown in 1831. Nuclei are roughly spherical in shape, and in animal cells, they are typically located in the central region of the cell (figure 5.10a). In some cells, a network of fine cytoplasmic filaments seems to cradle the nucleus in this position. The nucleus is the repository of the genetic information that directs the activities of a living eukaryotic cell. Most eukaryotic cells possess a single nucleus, although the cells of fungi and some other groups may have several to many nuclei. Mammalian erythrocytes (red blood cells) lose their nuclei when they mature. Many nuclei exhibit a dark-staining zone called the **nucleolus,** which is a region where intensive synthesis of ribosomal RNA is taking place.

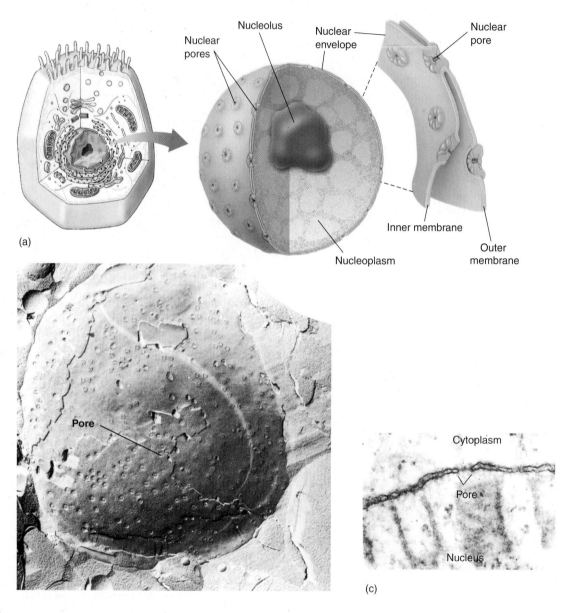

FIGURE 5.10

The nucleus. (*a*) The nucleus is composed of a double membrane, called a nuclear envelope, enclosing a fluid-filled interior containing the chromosomes. In cross section, the individual nuclear pores are seen to extend through the two membrane layers of the envelope; the dark material within the pore is protein, which acts to control access through the pore. (*b*) A freeze-fracture electron micrograph (see figure 6.6) of a cell nucleus, showing nuclear pores (9500×). (*c*) A transmission electron micrograph of the nuclear membrane, showing a nuclear pore.

The Nuclear Envelope: Getting In and Out

The surface of the nucleus is bounded by *two* phospholipid bilayer membranes, which together make up the **nuclear envelope** (see figure 5.10). The outer membrane of the nuclear envelope is continuous with the cytoplasm's interior membrane system, called the endoplasmic reticulum. Scattered over the surface of the nuclear envelope, like craters on the moon, are shallow depressions called **nuclear pores** (figure 5.10*b,c*). These pores form 50 to 80 nanometers apart at locations where the two membrane layers of the nuclear envelope pinch together. Rather than being empty, nuclear pores are filled with proteins that act as gatekeepers, permitting certain molecules to pass into and out of the nucleus. Passage is restricted primarily to two kinds of molecules: (1) proteins moving into the nucleus to be incorporated into nuclear structures or to catalyze nuclear activities; and (2) RNA and protein-RNA complexes formed in the nucleus and exported to the cytoplasm.

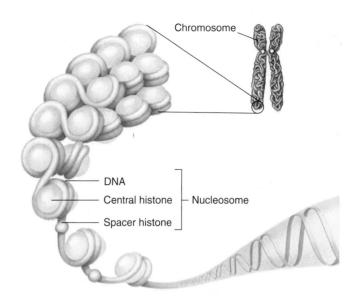

FIGURE 5.11
Nucleosomes. Each nucleosome is a region in which the DNA is wrapped tightly around a cluster of histone proteins.

The Chromosomes: Packaging the DNA

In both prokaryotes and eukaryotes, DNA contains the hereditary information specifying cell structure and function. However, unlike the circular DNA of prokaryotes, the DNA of eukaryotes is divided into several linear **chromosomes.** Except when a cell is dividing, its chromosomes are extended into threadlike strands, called **chromatin,** of DNA complexed with protein. This open arrangement allows proteins to attach to specific nucleotide sequences along the DNA and regulate gene expression. Without this access, DNA could not direct the day-to-day activities of the cell. During all cellular activities, the chromosomes are associated with packaging proteins called **histones.** When the cell is functioning normally, the DNA is loosely coiled around clusters of histones called **nucleosomes.** These structures resemble beads on a string (figure 5.11). Configuring chromosomes into a more extended form permits enzymes to make RNA copies of DNA. These RNA copies of the information in the DNA direct the synthesis of proteins. When a cell prepares to divide, the DNA coils up around the histones into a much more highly condensed form until the DNA is in a compact mass. Under a light microscope, these fully condensed chromosomes are readily seen in dividing cells as densely staining rods (figure 5.12). After cell division, eukaryotic chromosomes uncoil and can no longer be individually distinguished with a light microscope.

The nucleus of a eukaryotic cell contains the cell's genetic information and isolates it from the rest of the cell. A distinctive feature of eukaryotes is the organization of their DNA into complex chromosomes.

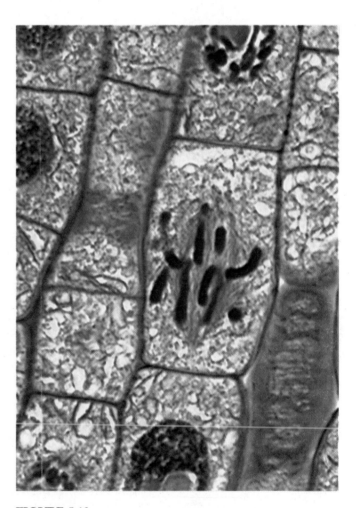

FIGURE 5.12
Eukaryotic chromosomes. These condensed chromosomes within an onion root tip are visible under the light microscope (500×).

The Endomembrane System

The Endoplasmic Reticulum: Compartmentalizing the Cell

The interior of a eukaryotic cell is packed with membranes (table 5.2) so thin that they are invisible under the low resolving power of light microscopes. This **endomembrane system** fills the cell, dividing it into compartments, channeling the passage of molecules through the interior of the cell, and providing surfaces for the synthesis of lipids and some proteins. The presence of these membranes in eu-karyotic cells constitutes one of the most fundamental distinctions between eukaryotes and prokaryotes.

The largest of the internal membranes is called the **endoplasmic reticulum (ER).** *Endoplasmic* means "within the cytoplasm," and *reticulum* is Latin for "a little net." Like the plasma membrane, the ER is composed of a lipid bilayer embedded with proteins. It weaves in sheets through the interior of the cell, creating a series of channels between its folds (figure 5.13). Of the many compartments in eukaryotic cells, the two largest are the inner region of the ER, called the **cisternal space,** and the region exterior to it, the **cytosol.**

Table 5.2 Eukaryotic Cell Structures and Their Functions

Structure		Description	Function
Cell wall		Outer layer of cellulose or chitin; or absent	Protection; support
Cytoskeleton		Network of protein filaments	Structural support; cell movement
Flagella (cilia)		Cellular extensions with 9 + 2 arrangement of pairs of microtubules	Motility or moving fluids over surfaces
Plasma membrane		Lipid bilayer with embedded proteins	Regulates what passes into and out of cell; cell-to-cell recognition
Endoplasmic reticulum (ER)		Network of internal membranes	Forms compartments and vesicles; participates in protein and lipid synthesis
Nucleus		Structure (usually spherical) that contains chromosomes and is surrounded by double membrane	Control center of cell; directs protein synthesis and cell reproduction
Golgi apparatus		Stacks of flattened vesicles	Packages proteins for export from cell; forms secretory vesicles
Lysosomes		Vesicles derived from Golgi apparatus that contain hydrolytic digestive enzymes	Digest worn-out organelles and cell debris; play role in cell death
Microbodies		Vesicles that are formed from incorporation of lipids and proteins and that contain oxidative and other enzymes	Isolate particular chemical activities from rest of cell
Mitochondria		Bacteria-like elements with double membrane	"Power plants" of the cell; sites of oxidative metabolism
Chloroplasts		Bacteria-like elements with membranes containing chlorophyll, a photosynthetic pigment	Sites of photosynthesis
Chromosomes		Long threads of DNA that form a complex with protein	Contain hereditary information
Nucleolus		Site of genes for rRNA synthesis	Assembles ribosomes
Ribosomes		Small, complex assemblies of protein and RNA, often bound to endoplasmic reticulum	Sites of protein synthesis

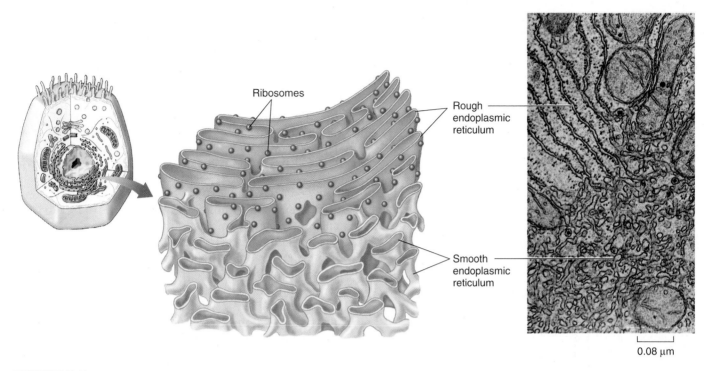

Ribosomes

Rough endoplasmic reticulum

Smooth endoplasmic reticulum

0.08 μm

FIGURE 5.13

The endoplasmic reticulum. Rough ER has ribosomes associated with only one side of the endomembrane; the other side is the boundary of a separate compartment within the cell into which the ribosomes extrude newly made proteins destined for secretion. Smooth endoplasmic reticulum has few to no bound ribosomes.

Rough ER: Manufacturing Proteins for Export. The ER surface regions that are devoted to protein synthesis are heavily studded with **ribosomes,** large molecular aggregates of protein and ribonucleic acid (RNA) that translate RNA copies of genes into protein. (We will examine ribosomes in detail later in this chapter.) Through the electron microscope, these ribosome-rich regions of the ER appear pebbly, like the surface of sandpaper, and they are therefore called **rough ER** (see figure 5.13).

The proteins synthesized on the surface of the rough ER are destined to be exported from the cell, sent to lysosomes or vacuoles, or embedded in the plasma membrane. Proteins to be exported contain special amino acid sequences called **signal sequences.** As a new protein is made by a free ribosome (one not attached to a membrane), the signal sequence of the growing polypeptide attaches to a recognition factor that carries the ribosome and its partially completed protein to a "docking site" on the surface of the ER. As the protein is assembled, it passes through the ER membrane into the interior ER compartment, the cisternal space, from which it is transported by vesicles to the Golgi apparatus (figure 5.14). The protein then travels within vesicles to the inner surface of the plasma membrane, where it is released to the outside of the cell.

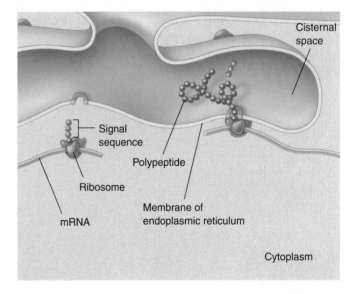

Cisternal space

Signal sequence

Polypeptide

Ribosome

mRNA

Membrane of endoplasmic reticulum

Cytoplasm

FIGURE 5.14

Signal sequences direct proteins to their destinations in the cell. In this example, a sequence of hydrophobic amino acids (the signal sequence) on a secretory protein attaches them (and the ribosomes making them) to the membrane of the ER. As the protein is synthesized, it passes into the cisternal space of the ER. The signal sequence is clipped off after the leading edge of the protein enters the cisternal space.

Smooth ER: Organizing Internal Activities. Regions of the ER with relatively few bound ribosomes are referred to as **smooth ER.** The membranes of the smooth ER contain many embedded enzymes. Enzymes anchored within the ER, for example, catalyze the synthesis of a variety of carbohydrates and lipids. In cells that carry out extensive lipid synthesis, such as those in the testes, intestine, and brain, smooth ER is particularly abundant. In the liver, the enzymes of the smooth ER are involved in the detoxification of drugs, including amphetamines, morphine, codeine, and phenobarbital.

Some vesicles form at the plasma membrane by budding inward, a process called *endocytosis* (see chapter 6). Some then move into the cytoplasm and fuse with the smooth endoplasmic reticulum. Others form secondary lysosomes or other interior vesicles.

The Golgi Apparatus: Delivery System of the Cell

At various locations within the endomembrane system, flattened stacks of membranes called **Golgi bodies** occur, often interconnected with one another. These structures are named for Camillo Golgi, the nineteenth-century Italian physician who first called attention to them. The number of Golgi bodies a cell contains ranges from 1 or a few in protists to 20 or more in animal cells and several hundred in plant cells. They are especially abundant in glandular cells, which manufacture and secrete substances. Collectively, the Golgi bodies are referred to as the **Golgi apparatus** (figure 5.15).

The Golgi apparatus functions in the collection, packaging, and distribution of molecules synthesized at one place in the cell and utilized at another location in the cell. A Golgi body has a front and a back, with distinctly different membrane compositions at the opposite ends. The front, or receiving end, is called the *cis* face, and is usually located near ER. Materials move to the *cis* face in transport vesicles that bud off the ER. These vesicles fuse with the *cis* face, emptying their contents into the interior, or lumen, of the Golgi apparatus. These ER-synthesized molecules then pass through the channels of the Golgi apparatus until they reach the back, or discharging end, called the *trans* face, where they are discharged in secretory vesicles (figure 5.16).

Proteins and lipids manufactured on the rough and smooth ER membranes are transported into the Golgi apparatus and modified as they pass through it. The most common alteration is the addition or modification of short sugar chains, forming a *glycoprotein* when sugars are complexed to a protein and a *glycolipid* when sugars are bound to a lipid. In many instances, enzymes in the Golgi apparatus modify existing glycoproteins and glycolipids made in the ER by cleaving a sugar from their sugar chain, or modifying one or more of the sugars.

The newly formed or altered glycoproteins and glycolipids collect at the ends of the Golgi bodies in flattened, stacked membrane folds called **cisternae** (Latin, "collecting vessels"). Periodically, the membranes of the cisternae push together, pinching off small, membrane-bounded secretory vesicles containing the glycoprotein and glycolipid molecules. These vesicles then move to other locations in the cell, distributing the newly synthesized molecules to their appropriate destinations.

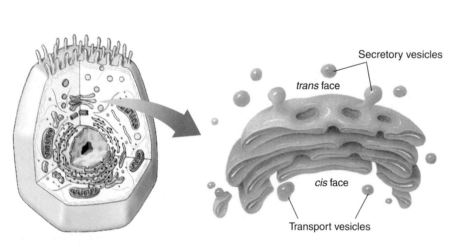
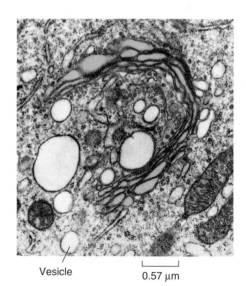

FIGURE 5.15
The Golgi apparatus. The Golgi apparatus is a smooth, concave, membranous structure located near the middle of the cell. It receives material for processing in transport vesicles on the *cis* face and sends the material packaged in secretory vesicles off the *trans* face. The substance in a vesicle could be for export out of the cell or for distribution to another region within the same cell.

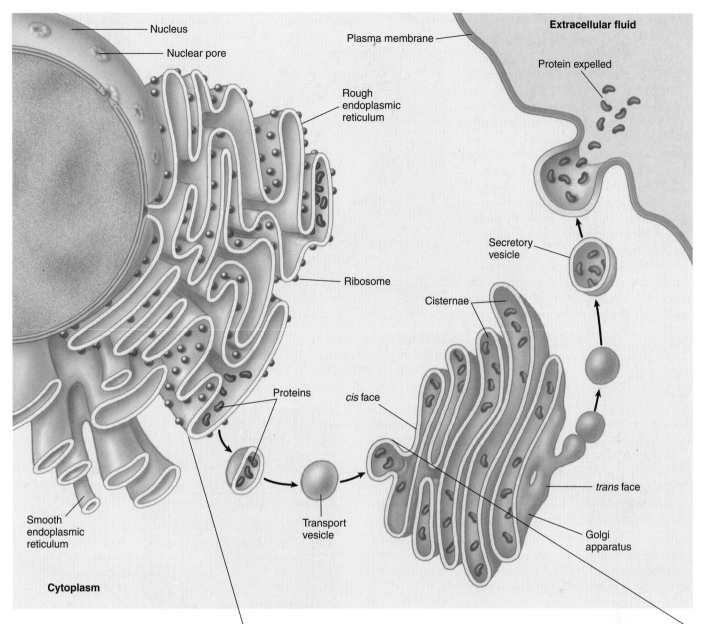

Nucleus

Nuclear pore

Extracellular fluid

Plasma membrane

Protein expelled

Rough endoplasmic reticulum

Ribosome

Secretory vesicle

Cisternae

Proteins

cis face

trans face

Transport vesicle

Golgi apparatus

Smooth endoplasmic reticulum

Cytoplasm

FIGURE 5.16

How proteins are transported within the cell. Proteins are manufactured at the ribosome and then released into the internal compartments of the rough ER. If the newly synthesized proteins are to be used at a distant location in or outside of the cell, they are transported within vesicles that bud off the rough ER and travel to the *cis* face, or receiving end, of the Golgi apparatus. There they are modified and packaged into secretory vesicles. The secretory vesicles then migrate from the *trans* face, or discharging end, of the Golgi apparatus to other locations in the cell, or they fuse with the plasma membrane, releasing their contents to the external cellular environment.

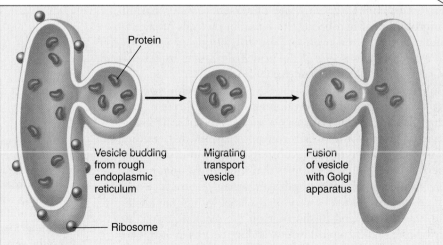

Protein

Vesicle budding from rough endoplasmic reticulum

Migrating transport vesicle

Fusion of vesicle with Golgi apparatus

Ribosome

Vesicles: Enzyme Storehouses

Lysosomes: Intracellular Digestion Centers. Lysosomes, membrane-bounded digestive vesicles, are also components of the endomembrane system that arise from the Golgi apparatus. They contain high levels of degrading enzymes, which catalyze the rapid breakdown of proteins, nucleic acids, lipids, and carbohydrates. Throughout the lives of eukaryotic cells, lysosomal enzymes break down old organelles, recycling their component molecules and making room for newly formed organelles. For example, mitochondria are replaced in some tissues every 10 days.

The digestive enzymes in lysosomes function best in an acidic environment. Lysosomes actively engaged in digestion keep their battery of hydrolytic enzymes (enzymes that catalyze the hydrolysis of molecules) fully active by pumping protons into their interiors and thereby maintaining a low internal pH. Lysosomes that are not functioning actively do not maintain an acidic internal pH and are called *primary lysosomes*. When a primary lysosome fuses with a food vesicle or other organelle, proton pumps in the lysosome are activated, its pH falls, and its arsenal of hydrolytic enzymes is activated; it is then called a *secondary lysosome*.

In addition to breaking down organelles and other structures within cells, lysosomes eliminate other cells that the cell has engulfed in a process called *phagocytosis*, a specific type of endocytosis (see chapter 6). When a white blood cell, for example, phagocytizes a passing pathogen, lysosomes fuse with the resulting "food vesicle," releasing their enzymes into the vesicle and degrading the material within (figure 5.17).

Microbodies. Eukaryotic cells contain a variety of enzyme-bearing, membrane-enclosed vesicles called **microbodies.** Microbodies are found in the cells of plants, animals, fungi, and protists. The distribution of enzymes into microbodies is one of the principal ways by which eukaryotic cells organize their metabolism.

While lysosomes bud from the endomembrane system, microbodies grow by incorporating lipids and protein, and then dividing. Plant cells have a special type of microbody called a **glyoxysome,** which contains enzymes that convert fats into carbohydrates. Another type of microbody, a **peroxisome,** contains enzymes that catalyze the removal of electrons and associated hydrogen atoms (figure 5.18). If these oxidative enzymes were not isolated within microbodies, they would tend to short-circuit the metabolism of the cytoplasm, which often involves adding hydrogen atoms to oxygen. The name *peroxisome* refers to the hydrogen perox-

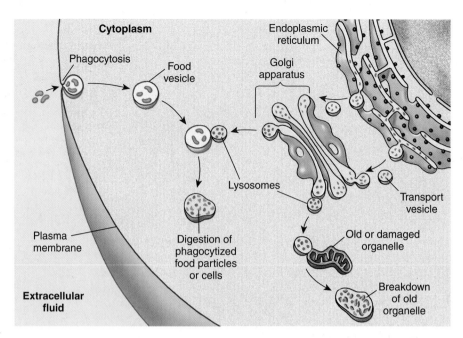

FIGURE 5.17
Lysosomes. Lysosomes contain hydrolytic enzymes that digest particles or cells taken into the cell by phagocytosis and break down old organelles.

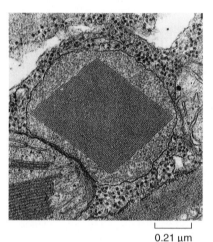

FIGURE 5.18
A peroxisome. Peroxisomes are spherical organelles that may contain a large, diamond-shaped crystal composed of protein. Peroxisomes contain digestive and detoxifying enzymes that produce hydrogen peroxide as a by-product.

0.21 µm

ide produced as a by-product of the activities of the oxidative enzymes in the microbody. Hydrogen peroxide is dangerous to cells because of its violent chemical reactivity. However, peroxisomes also contain the enzyme catalase, which breaks down hydrogen peroxide into harmless water and oxygen.

The endoplasmic reticulum (ER) is an extensive system of folded membranes that spatially organize the cell's biosynthetic activities. The Golgi apparatus collects, packages, modifies, and distributes molecules. Lysosomes and peroxisomes are vesicles that contain digestive and detoxifying enzymes. The isolation of these enzymes in vesicles protects the rest of the cell from the very reactive chemistry occurring inside.

Ribosomes: Sites of Protein Synthesis

Although the DNA in a cell's nucleus encodes the amino acid sequence of each protein in the cell, the proteins are not assembled there. A simple experiment demonstrates this: If a brief pulse of radioactive amino acid is administered to a cell, the radioactivity shows up associated with newly made protein, not in the nucleus, but in the cytoplasm. When investigators first carried out these experiments, they found that protein synthesis is associated with large RNA-protein complexes outside the nucleus called **ribosomes.**

Ribosomes are made up of several molecules of a special form of RNA called ribosomal RNA, or rRNA, bound within a complex of several dozen different proteins. Ribosomes are among the most complex molecular assemblies found in cells. Each ribosome is composed of two subunits (figure 5.19). The subunits join to form a functional ribosome only when they attach to another kind of RNA, called messenger RNA (mRNA), which is the transcribed copy of the coding information on the DNA in the nucleus. Ribosomes use the information in mRNA to direct the synthesis of a protein.

Proteins that function in the cytoplasm are made by free ribosomes that are not associated with organelle membranes, while proteins bound within membranes or destined for export from the cell are assembled by ribosomes bound to rough ER.

The Nucleolus Manufactures Ribosomal Subunits

When cells are synthesizing a large number of proteins, they must first make a large number of ribosomes. To facilitate this, many hundreds of copies of the genes encoding the ribosomal RNAs are clustered together on the chromosome. By transcribing RNA molecules from this cluster, the cell rapidly generates large numbers of the molecules needed to produce ribosomes.

The clusters of ribosomal RNA genes, the RNAs they produce, and the ribosomal proteins all combine within the nucleus during ribosome production. These areas where ribosomes are being assembled are easily visible within the nucleus as one or more dark-staining regions, called **nucleoli** (singular, *nucleolus;* figure 5.20). Nucleoli can be seen under the light microscope even when the chromosomes are extended, unlike the rest of the chromosomes, which are visible only when condensed.

Ribosomes are the sites of protein synthesis in the cytoplasm.

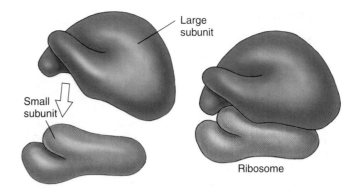

FIGURE 5.19

A ribosome. Ribosomes consist of a large and a small subunit composed of rRNA and protein. The individual subunits are synthesized in the nucleolus and then move through the nuclear pores to the cytoplasm, where they assemble to translate mRNA. Ribosomes serve as sites of protein synthesis.

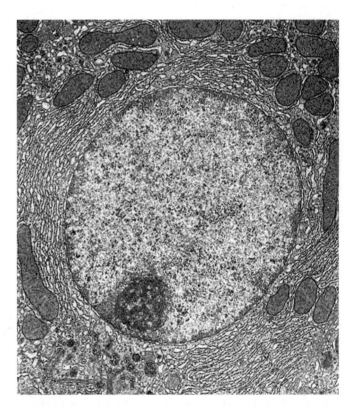

FIGURE 5.20

The nucleolus. This is the interior of a rat liver cell, magnified about 6000 times. A single large nucleus occupies the center of the micrograph. The electron-dense area in the lower left of the nucleus is the nucleolus, the area where the major components of the ribosomes are produced.

Organelles That Contain DNA

Among the most interesting cell organelles are those in addition to the nucleus that contain DNA.

Mitochondria: The Cell's Chemical Furnaces

Mitochondria (singular, *mitochondrion*) are typically tubular or sausage-shaped organelles about the size of bacteria and found in all types of eukaryotic cells (figure 5.21). Mitochondria are bounded by two membranes: a smooth outer membrane and an inner folded membrane with numerous contiguous layers called **cristae** (singular, *crista*). The cristae partition the mitochondrion into two compartments: a **matrix,** lying inside the inner membrane; and an outer compartment, or **intermembrane space,** lying between the two mitochondrial membranes. On the surface of the inner membrane, and also embedded within it, are proteins that carry out oxidative metabolism, the oxygen-requiring process by which energy in macromolecules is stored in ATP.

Mitochondria have their own DNA; this DNA contains several genes that produce proteins essential to the mitochondrion's role in oxidative metabolism. All of these genes are copied into RNA and used to make proteins within the mitochondrion. In this process, the mitochondria employ small RNA molecules and ribosomal components that the mitochondrial DNA also encodes. Thus, the mitochondrion, in many respects, acts as a cell within a cell, containing its own genetic information specifying proteins for its unique functions. The mitochondria are not fully autonomous, however, because most of the genes that produce the enzymes used in oxidative metabolism are located in the nucleus.

A eukaryotic cell does not produce brand-new mitochondria each time the cell divides. Instead, the mitochondria themselves divide in two, doubling in number, and these are partitioned between the new cells. Most of the components required for mitochondrial division are encoded by genes in the nucleus and translated into proteins by cytoplasmic ribosomes. Mitochondrial replication is, therefore, impossible without nuclear participation, and mitochondria thus cannot be grown in a cell-free culture.

Chloroplasts: Where Photosynthesis Takes Place

Plants and other eukaryotic organisms that carry out photosynthesis typically contain from one to several hundred **chloroplasts.** Chloroplasts bestow an obvious advantage on the organisms that possess them: They can manufacture their own food. Chloroplasts contain the

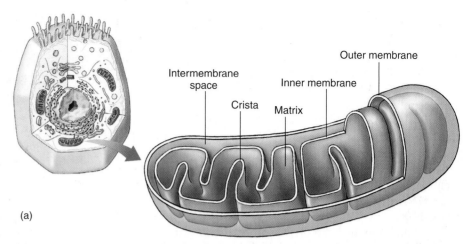

(a)

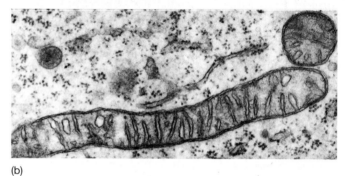

(b)

FIGURE 5.21
Mitochondria. (*a*) The inner membrane of a mitochondrion is shaped into folds called cristae, which greatly increase the surface area for oxidative metabolism. (*b*) Mitochondria in cross section and cut lengthwise (70,000×).

photosynthetic pigment chlorophyll that gives most plants their green color.

The chloroplast body is enclosed, like the mitochondrion, within two membranes, ones that resemble those of mitochondria (figure 5.22). However, chloroplasts are larger and more complex than mitochondria. In addition to the outer and inner membranes, which lie in close association with each other, chloroplasts have a closed compartment of stacked membranes called **grana** (singular, *granum*), which lie internal to the inner membrane. A chloroplast may contain a hundred or more grana, and each granum may contain from a few to several dozen disk-shaped structures called **thylakoids.** On the surface of the thylakoids are the light-capturing photosynthetic pigments, to be discussed in depth in chapter 10. Surrounding the thylakoid is a fluid matrix called the *stroma*.

Like mitochondria, chloroplasts contain DNA, but many of the genes that specify chloroplast components are also located in the nucleus. Some of the elements used in the photosynthetic process, including the specific protein components necessary to accomplish the reaction, are synthesized entirely within the chloroplast.

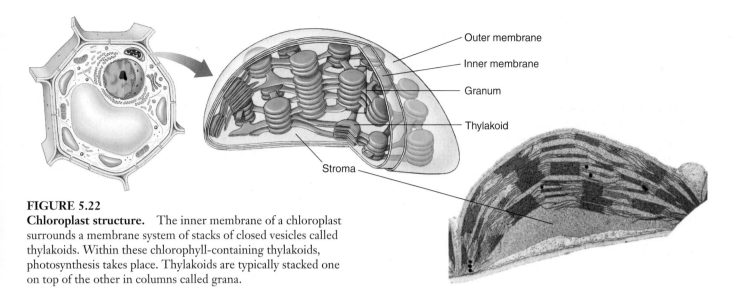

FIGURE 5.22
Chloroplast structure. The inner membrane of a chloroplast surrounds a membrane system of stacks of closed vesicles called thylakoids. Within these chlorophyll-containing thylakoids, photosynthesis takes place. Thylakoids are typically stacked one on top of the other in columns called grana.

Other DNA-containing organelles in plants, called leucoplasts, lack pigment and a complex internal structure. In root cells and some other plant cells, leucoplasts may serve as starch storage sites. A leucoplast that stores starch (amylose) is sometimes termed an **amyloplast.** These organelles—chloroplasts, leucoplasts, and amyloplasts—are collectively called **plastids.** All plastids are produced by the division of existing plastids.

Endosymbiosis

Symbiosis is a close relationship between organisms of different species that live together. As noted in chapter 4, the theory of **endosymbiosis** proposes that some of today's eukaryotic organelles evolved by a symbiosis in which one species of prokaryote was engulfed by and lived inside another species of prokaryote that was a precursor to eukaryotes (figure 5.23). According to the endosymbiont theory, the engulfed prokaryotes provided their hosts with certain advantages associated with their special metabolic abilities. Two key eukaryotic organelles are believed to be the descendants of these endosymbiotic prokaryotes: mitochondria, which are thought to have originated as bacteria capable of carrying out oxidative metabolism, and chloroplasts, which apparently arose from photosynthetic bacteria.

The endosymbiont theory is supported by a wealth of evidence. Both mitochondria and chloroplasts are surrounded by two membranes; the inner membrane probably evolved from the plasma membrane of the engulfed prokaryote, while the outer membrane is probably derived from the plasma membrane or endoplasmic reticulum of the host cell. Mitochondria are about the same size as most prokaryotes, and the cristae formed by their

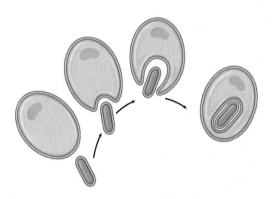

FIGURE 5.23
Endosymbiosis. This figure shows how a double membrane may have been created during the symbiotic origin of mitochondria or chloroplasts.

inner membranes resemble the folded membranes in various groups of bacteria. Mitochondrial ribosomes are also similar to prokaryotic ribosomes in size and structure. Both mitochondria and chloroplasts contain circular molecules of DNA similar to those in prokaryotes. Finally, mitochondria divide by simple fission, splitting in two just as prokaryotic cells do, and they apparently replicate and partition their DNA in much the same way as prokaryotes do.

Both mitochondria and chloroplasts have their own DNA, which contains specific genes related to some of their functions, but both depend on nuclear genes for other functions.

The Cytoskeleton: Interior Framework of the Cell

The cytoplasm of all eukaryotic cells is crisscrossed by a network of protein fibers that supports the shape of the cell and anchors organelles to fixed locations. This network, called the **cytoskeleton,** is a dynamic system, constantly forming and disassembling. Individual fibers form by **polymerization,** as identical protein subunits attract one another chemically and spontaneously assemble into long chains. Fibers disassemble in the same way, as one subunit after another breaks away from one end of the chain.

Eukaryotic cells may contain the following three types of cytoskeletal fibers, each formed from a different kind of subunit:

1. **Actin filaments.** Actin filaments are long fibers about 7 nanometers in diameter. Each filament is composed of two protein chains loosely twined together like two strands of pearls (figure 5.24). Each "pearl," or subunit, on the chains is the globular protein **actin.** Actin molecules spontaneously form these filaments, even in a test tube. Cells regulate the rate of actin polymerization through other proteins that act as switches, turning on polymerization when appropriate. Actin filaments are responsible for cellular movements such as contraction, crawling, "pinching" during division, and formation of cellular extensions.

2. **Microtubules.** Microtubules are hollow tubes about 25 nanometers in diameter, each composed of a ring of 13 protein protofilaments (see figure 5.24). Globular proteins consisting of dimers of alpha and beta *tubulin* subunits polymerize to form the 13 protofilaments. The protofilaments are arrayed side by side around a central core, giving the microtubule its characteristic tube shape. In many cells, microtubules form from nucleation centers near the center of the cell and radiate toward the periphery. They are in a constant state of flux, continually polymerizing and depolymerizing. The average half-life of a microtubule ranges from as long as 10 minutes in a nondividing animal cell to as short as 20 seconds in a dividing animal cell. The ends of the microtubule are designated as "+" (away from the nucleation center) or "−" (toward the nucleation center). Along with facilitating cellular movement, microtubules are responsible for moving materials within the cell itself. Special motor proteins, discussed on page 99, move cellular organelles around the cell on microtubular "tracks." *Kinesin* proteins move organelles toward the "+" end (toward the cell periphery), and *dyneins* move them toward the "−" end.

3. **Intermediate filaments.** The most durable element of the cytoskeleton in animal cells is a system of tough, fibrous protein molecules twined together in

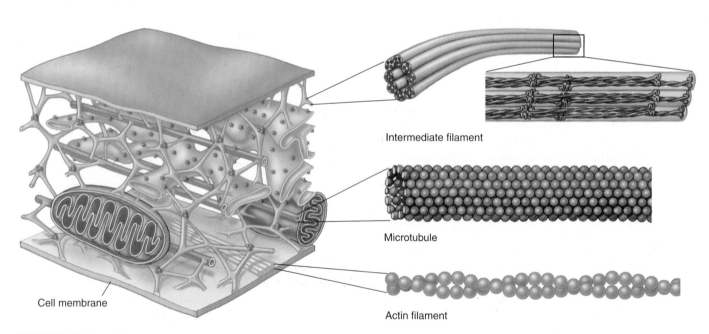

Cell membrane

Intermediate filament

Microtubule

Actin filament

FIGURE 5.24

Molecules that make up the cytoskeleton. *Actin filaments:* Actin filaments are made of two strands of the fibrous protein actin twisted together and usually occur in bundles. Actin filaments, though ubiquitous, are concentrated below the plasma membrane in bundles known as stress fibers, which may have a contractile function. *Microtubules:* Microtubules are composed of tubulin protein subunits arranged side by side to form a tube. Microtubules are comparatively stiff cytoskeletal elements that serve to organize metabolism and intracellular transport in the nondividing cell and stabilize cell structure. *Intermediate filaments:* Intermediate filaments are composed of overlapping staggered tetramers of protein. This molecular arrangement allows for a ropelike structure that imparts tremendous mechanical strength to the cell.

an overlapping arrangement (see figure 5.24). These fibers are characteristically 8 to 10 nanometers in diameter, intermediate in size between actin filaments and microtubules (which is why they are called intermediate filaments). Once formed, intermediate filaments are stable and usually do not break down. Intermediate filaments constitute a heterogeneous group of cytoskeletal fibers. The most common type, composed of protein subunits called *vimentin*, provides structural stability for many kinds of cells. *Keratin*, another class of intermediate filament, is found in epithelial cells (cells that line organs and body cavities) and associated structures such as hair and fingernails. The intermediate filaments of nerve cells are called *neurofilaments*.

Centrioles: Microtubule Assembly Centers

Centrioles are barrel-shaped organelles found in the cells of animals and most protists. They occur in pairs, usually located at right angles to each other near the nuclear membranes (figure 5.25); the region surrounding the pair in almost all animal cells is referred to as a **centrosome.** Centrioles help assemble the animal cell's microtubules. Cells of plants and fungi lack centrioles, and cell biologists are still in the process of characterizing their microtubule-organizing centers.

Moving Material Within the Cell

Actin filaments and microtubules often orchestrate their activities to affect cellular processes. For example, during cell reproduction (see chapter 11), newly replicated chromosomes move to opposite sides of a dividing cell because they are attached to shortening microtubules. Then, in animal cells, a belt of actin pinches the cell in two by contracting like a purse string. Muscle cells also use actin filaments to contract their cytoskeletons. The fluttering of an eyelash, the flight of an eagle, and the awkward crawling of a baby all depend on these cytoskeletal movements within muscle cells.

Not only is the cytoskeleton responsible for the cell's shape and movement, but it also provides a scaffold that holds certain enzymes and other macromolecules in defined areas of the cytoplasm. For example, many of the enzymes involved in cell metabolism bind to actin filaments; so do ribosomes. By moving and anchoring particular enzymes near one another, the cytoskeleton, like the endoplasmic reticulum, organizes the cell's activities.

Intracellular Molecular Motors. All eukaryotic cells must move materials from one place to another in the cytoplasm. Most cells use the endomembrane system as an intracellular highway; the Golgi apparatus packages materials into vesicles that move through the channels of the endoplasmic reticulum to the far reaches of the cell. However,

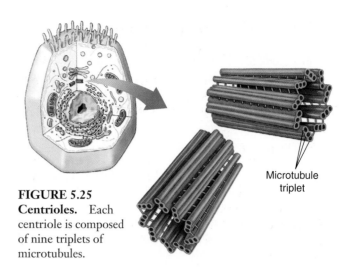

FIGURE 5.25 Centrioles. Each centriole is composed of nine triplets of microtubules.

Microtubule triplet

this highway is only effective over short distances. When a cell has to transport materials through a long extension, such as the axon of a nerve cell, passage through the highways is too slow. For these situations, eukaryotic cells have developed high-speed locomotives that run along microtubular tracks.

Four components are required: (1) a vesicle or organelle that is to be transported, (2) a motor molecule that provides the energy-driven motion, (3) a connector molecule that connects the vesicle to the motor molecule, and (4) microtubules on which the vesicle will ride like a train on a rail (figure 5.26). For example, embedded within the membranes of endoplasmic reticulum is a protein called kinectin that binds the ER vesicles to the motor protein **kinesin.** As nature's tiniest motors, these proteins literally pull the transport vesicles along the microtubular tracks. Kinesin uses ATP to power its movement toward the cell periphery, dragging the vesicle with it as it travels along the microtubule. Another set of vesicle proteins, called the dynactin complex, binds vesicles to the motor protein **dynein** (see figure 5.26),

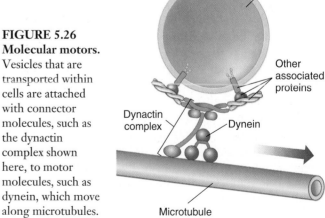

FIGURE 5.26 Molecular motors. Vesicles that are transported within cells are attached with connector molecules, such as the dynactin complex shown here, to motor molecules, such as dynein, which move along microtubules.

Vesicle

Other associated proteins

Dynactin complex

Dynein

Microtubule

which directs movement in the opposite direction, inward toward the cell's center. (Dynein is also involved in the movement of eukaryotic flagella, as discussed below.) The destination of a particular transport vesicle and its contents is thus determined by the nature of the linking protein embedded within the vesicle's membrane.

Cell Movement

Essentially all cell motion is tied to the movement of actin filaments, microtubules, or both. Intermediate filaments act as intracellular tendons, preventing excessive stretching of cells, and actin filaments play a major role in determining the shape of cells. Because actin filaments can form and dissolve so readily, they enable some cells to change shape quickly.

Some Cells Crawl. The arrangement of actin filaments within the cell cytoplasm allows cells to "crawl," *literally!* Crawling is a significant cellular phenomenon, essential to inflammation, clotting, wound healing, and the spread of cancer. White blood cells in particular exhibit this ability. Produced in the bone marrow, these cells are released into the circulatory system and then eventually crawl out of capillaries and into the tissues to destroy potential pathogens.

At the leading edge of a crawling cell, actin filaments rapidly polymerize, and their extension forces the edge of the cell forward. This extended region is stabilized when microtubules polymerize into the newly formed region. Forward movement of the cell overall is then achieved when myosin motors along the actin filaments contract, pulling the contents of the cell toward the newly extended front edge. Overall crawling of the cell takes place when these steps occur continuously, with a leading edge extending and stabilizing, and then motors contracting to pull the remaining cell contents along. Receptors on the cell surface can detect markers outside the cell and stimulate extension in specific directions, allowing cells to move toward particular targets.

Other Cells Swim: Flagella and Cilia. Earlier in this chapter, we described the structure of prokaryotic flagella. Eukaryotic cells have a *completely different* kind of flagellum, consisting of a circle of nine microtubule pairs surrounding two central microtubules; this arrangement is referred to as the **9 + 2 structure** (figure 5.27). As pairs of microtubules move past one another using arms composed of the motor protein dynein, the eukaryotic flagellum undulates rather than rotating. When examined carefully, each flagellum proves to be an outward projection of the cell's interior, containing cytoplasm and enclosed by the plasma membrane. The microtubules of the flagellum are derived from a **basal body,** situated just below the point where the flagellum protrudes from the surface of the cell.

The flagellum's complex microtubular apparatus evolved early in the history of eukaryotes. Although the

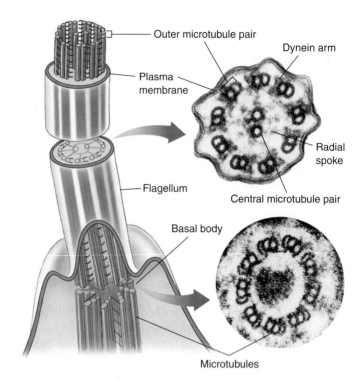

FIGURE 5.27
Flagella and cilia. A eukaryotic flagellum originates directly from a basal body. The flagellum has two microtubules in its core connected by radial spokes to an outer ring of nine paired microtubules with dynein arms. The basal body consists of nine microtubule triplets connected by short protein segments. The structure of cilia is similar to that of flagella, but cilia are usually shorter.

cells of many multicellular and some unicellular eukaryotes today no longer possess flagella and are nonmotile, an organization similar to the 9 + 2 arrangement of microtubules can still be found within them, in structures called **cilia** (singular, *cilium*). Cilia are short cellular projections that are often organized in rows. They are more numerous than flagella on the cell surface, but have the same internal structure. In many multicellular organisms, cilia carry out tasks far removed from their original function of propelling cells through water. In several kinds of vertebrate tissues, for example, the beating of rows of cilia moves water over the tissue surface. The sensory cells of the vertebrate ear also contain cilia; sound waves bend these cilia and provide the initial sensory input for hearing. Thus, the 9 + 2 structure of flagella and cilia appears to be a fundamental component of eukaryotic cells.

The three principal fibers of the cytoskeleton are actin filaments, microtubules, and intermediate filaments. These fibers interact to modulate cell shape and permit cell movement, and act to move materials within the cytoplasm.

5.4 Not all eukaryotic cells are the same.

Vacuoles and Cell Walls

Vacuoles: A Central Storage Compartment

Some kinds of eukaryotic cells contain unique organelles or structures. For example, the center of a plant cell usually contains a large **central vacuole** (figure 5.28) that stores water and other materials, such as sugars, ions, and pigments. The central vacuole also helps increase the surface area-to-volume ratio of the plant cell by applying pressure to the plasma membrane. The plasma membrane expands outward under this pressure, thereby increasing its surface area. Vacuoles are also found in some types of fungi and protists, and can have a variety of functions.

Cell Walls: Protection and Support

The cells of plants, fungi, and many types of protists have **cell walls,** which protect and support the cells. The cell walls of plants, protists, and fungi are chemically and structurally different from prokaryotic cell walls. In plants and protists, the cell walls are composed of fibers of the polysaccharide cellulose, while in fungi, the cell walls are composed of chitin. In plants, **primary walls** are laid down when the cell is still growing, and between the walls of adjacent cells is a sticky substance called the **middle lamella,** which glues the cells together (figure 5.29). Some plant cells produce strong **secondary walls,** which are deposited inside the primary walls of fully expanded cells.

> **Plant, fungi, and some protist cells store substances in a large central vacuole, and encase themselves within strong cell walls.**

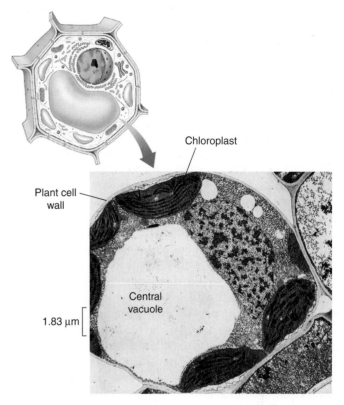

Chloroplast

Plant cell wall

Central vacuole

1.83 μm

FIGURE 5.28
The central vacuole. A plant's central vacuole stores dissolved substances and can expand in size to increase the surface area of a plant cell.

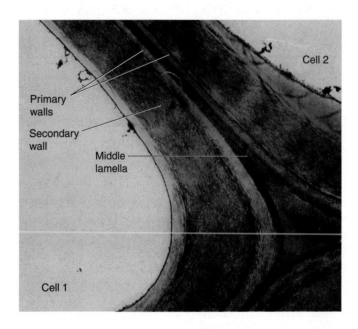

Cell 2

Primary walls

Secondary wall

Middle lamella

Cell 1

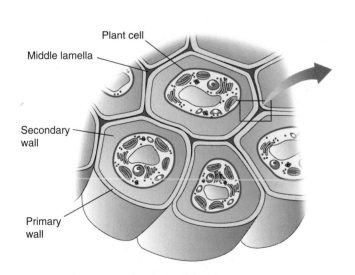

Plant cell

Middle lamella

Secondary wall

Primary wall

FIGURE 5.29
Cell walls in plants. Plant cell walls are thick, strong, and rigid. Primary cell walls are laid down when the cell is young. Thicker secondary cell walls may be added later when the cell is fully grown.

The Extracellular Matrix

As just discussed, many types of eukaryotic cells possess a cell wall exterior to the plasma membrane that protects the cell, maintains its shape, and prevents excessive water uptake. Animal cells are the great exception, lacking the cell walls that encase plants, fungi, and most protists. Instead, animal cells secrete an elaborate mixture of **glycoproteins** (proteins with short chains of sugars attached to them) into the space around them, forming the **extracellular matrix** (**ECM**) (figure 5.30). The fibrous protein collagen, the same protein in fingernails and hair, is abundant in the ECM. Strong fibers of collagen and another fibrous protein, elastin, are embedded within a complex web of other glycoproteins, called proteoglycans, that form a protective layer over the cell surface. The ECM is attached to the plasma membrane by a third kind of glycoprotein, **fibronectin**. Fibronectin molecules bind not only to ECM glycoproteins but also to proteins called **integrins,** which are an integral part of the plasma membrane. Integrins extend into the cytoplasm, where they are attached to the microfilaments of the cytoskeleton. Linking ECM and cytoskeleton, integrins allow the ECM to influence cell behavior in important ways, altering gene expression and cell migration patterns by a combination of mechanical and chemical signaling pathways. In this way, the ECM can help coordinate the behavior of all the cells in a particular tissue.

Table 5.3 compares and reviews the features of three types of cells.

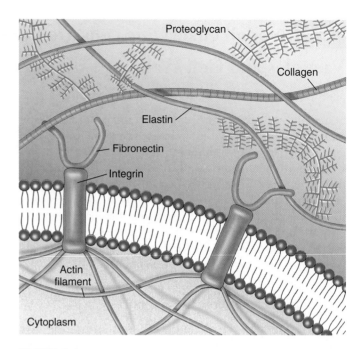

FIGURE 5.30
The extracellular matrix. Animal cells are surrounded by an extracellular matrix composed of various glycoproteins that give the cells support, strength, and resilience.

In animal cells, which lack a cell wall, the cytoskeleton is linked by integrin proteins to a web of glycoproteins called the extracellular matrix.

Table 5.3 A Comparison of Prokaryotic, Animal, and Plant Cells

	Prokaryote	Animal	Plant
EXTERIOR STRUCTURES			
Cell wall	Present (protein-polysaccharide)	Absent	Present (cellulose)
Cell membrane	Present	Present	Present
Flagella/cilia	May be present (single strand)	May be present	Absent except in sperm of a few species
INTERIOR STRUCTURES			
ER	Absent	Usually present	Usually present
Ribosomes	Present	Present	Present
Microtubules	Absent	Present	Present
Centrioles	Absent	Present	Absent
Golgi apparatus	Absent	Present	Present
Nucleus	Absent	Present	Present
Mitochondria	Absent	Present	Present
Chloroplasts	Absent	Absent	Present
Chromosomes	A single circle of DNA	Multiple; DNA-protein complex	Multiple; DNA-protein complex
Lysosomes	Absent	Usually present	Present
Vacuoles	Absent	Absent or small	Usually a large single vacuole

Concept Review

For interactive testing, visit the Online Learning Center with PowerWeb at www.mhhe.com/Raven7

5.1 All organisms are composed of cells.

Characteristics of Cells

- Prokaryotes have a single circular molecule of DNA, while eukaryotic cells have a nucleus, bounded by a nuclear membrane and containing DNA. (p. 80)
- Cytoplasm fills the interior of the cell, exclusive of the nucleus, and a plasma membrane encloses the cell and separates it from its surroundings. The plasma membrane contains several types of proteins that allow the cell to interact with its environment. (p. 80)
- Early cell theory included three principles: (1) All organisms are composed of one or more cells; (2) cells are the smallest living things; and (3) cells arise only by division of previously existing cells. (p. 81)
- Smaller cells are more advantageous than larger cells due to the limitations of communication in relation to surface area-to-volume ratios. (p. 81)

Visualizing Cells

- Resolution refers to the minimum distance two points can be apart and still be distinguished. (p. 82)
- Compound microscopes use magnifying lenses to achieve high magnification and clarity and to increase resolution. (p. 82)
- Other types of microscopes, including transmission electron microscopes and scanning electron microscopes, as well as staining procedures, are employed to view cells. (pp. 82–83)

5.2 Eukaryotic cells are more structurally complex than prokaryotic cells.

Prokaryotes: Simple Cells with Complex Physiology

- Most prokaryotic cells are encased by a strong cell wall and can be classified using the Gram staining procedure. (p. 84)
- Some prokaryotes use a flagellum for locomotion. (p. 84)
- Because prokaryotic cells lack membrane-bounded organelles, all cytoplasmic constituents have access to all areas of the cell. (p. 85)

Eukaryotes: Cells with Complex Interiors

- Eukaryotic cells have extensive internal compartmentalization, with multiple membrane-bounded organelles carrying out specific functions. (pp. 86–87)

5.3 Take a tour of a eukaryotic cell.

The Nucleus: Information Center for the Cell

- The nucleus is the repository for the genetic information that directs cell activities. (p. 88)
- The surface of the nucleus is bounded by two phospholipid bilayer membranes that form the nuclear envelope. (p. 89)
- DNA is organized into chromosomes located in the nucleus. (p. 89)

The Endomembrane System

- The endoplasmic reticulum is an extensive system of folded membranes that compartmentalizes the cell's interior. (p. 90)
- The Golgi apparatus collects, packages, modifies, and distributes molecules throughout the cell. (p. 92)
- Lysosomes contain digestive enzymes that catalyze the rapid breakdown of proteins, nucleic acids, and carbohydrates. (p. 94)

Ribosomes: Sites of Protein Synthesis

- Two ribosomal subunits join to form a ribosome when they attach to messenger RNA to direct protein synthesis. (p. 95)

Organelles That Contain DNA

- Mitochondria and chloroplasts have their own DNA specifying proteins for their unique functions. (p. 96)

The Cytoskeleton: Interior Framework of the Cell

- The cytoplasm of all eukaryotic cells is meshed by a network of protein fibers called the cytoskeleton, which supports the shape of the cell and anchors its organelles. (p. 98)
- Eukaryotic cells may contain three types of cytoskeletal fibers: actin filaments, microtubules, and intermediate filaments. (pp. 98–99)

5.4 Not all eukaryotic cells are the same.

Vacuoles and Cell Walls

- Plant cells usually contain a large central vacuole for storage, and are encased by thick cell walls and often even by strong secondary walls. (p. 101)

The Extracellular Matrix

- Animal cells lack the thick cell walls of plants, but secrete a mixture of glycoproteins to form an extracellular matrix, which helps coordinate the behavior of all the cells in a particular tissue. (p. 102)

Self Test

1. What type of microscope would you need to view a cellular structure that is 5 nm in size?
   a. a light microscope
   b. an electron microscope
   c. a compound microscope
   d. No microscope can resolve down to 5 nm.
2. Which of the following is *not* found in prokaryotic cells?
   a. ribosomes
   b. cell wall
   c. nucleus
   d. photosynthetic membranes
3. Which of the following statements is incorrect?
   a. DNA in the nucleus is usually coiled into chromosomes.
   b. The nucleolus is the site of ribosomal RNA synthesis.
   c. Some substances can pass into and out of the nucleus.
   d. Red blood cells cannot synthesize RNA.
4. Which of the following matches is *not* correct?
   a. ribosomes—rough ER
   b. protein synthesis—smooth ER
   c. rough ER—export of proteins out of cell
   d. smooth ER—cells in intestine
5. Which of the following is *not* produced by the Golgi apparatus?
   a. glycolipids
   b. glycoproteins
   c. liposomes
   d. secretory vesicles
6. What is the difference between a primary lysosome and a secondary lysosome?
   a. Primary lysosomes are larger than secondary lysosomes.
   b. Primary lysosomes are active, while secondary lysosomes are inactive.
   c. Primary lysosomes have a low pH, while secondary lysosomes have a high pH.
   d. Primary lysosomes have low levels of protons, while secondary lysosomes have high levels of protons.
7. Proteins that stay within the cell are produced
   a. on free ribosomes in the cytoplasm.
   b. in the nucleolus.
   c. on ribosomes attached to rough ER.
   d. on ribosomes and pass through the Golgi apparatus.
8. What do chloroplasts and mitochondria have in common?
   a. Both are present in animal cells.
   b. Both contain their own genetic material.
   c. Both are present in all eukaryotic cells.
   d. Both are present in plant cells.
9. Which of the following pairs is correctly matched?
   a. actin—MTOC
   b. intermediate fibers—protofilaments
   c. microtubules—"+"/"−" ends
   d. intermediate fibers—cellular movement
10. Which of the following eukaryotic organelles are believed to have evolved through endosymbiosis?
    a. nucleus and mitochondrion
    b. mitochondrion and chloroplast
    c. nucleus and endoplasmic reticulum
    d. chloroplast and endoplasmic reticulum

Test Your Visual Understanding

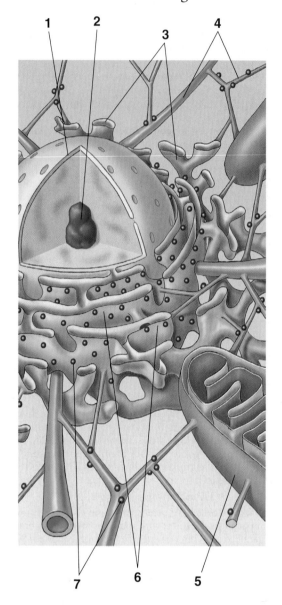

1. From the following list, match the correct label with its numbered leader line in the figure, and describe the function of the organelle in the eukaryotic cell.
   cytoskeleton   ribosomes
   mitochondrion   rough endoplasmic reticulum
   nuclear envelope   smooth endoplasmic reticulum
   nucleolus

Apply Your Knowledge

1. White blood cells (WBCs) circulate throughout the human body. Monocytes make up about 6% of total WBCs, and neutrophils make up about 65% of total WBCs. Monocytes have a diameter of 15 μm, and neutrophils have a diameter of 10 μm. Calculate the surface area and volume of each cell.

6

Membranes

Concept Outline

6.1 Biological membranes are fluid layers of lipid.

   The Phospholipid Bilayer. Cells are encased by membranes composed of a bilayer of phospholipid.

   The Lipid Bilayer Is Fluid. Because individual phospholipid molecules bind weakly to one another, the lipid bilayer of membranes is a fluid.

6.2 Proteins embedded within cell membranes determine their character.

   The Fluid Mosaic Model. A varied collection of proteins float within the lipid bilayer.

   Examining Cell Membranes. Visualizing a plasma membrane requires a powerful electron microscope.

   Kinds of Membrane Proteins. The proteins in a membrane function in support, transport, recognition, and reactions.

   Structure of Membrane Proteins. Membrane proteins contain nonpolar regions that span the lipid bilayer.

6.3 Passive transport across membranes moves down the concentration gradient.

   Diffusion. Random molecular motion results in a net movement of molecules to regions of lower concentration.

   Facilitated Diffusion. Passive movement across a membrane is often through specific carrier proteins.

   Osmosis. Water molecules also move down their concentration gradients toward the side of the membrane with more dissolved solute molecules.

6.4 Bulk transport utilizes endocytosis.

   Bulk Passage Into and Out of the Cell. To transport large particles, plasma membranes form vesicles.

6.5 Active transport across membranes requires energy.

   Active Transport. Cells transport molecules up a concentration gradient using ATP-powered carrier proteins. Active transport of ions also drives coupled uptake of other molecules up their concentration gradients.

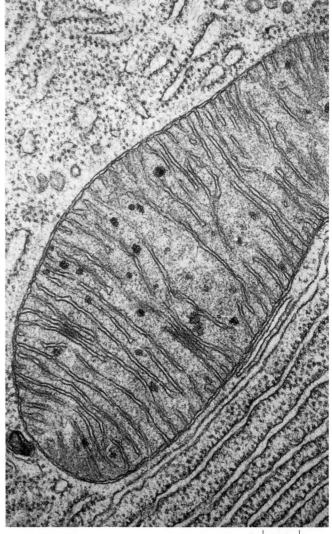

0.16 µm

FIGURE 6.1
Membranes within a human cell. Sheets of endoplasmic reticulum weave through the cell interior. The large oval is a mitochondrion, itself filled with extensive internal membranes.

A mong a cell's most important activities are its interactions with the environment, a give-and-take that never ceases. Without it, life could not persist. While living cells and eukaryotic organelles (figure 6.1) are encased within a lipid membrane through which few water-soluble substances can pass, the membrane contains protein passageways that permit specific substances to move into and out of the cell and allow the cell to exchange information with its environment. We call this delicate skin of lipids with embedded protein molecules a **plasma membrane.** This chapter examines the structure and function of this remarkable membrane.

6.1 Biological membranes are fluid layers of lipid.

The Phospholipid Bilayer

The membranes that encase all living cells are sheets of lipid only two molecules thick; more than 10,000 of these sheets piled on one another would just equal the thickness of this sheet of paper. The lipid layer that forms the foundation of a cell's membranes is composed of molecules called **phospholipids** (figure 6.2).

Phospholipids

Like the fat molecules you studied in chapter 3, a phospholipid has a backbone derived from a three-carbon molecule called glycerol. Attached to this backbone are fatty acids, long chains of carbon atoms ending in a carboxyl (—COOH) group. A fat molecule has three such chains, one attached to each carbon in the backbone; because these chains are nonpolar, they do not form hydrogen bonds with water, and the fat molecule is not water-soluble. A phospholipid, by contrast, has only two fatty acid chains attached to its backbone. The third carbon on the backbone is attached instead to a highly polar organic alcohol that readily forms hydrogen bonds with water. Because this alcohol is attached by a phosphate group, the molecule is called a *phospho*lipid.

One end of a phospholipid molecule is, therefore, strongly nonpolar (hydrophobic, or "water-fearing"), while the other end is strongly polar (hydrophilic, or "water-loving"). The two nonpolar fatty acids extend in one direction, roughly parallel to each other, and the polar alcohol group points in the other direction. Because of this structure, phospholipids are often diagrammed as a polar head with two dangling nonpolar tails, as in figure 6.2b.

Phospholipids Form Bilayer Sheets

What happens when a collection of phospholipid molecules is placed in water? The polar water molecules repel the long, nonpolar tails of the phospholipids as the water molecules seek partners for hydrogen bonding. Due to the polar nature of the water molecules, the nonpolar tails of the phospholipids end up packed closely together, sequestered as far as possible from water. Every phospholipid molecule orients its polar head toward water and its nonpolar tails away. When *two* layers form with the tails facing each other, no tails ever come in contact with water. The resulting structure is called a **phospholipid bilayer** (figure 6.3). Lipid bilayers form spontaneously, driven by the tendency of water molecules to form the maximum number of hydrogen bonds.

The nonpolar interior of a lipid bilayer impedes the passage of any water-soluble substances through the bilayer, just as a layer of oil impedes the passage of a drop

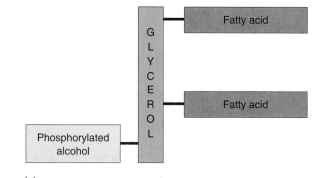

(a)

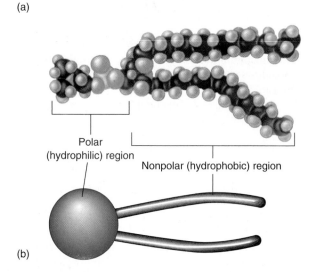

(b)

FIGURE 6.2
Phospholipid structure. (*a*) A phospholipid is a composite molecule similar to a triacylglycerol, except that only two fatty acids are bound to the glycerol backbone; a phosphorylated alcohol occupies the third position on the backbone. (*b*) Because the phosphorylated alcohol usually extends from one end of the molecule and the two fatty acid chains extend from the other, phospholipids are often diagrammed as a polar head with two nonpolar, hydrophobic tails.

of water ("oil and water do not mix"). This barrier to the passage of water-soluble substances is the key biological property of the lipid bilayer. In addition to the phospholipid molecules that make up the lipid bilayer, the membranes of every cell also contain proteins that extend through the lipid bilayer, providing passageways across the membrane.

The basic foundation of biological membranes is a lipid bilayer, which forms spontaneously. In such a layer, the nonpolar hydrophobic tails of phospholipid molecules point inward, forming a nonpolar barrier to water-soluble molecules.

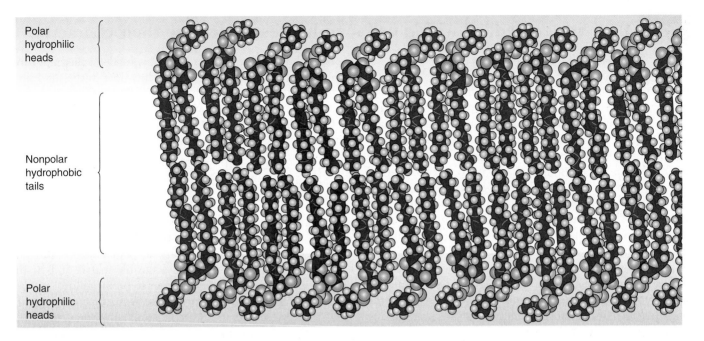

Polar
hydrophilic
heads

Nonpolar
hydrophobic
tails

Polar
hydrophilic
heads

FIGURE 6.3
A phospholipid bilayer. The basic structure of every plasma membrane is a double layer of lipid, in which phospholipids aggregate to form a bilayer with a nonpolar interior. The phospholipid tails do not align perfectly and have only a weak attraction to each other. Thus, the membrane is "fluid." Individual phospholipid molecules can move from one place to another in the membrane.

The Lipid Bilayer Is Fluid

A lipid bilayer is stable because water's affinity for hydrogen bonding never stops. Just as surface tension holds a soap bubble together, even though it is made of a liquid, so the hydrogen bonding of water holds a membrane together. But while water continually drives phospholipid molecules into this configuration, it does not locate specific phospholipid molecules relative to their neighbors in the bilayer. Because the interactions of phospholipids with each other are relatively weak, individual phospholipids and unanchored proteins are relatively free to move about within the membrane. This can be demonstrated vividly by fusing cells and watching their proteins reassort (figure 6.4).

Some membranes are more fluid than others, however. The tails of individual phospholipid molecules are attracted to one another when they line up close together. This causes the membrane to become less fluid, because aligned molecules must pull apart from one another before they can move about in the membrane. The greater the degree of alignment, the less fluid the membrane. Some phospholipid tails do not align well because they contain one or more double bonds between carbon atoms, introducing kinks in the tail and disrupting the van der Waals forces holding them together. Membranes containing such phospholipids are more fluid than membranes that lack them. Most membranes also contain steroid lipids such as cholesterol, which can either increase or decrease membrane fluidity, depending on temperature.

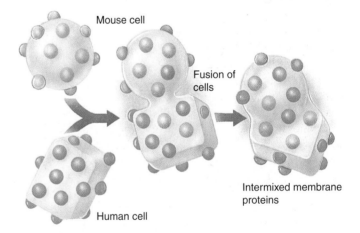

Mouse cell

Fusion of
cells

Intermixed membrane
proteins

Human cell

FIGURE 6.4
Protein movement in membranes. The way that proteins move about within membranes can be demonstrated easily by labeling the plasma membrane proteins of a mouse cell with fluorescent antibodies and then fusing that cell with a human cell. At first, all of the mouse proteins are located on the mouse side of the fused cell, and all of the human proteins are located on the human side of the fused cell. However, within an hour, the labeled and unlabeled proteins are intermixed throughout the hybrid cell's plasma membrane.

The lipid bilayer is liquid like a soap bubble, rather than solid like a rubber balloon.

6.2 Proteins embedded within cell membranes determine their character.

The Fluid Mosaic Model

A plasma membrane is composed of both lipids and globular proteins. For many years, biologists thought the protein covered the inner and outer surfaces of the phospholipid bilayer like a coat of paint. The widely accepted Davson-Danielli model, proposed in 1935, portrayed the membrane as a sandwich: a phospholipid bilayer between two layers of globular protein. This model, however, was not consistent with what researchers were learning in the 1960s about the structure of membrane proteins. Unlike most proteins found within cells, membrane proteins are not very soluble in water—they possess long stretches of nonpolar hydrophobic amino acids. If such proteins indeed coated the surface of the lipid bilayer, as the Davson-Danielli model suggested, then their nonpolar portions would separate the polar portions of the phospholipids from water, causing the bilayer to dissolve! Because this doesn't happen, there is clearly something wrong with the model.

In 1972, S. Singer and G. J. Nicolson revised the model in a simple but profound way: They proposed that the globular proteins are *inserted* into the lipid bilayer, with their nonpolar segments in contact with the nonpolar interior of the bilayer and their polar portions protruding out from the membrane surface. In this model, called the **fluid mosaic model,** a mosaic of proteins float in the fluid lipid bilayer like boats on a pond (figure 6.5).

Components of the Cell Membrane

A eukaryotic cell contains many membranes. While they are not all identical, they share the same fundamental architecture. Cell membranes are assembled from four components (table 6.1):

1. **Phospholipid bilayer.** Every cell membrane is composed of phospholipids in a bilayer. The other components of the membrane are embedded within the bilayer, which provides a flexible matrix and, at the same time, imposes a barrier to permeability.
2. **Transmembrane proteins.** A major component of every membrane is a collection of proteins that float on or in the lipid bilayer. These proteins provide passageways that allow substances and information to cross the membrane. Many membrane proteins are not fixed in position; they can move about, just as the phospholipid molecules do. Some membranes are crowded with proteins, while in others, the proteins are more sparsely distributed.

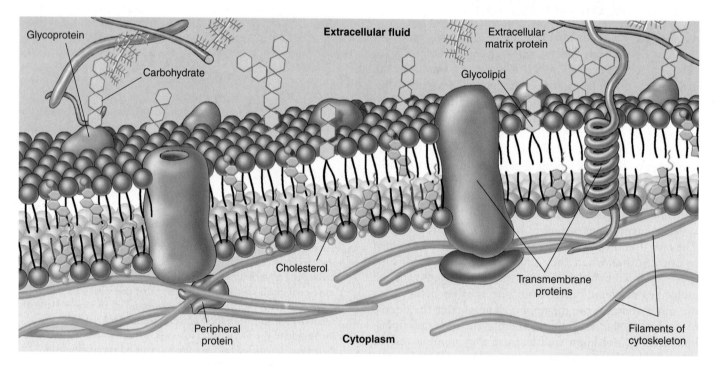

FIGURE 6.5
The fluid mosaic model of cell membranes.   A variety of proteins protrude through the plasma membrane of animal cells, and nonpolar regions of the proteins tether them to the membrane's nonpolar interior. Three important classes of membrane proteins are transport proteins, receptors, and cell surface markers. Carbohydrate chains are often bound to the extracellular portion of these proteins, as well as to the membrane phospholipids. These chains serve as distinctive identification tags, unique to particular cells.

Table 6.1 Components of the Cell Membrane

Component	Composition	Function	How It Works	Example
Phospholipid bilayer	Phospholipid molecules	Provides permeability barrier, matrix for proteins	Excludes water-soluble molecules from nonpolar interior of bilayer	Bilayer of cell is impermeable to water-soluble molecules, such as glucose
Transmembrane proteins	Carriers	Actively and passively transport molecules across membrane	"Escort" molecules through the membrane in a series of conformational changes	Glycophorin carrier for sugar transport; sodium-potassium pump
	Channels	Passively transport molecules across membrane	Create a tunnel that acts as a passage through membrane	Sodium and potassium channels in nerve cells
	Receptors	Transmit information into cell	Signal molecules bind to cell surface portion of the receptor protein; this alters the portion of the receptor protein within the cell, inducing activity	Specific receptors bind peptide hormones and neurotransmitters
Interior protein network	Spectrins	Determine shape of cell	Form supporting scaffold beneath membrane, anchored to both membrane and cytoskeleton	Red blood cell
	Clathrins	Anchor certain proteins to specific sites, especially on the exterior plasma membrane in receptor-mediated endocytosis	Proteins line coated pits and facilitate binding to specific molecules	Localization of low-density lipoprotein receptor within coated pits
Cell surface markers	Glycoproteins	"Self" recognition	Create a protein/carbohydrate chain shape characteristic of individual	Major histocompatibility complex protein recognized by immune system
	Glycolipid	Tissue recognition	Create a lipid/carbohydrate chain shape characteristic of tissue	A, B, O blood group markers

3. **Interior protein network.** Membranes are structurally supported by intracellular proteins that reinforce the membrane's shape. For example, a red blood cell has a characteristic biconcave shape because a scaffold of proteins called spectrin links proteins in the plasma membrane with actin filaments in the cell's cytoskeleton. Membranes use networks of other proteins to control the lateral movements of some key membrane proteins, anchoring them to specific sites.

4. **Cell surface markers.** Membrane sections assemble in the endoplasmic reticulum, transfer to the Golgi apparatus, and then are transported to the plasma membrane. The endoplasmic reticulum adds chains of sugar molecules to membrane proteins and lipids, creating a "sugar coating" called the glycocalyx that extends from the membrane on the outside of the cell. Different cell types exhibit different varieties of these glycoproteins and glycolipids on their surfaces, which act as cell identity markers.

Originally, it was believed the plasma membrane was uniform because of being highly fluid, with lipids and proteins free to diffuse rapidly in the plane of the membrane. However, in the last decade evidence has accumulated suggesting the plasma membrane is not homogeneous, and contains microdomains with distinct lipid and protein composition. One type of microdomain, the *lipid raft*, is heavily enriched in cholesterol and saturated fatty acids, and thus more tightly packed than the surrounding membrane. Lipid rafts appear to be involved in many important biological processes, including signal reception and cell movement. The structural proteins of replicating HIV viruses are targeted to lipid raft regions of the plasma membrane during virus assembly within infected cells.

The fluid mosaic model proposes that membrane proteins are embedded within the lipid bilayer. Membranes are composed of a lipid bilayer within which proteins are anchored. Plasma membranes are supported by a network of fibers and coated on the exterior with cell identity markers.

Examining Cell Membranes

Biologists examine the delicate, filmy structure of a cell membrane using electron microscopes that provide clear magnification to several thousand times. We discussed two types of electron microscopes in chapter 5: the transmission electron microscope (TEM) and the scanning electron microscope (SEM). When examining cell membranes with electron microscopy, specimens must be prepared for viewing.

In one method of preparing a specimen, the tissue of choice is embedded in a hard matrix, usually some sort of epoxy. The epoxy block is then cut with a microtome, a machine with a very sharp blade that makes incredibly thin slices. The knife moves up and down as the specimen advances toward it, causing transparent "epoxy shavings" less than 1 micrometer thick to peel away from the block of tissue. These shavings are placed on a grid, and a beam of electrons is directed through the grid with the TEM. At the high magnification an electron microscope provides, resolution is good enough to reveal the double layers of a membrane.

Freeze-fracturing a specimen is another way to visualize the inside of the membrane (figure 6.6). The tissue is embedded in a medium and quick-frozen with liquid nitrogen. The frozen tissue is then "tapped" with a knife, causing a crack between the phospholipid layers of membranes. Proteins, carbohydrates, pits, pores, channels, or any other structure affiliated with the membrane will pull apart (whole, usually) and stick with one side of the split membrane. A very thin coating of platinum is then evaporated onto the fractured surface, forming a replica or "cast" of the surface. Once the topography of the membrane has been preserved in the cast, the actual tissue is dissolved away, and the cast is examined with electron microscopy, creating a strikingly different view of the membrane.

> **Visualizing a plasma membrane requires a very powerful electron microscope. Freeze-fracturing reveals the interior of the lipid bilayer.**

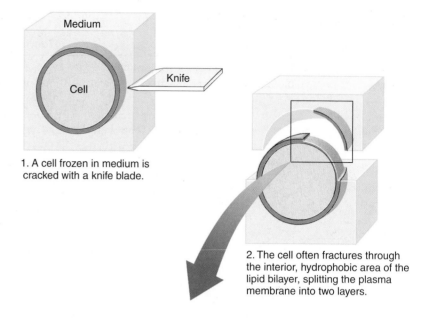

1. A cell frozen in medium is cracked with a knife blade.

2. The cell often fractures through the interior, hydrophobic area of the lipid bilayer, splitting the plasma membrane into two layers.

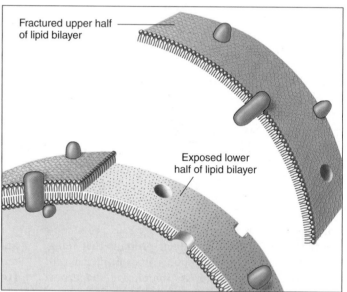

Fractured upper half of lipid bilayer

Exposed lower half of lipid bilayer

3. The plasma membrane separates such that proteins and other embedded membrane structures remain within one of the layers of the membrane.

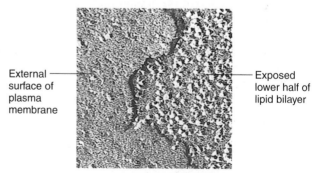

External surface of plasma membrane

Exposed lower half of lipid bilayer

4. The exposed membrane is coated with platinum which forms a replica of the membrane. The underlying membrane is dissolved away, and the replica is then viewed with electron microscopy.

FIGURE 6.6
Viewing a plasma membrane with freeze-fracture microscopy.

Kinds of Membrane Proteins

As we've seen, cell membranes are composed of a complex assembly of proteins enmeshed in a fluid array of phospholipid molecules. This enormously flexible design permits a broad range of interactions with the environment, some directly involving membrane proteins (figure 6.7). Although cells interact with their environment through their plasma membranes in many ways, we will focus on six key classes of membrane protein in this and the following chapter (chapter 7).

1. **Transporters.** Membranes are very selective, allowing only certain substances to enter or leave the cell, either through channels or carriers.
2. **Enzymes.** Cells carry out many chemical reactions on the interior surface of the plasma membrane, using enzymes attached to the membrane.
3. **Cell surface receptors.** Membranes are exquisitely sensitive to chemical messages, detecting them with receptor proteins on their surfaces.
4. **Cell surface identity markers.** Membranes carry cell surface markers that identify them to other cells. Most cell types carry their own ID tags, specific combinations of cell surface proteins characteristic of that cell type.
5. **Cell adhesion proteins.** Cells use specific proteins to glue themselves to one another. Some act by forming temporary interactions, while others form a more permanent bond.
6. **Attachments to the cytoskeleton.** Surface proteins that interact with other cells are often anchored to the cytoskeleton by linking proteins.

The many proteins embedded within a membrane carry out a host of functions, many of which are associated with transport of materials or information across the membrane.

Outside

Plasma membrane

Inside

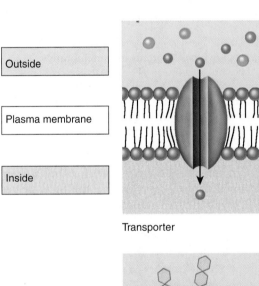

Transporter

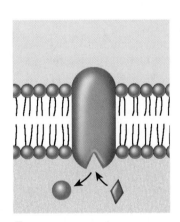

Enzyme

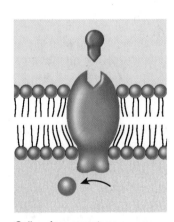

Cell surface receptor

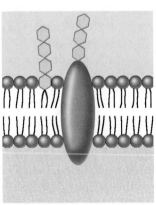

Cell surface identity marker

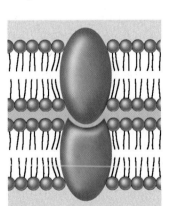

Cell adhesion

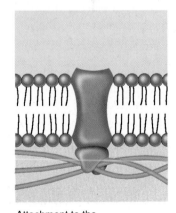

Attachment to the cytoskeleton

FIGURE 6.7
Functions of plasma membrane proteins. Membrane proteins act as transporters, enzymes, cell surface receptors, and cell surface markers, as well as aiding in cell-to-cell adhesion and securing the cytoskeleton.

Structure of Membrane Proteins

How can some proteins be anchored into particular positions on the cell membrane while others can move freely?

Anchoring Proteins in the Bilayer

Many membrane proteins are attached to the surface of the membrane by special molecules that associate with phospholipids and thereby anchor the protein to the membrane. Like a ship tied up to a floating dock, these proteins are free to move about on the surface of the membrane tethered to a phospholipid. These anchoring molecules are modified lipids that have (1) nonpolar regions that insert into the internal region of the lipid bilayer, and (2) chemical bonding domains that link directly to proteins.

In contrast, other proteins actually traverse the lipid bilayer. The part of the protein that extends through the lipid bilayer, in contact with the nonpolar interior, consists of nonpolar amino acid helices or β-pleated sheets (figure 6.8). Because water avoids nonpolar amino acids much as it does nonpolar lipid chains, the nonpolar portions of the protein are held within the interior of the lipid bilayer. Although the polar ends of the protein protrude from both sides of the membrane, the protein itself is locked into the membrane by its nonpolar segments. Any movement of the protein out of the membrane, in either direction, brings the nonpolar regions of the protein into contact with water, which "shoves" the protein back into the interior.

Extending Proteins Across the Bilayer

Cell membranes contain a variety of different **transmembrane proteins,** which differ in the way they traverse the lipid bilayer.

Single-Pass Anchors. A single nonpolar segment is adequate to anchor a protein in the membrane. For example, linking proteins of this sort attach the spectrin network of the cytoskeleton to the interior of the plasma membrane (figure 6.9). Many proteins that function as receptors for extracellular signals also have "single-pass" anchors that pass through the membrane only once. The portion of the receptor that extends out from the cell surface binds to specific hormones or other molecules, inducing changes at the other end of the protein in the cell's interior. In this way, information outside the cell is translated into action within the cell. The mechanisms of cell signaling will be addressed in detail in chapter 7.

Multiple-Pass Channels and Carriers. Other proteins have several helical segments that thread their way back and forth through the membrane, forming a channel like the hole in a doughnut. For example, bacteriorhodopsin is one of the key transmembrane proteins that carries out photosynthesis in bacteria. It contains seven nonpolar helical segments that traverse the membrane, forming a channel through which protons pass during the light-driven pumping of protons (figure 6.10). Other transmembrane proteins act as carriers to transport molecules across the membrane using different mechanisms (see section 6.5). All water-soluble molecules or ions that enter or leave the cell are either transported by carriers or pass through channels.

Pores. Some transmembrane proteins have extensive nonpolar regions with secondary configurations of β-pleated sheets instead of α helices. The β sheets form a characteristic motif, folding back and forth in a circle so the sheets come to be arranged like the staves of a barrel. This so-called β barrel, open on both ends, is a common feature of the porin class of proteins that are found within the outer membrane of some bacteria, where they allow molecules to pass through the membrane (figure 6.11).

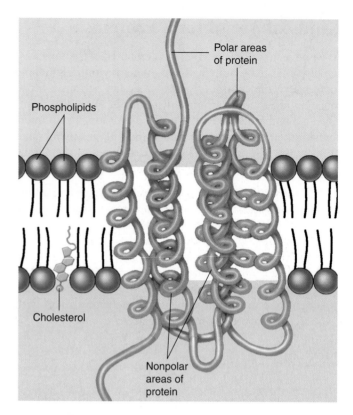

FIGURE 6.8
How nonpolar regions lock proteins into membranes.
A spiral helix of nonpolar amino acids (*red*) extends across the nonpolar lipid interior, while polar (*purple*) portions of the protein protrude out from the bilayer. The protein cannot move in or out because such a movement would drag nonpolar segments of the protein into contact with water.

Transmembrane proteins are anchored into the bilayer by their nonpolar segments. Some proteins pass through the bilayer only once; many channels pass back and forth through the bilayer repeatedly, creating a circular hole in the bilayer.

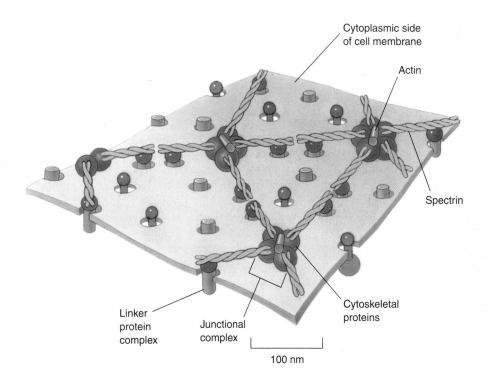

FIGURE 6.9
Linking proteins. Spectrin extends as a mesh anchored to the cytoplasmic side of a red blood cell plasma membrane. The spectrin protein is represented as a twisted dimer, attached to the membrane by special linker proteins. This cytoskeletal protein network confers resiliency to cells, such as red blood cells.

Cytoplasmic side of cell membrane

Actin

Spectrin

Cytoskeletal proteins

Junctional complex

Linker protein complex

100 nm

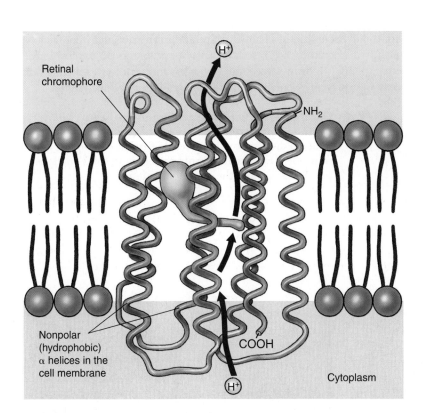

H⁺

Retinal chromophore

NH₂

Nonpolar (hydrophobic) α helices in the cell membrane

COOH

Cytoplasm

H⁺

FIGURE 6.10
A channel protein. This transmembrane protein mediates photosynthesis in the bacterium *Halobacterium halobium*. The protein traverses the membrane seven times with hydrophobic helical strands that are within the hydrophobic center of the lipid bilayer. The helical regions form a channel across the bilayer through which protons are pumped by the retinal chromophore (*green*).

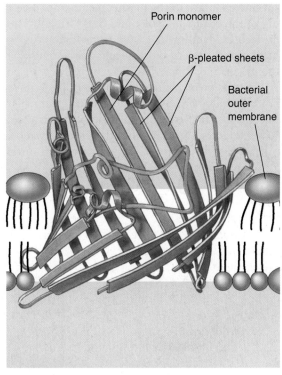

Porin monomer

β-pleated sheets

Bacterial outer membrane

FIGURE 6.11
A pore protein. The bacterial transmembrane protein porin creates large open tunnels called pores in the outer membrane of a bacterium. Sixteen strands of β-pleated sheets run antiparallel to each other, creating a β barrel in the bacterial outer cell membrane. The tunnel allows water and other materials to pass through the membrane.

6.3 Passive transport across membranes moves down the concentration gradient.

Diffusion

Molecules and ions dissolved in water are in constant motion, moving about randomly. This random motion causes a net movement of these substances from regions where their concentration is high to regions where their concentration is lower, a process called **diffusion** (figure 6.12). Net movement driven by diffusion will continue until the concentrations in all regions are the same. You can demonstrate diffusion by filling a jar to the brim with ink, capping it, placing it at the bottom of a bucket of water, and then carefully removing the cap. The ink molecules will slowly diffuse out from the jar until there is a uniform concentration in the bucket and the jar. This uniformity in the concentration of molecules is a type of equilibrium.

Membrane Transport Is Selective

Many molecules that cells require are polar and cannot pass through the nonpolar interior of the phospholipid bilayer. These molecules enter the cell by diffusion through specific channels in the plasma membrane. The inside lining of the channel is polar and thus "friendly" to the polar molecules, facilitating their transport across the membrane. Each type of biomolecule that is transported across the plasma membrane has its own type of transporter (that is, it has its own channel, which fits it like a glove and cannot be used by other molecules). Each channel is said to be selective for that type of molecule, and thus the cell is said to be **selectively permeable,** because only molecules admitted by the channels it possesses can enter it. The plasma membrane of a cell has many types of channels, each selective for a different type of molecule.

Diffusion of Ions Through Channels

Ions are solutes (substances dissolved in water) with an unequal number of protons and electrons. Those with an excess of protons are positively charged and called *cations*. Ions with more electrons than protons are negatively charged and called *anions*. Because they are charged, ions interact well with polar molecules such as water but are repelled by the nonpolar interior of a phospholipid bilayer. Therefore, ions cannot move between the cytoplasm of a cell and the extracellular fluid without the assistance of membrane transport proteins. **Ion channels** possess a hydrated interior that spans the membrane. Ions can diffuse through the channel in either direction without coming into contact with the hydrophobic tails of the phospholipids in the membrane. Two conditions determine the direction of net movement of the ions: their relative concentrations on either side of the membrane, and the voltage across the membrane (a topic we'll explore in chapter 45). Each type of channel is specific for a particular ion, such as calcium (Ca^{++}), sodium (Na^+), potassium (K^+), or chloride (Cl^-), or in some cases, for a few kinds of ions. Ion channels play an essential role in signaling by the nervous system.

> Diffusion is the net movement of substances to regions of lower concentration as a result of random motion. It tends to distribute substances uniformly. Membrane transport proteins allow only certain molecules and ions to diffuse through the plasma membrane.

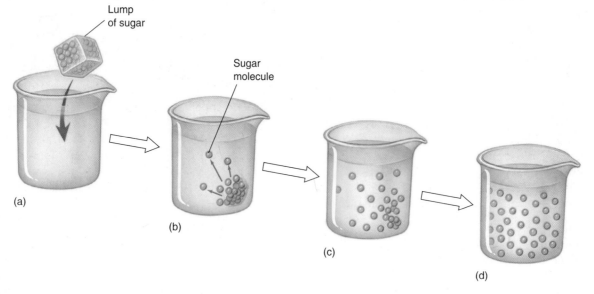

FIGURE 6.12
Diffusion. If a lump of sugar is dropped into a beaker of water (*a*), its molecules dissolve (*b*) and diffuse (*c*). Eventually, diffusion results in an even distribution of sugar molecules throughout the water (*d*).

Facilitated Diffusion

Carriers, another class of membrane proteins, transport ions as well as other solutes, such as sugars and amino acids, across the plasma membrane. Like channels, carriers are specific for a certain type of solute and can transport substances in either direction across the membrane. Unlike channels, however, they facilitate the movement of solutes across the membrane by physically binding to them on one side of the membrane and releasing them on the other. Again, the direction of the solute's net movement simply depends on its *concentration gradient* across the membrane. If the concentration is greater in the cytoplasm than outside the cell, the carrier will transport the molecule from the inside to the outside of the cell. This occurs because the solute is more likely to bind to the carrier on the cytoplasmic side of the membrane. If the concentration is greater in the extracellular fluid, the net movement will be from outside to inside. Thus, the net movement always occurs from areas of high concentration to low, just as it does in simple diffusion, but carriers facilitate the process. For this reason, this mechanism of transport is sometimes called **facilitated diffusion** (figure 6.13).

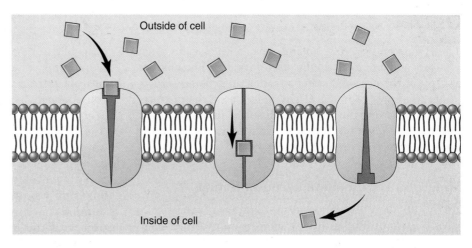

FIGURE 6.13
Facilitated diffusion is a carrier-mediated transport process. Molecules bind to a receptor on the extracellular side of the cell and are conducted through the plasma membrane by a membrane protein.

Facilitated Diffusion in Red Blood Cells

Several examples of facilitated diffusion by carrier proteins can be found in the plasma membranes of vertebrate red blood cells (RBCs). One RBC carrier protein, for example, transports a different molecule in each direction: Cl^- in one direction and bicarbonate ion (HCO_3^-) in the opposite direction. As you will learn in chapter 44, this carrier is important in transporting carbon dioxide in the blood.

A second important facilitated diffusion carrier in RBCs is the glucose transporter. Red blood cells keep their internal concentration of glucose low through a chemical trick: They immediately add a phosphate group to any entering glucose molecule, converting it to a highly charged glucose phosphate that cannot pass back across the membrane. This maintains a steep concentration gradient for unphosphorylated glucose, favoring its entry into the cell. The glucose transporter that carries glucose into the cell does not appear to form a channel in the membrane for the glucose to pass through. Instead, the transmembrane protein appears to bind the glucose and then flip its shape, dragging the glucose through the bilayer and releasing it on the inside of the plasma membrane. Once it releases the glucose, the glucose transporter reverts to its original shape. It is then available to bind the next glucose molecule that approaches the outside of the cell.

Saturation of Transport Through Carriers

A characteristic feature of transport through carriers is that its rate can become saturated. In other words, as the concentration gradient of a substance is progressively increased, its rate of transport will also increase to a certain point and then level off. Further increases in the gradient will produce no additional increase in rate. The explanation for this observation is that the membrane contains a limited number of carriers. When the concentration of the transported substance rises high enough, all of the carriers will be in use, and the capacity of the transport system will be saturated. In contrast, substances that move across the membrane by simple diffusion (diffusion through channels in the bilayer without the assistance of carriers) do not show saturation.

Facilitated diffusion provides the cell with a ready way to prevent the buildup of unwanted molecules within the cell or to take up needed molecules, such as sugars, that may be present outside the cell in high concentrations. Facilitated diffusion has three essential characteristics:

1. **It is specific.** A given carrier transports only certain molecules or ions.
2. **It is passive.** The direction of net movement is determined by the relative concentrations of the transported substance inside and outside the cell. It always moves from high concentration to low concentration.
3. **It saturates.** If all relevant protein carriers are in use, increases in the concentration gradient do not increase the transport rate.

Facilitated diffusion is the transport of molecules and ions across a membrane by specific carriers in the direction of lower concentration of those molecules or ions.

Osmosis

The cytoplasm of a cell contains ions and molecules, such as sugars and amino acids, dissolved in water. The mixture of these substances and water is called an *aqueous solution.* Water, the most common of the molecules in the mixture, is the **solvent,** and the substances dissolved in the water are **solutes.** The ability of water and solutes to diffuse across membranes has important consequences.

Molecules Diffuse down a Concentration Gradient

Both water and solutes diffuse from regions of high concentration to regions of low concentration; that is, they diffuse down their concentration gradients. When two regions are separated by a membrane, what happens depends on whether or not the solutes can pass freely through that membrane. Most solutes, including ions and sugars, are not lipid-soluble and, therefore, are unable to cross the lipid bilayer of the membrane.

Even water molecules, which are very polar, cannot readily cross a lipid bilayer. Water flow is facilitated by **aquaporins,** which are specialized channels for water. A simple experiment demonstrates this. If you place an amphibian egg in hypotonic spring water, where the solute concentration in the cell is higher than the surrounding water, it does not swell. If you then inject aquaporin mRNA into the egg, the channel proteins are expressed, and water will diffuse into the egg, causing it to swell.

Dissolved solutes interact with water molecules, which form hydration shells about the charged solute molecules. When a membrane separates two solutions with different concentrations of solutes, different concentrations of *free* water molecules exist on the two sides of the membrane. The side with higher solute concentration ties up more water molecules in hydration shells. As a consequence, free water molecules move down their concentration gradient, toward the higher solute concentration. This net water movement across a membrane toward a higher solute concentration by diffusion is called **osmosis** (figure 6.14).

The concentration of *all* solutes in a solution determines the **osmotic concentration** of the solution. If two solutions have unequal osmotic concentrations, the solution with the higher concentration is **hyperosmotic** (Greek *hyper,* "more than"), and the solution with the lower concentration is **hypoosmotic** (Greek *hypo,* "less than"). If the osmotic concentrations of two solutions are equal, the solutions are **isosmotic** (Greek *iso,* "the same").

In cells, a plasma membrane separates two aqueous solutions, one inside the cell (the cytoplasm) and one outside (the extracellular fluid). The direction of the net diffusion of water across this membrane is determined by the osmotic concentrations of the solutions on either side. For example, if the cytoplasm of a cell is *hypoosmotic* to the extracellular fluid, water diffuses out of the cell, toward the solution with the higher concentration of solutes (and, therefore, the lower concentration of unbound water molecules). This loss of water from the cytoplasm causes the cell to shrink until the osmotic concentrations of the cytoplasm and the extracellular fluid become equal.

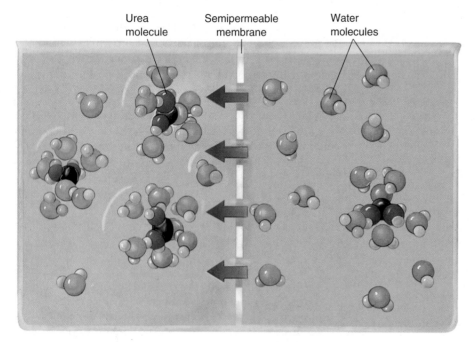

Urea molecule   Semipermeable membrane   Water molecules

FIGURE 6.14
Osmosis. Charged or polar substances are soluble in water because they form hydrogen bonds with water molecules clustered around them, forming hydration shells. When a polar solute (illustrated here with urea) is added to the solution on one side of a membrane, the water molecules that gather around each urea molecule are no longer free to diffuse across the membrane; in effect, the polar solute has reduced the concentration of free water molecules, creating a gradient. Water moves down the gradient by diffusion from the right to the left.

Osmotic Pressure

What happens if the cell's cytoplasm is *hyper*osmotic to the extracellular fluid? In this situation, water diffuses into the cell from the extracellular fluid, causing the cell to swell. The pressure of the cytoplasm pushing out against the cell membrane, or **hydrostatic pressure,** increases. On the other hand, the **osmotic pressure,** defined as the pressure that must be applied to stop the osmotic movement of water across a membrane, will also be at work (figure 6.15). If the membrane is strong enough, the cell reaches an equilibrium, at which the osmotic pressure, which tends to drive water into the cell, is exactly counterbalanced by the hydrostatic pressure, which tends to drive water back out of the cell. However, a plasma membrane by itself cannot withstand large internal pressures, and an isolated cell under such conditions would burst like an overinflated balloon. Accordingly, it is important for animal cells, which only have plasma membranes, to maintain isosmotic conditions. In contrast, the cells of prokaryotes, fungi, plants, and many protists are surrounded by strong cell walls, which can withstand high internal pressures without bursting.

Maintaining Osmotic Balance

Organisms have developed many solutions to the dilemma posed by being hyperosmotic to their environment.

Extrusion.   Some single-celled eukaryotes, such as the protist *Paramecium*, use organelles called contractile vacuoles to remove water. Each vacuole collects water from various parts of the cytoplasm and transports it to the central part of the vacuole, near the cell surface. The vacuole possesses a small pore that opens to the outside of the cell. By contracting rhythmically, the vacuole pumps out through this pore the water that is continuously seeping into the cell by osmosis.

Isosmotic Solutions.   Some organisms that live in the ocean adjust their internal concentration of solutes to match that of the surrounding seawater. Because they are isosmotic with respect to their environment, no net flow of water occurs into or out of these cells. Many terrestrial animals solve the problem in a similar way, by circulating a fluid through their bodies that bathes cells in an isosmotic solution. The blood in your body, for example, contains a high concentration of the protein albumin, which elevates the solute concentration of the blood to match that of your cells.

Turgor.   Most plant cells are hyperosmotic to their immediate environment, containing a high concentration of solutes in their central vacuoles. The resulting internal hydrostatic pressure, known as **turgor pressure,** presses the plasma membrane firmly against the interior of the cell

Hyperosmotic solution	Isosmotic solution	Hypoosmotic solution

| Shriveled cells | Normal cells | Cells swell and eventually burst |

Human red blood cells

| Cell body shrinks from cell wall | Flaccid cell | Normal turgid cell |

Plant cells

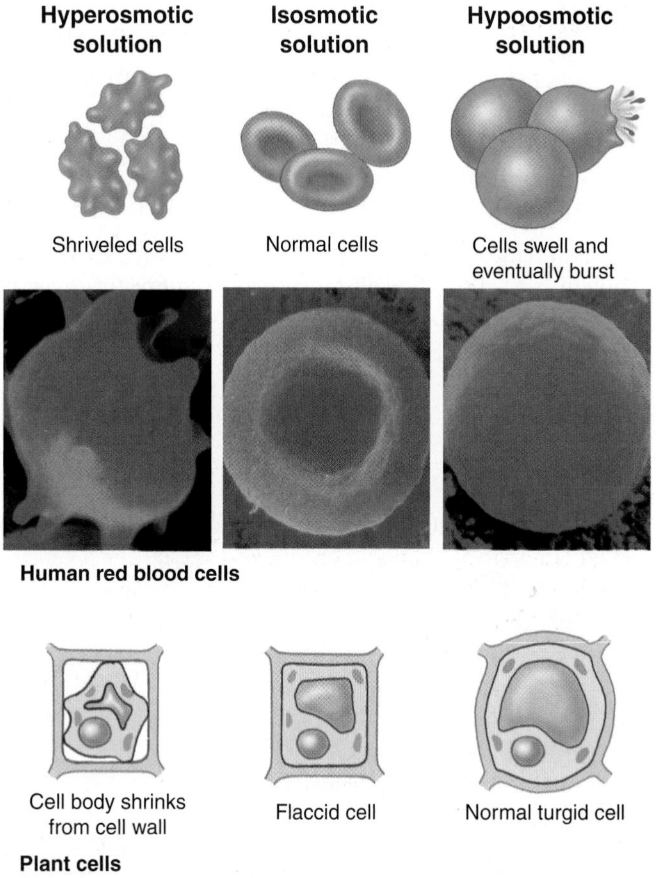

FIGURE 6.15
How solutes create osmotic pressure.   In a hyperosmotic solution, water moves out of the cell toward the higher concentration of solutes, causing the cell to shrivel. In an isosmotic solution, the concentration of solutes on either side of the plasma membrane is the same. Osmosis still occurs, but water diffuses into and out of the cell at the same rate, and the cell doesn't change size. In a hypoosmotic solution, the concentration of solutes is higher within the cell than outside, so the net movement of water is into the cell. A cell is an enclosed structure, and so as water enters the cell from a hypoosmotic solution, pressure is applied to the plasma membrane until the cell ruptures. This hydrostatic pressure is counterbalanced by osmotic pressure, the force required to stop the flow of water into the cell. Plant cells have strong cell walls that can apply adequate osmotic pressure to keep the cell from rupturing. This is not the case with animal cells.

wall, making the cell rigid. Most green plants depend on turgor pressure to maintain their shape, and thus they wilt when they lack sufficient water.

Osmosis is the diffusion of water, but not solutes, across a membrane.

6.4 Bulk transport utilizes endocytosis.

Bulk Passage Into and Out of the Cell

Endocytosis

The lipid nature of their plasma membranes raises a second problem for cells. The substances cells require for growth are for the most part large, polar molecules that cannot cross the hydrophobic barrier a lipid bilayer creates. How do organisms get these substances into their cells? One process is **endocytosis**, in which the plasma membrane envelops food particles. Cells use three major types of endocytosis: phagocytosis, pinocytosis, and receptor-mediated endocytosis (figure 6.16).

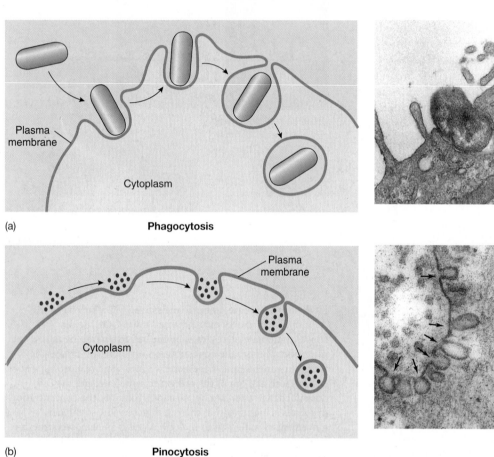

(a) **Phagocytosis**

(b) **Pinocytosis**

Coated pit
Target molecule
Clathrin
Receptor protein
Coated vesicle

(c) **Receptor-mediated endocytosis**

FIGURE 6.16
Endocytosis. Both (*a*) phagocytosis and (*b*) pinocytosis are forms of endocytosis. (*c*) In receptor-mediated endocytosis, cells have pits coated with the protein clathrin that initiate endocytosis when target molecules bind to receptor proteins in the plasma membrane. Photo inserts: (*a*) A TEM of phagocytosis of a bacterium, *Rickettsia tsutsugamushi*, by a mouse peritoneal mesothelial cell. The bacterium enters the host cell by phagocytosis and replicates in the cytoplasm. (*b*) A TEM of pinocytosis in a smooth muscle cell. (*c*) A coated pit appears in the plasma membrane of a developing egg cell, covered with a layer of proteins (80,000×). When an appropriate collection of molecules gathers in the coated pit, the pit deepens and will eventually seal off to form a vesicle.

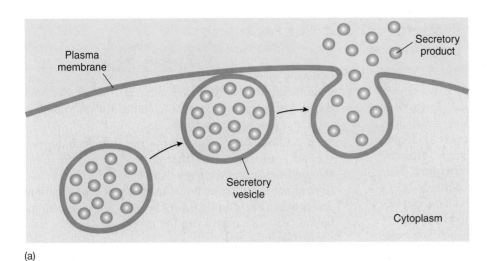
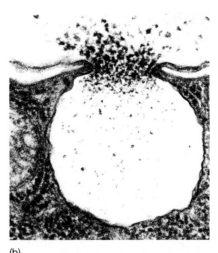

(a)   (b)

FIGURE 6.17

Exocytosis. (*a*) Proteins and other molecules are secreted from cells in small packets called vesicles, whose membranes fuse with the plasma membrane, releasing their contents to the cell surface. (*b*) A transmission electron micrograph showing exocytosis.

Phagocytosis and Pinocytosis. If the material the cell takes in is particulate (made up of discrete particles), such as an organism or some other fragment of organic matter (figure 6.16*a*), the process is called **phagocytosis** (Greek *phagein*, "to eat," + *cytos*, "cell"). If the material the cell takes in is liquid (figure 6.16*b*), it is called **pinocytosis** (Greek *pinein*, "to drink"). Pinocytosis is common among animal cells. Mammalian egg cells, for example, "nurse" from surrounding cells; the nearby cells secrete nutrients that the maturing egg cell takes up by pinocytosis. Virtually all eukaryotic cells constantly carry out these kinds of endocytosis, trapping particles and extracellular fluid in vesicles and ingesting them. Endocytosis rates vary from one cell type to another. They can be surprisingly high; some types of white blood cells ingest 25% of their cell volume each hour!

Receptor-Mediated Endocytosis. Specific molecules are often transported into eukaryotic cells through **receptor-mediated endocytosis.** Molecules to be transported first bind to specific receptors in the plasma membrane. The transport process is specific only to molecules having a shape that fits snugly into the receptor. The plasma membrane of a particular kind of cell contains a characteristic battery of receptor types, each for a different kind of molecule.

The portion of the receptor molecule inside the membrane is trapped in an indented pit coated with the protein clathrin. Each pit acts like a molecular mousetrap, closing over to form an internal vesicle when the right molecule enters the pit (figure 6.16*c*). The trigger that releases the trap is the binding of the properly fitted target molecule to a receptor embedded in the membrane of the pit. When binding occurs, the cell reacts by initiating endocytosis. The process is highly specific and very fast.

One type of molecule that is taken up by receptor-mediated endocytosis is called a low-density lipoprotein (LDL). LDL molecules bring cholesterol into the cell where it can be incorporated into membranes. Cholesterol plays a key role in determining the stiffness of the body's membranes. In the human genetic disease hypercholesterolemia, the receptors lack tails, so they are never caught in the clathrin-coated pits and, thus, never taken up by the cells. The cholesterol stays in the bloodstream of affected individuals, coating their arteries and leading to heart attacks.

It is important to understand that endocytosis in itself does not bring substances directly into the cytoplasm of a cell. The material taken in is still separated from the cytoplasm by the membrane of the vesicle.

Exocytosis

The reverse of endocytosis is **exocytosis,** the discharge of material from vesicles at the cell surface (figure 6.17). In plant cells, exocytosis is an important means of exporting the materials needed to construct the cell wall through the plasma membrane. Among protists, contractile vacuole discharge is a form of exocytosis. In animal cells, exocytosis provides a mechanism for secreting many hormones, neurotransmitters, digestive enzymes, and other substances.

> Cells import bulk materials by engulfing them with their plasma membranes in a process called endocytosis; similarly, they extrude or secrete material through exocytosis.

6.5 Active transport across membranes requires energy.

Active Transport

While diffusion, facilitated diffusion, and osmosis are passive transport processes that move materials down their concentration gradients, cells can also move substances across a cell membrane *up* their concentration gradients. This process requires the expenditure of energy, typically from ATP, and is therefore called **active transport**. Like facilitated diffusion, active transport involves highly selective protein carriers within the membrane. These carriers bind to the transported substance, which could be an ion or a simple molecule, such as a sugar, an amino acid, or a nucleotide to be used in the synthesis of DNA.

Active transport is one of the most important functions of any cell. It enables a cell to take up additional molecules of a substance that is already present in its cytoplasm in concentrations higher than in the extracellular fluid. Without active transport, for example, liver cells would be unable to accumulate glucose molecules from the blood plasma, because the glucose concentration is often higher inside the liver cells than it is in the plasma. Active transport also enables a cell to move substances from its cytoplasm to the extracellular fluid despite higher external concentrations.

The Sodium-Potassium Pump

The use of ATP in active transport may be direct or indirect. Let's first consider how ATP is used directly to move ions against their concentration gradient. More than one-third of all of the energy expended by an animal cell that is not actively dividing is used in the active transport of sodium (Na^+) and potassium (K^+) ions. Most animal cells have a low internal concentration of Na^+, relative to their surroundings, and a high internal concentration of K^+. They maintain these concentration differences by actively pumping Na^+ out of the cell and K^+ in. The remarkable protein that transports these two ions across the cell membrane is known as the **sodium-potassium pump** (figure 6.18). The cell obtains the energy it needs to operate the pump from adenosine triphosphate (ATP), a molecule we'll learn more about in chapter 8.

The important characteristic of the sodium-potassium pump is that it is an active transport process, transporting Na^+ and K^+ from areas of low concentration to areas of high concentration. This transport up concentration gradients is the opposite of the passive transport in diffusion; it is achieved only by the constant expenditure of metabolic energy. The sodium-potassium pump works through the following series of conformational changes in the transmembrane protein:

Step 1. Three sodium ions bind to the cytoplasmic side of the protein, causing the protein to change its conformation.

Step 2. In its new conformation, the protein binds a molecule of ATP and cleaves it into adenosine diphosphate and phosphate (ADP + P_i). ADP is released, but the phosphate group remains bound to the protein. The protein is now phosphorylated.

Step 3. The phosphorylation of the protein induces a second conformational change in the protein. This change translocates the three Na^+ across the membrane, so they now face the exterior. In this new conformation, the protein has a low affinity for Na^+, and the three bound Na^+ dissociate from the protein and diffuse into the extracellular fluid.

Step 4. The new conformation has a high affinity for K^+, two of which bind to the extracellular side of the protein as soon as it is free of the Na^+.

Step 5. The binding of the K^+ causes another conformational change in the protein, this time resulting in the dissociation of the bound phosphate group.

Step 6. Freed of the phosphate group, the protein reverts to its original conformation, exposing the two K^+ to the cytoplasm. This conformation has a low affinity for K^+, so the two bound K^+ dissociate from the protein and diffuse into the interior of the cell. The original conformation has a high affinity for Na^+; when these ions bind, they initiate another cycle.

In every cycle, three Na^+ leave the cell and two K^+ enter. The changes in protein conformation that occur during the cycle are rapid, enabling each carrier to transport as many as 300 Na^+ per second. The sodium-potassium pump appears to be ubiquitous in animal cells, although cells vary widely in the number of pump proteins they contain.

Coupled Transport

Some molecules are moved up their concentration gradient by using the energy stored in a gradient of a different molecule. In this process, the energy released as one molecule moves down its concentration gradient is captured and used to move a different molecule against its gradient. As we just learned, the energy stored in ATP molecules can be used to create a gradient of Na^+ and K^+ ions across the membrane. These gradients can then be used to power the transport of other molecules across the membrane.

For example, the transport of glucose across the membrane in animal cells requires energy for two reasons. First, glucose is a large polar molecule that cannot move through the membrane by itself, and second, the concentration of glucose inside the cell is frequently higher than the concentration outside the cell. The glucose transporter uses the Na^+ gradient produced by the

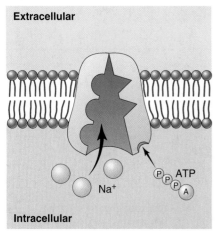

1. Protein in membrane binds intracellular sodium.

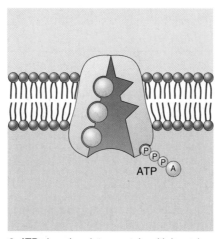

2. ATP phosphorylates protein with bound sodium.

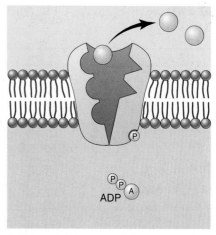

3. Phosphorylation causes conformational change in protein, allowing sodium to leave.

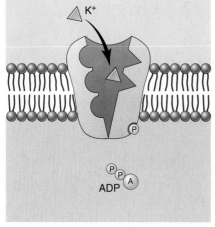

4. Extracellular potassium binds to exposed sites.

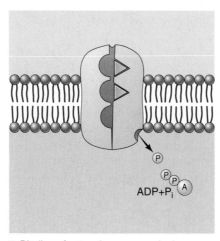

5. Binding of potassium causes dephosphorylation of protein.

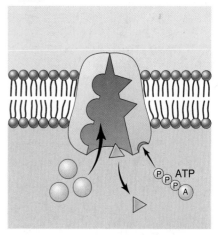

6. Dephosphorylation of protein triggers change back to original conformation, potassium moves into cell, and the cycle repeats.

FIGURE 6.18

The sodium-potassium pump. The protein carrier known as the sodium-potassium pump transports sodium (Na^+) and potassium (K^+) ions across the plasma membrane. For every three Na^+ transported out of the cell, two K^+ are transported into the cell. The sodium-potassium pump is fueled by ATP.

sodium-potassium pump as a source of energy to power the movement of glucose into the cell. In this system, both glucose and Na^+ bind to the transport protein, which allows Na^+ to pass down its concentration gradient, capturing the energy and using it to move glucose into the cell. In this kind of cotransport, both molecules are moving in the same direction across the membrane, and hence it is called *symport* (figure 6.19).

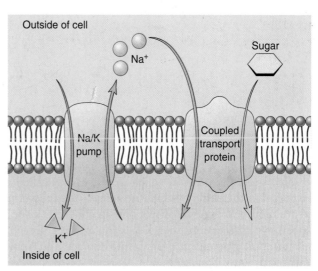

FIGURE 6.19

Coupled transport. A membrane protein transports sodium ions into the cell, down their concentration gradient, at the same time it transports a glucose molecule into the cell. The gradient driving the Na^+ entry is so great that sugar molecules can be brought in against their concentration gradient.

Table 6.2 Mechanisms for Transport Across Cell Membranes

Process	How It Works	Example
PASSIVE PROCESSES		
Diffusion		
Direct	Random molecular motion produces net migration of molecules toward region of lower concentration	Movement of oxygen into cells
Protein channel	Polar molecules pass through a protein channel	Movement of ions in or out of cell
Facilitated diffusion		
Protein carrier	Molecule binds to carrier protein in membrane and is transported across; net movement is toward region of lower concentration	Movement of glucose into cells
Osmosis		
Aquaporins	Diffusion of water across differentially permeable membrane	Movement of water into cells placed in a hypotonic solution
ACTIVE PROCESSES		
Endocytosis		
Membrane vesicle		
Phagocytosis	Particle is engulfed by membrane, which folds around it and forms a vesicle	Ingestion of bacteria by white blood cells
Pinocytosis	Fluid droplets are engulfed by membrane, which forms vesicles around them	"Nursing" of human egg cells
Receptor-mediated endocytosis	Endocytosis triggered by a specific receptor	Cholesterol uptake
Exocytosis		
Membrane vesicle	Vesicles fuse with plasma membrane and eject contents	Secretion of mucus
Active transport		
Protein carrier		
Na$^+$/K$^+$ pump	Carrier expends energy to export a substance across a membrane against its concentration gradient	Na$^+$ and K$^+$ against their concentration gradients
Coupled transport	Molecules are transported across a membrane against their concentration gradients by the cotransport of sodium ions or protons down their concentration gradients	Coupled uptake of glucose into cells against its concentration gradient

In a related process, called **countertransport,** the inward movement of Na$^+$ is coupled with the outward movement of another substance, such as Ca^{++} or H$^+$. As in cotransport, both Na$^+$ and the other substance bind to the same transport protein, which in this case is called an **antiport,** but they bind on opposite sides of the membrane and are moved in opposite directions. In countertransport, the cell uses the energy released as Na$^+$ moves down its concentration gradient into the cell to extrude a substance up its concentration gradient.

The mechanisms for transport across cell membranes are summarized in table 6.2.

Active transport moves a solute across a membrane up its concentration gradient, using protein carriers driven by the expenditure of chemical energy. Many molecules are cotransported into cells up their concentration gradients by coupling their movement to that of sodium ions or protons moving down their concentration gradients.

Concept Review

For interactive testing, visit the Online Learning Center with PowerWeb at www.mhhe.com/Raven7

6.1 Biological membranes are fluid layers of lipid.

The Phospholipid Bilayer

- Biological membranes are composed of phospholipid bilayers with hydrophilic, polar heads and hydrophobic, nonpolar, fatty acid tails. (p. 106)
- Phospholipid molecules orient their polar heads toward water and their nonpolar tails away from water. (p. 106)
- The nonpolar interior of a lipid bilayer impedes the passage of any water-soluble substances through the bilayer. (p. 106)

The Lipid Bilayer Is Fluid

- A lipid bilayer is stable because water's affinity for hydrogen bonding never stops, and the hydrogen bonding holds the membrane together in a liquid form. (p. 107)

6.2 Proteins embedded within cell membranes determine their character.

The Fluid Mosaic Model

- In the fluid mosaic model, a mosaic of proteins floats in the fluid lipid bilayer. (p. 108)
- Cell membranes are assembled from four components: phospholipid bilayer, transmembrane proteins, interior protein network, and cell surface markers. (pp. 108–109)

Examining Cell Membranes

- Electron microscopes must be used to visualize the structure of a cell membrane. (p. 110)
- Different procedures can be employed for preparing a specimen for viewing, including embedding tissue in a hard matrix and freeze-fracturing a specimen. (p. 110)

Kinds of Membrane Proteins

- Six key classes of membrane proteins include transporters, enzymes, cell surface receptors, cell surface identity markers, cell adhesion proteins, and attachments to the cytoskeleton. (p. 111)

Structure of Membrane Proteins

- Transmembrane proteins are anchored into the bilayer by nonpolar segments. (p. 112)
- Cell membranes contain a variety of different transmembrane proteins, including single-pass anchors, multiple-pass channels and carriers, and pores. (p. 112)

6.3 Passive transport across membranes moves down the concentration gradient.

Diffusion

- Random motion of molecules causes a net movement of substances from regions of high concentration to regions of lower concentration, and this movement continues until all regions exhibit the same concentration. (p. 114)
- Each transport protein in the plasma membrane is selectively permeable, and thus only allows certain molecules to diffuse through. (p. 114)

Facilitated Diffusion

- Facilitated diffusion occurs as molecules move from an area of higher concentration to an area of lower concentration via specific carriers. (p. 115)
- The essential characteristics of facilitated diffusion are specificity, passivity, and saturation. (p. 115)

Osmosis

- During osmosis, water moves across a membrane toward a solution with a higher solute concentration. (p. 116)
- The direction of net diffusion of water across the membrane is determined by the osmotic concentrations of the solutions on either side. (p. 116)
- Organisms have developed many solutions to being hyperosmotic to their environment, including extrusion, isomotic solutions, and turgor. (p. 117)

6.4 Bulk transport utilizes endocytosis.

Bulk Passage Into and Out of the Cell

- Endocytosis occurs when the plasma membrane envelops food particles and brings them into the cell interior. Three major forms of endocytosis are phagocytosis, pinocytosis, and receptor-mediated endocytosis. (pp. 118–119)
- Exocytosis refers to the discharge of materials from vesicles at the cell surface. (p. 119)

6.5 Active transport across membranes requires energy.

Active Transport

- Active transport is the movement of a solute across a membrane against its concentration gradient, requiring the use of protein carriers with the expenditure of ATP. (p. 120)
- Active transport involves highly selective membrane protein carriers. (p. 120)

Test Your Understanding

Self Test

1. Why is the phospholipid molecule so appropriate as the primary structural component of plasma membranes?
   a. Phospholipids are completely insoluble in water.
   b. Phospholipids form strong chemical bonds between the molecules, forming a stable structure.
   c. Phospholipids form a selectively permeable structure.
   d. Phospholipids form chemical bonds with membrane proteins that keep the proteins within the membrane.
2. Which increases the fluidity of the plasma membrane?
   a. having a large number of membrane proteins
   b. the tight alignment of phospholipids
   c. cholesterol present in the membrane
   d. double bonds between carbon atoms in the fatty acid tails
3. Which best describes the structure of a plasma membrane?
   a. proteins embedded within two layers of phospholipids
   b. phospholipids sandwiched between two layers of proteins
   c. proteins sandwiched between two layers of phospholipids
   d. a layer of proteins on top of a layer of phospholipids
4. What locks all transmembrane proteins in the bilayer?
   a. chemical bonds that form between the phospholipids and the proteins
   b. hydrophobic interactions between nonpolar amino acids of the proteins and the aqueous environments of the cell
   c. attachment to the cytoskeleton
   d. the addition of sugar molecules to the protein surface facing the external environment
5. The movement of sodium ions from an area of higher concentration to an area of lower concentration is called
   a. active transport.
   b. osmosis.
   c. diffusion.
   d. phagocytosis.
6. A cell placed in distilled water will
   a. shrivel up.
   b. swell.
   c. lose water.
   d. result in no net diffusion of water molecules.
7. Sucrose cannot pass through the membrane of a red blood cell (RBC), but water and glucose can. Which solution would cause the RBC to shrink the most?
   a. a hyperosmotic sucrose solution
   b. a hyperosmotic glucose solution
   c. a hypoosmotic sucrose solution
   d. a hypoosmotic glucose solution
8. Which of the following processes requires membrane proteins?
   a. exocytosis
   b. phagocytosis
   c. receptor-mediated endocytosis
   d. pinocytosis
9. Exocytosis involves
   a. the ingestion of large organic molecules or organisms.
   b. the use of ATP.
   c. the uptake of fluids from the environment.
   d. the discharge of materials from cellular vesicles.
10. Molecules that are transported into the cell *up* their concentration gradients do so by
   a. facilitated diffusion.
   b. osmosis.
   c. coupled transport.
   d. none of these.

Test Your Visual Understanding

1. In this figure of a transmembrane protein, what colored area of the protein contains nonpolar amino acids? What colored area contains polar amino acids? What colored area contains amino acids carrying a positive or negative charge?

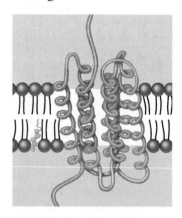

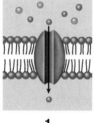 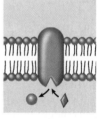 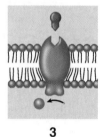

1   **2**   **3**

2. Match the function of the membrane protein with the appropriate numbered figure:
   a. cell surface receptor
   b. transporter
   c. cell surface identity marker
   d. enzyme

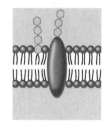

4

Apply Your Knowledge

1. If during the action of the sodium-potassium pump, 150 molecules of ATP are used, how many sodium ions are transported across the membrane?
2. If a cell's cytoplasm were hyperosmotic to the extracellular fluid, how would the concentration of solutes in the cytoplasm compare with that in the extracellular fluid? Assuming the membrane was permeable only to water, in which direction would water move?
3. Cholera, a disease caused by a bacterial infection of *Vibrio cholerae*, results in severe diarrhea leading to dehydration. A toxin released by the bacterium causes the release of chloride ions (Cl^-) from cells lining the small intestine and inhibits the uptake of sodium ions (Na^+) by these cells. Explain how this disruption of cellular ion concentrations would result in extreme dehydration.
4. Cystic fibrosis is a genetic disease that results in thick, mucous secretions that clog air passages in the lungs. Faulty chloride ion channels keep Cl^- and Na^+ in the cells that line the airways, increasing the intracellular ion concentrations. How does this cause the mucus in the airways to become thick?

7

Cell–Cell Interactions

Concept Outline

7.1 Cells signal one another with chemicals.

Receptor Proteins and Signaling Between Cells.
Receptor proteins embedded in the plasma membrane change shape when they bind specific signal molecules. The process of isolating and identifying these receptors has been aided by new techniques.

Types of Cell Signaling. Cell signaling can occur between adjacent cells, although chemical signals called hormones act over long distances.

7.2 Proteins in the cell and on its surface receive signals from other cells.

Intracellular Receptors. Some receptors are located within the cell cytoplasm. These receptors respond to lipid-soluble signals, such as steroid hormones.

Cell Surface Receptors. Many cell-to-cell signals are water-soluble and cannot penetrate membranes. Instead, the signals are received by transmembrane proteins protruding from the cell surface.

7.3 Follow the journey of information into the cell.

Initiating the Intracellular Signal. Cell surface receptors often use "second messengers" to transmit a signal to the cytoplasm.

Amplifying the Signal: Protein Kinase Cascades.
Surface receptors and second messengers amplify signals as they travel into the cell, often toward the cell nucleus.

7.4 Cell surface proteins mediate cell–cell interactions.

The Expression of Cell Identity. Cells possess on their surfaces a variety of tissue-specific identity markers that identify both the tissue and the individual.

Intercellular Adhesion. Cells attach themselves to one another with protein links. Some of the links are very strong, others more transient. The three categories of cell junctions are tight junctions, anchoring junctions, and communicating junctions. Each serves a different function in the cell.

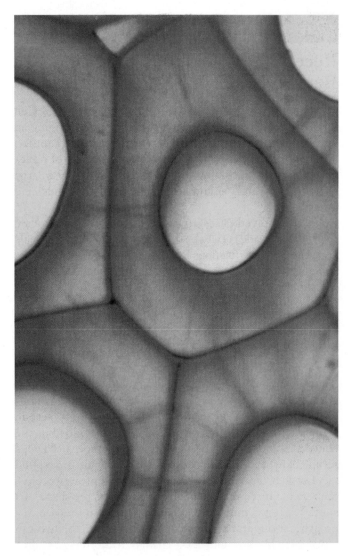

FIGURE 7.1
Persimmon cells in close contact with one another. These plant cells and all cells, no matter what their function, interact with their environment, including the cells around them.

Did you know that each of the 100 trillion cells of your body shares one key feature with the cells of tigers, bumblebees, and persimmons (figure 7.1)—a feature that most prokaryotes and protists lack? Your cells touch and communicate with one another. Sending and receiving a variety of chemical signals, they coordinate their behavior so that your body functions as an integrated whole, rather than as a massive collection of individual cells acting independently. The ability of cells to communicate with one another is the hallmark of multicellular organisms. In this chapter, we look in detail at how the cells of multicellular organisms interact with one another, first exploring how they signal one another with chemicals and then examining the ways in which their cell surfaces interact to organize tissues and body structures.

7.1 Cells signal one another with chemicals.

Receptor Proteins and Signaling Between Cells

Communication between cells is common in nature. Cell signaling occurs in all multicellular organisms, providing an indispensable mechanism for cells to influence one another. The cells of multicellular organisms use a variety of molecules as signals, including not only peptides, but also large proteins, individual amino acids, nucleotides, steroids, and other lipids.

Even dissolved gases are used as signals. Nitric oxide (NO) plays a wide variety of roles, including mediating male erections (Viagra functions by stimulating NO release).

Some of these molecules are attached to the surface of the signaling cell; others are secreted through the plasma membrane or released by exocytosis.

Cell Surface Receptors

Any given cell of a multicellular organism is exposed to a constant stream of signals. At any time, hundreds of different chemical signals may be in the environment surrounding the cell. However, each cell responds only to certain signals and ignores the rest, like a person following the conversation of one or two individuals in a noisy, crowded room. How does a cell "choose" which signals to respond to? Located on or within the cell are **receptor proteins,** each with a three-dimensional shape that fits the shape of a specific signal molecule. When a signal molecule approaches a receptor protein of the right shape, the two can bind. This binding induces a change in the receptor protein's shape, ultimately producing a response in the cell. Hence, a given cell responds to the signal molecules that fit the particular set of receptor proteins it possesses, and ignores those for which it lacks receptors.

The Hunt for Receptor Proteins

The characterization of receptor proteins has presented a very difficult technical problem because of their relative scarcity in the cell. Because these proteins may constitute less than 0.01% of the total mass of protein in a cell, purifying them is analogous to searching for a particular grain of sand in a sand dune! To make matters worse, any specific receptor may only be expressed in a small subset of cells in the body. However, two recent techniques—immunochemistry and molecular genetics—have enabled cell biologists to make rapid progress in this area.

Immunochemistry. The first method uses antibodies that recognize specific molecules that bind to receptors.

Using these tools, researchers can selectively purify specific receptor molecules from the complex mix of proteins found in a population of cells. Once purified, the specific chemistry and structure of the receptor molecules can be studied in detail.

Antibodies are special proteins produced by the immune system that, like receptors, bind very specifically to particular target molecules. The body naturally produces antibodies that bind to any foreign molecule entering the body as part of the immune defense. Each antibody-producing cell of the immune system makes its own unique antibody molecule with its own specific ability to bind to the target molecule. Researchers have taken advantage of this by isolating a single immune cell that is producing an antibody that recognizes a particular receptor or other target molecule. By growing a population of cells from that one starting cell (or cloning that cell), a monoclonal antibody can be purified that specifically binds to the target molecule.

Molecular Genetics. A second powerful approach has also allowed great advances in the study of the structure and function of receptors. Although receptors are commonly in low abundance and difficult to purify directly, the cellular defects caused by receptor malfunction can be very pronounced. Researchers have taken advantage of this by intentionally creating mutations in genes that cause significant cellular defects characteristic of receptor malfunction. The genes altered in these mutants are then purified using a variety of techniques, many of which exploit the wealth of information now available from the sequence of entire genomes of organisms (see chapter 17). By characterizing the structure of the genes, much can be learned about the structure and function of the proteins they encode. More importantly, by analyzing the sites within the gene that can be mutated to produce specific defects in the function of the receptor, the relationship between specific protein structures and their cellular functions can be determined.

Remarkably, these techniques have revealed that the enormous number of receptor proteins can be grouped into just a handful of "families" containing many related receptors. Later in this chapter, we will meet some of the members of these receptor families.

Cells in a multicellular organism communicate with others by releasing signal molecules that bind to receptor proteins on the surface of the other cells. Recent advances in protein isolation have yielded a wealth of information about the structure and function of these proteins.

Types of Cell Signaling

Cells can communicate through any of four basic mechanisms, depending primarily on the distance between the signaling and responding cells (figure 7.2). In addition to using these four basic mechanisms, some cells actually send signals to themselves, secreting signals that bind to specific receptors on their own plasma membranes. This process, called *autocrine signaling*, is thought to play an important role in reinforcing developmental changes.

Direct Contact

As we saw in chapter 6, the surface of a eukaryotic cell is richly populated with proteins, carbohydrates, and lipids attached to and extending outward from the plasma membrane. When cells are very close to one another, some of the molecules on the plasma membrane of one cell can be recognized by receptors on the plasma membrane of an adjacent cell. Many of the important interactions between cells in early development occur by means of *direct contact* between cell surfaces (figure 7.2a). We'll examine contact-dependent interactions more closely later in this chapter.

Paracrine Signaling

Signal molecules released by cells can diffuse through the extracellular fluid to other cells. If those molecules are taken up by neighboring cells, destroyed by extracellular enzymes, or quickly removed from the extracellular fluid in some other way, their influence is restricted to cells in the immediate vicinity of the releasing cell. Signals with such short-lived, local effects are called *paracrine signals* (figure 7.2b). Like direct contact, paracrine signaling plays an important role in early development, coordinating the activities of clusters of neighboring cells.

Endocrine Signaling

If a released signal molecule remains in the extracellular fluid, it may enter the organism's circulatory system and travel widely throughout the body. These longer-lived signal molecules, which may affect cells very distant from the releasing cell, are called **hormones,** and this type of intercellular communication is known as *endocrine signaling* (figure 7.2c).

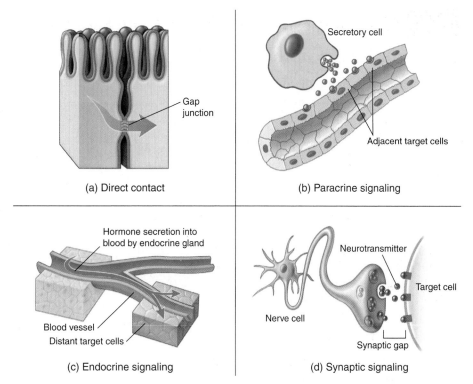

(a) Direct contact

(b) Paracrine signaling

(c) Endocrine signaling

(d) Synaptic signaling

FIGURE 7.2
Four kinds of cell signaling. Cells communicate in several ways. (*a*) Two cells in direct contact with each other may send signals across gap junctions. (*b*) In paracrine signaling, secretions from one cell have an effect only on cells in the immediate area. (*c*) In endocrine signaling, hormones are released into the organism's circulatory system, which carries them to the target cells. (*d*) Chemical synapse signaling involves transmission of signal molecules, called neurotransmitters, from a neuron over a small synaptic gap to the target cell.

Chapter 47 discusses endocrine signaling in detail. Both animals and plants use this signaling mechanism extensively.

Synaptic Signaling

In animals, the cells of the nervous system provide rapid communication with distant cells. Their signal molecules, **neurotransmitters,** do not travel to the distant cells through the circulatory system as hormones do. Rather, the long, fiberlike extensions of nerve cells release neurotransmitters from their tips very close to the target cells (figure 7.2d). The association of a neuron and its target cell is called a **chemical synapse,** and this type of intercellular communication is called *synaptic signaling*. While paracrine signals move through the fluid between cells, neurotransmitters cross the synaptic gap and persist only briefly. We will examine synaptic signaling more fully in chapter 45.

A cell can signal others by direct contact, by locally diffusing paracrine signals, by hormones that travel to distant cells, and by synaptic signaling between a nerve cell and other nerve or muscle cells.

7.2 Proteins in the cell and on its surface receive signals from other cells.

Intracellular Receptors

All cell signaling pathways share certain common elements, including a chemical signal that passes from one cell to another and a receptor that receives the signal in or on the target cell. We've looked at the sorts of signals that pass from one cell to another. Now let's consider the nature of the receptors that receive the signals. Table 7.1 summarizes the types of receptors we will discuss in this chapter.

Many cell signals are lipid-soluble or very small molecules that can readily pass across the plasma membrane of the target cell and into the cell, where they interact with a receptor inside. Some bind to protein receptors located in the cytoplasm; others pass across the nuclear membrane as well and bind to receptors within the nucleus. These **intracellular receptors** (figure 7.3) may trigger a variety of responses in the cell, depending on the receptor.

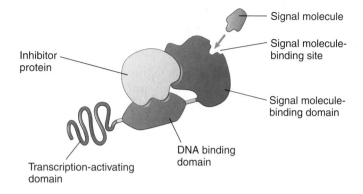

FIGURE 7.3
Basic structure of a gene-regulating intracellular receptor. These receptors are located within the cell and function in the reception of signals such as steroid hormones, vitamin D, and thyroid hormone.

Mechanism	Structure	Function	Example
Table 7.1 Cell-Communicating Mechanisms			
INTRACELLULAR RECEPTORS	No extracellular signal-binding site	Receives signals from lipid-soluble or noncharged, nonpolar small molecules	Receptors for NO, steroid hormone, vitamin D, and thyroid hormone
CELL SURFACE RECEPTORS			
Chemically gated ion channels	Multipass transmembrane protein forming a central pore	Molecular "gates" triggered chemically to open or close	Neurons
Enzymic receptors	Single-pass transmembrane protein	Binds signal extracellularly; catalyzes response intracellularly	Phosphorylation of protein kinases
G-protein-linked receptors	Seven-pass transmembrane protein with cytoplasmic binding site for G protein	Binding of signal to receptor causes GTP to bind a G protein; G protein, with attached GTP, detaches to deliver the signal inside the cell	Peptide hormones, rod cells in the eyes
PHYSICAL CONTACT WITH OTHER CELLS			
Surface markers	Variable; integral proteins or glycolipids in plasma membrane	Identify the cell	MHC complexes, blood groups, antibodies
Tight junctions	Tightly bound, leakproof, fibrous protein seal that surrounds cell	Organizing junction: holds cells together such that materials pass *through* but not *between* the cells	Junctions between epithelial cells in the gut
Desmosomes	Intermediate filaments of cytoskeleton linked to adjoining cells through cadherins	Anchoring junction: binds cells together	Epithelium
Adherens junctions	Transmembrane fibrous proteins	Anchoring junction: connects extracellular matrix to cytoskeleton	Tissues with high mechanical stress, such as the skin
Gap junctions	Six transmembrane connexon proteins creating a pore that connects cells	Communicating junction: allows passage of small molecules from cell to cell in a tissue	Excitable tissue such as heart muscle
Plasmodesmata	Cytoplasmic connections between gaps in adjoining plant cell walls	Communicating junction between plant cells	Plant tissues

Receptors That Act as Gene Regulators

Some intracellular receptors act as regulators of gene expression. Among them are the receptors for steroid hormones, such as cortisol, estrogen, and progesterone, as well as the receptors for a number of other small, lipid-soluble signal molecules, such as vitamin D and thyroid hormone. All of these receptors have similar structures; the genes that code for them may well be the evolutionary descendants of a single ancestral gene. Because of their structural similarities, they are all part of the *intracellular receptor superfamily*.

Each of these receptors has a binding site for DNA. In its inactive state, the receptor typically cannot bind DNA because an inhibitor protein occupies the binding site. When the signal molecule binds to another site on the receptor, the inhibitor is released, and the DNA-binding site is exposed (figure 7.4). The receptor then binds to specific nucleotide sequences on the DNA, which activates (or, in a few instances, suppresses) particular genes, usually located adjacent to the regulatory sites.

The lipid-soluble signal molecules that intracellular receptors recognize tend to persist in the blood far longer than water-soluble signals. Most water-soluble hormones break down within minutes, and neurotransmitters break down within seconds or even milliseconds. On the other hand, a steroid hormone such as cortisol or estrogen persists for hours.

The target cell's response to a lipid-soluble cell signal can vary enormously, depending on the nature of the cell. This is true even when different target cells have the same intracellular receptor, for two reasons: First, the binding site for the receptor on the target DNA differs from one cell type to another, so that different genes are affected when the signal-receptor complex binds to the DNA. Second, most eukaryotic genes have complex controls. We will discuss these controls in detail in chapter 18, but for now it is sufficient to note that several different regulatory proteins are usually involved in reading a eukaryotic gene. Thus, the intracellular receptor interacts with different signals in different tissues. Depending on the cell-specific controls operating in different tissues, the effect the intracellular receptor produces when it binds with DNA will vary.

Receptors That Act as Enzymes

Other intracellular receptors act as enzymes. A very interesting example is the receptor for the signal molecule nitric oxide (NO). The small gas molecule NO diffuses readily out of the cells where it is produced and passes directly into neighboring cells, where it binds to the enzyme guanylyl cyclase. Binding of NO activates the enzyme, enabling it to catalyze the synthesis of cyclic guanosine monophosphate (GMP), an intracellular messenger molecule that produces cell-specific responses such as the relaxation of smooth muscle cells.

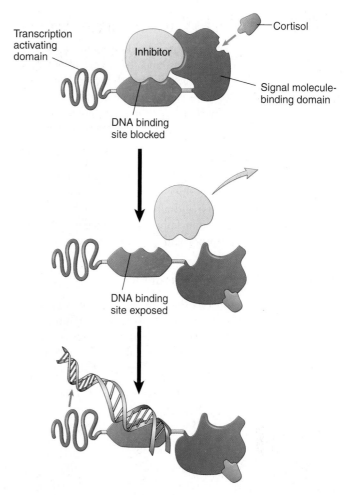

FIGURE 7.4
How intracellular receptors regulate gene transcription. In this model, the binding of the steroid hormone cortisol to a DNA regulatory protein causes it to alter its shape. The inhibitor is released, exposing the DNA-binding site of the regulatory protein. The DNA binds to the site, positioning a specific nucleotide sequence over the transcription-activating domain of the receptor and initiating transcription.

NO has only recently been recognized as a signal molecule in vertebrates. Already, however, a wide variety of roles have been documented. For example, when the brain sends a nerve signal relaxing the smooth muscle cells lining the walls of vertebrate blood vessels, the signal molecule acetylcholine released by the nerve cell near the muscle does not interact with the muscle cell directly. Instead, it causes nearby epithelial cells to produce NO, which then causes the smooth muscle to relax, allowing the vessel to expand and thereby increasing blood flow.

Many target cells possess intracellular receptors, which are activated by substances that pass through the plasma membrane.

Cell Surface Receptors

Most signal molecules are water-soluble, including neurotransmitters, peptide hormones, and the many proteins that multicellular organisms employ as growth factors during development. Water-soluble signals cannot diffuse through plasma membranes. Therefore, to trigger responses in cells, they must bind to receptor proteins on the outer surface of the cell. These **cell surface receptors** (figure 7.5) convert the extracellular signal to an intracellular one, responding to the binding of the signal molecule to the cell's outside by producing a change inside the cell. Most of a cell's receptors are cell surface receptors, and almost all of them belong to one of three receptor superfamilies: chemically gated ion channels, enzymic receptors, and G-protein-linked receptors.

Chemically Gated Ion Channels

Chemically gated ion channels are receptor proteins that ions pass through. The receptor proteins that bind many neurotransmitters have the same basic structure (figure 7.5a). Each is a "multipass" transmembrane protein, meaning that the chain of amino acids threads back and forth across the plasma membrane several times. In the center of the protein is a pore that connects the extracellular fluid with the cytoplasm. The pore is big enough for ions to pass through, so the protein functions as an **ion channel.** The channel is said to be chemically gated because it opens only when a chemical (the neurotransmitter) binds to it. The type of ion (sodium, potassium, calcium, or chloride, for example) that flows across the membrane when a chemically gated ion channel opens depends on the specific shape and charge structure of the channel.

Enzymic Receptors

Many cell surface receptors either act as enzymes or are directly linked to enzymes (figure 7.5b). When a signal molecule binds to the receptor, it activates the enzyme. In almost all cases, these enzymes are **protein kinases,** enzymes that add phosphate groups to proteins. Most enzymic receptors have the same general structure. Each is a single-pass transmembrane protein (the amino acid chain passes through the plasma membrane only once); the portion that binds the signal molecule lies outside the cell, and the portion that carries out the enzyme activity is exposed to the cytoplasm.

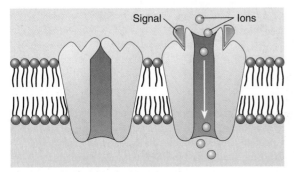

(a) Chemically gated ion channel

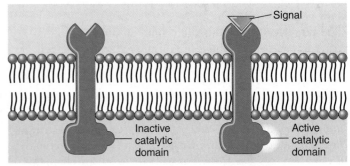

(b) Enzymic receptor

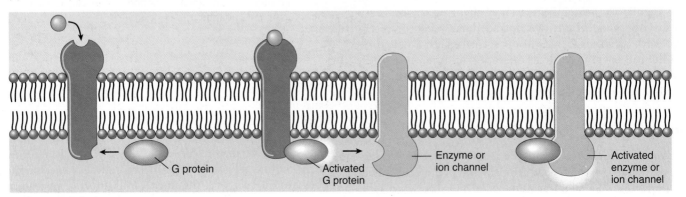

(c) G-protein-linked receptor

FIGURE 7.5
Cell surface receptors. (*a*) Chemically gated ion channels are multipass transmembrane proteins that form a pore in the plasma membrane. This pore is opened or closed by chemical signals. (*b*) Enzymic receptors are single-pass transmembrane proteins that bind the signal on the extracellular surface. A catalytic region on their cytoplasmic portion then initiates enzymatic activity inside the cell. (*c*) G-protein-linked receptors bind to the signal outside the cell and to G proteins inside the cell. The G protein then activates an enzyme or ion channel, mediating the passage of a signal from the cell's surface to its interior.

G-Protein-Linked Receptors

A third class of cell surface receptors acts indirectly on enzymes or ion channels in the plasma membrane with the aid of an assisting protein, called a *guanosine triphosphate* (GTP)-*binding protein*, or **G protein** (figure 7.5c). Receptors in this category use G proteins to mediate passage of the signal from the membrane surface into the cell interior.

How G-Protein-Linked Receptors Work. G proteins are mediators that initiate a diffusible signal in the cytoplasm. They form a transient link between the receptor on the cell surface and the signal pathway within the cytoplasm. Importantly, this signal has a relatively short duration. When a signal arrives, it finds the G protein nestled into the G-protein-linked receptor on the cytoplasmic side of the plasma membrane. Once the signal molecule binds to the receptor, the G-protein-linked receptor changes shape. This change in receptor shape twists the G protein, causing it to bind GTP. The G protein can now diffuse away from the receptor. The "activated" complex of a G protein with attached GTP is then free to initiate a number of events. However, this activation is short-lived, because GTP has a relatively short life span (seconds to minutes) before it is broken down into GDP + P_i. This elegant arrangement allows the G proteins to activate numerous pathways, but only in a transient manner. In order for a pathway to "stay on," there must be a continuous source of incoming extracellular signals. When the rate of external signal drops off, the pathway shuts down.

Scientists have identified more than 100 different G-protein-linked receptors, more than any other kind of cell surface receptor. These receptors mediate signaling, utilizing an incredible range of molecules, including peptide hormones, neurotransmitters, fatty acids, and amino acids. Despite this great variation in specificity, however, all G-protein-linked receptors whose amino acid sequences are known have a similar structure. They are almost certainly closely related in an evolutionary sense, probably arising from a single ancestral sequence. Each of these G-protein-linked receptors is a seven-pass transmembrane protein (figure 7.6); that is, a single polypeptide chain threads back and forth across the lipid bilayer seven times, creating a channel through the membrane.

Evolutionary Origin of G-Protein-Linked Receptors. As research revealed the structure of G-protein-linked receptors, an interesting pattern emerged. The same seven-pass structural motif is seen again and again: in sensory receptors such as the light-activated rhodopsin protein in the vertebrate eye, in the light-activated bacteriorhodopsin proton pump that plays a key role in bacterial photosynthesis, in the receptor that recognizes the yeast mating factor protein, and in many other sensory receptors. Vertebrate

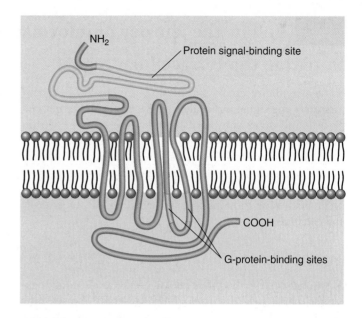

FIGURE 7.6
The G-protein-linked receptor. Each G-protein-linked receptor is a seven-pass transmembrane protein.

rhodopsin is in fact a G-protein-linked receptor and utilizes a G protein. Bacteriorhodopsin is not. The occurrence of the seven-pass structural motif in both, and in so many other G-protein-linked receptors, suggests that this motif is a very ancient one, and that G-protein-linked receptors may have evolved from sensory receptors of single-celled ancestors.

Discovery of G Proteins. Martin Rodbell of the National Institute of Environmental Health Sciences and Alfred Gilman of the University of Texas Southwestern Medical Center received the 1994 Nobel Prize for Medicine or Physiology for their work on G proteins. Rodbell and Gilman's work has proven to have significant ramifications. G proteins are involved in the mechanism employed by over half of all medicines in use today. Studying G proteins will vastly expand our understanding of how these medicines work. Furthermore, the investigation of G proteins should help elucidate how cells communicate in general and how they contribute to the overall physiology of an organism. As Gilman says, G proteins are "involved in everything from sex in yeast to cognition in humans."

> **Most receptors are located on the surface of the plasma membrane. Chemically gated ion channels open or close when signal molecules bind to the channel, allowing specific ions to diffuse through. Enzymic receptors typically activate intracellular proteins by phosphorylation. G-protein-linked receptors activate an intermediary protein, which then effects the intracellular change.**

7.3 Follow the journey of information into the cell.

Initiating the Intracellular Signal

Some enzymic receptors and most G-protein-linked receptors carry the signal molecule's message into the target cell by utilizing other substances to relay the message within the cytoplasm. These other substances, small molecules or ions called **second messengers**, alter the behavior of particular proteins by binding to them and changing their shape. The two most widely used second messengers are cyclic adenosine monophosphate (**cyclic AMP, or cAMP**) and calcium ions.

cAMP

All animal cells studied thus far use cAMP as a second messenger (chapter 47 discusses cAMP in detail). To see how cAMP typically works as a messenger, let's examine what happens when the hormone epinephrine binds to a particular type of G-protein-linked receptor called the β-adrenergic receptor in muscle cells (figure 7.7). When epinephrine binds with this receptor, it activates a G protein, which then stimulates the enzyme **adenylyl cyclase** to produce large amounts of cAMP within the cell (figure 7.8a). The cAMP then binds to and activates the enzyme α-kinase, which adds phosphates to specific proteins in the cell. The effect of this phosphorylation on cell function depends on the identity of the cell and the proteins that are phosphorylated. In muscle cells, for example, the α-kinase phosphorylates and thereby activates enzymes that stimulate the breakdown of glycogen into glucose and inhibit the synthesis of glycogen from glucose. More glucose is then available to the muscle cells.

Calcium

Calcium ions (Ca^{++}) serve widely as second messengers. Ca^{++} levels inside the cytoplasm are normally very low (less than 10^{-7} M), while outside the cell and in the endoplasmic reticulum, Ca^{++} levels are quite high (about 10^{-3} M). Chemically gated calcium channels in the endoplasmic reticulum membrane act as switches; when they open, Ca^{++} rushes into the cytoplasm and triggers proteins sensitive to Ca^{++} to initiate a variety of activities. The efflux of Ca^{++} from the endoplasmic reticulum causes skeletal muscle cells to contract and endocrine cells to secrete hormones.

Gated Ca^{++} channels are opened by a G-protein-linked receptor. In response to signals from other cells,

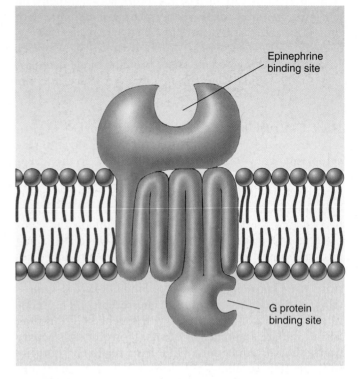

FIGURE 7.7

Structure of the β-adrenergic receptor. The receptor is a G-protein-linked molecule that, when it binds to an extracellular signal molecule, stimulates voluminous production of cAMP inside the cell, which then effects the cellular change.

the receptor activates its G protein, which in turn activates the enzyme *phospholipase C*. This enzyme catalyzes the production of *inositol trisphosphate* (IP_3) from phospholipids in the plasma membrane. The IP_3 molecules diffuse through the cytoplasm to the endoplasmic reticulum and bind to the Ca^{++} channels. This opens the channels and allows Ca^{++} to flow from the endoplasmic reticulum into the cytoplasm (figure 7.8b).

Ca^{++} initiates some cellular responses by binding to *calmodulin*, a 148-amino-acid cytoplasmic protein that contains four binding sites for Ca^{++} (figure 7.9). When four Ca^{++} ions are bound to calmodulin, the calmodulin/Ca^{++} complex binds to other proteins and activates them.

Cyclic AMP and Ca^{++} often behave as second messengers, relaying messages from receptors to target proteins.

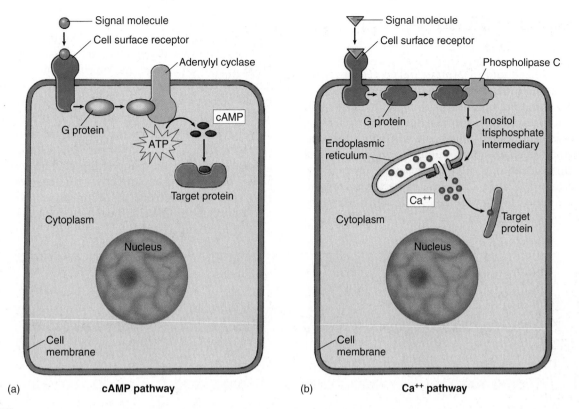

(a) **cAMP pathway**

(b) **Ca⁺⁺ pathway**

FIGURE 7.8

How second messengers work. (*a*) The cyclic AMP (cAMP) pathway. An extracellular receptor binds to a signal molecule and, through a G protein, activates the membrane-bounded enzyme adenylyl cyclase. This enzyme catalyzes the synthesis of cAMP, which binds to the target protein to initiate the cellular change. (*b*) The calcium (Ca^{++}) pathway. An extracellular receptor binds to another signal molecule and, through another G protein, activates the enzyme phospholipase C. This enzyme stimulates the production of inositol trisphosphate, which binds to and opens calcium channels in the membrane of the endoplasmic reticulum. Ca^{++} is released into the cytoplasm, effecting a change in the cell.

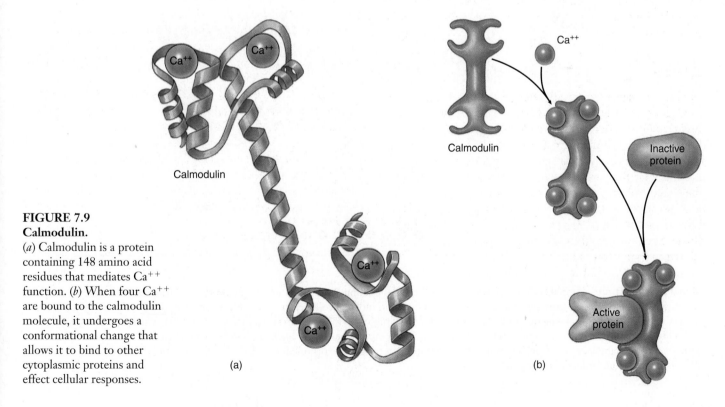

FIGURE 7.9
Calmodulin.
(*a*) Calmodulin is a protein containing 148 amino acid residues that mediates Ca^{++} function. (*b*) When four Ca^{++} are bound to the calmodulin molecule, it undergoes a conformational change that allows it to bind to other cytoplasmic proteins and effect cellular responses.

(a)

(b)

Amplifying the Signal: Protein Kinase Cascades

Both enzyme-linked and G-protein-linked receptors receive signals at the surface of the cell, but as we've seen, the target cell's response rarely takes place there. In most cases, the signals are relayed to the cytoplasm or the nucleus by second messengers, which influence the activity of one or more enzymes or genes and so alter the behavior of the cell. But most signaling molecules are found in such low concentrations that their diffusion across the cytoplasm would take a great deal of time unless the signal were amplified. Therefore, most enzyme-linked and G-protein-

linked receptors use a chain of other protein messengers to amplify the signal as it is being relayed to the nucleus.

How is the signal amplified? Imagine a relay race where, at the end of each stage, the finishing runner tags five new runners to start the next stage. The number of runners would increase dramatically as the race progresses: 1, then 5, 25, 125, and so on. The same sort of process takes place as a signal is passed from the cell surface to the cytoplasm or nucleus. First, the receptor activates a stage-one protein, almost always by phosphorylating it. The receptor either adds a phosphate group directly, or it activates a G protein that goes on to activate a second protein that does the phosphorylation. Once activated, each of these stage-one

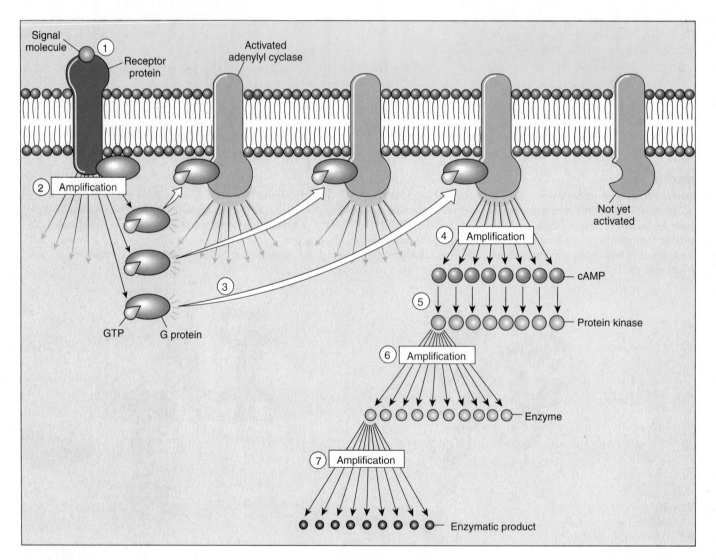

FIGURE 7.10

Signal amplification. Amplification at many stages of the cell-signaling process can ultimately produce a large response by the cell. One cell surface receptor (*1*), for example, may activate many G protein molecules (*2*), each of which activates a molecule of adenylyl cyclase (*3*), yielding an enormous production of cAMP (*4*). Each cAMP molecule in turn activates a protein kinase (*5*), which can phosphorylate and thereby activate several copies of a specific enzyme (*6*). Each of *those* enzymes can then catalyze many chemical reactions (*7*). Starting with 10^{-10} M of signaling molecule, one cell surface receptor can trigger the production of 10^{-6} M of one of the products, an amplification of four orders of magnitude.

proteins in turn activates a large number of stage-two proteins; then each of them activates a large number of stage-three proteins; and so on (figure 7.10). A single cell surface receptor can thus stimulate a cascade of protein kinases to amplify the signal and cause a strong response inside the cell to a weak signal outside.

This kind of chain reaction or signaling cascade is common after receptors are triggered. It can involve several different proteins, but commonly involves G proteins, phospholipase C, IP_3, calcium, and sometimes protein kinases. Although each receptor-triggered signal transduction event may employ some of the same intermediate molecules, they ultimately activate particular sets of targets to cause a cellular response that is unique and specific to the particular receptor triggered.

The Vision Amplification Cascade

Let's trace a protein amplification cascade to see exactly how it works. In vision, a single, light-activated rhodopsin (a G-protein-linked receptor) activates hundreds of molecules of the G protein transducin in the first stage of the relay. In the second stage, each transducin causes an enzyme to modify thousands of molecules of a special inside-the-cell messenger called cyclic GMP. (We will discuss cyclic GMP in more detail later.) In about 1 second, a single rhodopsin signal passing through this two-step cascade splits more than 10^5 (100,000) cyclic GMP molecules (figure 7.11). The rod cells of humans are sufficiently sensitive to detect brief flashes of just 5 photons.

The Cell Division Amplification Cascade

The general principle of signal amplification occurs in many cell types and for many receptor types. However, the specific mechanisms of how that amplification is achieved vary considerably. Cell division, for example, is regulated by integrating many signals from outside the cell. One of the receptors involved acts as a protein kinase. The receptor responds to growth-promoting signals by phosphorylating an intracellular protein kinase called ras, which then activates a series of interacting phosphorylation cascades, some with five or more stages. If the ras protein becomes hyperactive for any reason, the cell acts as if it is being constantly stimulated to divide. Ras proteins were first discovered in cancer cells. A mutation of the gene that encodes ras had caused it to become hyperactive, resulting in unrestrained cell proliferation. Almost one-third of human cancers have such a mutation in a *ras* gene.

A small number of surface receptors can generate a vast intracellular response as each stage of the pathway amplifies the next.

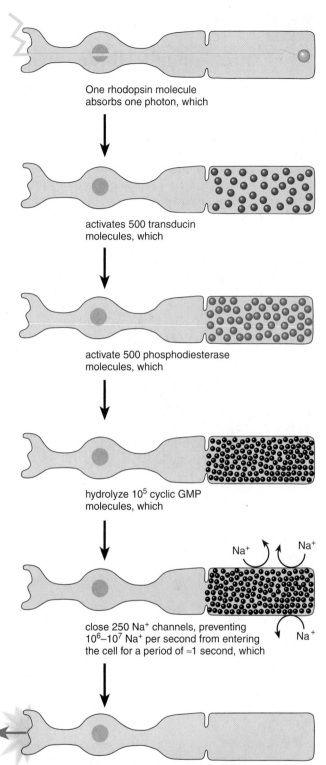

One rhodopsin molecule absorbs one photon, which

activates 500 transducin molecules, which

activate 500 phosphodiesterase molecules, which

hydrolyze 10^5 cyclic GMP molecules, which

close 250 Na$^+$ channels, preventing 10^6–10^7 Na$^+$ per second from entering the cell for a period of ≈1 second, which

hyperpolarizes the rod cell membrane by 1 mV, sending a visual signal to the brain.

FIGURE 7.11

The role of signal amplification in vision.   In this vertebrate light-detecting rod cell, *one* single rhodopsin pigment molecule, when excited by a photon, ultimately yields *100,000* split cyclic GMP molecules, which will then effect a change in the membrane of the rod cell that is interpreted by the organism as a visual event.

Cell surface proteins mediate cell–cell interactions.

The Expression of Cell Identity

With the exception of a few primitive types of organisms, the hallmark of multicellular life is the development of highly specialized groups of cells called **tissues,** such as blood and muscle. Remarkably, each cell within a tissue performs the functions of that tissue and no other, even though all cells of the body are derived from a single fertilized cell and contain the same genetic information—all of the genes found in the genome. However, as an organism develops, the cells acquire their specific identities by carefully controlling the *expression* of those genes, turning on the specific set of genes that encode the particular functions of each cell type. How do cells sense where they are, and how do they "know" which type of tissue they belong to?

Tissue-Specific Identity Markers

One key set of genes function to mark the surfaces of cells to identify them as being of a particular type. When cells make contact, they "read" each others' cell surface markers and react accordingly. When cells that are part of the same tissue type recognize each other, they frequently respond by forming connections between their cell surfaces to better coordinate their functions.

Glycolipids. Most tissue-specific cell surface markers are glycolipids, lipids with carbohydrate heads. The glycolipids on the surface of red blood cells are also responsible for the differences among A, B, and O blood types.

MHC Proteins. One important example of the function of cell surface markers is the recognition of "self" and "nonself" cells by the immune system. This is an important function for multicellular organisms, where the ability to distinguish between the cells of the organism and foreign cells is important.

The immune system of vertebrates uses a particular set of markers to distinguish self from nonself cells, encoded by genes of the *major histocompatibility complex* (MHC). These genes encode a set of cell surface proteins that are unique to the individual and serve as effective identity tags. The MHC proteins and other self-identifying markers are single-pass proteins anchored in the plasma membrane, and many of them are members of a large superfamily of receptors, the immunoglobulins (figure 7.12). Cells of the immune system continually inspect other cells they encounter in the body, triggering the destruction of cells that display foreign or nonself identity markers.

Every cell contains a specific array of marker proteins on its surface. These markers identify each type of cell in a very precise way.

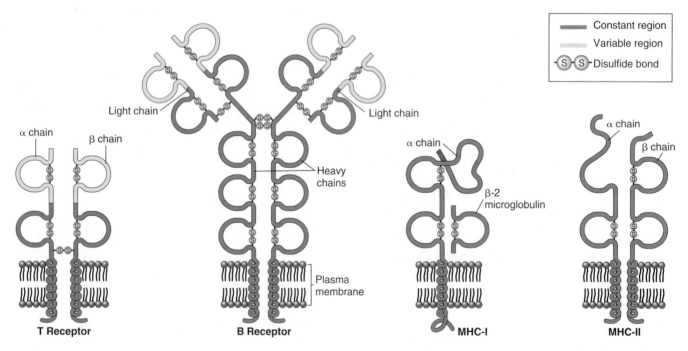

FIGURE 7.12
Structure of the immunoglobulin family of cell surface marker proteins. T and B cell receptors help mediate the immune response in organisms by recognizing and binding to foreign cell markers. MHC antigens label cells as "self," so that the immune system attacks only invading entities, such as bacteria, viruses, and often the cells of transplanted organs that have different MHC antigens.

Intercellular Adhesion

Not all physical contacts between cells in a multicellular organism are fleeting touches. In fact, most cells are in physical contact with other cells at all times, usually as members of organized tissues such as those in a leaf, or in your lungs, heart, or gut. These cells and the mass of other cells clustered around them form long-lasting or permanent connections with each other called **cell junctions** (figure 7.13). The nature of the physical connections between the cells of a tissue in large measure determines what the tissue is like. Indeed, a tissue's proper functioning often depends critically upon how the individual cells are arranged within it. Just as a house cannot maintain its structure without nails and cement, so a tissue cannot maintain its characteristic architecture without the appropriate cell junctions.

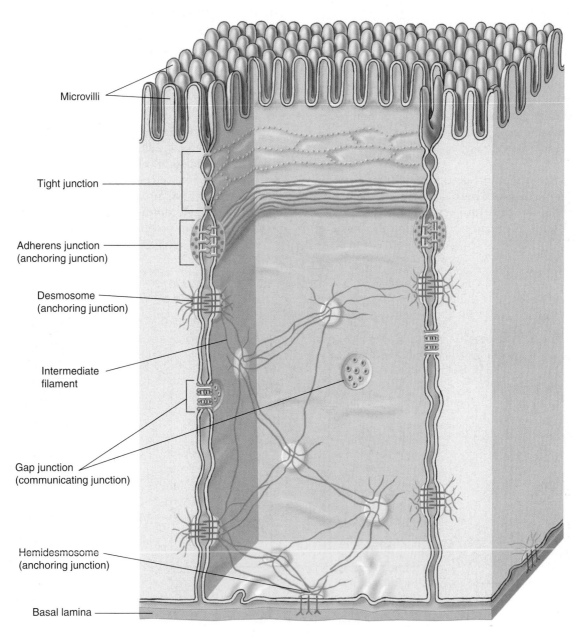

Microvilli

Tight junction

Adherens junction
(anchoring junction)

Desmosome
(anchoring junction)

Intermediate
filament

Gap junction
(communicating junction)

Hemidesmosome
(anchoring junction)

Basal lamina

FIGURE 7.13
An overview of cell junction types. Here, gut epithelial cells illustrate the comparative structures and locations of common cell junctions.

Cell junctions are divided into three categories, based upon their functions: tight junctions, anchoring junctions, and communicating junctions (figure 7.14).

Tight Junctions

Sometimes called occluding junctions, tight junctions connect the plasma membranes of adjacent cells in a sheet, preventing small molecules from leaking between the cells. This allows the sheet of cells to act as a wall within the organ, keeping molecules on one side or the other.

Creating Sheets of Cells. The cells that line an animal's digestive tract are organized in a sheet only one cell thick. One surface of the sheet faces the inside of the tract, and the other faces the extracellular space where blood vessels are located. Tight junctions encircle each cell in the sheet, like a belt cinched around a person's waist. The junctions between neighboring cells are so securely attached that there is no space between them for leakage. Hence, nutrients absorbed from the food in the digestive tract must pass directly through the cells in the sheet to enter the blood because they cannot pass through spaces between cells.

Partitioning the Sheet. The tight junctions between the cells lining the digestive tract also partition the plasma membranes of these cells into separate compartments. Transport proteins in the membrane facing the inside of

the tract carry nutrients from that side to the cytoplasm of the cells. Other proteins, located in the membrane on the opposite side of the cells, transport those nutrients from the cytoplasm to the extracellular fluid, where they can enter the blood. For the sheet to absorb nutrients properly, these proteins must remain in the correct locations within the fluid membrane. Tight junctions effectively segregate the proteins on opposite sides of the sheet, preventing them from drifting within the membrane from one side of the sheet to the other. When tight junctions are experimentally disrupted, just this sort of migration occurs.

Anchoring Junctions

Anchoring junctions mechanically attach the cytoskeleton of a cell to the cytoskeletons of other cells or to the extracellular matrix. These junctions are most common in tissues subject to mechanical stress, such as muscle and skin epithelium.

Cadherin and Intermediate Filaments: Desmosomes. Anchoring junctions called **desmosomes** connect the cytoskeletons of adjacent cells (figure 7.15), while *hemidesmosomes* anchor epithelial cells to a basement membrane. Proteins called **cadherins,** most of which are single-pass transmembrane glycoproteins, create the critical link. A variety of attachment proteins link the short cytoplasmic end

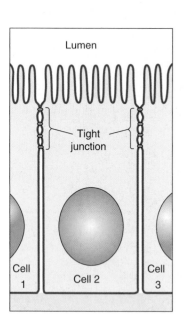 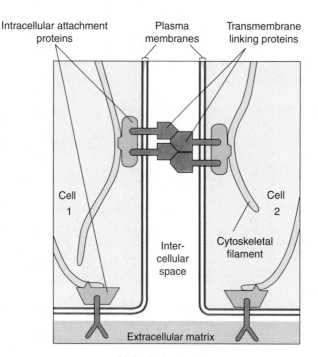 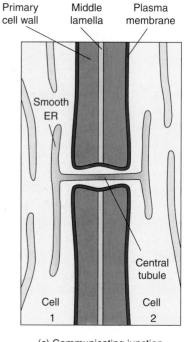

(a) Tight junction   (b) Anchoring junction   (c) Communicating junction

FIGURE 7.14
The three types of cell junctions. These models represent current thinking on how the structures of the three major types of cell junctions facilitate their function: (*a*) tight junction; (*b*) anchoring junction; (*c*) communicating junction.

of a cadherin to the intermediate filaments in the cytoskeleton. The other end of the cadherin molecule projects outward from the plasma membrane, joining directly with a cadherin protruding from an adjacent cell in a firm handshake, binding the cells together.

Connections between proteins tethered to the intermediate filaments are much more secure than connections between free-floating membrane proteins. As described in chapter 6, proteins are suspended within the membrane by relatively weak interactions between the nonpolar portions of the protein and the membrane lipids. It would not take much force to pull an untethered protein completely out of the membrane.

Cadherin and Actin Filaments. Cadherins can also connect the actin frameworks of cells in cadherin-mediated junctions (figure 7.16). When they do, they form less stable links between cells than when they connect intermediate filaments. Many kinds of actin-linking cadherins occur in different tissues. During vertebrate development, the migration of neurons in the embryo is associated with changes in the type of cadherin expressed on their plasma membranes. This suggests that gene-controlled changes in cadherin expression may provide the migrating cells with a "roadmap" to their destination.

Integrin-Mediated Links. Anchoring junctions called **adherens junctions** connect the actin filaments of one cell with those of neighboring cells or with the extracellular matrix. The linking proteins in these junctions are members of a large superfamily of cell surface receptors called integrins that bind to a protein component of the extracellular matrix. There appear to be many different kinds of integrins (cell biologists have identified 20), each with a slightly different-shaped binding domain. The exact component of the matrix that a given cell binds to depends on which combination of integrins that cell has in its plasma membrane.

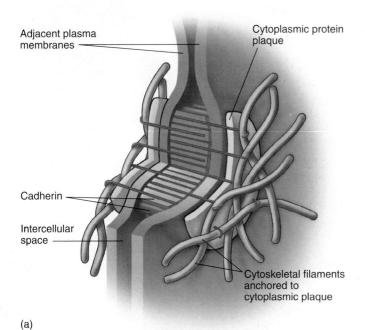

Adjacent plasma membranes

Cytoplasmic protein plaque

Cadherin

Intercellular space

Cytoskeletal filaments anchored to cytoplasmic plaque

(a)

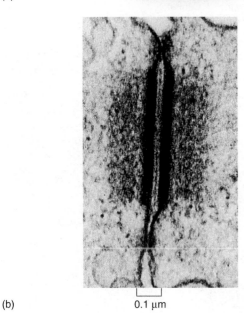

(b)

0.1 μm

FIGURE 7.15
Desmosomes. (*a*) Cadherin proteins create the adhering link between adjoining cells. (*b*) A desmosome anchors adjacent cells to one another.

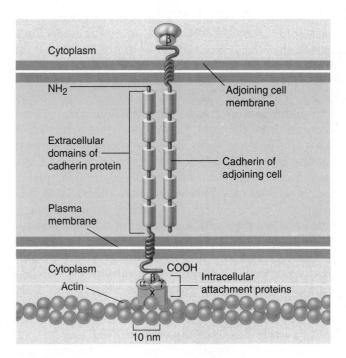

Cytoplasm

NH₂

Extracellular domains of cadherin protein

Plasma membrane

Cytoplasm

Actin

Adjoining cell membrane

Cadherin of adjoining cell

COOH

Intracellular attachment proteins

10 nm

FIGURE 7.16
A cadherin-mediated junction. The cadherin molecule is anchored to actin in the cytoskeleton and passes through the membrane to interact with the cadherin of an adjoining cell.

Communicating Junctions

Many cells communicate with adjacent cells through direct connections called **communicating junctions.** In these junctions, a chemical or electrical signal passes directly from one cell to an adjacent one. Communicating junctions establish direct physical connections that link the cytoplasms of two cells together, permitting small molecules or ions to pass from one to the other. In animals, these direct communication channels between cells are called gap junctions, and in plants, plasmodesmata.

Gap Junctions in Animals. Gap junctions are composed of structures called connexons, complexes of six identical transmembrane proteins (figure 7.17). The proteins in a connexon are arranged in a circle to create a channel through the plasma membrane that protrudes several nanometers from the cell surface. A gap junction forms when the connexons of two cells align perfectly, creating an open channel that spans the plasma membranes of both cells. Gap junctions provide passageways large enough to permit small substances, such as simple sugars and amino acids, to pass from the cytoplasm of one cell to that of the next, yet small enough to prevent the passage of larger molecules such as proteins. The connexons hold the plasma membranes of the paired cells about 4 nanometers apart, in marked contrast to the more-or-less direct contact between the lipid bilayers in a tight junction.

Gap junction channels are dynamic structures that can open or close in response to a variety of factors, including Ca^{++} and H^+ ions. This gating serves at least one important function. When a cell is damaged, its plasma membrane often becomes leaky. Ions in high concentrations outside the cell, such as Ca^{++}, flow into the damaged cell and shut its gap junction channels. This isolates the cell and so prevents the damage from spreading to other cells.

Plasmodesmata in Plants. In plants, cell walls separate every cell from all others. Cell–cell junctions occur only at holes or gaps in the walls, where the plasma membranes of adjacent cells can come into contact with each other. Cytoplasmic connections that form across the touching plasma membranes are called **plasmodesmata** (figure 7.18). The majority of living cells within a higher plant are connected with their neighbors by these junctions. Plasmodesmata function much like gap junctions in animal cells, although their structure is more complex. Unlike gap junctions, plasmodesmata are lined with plasma membrane and contain a central tubule that connects the endoplasmic reticula of the two cells.

Cells attach themselves to one another with long-lasting bonds called cell junctions. Tight junctions connect the plasma membranes of adjacent cells in sheets. Anchoring junctions attach to the cells' cytoskeletons. Communicating junctions permit passage of substances between cells.

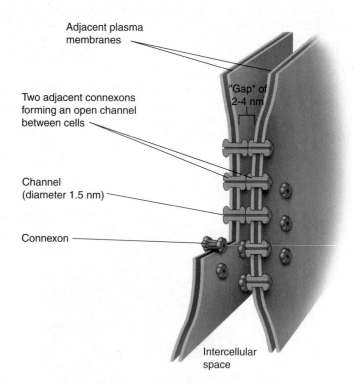

FIGURE 7.17
Gap junctions. Connexons in gap junctions create passageways that connect the cytoplasms of adjoining cells. Gap junctions readily allow the passage of small molecules and ions required for rapid communication (such as in heart tissue), but do not allow the passage of larger molecules such as proteins.

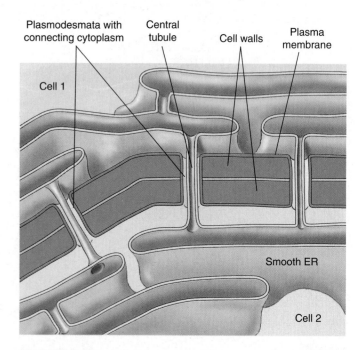

FIGURE 7.18
Plasmodesmata. Plant cells can communicate through specialized openings in their cell walls, called plasmodesmata, where the cytoplasms of adjoining cells are connected (also see figure 7.14c).

Concept Review

7.1 Cells signal one another with chemicals.

Receptor Proteins and Signaling Between Cells

- Receptor proteins are located on or within the cell, and have three-dimensional shapes that fit the shape of specific signal molecules. (p. 126)
- Binding of the signal molecule with the receptor protein induces a change in the protein's shape and produces a cellular response. (p. 126)
- Immunochemistry and molecular genetics are being used to locate and characterize receptor proteins. (p. 126)

Types of Cell Signaling

- Cells can communicate through any of four basic mechanisms: direct contact, paracrine signaling, endocrine signaling, or synaptic signaling. (p. 127)

7.2 Proteins in the cell and on its surface receive signals from other cells.

Intracellular Receptors

- All cell-signaling pathways share certain common elements, including a chemical signal that passes from one cell to another and a receptor that receives the signal in or on the target cell. (p. 128)
- Intracellular receptors may trigger a variety of responses in the cell, dependent on the receptor. (pp. 128–129)

Cell Surface Receptors

- Cell surface receptors convert the extracellular signal to an intercellular one, responding to the binding of the signal molecule to the cell's outside by producing a change inside the cell. (p. 130)
- Many cell surface receptors either act as enzymes or are directly linked to enzymes. (p. 130)
- G-protein-linked receptors activate an intermediary protein, which then effects the intercellular change. (p. 131)

7.3 Follow the journey of information into the cell.

Initiating the Intracellular Signal

- Second messengers, such as cAMP and calcium ions, relay messages from receptors to target proteins. (p. 132)

Amplifying the Signal: Protein Kinase Cascades

- Some surface receptors generate large intracellular responses because each stage of the pathway amplifies the next, causing a cascading effect. (pp. 134–135)

7.4 Cell surface proteins mediate cell–cell interactions.

The Expression of Cell Identity

- As an organism develops, its cells acquire their specific identities by controlling gene expression, turning on the specific set of genes that encode the particular functions of each cell type. (p. 136)
- Every cell contains surface marker proteins that uniquely identify each cell type. (p. 136)

Intercellular Adhesion

- Cells attach to one another using cell junctions. (p. 137)
- Tight junctions connect the plasma membranes of adjacent cells in a sheet. (p. 138)
- Anchoring junctions mechanically attach the cytoskeleton of a cell to the cytoskeletons of other cells or to the extracellular matrix. (p. 138)
- Communication junctions allow communication with adjacent cells through direct connections. (p. 140)

Self Test

1. Which of the following techniques has recently aided the study of receptor proteins?
   a. protein purification
   b. monoclonal antibodies
   c. isolation of cell signal molecules
   d. all of these
2. Which of the following describes autocrine signaling?
   a. Signal molecules released by cells diffuse through the extracellular fluid to other cells.
   b. Signal molecules enter the organism's circulatory system and travel throughout the body.
   c. Signal molecules are released from a cell and bind to receptors on its own plasma membrane.
   d. Signal molecules are released into a narrow space between cells called a synapse.
3. Intracellular receptors usually bind
   a. water-soluble signals.
   b. large molecules that act as signals.
   c. signals on the cell surface.
   d. lipid-soluble signals.
4. Which of the following is *not* a type of cell surface receptor?
   a. chemically gated ion channels
   b. intracellular receptors
   c. enzymic receptors
   d. G-protein-linked receptors
5. Which of the following is *not* a second messenger?
   a. adenylyl cyclase
   b. cyclic adenosine monophosphate
   c. calcium ions
   d. cAMP
6. The amplification of a cellular signal requires all but which of the following?
   a. a second messenger
   b. DNA
   c. a signal molecule
   d. a cascade of protein kinases
7. MHC proteins are
   a. molecules that determine a person's blood type.
   b. large molecules that pass through the membrane many times.
   c. identity markers present on the surface of an individual's cells.
   d. different for each type of tissue in the body.
8. Sheets of cells are formed from which type of cell junctions?
   a. tight junctions
   b. anchoring junctions
   c. communication junctions
   d. none of these
9. Cadherin can be found in which of the following?
   a. tight junctions
   b. anchoring junctions
   c. communication junctions
   d. adherens junctions
10. Plasmodesmata are a type of
    a. gap junction.
    b. anchoring junction.
    c. communicating junction.
    d. tight junction.

Test Your Visual Understanding

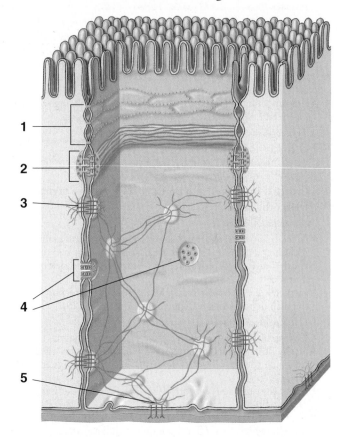

1. Match the following labels with the appropriate structures in the figure, and explain where each type of junction is found.
   adherens junction
   desmosome
   gap junctions
   hemidesmosome
   tight junction

Apply Your Knowledge

1. In paracrine signaling, the signal molecule is destroyed by enzymes in 6 milliseconds. The diffusion rate of the signal through the extracellular fluid is 2 nanometer/1 millisecond. How long will the signal last in the extracellular fluid, and what is the farthest distance a target cell can be from the releasing cell to be affected by the signal?
2. At first glance, the signaling systems that involve cell surface receptors may appear rather complex and indirect, with their use of G proteins, second messengers, and often multiple stages of enzymes. What are the advantages of such seemingly complex response systems?
3. *Shigella flexneri* is one of several species of bacteria that cause shigellosis, or bacillary dysentery. Recent evidence has shown that *S. flexneri* cannot spread between the epithelial cells of the intestines without the expression of cadherin by those cells. Why do you suppose it cannot?

8

Energy and Metabolism

Concept Outline

8.1 The laws of thermodynamics describe how energy changes.

The Flow of Energy in Living Things. Potential energy is present in the electrons of atoms, and so can be transferred from one molecule to another.

The Laws of Thermodynamics. Energy is never lost, but as it is transferred, more and more of it dissipates as heat, a disordered form of energy.

Free Energy. In a chemical reaction, the energy released or supplied is the difference between the bond energies of the reactants and the products, corrected for disorder.

Activation Energy. To start a chemical reaction, an input of energy is required to destabilize existing chemical bonds. Catalysts lower this activation energy, allowing exergonic reactions to proceed faster.

8.2 Enzymes are biological catalysts.

Enzymes. Globular proteins called enzymes catalyze chemical reactions within cells.

How Enzymes Work. Enzymes have sites on their surface shaped to fit their substrates snugly, forcing chemically reactive groups close enough to facilitate a reaction.

Enzymes Take Many Forms. Some enzymes are associated in complex groups; others are not even proteins.

Factors Affecting Enzyme Activity. Each enzyme works most efficiently at its optimal temperature and pH. Metal ions or other substances often help enzymes carry out catalysis.

8.3 ATP is the energy currency of life.

ATP. Cells store and release energy from the phosphate bonds of ATP, the energy currency of the cell.

8.4 Metabolism is the chemical life of a cell.

Biochemical Pathways: The Organizational Units of Metabolism. Biochemical pathways, where the product of one reaction becomes the substrate for the next, are the organizational units of metabolism.

FIGURE 8.1
Lion at lunch. Energy that this lion extracts from its meal of giraffe will be used to power its roar, fuel its running, and build a bigger lion.

Life can be viewed as a constant flow of energy, channeled by organisms to do the work of living. Each of the significant properties by which we define life—order, growth, reproduction, responsiveness, and internal regulation—requires a constant supply of energy (figure 8.1). Deprived of a source of energy, life stops. Therefore, a comprehensive study of life would be impossible without discussing *bioenergetics*, the analysis of how energy powers the activities of living systems. In this chapter, we focus on energy—what it is and how organisms capture, store, and use it.

The Flow of Energy in Living Things

Energy is defined as the capacity to do work. It can be considered to exist in two states (figure 8.2). **Kinetic energy** is the energy of motion. Moving objects perform work by causing other matter to move. **Potential energy** is stored energy. Objects that are not actively moving but have the capacity to do so possess potential energy. A boulder perched on a hilltop has potential energy; as it begins to roll downhill, some of its potential energy is converted into kinetic energy. Much of the work that living organisms carry out involves transforming potential energy to kinetic energy.

Energy can take many forms: mechanical energy, heat, sound, electric current, light, or radioactive radiation. Because it can exist in so many forms, there are many ways to measure energy. The most convenient is in terms of heat, because all other forms of energy can be converted into heat. In fact, the study of energy is called **thermodynamics,** meaning "heat changes." The unit of heat most commonly employed in biology is the **kilocalorie** (kcal). One kilocalorie is equal to 1000 calories (cal), and one calorie is the heat required to raise the temperature of one gram of water one degree Celsius (°C). (It is important not to confuse calories with a term related to diets and nutrition, the Calorie with a capital C, which is actually another term for kilocalorie.) Another energy unit, often used in physics, is the **joule;** one joule equals 0.239 cal.

Oxidation-Reduction

Energy flows into the biological world from the sun, which shines a constant beam of light on the earth. It is estimated that the sun provides the earth with more than 13×10^{23} calories per year, or 40 million billion calories per second! Plants, algae, and certain kinds of bacteria capture a fraction of this energy through photosynthesis. In photosynthesis, energy garnered from sunlight is used to combine small molecules (water and carbon dioxide) into more complex molecules (sugars). The energy is stored as potential energy in the covalent bonds between atoms in the sugar molecules. Recall from chapter 2 that an atom consists of a central nucleus surrounded by one or more orbiting electrons, and a covalent bond forms when two atomic nuclei share valence electrons. Breaking such a bond requires energy to pull the nuclei apart. Indeed, the strength of a covalent bond is measured by the amount of energy required to break it. For example, it takes 98.8 kcal to break one mole (6.023×10^{23}) of carbon-hydrogen (C—H) bonds.

During a chemical reaction, the energy stored in chemical bonds may transfer to new bonds. In some of these reactions, electrons actually pass from one atom or molecule to another. When an atom or molecule loses an electron, it is said to be oxidized, and the process by which this occurs is called **oxidation.** The name reflects the fact that in biological systems oxygen, which attracts electrons strongly, is the most common electron accep-

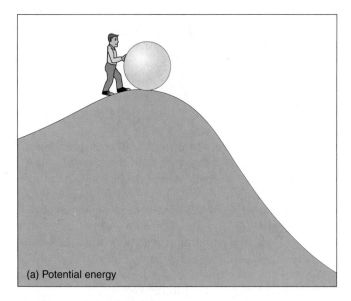

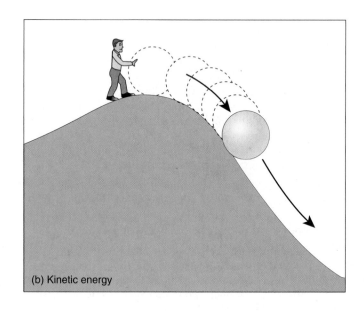

(a) Potential energy

(b) Kinetic energy

FIGURE 8.2

Potential and kinetic energy.   (*a*) Objects that have the capacity to move but are not moving have potential energy. The energy required to move the ball up the hill is stored as potential energy. (*b*) Objects that are in motion have kinetic energy. The stored energy is released as kinetic energy as the ball rolls down the hill.

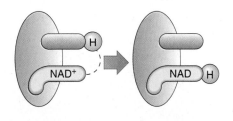

1. Enzymes that harvest hydrogen atoms have a binding site for NAD$^+$ located near another binding site. NAD$^+$ and an energy-rich molecule bind to the enzyme.

2. In an oxidation-reduction reaction, a hydrogen atom is transferred to NAD$^+$, forming NADH.

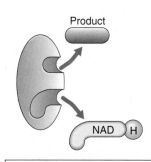

3. NADH then diffuses away and is available to other molecules.

FIGURE 8.3
Oxidation-reduction reactions often employ cofactors. Cells use a chemical cofactor called NAD$^+$ to carry out oxidation-reduction reactions. Energetic electrons are often paired with a proton as a hydrogen atom. Molecules that gain energetic electrons are said to be reduced, while ones that lose energetic electrons are said to be oxidized. NAD$^+$ oxidizes energy-rich molecules by acquiring their hydrogens (in the figure, this proceeds 1→2→3) and then reduces other molecules by giving the hydrogens to them (in the figure, this proceeds 3→2→1).

tor. Conversely, when an atom or molecule gains an electron, it is said to be reduced, and the process is called **reduction.** Oxidation and reduction always take place together, because every electron that is lost by an atom through oxidation is gained by some other atom through reduction. Therefore, chemical reactions of this sort are called **oxidation-reduction (redox) reactions** (figure 8.3). Energy is transferred from one molecule to another via redox reactions. The reduced form of a molecule thus has a higher level of energy than the oxidized form (figure 8.4). The ability of organisms to store energy in molecules by transferring electrons to them is referred to as *reducing power*, and is a fundamental property of living systems.

Oxidation-reduction reactions play a key role in the flow of energy through biological systems because the electrons that pass from one atom to another carry energy with them. The amount of energy an electron possesses depends on how far it is from the nucleus and how strongly the nucleus attracts it. Generally, electrons that are close to the nucleus and held tightly by it have less energy than electrons that are farther from the nucleus and held loosely by it. Light (and other forms of energy) can add energy to an electron and boost it to a higher energy level. When this electron departs from one atom (oxidation) and moves to another (reduction), the electron's added energy is transferred with it, and the electron orbits the second atom's nucleus at the higher en-

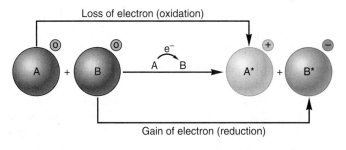

Loss of electron (oxidation)

Gain of electron (reduction)

Low energy
High energy

FIGURE 8.4
Redox reactions. Oxidation is the loss of an electron; reduction is the gain of an electron. In this example, the charges of molecules A and B are shown in small circles to the upper right of each molecule. Molecule A loses energy as it loses an electron, while molecule B gains energy as it gains an electron.

ergy level. The added energy is stored as potential chemical energy that the atom can later release when the electron returns to its original energy level.

Energy is the capacity to do work, either actively (kinetic energy) or stored for later use (potential energy). Energy is transferred with electrons. Oxidation is the loss of an electron; reduction is the gain of one.

Chapter 8 Energy and Metabolism **145**

The Laws of Thermodynamics

All activities of living organisms—running, thinking, singing, reading these words—involve changes in energy. A set of two universal laws we call the laws of thermodynamics govern all energy changes in the universe, from nuclear reactions to the buzzing of a bee.

The First Law of Thermodynamics

The **First Law of Thermodynamics** concerns the amount of energy in the universe. It states that energy cannot be created or destroyed; it can only change from one form to another (from potential to kinetic, for example). The total amount of energy in the universe remains constant.

The lion eating a giraffe in figure 8.1 is in the process of acquiring energy. Rather than creating new energy or capturing the energy in sunlight, the lion is merely transferring some of the potential energy stored in the giraffe's tissues to its own body (just as the giraffe obtained the potential energy stored in the plants it ate while it was alive). Within any living organism, this chemical potential energy can be shifted to other molecules and stored in different chemical bonds, or it can convert into other forms, such as kinetic energy, light, or electricity. During each conversion, some of the energy dissipates into the environment as **heat**, a measure of the random motion of molecules (and, hence, a measure of one form of kinetic energy). Energy continuously flows through the biological world in one direction, with new energy from the sun constantly entering the system to replace the energy dissipated as heat.

Heat can be harnessed to do work only when there is a heat gradient—that is, a temperature difference between two areas (this is how a steam engine functions). Cells are too small to maintain significant internal temperature differences, so heat energy is incapable of doing the work of cells. Thus, although the total amount of energy in the universe remains constant, the energy available to do work decreases as more of it progressively dissipates as heat.

The Second Law of Thermodynamics

The **Second Law of Thermodynamics** concerns the transformation of potential energy into heat, or random molecular motion. It states that the disorder (more for-

Disorder happens "spontaneously"

Organization requires energy

FIGURE 8.5
Entropy in action. As time elapses, a child's room becomes more disorganized. It takes effort to clean it up.

mally called *entropy*) in the universe is continuously increasing. Put simply, disorder is more likely than order. For example, it is much more likely that a column of bricks will tumble over than that a pile of bricks will arrange themselves spontaneously to form a column. In general, energy transformations proceed spontaneously to convert matter from a more ordered, less stable form to a less ordered, more stable form (figure 8.5).

Entropy

Entropy is a measure of the disorder of a system, so the Second Law of Thermodynamics can also be stated simply as "entropy increases." When the universe formed, it held all the potential energy it will ever have. It has become progressively more disordered ever since, with every energy exchange increasing the amount of entropy.

The First Law of Thermodynamics states that energy cannot be created or destroyed; it can only undergo conversion from one form to another. The Second Law of Thermodynamics states that disorder (entropy) in the universe is increasing. As energy is used, more and more of it is converted to heat, the energy of random molecular motion.

Free Energy

It takes energy to break the chemical bonds that hold the atoms in a molecule together. Heat energy, because it increases atomic motion, makes it easier for the atoms to pull apart. Both chemical bonding and heat have a significant influence on a molecule, the former reducing disorder and the latter increasing it. The net effect, the amount of energy actually available to break and subsequently form other chemical bonds, is called the **free energy** of that molecule. In a more general sense, free energy is defined as the energy available to do work in any system. In a molecule within a cell, where pressure and volume usually do not change, the free energy is denoted by the symbol G (for "Gibbs' free energy," which limits the system being considered to the cell). G is equal to the energy contained in a molecule's chemical bonds (called *enthalpy* and designated **H**) minus the energy unavailable because of disorder (called *entropy* and symbolized as **S**) times the absolute temperature, **T,** in degrees Kelvin (K = °C + 273):

$$G = H - TS$$

Chemical reactions break some bonds in the reactants and form new bonds in the products. Consequently, reactions can produce changes in free energy. When a chemical reaction occurs under conditions of constant temperature, pressure, and volume—as do most biological reactions—the change Δ in free energy (ΔG) is simply:

$$\Delta G = \Delta H - T\Delta S$$

The change in free energy, or ΔG, is a fundamental property of chemical reactions. In some reactions, the ΔG is positive. This means that the products of the reaction contain *more* free energy than the reactants; the bond energy (H) is higher or the disorder (S) in the system is lower. Such reactions do not proceed spontaneously because they require an input of energy. Any reaction that requires an input of energy is said to be **endergonic** ("inward energy").

In other reactions, the ΔG is negative. The products of the reaction contain less free energy than the reactants; either the bond energy is lower or the disorder is higher, or both. Such reactions tend to proceed spontaneously. Any chemical reaction tends to proceed spontaneously if the difference in disorder (TΔS) is *greater* than the difference in bond energies between reactants and products (ΔH). Note that spontaneous does not mean the same thing as instantaneous. A spontaneous reaction may proceed very slowly. These reactions release the excess free energy as heat and are thus said to be **exergonic** ("outward energy"). Figure 8.6 sums up endergonic and exergonic reactions.

Free energy is the energy available to do work. Within cells, the change in free energy (ΔG) is the difference in bond energies between reactants and products (ΔH), minus any change in the degree of disorder of the system (TΔS). Any reaction whose products contain less free energy than the reactants (ΔG is negative) tends to proceed spontaneously.

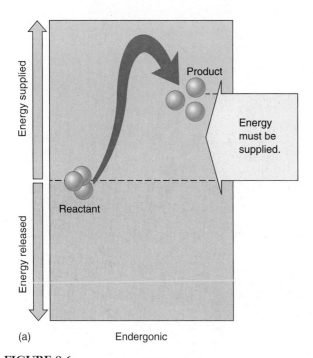

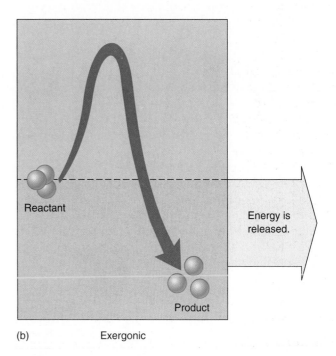

FIGURE 8.6
Energy in chemical reactions. (*a*) In an endergonic reaction, the products of the reaction contain more energy than the reactants, and the extra energy must be supplied for the reaction to proceed. (*b*) In an exergonic reaction, the products contain less energy than the reactants, and the excess energy is released.

Activation Energy

If all chemical reactions that release free energy tend to occur spontaneously, why haven't all such reactions already occurred? One reason is that most reactions require an input of energy to get started. Before it is possible to form new chemical bonds, even bonds that contain less energy, it is first necessary to break the existing bonds, and that takes energy. The extra energy required to destabilize existing chemical bonds and initiate a chemical reaction is called **activation energy** (figure 8.7*a*).

The rate of an exergonic reaction depends on the activation energy required for the reaction to begin. Reactions with larger activation energies tend to proceed more slowly because fewer molecules succeed in overcoming the initial energy hurdle. Activation energies are not constant, however. Stressing particular chemical bonds can make them easier to break. The process of influencing chemical bonds in a way that lowers the activation energy needed to initiate a reaction is called **catalysis,** and substances that accomplish this are known as catalysts (figure 8.7*b*).

Catalysts cannot violate the basic laws of thermodynamics; they cannot, for example, make an endergonic reaction proceed spontaneously. By reducing the activation energy, a catalyst accelerates both the forward and the reverse reactions by exactly the same amount. Hence, it does not alter the proportion of reactant ultimately converted into product.

To grasp this, imagine a bowling ball resting in a shallow depression on the side of a hill. Only a narrow rim of dirt below the ball prevents it from rolling down the hill. Now imagine digging away that rim of dirt. If you remove enough dirt from below the ball, it will start to roll down the hill—but removing dirt from below the ball will *never* cause the ball to roll up the hill. Removing the lip of dirt simply allows the ball to move freely; gravity determines the direction it then travels. Lowering the resistance to the ball's movement will promote the movement dictated by its position on the hill.

Similarly, the direction in which a chemical reaction proceeds is determined solely by the difference in free energy between reactants and products. Like digging away the soil below the bowling ball on the hill, catalysts reduce the energy barrier that is preventing the reaction from proceeding. Catalysts don't favor endergonic reactions any more than digging makes the hypothetical bowling ball roll uphill. Only exergonic reactions can proceed spontaneously, and catalysts cannot change that. What catalysts *can* do is make a reaction proceed much faster.

The rate of a reaction depends on the activation energy necessary to initiate it. Catalysts reduce the activation energy and so increase the rates of reactions, although they do not change the final proportions of reactants and products.

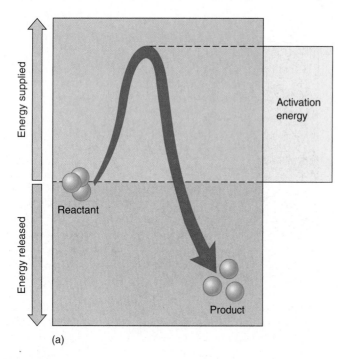

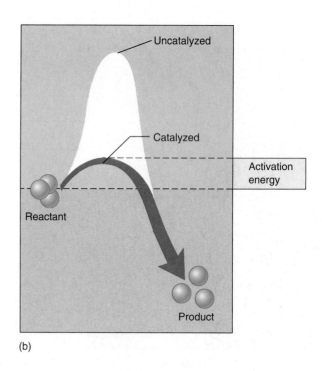

FIGURE 8.7
Activation energy and catalysis. (*a*) Exergonic reactions do not necessarily proceed rapidly because energy must be supplied to destabilize existing chemical bonds. This extra energy is the activation energy for the reaction. (*b*) Catalysts accelerate particular reactions by lowering the amount of activation energy required to initiate the reaction. Catalysts do not change the free energy of the reactants or products and therefore do not alter the free energy change produced by the reaction.

8.2 Enzymes are biological catalysts.

Enzymes

The chemical reactions within living organisms are regulated by controlling the points at which catalysis takes place. Life itself, therefore, is regulated by catalysts. The agents that carry out most of the catalysis in living organisms are proteins called **enzymes.** (There is increasing evidence that some types of biological catalysis are carried out by RNA molecules.) The unique three-dimensional shape of an enzyme enables it to stabilize a temporary association between **substrates,** the molecules that will undergo the reaction. By bringing two substrates together in the correct orientation, or by stressing particular chemical bonds of a substrate, an enzyme lowers the activation energy required for new bonds to form. The reaction thus proceeds much more quickly than it would without the enzyme. Because the enzyme itself is not changed or consumed in the reaction, only a small amount of an enzyme is needed, and it can be used over and over.

As an example of how an enzyme works, let's consider the reaction of carbon dioxide and water to form carbonic acid. This important enzyme-catalyzed reaction occurs in vertebrate red blood cells:

$$CO_2 + H_2O \rightleftharpoons H_2CO_3$$

carbon water   carbonic
dioxide   acid

This reaction may proceed in either direction, but because it has a large activation energy, the reaction is very slow in the absence of an enzyme: Perhaps 200 molecules of carbonic acid form in an hour in a cell. Reactions that proceed this slowly are of little use to a cell. Cells overcome this problem by employing an enzyme within their cytoplasm called *carbonic anhydrase* (enzyme names usually end in "–ase"). Under the same conditions, but in the presence of carbonic anhydrase, an estimated 600,000 molecules of carbonic acid form every *second!* Thus, the enzyme increases the reaction rate more than one million times. (See Box: Catalysis: A Closer Look at Carbonic Anhydrase.)

Thousands of different kinds of enzymes are known, each catalyzing one or a few specific chemical reactions. By facilitating particular chemical reactions, the enzymes in a cell determine the course of metabolism—the collection of all chemical reactions—in that cell. Different types of cells contain different sets of enzymes, and this difference contributes to structural and functional variations among cell types. For example, the chemical reactions taking place within a red blood cell differ from those that occur within a nerve cell, in part because the cytoplasm and membranes of red blood cells and nerve cells contain different arrays of enzymes.

Cells use proteins called enzymes as catalysts to lower activation energies.

Catalysis: A Closer Look at Carbonic Anhydrase

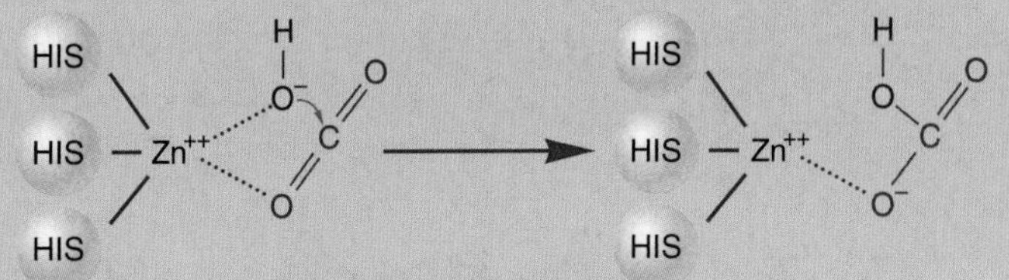

FIGURE 8.A

One of the most rapidly acting enzymes in the human body is carbonic anhydrase, which plays a key role in blood by converting dissolved carbon dioxide (CO_2) into carbonic acid, which dissociates into bicarbonate and hydrogen ions:

$$CO_2 + H_2O \rightarrow H_2CO_3 \rightarrow HCO_3^- + H^+$$

Fully 70% of the CO_2 transported by the blood is transported as bicarbonate ion. This reaction is exergonic and can proceed spontaneously, but its energy of activation is significant, so that little conversion to bicarbonate occurs spontaneously. In the presence of the enzyme carbonic anhydrase, however, the rate of the reaction accelerates by a factor of more than one million!

How does carbonic anhydrase catalyze this reaction so effectively? The active site of the enzyme is a deep cleft traversing the enzyme, as if it had been cut with the blade of an ax. Deep within the cleft, some 1.5 nanometers from the surface, are three histidines, their imidazole (nitrogen ring) groups all pointed at the same place in the center of the cleft. Together, they hold a zinc ion (Zn^{++}) firmly in position. This zinc ion will be the cutting blade of the catalytic process.

Here is how the zinc catalyzes the reaction: Immediately adjacent to the position of the zinc atom in the cleft are a group of amino acids that recognize and bind CO_2. When the CO_2 binds to this site, it interacts with the Zn^{++} in the cleft, orienting in the

plane of the cleft. Meanwhile, water bound to the zinc is rapidly converted to hydroxide ion. This hydroxide ion is now precisely positioned to attack the CO_2. When it does so, HCO_3^- is formed—and the enzyme is unchanged (figure 8.A).

Carbonic anhydrase is an effective catalyst because it brings its two substrates into close proximity and optimizes their orientation for reaction. Other enzymes use other mechanisms. Many, for example, use charged amino acids to polarize substrates or electronegative amino acids to stress particular bonds. Whatever the details of the reaction, however, the precise positioning of substrates achieved by the particular shape of the enzyme always plays a key role.

How Enzymes Work

Most enzymes are globular proteins with one or more pockets or clefts on their surface called **active sites** (figure 8.8). Substrates bind to the enzyme at these active sites, forming an **enzyme-substrate complex** (figure 8.9). For catalysis to occur within the complex, a substrate molecule must fit precisely into an active site. When that happens, amino acid side groups of the enzyme end up in close proximity to certain bonds of the substrate. These side groups interact chemically with the substrate, usually stressing or distorting a particular bond and consequently lowering the activation energy needed to break the bond. Once the bonds of the substrates are broken, or new bonds are formed, the substrates have been converted to products. These products then dissociate from the enzyme.

Proteins are not rigid. The binding of a substrate induces the enzyme to adjust its shape slightly, leading to a better *induced fit* between enzyme and substrate. This interaction may also facilitate the binding of other substrates; in such cases, one substrate "activates" the enzyme to receive other substrates.

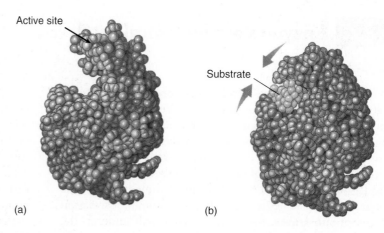

(a)   (b)

FIGURE 8.8

The active site of the enzyme lysozyme. (*a*) The active site, a groove running through lysozyme, fits the shape of the polysaccharide (a chain of sugars) that makes up bacterial cell walls. (*b*) When such a chain of sugars, indicated in yellow, slides into the groove, its entry induces the protein to alter its shape slightly and embrace the substrate more intimately. This induced fit positions a glutamic acid residue in the protein next to the bond between two adjacent sugars, and the glutamic acid "steals" an electron from the bond, causing it to break.

> Enzymes are specific in their choice of substrates. This specificity is due to the active site of the enzyme, which is shaped so that only a certain substrate molecule will fit into it.

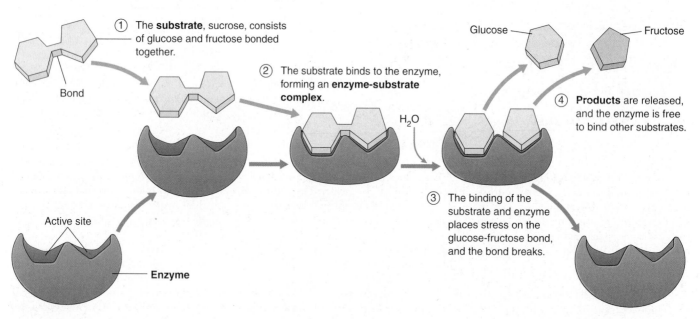

① The **substrate**, sucrose, consists of glucose and fructose bonded together.

Bond

Active site

Enzyme

② The substrate binds to the enzyme, forming an **enzyme-substrate complex.**

H_2O

③ The binding of the substrate and enzyme places stress on the glucose-fructose bond, and the bond breaks.

Glucose

Fructose

④ **Products** are released, and the enzyme is free to bind other substrates.

FIGURE 8.9

The catalytic cycle of an enzyme. Enzymes increase the speed with which chemical reactions occur, but they are not altered themselves as they do so. In the reaction illustrated here, the enzyme sucrase is splitting the sugar sucrose (present in most candy) into two simpler sugars: glucose and fructose. (*1*) First, the sucrose substrate binds to the active site of the enzyme, fitting into a depression in the enzyme surface. (*2*) The binding of sucrose to the active site forms an enzyme-substrate complex and induces the sucrase molecule to alter its shape, fitting more tightly around the sucrose. (*3*) Amino acid residues in the active site, now in close proximity to the bond between the glucose and fructose components of sucrose, break the bond. (*4*) The enzyme releases the resulting glucose and fructose fragments, the products of the reaction, and is then ready to bind another molecule of sucrose and run through the catalytic cycle once again. This cycle is often summarized by the equation: $\mathbf{E} + \mathbf{S} \leftrightarrow [\mathbf{ES}] \leftrightarrow \mathbf{E} + \mathbf{P}$, where E = enzyme, S = substrate, ES = enzyme-substrate complex, and P = products.

Enzymes Take Many Forms

While many enzymes are suspended in the cytoplasm of cells, not attached to any structure, other enzymes function as integral parts of cell membranes and organelles.

Multienzyme Complexes

Often in cells, the several enzymes catalyzing the different steps of a sequence of reactions are associated with one another in noncovalently bonded assemblies called *multienzyme complexes*. The bacterial pyruvate dehydrogenase multienzyme complex seen in figure 8.10 contains enzymes that carry out three sequential reactions in oxidative metabolism. Each complex has multiple copies of each of the three enzymes—60 protein subunits in all. The many subunits work in concert, like a tiny factory.

Multienzyme complexes offer the following significant advantages in catalytic efficiency:

1. The rate of any enzyme reaction is limited by the frequency with which the enzyme collides with its substrate. If a series of sequential reactions occurs within a multienzyme complex, the product of one reaction can be delivered to the next enzyme without releasing it to diffuse away.
2. Because the reacting substrate never leaves the complex during its passage through the series of reactions, the possibility of unwanted side reactions is eliminated.
3. All of the reactions that take place within the multienzyme complex can be controlled as a unit.

In addition to pyruvate dehydrogenase, which controls entry to the Krebs cycle (see chapter 9), several other key processes in the cell are catalyzed by multienzyme complexes. One well-studied system is the fatty acid synthetase complex that catalyzes the synthesis of fatty acids from two-carbon precursors. There are seven different enzymes in this multienzyme complex, and the reaction intermediates remain associated with the complex for the entire series of reactions.

Not All Biological Catalysts Are Proteins

Until a few years ago, most biology textbooks contained statements such as "Proteins called enzymes are the catalysts of biological systems." We can no longer make that statement without qualification. As discussed in chapter 4, Tom Cech and his colleagues at the University of Col-

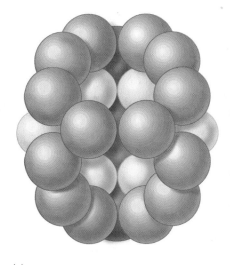

(a)

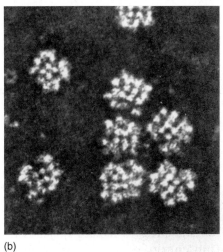

(b)

FIGURE 8.10
The enzyme pyruvate dehydrogenase. The enzyme (model, *a*) that carries out the oxidation of pyruvate is one of the most complex enzymes known. It has 60 protein subunits, many of which can be seen in (*b*) the electron micrograph (200,000×).

orado reported in 1981 that certain reactions involving RNA molecules appear to be catalyzed in cells by RNA itself, rather than by enzymes. This initial observation has been corroborated by additional examples of RNA catalysis in the last few years. Like enzymes, these RNA catalysts, which are loosely called "ribozymes," greatly accelerate the rate of particular biochemical reactions and show extraordinary specificity with respect to the substrates on which they act.

There appear to be at least two sorts of ribozymes. Some ribozymes have folded structures and catalyze reactions on themselves, a process called *intra*molecular catalysis. Other ribozymes act on other molecules without themselves being changed, a process called *inter*molecular catalysis. Many important cellular reactions involve small RNA molecules, including reactions that remove unnecessary sections from RNA copies of genes, that prepare ribosomes for protein synthesis, and that facilitate the replication of DNA within mitochondria. In all of these cases, the possibility of RNA catalysis is being actively investigated. It seems likely, particularly in the complex process of photosynthesis, that both enzymes and RNA play important catalytic roles.

The ability of RNA, an informational molecule, to act as a catalyst has stirred great excitement among biologists because it appears to provide a potential answer to the question posed in chapter 4: Which came first, the protein or the nucleic acid? It now seems at least possible that RNA may have evolved first and catalyzed the formation of the first proteins.

Not all biological catalysts float freely in the cytoplasm. Some are part of other structures, and others are not even proteins.

Factors Affecting Enzyme Activity

The rate of an enzyme-catalyzed reaction is affected by the concentrations of both the substrate and the enzyme that works on it. In addition, any chemical or physical factor that alters the enzyme's three-dimensional shape—such as temperature, pH, salt concentration, and the binding of specific regulatory molecules—can affect the enzyme's ability to catalyze the reaction.

Temperature

Increasing the temperature of an uncatalyzed reaction will increase its rate because the additional heat represents an increase in random molecular movement. This can add stress to molecular bonds and affect the activation energy of a reaction. The rate of an enzyme-catalyzed reaction also increases with temperature, but only up to a point called the *optimum temperature* (figure 8.11*a*). Below this temperature, the hydrogen bonds and hydrophobic interactions that determine the enzyme's shape are not flexible enough to permit the induced fit that is optimum for catalysis. Above the optimum temperature, these forces are too weak to maintain the enzyme's shape against the increased random movement of the atoms in the enzyme. At these higher temperatures, the enzyme denatures, as we described in chapter 3. Most human enzymes have temperature optima between 35°C and 40°C, a range that includes normal body temperature. Prokaryotes that live in hot springs have more stable enzymes (that is, enzymes held together more strongly), so the temperature optima for those enzymes can be 70°C or higher.

pH

Ionic interactions between oppositely charged amino acid residues, such as glutamic acid (−) and lysine (+), also hold enzymes together. These interactions are sensitive to the hydrogen ion concentration of the fluid the enzyme is dissolved in, because changing that concentration shifts the balance between positively and negatively charged amino acid residues. For this reason, most enzymes have an *optimum pH* that usually ranges from pH 6 to 8. Those enzymes able to function in very acidic environments are proteins that maintain their three-dimensional shape even in the presence of high levels of hydrogen ion. The enzyme pepsin, for example, digests proteins in the stomach at pH 2, a very acidic level (figure 8.11*b*).

Inhibitors and Activators

Enzyme activity is sensitive to the presence of specific substances that bind to the enzyme and cause changes in its shape. Through these substances, a cell is able to regulate which of its enzymes are active and which are inactive at a

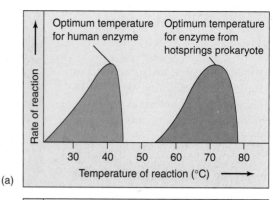

(a)

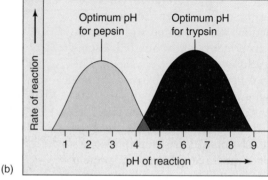

(b)

FIGURE 8.11
Enzyme sensitivity to the environment. The activity of an enzyme is influenced by both (*a*) temperature and (*b*) pH. Most human enzymes, such as the protein-degrading enzyme trypsin, work best at temperatures of about 40°C and within a pH range of 6 to 8.

particular time. This allows the cell to increase its efficiency and to control changes in its characteristics during development. A substance that binds to an enzyme and *decreases* its activity is called an **inhibitor.** Very often, the end product of a biochemical pathway acts as an inhibitor of an early reaction in the pathway, a process called *feedback inhibition* (to be discussed later).

Enzyme inhibition occurs in two ways: **Competitive inhibitors** compete with the substrate for the same active site, displacing a percentage of substrate molecules from the enzymes; **noncompetitive inhibitors** bind to the enzyme in a location other than the active site, changing the shape of the enzyme and making it unable to bind to the substrate (figure 8.12). Most noncompetitive inhibitors bind to a specific portion of the enzyme called an **allosteric site** (Greek *allos,* "other," + *steros,* "form"). These sites serve as chemical on/off switches; the binding of a substance to the site can switch the enzyme between its active and inactive configurations. A substance that binds to an allosteric site and reduces enzyme activity is called an **allosteric inhibitor** (figure 8.12*b*). Alternatively, **activators** bind to allosteric sites and keep the enzymes in their active configurations, thereby *increasing* enzyme activity.

Enzyme Cofactors

Enzyme function is often assisted by additional chemical components known as **cofactors.** For example, the active sites of many enzymes contain metal ions that help draw electrons away from substrate molecules. Zinc is used by some enzymes, such as protein-digesting carboxypeptidase, to draw electrons away from their position in covalent bonds, making the bonds less stable and easier to break. Other elements, such as molybdenum and manganese, are also used as cofactors. Like zinc, these substances are required in the diet in small amounts. When the cofactor is a nonprotein organic molecule, it is called a **coenzyme.** Many vitamins are parts of coenzymes.

In numerous oxidation-reduction reactions that are catalyzed by enzymes, the electrons pass in pairs from the active site of the enzyme to a coenzyme that serves as the electron acceptor. The coenzyme then transfers the electrons to a different enzyme, which releases them (and the energy they bear) to the substrates in another reaction. Often, the electrons combine with protons (H^+) as hydrogen atoms. In this way, coenzymes shuttle energy in the form of hydrogen atoms from one enzyme to another in a cell.

One of the most important coenzymes is the hydrogen acceptor **nicotinamide adenine dinucleotide (NAD^+)** (figure 8.13). The NAD^+ molecule is composed of two nucleotides bound together. As you may recall from chapter 3, a nucleotide is a five-carbon sugar with one or more phosphate groups attached to one end and an organic base attached to the other end. The two nucleotides that make up NAD^+, nicotinamide monophosphate (NMP) and adenine monophosphate (AMP), are joined head-to-head by their phosphate groups. The two nucleotides serve different functions in the NAD^+ molecule: AMP acts as the core, providing a shape recognized by many enzymes; NMP is the active part of the molecule, contributing a site that is readily reduced (that is, easily accepts electrons).

When NAD^+ acquires an electron and a hydrogen atom (actually, two electrons and a proton) from the active site of an enzyme, it is reduced to NADH. The NADH molecule now carries the two energetic electrons and the proton, and can supply them to other molecules and reduce them. The oxidation of energy-containing molecules, which provides energy to cells, involves stripping electrons from those molecules and donating them to NAD^+. As we'll see, much of the energy of NADH is transferred to other molecules.

Enzymes have an optimum temperature and pH, at which the enzyme functions most effectively. Inhibitors decrease enzyme activity, while activators increase it. The activity of enzymes is often facilitated by cofactors, which can be metal ions or other substances. Cofactors that are nonprotein organic molecules are called coenzymes.

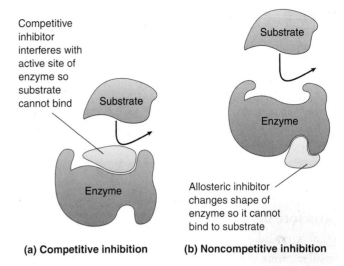

(a) Competitive inhibition   **(b) Noncompetitive inhibition**

FIGURE 8.12
How enzymes can be inhibited. (*a*) In competitive inhibition, the inhibitor interferes with the active site of the enzyme. (*b*) In noncompetitive inhibition, the inhibitor binds to the enzyme at a place away from the active site, effecting a conformational change in the enzyme so that it can no longer bind to its substrate.

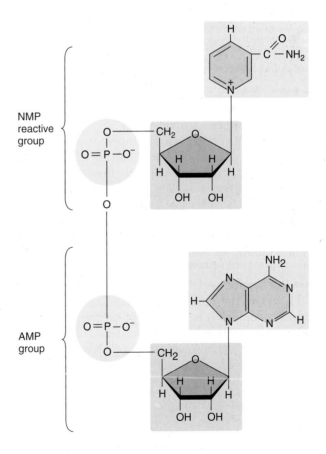

FIGURE 8.13
The chemical structure of nicotinamide adenine dinucleotide (NAD^+). This key coenzyme is composed of two nucleotides, NMP and AMP, bound together.

8.3 ATP is the energy currency of life.

ATP

The chief energy currency all cells use is a molecule called **adenosine triphosphate (ATP).** Cells use their supply of ATP to power almost every energy-requiring process they carry out, from making sugars, to supplying activation energy for chemical reactions, to actively transporting substances across membranes, to moving through their environment and growing.

Structure of the ATP Molecule

Each ATP molecule is a nucleotide composed of three smaller components (figure 8.14). The first component is a five-carbon sugar, ribose, which serves as the backbone to which the other two subunits are attached. The second component is adenine, an organic molecule composed of two carbon-nitrogen rings. Each of the nitrogen atoms in the ring has an unshared pair of electrons and weakly attracts hydrogen ions. Adenine, therefore, acts chemically as a base and is usually referred to as a nitrogenous base (it is one of the four nitrogenous bases found in DNA and RNA). The third component of ATP is a triphosphate group (a chain of three phosphates).

How ATP Stores Energy

The key to how ATP stores energy lies in its triphosphate group. Phosphate groups are highly negatively charged, so they repel one another strongly. Because of this electrostatic repulsion, the covalent bonds joining the phosphates are unstable. The molecule is often referred to as a "coiled spring," the phosphates straining away from one another.

The unstable bonds holding the phosphates together in the ATP molecule have a low activation energy and are easily broken. When they break, they can transfer a considerable amount of energy. In most reactions involving ATP, only the outermost high-energy phosphate bond is hydrolyzed, cleaving off the phosphate group on the end. When this happens, ATP becomes **adenosine diphosphate (ADP),** and energy equal to 7.3 kcal/mole is released under standard conditions. The liberated phosphate group usually attaches temporarily to some intermediate molecule. When that molecule is dephosphorylated, the phosphate group is released as inorganic phosphate (P_i).

How ATP Powers Energy-Requiring Reactions

Cells use ATP to drive endergonic reactions. Such reactions do not proceed spontaneously, because their products possess more free energy than their reactants. However, if the cleavage of ATP's terminal high-energy bond releases more energy than the other reaction consumes, the two re-

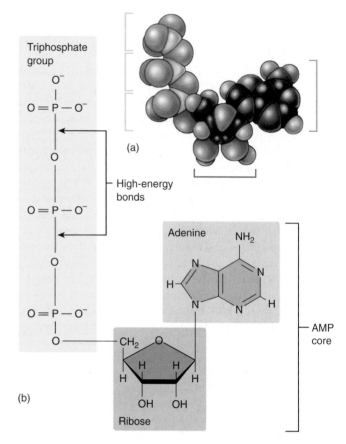

FIGURE 8.14
The ATP molecule. (*a*) The model and (*b*) the structural diagram both show that, like NAD^+, ATP has a core of AMP. In ATP, the reactive group added to the end of the AMP phosphate group is not another nucleotide but rather a chain of two phosphate groups.

actions can be coupled so that the energy released by the hydrolysis of ATP can be used to supply the second endergonic reaction with the energy it needs. Together, these reactions have a net release of energy and are therefore exergonic and will proceed spontaneously. Because almost all endergonic reactions require less energy than is released by the cleavage of ATP, ATP can provide most of the energy a cell needs.

The same feature that makes ATP an effective energy donor—the instability of its phosphate bonds—precludes it from being a good long-term energy storage molecule. Fats and carbohydrates serve that function better. Most cells do not maintain large stockpiles of ATP. Instead, they typically have only a few seconds' supply of ATP at any given time, and they continually produce more from ADP and P_i.

The instability of its phosphate bonds make ATP an excellent energy donor.

Metabolism is the chemical life of a cell.

Biochemical Pathways: The Organizational Units of Metabolism

Living chemistry, the total of all chemical reactions carried out by an organism, is called **metabolism** (Greek *metabole*, "change"). Those reactions that expend energy to make or transform chemical bonds are called *anabolic* reactions, or **anabolism.** Reactions that harvest energy when chemical bonds are broken are called *catabolic* reactions, or **catabolism.**

Organisms contain thousands of different kinds of enzymes that catalyze a bewildering variety of reactions. Many of these reactions in a cell occur in sequences called **biochemical pathways.** In such pathways, the product of one reaction becomes the substrate for the next (figure 8.15). Biochemical pathways are the organizational units of metabolism, the elements an organism controls to achieve coherent metabolic activity. Most sequential enzyme steps in biochemical pathways take place in specific compartments of the cell; for example, the steps of the krebs cycle (chapter 9), occur inside mitochondria. By determining where many of the enzymes that catalyze these steps are located, we can "map out" a model of metabolic processes in the cell.

How Biochemical Pathways Evolved

In the earliest cells, the first biochemical processes probably involved energy-rich molecules scavenged from the environment. Most of the molecules necessary for these processes are thought to have existed in the "organic soup" of the early oceans. The first catalyzed reactions were probably simple, one-step reactions that brought these molecules together in various combinations. Eventually, the energy-rich molecules became depleted in the external environment, and only organisms that had evolved some means of making those molecules from other substances in the environment could survive. Thus, a hypothetical reaction,

$$\begin{matrix} F \\ + \rightarrow H \\ G \end{matrix}$$

where two energy-rich molecules (F and G) react to produce compound H and release energy, became more complex when the supply of F in the environment ran out. A new reaction was added in which the depleted molecule, F, is made from another molecule, E, which was also present in the environment:

$$\begin{matrix} E \rightarrow F \\ + \rightarrow H \\ G \end{matrix}$$

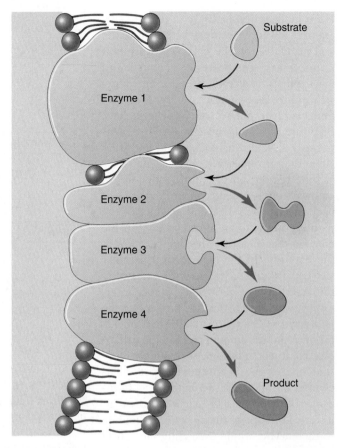

FIGURE 8.15
A biochemical pathway. The original substrate is acted on by enzyme 1, changing the substrate to a new form recognized by enzyme 2. Each enzyme in the pathway acts on the product of the previous stage.

When the supply of E in turn became depleted, organisms that were able to make it from some other available precursor, D, survived. When D became depleted, those organisms in turn were replaced by ones able to synthesize D from another molecule, C:

$$\begin{matrix} C \rightarrow D \rightarrow E \rightarrow F \\ + \rightarrow H \\ G \end{matrix}$$

This hypothetical biochemical pathway would have evolved slowly through time, with the final reactions in the pathway evolving first and earlier reactions evolving later. Looking at the pathway now, we would say that the organism, starting with compound C, is able to synthesize H by means of a series of steps. This is how the biochemical pathways within organisms are thought to have evolved—not all at once, but one step at a time, backward.

How Biochemical Pathways Are Regulated

For a biochemical pathway to operate efficiently, its activity must be coordinated and regulated by the cell. Not only is it unnecessary to synthesize a compound when plenty is already present, but doing so would waste energy and raw materials that could be put to use elsewhere. It is, therefore, advantageous for a cell to temporarily shut down biochemical pathways when their products are not needed.

The regulation of simple biochemical pathways often depends on an elegant feedback mechanism: The end product of the pathway binds to an allosteric site on the enzyme that catalyzes the first reaction in the pathway. In the hypothetical pathway we just described, the enzyme catalyzing the reaction C → D would possess an allosteric site for H, the end product of the pathway. As the pathway churned out its product and the amount of H in the cell increased, it would become increasingly likely that one of the H molecules would encounter the allosteric site on the C → D enzyme. If the product H functioned as an allosteric inhibitor of the enzyme, its binding to the enzyme would essentially shut down the reaction C → D. Shutting down this reaction, the first reaction in the pathway, effectively shuts down the whole pathway. Hence, as the cell produces increasing quantities of the product H, it automatically inhibits its ability to produce more. This mode of regulation is called **feedback inhibition** (figure 8.16).

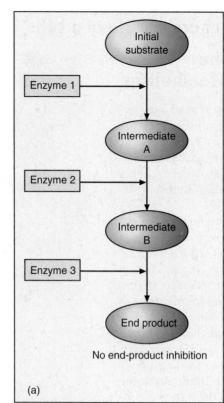 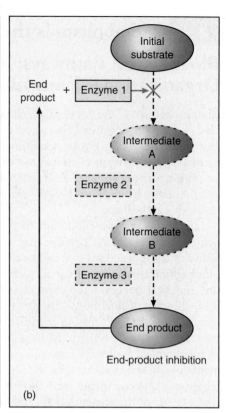

FIGURE 8.16

Feedback inhibition. (*a*) A biochemical pathway with no feedback inhibition. (*b*) A biochemical pathway in which the final end product becomes the allosteric inhibitor for the first enzyme in the pathway. In other words, the formation of the pathway's final end product stops the pathway.

A biochemical pathway is an organized series of reactions, often regulated as a unit.

A Vocabulary of Metabolism

activation energy   The energy required to destabilize chemical bonds and to initiate a chemical reaction.

catalysis   Acceleration of the rate of a chemical reaction by lowering the activation energy.

coenzyme   A nonprotein organic molecule that plays an accessory role in enzyme-catalyzed reactions, often by acting as a donor or acceptor of electrons. NAD^+ is a coenzyme.

endergonic reaction   A chemical reaction to which energy from an outside source must be added before the reaction proceeds; the opposite of an exergonic reaction.

entropy   A measure of the randomness or disorder of a system. In cells, it is a measure of how much energy has become so dispersed (usually as evenly distributed heat) that it is no longer available to do work.

exergonic reaction   An energy-yielding chemical reaction. Exergonic reactions tend to proceed spontaneously, although activation energy is required to initiate them.

free energy   Energy available to do work.

kilocalorie   1000 calories. A calorie is the heat required to raise the temperature of 1 gram of water by 1°C.

metabolism   The sum of all chemical processes occurring within a living cell or organism.

oxidation   The loss of an electron by an atom or molecule. It occurs simultaneously with reduction of some other atom or molecule because an electron that is lost by one is gained by another.

reduction   The gain of an electron by an atom or molecule. Oxidation-reduction reactions are an important means of energy transfer within living systems.

substrate   A molecule on which an enzyme acts; the initial reactant in an enzyme-catalyzed reaction.

Concept Review

For interactive testing, visit the Online Learning Center with PowerWeb at www.mhhe.com/Raven7

8.1 **The laws of thermodynamics describe how energy changes.**

The Flow of Energy in Living Things

- Energy is the capacity to do work. Kinetic energy is energy of motion, while potential energy is stored energy. (p. 144)
- Oxidation is the process whereby an atom or molecule loses an electron, and reduction is the process whereby an atom or molecule gains an electron. Oxidation-reduction reactions always take place together. (p. 145)

The Laws of Thermodynamics

- The First Law of Thermodynamics states that energy cannot be created or destroyed; it can only change from one form to another. (p. 146)
- During each energy conversion, some energy dissipates into the environment as heat. (p. 146)
- The Second Law of Thermodynamics states that disorder (entropy) in the universe is continuously increasing. (p. 146)
- Entropy is always increasing because, as more energy is used, more energy is converted to heat. (p. 146)

Free Energy

- The amount of energy available to break and form chemical bonds is referred to as a molecule's free energy. (p. 147)
- Reactions in which the products contain less free energy than the reactants tend to proceed spontaneously. (p. 147)

Activation Energy

- Activation energy is the extra energy required to destabilize existing chemical bonds. (p. 148)
- Catalysts increase the rate of reaction by lowering the activation energy, but do not alter the proportions of the reactants or products. (p. 148)

8.2 **Enzymes are biological catalysts.**

Enzymes

- Cells use proteins (enzymes) to lower activation energies. Thousands of different enzymes are known, each catalyzing one or a few specific chemical reactions. (p. 149)

How Enzymes Work

- Most enzymes are globular proteins with one or more active sites where specific substrates can bind, forming an enzyme-substrate complex. (p. 150)

Enzymes Take Many Forms

- Multienzyme complexes are noncovalently bonded assemblies of enzymes that catalyze different steps of a reaction sequence, and can offer significant advantages in catalytic efficiency. (p. 151)
- Not all biological catalysts are proteins; for example, certain reactions can be catalyzed by ribosomes. (p. 151)

Factors Affecting Enzyme Activity

- Several factors can affect the activity of an enzyme, including temperature, pH, inhibitors and activators, and enzyme cofactors. (pp. 152–153)

8.3 **ATP is the energy currency of life.**

ATP

- The chief energy currency all cells use is a molecule called adenosine triphosphate (ATP). (p. 154)
- Because of electrostatic repulsion, the covalent bonds between phosphates are unstable; thus, ATP is an excellent energy donor. (p. 154)

8.4 **Metabolism is the chemical life of a cell.**

Biochemical Pathways: The Organizational Units of Metabolism

- Metabolism refers to the total of all chemical reactions carried out by an organism. (p. 155)
- Biochemical pathways occur when the product of one reaction becomes the substrate for the next reaction. (p. 155)
- To operate efficiently, a biochemical pathway must be coordinated and regulated by the cell, and regulation often takes place as feedback inhibition. (p. 156)

Self Test

1. An atom gains energy when
   a. an electron is lost from it.
   b. it undergoes oxidation.
   c. it undergoes reduction.
   d. it undergoes an oxidation-reduction reaction.

2. Which of the following is concerned with the amount of energy in the universe?
   a. the First Law of Thermodynamics
   b. the Second Law of Thermodynamics
   c. thermodynamics
   d. entropy

3. In a chemical reaction, if ΔG is negative, it means that
   a. the products contain more free energy than the reactants.
   b. an input of energy is required to break the bonds.
   c. the reaction will proceed spontaneously.
   d. the reaction is endergonic.

4. A catalyst
   a. allows an endergonic reaction to proceed more quickly.
   b. increases the activation energy so that a reaction can proceed more quickly.
   c. lowers the amount of energy needed for a reaction to proceed.
   d. is required for an exergonic reaction to occur.

5. Which of the following statements about enzymes is false?
   a. Enzymes are catalysts within cells.
   b. All the cells of an organism contain the same enzymes.
   c. Enzymes bring substances together so that they undergo a reaction.
   d. Enzymes lower the activation energy of spontaneous reactions in the cell.

6. A multienzyme complex contains
   a. many copies of just one enzyme.
   b. one enzyme and its substrate.
   c. enzymes that catalyze a series of reactions.
   d. side reactions on a substrate.

7. Which of the following has *no* effect on the rate of enzyme-catalyzed reactions?
   a. temperature
   b. pH
   c. concentration of substrate
   d. none of these

8. How is ATP used in the cell to produce cellular energy?
   a. ATP provides energy to drive exergonic reactions.
   b. ATP hydrolysis is coupled to endergonic reactions.
   c. A liberated phosphate group attaches to another molecule, which generates energy.
   d. ATP stores energy by the repulsion of the negatively charged phosphates.

9. Anabolic reactions are reactions that
   a. break chemical bonds.
   b. make chemical bonds.
   c. harvest energy.
   d. occur in a sequence.

10. How is a biochemical pathway regulated?
    a. The product of one reaction becomes the substrate for the next.
    b. The end product replaces the initial substrate in the pathway.
    c. The end product inhibits the first enzyme in the pathway by binding to an allosteric site.
    d. All of these are correct.

Test Your Visual Understanding

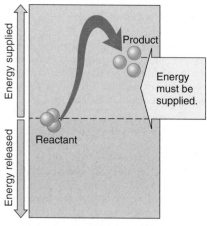

Panel a

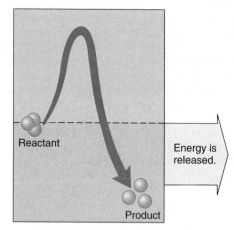

Panel b

1. In which of these two reactions would the change in free energy, ΔG, be positive? Would the product of this reaction have more or less free energy compared to the reactant? Name this type of reaction.

2. In which of these two reactions could a catalyst help speed the reaction? How would the catalyst speed the reaction? Name this type of reaction.

Apply Your Knowledge

1. In a biochemical pathway, three ATP molecules are hydrolyzed. The endergonic reactions in the pathway require a total of 17.3 kcal/mole of energy to drive the reactions of the pathway. What is the overall change in free energy of the biochemical pathway? Is the overall pathway endergonic or exergonic?

2. Oxidation-reduction reactions can involve a wide variety of molecules. Why are those involving hydrogen and oxygen of paramount importance in biological systems?

3. Almost no sunlight penetrates the deep ocean. However, many fish that live there attract prey and potential mates by producing their own light. Where does that light come from? Does its generation require energy?

9
How Cells Harvest Energy

Concept Outline

9.1 Cells harvest the energy in chemical bonds.

Using Chemical Energy to Drive Metabolism. The energy in C—H, C—O, and other chemical bonds can be captured and used to fuel the synthesis of ATP.

9.2 Cellular respiration oxidizes food molecules.

An Overview of Glucose Catabolism. The chemical energy in sugar is harvested by both substrate-level phosphorylation and aerobic respiration.

Stage One: Glycolysis. The 10 reactions of glycolysis capture energy from glucose by reshuffling the bonds.

Stage Two: The Oxidation of Pyruvate. Pyruvate, the product of glycolysis, is oxidized to acetyl-CoA.

Stage Three: The Krebs Cycle. In a series of reactions, electrons are stripped from acetyl-CoA.

Harvesting Energy by Extracting Electrons. A series of oxidation-reduction reactions strip electrons from glucose and use the energy of these electrons to power the synthesis of ATP.

Stage Four: The Electron Transport Chain. The electrons harvested from glucose pass through a chain of membrane proteins that use the energy to pump protons, driving the synthesis of ATP by ATP synthase.

Summarizing the Yield of Aerobic Respiration. The oxidation of glucose by aerobic respiration in eukaryotes produces nearly three dozen ATP molecules.

Regulating Aerobic Respiration. High levels of ATP tend to shut down cellular respiration by inhibiting key reactions.

9.3 Catabolism of proteins and fats can yield considerable energy.

Glucose Is Not the Only Source of Energy. Proteins and fats are dismantled and the products fed into cellular respiration.

9.4 Cells can metabolize food without oxygen.

Fermentation. Fermentation allows continued metabolism in the absence of oxygen by donating the electrons harvested in glycolysis to organic molecules.

9.5 The stages of cellular respiration evolved over time.

The Evolution of Metabolism. The major metabolic processes evolved over a long period, building on what had come before.

FIGURE 9.1
Harvesting chemical energy. Organisms such as these harvest mice depend on the energy stored in the chemical bonds of the food they eat to power their life processes.

Life is driven by energy. All the activities organisms carry out—the swimming of bacteria, the purring of a cat, your thinking about these words—use energy. In this chapter, we discuss the processes all cells use to derive chemical energy from organic molecules and to convert that energy to ATP. Then, in chapter 10, we will examine photosynthesis, which uses light energy rather than chemical energy. The reason we consider the conversion of chemical energy to ATP first is that all organisms, both photosynthesizers and the organisms that feed on them (including the field mice in figure 9.1), are capable of harvesting energy from chemical bonds. However, as you will see, this process and photosynthesis have much in common.

159

Cells harvest the energy in chemical bonds.

Using Chemical Energy to Drive Metabolism

Plants, algae, and some bacteria harvest the energy of sunlight through photosynthesis, converting radiant energy into chemical energy. These organisms, along with a few others that use chemical energy in a similar way, are called **autotrophs** ("self-feeders"). All other organisms live on the energy autotrophs produce, using them as food, and are called **heterotrophs** ("fed by others"). At least 95% of the kinds of organisms on earth—all animals and fungi, and most protists and prokaryotes—are heterotrophs.

Where is the chemical energy in food, and how do heterotrophs harvest it to carry out the many tasks of living? Most foods contain a variety of carbohydrates, proteins, and fats, all rich in energy-laden chemical bonds. Carbohydrates and fats, for example, possess many carbon-hydrogen (C—H), as well as carbon-oxygen (C—O) bonds. The job of extracting energy from this complex organic mixture is tackled in stages. First, enzymes break down the large molecules into smaller ones, a process called **digestion.** Then, other enzymes dismantle these fragments a little at a time, harvesting energy from C—H and other chemical bonds at each stage. This process is called **catabolism.**

Cellular Respiration

The energy of a chemical bond is contained in the potential energy of the electrons that make up the bond. Cells harvest this energy by breaking the bonds and shifting the electrons from one molecule to the next. During each transfer, the electrons lose some of their energy. Some of this energy may be captured and used to make ATP or form other chemical bonds; the rest is lost as heat. At the end of this process, the high-energy electrons from the initial chemical bonds have lost much of their energy, and these depleted electrons are transferred to a final electron acceptor. When this acceptor molecule is oxygen, the process is called **aerobic respiration.** When the final electron acceptor is an inorganic molecule other than oxygen, the process is called **anaerobic respiration,** and when it is an organic molecule, the process is called **fermentation.**

Chemically, there is little difference between the catabolism of carbohydrates in a cell and the burning of wood in a fireplace. In both instances, the reactants are carbohydrates and oxygen, and the products are carbon dioxide, water, and energy:

$$C_6H_{12}O_6 + 6\ O_2 \rightarrow 6\ CO_2 + 6\ H_2O + \text{energy (heat or ATP)}$$

The change in free energy in this reaction is −720 kilocalories (−3012 kilojoules) per mole of glucose under the conditions found within a cell. (The traditional value of −686 kilocalories, or −2870 kJ, per mole refers to standard conditions—room temperature, one atmosphere of pressure, etc.) This change in free energy results largely from the breaking of the six C—H bonds in the glucose molecule. The negative sign indicates that the products possess *less* free energy than the reactants. The same amount of energy is released whether glucose is catabolized or burned, but when it is burned, most of the energy is released as heat. This heat cannot be used to perform work in cells. The key to a cell's ability to harvest useful energy through the catabolism of food molecules such as glucose is its conversion of a portion of the energy into a more useful form. Cells do this by using some of the energy to drive the production of ATP, a molecule that can power cellular activities.

The ATP Molecule

Adenosine triphosphate (ATP) is the energy currency of the cell, the molecule that transfers the energy captured during cellular respiration to the many sites that use energy in the cell. How is ATP able to transfer energy so readily? Recall from chapter 8 that ATP is composed of a sugar (ribose) bound to an organic base (adenine) and a chain of three phosphates (a triphosphate group). As shown in figure 9.2, each phosphate group is negatively charged. Because like charges repel each other, the linked phosphate groups push against each other and stress the bond that holds them together. The linked phosphates store the energy of their electrostatic repulsion in the bonds that hold the phosphates together. Transferring a phosphate group to another molecule relaxes the electrostatic spring of ATP, at the same time cocking the spring of the molecule that is phosphorylated. This molecule can then use the energy to undergo some change that requires work.

How Cells Use ATP

Cells use ATP to do most of those activities that require work. One of the most obvious is movement. Tiny fibers within muscle cells pull against one another when muscles contract. Mitochondria move a meter or more along the narrow nerve cells that connect your feet with your spine. Chromosomes are pulled by microtubules during cell division. All of these movements by cells require the expenditure of ATP energy.

A second major way cells use ATP is to drive endergonic reactions. Many of the synthetic activities of the cell are endergonic, because building molecules takes energy. The chemical bonds of the products of these reactions contain more energy, or are more organized, than the reactants. The reaction can't proceed until that extra energy is supplied to the reaction. It is ATP that provides this needed energy.

How ATP Drives Endergonic Reactions

How does ATP drive an endergonic reaction? The enzyme that catalyzes the endergonic reaction has two binding sites on its surface, one for the reactant and another for ATP. The ATP site splits the ATP molecule, liberating over 7 kilocalories (30 kJ) of chemical energy. This energy pushes the reactant at the second site "uphill," driving the endergonic reaction.

When the splitting of ATP molecules drives an energy-requiring reaction in a cell, the two parts of the reaction—ATP-splitting and endergonic—take place in concert. In some cases, the two parts both occur on the surface of the same enzyme; they are physically linked, or "coupled." In other cases, a high-energy phosphate from ATP attaches to the protein catalyzing the endergonic process, activating it. Coupling energy-requiring reactions to the splitting of ATP in this way is one of the key tools cells use to manage energy.

ATP Synthase: A Micro-motor Responsible for the Production of Most ATP

Because of its ability to supply energy for such a broad range of metabolic reactions, ATP is in very high demand in cells and must be synthesized in very large quantities. Although the enzymes that break down fats and sugars help produce a few molecules of ATP directly from the energy supplied by the high-energy bonds of the substrates, the vast majority of ATP produced in the cell is made by the enzyme **ATP synthase,** one of the most important enzymes in all living systems.

ATP synthase catalyzes the synthesis of ATP by using the energy stored in a gradient of protons that is built across membranes, either the plasma membrane of prokaryotes or the inner membranes of mitochondria or chloroplasts. This proton gradient is produced by pumping protons across the membrane using the energy from a series of redox reactions. The electrons driving these reactions are extracted from the high-energy molecules broken down during catabolism, or are energized by light striking chlorophyll in photosynthesis.

ATP synthase uses a fascinating molecular mechanism to perform ATP synthesis (figure 9.3). The enzyme is embedded in the membrane and provides a channel through which protons can move across the membrane down their concentration gradient. As they do so, the energy they release causes components of the enzyme complex to rotate. The mechanical energy of this rotation is then converted to the chemical bond that holds the third phosphate on ATP. Thus, the synthesis of ATP is achieved by a tiny rotary motor whose rotation is driven directly by a gradient of protons. Much of the metabolic machinery a cell uses to break down glucose or other energy-rich molecules is devoted to harvesting the high-energy electrons that supply the power needed to pump protons and create this gradient.

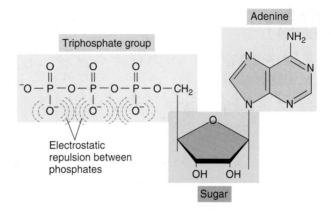

FIGURE 9.2
Structure of the ATP molecule.

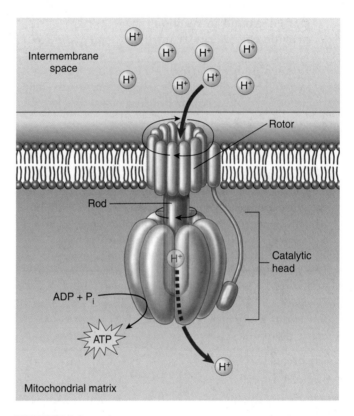

FIGURE 9.3
ATP synthase at work. Protons move across the membrane down their concentration gradient. The energy released causes the rotor and rod structures to rotate. This mechanical energy is converted to chemical energy with the formation of ATP.

The catabolism of glucose into carbon dioxide and water in living organisms releases about 720 kilocalories (3012 kJ) of energy per mole of glucose. This energy is captured in ATP, which stores the energy by linking charged phosphate groups near one another. When the phosphate bonds in ATP are hydrolyzed, energy is released and available to do work.

9.2 Cellular respiration oxidizes food molecules.

An Overview of Glucose Catabolism

Cells are able to make ATP from the catabolism of organic molecules in two different ways.

1. In **substrate-level phosphorylation**, ATP is formed by transferring a phosphate group directly to ADP from a phosphate-bearing intermediate (figure 9.4). During **glycolysis**, discussed below, the chemical bonds of glucose are shifted around in reactions that provide the energy required to form ATP.

2. In **aerobic respiration**, ATP is synthesized by ATP synthase powered by a proton gradient formed from electrons harvested from organic molecules. These electrons, with their energy depleted, are then donated to oxygen gas. Eukaryotes and aerobic prokaryotes produce the vast majority of their ATP this way.

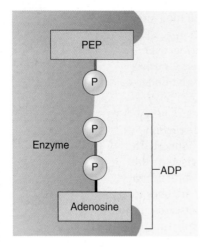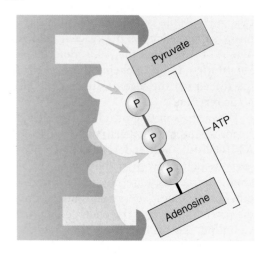

FIGURE 9.4

Substrate-level phosphorylation. Some molecules, such as phosphoenolpyruvate (PEP), possess a high-energy phosphate bond similar to the bonds in ATP. When PEP's phosphate group is transferred enzymatically to ADP, the energy in the bond is conserved, and ATP is created.

In most organisms, these two processes are combined. To harvest energy to make ATP from the sugar glucose in the presence of oxygen, the cell carries out a complex series of enzyme-catalyzed reactions that occur in four stages: The first stage captures energy by substrate-level phosphorylation through glycolysis; the next three stages carry out aerobic respiration by oxidizing the end product of glycolysis. In this section, we first provide an overview of these stages; then we return to each topic for an in-depth examination.

Glycolysis

Stage One: Glycolysis. The first stage of extracting energy from glucose is a 10-reaction biochemical pathway called glycolysis that produces ATP by substrate-level phosphorylation. The enzymes that catalyze the glycolytic reactions are in the cytoplasm of the cell, not bound to any membrane or organelle. Two ATP molecules are used up early in the pathway, and four ATP molecules are formed by substrate-level phosphorylation. This yields a net of two ATP molecules for each molecule of glucose catabolized. In addition, four electrons are harvested from the chemical bonds of glucose and carried by NADH so they can be used to form ATP by aerobic respiration. Still, the total yield of ATP is small. When the glycolytic process is completed, the two molecules of pyruvate that are formed still contain most of the energy the original glucose molecule held.

Aerobic Respiration

Stage Two: Pyruvate Oxidation. In the second stage, pyruvate, the end product of glycolysis, is converted into carbon dioxide and a two-carbon molecule called acetyl-CoA. For each molecule of pyruvate converted, one molecule of NAD^+ is reduced to NADH, again to carry electrons that can be used to make ATP.

Stage Three: The Krebs Cycle. The third stage introduces this acetyl-CoA into a cycle of nine reactions called the Krebs cycle, named after the British biochemist Sir Hans Krebs, who discovered it. (The Krebs cycle is also called the *citric acid cycle*, for the citric acid, or citrate, formed in its first step, and less commonly, the *tricarboxylic acid cycle*, because citrate has three carboxyl groups.) In the Krebs cycle, two more ATP molecules are extracted by substrate-level phosphorylation, and a large number of electrons are removed by the reduction of NAD^+ to NADH.

Stage Four: The Electron Transport Chain. In the fourth stage, the energetic electrons carried by NADH are transferred to a series of electron carriers that progressively extract the energy the electrons possess and use it to pump protons across a membrane. The resulting proton gradient is used by ATP synthase to produce ATP.

Pyruvate oxidation, the reactions of the Krebs cycle, and ATP production by electron transport chains occur within many forms of prokaryotes and inside the mitochondria of all eukaryotes. Recall from chapter 5 that mitochondria are thought to have evolved from bacteria. Figure 9.5 provides an overview of aerobic respiration.

Anaerobic Respiration

As just noted, in the presence of oxygen, cells can respire aerobically, using oxygen to accept the electrons harvested from food molecules. But even when no oxygen is present to accept the electrons, some organisms can still respire *anaerobically*, using inorganic molecules to accept the electrons. For example, many prokaryotes use sulfur, nitrate, or other inorganic compounds as the electron acceptor in place of oxygen.

Methanogens.
Among the heterotrophs that practice anaerobic respiration are primitive archaebacteria such as the thermophiles and methanogens discussed in chapter 4. Methanogens use carbon dioxide (CO_2) as the electron acceptor, reducing CO_2 to CH_4 (methane) with the hydrogens derived from organic molecules produced by other organisms.

Sulfur Bacteria.
Evidence of a second anaerobic respiratory process among primitive bacteria is seen in a group of rocks about 2.7 billion years old, known as the Woman River iron formation. Organic material in these rocks is enriched for the light isotope of sulfur, ^{32}S, relative to the heavier isotope, ^{34}S. No known geochemical process produces such enrichment, but biological sulfur reduction does, in a process still carried out today by certain primitive prokaryotes. In this sulfate respiration, the prokaryotes derive energy from the reduction of inorganic sulfates (SO_4) to hydrogen sulfide (H_2S). The hydrogen atoms are obtained from organic molecules other organisms produce. These prokaryotes thus do the same thing methanogens do, but they use SO_4 as the oxidizing (that is, electron-accepting) agent in place of CO_2.

The sulfate reducers set the stage for the evolution of photosynthesis, creating an environment rich in H_2S. As discussed in chapter 10, the first form of photosynthesis obtained hydrogens from H_2S using the energy of sunlight.

> In aerobic respiration, the cell harvests energy from glucose molecules in a sequence of four major pathways: glycolysis, pyruvate oxidation, the Krebs cycle, and the electron transport chain. Oxygen is the final electron acceptor. Anaerobic respiration donates the harvested electrons to other inorganic compounds.

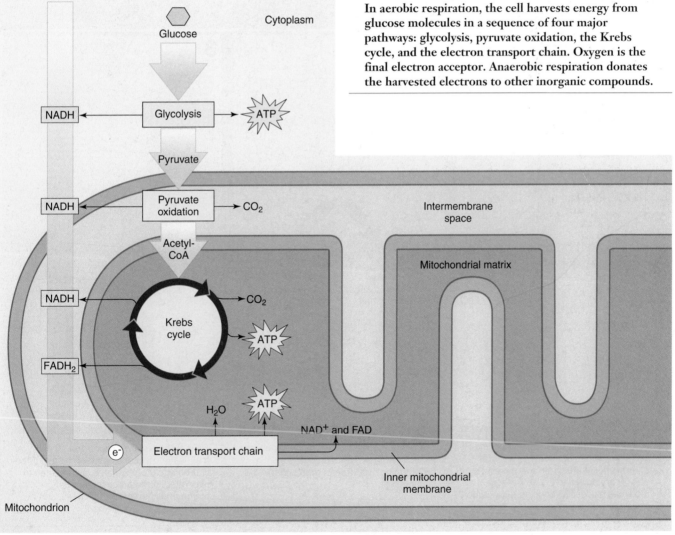

FIGURE 9.5
An overview of aerobic respiration.

Stage One: Glycolysis

The metabolism of primitive organisms focused on glucose. Glucose molecules can be dismantled in many ways, but primitive organisms evolved a glucose-catabolizing process that releases enough free energy to drive the synthesis of ATP in coupled reactions. This process, called glycolysis, occurs in the cytoplasm and involves a sequence of 10 reactions that convert glucose into 2 three-carbon molecules of pyruvate (figure 9.6). For each molecule of glucose that passes through this transformation, the cell nets two ATP molecules by substrate-level phosphorylation.

Priming

The first half of glycolysis consists of five sequential reactions that convert one molecule of glucose into two molecules of the three-carbon compound glyceraldehyde 3-phosphate (G3P). These reactions demand the expenditure of ATP, so they are an energy-requiring process.

Step A: Glucose priming. Three reactions "prime" glucose by changing it into a compound that can be cleaved readily into 2 three-carbon phosphorylated molecules. Two of these reactions require the cleavage of ATP, so this step requires the cell to use 2 ATP molecules.

Step B: Cleavage and rearrangement. In the first of the remaining pair of reactions, the six-carbon product of step A is split into 2 three-carbon molecules. One is G3P, and the other is then converted to G3P by the second reaction (figure 9.7).

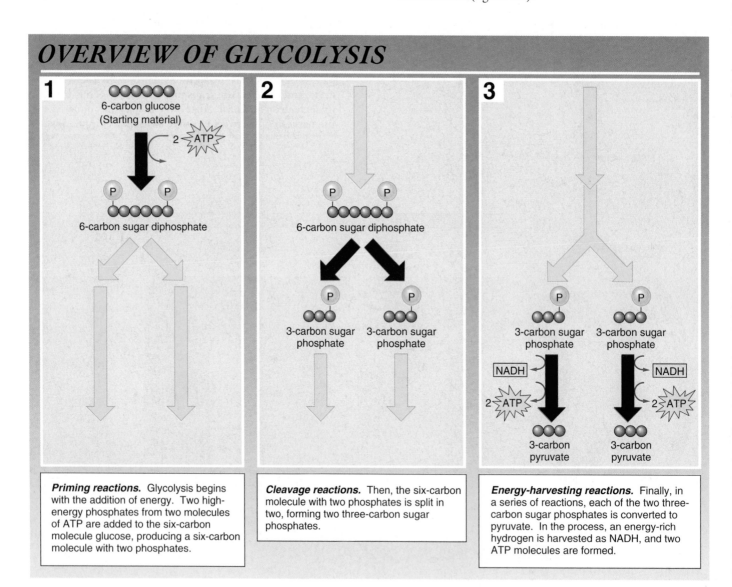

OVERVIEW OF GLYCOLYSIS

1

6-carbon glucose
(Starting material)

2 ATP

P   P

6-carbon sugar diphosphate

Priming reactions. Glycolysis begins with the addition of energy. Two high-energy phosphates from two molecules of ATP are added to the six-carbon molecule glucose, producing a six-carbon molecule with two phosphates.

2

P   P

6-carbon sugar diphosphate

P   P

3-carbon sugar phosphate   3-carbon sugar phosphate

Cleavage reactions. Then, the six-carbon molecule with two phosphates is split in two, forming two three-carbon sugar phosphates.

3

P   P

3-carbon sugar phosphate   3-carbon sugar phosphate

NADH   NADH

2 ATP   2 ATP

3-carbon pyruvate   3-carbon pyruvate

Energy-harvesting reactions. Finally, in a series of reactions, each of the two three-carbon sugar phosphates is converted to pyruvate. In the process, an energy-rich hydrogen is harvested as NADH, and two ATP molecules are formed.

FIGURE 9.6
How glycolysis works.

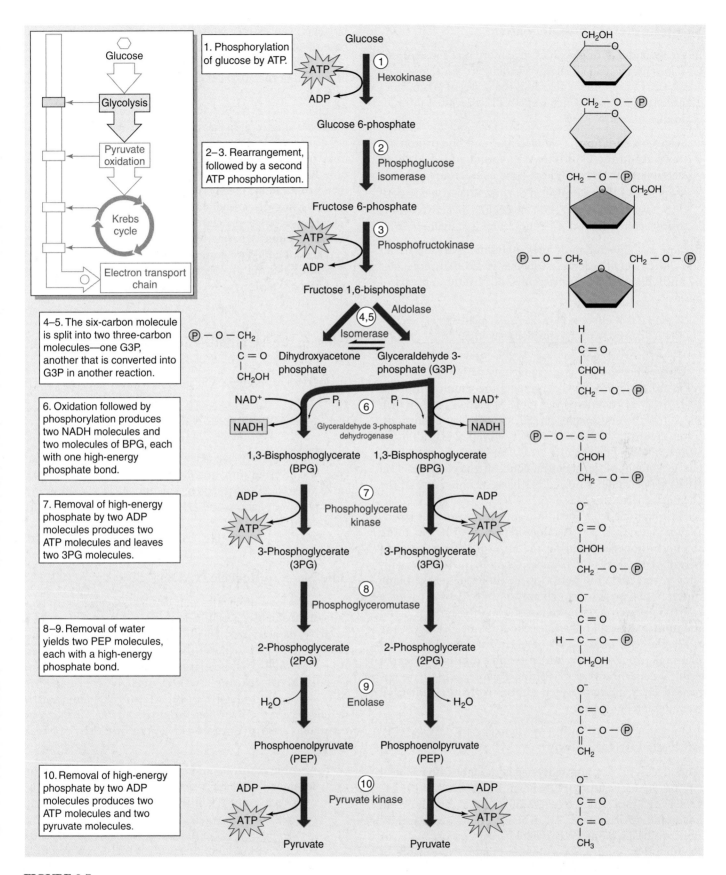

FIGURE 9.7
The glycolytic pathway. The first five reactions convert a molecule of glucose into two molecules of G3P. The second five reactions convert G3P into pyruvate.

Substrate-Level Phosphorylation

In the second half of glycolysis, five more reactions convert G3P into pyruvate in an energy-yielding process that generates ATP. Overall, then, glycolysis is a series of 10 enzyme-catalyzed reactions in which some ATP is invested in order to produce more.

Step C: Oxidation. Two electrons and one proton are transferred from G3P to NAD^+, forming NADH. Both electrons in the new covalent bond come from G3P.

Step D: ATP generation. Four reactions convert G3P into another three-carbon molecule, pyruvate. This process generates two ATP molecules (see figure 9.4).

Because each glucose molecule is split into two G3P molecules, the overall reaction sequence yields two molecules of ATP, as well as two molecules of NADH and two of pyruvate:

$$\begin{array}{l} 4\text{ ATP (2 ATP for each of the 2 G3P molecules in step D)} \\ \underline{-\ 2\text{ ATP (used in the two reactions in step A)}} \\ 2\text{ ATP} \end{array}$$

Under the nonstandard conditions within a cell, each ATP molecule produced represents the capture of about 12 kcal (50 kJ) of energy per mole of glucose, rather than the 7.3 kcal traditionally quoted for standard conditions. This means glycolysis harvests about 24 kcal/mole (100 kJ/mole). This is not a great deal of energy. The total energy content of the chemical bonds of glucose is 686 kcal (2870 kJ) per mole, so glycolysis harvests only 3.5% of the chemical energy of glucose.

Although far from ideal in terms of the amount of energy it releases, glycolysis does generate ATP. For more than a billion years during the anaerobic first stages of life on earth, it was the primary way heterotrophic organisms generated ATP from organic molecules. Like many biochemical pathways, glycolysis is believed to have evolved backward, with the last steps in the process being the most ancient. Thus, the second half of glycolysis, the ATP-yielding breakdown of G3P, may have been the original process early heterotrophs used to generate ATP. The synthesis of G3P from glucose would have appeared later, perhaps when alternative sources of G3P were depleted.

All Cells Use Glycolysis

The glycolytic reaction sequence is thought to have been among the earliest of all biochemical processes to evolve. It uses no molecular oxygen and occurs readily in an anaerobic environment. All of its reactions occur free in the cytoplasm; none is associated with any organelle or membrane structure. Every living creature is capable of carrying out glycolysis. Most present-day organisms, however, can extract considerably more energy from glucose through aerobic respiration.

Why does glycolysis take place even now, since its energy yield in the absence of oxygen is comparatively so paltry? The answer is that evolution is an incremental process: Change occurs by improving on past successes. In catabolic metabolism, glycolysis satisfied the one essential evolutionary criterion—it was an improvement. Cells that could not carry out glycolysis were at a competitive disadvantage, and only cells capable of glycolysis survived. Later improvements in catabolic metabolism built on this success. Glycolysis was not discarded during the course of evolution; rather, it served as the starting point for the further extraction of chemical energy. Metabolism evolved as one layer of reactions added to another. Nearly every present-day organism carries out glycolysis, as a metabolic memory of its evolutionary past.

Closing the Metabolic Circle: The Regeneration of NAD^+

Inspect for a moment the net reaction of the glycolytic sequence:

$$\text{glucose} + 2\text{ ADP} + 2\text{ P}_i + 2\text{ NAD}^+ \rightarrow$$
$$2\text{ pyruvate} + 2\text{ ATP} + 2\text{ NADH} + 2\text{ H}^+ + 2\text{ H}_2\text{O}$$

You can see that three changes occur in glycolysis: (1) Glucose is converted into two molecules of pyruvate; (2) two molecules of ADP are converted into ATP via substrate-level phosphorylation; and (3) two molecules of NAD^+ are reduced to NADH.

The Need to Recycle NADH

As long as food molecules that can be converted into glucose are available, a cell can continually churn out ATP to drive its activities. In doing so, however, it accumulates NADH and depletes the pool of NAD^+ molecules. A cell does not contain a large amount of NAD^+, and for glycolysis to continue, NADH must be recycled into NAD^+. Some molecule other than NAD^+ must ultimately accept the hydrogen atom taken from G3P and be reduced. Two processes can carry out this key task (figure 9.8):

1. **Aerobic respiration.** Oxygen is an excellent electron acceptor. Through a series of electron transfers, the hydrogen atom taken from G3P can be donated to oxygen, forming water. This is what happens in the cells of eukaryotes in the presence of oxygen. Because air is rich in oxygen, this process is also referred to as aerobic metabolism.

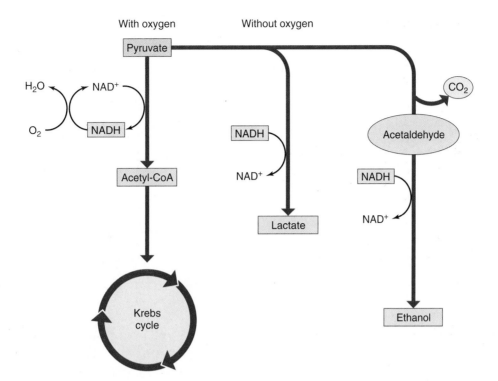

FIGURE 9.8
What happens to pyruvate, the product of glycolysis? In the presence of oxygen, pyruvate is oxidized to acetyl-CoA, which enters the Krebs cycle. In the absence of oxygen, pyruvate is instead reduced, accepting the electrons extracted during glycolysis and carried by NADH. When pyruvate is reduced directly, as in muscle cells, the product is lactate. When carbon dioxide is first removed from pyruvate and the product, acetaldehyde, is then reduced, as in yeast cells, the product is ethanol.

2. **Fermentation.** When oxygen is unavailable, an organic molecule, such as acetaldehyde in wine fermentation, can accept the hydrogen atom instead (figure 9.9). Such fermentation plays an important role in the metabolism of most organisms, even those capable of aerobic respiration.

The fate of the pyruvate that is produced by glycolysis depends upon which of these two processes takes place. The aerobic respiration path starts with the oxidation of pyruvate to a molecule called acetyl-CoA, which is then further oxidized in a series of reactions called the Krebs cycle. The fermentation path, by contrast, involves the reduction of all or part of pyruvate. We will start by examining aerobic respiration (stages two through four), and then look briefly at fermentation.

Glycolysis generates a small amount of ATP by reshuffling the bonds of glucose molecules. In glycolysis, two molecules of NAD^+ are reduced to NADH. NAD^+ must be regenerated for glycolysis to continue unabated.

FIGURE 9.9
How wine is made. The conversion of pyruvate to ethanol takes place naturally in grapes left to ferment on vines, as well as in fermentation vats of crushed grapes. Yeasts carry out the process, but when their conversion increases the ethanol concentration to about 12%, the toxic effects of the alcohol kill the yeast cells. What is left is wine.

Stage Two: The Oxidation of Pyruvate

In the presence of oxygen, the oxidation of glucose that begins in glycolysis continues where glycolysis leaves off—with pyruvate. In eukaryotic organisms, the extraction of additional energy from pyruvate takes place exclusively inside mitochondria. The cell harvests pyruvate's considerable energy in two steps: first, by oxidizing pyruvate to form acetyl-CoA, and then by oxidizing acetyl-CoA in the Krebs cycle.

Producing Acetyl-CoA

Pyruvate is oxidized in a single "decarboxylation" reaction that cleaves off one of pyruvate's three carbons. This carbon then departs as CO_2 (figure 9.10, *top*), a waste molecule that must be "exhaled" from the organism. This reaction produces a two-carbon fragment called an acetyl group, as well as a pair of electrons and their associated hydrogen, which reduce NAD^+ to NADH. The reaction is complex, involving three intermediate stages, and is catalyzed within mitochondria by a *multienzyme complex*. As chapter 8 noted, such a complex organizes a series of enzymatic steps so that the chemical intermediates do not diffuse away or undergo other reactions. Within the complex, component polypeptides pass the substrates from one enzyme to the next, without releasing them. *Pyruvate dehydrogenase*, the complex of enzymes that removes CO_2 from pyruvate, is one of the largest enzymes known; it contains 60 subunits! In the course of the reaction, the acetyl group removed from pyruvate combines with a cofactor called coenzyme A (CoA), forming a compound known as **acetyl-CoA**:

$$\text{pyruvate} + NAD^+ + \text{CoA} \rightarrow \text{acetyl-CoA} + NADH + CO_2$$

This reaction produces a molecule of NADH, which is later used to produce ATP. Of far greater significance than the reduction of NAD^+ to NADH, however, is the production of acetyl-CoA (figure 9.10, *bottom*). Acetyl-CoA is important because so many different metabolic processes generate it. Not only does the oxidation of pyruvate, an intermediate in carbohydrate catabolism, produce it, but the metabolic breakdown of proteins, fats, and other lipids also generates acetyl-CoA. Indeed, almost all molecules catabolized for energy are converted into acetyl-CoA. Acetyl-CoA is then channeled into fat synthesis or into ATP production, depending on the organism's energy requirements. Acetyl-CoA is a key point for the many catabolic processes of the eukaryotic cell.

Using Acetyl-CoA

Although the cell forms acetyl-CoA in many ways, only a limited number of processes use acetyl-CoA. Most of it is either directed toward energy storage (lipid synthesis, for example) or oxidized in the Krebs cycle to produce ATP.

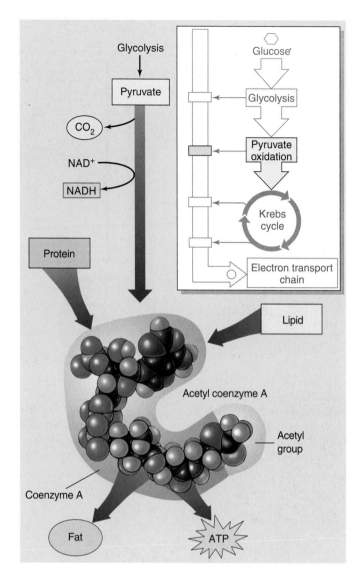

FIGURE 9.10
The oxidation of pyruvate. This complex reaction involves the reduction of NAD^+ to NADH and is thus a significant source of metabolic energy. Its product, acetyl-CoA, is the starting material for the Krebs cycle. Almost all molecules that are catabolized for energy are converted into acetyl-CoA, which is then channeled into fat synthesis or into ATP production.

Which of these two options is taken depends on the level of ATP in the cell. When ATP levels are high, the oxidative pathway is inhibited, and acetyl-CoA is channeled into fatty acid synthesis. This explains why many animals (humans included) develop fat reserves when they consume more food than their bodies require. Alternatively, when ATP levels are low, the oxidative pathway is stimulated, and acetyl-CoA flows into energy-producing oxidative metabolism.

In the second energy-harvesting stage of glucose catabolism, pyruvate is decarboxylated, yielding acetyl-CoA, NADH, and CO_2. This process occurs within the mitochondrion.

Stage Three: The Krebs Cycle

After glycolysis catabolizes glucose to produce pyruvate, and pyruvate is oxidized to form acetyl-CoA, the third stage of extracting energy from glucose begins. In this third stage, acetyl-CoA is oxidized in a series of nine reactions called the **Krebs cycle**. These reactions occur in the matrix of mitochondria. In this cycle, the two-carbon acetyl group of acetyl-CoA combines with a four-carbon molecule called oxaloacetate (figure 9.11). The resulting six-carbon molecule then goes through a sequence of electron-yielding oxidation reactions, during which two CO_2 molecules split off, restoring oxaloacetate. The oxaloacetate is then recycled to bind to another acetyl group. In each turn of the cycle, a new acetyl group replaces the two CO_2 molecules lost, and more electrons are extracted to drive *proton pumps* that generate ATP.

Overview of the Krebs Cycle

The nine reactions of the Krebs cycle occur in two steps:

Step A: Priming. Three reactions prepare the six-carbon molecule for energy extraction. First, acetyl-CoA joins the cycle, and then chemical groups are rearranged.

Step B: Energy extraction. Four of the six reactions in this step are oxidations in which electrons are removed, and one generates an ATP equivalent directly by substrate-level phosphorylation.

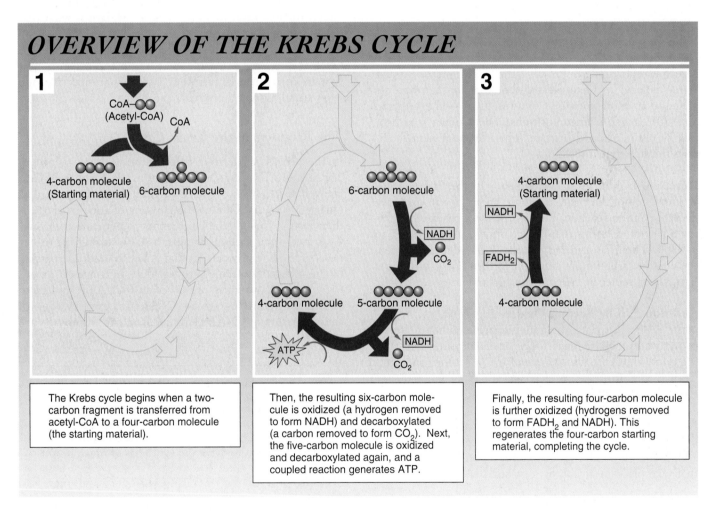

OVERVIEW OF THE KREBS CYCLE

1
CoA—◯◯
(Acetyl-CoA)   CoA
4-carbon molecule (Starting material)   6-carbon molecule

The Krebs cycle begins when a two-carbon fragment is transferred from acetyl-CoA to a four-carbon molecule (the starting material).

2
6-carbon molecule
NADH
CO_2
4-carbon molecule   5-carbon molecule
ATP   NADH
CO_2

Then, the resulting six-carbon molecule is oxidized (a hydrogen removed to form NADH) and decarboxylated (a carbon removed to form CO_2). Next, the five-carbon molecule is oxidized and decarboxylated again, and a coupled reaction generates ATP.

3
4-carbon molecule (Starting material)
NADH
FADH$_2$
4-carbon molecule

Finally, the resulting four-carbon molecule is further oxidized (hydrogens removed to form FADH$_2$ and NADH). This regenerates the four-carbon starting material, completing the cycle.

FIGURE 9.11
How the Krebs cycle works.

The Reactions of the Krebs Cycle

The Krebs cycle consists of nine sequential reactions that cells use to extract energetic electrons and drive the synthesis of ATP (figure 9.12). A two-carbon group from acetyl-CoA enters the cycle at the beginning, and two CO_2 molecules and eight electrons are given off during the cycle.

Reaction 1: Condensation. The two-carbon group from acetyl-CoA joins with a four-carbon molecule, oxaloacetate, to form a six-carbon molecule, *citrate*. This condensation reaction is irreversible, committing the two-carbon acetyl group to the Krebs cycle. The reaction is inhibited when the cell's ATP concentration is high and stimulated when it is low. Hence, when the cell possesses ample amounts of ATP, the Krebs cycle shuts down, and acetyl-CoA is channeled into fat synthesis.

Reactions 2 and 3: Isomerization. Before the oxidation reactions can begin, the hydroxyl (—OH) group of citrate must be repositioned. This is done in two steps: First, a water molecule is removed from one carbon; then water is added to a different carbon. As a result, an —H group and an —OH group change positions. The product is an isomer of citrate called isocitrate. This rearrangement facilitates the subsequent reactions.

Reaction 4: The First Oxidation. In the first energy-yielding step of the cycle, isocitrate undergoes an oxidative decarboxylation reaction. First, isocitrate is oxidized, yielding a pair of electrons that reduce a molecule of NAD^+ to NADH. Then the oxidized intermediate is decarboxylated; the central carbon atom splits off to form CO_2, yielding a five-carbon molecule called α-ketoglutarate.

Reaction 5: The Second Oxidation. Next, α-ketoglutarate is decarboxylated by a multienzyme complex similar to pyruvate dehydrogenase. The succinyl group left after the removal of CO_2 joins to coenzyme A, forming succinyl-CoA. In the process, two electrons are extracted, and they reduce another molecule of NAD^+ to NADH.

Reaction 6: Substrate-Level Phosphorylation. The linkage between the four-carbon succinyl group and CoA is a high-energy bond. In a coupled reaction similar to those that take place in glycolysis, this bond is cleaved, and the energy released drives the phosphorylation of guanosine diphosphate (GDP), forming guanosine triphosphate (GTP). GTP is readily converted into ATP, and the four-carbon fragment that remains is called succinate.

Reaction 7: The Third Oxidation. Next, succinate is oxidized to fumarate. The free energy change in this reaction is not large enough to reduce NAD^+. Instead, flavin adenine dinucleotide (FAD) is the electron acceptor. Unlike NAD^+, FAD is not free to diffuse within the mitochondrion; it is an integral part of the inner mitochondrial membrane. Its reduced form, $FADH_2$, can only contribute electrons to the electron transport chain in the membrane.

Reactions 8 and 9: Regeneration of Oxaloacetate. In the final two reactions of the cycle, a water molecule is added to fumarate, forming malate. Malate is then oxidized, yielding a four-carbon molecule of oxaloacetate and two electrons that reduce a molecule of NAD^+ to NADH. Oxaloacetate, the molecule that began the cycle, is now free to combine with another two-carbon acetyl group from acetyl-CoA and reinitiate the cycle.

The Products of the Krebs Cycle

In the process of aerobic respiration, glucose is entirely consumed. The six-carbon glucose molecule is first cleaved into a pair of three-carbon pyruvate molecules during glycolysis. One of the carbons of each pyruvate is then lost as CO_2 in the conversion of pyruvate to acetyl-CoA; two other carbons are lost as CO_2 during the oxidations of the Krebs cycle. All that is left to mark the passing of the glucose molecule into six CO_2 molecules is its energy, some of which is preserved in four ATP molecules and in the reduced state of 12 electron carriers. Ten of these carriers are NADH molecules; the other two are $FADH_2$.

> **The Krebs cycle generates two ATP molecules per molecule of glucose, the same number generated by glycolysis. More importantly, the Krebs cycle and the oxidation of pyruvate harvest many energized electrons, which can be directed to the electron transport chain to drive the synthesis of much more ATP.**

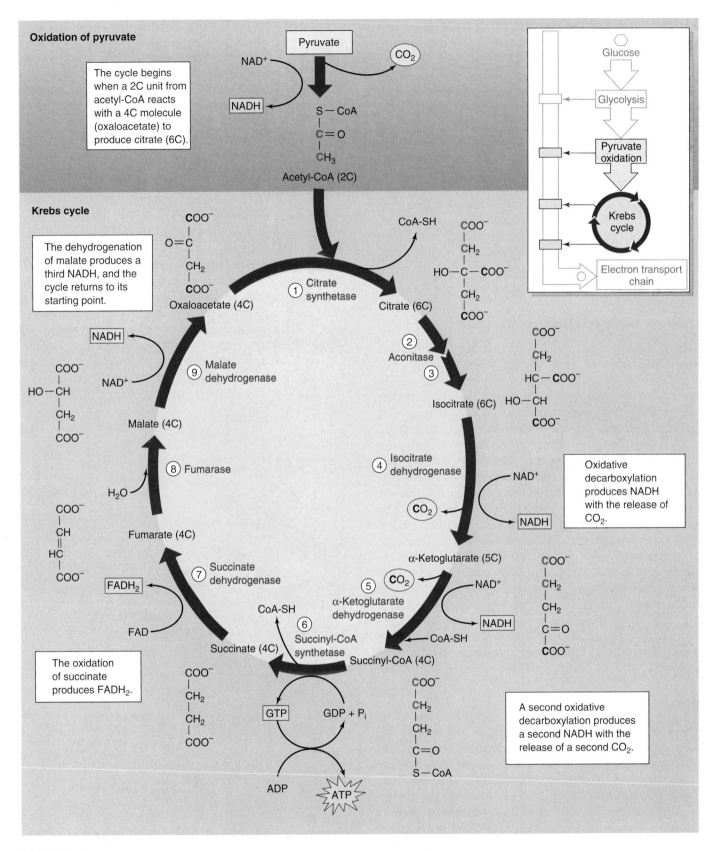

Oxidation of pyruvate

The cycle begins when a 2C unit from acetyl-CoA reacts with a 4C molecule (oxaloacetate) to produce citrate (6C).

Krebs cycle

The dehydrogenation of malate produces a third NADH, and the cycle returns to its starting point.

Oxaloacetate (4C)

① Citrate synthetase

Citrate (6C)

② Aconitase

③

Isocitrate (6C)

⑨ Malate dehydrogenase

Malate (4C)

⑧ Fumarase

Fumarate (4C)

⑦ Succinate dehydrogenase

Succinate (4C)

⑥ Succinyl-CoA synthetase

Succinyl-CoA (4C)

⑤ α-Ketoglutarate dehydrogenase

α-Ketoglutarate (5C)

④ Isocitrate dehydrogenase

Oxidative decarboxylation produces NADH with the release of CO_2.

A second oxidative decarboxylation produces a second NADH with the release of a second CO_2.

The oxidation of succinate produces $FADH_2$.

Glucose
Glycolysis
Pyruvate oxidation
Krebs cycle
Electron transport chain

FIGURE 9.12
The Krebs cycle.   This series of reactions takes place within the matrix of the mitochondrion. For the complete breakdown of a molecule of glucose, the two molecules of acetyl-CoA produced by glycolysis and pyruvate oxidation each have to make a trip around the Krebs cycle. Follow the different carbons through the cycle, and notice the changes that occur in the carbon skeletons of the molecules as they proceed through the cycle.

rvesting Energy by Extracting Electrons

To understand how cells direct some of the energy released during glucose catabolism into ATP production, we need to take a closer look at the electrons in the C—H bonds of the glucose molecule. We stated in chapter 8 that when an electron is removed from one atom and donated to another, the electron's potential energy of position is also transferred. In this process, the atom that receives the electron is reduced. We spoke of reduction in an all-or-none fashion, as if it involved the complete transfer of an electron from one atom to another. Often this is just what happens. However, sometimes a reduction simply changes the *degree of sharing* within a covalent bond. Let us now revisit that discussion and consider what happens when the transfer of electrons is incomplete.

A Closer Look at Oxidation-Reduction

The catabolism of glucose is an oxidation-reduction reaction. The covalent electrons in the C—H bonds of glucose are shared approximately equally between the C and H atoms because carbon and hydrogen nuclei have about the same affinity for valence electrons (that is, they exhibit similar *electronegativity*). However, when the carbon atoms of glucose react with oxygen to form carbon dioxide, the electrons in the new covalent bonds take a different position. Instead of being shared equally, the electrons that were associated with the carbon atoms in glucose shift far toward the oxygen atoms in CO_2 because oxygen is very electronegative. Since these electrons are pulled farther from the carbon atoms, the carbon atoms of glucose have been oxidized (loss of electrons) and the oxygen atoms reduced (gain of electrons). Similarly, when the hydrogen atoms of glucose combine with oxygen atoms to form water, the oxygen atoms draw the shared electrons strongly toward them; again, oxygen is reduced and glucose is oxidized. In this reaction, oxygen is an oxidizing (electron-attracting) agent because it oxidizes the atoms of glucose.

Releasing Energy

The key to understanding the oxidation of glucose is to focus on the energy of the shared electrons. In a covalent bond, energy must be added to remove an electron from an atom, just as energy must be used to roll a boulder up a hill. The more electronegative the atom, the steeper the energy hill that must be climbed to pull an electron away from it. However, energy is released when an electron is shifted away from a less electronegative atom and *closer* to a more electronegative atom, just as energy is released when a boulder is allowed to roll down a hill. In the catabolism of glucose, energy is released when glucose is oxidized, as electrons relocate closer to oxygen (figure 9.13).

Glucose is an energy-rich food because it has an abundance of C—H bonds. Viewed in terms of oxidation-reduction, glucose possesses a wealth of electrons held far from their atoms, all with the potential to move closer toward oxygen. In oxidative respiration, energy is released not simply because the hydrogen atoms of the C—H bonds are transferred from glucose to oxygen, but because the positions of the valence electrons shift. This shift releases energy that can be used to make ATP.

Reducing Power

Just as electrons are donated to NAD^+, reducing it to NADH and allowing it to carry the associated energy, NADH can donate its electrons to other molecules, reducing them and being oxidized itself. This ability to supply high-energy electrons is critical to the biosynthesis of many organic molecules, including fats and sugars. In animals, when ATP is plentiful, the reducing power of the accumulated NADH is diverted to supplying fatty acid precursors with high-energy electrons, reducing them to form fats and storing the energy of the electrons.

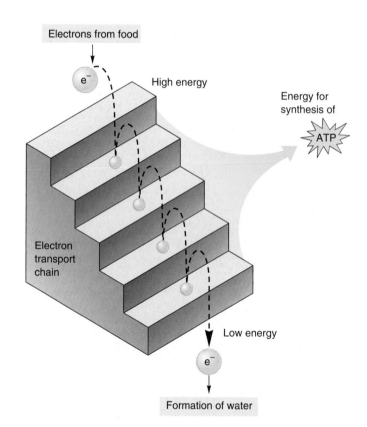

FIGURE 9.13
How electron transport works. This diagram shows how ATP is generated when electrons transfer from one energy level to another. Rather than releasing a single explosive burst of energy, electrons "fall" to lower and lower energy levels in steps, releasing stored energy with each fall as they tumble to the lowest (most electronegative) electron acceptor.

Harvesting the Energy in Stages

It is generally true that the larger the release of energy in any single step, the more of that energy is released as heat (random molecular motion) and the less is available to be channeled into more useful paths. In the combustion of gasoline, the same amount of energy is released whether all of the gasoline in a car's gas tank explodes at once or whether the gasoline burns in a series of very small explosions inside the cylinders. By releasing the energy in gasoline a little at a time, the harvesting efficiency is greater, and more of the energy can be used to push the pistons and move the car.

The same principle applies to the oxidation of glucose inside a cell. If all of the hydrogens were transferred to oxygen in one explosive step, releasing all of the free energy at once, the cell would recover very little of that energy in a useful form. Instead, cells burn their fuel much as a car does, a little at a time. The six hydrogens in the C—H bonds of glucose are stripped off in stages in the series of enzyme-catalyzed reactions collectively referred to as glycolysis and the Krebs cycle. We have had a great deal to say about these reactions already in this chapter. Recall that the hydrogens are removed by transferring them to a coenzyme carrier, NAD$^+$ (figure 9.14). As discussed in chapter 8, NAD$^+$ is a very versatile electron acceptor, shuttling energy-bearing electrons throughout the cell. In harvesting the energy of glucose, NAD$^+$ acts as the primary electron acceptor.

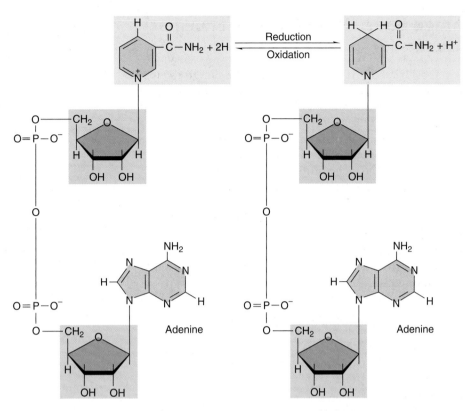

NAD$^+$: oxidized form of nicotinamide   **NADH: reduced form of nicotinamide**

FIGURE 9.14
NAD$^+$ and NADH.   This dinucleotide serves as an "electron shuttle" during cellular respiration. NAD$^+$ accepts electrons from catabolized macromolecules and is reduced to NADH.

Following the Electrons

As you examine these reactions, try not to become confused by the changes in electrical charge. Always *follow the electrons*. Enzymes extract two hydrogens—that is, two electrons and two protons—from glucose and transfer both electrons and one of the protons to NAD$^+$. The other proton is released as a hydrogen ion, H$^+$, into the surrounding solution. This transfer converts NAD$^+$ into NADH; that is, two negative electrons and one positive proton are added to one positively charged NAD$^+$ to form NADH, which is electrically neutral.

Energy captured by NADH is not harvested all at once. Instead of being transferred directly to oxygen, the two electrons carried by NADH are passed along the **electron transport chain.** This chain consists of a series of molecules, mostly proteins, embedded within the inner membranes of mitochondria. NADH delivers electrons to the top of the electron transport chain, and oxygen captures them at the bottom. The oxygen then joins with hydrogen ions to form water. At each step in the chain, the electrons move to a slightly more electronegative carrier, and their positions shift slightly. Thus, the electrons move *down* an energy gradient. The entire process releases a total of 53 kcal/mole (222 kJ/mole) under standard conditions. The transfer of electrons along this chain allows the energy to be extracted gradually. Next, we will discuss how this energy is put to work to drive the production of ATP.

The catabolism of glucose involves a series of oxidation-reduction reactions that release energy by repositioning electrons closer to oxygen atoms. Energy is thus harvested from glucose molecules in gradual steps, using NAD$^+$ as an electron carrier.

Stage Four: The Electron Transport Chain

The NADH and FADH$_2$ molecules formed during the first three stages of aerobic respiration each contain a pair of electrons that were gained when NAD$^+$ and FAD were reduced. The NADH molecules carry their electrons to the inner mitochondrial membrane, where they transfer the electrons to a series of membrane-associated proteins collectively called the **electron transport chain.**

Moving Electrons Through the Electron Transport Chain

The first of the proteins to receive the electrons is a complex, membrane-embedded enzyme called **NADH dehydrogenase.** A carrier called ubiquinone then passes the electrons to a protein-cytochrome complex called the *bc$_1$ complex.* This complex, along with others in the chain, operates as a proton pump, driving a proton out across the membrane (figure 9.15).

The electron is then carried by another carrier, *cytochrome c,* to the cytochrome oxidase complex. This complex uses four such electrons to reduce a molecule of oxygen. Each oxygen then combines with two hydrogen ions to form water:

$$O_2 + 4\,H^+ + 4\,e^- \rightarrow 2\,H_2O$$

While NADH contributes its electrons to the first protein of the electron transport chain, NADH dehydrogenase, FADH$_2$, which is always attached to the inner mitochondrial membrane, feeds its electrons into the electron transport chain later, to ubiquinone.

It is the availability of a plentiful electron acceptor (often oxygen) that makes oxidative respiration possible. As we'll see in chapter 10, the electron transport chain used in aerobic respiration is similar to, and may well have evolved from, the chain employed in aerobic photosynthesis.

Building an Electrochemical Gradient

In eukaryotes, aerobic metabolism takes place within the mitochondria present in virtually all cells. The internal compartment, or matrix, of a mitochondrion contains the enzymes that carry out the reactions of the Krebs cycle. As

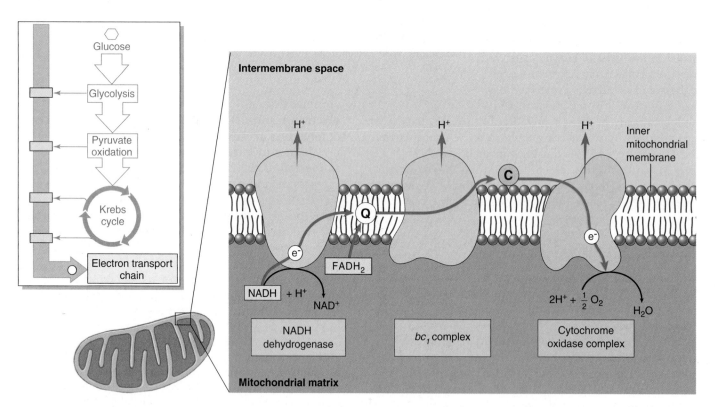

FIGURE 9.15
The electron transport chain.   High-energy electrons harvested from catabolized molecules are transported (*red arrows*) by mobile electron carriers (ubiquinone, marked Q, and cytochrome *c*, marked C) along a chain of membrane proteins. Three proteins use portions of the electrons' energy to pump protons (*blue arrows*) out of the matrix and into the intermembrane space. The electrons are finally donated to oxygen to form water.

the electrons harvested by oxidative respiration are passed along the electron transport chain, the energy they release transports protons out of the matrix and into the outer compartment, sometimes called the intermembrane space. Three transmembrane proteins in the inner mitochondrial membrane (see figure 9.15) actually accomplish the transport. The flow of excited electrons induces a change in the shape of these pump proteins, which causes them to transport protons across the membrane. The electrons contributed by NADH activate all three of these proton pumps, while those contributed by $FADH_2$ activate only two.

Producing ATP: Chemiosmosis

As the proton concentration in the intermembrane space rises above that in the matrix, the matrix becomes slightly negative in charge. This internal negativity attracts the positively charged protons and induces them to reenter the matrix. The higher outer concentration tends to drive protons back in by diffusion; since membranes are relatively impermeable to ions, most of the protons that reenter the matrix pass through ATP synthase, which uses the energy of the gradient to catalyze the synthesis of ATP from ADP and P_i. Thus, the vast majority of the energy originally contained in glucose is ultimately used to generate a gradient of protons in the inner membrane of the mitochondrion. This gradient is then used to drive the synthesis of ATP, which is transported by facilitated diffusion to the many places in the cell where enzymes require energy to drive their reactions. Because the chemical formation of ATP is driven by a diffusion force similar to osmosis, this process is referred to as **chemiosmosis** (figure 9.16).

Thus, the electron transport chain uses electrons harvested in aerobic respiration to pump a large number of protons across the inner mitochondrial membrane. Their subsequent reentry into the mitochondrial matrix through ATP synthase drives the synthesis of ATP by chemiosmosis. Figure 9.17 summarizes the overall process.

The electron transport chain is a series of five membrane-associated proteins. Electrons delivered by NADH and $FADH_2$ are passed from protein to protein along the chain, like a baton in a relay race. These electrons are used to pump protons out of the mitochondrial matrix via the electron transport chain. The return of the protons into the matrix generates ATP.

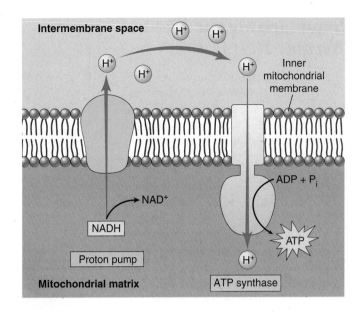

FIGURE 9.16
Chemiosmosis. NADH transports high-energy electrons harvested from the catabolism of macromolecules to "proton pumps" that use the energy to pump protons out of the mitochondrial matrix. As a result, the concentration of protons in the intermembrane space rises, inducing protons to diffuse back into the matrix. Many of the protons pass through ATP synthase, which couples the reentry of protons to the production of ATP.

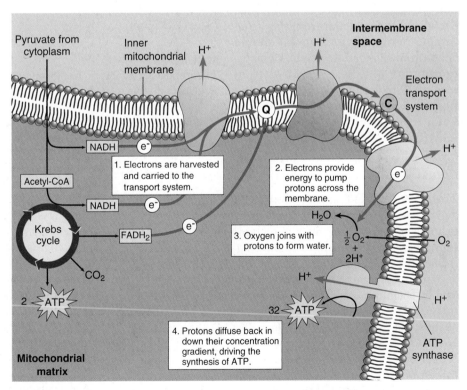

FIGURE 9.17
ATP generation during the Krebs cycle and electron transport chain. This process begins with pyruvate, the product of glycolysis, and ends with the synthesis of ATP.

Summarizing the Yield of Aerobic Respiration

How much metabolic energy does a cell actually gain from the electrons harvested from a molecule of glucose, using the electron transport chain to produce ATP by chemiosmosis?

Theoretical Yield

The chemiosmotic model suggests that one ATP molecule is generated for each proton pump activated by the electron transport chain. As the electrons from NADH activate three pumps and those from $FADH_2$ activate two, we would expect each molecule of NADH and $FADH_2$ to generate three and two ATP molecules, respectively. However, because eukaryotic cells carry out glycolysis in their cytoplasm and the Krebs cycle within their mitochondria, they must transport the two molecules of NADH produced during glycolysis across the mitochondrial membranes, which requires one ATP per molecule of NADH. Thus, the net ATP production is decreased by two. Therefore, the overall ATP production resulting from aerobic respiration *theoretically* should be: 4 (from substrate-level phosphorylation during glycolysis) + 30 (3 from each of 10 molecules of NADH) + 4 (2 from each of 2 molecules of $FADH_2$) − 2 (for transport of glycolytic NADH) = 36 molecules of ATP (figure 9.18).

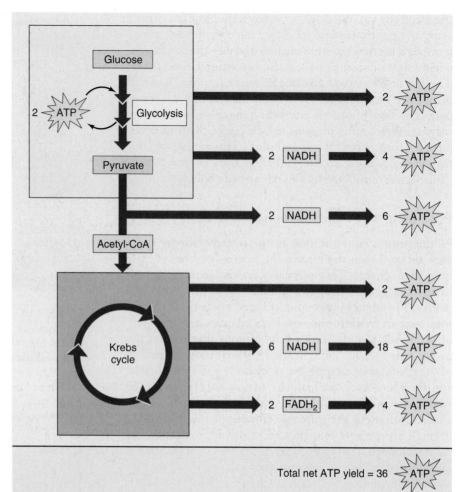

FIGURE 9.18
Theoretical ATP yield. The theoretical yield of ATP harvested from glucose by aerobic respiration totals 36 molecules.

Actual Yield

The amount of ATP *actually* produced in a eukaryotic cell during aerobic respiration is somewhat lower than 36, for two reasons. First, the inner mitochondrial membrane is somewhat "leaky" to protons, allowing some of them to reenter the matrix without passing through ATP synthase. Second, mitochondria often use the proton gradient generated by chemiosmosis for purposes other than ATP synthesis (such as transporting pyruvate into the matrix). Consequently, the actual measured values of ATP generated by NADH and $FADH_2$ are closer to 2.5 for each NADH and 1.5 for each $FADH_2$. With these corrections, the overall harvest of ATP from a molecule of glucose in a eukaryotic cell is closer to: 4 (from substrate-level phosphorylation) + 25 (2.5 from each of 10 molecules of NADH) + 3 (1.5 from each of 2 molecules of $FADH_2$) − 2 (transport of glycolytic NADH) = 30 molecules of ATP.

The catabolism of glucose by aerobic respiration, in contrast to glycolysis, is quite efficient. Aerobic respiration in a eukaryotic cell harvests about $(7.3 \times 30)/686 = 32\%$ of the energy available in glucose. (By comparison, a typical car converts only about 25% of the energy in gasoline into useful energy.)

The higher efficiency of aerobic respiration was one of the key factors that fostered the evolution of heterotrophs. With this mechanism for producing ATP, it became feasible for nonphotosynthetic organisms to derive metabolic energy exclusively from the oxidative breakdown of other organisms. As long as some organisms captured energy by photosynthesis, others could exist solely by feeding on them.

> Aerobic respiration produces approximately 30 molecules of ATP from each molecule of glucose in a eukaryotic cell. This represents about one-third of the energy in the chemical bonds of glucose.

Regulating Aerobic Respiration

When cells possess plentiful amounts of ATP, the key reactions of glycolysis, the Krebs cycle, and fatty acid breakdown are inhibited, slowing ATP production. The regulation of these biochemical pathways by the level of ATP is an example of feedback inhibition. Conversely, when ATP levels in the cell are low, ADP levels are high, and ADP activates enzymes in the pathways of carbohydrate catabolism to stimulate the production of more ATP.

Control of glucose catabolism occurs at two key points in the catabolic pathway (figure 9.19). The control point in glycolysis is the enzyme phosphofructokinase, which catalyzes the conversion of fructose phosphate to fructose bisphosphate. This is the first reaction of glycolysis that is not readily reversible, committing the substrate to the glycolytic sequence. High levels of ADP relative to ATP (implying a need to convert more ADP to ATP) stimulate phosphofructokinase, committing more sugar to the catabolic pathway; so do low levels of citrate (implying the Krebs cycle is not running at full tilt and needs more input). The main control point in the oxidation of pyruvate occurs at the committing step in the Krebs cycle with the enzyme pyruvate decarboxylase. This enzyme is inhibited by high levels of NADH, a key product of the Krebs cycle, implying that no more is needed.

Another control point in the Krebs cycle is the enzyme citrate synthetase, which catalyzes the first reaction, the conversion of oxaloacetate and acetyl-CoA into citrate. High levels of ATP inhibit citrate synthetase (as well as pyruvate decarboxylase and two other Krebs cycle enzymes), shutting down the catabolic pathway.

Relative levels of ADP and ATP regulate the catabolism of glucose at key committing reactions.

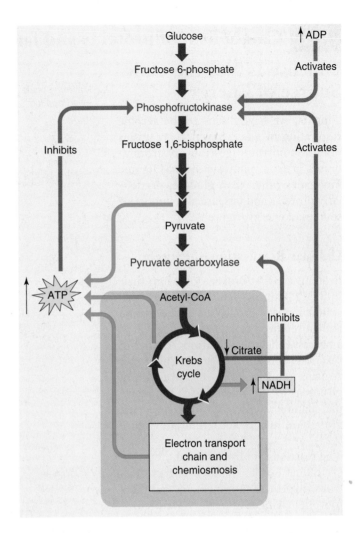

FIGURE 9.19
Control of glucose catabolism. The relative levels of ADP and ATP control the catabolic pathway at two key points: the committing reactions of glycolysis and the Krebs cycle.

A Vocabulary of ATP Generation

aerobic respiration The portion of cellular respiration that requires oxygen as an electron acceptor; it includes pyruvate oxidation, the Krebs cycle, and the electron transport chain.

anaerobic respiration Cellular respiration in which inorganic electron acceptors other than oxygen are used; it includes glycolysis.

cellular respiration The anaerobic oxidation of organic molecules to produce ATP in which the final electron acceptor is organic; it includes aerobic and anaerobic respiration.

chemiosmosis The passage of high-energy electrons along the electron transport chain, which is coupled to the pumping of protons across a membrane and the return of protons to the original side of the membrane through ATP synthase, driving the production of ATP.

fermentation Alternative ATP-producing pathway performed by some cells in the absence of oxygen, in which the final electron acceptor is an organic molecule.

oxidation The loss of an electron. In cellular respiration, high-energy electrons are stripped from food molecules, oxidizing them. Any oxidation must be accompanied by a corresponding reduction.

photosynthesis The chemiosmotic generation of ATP and complex organic molecules powered by the energy derived from light.

substrate-level phosphorylation The generation of ATP by the direct transfer of a phosphate group to ADP from another phosphorylated molecule.

9.3 Catabolism of proteins and fats can yield considerable energy.

Glucose Is Not the Only Source of Energy

Thus far we have discussed the aerobic respiration of glucose, which organisms obtain from the digestion of carbohydrates or from photosynthesis. Organic molecules other than glucose, particularly proteins and fats, are also important sources of energy (figure 9.20).

Cellular Respiration of Protein

Proteins are first broken down into their individual amino acids. The nitrogen-containing side group (the amino group) is then removed from each amino acid in a process called **deamination**. A series of reactions convert the carbon chain that remains into a molecule that takes part in glycolysis or the Krebs cycle. For example, alanine is converted into pyruvate, glutamate into α-ketoglutarate (figure 9.21), and aspartate into oxaloacetate. The reactions of glycolysis and the Krebs cycle then extract the high-energy electrons from these molecules and put them to work making ATP.

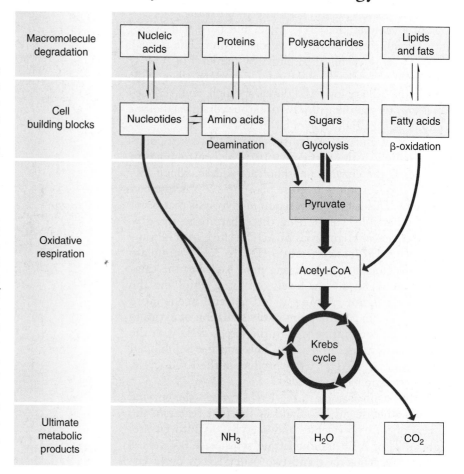

FIGURE 9.20
How cells extract chemical energy. All eukaryotes and many prokaryotes extract energy from organic molecules by oxidizing them. The first stage of this process, breaking down macromolecules into their constituent parts, yields little energy. The second stage, oxidative or aerobic respiration, extracts energy, primarily in the form of high-energy electrons, and produces water and carbon dioxide.

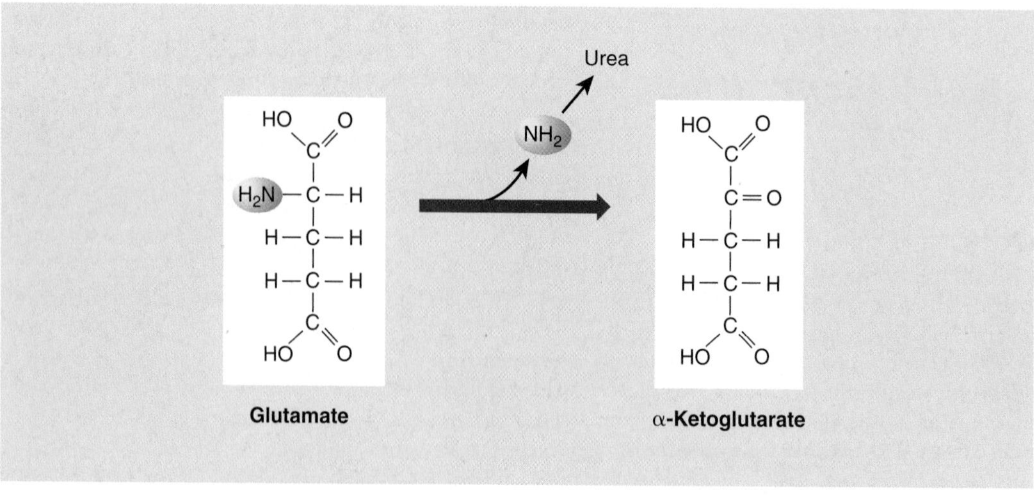

FIGURE 9.21
Deamination. After proteins are broken down into their amino acid constituents, the amino groups are removed from the amino acids to form molecules that participate in glycolysis and the Krebs cycle. For example, the amino acid glutamate becomes α-ketoglutarate, a Krebs cycle molecule, when it loses its amino group.

Cellular Respiration of Fat

Fats are broken down into fatty acids plus glycerol. The tails of fatty acids typically have 16 or more —CH₂ links, and the many hydrogen atoms in these long tails provide a rich harvest of energy. Fatty acids are oxidized in the matrix of the mitochondrion. Enzymes there remove the two-carbon acetyl groups from the end of each fatty acid tail until the entire fatty acid is converted into acetyl groups (figure 9.22). Each acetyl group then combines with coenzyme A to form acetyl-CoA. This process is known as **β oxidation.**

How much ATP does the catabolism of fatty acids produce? Let's compare a hypothetical six-carbon fatty acid with the six-carbon glucose molecule, which we've said yields about 30 molecules of ATP in a eukaryotic cell. Two rounds of β oxidation would convert the fatty acid into three molecules of acetyl-CoA. Each round requires one molecule of ATP to prime the process, but it also produces one molecule of NADH and one of FADH₂. These molecules together yield four molecules of ATP (assuming 2.5 ATPs per NADH and 1.5 ATPs per FADH₂). The oxidation of each acetyl-CoA in the Krebs cycle ultimately produces an additional 10 molecules of ATP. Overall, then, the ATP yield of a six-carbon fatty acid would be approximately: 8 (from two rounds of β oxidation) − 2 (for priming those two rounds) + 30 (from oxidizing the three acetyl-CoAs) = 36 molecules of ATP. Therefore, the respiration of a six-carbon fatty acid yields 20% more ATP than the respiration of glucose. Moreover, a fatty acid of that size would weigh less than two-thirds as much as glucose, so a gram of fatty acid contains more than twice as many kilocalories as a gram of glucose. That is why fat is a storage molecule for excess energy in many types of animals. If excess energy were stored instead as carbohydrate, as it is in plants, animal bodies would be much bulkier.

Proteins, fats, and other molecules are also metabolized for energy. The amino acids of proteins are first deaminated, and fats undergo a process called β oxidation.

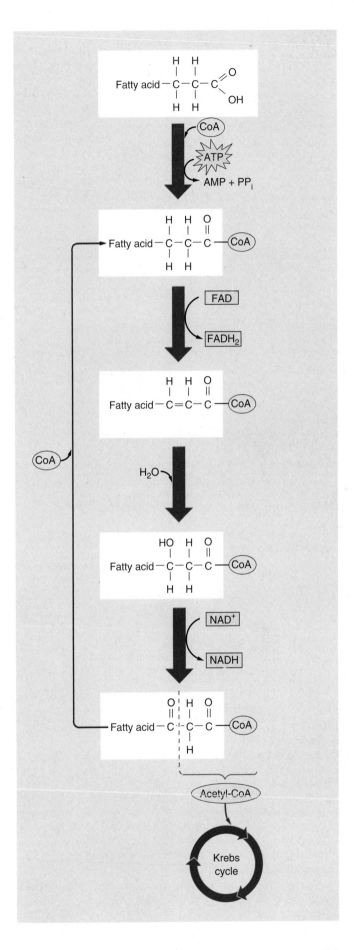

FIGURE 9.22

β oxidation. Through a series of reactions known as β oxidation, the last two carbons in a fatty acid tail combine with coenzyme A to form acetyl-CoA, which enters the Krebs cycle. The fatty acid, now two carbons shorter, enters the pathway again and keeps reentering until all its carbons have been used to form acetyl-CoA molecules. Each round of β oxidation uses one molecule of ATP and generates one molecule each of FADH₂ and NADH.

Metabolic Efficiency and the Length of Food Chains

It has been estimated that a heterotroph limited to glycolysis captures only 3.5% of the energy in the autotrophs it consumes as food. Hence, any other heterotrophs that consume the first heterotroph will capture through glycolysis 3.5% of the energy in it, or 0.12% of the energy available in the original autotrophs. A very large base of autotrophs would thus be needed to support a small number of heterotrophs.

When organisms became able to extract energy from organic molecules by oxidative metabolism, this constraint became far less severe, because the efficiency of oxidative respiration is estimated to be about 32%. This increased efficiency results in the transmission of much more energy from one trophic level to another than does glycolysis. (A trophic level is a step in the movement of energy through an ecosystem.) The efficiency of oxidative metabolism has made possible the evolution of food chains in which autotrophs are consumed by heterotrophs, which are consumed by other heterotrophs, and so on. You will read more about food chains in chapter 55.

Even with oxidative metabolism, approximately two-thirds of the available energy is lost at each trophic level, and that limits how long a food chain can be. Most food chains, like the one illustrated in figure 9.A, involve only three or rarely four trophic levels. Too much energy is lost at each transfer to allow chains to be much longer than that. For example, it would be impossible for a large human population to subsist by eating lions captured from the Serengeti Plain of Africa; the amount of grass available there would not support enough zebras and other herbivores to maintain the number of lions needed to feed the human population. Thus, the ecological complexity of our world is fixed in a fundamental way by the chemistry of oxidative respiration.

Stage 1: Photosynthesizers

Stage 2: Herbivores

Stage 3: Carnivore

Stage 4: Scavengers

Stage 5: Refuse utilizers

FIGURE 9.A
A food chain in the savannas, or open grasslands, of East Africa. Stage 1: *Photosynthesizer.* The grass under this yellow fever tree grows actively during the hot, rainy season, capturing the energy of the sun and storing it in molecules of glucose, which are then converted into starch and stored in the grass. Stage 2: *Herbivore.* This mother zebra and her baby consume the grass and transfer some of its stored energy into their own bodies. Stage 3: *Carnivore.* The lion feeds on zebras and other animals, capturing part of their stored energy and storing it in its own body. Stage 4: *Scavenger.* This hyena and the vultures occupy the same level in the food chain as the lion. They are also consuming the body of the dead zebra, which has been abandoned by the lion. Stage 5: *Refuse utilizer.* These butterflies, mostly *Precis octavia*, are feeding on the material left in the hyena's dung after the food the hyena consumed had passed through its digestive tract. At each of these four levels, only about one-third or less of the energy present is used by the recipient.

Cells can metabolize food without oxygen.

Fermentation

In the absence of oxygen, aerobic metabolism cannot occur, and cells must rely exclusively on glycolysis to produce ATP. Under these conditions, the hydrogen atoms generated by glycolysis are donated to organic molecules in a process called **fermentation** that recycles NAD^+, the electron acceptor required for glycolysis to proceed.

Bacteria carry out more than a dozen kinds of fermentations, all using some form of organic molecule to accept the hydrogen atom from NADH and thus recycle NAD^+:

organic molecule → reduced organic molecule
+ NADH   + NAD^+

Often the reduced organic compound is an organic acid—such as acetic acid, butyric acid, propionic acid, or lactic acid—or an alcohol.

Ethanol Fermentation

Eukaryotic cells are capable of only a few types of fermentation. In one type, which occurs in single-celled fungi called yeast, the molecule that accepts hydrogen from NADH is pyruvate, the end product of glycolysis itself. Yeast enzymes remove a terminal CO_2 group from pyruvate through decarboxylation, producing a two-carbon molecule called acetaldehyde. The CO_2 released causes bread made with yeast to rise, while bread made without yeast (unleavened bread) does not. The acetaldehyde accepts a hydrogen atom from NADH, producing NAD^+ and ethanol (ethyl alcohol) (figure 9.23). This particular type of fermentation is of great interest to humans, since it is the source of the ethanol in wine and beer. Ethanol is a by-product of fermentation that is actually toxic to yeast; as it approaches a concentration of about 12%, it begins to kill the yeast. That explains why naturally fermented wine contains only about 12% ethanol.

Lactic Acid Fermentation

Most animal cells regenerate NAD^+ without decarboxylation. Muscle cells, for example, use an enzyme called lactate dehydrogenase to transfer a hydrogen atom from NADH back to the pyruvate that is produced by glycolysis. This reaction converts pyruvate into lactic acid and regenerates NAD^+ from NADH. It therefore closes the metabolic circle, allowing glycolysis to continue as long as glucose is available. Circulating blood removes excess lactate (the ionized form of lactic acid) from muscles, but when removal cannot keep pace with production, the accumulating lactic acid interferes with muscle function and contributes to muscle fatigue.

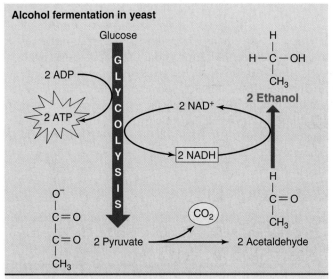

Alcohol fermentation in yeast

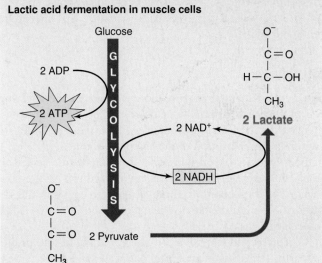

Lactic acid fermentation in muscle cells

FIGURE 9.23
Fermentation. Yeasts carry out the conversion of pyruvate to ethanol (see also figure 9.9). Muscle cells convert pyruvate into lactate, which is less toxic than ethanol. However, lactate is still toxic enough to produce a painful sensation in muscles during heavy exercise, when oxygen in the muscles is depleted.

In fermentation, which occurs in the absence of oxygen, the electrons that result from the glycolytic breakdown of glucose are donated to an organic molecule, regenerating NAD^+ from NADH.

9.5 The stages of cellular respiration evolved over time.

The Evolution of Metabolism

We talk about cellular respiration as a continuous series of stages, but it is important to note that these stages evolved over time. Metabolism has changed a great deal as life on earth has evolved. This has been particularly true of the re- actions organisms use to capture energy from the sun to build organic molecules (anabolism) and then break down organic molecules to obtain energy (catabolism). These processes evolved in concert with each other.

Degradation

The most primitive forms of life are thought to have ob- tained chemical energy by degrading, or breaking down, organic molecules that were abiotically produced.

The first major event in the evolution of metabolism was the origin of the ability to harness chemical bond energy. At an early stage, organisms began to store this energy in the bonds of ATP.

Glycolysis

The second major event in the evolution of metabolism was glycolysis, the initial breakdown of glucose. As proteins evolved diverse catalytic functions, it became possible to capture a larger fraction of the chemical bond energy in or- ganic molecules by breaking chemical bonds in a series of steps. For example, the progressive breakdown of the six- carbon sugar glucose into three-carbon molecules in the 10 steps of glycolysis results in the net production of two ATP molecules.

Glycolysis undoubtedly evolved early in the history of life on earth, since this biochemical pathway has been retained by all living organisms. It is a chemical process that does not appear to have changed for well over 2 billion years.

Anaerobic Photosynthesis

The third major event in the evolution of metabolism was anaerobic photosynthesis. Early in the history of life, some organisms evolved a different way of generating ATP, called photosynthesis. Instead of obtaining energy for ATP synthesis by reshuffling chemical bonds, as in glycolysis, these organisms developed the ability to use light to pump protons out of their cells, and to use the resulting proton gradient to power the production of ATP through chemiosmosis.

Photosynthesis evolved in the absence of oxygen and works well without it. Dissolved H_2S, present in the oceans beneath an atmosphere free of oxygen gas, served as a ready source of hydrogen atoms for building organic molecules. Free sulfur was produced as a by-product of this reaction.

Oxygen-Forming Photosynthesis

The substitution of H_2O for H_2S in photosynthesis was the fourth major event in the history of metabolism. Oxygen- forming photosynthesis employs H_2O rather than H_2S as a source of hydrogen atoms and their associated electrons. Because it garners its hydrogen atoms from reduced oxygen rather than from reduced sulfur, it generates oxygen gas rather than free sulfur.

More than 2 billion years ago, small cells capable of car- rying out this oxygen-forming photosynthesis, such as cyanobacteria, became the dominant forms of life on earth. Oxygen gas began to accumulate in the atmosphere. This was the beginning of a great transition that changed condi- tions on earth permanently. Our atmosphere is now 20.9% oxygen, every molecule of which is derived from an oxygen-forming photosynthetic reaction.

Nitrogen Fixation

Nitrogen fixation was the fifth major step in the evolution of metabolism. Proteins and nucleic acids cannot be syn- thesized from the products of photosynthesis because both of these biologically critical molecules contain nitrogen. Obtaining nitrogen atoms from N_2 gas, a process called *ni- trogen fixation*, requires breaking an N≡N triple bond. This important reaction evolved in the hydrogen-rich at- mosphere of the early earth, where no oxygen was present. Oxygen acts as a poison to nitrogen fixation, which today occurs only in oxygen-free environments or in oxygen-free compartments within certain prokaryotes.

Aerobic Respiration

Aerobic respiration is the sixth and final event in the his- tory of metabolism. This cellular process harvests energy by stripping energetic electrons from organic molecules. Aerobic respiration employs the same kind of proton pumps as photosynthesis, and is thought to have evolved as a modification of the basic photosynthetic machinery.

Biologists think that the ability to carry out photosyn- thesis without H_2S first evolved among purple nonsulfur bacteria, which obtain their hydrogens from organic com- pounds instead. It was perhaps inevitable that among the descendants of these respiring photosynthetic bacteria, some would eventually do without photosynthesis entirely, subsisting only on the energy and hydrogens derived from the breakdown of organic molecules. The mitochondria within all eukaryotic cells are thought to be descendants of these bacteria.

Six major innovations highlight the evolution of metabolism as we know it today.

Concept Review

For interactive testing, visit the Online Learning Center with PowerWeb at www.mhhe.com/Raven7

9.1 Cells harvest the energy in chemical bonds.

Using Chemical Energy to Drive Metabolism

- Autotrophs produce their own chemical energy, while heterotrophs live on the energy autotrophs produce. (p. 160)
- The energy of a chemical bond is contained in the potential energy of the electrons that make up the bond. (p. 160)
- Cells use some of the energy gained by catabolizing food to drive ATP production. (p. 160)
- ATP stores energy by linking charged phosphate groups near one another. (p. 160)
- Cells use ATP to facilitate movement and to drive endergonic reactions. (p. 160)
- The vast majority of ATP produced in the cell is made by ATP synthase. (p. 161)

9.2 Cellular respiration oxidizes food molecules.

An Overview of Glucose Catabolism

- Cells can make ATP from the catabolism of organic molecules two ways: substrate-level phosphorylation and aerobic respiration. (p. 162)
- In many organisms, cells harvest energy from glucose molecules in a sequence of four pathways: glycolysis, pyruvate oxidation, the Krebs cycle, and the electron transport chain. (pp. 162–163)
- Anaerobic respiration donates harvested electrons to inorganic compounds other than oxygen. (p. 163)

Stage One: Glycolysis

- Glycolysis generates ATP by shuffling the bonds in glucose molecules. Two molecules of NAD^+ are reduced to NADH. (p. 164)
- NAD^+ must be regenerated for glycolysis to continue. (pp. 166–167)

Stage Two: The Oxidation of Pyruvate

- Pyruvate is decarboxylated within the mitochondrion, yielding acetyl-CoA, NADH, and CO_2. (p. 168)

Stage Three: The Krebs Cycle

- The Krebs cycle is a series of nine reactions that oxidize acetyl-CoA in the matrix of a mitochondrion. (p. 169)
- The Krebs cycle yields two molecules of ATP per molecule of glucose. (p. 170)

Harvesting Energy by Extracting Electrons

- Glucose catabolism involves a series of oxidation-reduction reactions that release energy by repositioning electrons closer to oxygen atoms. (pp. 172–173)
- Energy is harvested in gradual steps, using NAD^+ as an electron carrier. (p. 172)

Stage Four: The Electron Transport Chain

- The electron transport chain is a series of membrane-associated proteins. Electrons delivered by NADH and $FADH_2$ are passed along the protein chain, and are used to pump protons out of the mitochondrial matrix via the electron transport chain. The return of protons into the matrix through ATP synthase generates ATP. (p. 175)

Summarizing the Yield of Aerobic Respiration

- The theoretical yield of aerobic respiration is 36 molecules of ATP, while the actual yield is around 30 molecules of ATP. (p. 176)
- Aerobic respiration harvests about 32% of the energy available in glucose. (p. 176)

Regulating Aerobic Respiration

- Relative levels of ADP and ATP control the catabolic pathway at the committing reactions of glycolysis and the Krebs cycle. (p. 177)

9.3 Catabolism of proteins and fats can yield considerable energy.

Cellular Respiration of Protein

- Proteins are first broken down into their individual amino acids, and then the amino group is removed from each amino acid by deamination. (p. 178)
- Glycolysis and the Krebs cycle then extract high-energy electrons from the molecules and use them in producing ATP. (p. 178)

Cellular Respiration of Fat

- Fats are metabolized for energy by β oxidation. (p. 179)

9.4 Cells can metabolize food without oxygen.

Fermentation

- Fermentation occurs in the absence of oxygen as electrons from the glycolytic breakdown of glucose are donated to an organic molecule, regenerating NAD^+ from NADH. (p. 181)

9.5 The stages of cellular respiration evolved over time.

The Evolution of Metabolism

- The six major innovations of metabolism are degradation, glycolysis, anaerobic photosynthesis, oxygen-forming photosynthesis, nitrogen fixation, and aerobic respiration. (p. 182)

Self Test

1. In cellular respiration, energy-depleted electrons are donated to an inorganic molecule. In fermentation, what molecule accepts these electrons?
   a. oxygen
   b. an organic molecule
   c. sulfur
   d. an inorganic molecule other than O_2

2. Which of the following is *not* a stage of aerobic respiration?
   a. glycolysis
   b. pyruvate oxidation
   c. the Krebs cycle
   d. the electron transport chain

3. Which steps in glycolysis require the input of energy?
   a. the glucose priming steps
   b. the phosphorylation of glucose
   c. the phosphorylation of fructose 6-phosphate
   d. All of these steps require the input of energy.

4. Pyruvate dehydrogenase is a multienzyme complex that catalyzes a series of reactions. Which of the following is *not* carried out by pyruvate dehydrogenase?
   a. a decarboxylation reaction
   b. the production of ATP
   c. producing an acetyl group from pyruvate
   d. combining the acetyl group with a cofactor

5. How many molecules of CO_2 are produced for each molecule of glucose that passes through glycolysis and the Krebs cycle?
   a. 2
   b. 3
   c. 6
   d. 7

6. The electrons generated from the Krebs cycle are transferred to _____ and then are shuttled to _____.
   a. NAD^+ / oxygen
   b. NAD^+ / electron transport chain
   c. NADH / oxygen
   d. NADH / electron transport chain

7. The electron transport chain pumps protons
   a. out of the mitochondrial matrix.
   b. out of the intermembrane space and into the matrix.
   c. out of the mitochondrion and into the cytoplasm.
   d. out of the cytoplasm and into the mitochondrion.

8. What process of cellular respiration generates the most ATP?
   a. glycolysis
   b. oxidation of pyruvate
   c. Krebs cycle
   d. chemiosmosis

9. Oxidizing which of the following substances yields the most energy?
   a. proteins
   b. glucose
   c. fatty acids
   d. water

10. The final electron acceptor in lactic acid fermentation is
    a. pyruvate.
    b. NAD^+.
    c. lactic acid.
    d. O_2.

Test Your Visual Understanding

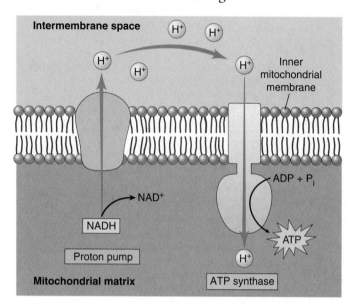

Answer the following questions related to this figure, which shows the process of chemiosmosis.

1. In the figure, proton pumps are transporting hydrogen ions across the membrane. What is the driving force of this pump? What type of membrane transport is the proton pump? Explain.

2. Why is this process called chemiosmosis? What is the force driving the synthesis of ATP? How could the process of ATP synthesis be inhibited or shut down?

Apply Your Knowledge

1. How much energy would be generated in the cells of a person who consumed a diet of pyruvate instead of glucose? Calculate the energy generated on a per molecule basis.

2. As explained in chapter 5, mitochondria are thought to have evolved from bacteria that were engulfed by and lived symbiotically within early eukaryotic cells. Why haven't present-day eukaryotic cells dispensed with mitochondria, placing all of the mitochondrial genes in the nucleus and carrying out all of the metabolic functions of the mitochondria within the cytoplasm?

3. Why do plants typically store their excess energy as carbohydrates rather than fat?

4. If you poke a hole in a mitochondrion, can it still perform oxidative respiration? Can fragments of mitochondria perform oxidative respiration?

10

Photosynthesis

Concept Outline

FIGURE 10.1
Capturing energy. These sunflower plants, growing vigorously in the August sun, are capturing light energy for conversion into chemical energy through photosynthesis.

The rich diversity of life that covers our earth would be impossible without photosynthesis. Every oxygen atom in the air we breathe was once part of a water molecule, liberated by photosynthesis. All the energy released by the burning of coal, firewood, gasoline, and natural gas, and by our bodies' burning of all the food we eat—directly or indirectly—has been captured from sunlight by photosynthesis. It is vitally important, then, that we understand photosynthesis. Research may enable us to improve crop yields and land use, important goals in an increasingly crowded world. In chapter 9, we described how cells extract chemical energy from food molecules and use that energy to power their activities. In this chapter, we examine photosynthesis, the process by which organisms capture energy from sunlight and use it to build food molecules that are rich in chemical energy (figure 10.1).

10.1 What is photosynthesis?

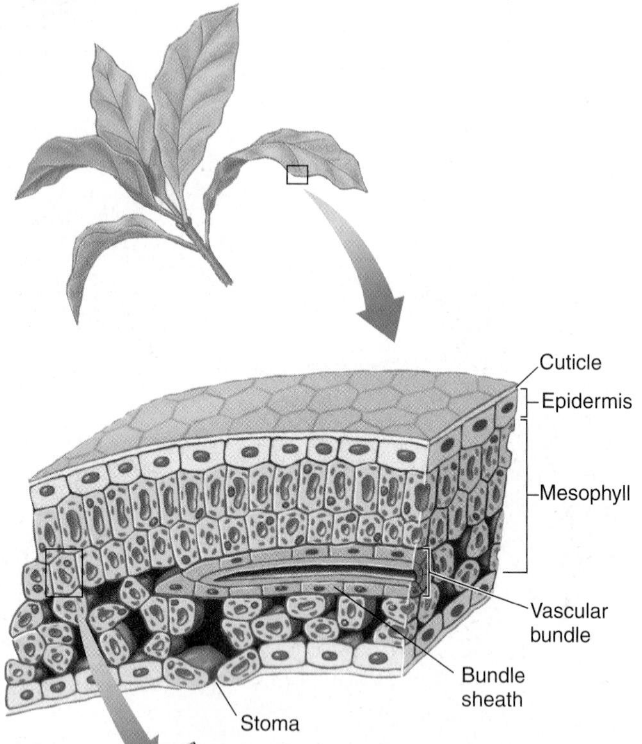

Cuticle
Epidermis
Mesophyll
Vascular bundle
Bundle sheath
Stoma

Chloroplasts
Nucleus
Vacuole
Cell wall

Inner membrane
Outer membrane
Granum
Stroma
Thylakoid

The Chloroplast as a Photosynthetic Machine

Life is powered by sunshine. The energy used by most living cells comes ultimately from the sun, captured by plants, algae, and bacteria through the process of photosynthesis. The diversity of life is only possible because our planet is awash in energy streaming earthward from the sun. Each day, the radiant energy that reaches the earth equals about 1 million Hiroshima-sized atomic bombs. Photosynthesis captures about 1% of this huge supply of energy (an amount equal to 10,000 Hiroshima bombs), and uses it to provide the energy that drives all life.

The Photosynthetic Process: A Summary

Photosynthesis occurs in many kinds of bacteria and algae, as well as in the leaves and sometimes the stems of green plants. Figure 10.2 describes the levels of organization in a plant leaf. Recall from chapter 5 that the cells of plant leaves contain organelles called chloroplasts, which carry out the photosynthetic process. No other structure in a

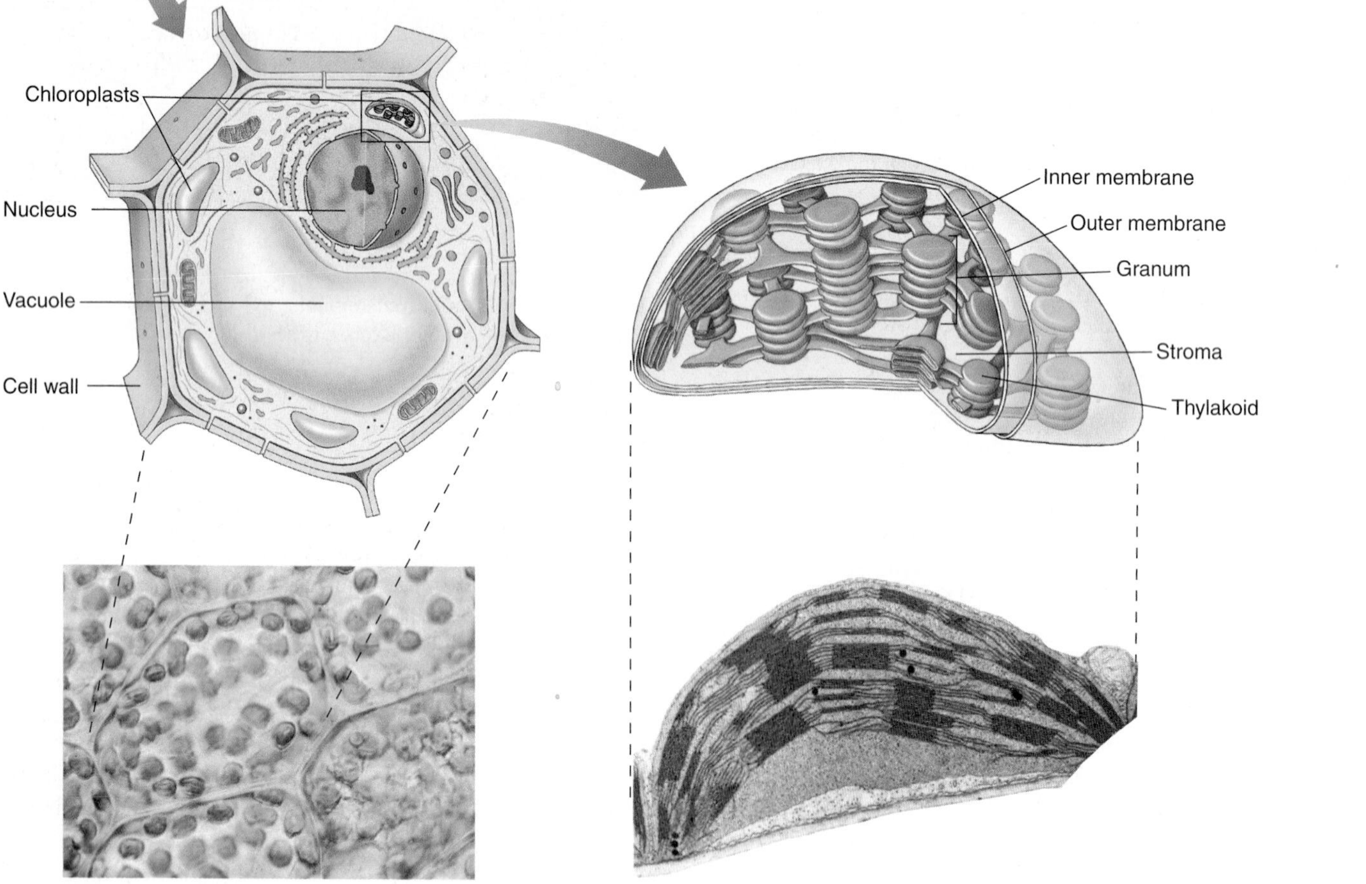

FIGURE 10.2
Journey into a leaf. A plant leaf possesses a thick layer of cells (the mesophyll) rich in chloroplasts. The flattened thylakoids in the chloroplast are stacked into columns called grana. The light-dependent reactions take place on the thylakoid membrane and generate the

plant cell is able to carry out photosynthesis. Photosynthesis takes place in three stages: (1) capturing energy from sunlight; (2) using the energy to make ATP and reducing power in the form of a compound called NADPH; and (3) using the ATP and NADPH to power the synthesis of organic molecules from CO_2 in the air (carbon fixation).

The first two stages can only take place in the presence of light and are commonly called the **light-dependent reactions.** The third stage, the formation of organic molecules from atmospheric CO_2, occurs in a process called the **Calvin cycle.** As long as ATP and NADPH are available, the Calvin cycle can occur either in the presence or in the absence of light, and so the reactions of the Calvin cycle are also called the **light-independent reactions.**

The following simple equation summarizes the overall process of photosynthesis:

$$6\ CO_2 + 12\ H_2O + light \rightarrow C_6H_{12}O_6 + 6\ H_2O + 6\ O_2$$

carbon   water   glucose   water   oxygen
dioxide

Inside the Chloroplast

The internal membranes of chloroplasts, called *thylakoids*, are organized into sacs stacked on one another in columns called **grana** (singular, *granum*). The thylakoid membranes house the photosynthetic pigments for capturing light energy and the machinery to make ATP. Surrounding the thylakoid membrane system is a semiliquid substance called *stroma*. The stroma houses the enzymes needed to assemble

organic molecules from CO_2 using energy from ATP and reducing power in NADPH. In the membranes of thylakoids, photosynthetic pigments are clustered together to form a **photosystem.**

Each pigment molecule within the photosystem is capable of capturing *photons*, which are packets of energy. A lattice of proteins holds the pigments in close contact with one another. When light of a proper wavelength strikes a pigment molecule in the photosystem, the resulting excitation passes from one chlorophyll molecule to another. The excited electron is not transferred physically— rather, its *energy* passes from one molecule to another. A crude analogy to this form of energy transfer is the initial "break" in a game of pool. If the cue ball squarely hits the point of the triangular array of 15 pool balls, the two balls at the far corners of the triangle fly off, but none of the central balls move. The energy passes through the central balls to the most distant ones.

Eventually, the energy arrives at a key chlorophyll molecule that is touching a membrane-bound protein. The energy is transferred as an excited electron to that protein, which passes it on to a series of other membrane proteins that put the energy to work making ATP and NADPH and building organic molecules. The photosystem thus acts as a large antenna, gathering the light energy harvested by many individual pigment molecules.

The light-dependent reactions of photosynthesis take place within thylakoid membranes within chloroplasts in leaf cells.

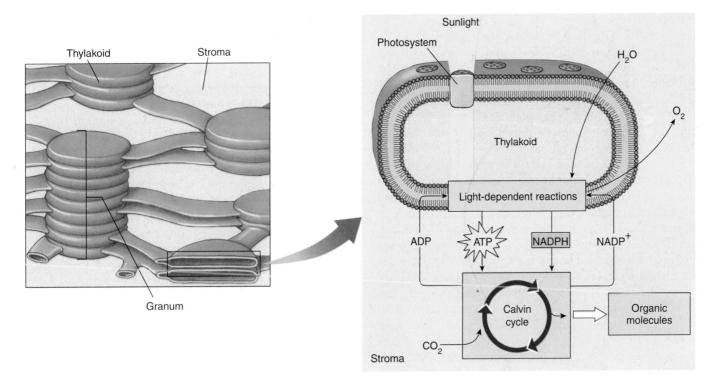

FIGURE 10.2 (continued)
ATP and NADPH that fuel the Calvin cycle. The fluid interior matrix of a chloroplast, the stroma, contains the enzymes that carry out the Calvin cycle.

10.2 Learning about photosynthesis: an experimental journey.

The Role of Soil and Water

The story of how we learned about photosynthesis is one of the most interesting in science and serves as a good introduction to this complex process. It begins over 300 years ago, with a simple but carefully designed experiment by a Belgian doctor, Jan Baptista van Helmont (1577–1644). From the time of the Greeks, plants were thought to obtain their food from the soil, literally sucking it up with their roots; van Helmont thought of a simple way to test this idea. He planted a small willow tree in a pot of soil, after first weighing the tree and the soil. The tree grew in the pot for several years, during which time van Helmont added only water. At the end of five years, the tree was much larger, its weight having increased by 74.4 kilograms. However, the soil in the pot weighed only 57 grams less than it had five years earlier, so obviously all of this added mass could not have come from the soil. With this experiment, van Helmont demonstrated that the substance of the plant was not produced only from the soil. He incorrectly concluded that the water he had been adding mainly accounted for the plant's increased mass.

A hundred years passed before the story became clearer. The key clue was provided by the English scientist Joseph Priestly, in his pioneering studies of the properties of air. On the 17th of August, 1771, Priestly "accidentally hit upon a method of restoring air that had been injured by the burning of candles." He "put a [living] sprig of mint into air in which a wax candle had burnt out and found that, on the 27th of the same month, another candle could be burned in this same air." Somehow, the vegetation seemed to have restored the air! Priestly found that while a mouse could not breathe candle-exhausted air, air "restored" by vegetation was not "at all inconvenient to a mouse." The key clue was that *living vegetation adds something to the air.*

How does vegetation "restore" air? Twenty-five years later, the Dutch physician Jan Ingenhousz solved the puzzle. Working over several years, Ingenhousz reproduced and significantly extended Priestly's results, demonstrating that air was restored only in the presence of sunlight and only by a plant's green leaves, not by its roots. He proposed that the green parts of the plant carry out a process (which we now call photosynthesis) that uses sunlight to split carbon dioxide (CO_2) into carbon and oxygen. He suggested that the oxygen was released as O_2 gas into the air, while the carbon atom combined with water to form carbohydrates. His proposal was a good guess, even though the latter step was subsequently modified. Chemists later found that the proportions of carbon, oxygen, and hydrogen atoms in carbohydrates are indeed about one atom of carbon per molecule of water (as the term *carbohydrate* indicates). A Swiss botanist found in 1804 that water was a necessary reactant. By the end of that century, the overall reaction for photosynthesis could be written as:

$$CO_2 + H_2O + \text{light energy} \rightarrow (CH_2O) + O_2$$

It turns out, however, that there's more to it than that. When researchers began to examine the process in more detail in the twentieth century, the role of light proved to be unexpectedly complex.

Van Helmont showed that soil did not add mass to a growing plant. Priestly, Ingenhousz, and others then worked out the basic chemical reaction.

Discovery of the Light-Independent Reactions

Ingenhousz's early equation for photosynthesis includes one factor we have not discussed: light energy. What role does light play in photosynthesis? At the beginning of the twentieth century, the English plant physiologist F. F. Blackman began to question the role of light in photosynthesis. In 1905, he came to the startling conclusion that photosynthesis is in fact a multistage process, only one of which uses light directly.

Blackman measured the effects of different light intensities, CO_2 concentrations, and temperatures on photosynthesis. As long as light intensity was relatively low, he found photosynthesis could be accelerated by increasing the amount of light, but not by increasing the temperature or CO_2 concentration (figure 10.3). At high light intensities, however, an increase in temperature or CO_2 concentration greatly accelerated photosynthesis. Blackman concluded that photosynthesis consists of an initial set of what he called "light" reactions, that are largely independent of temperature, and a second set of "dark" reactions (also called light-independent reactions), that seemed to be independent of light but limited by CO_2. Do not be confused by Blackman's labels—the so-called "dark" reactions occur in the light (in fact, they require the products of the light-dependent reactions); their name simply indicates that light is not *directly* involved in those reactions.

Blackman found that increased temperature speeds the rate of the dark carbon-reducing reactions, but only up to about 35°C. Higher temperatures cause the rate to fall off rapidly. Because 35°C is the temperature at which many plant enzymes begin to be denatured (the hydrogen bonds that hold an enzyme in its particular catalytic shape begin to be disrupted), Blackman concluded that enzymes must carry out the light-independent reactions.

Blackman showed that capturing photosynthetic energy requires sunlight, while building organic molecules does not.

The Role of Light and Reducing Power

The role of light in the so-called light-dependent and light-independent reactions was worked out in the 1930s by C. B. van Niel, then a graduate student at Stanford University. While studying photosynthesis in bacteria, van Niel discovered that purple sulfur bacteria do not release oxygen during photosynthesis; instead, they convert hydrogen sulfide (H_2S) into globules of pure elemental sulfur that accumulate inside themselves. The process van Niel observed was:

$$CO_2 + 2\ H_2S + \text{light energy} \rightarrow (CH_2O) + H_2O + 2\ S$$

The striking parallel between this equation and Ingenhousz's equation led van Niel to propose that the generalized process of photosynthesis is in fact:

$$CO_2 + 2\ H_2A + \text{light energy} \rightarrow (CH_2O) + H_2O + 2\ A$$

In this equation, the substance H_2A serves as an electron donor. In photosynthesis performed by green plants, H_2A is water, while among purple sulfur bacteria, H_2A is hydrogen sulfide. The product, A, comes from the splitting of H_2A. Therefore, the O_2 produced during green plant photosynthesis results from splitting water, not carbon dioxide.

When isotopes came into common use in biology in the early 1950s, it became possible to test van Niel's revolutionary proposal. Investigators examined photosynthesis in green plants supplied with ^{18}O water; they found that the ^{18}O label ended up in oxygen gas rather than in carbohydrate, just as van Niel had predicted:

$$CO_2 + 2\ H_2{}^{18}O + \text{light energy} \rightarrow (CH_2O) + H_2O + {}^{18}O_2$$

In algae and green plants, the carbohydrate typically produced by photosynthesis is the sugar glucose, which has six carbons. The complete balanced equation for photosynthesis in these organisms thus becomes:

$$6\ CO_2 + 12\ H_2O + \text{light energy} \rightarrow C_6H_{12}O_6 + 6\ O_2 + 6\ H_2O$$

We now know that the light-dependent reactions of photosynthesis use the energy of light to reduce NADP (an electron carrier molecule) to NADPH and to manufacture ATP. The NADPH and ATP from this stage of photosynthesis are then used in a subsequent stage, the Calvin cycle, to reduce the carbon in carbon dioxide and form a simple sugar whose carbon skeleton can be used to synthesize other organic molecules.

In his pioneering work on the light-dependent reactions, van Niel had further proposed that the reducing power (H^+) generated by the splitting of water was used to convert CO_2 into organic matter in a process he called **carbon fixation**. This reducing power supplies high-energy electrons from NADPH to replace the low-energy electrons in the C—O bonds of carbon dioxide. These high-energy electrons form the C—H bonds of the newly synthesized organic molecules.

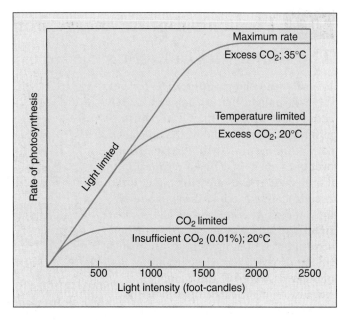

FIGURE 10.3

Discovery of the light-independent reactions. Blackman measured photosynthesis rates under differing light intensities, CO_2 concentrations, and temperatures. As this graph shows, light is the limiting factor at low light intensities, while temperature and CO_2 concentration are the limiting factors at higher light intensities. **Blackman found that increasing light intensity above 2000 foot-candles did not lead to any further increase in the rate of photosynthesis. Can you suggest a hypothesis that would explain why?**

In the 1950s, Robin Hill demonstrated that van Niel was indeed right, and that light energy could be used to generate reducing power. Chloroplasts isolated from leaf cells were able to reduce a dye and release oxygen in response to light. Later experiments showed that the electrons released from water were transferred to NADP$^+$. Arnon and co-workers showed that illuminated chloroplasts deprived of CO_2 accumulate ATP. If CO_2 is then introduced, neither ATP nor NADPH accumulate, and the CO_2 is assimilated into organic molecules. These experiments are important for three reasons: First, they firmly demonstrate that photosynthesis occurs only within chloroplasts. Second, they show that the light-dependent reactions use light energy to reduce NADP$^+$ and to manufacture ATP. Third, they confirm that the ATP and NADPH from this early stage of photosynthesis are then used in the later light-independent reactions to reduce carbon dioxide, forming simple sugars.

Van Niel discovered that photosynthesis splits water molecules, incorporating the carbon atoms of carbon dioxide gas and the hydrogen atoms of water into organic molecules and leaving oxygen gas.
Hill showed that plants can use light energy to generate reducing power. The incorporation of carbon dioxide into organic molecules in the light-independent reactions is called carbon fixation.

10.3 Pigments capture energy from sunlight.

The Biophysics of Light

Where is the energy in light? What is there in sunlight that a plant can use to reduce carbon dioxide? This is the mystery of photosynthesis, the one factor fundamentally different from processes such as cellular respiration. To answer these questions, we will need to consider the physical nature of light. James Clerk Maxwell had theorized that light was an electromagnetic wave—that is, that light moves through the air as oscillating electric and magnetic fields. Proof of this came in a curious experiment carried out in a laboratory in Germany in 1887. A young physicist, Heinrich Hertz, was attempting to verify a highly mathematical theory that predicted the existence of electromagnetic waves. To see whether such waves existed, Hertz designed a clever experiment. On one side of a room, he constructed a powerful spark generator that consisted of two large, shiny metal spheres standing near each other on tall, slender rods. When a very high static electrical charge was built up on one sphere, sparks jumped across to the other sphere.

After constructing this device, Hertz set out to investigate whether the sparking would create invisible electromagnetic waves, so-called radio waves, as predicted by the mathematical theory. On the other side of the room, he placed the world's first radio receiver, a thin metal hoop on an insulating stand. There was a small gap at the bottom of the hoop, so that the hoop did not quite form a complete circle. When Hertz turned on the spark generator across the room, he saw tiny sparks passing across the gap in the hoop! This was the first demonstration of radio waves. But Hertz noted another curious phenomenon. When UV light was shining across the gap on the hoop, the sparks were produced more readily. This unexpected facilitation, called the *photoelectric effect*, puzzled investigators for many years.

The photoelectric effect was finally explained using a concept proposed by Max Planck in 1901. Planck developed an equation that predicted the blackbody radiation curve based upon the assumption that light and other forms of radiation behave as units of energy called *photons*. In 1905, Albert Einstein explained the photoelectric effect utilizing the photon concept. Ultraviolet light has photons of sufficient energy that, when they fell on the loop, electrons were ejected from the metal surface. The photons had transferred their energy to the electrons, literally blasting them from the ends of the hoop and thus facilitating the

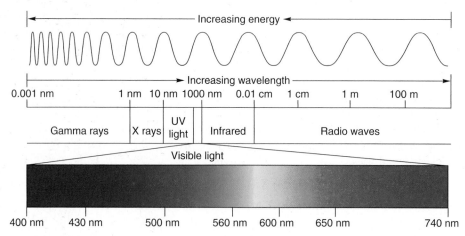

FIGURE 10.4
The electromagnetic spectrum. Light is a form of electromagnetic energy conveniently thought of as a wave. The shorter the wavelength of light, the greater its energy. Visible light represents only a small part of the electromagnetic spectrum between 400 and 740 nanometers.

passage of the electric spark induced by the radio waves. Visible wavelengths of light were unable to remove the electrons because their photons did not have enough energy to free the electrons from the metal surface at the ends of the hoop.

The Energy in Photons

Not all photons possess the same amount of energy. Instead, the energy content of a photon is inversely proportional to the wavelength of the light: Short-wavelength light contains photons of higher energy than long-wavelength light (figure 10.4). X rays, which contain a great deal of energy, have very short wavelengths—much shorter than those of visible light, making them ideal for use in high-resolution microscopes.

Hertz had noted that the strength of the photoelectric effect depends on the wavelength of light; that is, short wavelengths are much more effective than long ones in producing the photoelectric effect. Einstein's theory of the photoelectric effect provides an explanation: Sunlight contains photons of many different energy levels, only some of which our eyes perceive as visible light. The highest energy photons, at the short-wavelength end of the electromagnetic spectrum (see figure 10.4), are gamma rays, with wavelengths of less than 1 nanometer; the lowest energy photons, with wavelengths of up to thousands of meters, are radio waves. Within the visible portion of the spectrum, violet light has the shortest wavelength and the most energetic photons, and red light has the longest wavelength and the least energetic photons.

Ultraviolet Light

The sunlight that reaches the earth's surface contains a significant amount of ultraviolet (UV) light, which, because of its shorter wavelength, possesses considerably more energy than visible light. UV light is thought to have been an important source of energy on the primitive earth when life originated. Today's atmosphere contains ozone (derived from oxygen gas), which absorbs most of the UV photons in sunlight, but a considerable amount of UV light still manages to penetrate the atmosphere. This UV light is a potent force in disrupting the bonds of DNA, causing mutations that can lead to skin cancer. As we will describe in chapter 56, loss of atmospheric ozone due to human activities threatens to cause an enormous jump in the incidence of human skin cancers throughout the world.

Absorption Spectra and Pigments

How does a molecule "capture" the energy of light? A photon can be envisioned as a very fast-moving packet of energy. When it strikes a molecule, its energy is either lost as heat or absorbed by the electrons of the molecule, boosting those electrons into higher energy levels. Whether or not the photon's energy is absorbed depends on how much energy it carries (defined by its wavelength) and also on the chemical nature of the molecule it hits. As we saw in chapter 2, electrons occupy discrete energy levels in their orbits around atomic nuclei. To boost an electron into a different energy level requires just the right amount of energy, just as reaching the next rung on a ladder requires you to raise your foot just the right distance. A specific atom can, therefore, absorb only certain photons of light—namely, those that correspond to the atom's available electron energy levels. As a result, each molecule has a characteristic **absorption spectrum,** the range and efficiency of photons it is capable of absorbing.

Molecules that are good absorbers of light in the visible range are called **pigments.** Organisms have evolved a variety of different pigments, but only two general types are used in green plant photosynthesis: carotenoids and chlorophylls. Chlorophylls absorb photons within narrow energy ranges. Two kinds of chlorophyll in plants, chlorophylls *a* and *b*, preferentially absorb violet-blue and red light (figure 10.5). Neither of these pigments absorbs photons with wavelengths between about 500 and 600 nanometers, and light of these wavelengths is, therefore, reflected by plants.

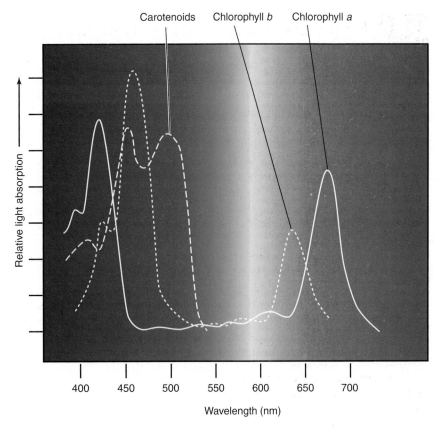

FIGURE 10.5

The absorption spectrum of chlorophyll and carotenoids. The peaks represent wavelengths of sunlight absorbed by the two common forms of photosynthetic pigment, chlorophylls *a* and *b*, and by the carotenoids. Chlorophylls absorb predominantly violet-blue and red light in two narrow bands of the spectrum and reflect green light in the middle of the spectrum. Carotenoids absorb mostly blue and green light and reflect orange and yellow light.

When these photons are subsequently absorbed by the pigment in our eyes, we perceive them as green.

Chlorophyll *a* is the main photosynthetic pigment and the only pigment that can act directly to convert light energy to chemical energy. However, chlorophyll *b*, acting as an *accessory* or secondary light-absorbing pigment, complements and adds to the light absorption of chlorophyll *a*. Chlorophyll *b* has an absorption spectrum shifted toward the green wavelengths. Therefore, chlorophyll *b* can absorb photons that chlorophyll *a* cannot, greatly increasing the proportion of the photons in sunlight that plants can harvest. An important group of accessory pigments, the carotenoids, assist in photosynthesis by capturing energy from light composed of wavelengths that are not efficiently absorbed by either chlorophyll.

In photosynthesis, photons of light are absorbed by pigments; the wavelength of light absorbed depends upon the specific pigment.

Chlorophylls and Carotenoids

Chlorophylls absorb photons by means of an excitation process analogous to the photoelectric effect. These pigments contain a complex ring structure, called a porphyrin ring, with alternating single and double bonds. At the center of the ring is a magnesium atom. Photons absorbed by the pigment molecule excite electrons in the ring, which are then channeled away through the alternating carbon-bond system. Several small side groups attached to the outside of the ring alter the absorption properties of the molecule in different kinds of chlorophyll (figure 10.6). The precise absorption spectrum is also influenced by the local microenvironment created by the association of chlorophyll with specific proteins.

Once Ingenhousz demonstrated that only the green parts of plants can "restore" air, researchers suspected chlorophyll was the primary pigment that plants employ to absorb light in photosynthesis. Experiments conducted in the 1800s clearly verified this suspicion. One such experiment, performed by T. W. Englemann in 1882 (figure 10.7), serves as a particularly elegant example, simple in design and clear in outcome. Englemann set out to characterize the **action spectrum** of photosynthesis—that is, the relative effectiveness of different wavelengths of light in promoting photosynthesis. He carried out the entire experiment utilizing a single slide mounted on a microscope. To obtain different wavelengths of light, he placed a prism under his microscope, splitting the light that illuminated the slide into a spectrum of colors. He then arranged a filament of green algal cells across the spectrum, so that different parts of the filament were illuminated with different wavelengths, and allowed the algae to carry out photosynthesis. To assess how fast photosynthesis was proceeding, Englemann chose to monitor the rate of oxygen production. Lacking a mass spectrometer and other modern instruments, he added aerotactic (oxygen-seeking) bacteria to the slide; he knew they would gather along the filament at locations where oxygen was being produced. He found that the bacteria accumulated in areas illuminated by red and violet light, the two colors most strongly absorbed by chlorophyll.

All plants, algae, and cyanobacteria use chlorophyll *a* as their primary pigments. It is reasonable to ask why these photosynthetic organisms do not use a pigment like retinal (the pigment in our eyes), which has a broad absorption spectrum that covers the range of 500 to 600 nanometers. The most likely hypothesis involves *photoefficiency*. Although retinal absorbs a broad range of wavelengths, it does so with relatively low efficiency. Chlorophyll, in contrast, absorbs in only two narrow bands, but does so with high efficiency. Therefore, plants and most other photosynthetic organisms achieve far higher overall photon capture rates with chlorophyll than with other pigments.

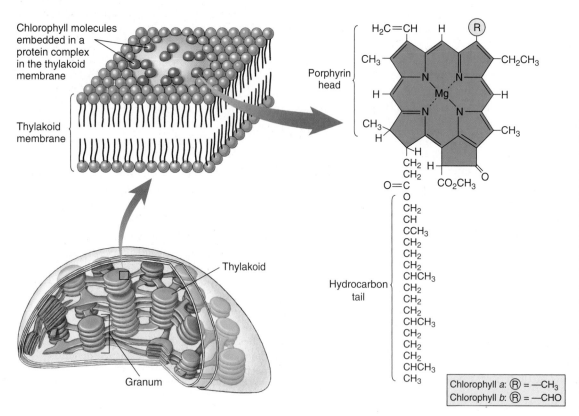

FIGURE 10.6 Chlorophyll. Chlorophyll molecules consist of a porphyrin head and a hydrocarbon tail that anchors the pigment molecule to hydrophobic regions of proteins embedded within the membranes of thylakoids. The only difference between the two chlorophyll molecules is the substitution of a —CHO (aldehyde) group in chlorophyll *b* for a —CH₃ (methyl) group in chlorophyll *a*.

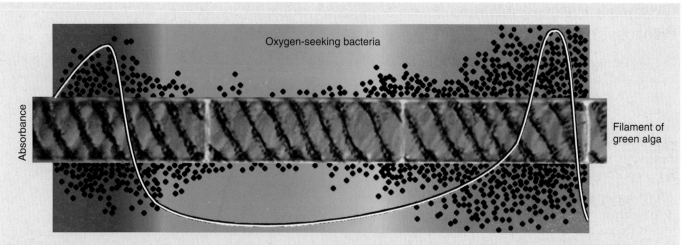

Oxygen-seeking bacteria

Absorbance

Filament of green alga

T.W. Englemann revealed the action spectrum of photosynthesis in the filamentous alga *Spirogyra* in 1882. Englemann used the rate of oxygen production to measure the rate of photosynthesis. As his oxygen indicator, he chose bacteria that are attracted by oxygen. In place of the mirror and diaphragm usually used to illuminate objects under view in his microscope, he substituted a "microspectral apparatus," which, as its name implies, produced a tiny spectrum of colors that it projected upon the slide under the microscope. Then he arranged a filament of algal cells parallel to the spread of the spectrum. The oxygen-seeking bacteria congregated mostly in the areas where the violet and red wavelengths fell upon the algal filament.

FIGURE 10.7

Constructing an action spectrum for photosynthesis. As you can see, the action spectrum for photosynthesis that Englemann revealed in his experiment parallels the absorption spectrum of chlorophyll (see figure 10.5).

Oak leaf in summer

Oak leaf in autumn

FIGURE 10.8

Fall colors are produced by carotenoids and other accessory pigments. During the spring and summer, chlorophyll in leaves masks the presence of carotenoids and other accessory pigments. When cool fall temperatures cause leaves to cease manufacturing chlorophyll, the chlorophyll is no longer present to reflect green light, and the leaves reflect the orange and yellow light that carotenoids and other pigments do not absorb.

Carotenoids consist of carbon rings linked to chains with alternating single and double bonds. They can absorb photons with a wide range of energies, although they are not always highly efficient in transferring this energy. Carotenoids assist in photosynthesis by capturing energy from light composed of wavelengths that are not efficiently absorbed by chlorophylls (figure 10.8; see figure 10.5).

A typical carotenoid is β-carotene, whose two carbon rings are connected by a chain of 18 carbon atoms with alternating single and double bonds. Splitting a molecule of β-carotene into equal halves produces two molecules of vitamin A. Oxidation of vitamin A produces retinal, the pigment used in vertebrate vision. This explains why carrots, which are rich in β-carotene, enhance vision.

A pigment is a molecule that absorbs light. The wavelengths absorbed by a particular pigment depend on the available energy levels to which light-excited electrons can be boosted in the pigment.

Organizing Pigments into Photosystems

The light-dependent reactions of photosynthesis occur in membranes. In photosynthetic bacteria, the plasma membrane itself is the photosynthetic membrane. In plants and algae, by contrast, photosynthesis is carried out by organelles that are the evolutionary descendants of photosynthetic bacteria, chloroplasts—in other words, the photosynthetic membranes exist *within* the chloroplasts. The light-dependent reactions take place in four stages:

1. **Primary photoevent.** A photon of light is captured by a pigment. The result of this primary photoevent is the excitation of an electron within the pigment.
2. **Charge separation.** This excitation energy is transferred to a specialized chlorophyll pigment termed a **reaction center,** which reacts by transferring an energetic electron to an acceptor molecule, thus initiating electron transport.
3. **Electron transport.** The excited electron is shuttled along a series of electron-carrier molecules embedded within the photosynthetic membrane. Several of them react by transporting protons across the membrane, generating a gradient of proton concentration. Its arrival at the pump induces the transport of a proton across the membrane. The electron is then passed to an acceptor.
4. **Chemiosmosis.** The protons that accumulate on one side of the membrane now flow back across the membrane through ATP synthase where chemiosmotic synthesis of ATP takes place, just as it does in aerobic respiration.

Discovery of Photosystems

One way to study how pigments absorb light is to measure the dependence of the output of photosynthesis on the intensity of illumination—that is, how much photosynthesis is produced by how much light. When experiments of this sort are done on plants, they show that the output of photosynthesis increases linearly at low intensities but lessens at higher intensities, finally saturating at high-intensity light. Saturation occurs because all of the light-absorbing capacity of the plant is in use; additional light doesn't increase the output because there is nothing to absorb the added photons.

It is tempting to think that at saturation, all of a plant's pigment molecules are in use. In 1932, plant physiologists Robert Emerson and William Arnold set out to test this hypothesis using an organism in which they could measure both the number of chlorophyll molecules and the output of photosynthesis. In their experiment, they measured the oxygen yield of photosynthesis when a culture of *Chlorella* (a unicellular green alga) was exposed to very brief light flashes lasting only a few microseconds. Assuming the hy-

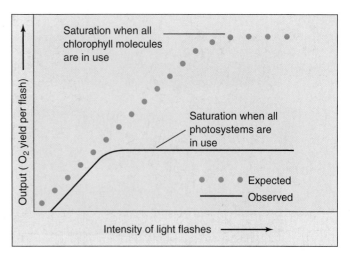

FIGURE 10.9
Emerson and Arnold's experiment. When photosynthetic saturation is achieved, further increases in intensity cause no increase in output.
Under what experimental conditions would you expect the saturation levels for a given number of chlorophyll molecules to be higher?

pothesis of pigment saturation to be correct, they expected to find that as they increased the intensity of the flashes, the yield per flash would increase, until each chlorophyll molecule absorbed a photon, which would then be used in the light-dependent reactions, producing a molecule of O_2.

Unexpectedly, this is not what happened. Instead, saturation was achieved much earlier, with only one molecule of O_2 per 2500 chlorophyll molecules (figure 10.9)! This led Emerson and Arnold to conclude that light is absorbed not by independent pigment molecules, but rather by clusters of chlorophyll and accessory pigment molecules, which have come to be called *photosystems* (described in more detail below). Light is absorbed by any one of the hundreds of pigment molecules in a photosystem, which transfer their excitation energy to one with a lower energy level than the others. This reaction center of the photosystem acts as an energy sink, trapping the excitation energy. It was the saturation of these reaction centers, not individual molecules, that was observed by Emerson and Arnold.

Architecture of a Photosystem

In chloroplasts and all but the most primitive prokaryotes, light is captured by **photosystems.** Each photosystem is a network of chlorophyll *a* molecules, accessory pigments, and associated proteins held within a protein matrix on the surface of the photosynthetic membrane. Like a magnifying glass focusing light on a precise point, a photosystem channels the excitation energy gathered by any one of its pigment molecules to a specific molecule, the reaction center chlorophyll. This molecule then passes the energy out of the photosystem so it can be put to work driving the synthesis of ATP and molecules.

A photosystem thus consists of two closely linked components: (1) an *antenna complex* of hundreds of pigment molecules that gather photons and feed the captured light energy to the reaction center; and (2) a *reaction center*, consisting of one or more chlorophyll *a* molecules in a matrix of protein, that passes the energy out of the photosystem.

The Antenna Complex. The antenna complex captures photons from sunlight (figure 10.10). In chloroplasts, the antenna complex is a web of chlorophyll molecules linked together and held tightly on the thylakoid membrane by a matrix of proteins. Varying amounts of carotenoid accessory pigments may also be present. The protein matrix serves as a sort of scaffold, holding individual pigment molecules in orientations that are optimal for energy transfer. The excitation energy resulting from the absorption of a photon passes from one pigment molecule to an adjacent molecule on its way to the reaction center. After the transfer, the excited electron in each molecule returns to the low-energy level it had before the photon was absorbed. Consequently, it is energy, not the excited electrons themselves, that passes from one pigment molecule to the next. The antenna complex funnels the energy from many electrons to the reaction center.

The Reaction Center. The reaction center is a transmembrane protein-pigment complex. In the reaction center of purple photosynthetic bacteria, which is simpler than in chloroplasts but better understood, a pair of chlorophyll *a* molecules acts as a trap for photon energy, passing an excited electron to an acceptor precisely positioned as its neighbor. Note that here the excited electron itself is transferred, not just the energy as we saw in pigment-pigment transfers. This allows the photon excitation to move away from the chlorophylls and is the key conversion of light to chemical energy.

Figure 10.11 shows the transfer of energy from the reaction center to the primary electron acceptor. By energizing an electron of the reaction center chlorophyll, light creates a strong electron donor where none existed before. The chlorophyll transfers the energized electron to the primary acceptor, a molecule of quinone, reducing the quinone and converting it to a strong electron donor. A weak electron donor then donates a low-energy electron to the chlorophyll, restoring it to its original condition. In plant chloroplasts, water serves as this weak electron

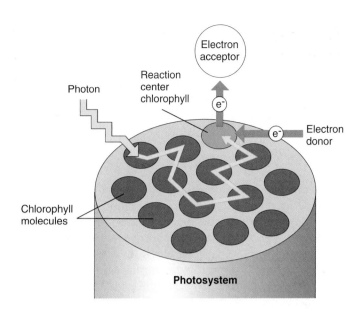

FIGURE 10.10

How the antenna complex works. When light of the proper wavelength strikes any pigment molecule within a photosystem, the light is absorbed by that pigment molecule. The excitation energy is then transferred from one molecule to another within the cluster of pigment molecules until it encounters the reaction center chlorophyll *a*. When excitation energy reaches the reaction center chlorophyll, electron transfer is initiated.

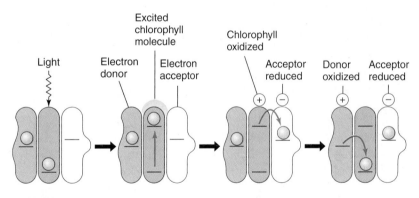

FIGURE 10.11

Converting light to chemical energy. The reaction center chlorophyll donates a light-energized electron to the primary electron acceptor, reducing it. The oxidized chlorophyll then fills its electron "hole" by oxidizing a donor molecule.

donor. When water is oxidized this way, oxygen is released along with two protons.

> **Photosystems contain pigments that capture photon energy from light. The pigments transfer the energy to reaction centers. There, the energy excites electrons, which are channeled away to do chemical work.**

How Photosystems Convert Light to Chemical Energy

Bacteria Use a Single Photosystem

Photosynthetic pigment arrays are thought to have evolved more than 2 billion years ago in bacteria similar to the sulfur bacteria alive today. A two-stage process takes place within bacterial photosystems.

1. Electron is joined with a proton to make hydrogen.

In these bacteria, the absorption of a photon of light at a peak absorption of 870 nanometers (near infrared, not visible to the human eye) by the photosystem results in the transmission of an energetic electron along an electron transport chain, eventually combining with a proton to form a hydrogen atom. In the sulfur bacteria, the electron is extracted from hydrogen sulfide, leaving elemental sulfur and protons as by-products. In bacteria that evolved later, as well as in plants and algae, the electron comes from water, producing oxygen and protons as by-products.

2. Electron is recycled to chlorophyll.

The ejection of an electron from the bacterial reaction center leaves it short one electron. Before the photosystem of the sulfur bacteria can function again, an electron must be returned. These bacteria channel the electron back to the pigment through an electron transport system similar to the one described in chapter 9; the electron's passage drives a proton pump that promotes the chemiosmotic synthesis of ATP. One molecule of ATP is produced for every three electrons that follow this path. Viewed overall (figure 10.12), the path of the electron is thus a circle. Chemists therefore call the electron transfer process leading to ATP formation **cyclic photophosphorylation.** Note, however, that the electron that left the P_{870} reaction center was a high-energy electron, boosted by the absorption of a photon of light, while the electron that returns has only as much energy as it had before the photon was absorbed. The difference in the energy of that electron is the photosynthetic payoff, the energy that drives the proton pump.

For more than a billion years, cyclic photophosphorylation was the only form of photosynthetic light-dependent reaction that organisms used. However, its major limitation is that it is geared only toward energy production, not toward biosynthesis. Most photosynthetic organisms incorporate atmospheric carbon dioxide into carbohydrates. Because the carbohydrate molecules are more reduced (have more hydrogen atoms) than carbon dioxide, a source of reducing power (that is, hydrogens) must be provided. Cyclic photophosphorylation does not do this. The hydrogen atoms extracted from H_2S are used as a source of protons, and are not available to join to carbon. Thus, bacteria that are restricted to this process must scavenge hydrogens from other sources, an inefficient undertaking.

Plants Use Two Photosystems

After the sulfur bacteria appeared, other kinds of bacteria evolved an improved version of the photosystem that overcame the limitation of cyclic photophosphorylation in a neat and simple way: A second, more powerful photosystem using another arrangement of chlorophyll *a* was combined with the original.

In this second photosystem, called **photosystem II,** molecules of chlorophyll *a* are arranged with a different geometry, so that more shorter-wavelength, higher-energy photons are absorbed than in the ancestral photosystem, which is called **photosystem I.** As in the ancestral photosystem, energy is transmitted from one pigment molecule to another within the antenna complex of these photosystems until it reaches the reaction center, a particular pigment molecule positioned near a strong membrane-bound electron acceptor. In photosystem II, the absorption peak (that is, the wavelength of light most strongly absorbed) of the pigments is approximately 680 nanometers; therefore, the reaction center pigment is called P_{680}. The absorption peak of photosystem I pigments in plants is 700 nanometers, so its reaction center pigment is called P_{700}. Working together, the two photosystems carry out a noncyclic electron transfer.

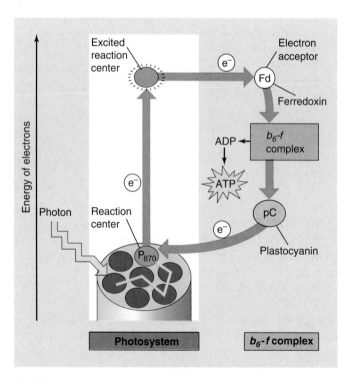

FIGURE 10.12
The path of an electron in sulfur bacteria. When a light-energized electron is ejected from the photosystem reaction center (P_{870}), it passes in a circle, eventually returning to the photosystem from which it was ejected.

When the rate of photosynthesis was measured using two light beams of different wavelengths (one red and the other far-red), the rate was greater than the sum of the rates using individual beams of red and far-red light (figure 10.13). This surprising result, called the **enhancement effect,** can be explained by a mechanism involving two photosystems acting in series (that is, one after the other), one of which absorbs preferentially in the red, the other in the far-red.

The use of two photosystems solves the problem of obtaining reducing power in a simple and direct way, by harnessing the energy of two photosystems. The scheme shown in figure 10.14, called a *Z diagram*, illustrates the two electron-energizing steps, one catalyzed by each photosystem. The electrons originate from water, which holds onto its electrons very tightly (redox potential = +820 mV), and end up in NADPH, which holds its electrons much more loosely (redox potential = −320 mV).

In sulfur bacteria, excited electrons ejected from the reaction center travel a circular path, driving a proton pump and then returning to their original photosystem. Plants employ two photosystems in series, which generates power to reduce $NADP^+$ to NADPH with enough left over to make ATP.

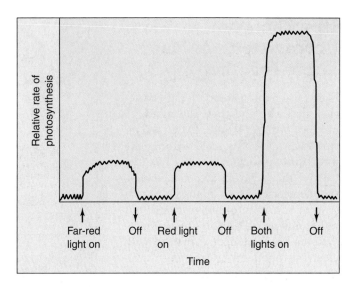

FIGURE 10.13
The enhancement effect. The rate of photosynthesis when red and far-red light are provided together is greater than the sum of the rates when each wavelength is provided individually. This result baffled researchers in the 1950s. Today, it provides the key evidence that photosynthesis is carried out by two photochemical systems with slightly different wavelength optima.
What would you conclude if "both lights on" did not change the relative rate of photosynthesis?

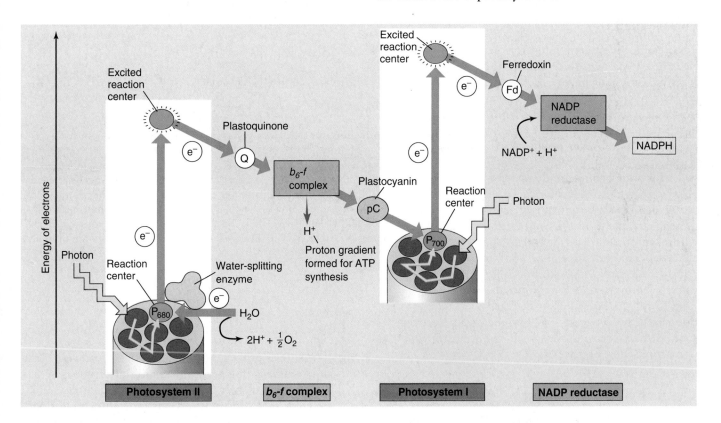

FIGURE 10.14
A Z diagram of photosystems I and II. Two photosystems work sequentially. First, a photon of light ejects a high-energy electron from photosystem II; that electron is used to pump a proton across the membrane, contributing chemiosmotically to the production of a molecule of ATP. The ejected electron then passes along a chain of cytochromes to photosystem I. When photosystem I absorbs a photon of light, it ejects a high-energy electron used to drive the formation of NADPH.

How the Two Photosystems of Plants Work Together

Plants use photosystems II and I in series, first one and then the other, to produce both ATP and NADPH. This two-stage process is called **noncyclic photophosphorylation,** because the path of the electrons is not a circle—the electrons ejected from the photosystems do not return to them, but rather end up in NADPH. The photosystems are replenished instead with electrons obtained by splitting water. Photosystem II acts first. High-energy electrons generated by photosystem II are used to synthesize ATP and then passed to photosystem I to drive the production of NADPH. For every pair of electrons obtained from water, one molecule of NADPH and slightly more than one molecule of ATP are produced.

Photosystem II

The reaction center of photosystem II, called P_{680}, closely resembles the reaction center of purple bacteria. It consists of more than 10 transmembrane protein subunits. The light-harvesting antenna complex consists of some 250 molecules of chlorophyll *a* and accessory pigments bound to several protein chains. In photosystem II, the oxygen atoms of two water molecules bind to a cluster of manganese atoms that are embedded within an enzyme and bound to the reaction center. In a way that is poorly understood, this enzyme splits water, removing electrons one at a time to fill the holes left in the reaction center by departure of light-energized electrons. As soon as four electrons have been removed from the two water molecules, O_2 is released.

The Path to Photosystem I

The primary electron acceptor for the light-energized electrons leaving photosystem II is a quinone molecule, as it was in the bacterial photosystem described earlier. The reduced quinone that results (*plastoquinone*, symbolized Q) is a strong electron donor; it passes the excited

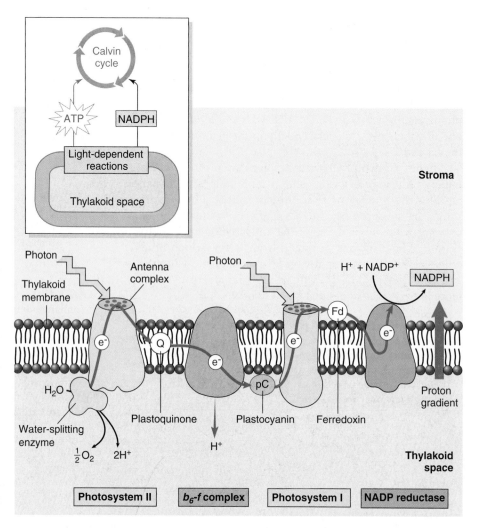

FIGURE 10.15

The photosynthetic electron transport system.   When a photon of light strikes a pigment molecule in photosystem II, it excites an electron. This electron is coupled to a proton stripped from water by an enzyme and is passed along a chain of membrane-bound cytochrome electron carriers (*red arrow*). When water is split, oxygen is released from the cell, and the hydrogen ions remain in the thylakoid space. At the proton pump (*b_6-f* complex), the energy supplied by the photon is used to transport a proton across the membrane into the thylakoid space. The concentration of hydrogen ions within the thylakoid thus increases further. When photosystem I absorbs another photon of light, its pigment passes a second high-energy electron to a reduction complex, which generates NADPH.

electron to a proton pump called the *b_6-f complex* embedded within the thylakoid membrane (figure 10.15). This complex closely resembles the *bc_1 complex* in the respiratory electron transport chain of mitochondria discussed in chapter 9. Arrival of the energetic electron causes the *b_6-f* complex to pump a proton into the thylakoid space. A small, copper-containing protein called *plastocyanin* (symbolized pC) then carries the electron to photosystem I.

Making ATP: Chemiosmosis

Each thylakoid is a closed compartment into which protons are pumped from the stroma by the b_6-f complex. The splitting of water also produces added protons that contribute to the gradient. The thylakoid membrane is impermeable to protons, so protons cross back out almost exclusively via the channels provided by *ATP synthases*. These channels protrude like knobs on the external surface of the thylakoid membrane. As protons pass out of the thylakoid through the ATP synthase channel, ADP is phosphorylated to ATP and released into the stroma, the fluid matrix inside the chloroplast (figure 10.16). The stroma contains the enzymes that catalyze the reactions of carbon fixation.

Photosystem I

The reaction center of photosystem I, called P_{700}, is a transmembrane complex consisting of at least 13 protein subunits. Energy is fed to it by an antenna complex consisting of 130 chlorophyll *a* and accessory pigment molecules. Photosystem I accepts an electron from plastocyanin into the hole created by the exit of a light-energized electron. This arriving electron has by no means lost all of its light-excited energy; almost half remains. Thus, the absorption of a photon of light energy by photosystem I boosts the electron leaving the reaction center to a very high energy level. Unlike photosystem II and the bacterial photosystem, photosystem I does not rely on quinones as electron acceptors. Instead, it passes electrons to an iron-sulfur protein called *ferredoxin* (Fd).

Making NADPH

Photosystem I passes electrons to ferredoxin on the stromal side of the membrane (outside the thylakoid). The reduced ferredoxin carries a very-high-potential electron. Two of them, from two molecules of reduced ferredoxin, are then donated to a molecule of $NADP^+$ to form NADPH. The reaction is catalyzed by the membrane-bound enzyme *NADP reductase*. Because the reaction occurs on the stromal side of the membrane and involves the uptake of a proton in forming NADPH, it contributes further to the proton gradient established during photosynthetic electron transport.

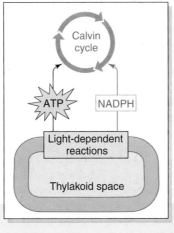

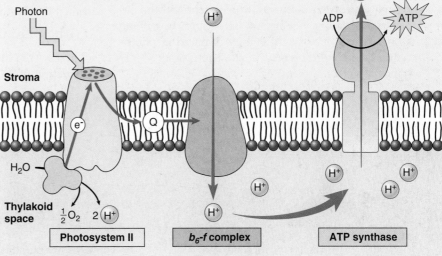

FIGURE 10.16

Chemiosmosis in a chloroplast. The b_6-f complex embedded in the thylakoid membrane pumps protons into the interior of the thylakoid. ATP is produced on the outside surface of the membrane (stroma side), as protons diffuse back out of the thylakoid through ATP synthase channels.

Making More ATP

The passage of an electron from water to NADPH in the noncyclic photophosphorylation described previously generates one molecule of NADPH and slightly more than one molecule of ATP. However, as you will learn later in this chapter, building organic molecules takes more energy than that—it takes one-and-a-half ATP molecules per NADPH molecule to fix carbon. To produce the extra ATP, many plant species are capable of short-circuiting photosystem I, switching photosynthesis into a *cyclic photophosphorylation* mode, so that the light-excited electron leaving photosystem I is used to make ATP instead of NADPH. The energetic electron is simply passed back to the b_6-f complex rather than passing on to $NADP^+$. The b_6-f complex pumps out a proton, adding to the proton gradient driving the chemiosmotic synthesis of ATP. The relative proportions of cyclic and noncyclic photophosphorylation in these plants determine the relative amounts of ATP and NADPH available for building organic molecules.

The electrons that photosynthesis strips from water molecules provide the energy to form ATP and NADPH. The residual oxygen atoms of the water molecules combine to form oxygen gas.

10.4 Cells use the energy and reducing power captured by the light-dependent reactions to make organic molecules.

The Calvin Cycle

Photosynthesis is a way of making organic molecules from carbon dioxide. These molecules contain many C—H bonds and are highly reduced compared with CO_2. To build them, cells use raw materials provided by light-dependent reactions:

1. **Energy.** ATP (provided by cyclic and noncyclic photophosphorylation) drives the endergonic reactions.
2. **Reducing power.** NADPH (provided by photosystem I) provides a source of hydrogens and the energetic electrons needed to bind them to carbon atoms. Much of the light energy captured in photosynthesis ends up invested in the energy-rich C—H bonds of sugars.

Discovering the Calvin Cycle

Nearly 100 years ago, Blackman concluded that, because of its temperature dependence, photosynthesis might involve enzyme-catalyzed reactions. These reactions form a cycle of enzyme-catalyzed steps similar to the Krebs cycle. This cycle of reactions is called the **Calvin cycle,** after its discoverer, Melvin Calvin of the University of California, Berkeley. As shown by the overview in figure 10.17, the cycle begins when CO_2 binds RuBP to form PGA. Because PGA contains three carbon atoms, this process is also called **C_3 photosynthesis.**

Carbon Fixation

The key step in the Calvin cycle—the event that makes the reduction of CO_2 possible—is the attachment of CO_2 to a

THE CALVIN CYCLE

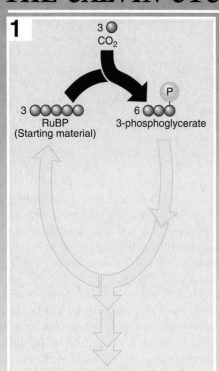

1

The Calvin cycle begins when a carbon atom from a CO_2 molecule is added to a five-carbon molecule (the starting material). The resulting six-carbon molecule is unstable and immediately splits into three-carbon molecules.

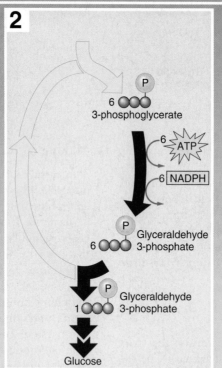

2

Then, through a series of reactions, energy from ATP and hydrogens from NADPH (the products of the light-dependent reactions) are added to the three-carbon molecules. The now-reduced three-carbon molecules either combine to make glucose or are used to make other molecules.

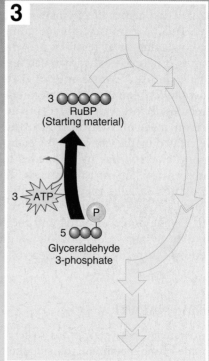

3

Most of the reduced three-carbon molecules are used to regenerate the five-carbon starting material, thus completing the cycle.

FIGURE 10.17
How the Calvin cycle works.

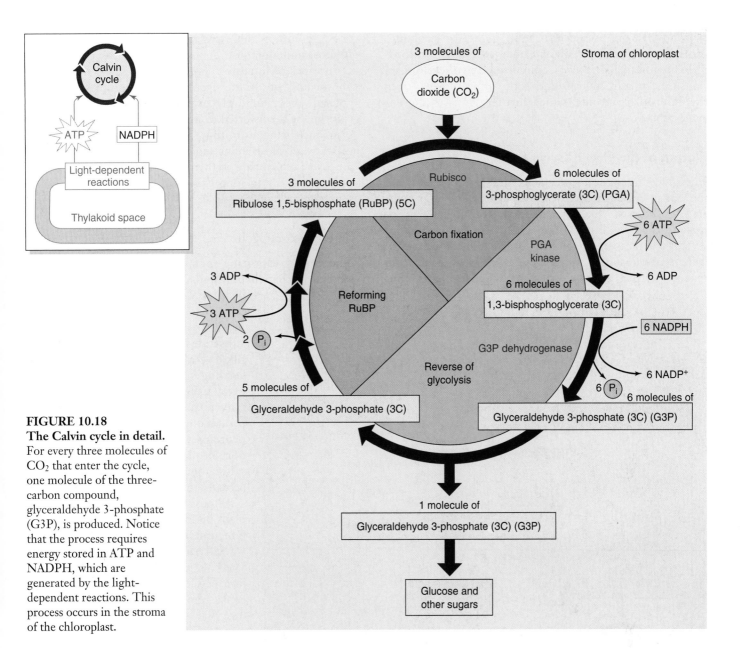

FIGURE 10.18
The Calvin cycle in detail.
For every three molecules of CO_2 that enter the cycle, one molecule of the three-carbon compound, glyceraldehyde 3-phosphate (G3P), is produced. Notice that the process requires energy stored in ATP and NADPH, which are generated by the light-dependent reactions. This process occurs in the stroma of the chloroplast.

very special organic molecule. Photosynthetic cells produce this molecule by reassembling the bonds of two intermediates in glycolysis, fructose 6-phosphate and glyceraldehyde 3-phosphate, to form the energy-rich five-carbon sugar *ribulose 1,5-bisphosphate (RuBP)*.

CO_2 binds to RuBP in the key process called **carbon fixation,** forming two three-carbon molecules of *3-phosphoglycerate (PGA)*. The enzyme that carries out this reaction, *ribulose bisphosphate carboxylase/oxygenase* (usually abbreviated *rubisco*) is a very large, four-subunit enzyme present in the chloroplast stroma.

Reactions of the Calvin Cycle

In a series of reactions, three molecules of CO_2 are fixed by rubisco to produce six molecules of PGA (containing $6 \times 3 = 18$ carbon atoms in all, three from CO_2 and 15 from

RuBP). The 18 carbon atoms then undergo a cycle of reactions that regenerates the three molecules of RuBP used in the initial step (containing $3 \times 5 = 15$ carbon atoms). This leaves one molecule of glyceraldehyde 3-phosphate (three carbon atoms) as the net gain.

The net equation of the Calvin cycle is:

$3 \, CO_2 + 9 \, ATP + 6 \, NADPH + \text{water} \rightarrow$
   glyceraldehyde 3-phosphate $+ 8 \, P_i + 9 \, ADP + 6 \, NADP^+$

With three full turns of the cycle, three molecules of carbon dioxide enter, a molecule of glyceraldehyde 3-phosphate (G3P) is produced, and three molecules of RuBP are regenerated (figure 10.18).

We now know that light is required *indirectly* for different segments of the CO_2 reduction reactions. Five of the Calvin cycle enzymes—including rubisco—are light activated; that is, they become functional or operate more efficiently in the presence of light. Light also promotes transport of three-carbon intermediates across chloroplast membranes that are required for Calvin cycle reactions. And finally, light promotes the influx of Mg^{++} into the chloroplast stroma, which further activates the enzyme rubisco.

Output of the Calvin Cycle

The glyceraldehyde 3-phosphate that is the product of the Calvin cycle is a three-carbon sugar, a key intermediate in glycolysis. Much of it is exported from the chloroplast to the cytoplasm of the cell, where the reversal of several reactions in glycolysis allows it to be converted to fructose 6-phosphate and glucose 1-phosphate, and from that to sucrose, a major transport sugar in plants. (Sucrose, common table sugar, is a disaccharide made of fructose and glucose.)

In times of intensive photosynthesis, glyceraldehyde 3-phosphate levels in the stroma of the chloroplast rise. As a consequence, some glyceraldehyde 3-phosphate in the chloroplast is converted to glucose 1-phosphate, in a set of reactions analogous to those done in the cytoplasm, by reversing several reactions similar to those of glycolysis. The glucose 1-phosphate is then combined into an insoluble polymer, forming long chains of starch stored as bulky starch grains in chloroplasts.

The Energy Cycle

The energy-capturing metabolisms of the chloroplasts studied in this chapter and the mitochondria studied in chapter 9 are intimately related (figure 10.19). Photosynthesis uses the products of respiration as its starting substrates, and respiration uses the products of photosynthesis as its starting substrates. The Calvin cycle even uses part of the ancient glycolytic pathway, run in reverse, to produce glucose. Also, the principal proteins involved in electron transport in plants are related to those in mitochondria, and in many cases are actually the same.

Photosynthesis is but one aspect of plant biology, although it is an important one. In chapters 35 through 41, we will examine plants in more detail. We have treated photosynthesis here, in a part devoted to cell biology, because photosynthesis arose long before plants did, and all organisms depend directly or indirectly on photosynthesis for the energy that powers their lives.

Plants incorporate carbon dioxide into sugars by means of the Calvin cycle, which is driven by the ATP and NADPH produced in the light-dependent reactions. The Calvin cycle essentially reverses the breakdown of such molecules that occurs in the mitochondrion, forming a cycle in which energy enters from the sun and leaves as heat and work.

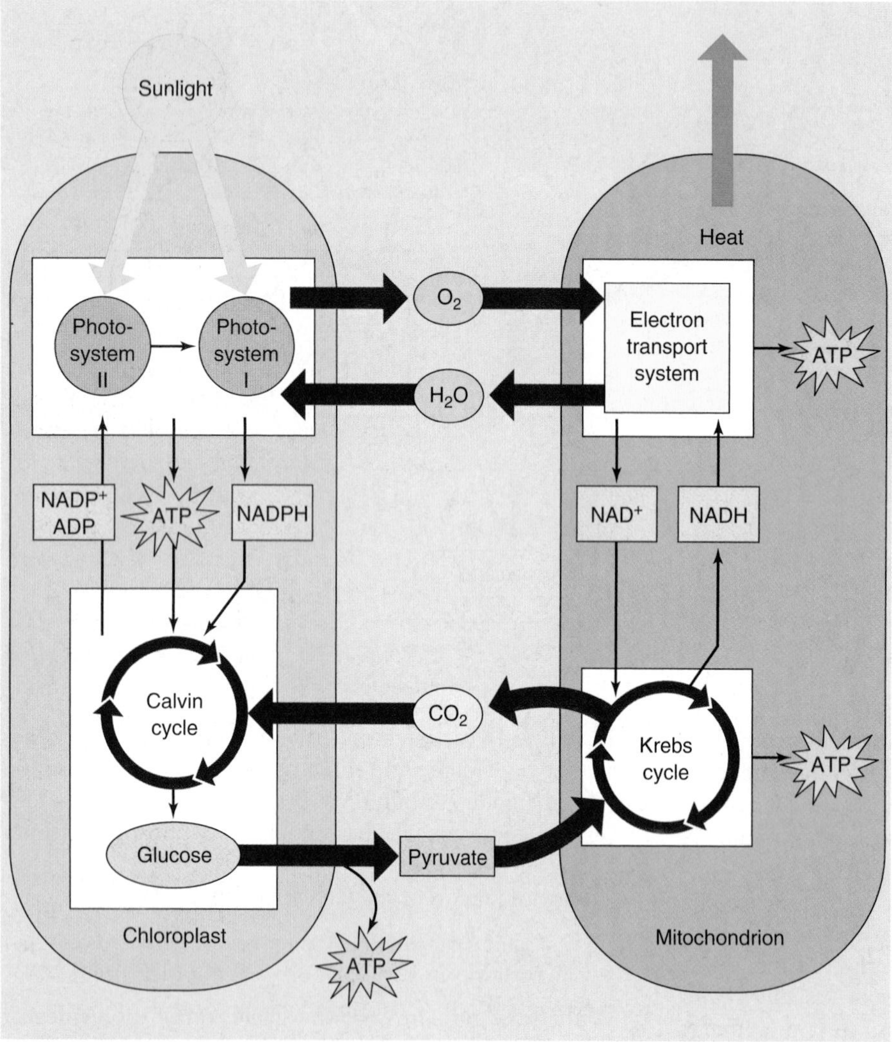

FIGURE 10.19
Chloroplasts and mitochondria: Completing an energy cycle.
Water and oxygen gas cycle between chloroplasts and mitochondria within a plant cell, as do glucose and CO_2. Cells with chloroplasts require an outside source of CO_2 and water and generate glucose and oxygen. Cells without chloroplasts, such as animal cells, require an outside source of glucose and oxygen and generate CO_2 and water.

Photorespiration

Evolution does not necessarily result in optimum solutions. Rather, it favors workable solutions that can be derived from others that already exist. Photosynthesis is no exception. Rubisco, the enzyme that catalyzes the key carbon-fixing reaction of photosynthesis, provides a decidedly suboptimal solution. This enzyme has a second enzyme activity that interferes with the Calvin cycle, *oxidizing* RuBP. In this process, called **photorespiration,** O_2 is incorporated into RuBP, which undergoes additional reactions that actually release CO_2. Hence, photorespiration releases CO_2—essentially undoing the Calvin cycle, which reduces CO_2 to carbohydrate.

The carboxylation and oxidation of RuBP are catalyzed at the same active site on rubisco, and CO_2 and O_2 compete with each other at this site. Under normal conditions at 25°C, the rate of the carboxylation reaction is four times that of the oxidation reaction, meaning that 20% of photosynthetically fixed carbon is lost to photorespiration. This loss rises substantially as temperature increases, because under hot, arid conditions, specialized openings in the leaf called *stomata* (singular, *stoma*) close to conserve water. This closing also cuts off the supply of CO_2 entering the leaf and does not allow O_2 produced by photosynthesis to exit the leaf (figure 10.20). As a result, the low CO_2 and high O_2 conditions within the leaf favor photorespiration.

Plants that fix carbon using only C_3 photosynthesis (the Calvin cycle) are called C_3 plants. In C_3 photosynthesis, RuBP is carboxylated to form a three-carbon compound via the activity of rubisco. Other plants use **C_4 photosynthesis,** in which phosphoenolpyruvate, or PEP, is carboxylated to form a four-carbon compound using the enzyme PEP carboxylase. This enzyme has no oxidation activity, and thus no photorespiration. Furthermore, PEP carboxylase has a much greater affinity for CO_2 than does rubisco. In the C_4 pathway, the four-carbon compound undergoes further modification, only to be decarboxylated. The CO_2 that is released is then captured by rubisco and drawn into the Calvin cycle. Because an organic compound is donating the CO_2, the effective concentration of CO_2 relative to O_2 is increased, and photorespiration is minimized.

The reduction in the yield of fixed carbon as a result of photorespiration is not trivial. C_3 plants lose between 25 and 50% of their photosynthetically fixed carbon in this way. The rate depends largely upon the temperature. In tropical climates, especially those in which the temperature is often above 28°C, the problem is severe, and it has a major impact on tropical agriculture.

The C_4 Pathway

Plants adapted to warmer environments have evolved two principal ways that use the C_4 pathway to deal with the problem of losing large amounts of fixed carbon. In one ap-

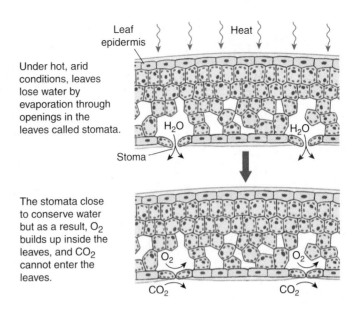

FIGURE 10.20
Conditions favoring photorespiration. In hot, arid environments, stomata close to conserve water, which also prevents CO_2 from entering the leaf and O_2 from exiting the leaf. The high O_2/low CO_2 conditions favor photorespiration.

proach, plants conduct C_4 photosynthesis in the mesophyll cells and the Calvin cycle in the bundle-sheath cells. This creates high local levels of CO_2 to favor the carboxylation reaction of rubisco. These plants, called C_4 plants, include corn, sugarcane, sorghum, and a number of other grasses. In the C_4 pathway, the three-carbon metabolite phosphoenolpyruvate is carboxylated to form the four-carbon molecule oxaloacetate, which is the first product of CO_2 fixation (figure 10.21*a*). In C_4 plants, oxaloacetate is in turn converted into the intermediate malate, which is transported to an adjacent bundle-sheath cell. Inside the bundle-sheath cell, malate is decarboxylated to produce pyruvate, releasing CO_2. Because bundle-sheath cells are impermeable to CO_2, the CO_2 is retained within them in high concentrations. Pyruvate returns to the mesophyll cell, where two of the high-energy bonds in an ATP molecule are split to convert the pyruvate back into phosphoenolpyruvate, thus completing the cycle.

The enzymes that carry out the Calvin cycle in a C_4 plant are located within the bundle-sheath cells, where the increased CO_2 concentration decreases photorespiration. Because each CO_2 molecule is transported into the bundle-sheath cells at a cost of two high-energy ATP bonds, and because six carbons must be fixed to form a molecule of glucose, 12 additional molecules of ATP are required to form a molecule of glucose. In C_4 photosynthesis, the energetic cost of forming glucose is almost twice that of C_3 photosynthesis: 30 molecules of ATP versus 18. Nevertheless, C_4 photosynthesis is advantageous in a hot climate; otherwise, photorespiration would remove more than half of the carbon fixed.

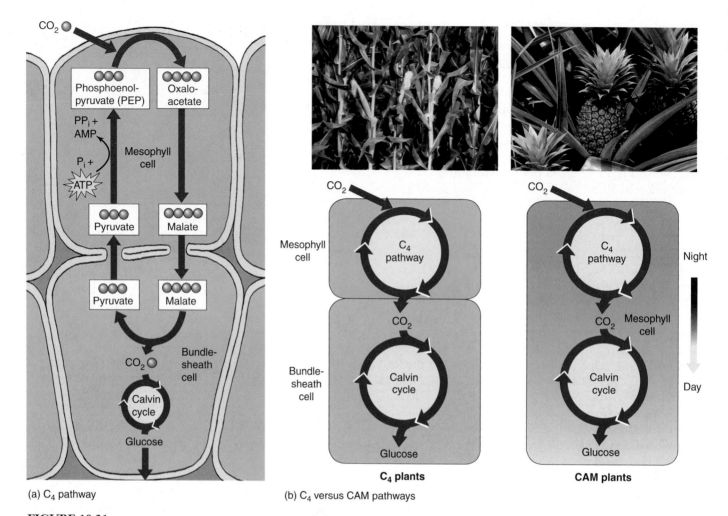

(a) C₄ pathway

(b) C₄ versus CAM pathways

FIGURE 10.21
Carbon fixation in C$_4$ plants. (*a*) This process is called the C$_4$ pathway because the first molecule formed, oxaloacetate, is a molecule containing four carbons. (*b*) A comparison of C$_4$ and CAM plants. Both C$_4$ and CAM plants utilize the C$_4$ and the C$_3$ pathways. In C$_4$ plants, the pathways are separated spatially: The C$_4$ pathway takes place in the mesophyll cells, and the C$_3$ pathway in the bundle-sheath cells. In CAM plants, the two pathways are separated temporally: The C$_4$ pathway is utilized at night, and the C$_3$ pathway during the day.

The Crassulacean Acid Pathway

A second strategy to decrease photorespiration in hot regions has been adopted by many succulent (water-storing) plants, such as cacti, pineapples, and some members of about two dozen other plant groups. This mode of initial carbon fixation is called **crassulacean acid metabolism (CAM),** after the plant family Crassulaceae (the stonecrops, or hens-and-chicks), in which it was first discovered. In these plants, the stomata, openings in leaves through which CO$_2$ enters and water vapor is lost, open during the night and close during the day. This pattern of stomatal opening and closing is the reverse of that in most plants. CAM plants open stomata at night and initially fix CO$_2$ into organic compounds using the C$_4$ pathway. These organic compounds accumulate throughout the night and are decarboxylated during the day to yield high levels of CO$_2$. In the day, these high levels of CO$_2$ drive the Calvin cycle and minimize photorespiration. Like C$_4$ plants, CAM plants use both C$_4$ and C$_3$ pathways. They differ from C$_4$ plants in that they use the C$_4$ pathway at night and the C$_3$ pathway during the day *within the same cells.* In C$_4$ plants, the two pathways take place in different cells (figure 10.21*b*).

Photorespiration results in decreased yields of photosynthesis. C$_4$ and CAM plants circumvent this problem through modifications of leaf architecture and photosynthetic chemistry that locally increase CO$_2$ concentrations. C$_4$ plants isolate CO$_2$ production spatially, while CAM plants isolate it temporally.

For interactive testing, visit the Online Learning Center with PowerWeb at www.mhhe.com/Raven7

10.1 What is photosynthesis?

The Chloroplast as a Photosynthetic Machine

- The equation for photosynthesis is: carbon dioxide + water + light yields glucose + water + oxygen. (p. 187)
- The light-dependent reactions occur within thylakoid membranes within chloroplasts. (p. 187)

10.2 Learning about photosynthesis: An experimental journey.

The Role of Soil and Water

- Jan Baptista van Helmont demonstrated that soil did not add mass to a growing plant, while Priestly, Ingenhousz, and other chemists worked out the basic formula for photosynthesis: carbon dioxide + water + light yields sugar and oxygen. (p. 188)

Discovery of the Light-Independent Reactions

- In the early 1900s, Blackman showed that capturing photosynthetic energy requires the input of sunlight, but building organic molecules does not. (p. 188)

The Role of Light and Reducing Power

- Van Niel discovered that photosynthesis splits water molecules, incorporating the carbon atoms of carbon dioxide gas and the hydrogen atoms of water into organic molecules and oxygen gas. (p. 189)
- Hill showed that plants can use light energy to generate reducing power. (p. 189)
- Carbon fixation refers to the incorporation of carbon dioxide into organic molecules in the light-independent reactions. (p. 189)

10.3 Pigments capture energy from sunlight.

The Biophysics of Light

- Short-wavelength light contains photons of higher energy than long-wavelength light. (p. 190)
- Sunlight reaching the earth's surface contains a significant amount of ultraviolet light, which possesses considerably more energy than visible light. (p. 191)
- In photosynthesis, photons are absorbed by plant pigments, and specific pigments absorb specific wavelengths. (p. 191)
- Chlorophyll *a* is the main photosynthetic pigment, although chlorophyll *b* and carotenoids also play important roles. (p. 191)

Chlorophylls and Carotenoids

- Chlorophylls absorb photons by means of an excitation process. (p. 192)
- The wavelengths absorbed by a pigment depend on the available energy level to which light-excited electrons can be boosted. (p. 193)

Organizing Pigments into Photosystems

- The light-dependent reactions take place in four stages: primary photoevent, charge separation, electron transport, and chemiosmosis. (p. 194)
- Pigments within photosystems transfer energy to reaction centers where the energy excites electrons that are channeled to perform chemical work. (p. 195)

How Photosystems Convert Light to Chemical Energy

- Plants employ two photosystems in series, which generate power to reduce $NADP^+$ to NADPH. (p. 196)

How the Two Photosystems of Plants Work Together

- Plants use photosystems II and I in series (noncyclic phosphorylation). (p. 198)
- High-energy electrons generated by photosystem II are used to synthesize ATP and then passed to photosystem I to drive the production of NADPH. (p. 199)

10.4 Cells use the energy and reducing power captured by the light-dependent reactions to make organic molecules.

The Calvin Cycle

- The Calvin cycle is also known as C_3 photosynthesis. (p. 200)
- CO_2 binds to RuBP in carbon fixation, forming two three-carbon molecules of PGA. (pp. 200–201)
- Plants incorporate carbon dioxide into sugars in the Calvin cycle, which is driven by the ATP and NADPH produced in the light-dependent reactions. (p. 202)

Photorespiration

- Photorespiration releases CO_2 and results in decreased yields of photosynthesis. (p. 203)
- C_4 photosynthesis circumvents photorespiration by creating high local levels of CO_2 in bundle sheath cells. (p. 203)
- CAM plants isolate CO_2 temporally by opening stomata at night instead of during the day. (p. 204)

Self Test

1. Within chloroplasts, the semiliquid matrix in which the Calvin cycle occurs is called
   a. stroma.
   b. thylakoids.
   c. grana.
   d. photosystem.
2. Visible light occupies what part of the electromagnetic spectrum?
   a. the entire spectrum
   b. the entire upper half (with longer wavelengths)
   c. a small portion in the middle
   d. the entire lower half (with shorter wavelengths)
3. The colors of light that are most effective for photosynthesis are
   a. red, blue, and violet.
   b. green, yellow, and orange.
   c. infrared and ultraviolet.
   d. All colors of light are equally effective.
4. A photosystem consists of
   a. a group of chlorophyll molecules, all of which contribute excited electrons to the synthesis of ATP.
   b. a pair of chlorophyll a molecules.
   c. a group of chlorophyll molecules held together by proteins.
   d. a group of chlorophyll molecules that funnels light energy toward a single chlorophyll b molecule.
5. Which photosystem is believed to have evolved first?
   a. photosystem I
   b. photosystem II
   c. cyclic photophosphorylation
   d. All photosystems evolved at the same time, but in different organisms.
6. Oxygen is produced during photosynthesis when
   a. the carbon is removed from carbon dioxide to make carbohydrates.
   b. hydrogen from water is added to carbon dioxide to make carbohydrates.
   c. water molecules are split to provide electrons for photosystem I.
   d. water molecules are split to provide electrons for photosystem II.
7. During photosynthesis, ATP molecules are generated by
   a. the Calvin cycle.
   b. chemiosmosis.
   c. the electron transport chain.
   d. light striking the chlorophyll molecules.
8. The overall purpose of the Calvin cycle is to
   a. generate molecules of ATP.
   b. generate NADPH.
   c. give off oxygen for animal use.
   d. build organic (carbon) molecules.
9. The final product of the Calvin cycle is
   a. RuBP.
   b. G3P.
   c. glucose.
   d. PGA.
10. C_4 photosynthesis is an adaptation to hot, dry conditions in which
    a. CO_2 is fixed and stored in the leaf.
    b. water is stored in the stem.
    c. oxygen is stored in the root.
    d. light energy is stored in chloroplasts.

Test Your Visual Understanding

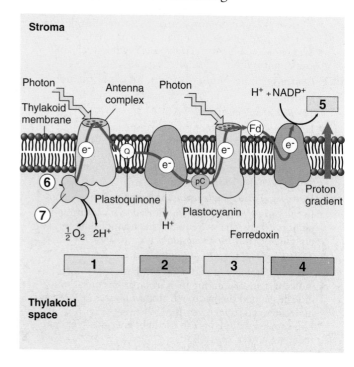

1. Match the following labels with their appropriate location on the figure.
   a. b_6-f complex
   b. H_2O
   c. NADP reductase
   d. NADPH
   e. photosystem I
   f. photosystem II
   g. water-splitting enzyme

Apply Your Knowledge

1. To reduce six molecules of carbon dioxide to glucose via photosynthesis, how many molecules of NADPH and ATP are required?
2. What is the advantage of having many pigment molecules in each photosystem but only one reaction center chlorophyll? In other words, why not couple every pigment molecule directly to an electron acceptor?
3. The two photosystems, P_{680} and P_{700}, of cyanobacteria, algae, and plants yield an oxidant capable of cleaving water. How might the subsequent evolution of cellular respiration have been different if the two photosystems had not evolved, and all photosynthetic organisms were restricted to the cyclic photophosphorylation used by the sulfur bacteria?
4. In theory, a plant kept in total darkness could still manufacture glucose, if it were supplied with which molecules?

11

How Cells Divide

Concept Outline

11.1 Prokaryotes divide far more simply than do eukaryotes.

   Cell Division in Prokaryotes. Prokaryotic cells divide by splitting in two.

11.2 The chromosomes of eukaryotes are highly ordered structures.

   Discovery of Chromosomes. All eukaryotic cells contain chromosomes, but different organisms possess differing numbers of chromosomes.

   The Structure of Eukaryotic Chromosomes. Proteins play an important role in packaging DNA in chromosomes.

11.3 Mitosis is a key phase of the cell cycle.

   The Cell Cycle. The cell cycle consists of three growth phases, a nuclear division phase, and a cytoplasmic division stage.

   Interphase: Preparing for Mitosis. In interphase, the cell grows, replicates its DNA, and prepares for cell division.

   Mitosis. In prophase, the chromosomes condense, and microtubules attach sister chromosomes to opposite poles of the cell. In metaphase, the chromosomes align along the center of the cell. In anaphase, the chromosomes separate; in telophase, the spindle dissipates and the nuclear envelope re-forms.

   Cytokinesis. In cytokinesis, the cytoplasm separates into two roughly equal halves.

11.4 The cell cycle is carefully controlled.

   General Strategies of Cell Cycle Control. At three points in the cell cycle, feedback from the cell determines whether the cycle will continue.

   Molecular Mechanisms of Cell Cycle Control. Special proteins regulate the checkpoints of the cell cycle.

   Cancer and the Control of Cell Proliferation. Cancer results from damage to genes encoding proteins that regulate the cell division cycle.

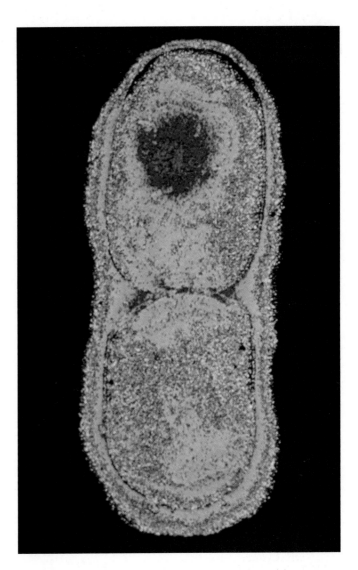

FIGURE 11.1
Cell division in prokaryotes. It's hard to imagine fecal coliform bacteria as being beautiful, but here is *Escherichia coli*, inhabitant of the large intestine and the biotechnology lab, spectacularly caught in the act of fission.

All species of organisms—bacteria, alligators, the weeds in a lawn—grow and reproduce. From the smallest creature to the largest, all species produce offspring like themselves and pass on the hereditary information that makes them what they are. In this chapter, we examine how cells divide and reproduce (figure 11.1). The mechanism of cell reproduction and its biological consequences have changed significantly during the evolution of life on earth. The process is complex in eukaryotes, involving both the replication of chromosomes and their separation into daughter cells. Much of what we are learning about the causes of cancer relates to how cells control this process, and in particular their propensity to divide, a mechanism that in broad outline remains the same in all eukaryotes.

11.1 Prokaryotes divide far more simply than do eukaryotes.

Cell Division in Prokaryotes

The end result of cell division in both prokaryotic and eukaryotic cells is two daughter cells, each with the same genetic information as the original cell. The differences between these two basic cell types lead to large differences in how this process occurs. Despite these differences, the essentials of the process are the same: duplication and segregation of genetic information into daughter cells, and division of cellular contents. We will begin by looking at the simpler process, which occurs in prokaryotes: division by **binary fission.**

Most prokaryotes have a genome made up of a single, circular DNA molecule. Despite its apparent simplicity, the DNA molecule of the bacterium *Escherichia coli* is actually on the order of 500 times longer than the cell itself! Thus, this "simple" structure is actually exquisitely packaged to fit into the cell. Although not found in a nucleus, the DNA is in a compacted form called a *nucleoid* that is distinct from the cytoplasm around it.

For many years, it was believed that the *E. coli* DNA molecule was passively segregated by attachment to the membrane and growth of the membrane as the cell elongates. More recently, a more complex picture is emerging that involves both active partitioning of the DNA and formation of a septum that divides the elongated cell in half. Although the details differ, species as different as *E. coli* and *Bacillus subtilis* both exhibit active partitioning of the newly replicated DNA molecules during the division process. This requires both specific sites on the chromosomes and a number of proteins actively involved in the process.

Binary fission begins with the replication of the prokaryotic DNA at a specific site—the origin of replication (see chapter 15)—and proceeds bidirectionally around the circular DNA to a specific site of termination (figure 11.2). Growth of the cell results in elongation, and the newly replicated DNA molecules are actively partitioned to one-fourth and three-quarters of the cell length. This process requires sequences near the origin of replication and results in these sequences being attached to the membrane. The cell itself is partitioned by the growth of new membrane and cell material called a septum (see figure 11.2). This process of septation is complex and under control of the cell as well.

The site of septation is usually the midpoint of the cell and begins with the formation of a ring composed of the molecule FtsZ (figure 11.3). This then results in the accumulation of a number of other proteins, including ones embedded in the membrane. The exact mechanism of septation is not known, but this structure grows inward radially until the cells pinch off into new cells.

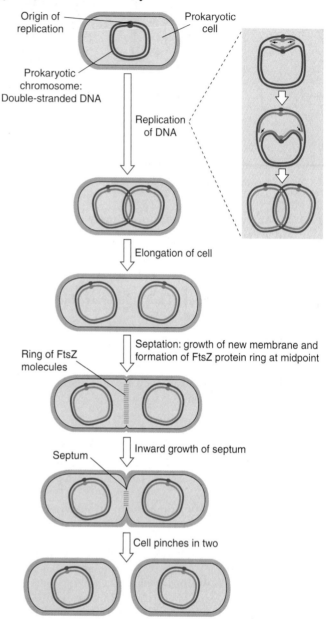

FIGURE 11.2

Binary fission. Prior to cell division, the prokaryotic DNA molecule replicates. The replication of the double-stranded, circular DNA molecule *(blue)* that constitutes the genome of a prokaryote begins at a specific site, called the origin of replication. The replication enzymes move out in both directions from that site and make copies *(red)* of each strand in the DNA duplex. The enzymes continue until they meet at another specific site, the terminus of replication. After the DNA is replicated, the cell elongates, and the DNA is partitioned in the cell. Septation then begins, in which new cell membrane material begins to grow and form a septum at approximately the midpoint of the cell. A protein molecule called FtsZ facilitates this process. When the septum is complete, the cell pinches in two, and two daughter cells are formed, each containing a prokaryotic DNA molecule.

FIGURE 11.3

The FtsZ protein. In these dividing *E. coli* bacteria, the FtsZ protein is fluorescent, and its location during binary fission can be seen. The protein assembles into a ring at approximately the midpoint of the cell, where it facilitates septation and cell division. Bacteria in which the *ftsZ* gene is mutated cannot divide.

The FtsZ molecule is interesting for a number of reasons. It is highly conserved evolutionarily, having been identified in most prokaryotes, including archaebacteria. It shows some small similarity to eukaryotic tubulin and can form filaments and rings. Recent 3-D crystals show similarity to tubulin as well. It is interesting to speculate that the elaborate spindle found in eukaryotic division may be related to this simple prokaryotic precursor (figure 11.4).

The evolution of eukaryotic cells led to much more complex genomes composed of multiple linear chromosomes housed in a membrane-bounded nucleus. These chromosomes contain even more DNA, and thus pose packaging problems that are solved by DNA being complexed with protein and packaged into functionally distinct chromosomes. This creates more challenges both for the replication of the genome and for its accurate segregation during cell division. The process that evolved to accomplish this segregation of chromosomes is called mitosis.

> **Prokaryotes divide by binary fission. Fission begins in the middle of the cell. An active partitioning process ensures that one genome will end up in each daughter cell.**

FIGURE 11.4

A comparison of protein assemblies during cell division among different organisms. The prokaryotic protein FtsZ has a structure that is similar to that of the eukaryotic protein tubulin. Tubulin is the protein component of microtubules, which are fibers that play an important role in eukaryotic cell division.

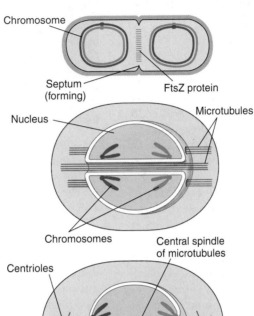
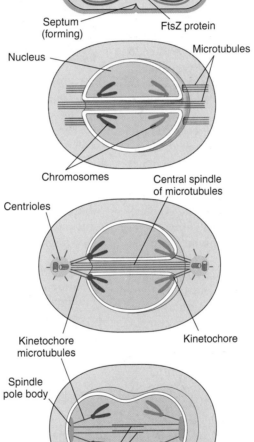
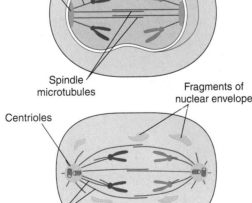

Prokaryotes
No nucleus; single circular chromosome. After DNA is replicated, it is partitioned in the cell. After cell elongation, FtsZ protein assembles into a ring and facilitates septation and cell division.

Some protists
Nucleus present and nuclear envelope remains intact during cell division. Chromosomes linear. Fibers called microtubules, composed of the protein tubulin, pass through tunnels in the nuclear membrane and set up an axis for separation of replicated chromosomes, and cell division.

Other protists
A spindle of microtubules forms between two pairs of centrioles at opposite ends of the cell. The spindle passes through one tunnel in the intact nuclear envelope. Kinetochore microtubules form between kinetochores on the chromosomes and the spindle poles and pull the chromosomes to each pole.

Yeasts
Nuclear envelope remains intact; spindle microtubules form inside the nucleus between spindle pole bodies. A single kinetochore microtubule attaches to each chromosome and pulls each to a pole.

Animals
Spindle microtubules begin to form between centrioles outside of nucleus. As these centrioles move to the poles, the nuclear envelope breaks down, and kinetochore microtubules attach kinetochores of chromosomes to spindle poles. Polar microtubules extend toward the center of the cell and overlap.

11.2 The chromosomes of eukaryotes are highly ordered structures.

Discovery of Chromosomes

Chromosomes were first observed by the German embryologist Walther Fleming in 1882, while he was examining the rapidly dividing cells of salamander larvae. When Fleming looked at the cells through what would now be a rather primitive light microscope, he saw minute threads within their nuclei that appeared to be dividing lengthwise. Fleming called their division **mitosis**, based on the Greek word *mitos*, meaning "thread."

Chromosome Number

Since their initial discovery, chromosomes have been found in the cells of all eukaryotes examined. Their number may vary enormously from one species to another. A few kinds of organisms have only 1 pair of chromosomes, while some ferns have more than 500 pairs (table 11.1). Most eukaryotes have between 10 and 50 chromosomes in their body cells.

Human cells each have 46 chromosomes, consisting of 23 nearly identical pairs (figure 11.5). Each of these 46 chromosomes contains hundreds or thousands of genes that play important roles in determining how a person's body develops and functions. For this reason, possession of all the chromosomes is essential to survival. Humans missing even one chromosome, a condition called *monosomy*, do not survive embryonic development in most cases. Nor does the human embryo develop properly with an extra copy of any one chromosome, a condition called *trisomy*. For all but a few of the smallest chromosomes, trisomy is fatal, and even in those few cases, serious problems result. For example, individuals with an extra copy of

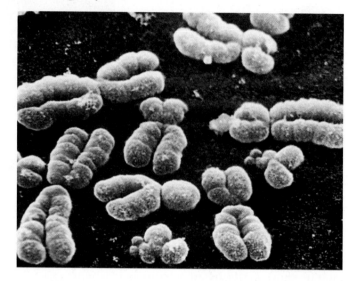

FIGURE 11.5

Human chromosomes. This photograph (950×) shows human chromosomes as they appear immediately before nuclear division. Each DNA molecule has already replicated, forming identical copies held together by a constriction called the centromere.

the very small chromosome 21 develop more slowly than normal and are mentally retarded, a condition called Down syndrome.

> All eukaryotic cells store their hereditary information in chromosomes, but different species utilize very different numbers of chromosomes to store this information.

Table 11.1 Chromosome Number in Selected Eukaryotes

Group	Total Number of Chromosomes	Group	Total Number of Chromosomes	Group	Total Number of Chromosomes
FUNGI		**PLANTS**		**VERTEBRATES**	
Neurospora (haploid)	7	*Haplopappus gracilis*	2	Opossum	22
Saccharomyces (a yeast)	16	Garden pea	14	Frog	26
		Corn	20	Mouse	40
INSECTS		Bread wheat	42	Human	46
Mosquito	6	Sugarcane	80	Chimpanzee	48
Drosophila	8	Horsetail	216	Horse	64
Honeybee	32	Adder's tongue fern	1262	Chicken	78
Silkworm	56			Dog	78

The Structure of Eukaryotic Chromosomes

In the century since chromosomes were discovered, we have learned a great deal about their structure and composition.

Composition of Chromatin

Chromosomes are composed of **chromatin,** a complex of DNA and protein; most chromosomes are about 40% DNA and 60% protein. A significant amount of RNA is also associated with chromosomes because chromosomes are the sites of RNA synthesis. The DNA of a chromosome is one very long, double-stranded fiber that extends unbroken through the entire length of the chromosome. A typical human chromosome contains about 140 million (1.4×10^8) nucleotides in its DNA. The amount of information one chromosome contains would fill about 280 printed books of 1000 pages each, if each nucleotide corresponded to a "word" and each page had about 500 words on it. Furthermore, if the strand of DNA from a single chromosome were laid out in a straight line, it would be about 5 centimeters (2 inches) long. Fitting such a strand into a nucleus is like cramming a string the length of a football field into a baseball—and that's only 1 of 46 chromosomes! In the cell, however, the DNA is coiled, allowing it to fit into a much smaller space than would otherwise be possible.

Chromosome Coiling

How can this long DNA fiber coil so tightly? If we gently disrupt a eukaryotic nucleus and examine the DNA with an electron microscope, we find that it resembles a string of beads (figure 11.6). Every 200 nucleotides, the DNA duplex is coiled around a core of eight histone proteins, forming a complex known as a **nucleosome.** Unlike most proteins, which have an overall negative charge, histones are positively charged, due to an abundance of the basic amino acids arginine and lysine. Thus, they are strongly attracted to the negatively charged phosphate groups of the DNA, and the histone cores act as "magnetic forms" that promote and guide the coiling of the DNA.

Further coiling occurs when the string of nucleosomes wraps up into higher-order coils called solenoids. This 30-nm solenoid forms the basis for interphase chromatin and is the starting point for the further compaction that occurs during mitosis. Chromatin appears to have some organization in the interphase nucleus, but it is not well understood. Further, geneticists have recognized for years that there are domains of chromatin that are not expressed, called **heterochromatin,** and domains of chromatin that are expressed, called **euchromatin.** During mitosis, the chromatin in the solenoid is further arranged around a scaffold of protein that is assembled at this time. The exact nature of this compaction is not known, but it includes radial looping of the solenoid about the protein scaffold.

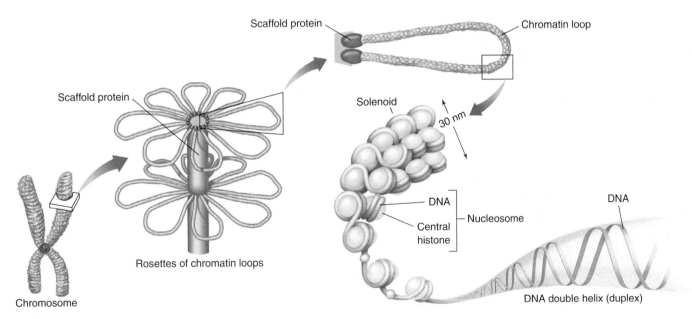

FIGURE 11.6
Levels of eukaryotic chromosomal organization. Nucleotides assemble into long double strands of DNA molecules. These strands require further packaging to fit into the cell nucleus. The DNA duplex is tightly bound to and wound around proteins called histones. The DNA-wrapped histones are called nucleosomes. The nucleosomes are further coiled into a solenoid. This solenoid is then organized into looped domains. The final organization of the chromosome is not known, but it appears to involve further radial looping into rosettes around a preexisting scaffolding of protein. The arrangement illustrated here is one of many possibilities.

Chromosome Karyotypes

Chromosomes may differ widely among species and sometimes even among individuals of the same species. They vary in size, staining properties, the location of the centromere (a constriction found on all chromosomes), the relative length of the two arms on either side of the centromere, and the positions of constricted regions along the arms. The particular array of chromosomes an individual possesses is called its **karyotype.** The karyotype in figure 11.7 shows marked differences in the chromosomes of humans.

When defining the number of chromosomes in a species, geneticists count the **haploid (n)** number of chromosomes. This refers to one complete set of chromosomes necessary to define an organism. For humans and many other species, the normal number of chromosomes in a cell is called the **diploid (2n)** number, which is twice the haploid number. This is a reflection of the equal genetic contribution that parents make to human offspring: one complete set of chromosomes from mom and one from dad. We refer to the maternal and paternal chromosomes as being **homologous,** and say that each chromosome has two **homologues.** So you received 23 chromosomes from your mother and 23 chromosomes from your father to produce the diploid number of 46. The timing of the events of meiosis and syngamy can lead to a variety of life cycles with varying amounts of time spent as haploid and diploid (see figure 12.3 and chapters 28, 29, and 30). Humans have a diploid dominant life cycle, meaning that the cells in our bodies that are haploid are gametes found in the gonads.

Chromosomes as seen in a karyotype are only present for a brief period during cell division. Prior to replicating, each chromosome has one **centromere,** a condensed area found on all eukaryotic chromosomes. After replication, each chromosome has two sister **chromatids** that appear to share a common centromere (figure 11.8). At the molecular level, this is probably not true, because centromeric DNA is not late-replicating. However, we can still call this one chromosome because, by convention, we count chromosomes by counting centromeres. Thus, entering division, a human cell has 46 chromosomes composed of 92 chromatids and 46 centromeres.

Eukaryotic genomes are larger and more complex than those of prokaryotes. Eukaryotic DNA is packaged tightly into chromosomes, enabling it to fit inside cells. Haploid cells contain one set of chromosomes, while diploid cells contain two sets.

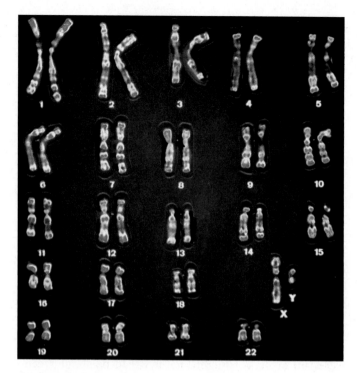

FIGURE 11.7
A human karyotype. The individual chromosomes that make up the 23 pairs differ widely in size and in centromere position. In this preparation, the chromosomes have been specifically stained to indicate further differences in their composition and to distinguish them clearly from one another.

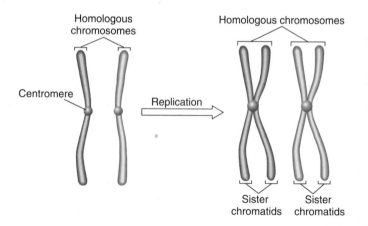

FIGURE 11.8
The difference between homologous chromosomes and sister chromatids. Homologous chromosomes are a pair of the same chromosome—say, chromosome number 16. Sister chromatids are the two replicas of a single chromosome held together by the centromeres after DNA replication.

11.3 Mitosis is a key phase of the cell cycle.

The Cell Cycle

The increased size and more complex organization of eukaryotic genomes over those of prokaryotes required radical changes in the process by which the two replicas of the genome are partitioned into the daughter cells during cell division. This division process is diagrammed as a **cell cycle** consisting of five phases (figure 11.9).

The Five Phases

G_1 is the primary growth phase of the cell. For many organisms, this encompasses the major portion of the cell's life span. **S** is the phase in which the cell synthesizes a replica of the genome. G_2 is the second growth phase, in which preparations are made for genomic separation. During this phase, mitochondria and other organelles replicate, chromosomes condense, and microtubules begin to assemble at a spindle. G_1, S, and G_2 together constitute **interphase,** the portion of the cell cycle between cell divisions.

M (for **mitosis**) is the phase of the cell cycle in which the microtubular apparatus assembles, binds to the chromosomes, and moves the sister chromatids apart. Mitosis is the essential step in the separation of the two daughter genomes. In this section, we will discuss mitosis as it occurs in animals and plants, where the process does not vary much; it is somewhat different among fungi and some protists. Although mitosis is a continuous process, it is traditionally subdivided into four stages: prophase, metaphase, anaphase, and telophase.

C (for **cytokinesis**) is the phase of the cell cycle when the cytoplasm divides, creating two daughter cells. In animal cells, the microtubule spindle helps position a contracting ring of actin that constricts like a drawstring to pinch the cell in two. In cells with a cell wall, such as plant cells, a plate forms between the dividing cells.

Duration of the Cell Cycle

The time it takes to complete a cell cycle varies greatly among organisms. Cells in growing animal embryos can complete their cell cycle in under 20 minutes; the shortest known animal nuclear division cycles occur in fruit fly embryos (8 minutes). Cells such as these simply divide their nuclei as quickly as they can replicate their DNA, without cell growth. Half of the cycle is taken up by S, half by M, and essentially none by G_1 or G_2. Because mature cells require time to grow, most of their cycles are much longer

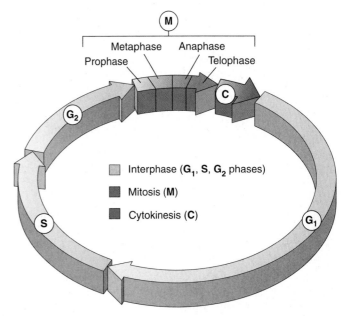

FIGURE 11.9
The cell cycle. This circle represents the 22-hour cell cycle in human cells growing in culture. G_1 represents the primary growth phase of the cell cycle, S the phase during which a replica of the genome is synthesized, and G_2 the second growth phase.

than those of embryonic tissue. Typically, a dividing mammalian cell completes its cell cycle in about 24 hours, but some cells, such as certain cells in the human liver, have cell cycles lasting more than a year. During the cycle, growth occurs throughout the G_1 and G_2 phases (referred to as "gap" phases because they separate S from M), as well as during the S phase. The M phase takes only about an hour, a small fraction of the entire cycle.

Most of the variation in the length of the cell cycle from one organism or tissue to the next occurs in the G_1 phase. Cells often pause in G_1 before DNA replication and enter a resting state called the **G_0 phase;** they may remain in this phase for days to years before resuming cell division. At any given time, most of the cells in an animal's body are in G_0 phase. Some, such as muscle and nerve cells, remain there permanently; others, such as liver cells, can resume G_1 phase in response to factors released during injury.

Most eukaryotic cells repeat a process of growth and division referred to as the cell cycle. The cycle can vary in length from a few minutes to several years.

Interphase: Preparing for Mitosis

The events that occur during interphase—the G_1, S, and G_2 phases—are very important for the successful completion of mitosis. During G_1, cells undergo the major portion of their growth. During the S phase, each chromosome replicates to produce two sister chromatids, which remain attached to each other at the **centromere.** The centromere is a point of constriction on the chromosome, containing a specific DNA sequence to which is bound a disk of protein called a **kinetochore.** This disk functions as an attachment site for fibers called microtubules that assist in cell division (figure 11.10). Each chromosome's centromere is located at a characteristic site.

The cell grows throughout interphase. The G_1 and G_2 segments of interphase are periods of active growth, during which proteins are synthesized and cell organelles produced. The cell's DNA replicates only during the S phase of the cell cycle.

After the chromosomes have replicated in S phase, they remain fully extended and uncoiled. This makes them invisible under the light microscope. In G_2 phase, they begin the long process of **condensation,** coiling ever more tightly. Special *motor proteins* are involved in the rapid final condensation of the chromosomes that occurs early in mitosis. Also during G_2 phase, the cells begin to assemble the machinery they will later use to move the chromosomes to opposite poles of the cell. In animal cells, a pair of microtubule-organizing centers called **centrioles** replicate. All eukaryotic cells undertake an extensive synthesis of *tubulin,* the protein of which microtubules are formed.

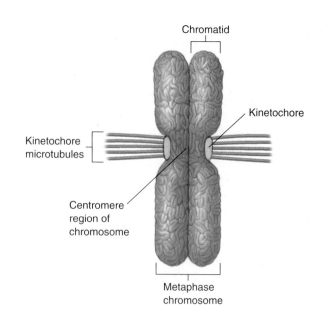

FIGURE 11.10
Kinetochores. In a metaphase chromosome, kinetochore microtubules are anchored to proteins at the centromere.

Interphase is that portion of the cell cycle in which the chromosomes are invisible under the light microscope because they are not yet condensed. It includes the G_1, S, and G_2 phases. In the G_2 phase, the cell mobilizes its resources for cell division.

A Vocabulary of Cell Division

binary fission Reproduction of a cell by division into two equal or nearly equal parts. Prokaryotes divide by binary fission.

centromere A constricted region of a chromosome about 220 nucleotides in length, composed of highly repeated DNA sequences. During mitosis, the centromere joins the two sister chromatids and is the site to which the kinetochores are attached.

chromatid One of the two copies of a replicated chromosome, joined by a single centromere to the other strand.

chromatin The complex of DNA and proteins of which eukaryotic chromosomes are composed.

chromosome The structure within cells that contains the genes. In eukaryotes, it consists of a single linear DNA molecule associated with proteins. The DNA replicates during S phase, and the replicas separate during M phase.

cytokinesis Division of the cytoplasm of a cell after nuclear division.

euchromatin The portion of a chromosome that is extended except during cell division, and from which RNA is transcribed.

heterochromatin The portion of a chromosome that remains permanently condensed and, therefore, is not transcribed into RNA. Most centromere regions are heterochromatic.

homologues Homologous chromosomes; in diploid cells, one of a pair of chromosomes that carry equivalent genes.

kinetochore A disk of protein bound to the centromere and attached to microtubules during mitosis, linking each chromatid to the spindle apparatus.

microtubule A hollow cylinder, about 25 nanometers in diameter, composed of subunits of the protein tubulin. Microtubules lengthen by the addition of tubulin subunits to their end(s) and shorten by the removal of subunits.

mitosis Nuclear division in which replicated chromosomes separate to form two genetically identical daughter nuclei. When accompanied by cytokinesis, it produces two identical daughter cells.

nucleosome The basic packaging unit of eukaryotic chromosomes, in which the DNA molecule is wound around a cluster of histone proteins. Chromatin is composed of long strings of nucleosomes that resemble beads on a string.

Mitosis

Prophase: Formation of the Mitotic Apparatus

When the chromosome condensation initiated in G_2 phase reaches the point at which individual condensed chromosomes first become visible with the light microscope, the first stage of mitosis, **prophase,** has begun. The condensation process continues throughout prophase; consequently, some chromosomes that start prophase as minute threads appear quite bulky before its conclusion. Ribosomal RNA synthesis ceases when the portion of the chromosome bearing the rRNA genes is condensed.

Assembling the Spindle Apparatus. The assembly of the microtubular apparatus that will later separate the sister chromatids also continues during prophase. In animal cells, the two centriole pairs formed during G_2 phase begin to move apart early in prophase, forming between them an axis of microtubules referred to as spindle fibers. By the time the centrioles reach the opposite poles of the cell, they have established a bridge of microtubules called the spindle apparatus between them. In plant cells, a similar bridge of microtubular fibers forms between opposite poles of the cell, although centrioles are absent in plant cells.

During the formation of the spindle apparatus, the nuclear envelope breaks down, and the endoplasmic reticulum reabsorbs its components. At this point, the microtubular spindle fibers extend completely across the cell, from one pole to the other. Their orientation determines the plane in which the cell will subsequently divide, through the center of the cell at right angles to the spindle apparatus.

In animal cell mitosis, the centrioles extend a radial array of microtubules toward the plasma membrane when they reach the poles of the cell. This arrangement of microtubules is called an **aster.** Although the aster's function is not fully understood, it probably braces the centrioles against the membrane and stiffens the point of microtubular attachment during the retraction of the spindle. Plant cells, which have rigid cell walls, do not form asters.

Linking Sister Chromatids to Opposite Poles. Each chromosome possesses two kinetochores, one attached to the centromere region of each sister chromatid (see figure 11.10). As prophase continues, a second group of microtubules appears to grow from the poles of the cell toward the centromeres. These microtubules connect the kinetochores on each pair of sister chromatids to the two poles of the spindle. Because microtubules extending from the two poles attach to opposite sides of the centromere, they attach one sister chromatid to one pole and the other sister chromatid to the other pole. This arrangement is absolutely critical to the process of mitosis; any mistakes in microtubule positioning can be disas-

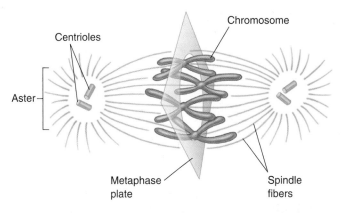

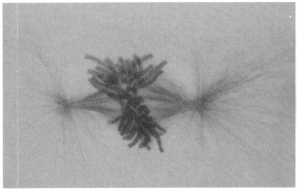

FIGURE 11.11
Metaphase. In metaphase, the chromosomes array themselves in a circle around the spindle midpoint.

trous. For example, the attachment of the two sides of a centromere to the same pole leads to a failure of the sister chromatids to separate, so that they end up in the same daughter cell.

Metaphase: Alignment of the Centromeres

In the second stage of mitosis, **metaphase,** the chromosomes align in the center of the cell. When viewed with a light microscope, the chromosomes appear to array themselves in a circle along the inner circumference of the cell, just as the equator girdles the earth (figure 11.11). An imaginary plane perpendicular to the axis of the spindle that passes through this circle is called the *metaphase plate.* The metaphase plate is not an actual structure, but rather an indication of the future axis of cell division. Positioned by the microtubules attached to the kinetochores of their centromeres, all of the chromosomes line up on the metaphase plate. At this point, which marks the end of metaphase, their centromeres are neatly arrayed in a circle, equidistant from the two poles of the cell, with microtubules extending back toward the opposite poles of the cell. Because of its shape, this arrangement is called a spindle.

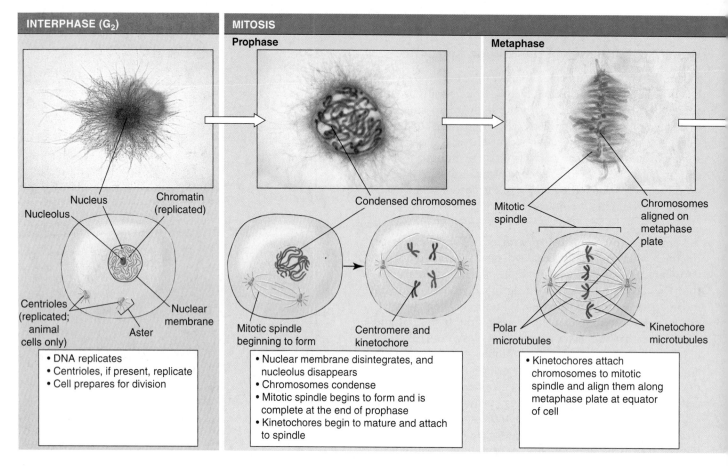

Prophase

Metaphase

Nucleus
Chromatin (replicated)
Nucleolus
Condensed chromosomes
Mitotic spindle
Chromosomes aligned on metaphase plate

Centrioles (replicated; animal cells only)
Nuclear membrane
Aster
Mitotic spindle beginning to form
Centromere and kinetochore
Polar microtubules
Kinetochore microtubules

- DNA replicates
- Centrioles, if present, replicate
- Cell prepares for division

- Nuclear membrane disintegrates, and nucleolus disappears
- Chromosomes condense
- Mitotic spindle begins to form and is complete at the end of prophase
- Kinetochores begin to mature and attach to spindle

- Kinetochores attach chromosomes to mitotic spindle and align them along metaphase plate at equator of cell

FIGURE 11.12

Mitosis and cytokinesis in plants (photos) and in animals (drawings). Mitosis (separation of the two genomes) occurs in four stages—prophase, metaphase, anaphase, and telophase—and is followed by cytokinesis (division into two separate cells). In the photos, the chromosomes of the African blood lily, *Haemanthus katharinae*, are stained blue, and microtubules are stained red.

Anaphase and Telophase: Separation of the Chromatids and Reformation of the Nuclei

Of all the stages of mitosis, shown in figure 11.12, **anaphase** is the shortest and the most beautiful to watch. It starts when the centromeres divide. Each centromere splits in two, freeing the two sister chromatids from each other.

To this point in mitosis, sister chromatids have been held together by a complex of proteins called cohesin. In yeasts, cleavage of the cohesin subunit Scc1p by a *separase* proteolytic enzyme is thought to be the event that actually releases the chromatids for migration to daughter cells in anaphase. In vertebrate cells, most of the cohesin dissociates from the chromosomes before the onset of metaphase. However, a small amount remains, locking the two chromatids together. Of particular importance is a subunit dubbed SCC1. Cleavage

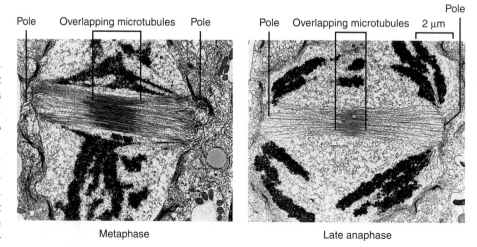

Pole   Overlapping microtubules   Pole
Pole   Overlapping microtubules   2 μm   Pole

Metaphase

Late anaphase

FIGURE 11.13

Microtubules slide past each other as the chromosomes separate. In these electron micrographs of dividing diatoms, the overlap of the microtubules lessens markedly during spindle elongation as the cell passes from metaphase to anaphase.

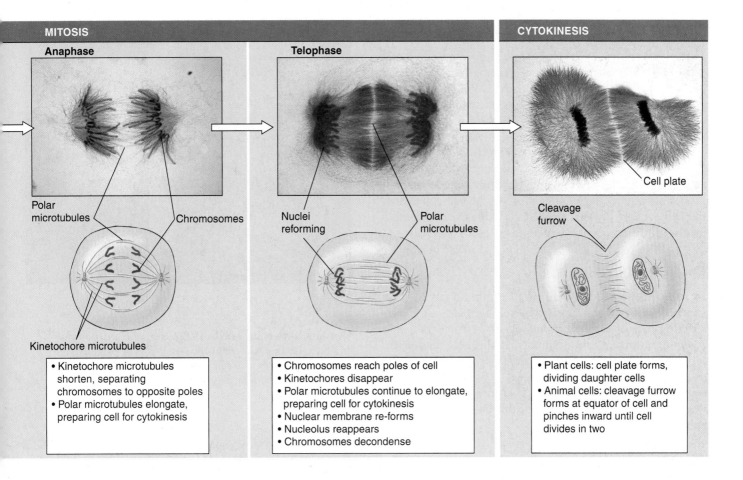

Anaphase

Polar microtubules
Chromosomes
Kinetochore microtubules

- Kinetochore microtubules shorten, separating chromosomes to opposite poles
- Polar microtubules elongate, preparing cell for cytokinesis

Telophase

Nuclei reforming
Polar microtubules

- Chromosomes reach poles of cell
- Kinetochores disappear
- Polar microtubules continue to elongate, preparing cell for cytokinesis
- Nuclear membrane re-forms
- Nucleolus reappears
- Chromosomes decondense

Cell plate

Cleavage furrow

- Plant cells: cell plate forms, dividing daughter cells
- Animal cells: cleavage furrow forms at equator of cell and pinches inward until cell divides in two

of the SCC1 cohesin subunit by separase enzymes is essential for centromere division and sister chromatid separation. The centromeres of all the chromosomes divide simultaneously. While the mechanism that achieves this synchrony is not well understood, it likely involves the timing of separase activation by a control element called the anaphase-promoting complex (APC), to be described in more detail later.

Freed from each other, the sister chromatids are pulled rapidly toward the poles to which their kinetochores are attached. In the process, two forms of movement take place simultaneously, each driven by microtubules.

First, *the poles move apart* as microtubular spindle fibers physically anchored to opposite poles slide past each other, away from the center of the cell (figure 11.13). Because another group of microtubules attach the chromosomes to the poles, the chromosomes move apart too. If a flexible membrane surrounds the cell, it becomes visibly elongated.

Second, *the centromeres move toward the poles* as the microtubules that connect them to the poles shorten. This shortening process is not a contraction; the microtubules do not get any thicker. Instead, tubulin subunits are removed from the kinetochore ends of the microtubules by the organizing center. As more subunits are removed, the chromatid-bearing microtubules are progressively disassembled, and the chromatids are pulled ever closer to the poles of the cell.

When the sister chromatids separate in anaphase, the accurate partitioning of the replicated genome—the essential element of mitosis—is complete. In **telophase,** the spindle apparatus disassembles as the microtubules are broken down into tubulin monomers that can be used to construct the cytoskeletons of the daughter cells. A nuclear envelope forms around each set of sister chromatids, which can now be called chromosomes because each has its own centromere. The chromosomes soon begin to uncoil into the more extended form that permits gene expression. One of the early group of genes expressed are the rRNA genes, resulting in the reappearance of the nucleolus.

During prophase, microtubules attach the centromeres joining pairs of sister chromatids to opposite poles of the spindle apparatus. During metaphase, each chromosome is drawn to a ring along the inner circumference of the cell by the microtubules extending from the centromere to the two poles of the spindle apparatus. During anaphase, the poles of the cell are pushed apart by microtubular sliding, and the sister chromatids are drawn to opposite poles by the shortening of the microtubules attached to them. During telophase, the spindle is disassembled, nuclear envelopes are reestablished, and the normal expression of genes present in the chromosomes is reinitiated.

Cytokinesis

Mitosis is complete at the end of telophase. The eukaryotic cell has partitioned its replicated genome into two nuclei positioned at opposite ends of the cell. While mitosis was going on, the cytoplasmic organelles, including mitochondria and chloroplasts (if present), were reassorted to areas that will separate and become the daughter cells. The replication of organelles takes place before cytokinesis, often in the S or G_2 phase. Cell division is still not complete at the end of mitosis, however, because the division of the cell proper has not yet begun. The phase of the cell cycle when the cell actually divides is called **cytokinesis.** It generally involves the cleavage of the cell into roughly equal halves.

Cytokinesis in Animal Cells

In animal cells and the cells of all other eukaryotes that lack cell walls, cytokinesis is achieved by means of a constricting belt of actin filaments. As these filaments slide past one another, the diameter of the belt decreases, pinching the cell and creating a *cleavage furrow* around the cell's circumference (figure 11.14*a*). As constriction proceeds, the furrow deepens until it eventually slices all the way into the center of the cell. At this point, the cell is divided in two (figure 11.14*b*).

Cytokinesis in Plant Cells

Plant cell walls are far too rigid to be squeezed in two by actin filaments. Instead, these cells assemble membrane components in their interior, at right angles to the spindle apparatus. This expanding membrane partition, called a **cell plate,** continues to grow outward until it reaches the interior surface of the plasma membrane and fuses with it, effectively dividing the cell in two (figure 11.15). Cellulose is then laid down on the new membranes, creating two new cell walls. The space between the daughter cells becomes impregnated with pectins and is called a **middle lamella.**

Cytokinesis in Fungi and Protists

In fungi and some groups of protists, the nuclear membrane does not dissolve, and as a result, all the events of mitosis occur entirely *within* the nucleus. Only after mitosis is complete in these organisms does the nucleus divide into two daughter nuclei; then, during cytokinesis, one nucleus goes to each daughter cell. This separate nuclear division phase of the cell cycle does not occur in plants, animals, or most protists.

After cytokinesis in any eukaryotic cell, the two daughter cells contain all the components of a complete cell. While mitosis ensures that both daughter cells contain a full complement of chromosomes, no similar mechanism ensures that organelles such as mitochondria and chloroplasts are

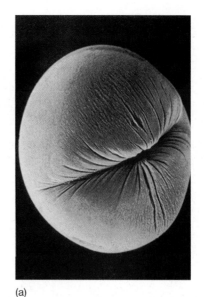
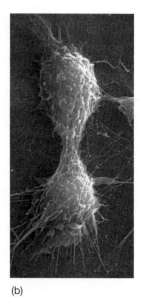

(a) (b)

FIGURE 11.14
Cytokinesis in animal cells. (*a*) A cleavage furrow forms around a dividing sea urchin egg (30×). (*b*) The completion of cytokinesis in an animal cell. The two daughter cells are still joined by a thin band of cytoplasm occupied largely by microtubules.

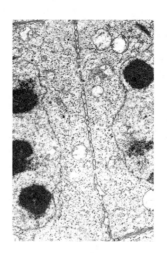
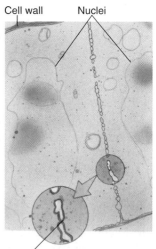

Cell wall   Nuclei

Vesicles containing membrane components fusing to form cell plate

FIGURE 11.15
Cytokinesis in plant cells. In this photograph and companion drawing, a cell plate is forming between daughter nuclei. Once the plate is complete, there will be two cells.

distributed equally between the daughter cells. However, as long as some of each organelle are present in each cell, the organelles can replicate to reach the number appropriate for that cell.

Cytokinesis is the physical division of the cytoplasm of a eukaryotic cell into two daughter cells.

11.4 The cell cycle is carefully controlled.

General Strategies of Cell Cycle Control

Our knowledge of how the cell cycle is controlled, while still incomplete, has grown enormously in the past 20 years. Our current view integrates two basic concepts: First, there are two irreversible points in the cell cycle—the replication of genetic material and the separation of the sister chromatids. Second, the cell cycle can be put on hold at specific points called *checkpoints*. At any of these checkpoints, the process is assayed for accuracy and can be halted if there are errors. This leads to extremely high fidelity overall for the entire process. The checkpoint organization of the cell cycle also makes it responsive to both the internal state of the cell, including nutritional state and integrity of genetic material, and to signals from the environment, which are integrated at major checkpoints.

Architecture of the Control System

A cell uses three main checkpoints to both assess the internal state of the cell and integrate external signals (figure 11.16): G_1/S, G_2/M, and late metaphase (the spindle checkpoint). Passage through these checkpoints is controlled by kinase enzymes composed of an enzymatic subunit partnered with a protein called cyclin. These enzymes are thus called **cyclin-dependent kinases,** or **Cdk's.** Cyclins are produced and degraded cyclically with mitosis, but it is the activity of Cdk's that drives the cell cycle.

G_1/S Checkpoint

The G_1/S checkpoint is the primary point at which the cell "decides" whether or not to divide. This checkpoint is therefore the primary point at which external signals can influence events of the cycle. It is the phase during which growth factors (discussed later in this chapter) affect the cycle. It is also the phase that links cell division to cell growth and nutrition. In yeast systems, where the majority of the genetic analysis of the cell cycle has been performed, this checkpoint is called "start." In animals, it is called the restriction point (R point). In all systems, once a cell has made this irreversible commitment to replicate its genome, it has committed to divide. The decision made at this checkpoint depends on external signals such as growth factors and internal signals such as nutritional state; it also depends on the genome being intact. Damage to DNA can halt the cycle at this point, as can starvation conditions or lack of growth factors.

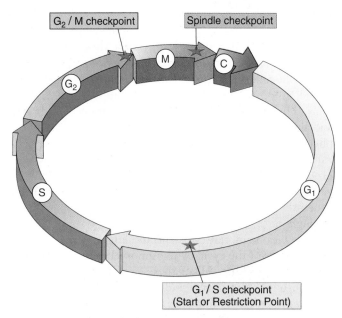

FIGURE 11.16
Control of the cell cycle. Cells use a centralized control system to check whether proper conditions have been achieved before passing three key checkpoints in the cell cycle.

G_2/M Checkpoint

The G_2/M checkpoint has received a large amount of attention due to its complexity and its importance as the stimulus for the events of mitosis. Historically, Cdk's active at this checkpoint were first identified as a class of substance that could be added to frog oocytes arrested in G_2 of meiosis to relieve this G_2 arrest. Because this was a normal step in the maturation of frog eggs, this substance was called "maturation promoting factor," a term that has now evolved into *M-phase-promoting factor* (MPF).

Passage through this checkpoint represents the commitment to mitosis. This checkpoint assesses the success of DNA replication and can stall the cycle if DNA has not been accurately replicated. DNA-damaging agents result in arrest at this checkpoint as well as at the G_1/S checkpoint.

Spindle Checkpoint

The spindle checkpoint ensures that all of the chromosomes are attached to the spindle in preparation for anaphase. The second irreversible step in the cycle is the separation of chromosomes during anaphase, and it is critical that they are properly arrayed at the metaphase plate.

The cell cycle is controlled at three checkpoints.

Molecular Mechanisms of Cell Cycle Control

On the previous page, we provided an overview of cell cycle control strategies; now we examine those processes in detail. The primary molecular mechanism of cell cycle control is phosphorylation, the addition of a phosphate group to the amino acids serine, threonine, and tyrosine in proteins. The enzymes that add phosphates are called kinases, and the enzymes that remove phosphates are called phosphatases. The phosphorylation of a protein can either activate or inactivate it, depending on the protein. Similarly, a protein that is inactivated by phosphorylation will be activated by dephosphorylation. The control of cellular processes by phosphorylation will be a continuing theme throughout this book.

The actions of cell-cycle-specific kinases drive the different stages of the cycle by phosphorylating a wide variety of cellular proteins. All of these protein targets are not known, but the importance of cell cycle kinases is clear. The enzymes that accomplish this phosphorylation are the cyclin-dependent kinases (Cdk's), consisting of an enzymatic subunit partnered with the protein cyclin (figure 11.17). Cyclins are proteins that display characteristic patterns of synthesis and degradation that coincide with phases of the cell cycle. The Cdk's are only active when the kinase is combined with the cyclin. The most important cell cycle kinase was identified in fission yeast and named cdc2. This Cdk can partner with different cyclins at different points in the cell cycle. Thus, the signal to begin comes from this Cdk combined with one cyclin, and the signal for the initiation of mitosis comes from this Cdk combined with a different cyclin.

The important question is: What controls the activity of the Cdk's during the cycle? For many years, a common view was that cyclins drove the cell cycle—that is, the periodic synthesis and destruction of cyclins acted as a clock for the cell cycle. More recently, it has become clear that the cdc2 kinase is controlled by phosphorylation: Phosphorylation at one site activates cdc2, and phosphorylation of cdc2 at another site inactivates it (see figure 11.17). Full activation of the cdc2 kinase requires complexing with the cyclin, and an appropriate pattern of phosphorylation. The exact molecular mechanisms of Cdk control are still an area of active investigation, but the main points are clear.

G₁/S Checkpoint

The G₁/S checkpoint integrates a number of signals, both internal and external (figure 11.18). The internal signals include the nutritional state of the cell and the size of the cell. The external signals include factors that promote cell growth and division (see the discussion of tumor suppressors later in this chapter). In mammalian cells, this is the checkpoint that involves the action of retinoblastoma pro-

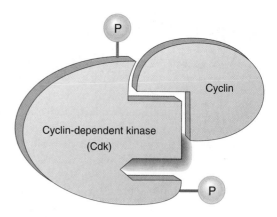

FIGURE 11.17

A complex of two proteins triggers passage through cell cycle checkpoints. Cdk is a protein kinase that activates numerous cell proteins by phosphorylating them. Cyclin is a regulatory protein required to activate Cdk. The activity of Cdk is also controlled by the pattern of phosphorylation: phosphorylation at one site (represented by the red site) inactivates the Cdk, and phosphorylation at another site (represented by the green site) activates the Cdk.

tein (discussed later). In yeasts, the molecular nature of the signal for this checkpoint (called "start") appears to be the accumulation of an S-phase-specific cyclin. This complexes with the cdc2 kinase to form the G₁/S Cdk. This Cdk phosphorylates a number of targets to result in the synthesis of proteins necessary for S phase.

G₂/M Checkpoint

The G₂/M checkpoint integrates a number of signals to trigger the commitment to mitosis. The Cdk that acts at this checkpoint, MPF, has been extensively analyzed in a number of different experimental systems. The control of MPF is sensitive to agents that disrupt or delay replication and to agents that damage DNA. It was once thought that MPF was controlled by the level of the M-phase-specific cyclins, but it has now become clear that this is not the case. While M-phase cyclin is necessary for MPF function, activity is controlled by inhibitory phosphorylation of cdc2. The critical signal in this process is the removal of inhibitor phosphates. This forms a molecular switch based on positive feedback as the active MPF further activates its own activating phosphatase. The checkpoint assesses the balance of kinases that add phosphates with the phosphatase that removes them. Damage to DNA acts through a complex pathway that includes damage sensing and response to tip the balance toward the inhibitory phosphorylation of MPF. In animal cells, the damage-sensing pathway involves the p53 protein, which has been found to be mutated in a variety of human cancers (see discussion of tumor suppressors later in this chapter).

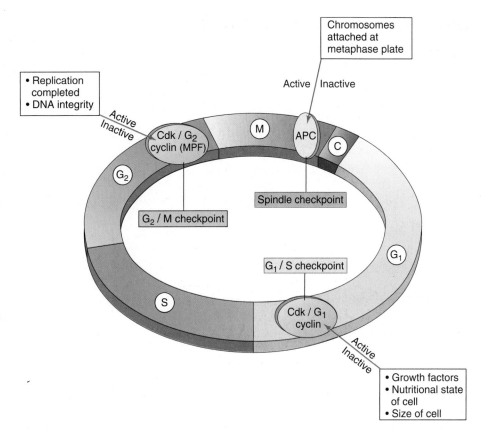

FIGURE 11.18
Checkpoints of the cell cycle. The cell cycle is controlled through three main checkpoints. These integrate internal and external signals to control progress through the cycle. These inputs control the state of two Cdk/cyclin complexes and the anaphase-promoting complex (APC). The arrows represent inputs, which can be complex networks, such as the signal transduction cascade seen in growth factor signaling (see figure 11.19).

Spindle Checkpoint

The spindle checkpoint is before the second irreversible step in the cell cycle, the anaphase separation of sister chromatids. This checkpoint ensures that all chromosomes are present at the metaphase plate and aligned with the centromeres of each sister chromatid oriented to opposite poles. The molecular details of the nature of this checkpoint are not clear, but both the presence of all chromosomes and the tension between opposite poles appear to be important. The signal is transmitted through the anaphase-promoting complex (APC). The sister chromatids at metaphase are still held together by a protein complex called cohesin. The APC acts by removing inhibitors of a protease that destroys the cohesin complex. This simultaneous removal of sister chromatid cohesin results in the separation of sister chromatids during anaphase.

Controlling the Cell Cycle in Multicellular Eukaryotes

Much of the work that led to our current understanding of cell cycle control came out of studies on unicellular fungi: budding yeast and fission yeast. Amazingly, most of these findings also hold true for animal cells, which have much greater constraints on growth due to their or-

ganization into tissues and organs. The major difference between animals and eukaryotes such as fungi and protists is twofold: First, multiple Cdk's control the cycle as opposed to the single Cdk in yeasts, and second, animal cells respond to a greater variety of external signals than yeasts, which primarily respond to signals necessary for mating.

The cells of multicellular eukaryotes are not free to make individual "decisions" about cell division, as yeast cells are. A multicellular body's organization cannot be maintained without severely limiting cell proliferation so that only certain cells divide, and only at appropriate times. The way cells inhibit individual growth of other cells is apparent in mammalian cells growing in tissue culture: A single layer of cells expands over a culture plate until the growing border of cells comes into contact with neighboring cells, and then the cells stop dividing. If a sector of cells is cleared away, neighboring cells rapidly refill that sector and then stop dividing again. How are cells able to sense the density of the cell culture around them? Each growing cell apparently binds minute amounts of positive regulatory signals called **growth factors,** proteins that stimulate cell division (such as MPF). When neighboring cells have used up what little growth factor is present, not enough is left to trigger cell division in any one cell.

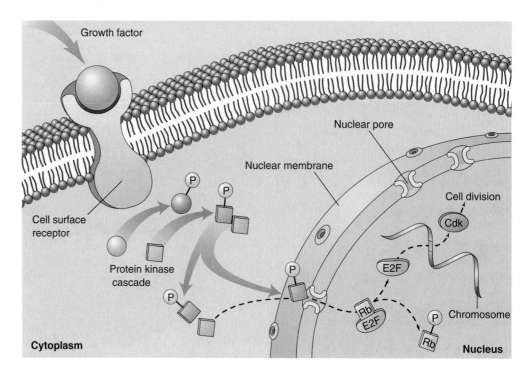

FIGURE 11.19
The cell proliferation-signaling pathway. Binding of a growth factor sets in motion a cascading intracellular signaling pathway (described in chapter 7), which activates nuclear regulatory proteins that trigger cell division. In this example, when the nuclear protein Rb is phosphorylated, another nuclear protein (the transcription factor E2F) is released and is then able to stimulate the production of Cdk proteins.

Growth Factors and the Cell Cycle

As you may recall from chapter 7 (cell–cell interactions), growth factors work by triggering intracellular signaling systems. Fibroblasts, for example, possess numerous receptors on their plasma membranes for one of the first growth factors to be identified, platelet-derived growth factor (PDGF). When PDGF binds to a membrane receptor, it initiates an amplifying chain of internal cell signals that stimulates cell division.

PDGF was discovered when investigators found that fibroblasts would grow and divide in tissue culture only if the growth medium contained blood serum (the liquid that remains after blood clots); blood plasma (blood from which the cells have been removed without clotting) would not work. The researchers hypothesized that platelets in the blood clots were releasing into the serum one or more factors required for fibroblast growth. Eventually, they isolated such a factor and named it PDGF.

Growth factors such as PDGF override cellular controls that otherwise inhibit cell division. When a tissue is injured, a blood clot forms, and the release of PDGF triggers neighboring cells to divide, helping to heal the wound. Only a tiny amount of PDGF (approximately 10^{-10} M) is required to stimulate cell division.

Characteristics of Growth Factors. Over 50 different proteins that function as growth factors have been isolated, and more undoubtedly exist. A specific cell surface receptor "recognizes" each growth factor, its shape fitting that growth factor precisely. When the growth factor binds with its receptor, the receptor reacts by triggering events within the cell (figure 11.19). The cellular selectivity of a particular growth factor depends upon which target cells bear its unique receptor. Some growth factors, such as PDGF and epidermal growth factor (EGF), affect a broad range of cell types, while others affect only specific types. For example, nerve growth factor (NGF) promotes the growth of certain classes of neurons, and erythropoietin triggers cell division in red blood cell precursors. Most animal cells need a combination of several different growth factors to overcome the various controls that inhibit cell division.

The G₀ Phase. If cells are deprived of appropriate growth factors, they stop at the G_1 checkpoint of the cell cycle. With their growth and division arrested, they remain in the G_0 phase, as we discussed earlier. This nongrowing state is distinct from the interphase stages of the cell cycle, G_1, S, and G_2.

The ability to enter G_0 accounts for the incredible diversity seen in the length of the cell cycle among different tissues. Epithelial cells lining the human gut divide more than twice a day, constantly renewing the lining of the digestive tract. By contrast, liver cells divide only once every year or two, spending most of their time in the G_0 phase. Mature neurons and muscle cells usually never leave G_0.

Two groups of proteins, cyclins and Cdk's, interact to regulate the cell cycle. Cells also receive protein signals called growth factors that affect cell division.

Cancer and the Control of Cell Proliferation

The unrestrained, uncontrolled growth of cells, called cancer, is addressed more fully in chapter 20. However, cancer certainly deserves mention in this chapter because it is essentially a disease of cell division—a failure of cell division *control*. Recent work has identified one of the culprits. Working independently, cancer scientists have repeatedly identified what has proven to be the same gene! Officially dubbed *p53* (researchers italicize the gene symbol to differentiate it from the protein), this gene plays a key role in the G₁ checkpoint of cell division. The gene's product, the p53 protein, monitors the integrity of DNA, checking that it is undamaged. If the p53 protein detects damaged DNA, it halts cell division and stimulates the activity of special enzymes to repair the damage. Once the DNA has been repaired, *p53* allows cell division to continue. In cases where the DNA damage is irreparable, *p53* then directs the cell to kill itself, activating an apoptosis (cell suicide) program (see chapter 19 for a discussion of apoptosis).

By halting division in damaged cells, *p53* prevents the development of many mutated cells, and it is therefore considered a tumor-suppressor gene (even though its activities are not limited to cancer prevention). Scientists have found that *p53* is entirely absent or damaged beyond use in the majority of cancerous cells they have examined. It is precisely because *p53* is nonfunctional that these cancer cells are able to repeatedly undergo cell division without being halted at the G₁ checkpoint (figure 11.20). To test this, scientists administered healthy p53 protein to rapidly dividing cancer cells in a petri dish: The cells soon ceased dividing and died.

Scientists at Johns Hopkins University School of Medicine have further reported that cigarette smoke causes mutations in the *p53* gene. This study, published in 1995, reinforced the strong link between smoking and cancer described in chapter 20.

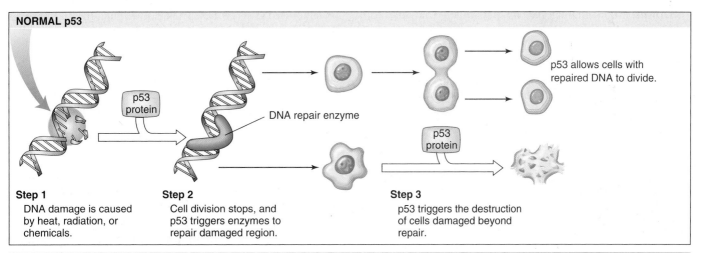

NORMAL p53

p53 protein

DNA repair enzyme

p53 allows cells with repaired DNA to divide.

p53 protein

Step 1
DNA damage is caused by heat, radiation, or chemicals.

Step 2
Cell division stops, and p53 triggers enzymes to repair damaged region.

Step 3
p53 triggers the destruction of cells damaged beyond repair.

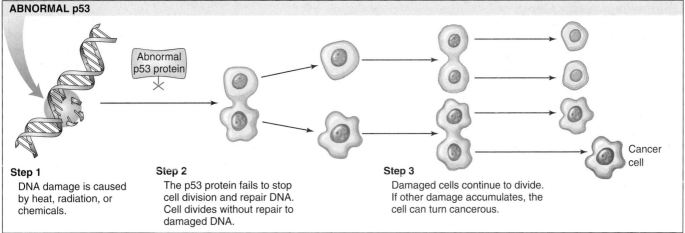

ABNORMAL p53

Abnormal p53 protein

Cancer cell

Step 1
DNA damage is caused by heat, radiation, or chemicals.

Step 2
The p53 protein fails to stop cell division and repair DNA. Cell divides without repair to damaged DNA.

Step 3
Damaged cells continue to divide. If other damage accumulates, the cell can turn cancerous.

FIGURE 11.20
Cell division and p53 protein. Normal p53 protein monitors DNA, destroying cells that have irreparable damage to their DNA. Abnormal p53 protein fails to stop cell division and repair DNA. As damaged cells proliferate, cancer develops.

Growth Factors and Cancer

The disease we call cancer is actually many different diseases, depending on the tissue affected. The common theme in all cases is the loss of control over the cell cycle.

Proto-oncogenes

Research has identified numerous so-called *oncogenes,* genes that can, when introduced into a cell, cause it to become a cancer cell. This identification then led to the discovery of **proto-oncogenes,** which are the normal cellular genes that become oncogenes when mutated. These proto-oncogenes interact with growth factors in several ways: They can be the receptors for growth factors, and they can be any of the many factors downstream of the signaling by the growth factor receptors. A receptor for a growth factor that is mutated so that it is permanently "on" leaves the cell no longer dependent on the growth factor. This is analogous to a light switch that is stuck on: The light will always be on. PDGF and EGF receptors both fall into the category of proto-oncogenes.

Downstream signaling genes include many that code for proteins acting between the initial growth factor/receptor interaction and the ultimate action of the Cdk's that control the cell cycle. Examples of this category include proteins involved in signal transduction pathways, such as ras, and the transcription factors myc, fos, and jun at the end of such pathways. All of these factors are produced as an early response to growth factor signaling, and themselves stimulate the production of a number of important proteins, including cyclins and Cdk's.

The number of proto-oncogenes identified has grown to more than 50 over the years. This line of research connects our understanding of cancer with our understanding of the molecular mechanisms governing cell cycle control.

Tumor-Suppressor Genes

After the discovery of proto-oncogenes, a second category of genes related to cancer was identified: tumor-suppressor genes. These very interesting genes normally act to inhibit the cell cycle; thus, loss of function of both copies of such genes can lead to cancer by releasing the cell cycle from inhibition. This is in contrast to the proto-oncogenes, in which only one mutant copy is necessary to produce the cancerous phenotype. That is, the proto-oncogenes act in a dominant fashion while tumor suppressors act in a recessive fashion (see chapter 13 for a discussion of dominant and recessive). The first such gene identified was the retinoblastoma susceptibility gene (*Rb*), which predisposes individuals for a rare form of cancer that affects the retina of the eye. Despite the recessive nature of *Rb* at the level of the cell, it is inherited as a dominant in families. This is because inheriting a single mutant copy of *Rb* means that you only have one "good"

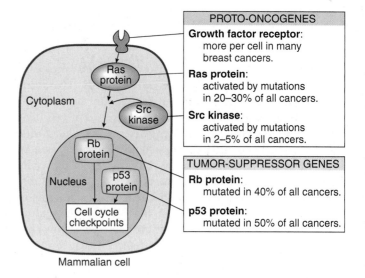

FIGURE 11.21
Key proteins associated with human cancers. Mutations in genes encoding key components of the cell division-signaling pathway are responsible for many cancers. Among them are proto-oncogenes encoding growth factor receptors, protein relay switches such as ras protein, and kinase enzymes such as src, that aid ras function. Mutations that disrupt tumor-suppressor proteins, such as Rb and p53, also foster cancer development.

copy left, and during the hundreds of thousands of divisions that occur to produce the retina, any error that results in loss or damage to the remaining good copy leads to a cancerous cell. A single cancerous cell in the retina then leads to the formation of a retinoblastoma.

The role of the Rb protein in the cell cycle is to integrate signals from growth factors. The Rb protein is called a "pocket protein" because it has binding pockets for other proteins. Its role is therefore to bind important regulatory proteins and prevent them from stimulating the production of the necessary cell cycle proteins discussed previously, such as cyclins or Cdk's. The binding of Rb to other proteins is controlled by phosphorylation such that when it is dephosphorylated, it can bind a variety of regulatory proteins, but loses this binding capacity when phosphorylated. The action of growth factors activates signaling pathways that result in the phosphorylation of Rb protein by a Cdk. This then brings us full circle, because the phosphorylation of Rb releases regulatory proteins that result in the production of S-phase cyclins that are necessary for the action of the Cdk required to pass the G_1/S boundary.

Figure 11.21 summarizes the types of genes that can cause cancer when mutated.

Some proto-oncogenes accelerate the cell cycle by promoting cyclins and Cdk's. Others, tumor-suppressor genes, suppress it by inhibiting their action. Cancer can result if mutations occur in either type of growth factor gene.

Concept Review

For interactive testing, visit the Online Learning Center with PowerWeb at www.mhhe.com/Raven7

11.1 Prokaryotes divide far more simply than do eukaryotes.

Cell Division in Prokaryotes

- Most prokaryotes have a genome made up of a single, circular DNA molecule, and replicate via binary fusion. (p. 208)
- Binary fusion begins with DNA replication, which starts at the origin site and proceeds bidirectionally around the circular DNA to a specific site of termination. (p. 208)
- The evolution of eukaryotic cells led to much more complex genomes and, thus, new and different ways to replicate and segregate the genome during cell division. (p. 209)

11.2 The chromosomes of eukaryotes are highly ordered structures.

Discovery of Chromosomes

- Chromosomes were first discovered in 1882 by Walther Fleming. (p. 210)
- The number of chromosomes varies from one species to another. Humans have 23 nearly identical pairs for a total of 46 chromosomes. (p. 210)

The Structure of Eukaryotic Chromosomes

- The DNA is a very long, double-stranded fiber extending unbroken through the entire length of the chromosome. A typical human chromosome contains about 140 million nucleotides. (p. 211)
- Every 200 nucleotides, the DNA duplex is coiled around a core of eight histone proteins, forming a nucleosome. (p. 211)
- The particular array of chromosomes an individual possesses is its karyotype. (p. 212)
- The number of different chromosomes a species contains is known as its haploid (n) number, and is considered one complete set of chromosomes. (p. 212)
- Humans are diploid, with homologues coming from both the maternal and paternal lineages. (p. 212)

11.3 Mitosis is a key phase of the cell cycle.

The Cell Cycle

- The typical cell cycle consists of five phases: G_1 is the primary growth phase; S is the replication phase; and G_2 is the second growth phase. (G_1, S, and G_2 combined constitute interphase.) M (mitosis) is the phase during which the microtubular apparatus separates sister chromatids, and C (cytokinesis) is the phase during which the cytoplasm divides, creating two daughter cells. (p. 213)
- The duration of the cell cycle varies from species to species, and can range from about 8 minutes to over a year. (p. 213)

Interphase: Preparing for Mitosis

- The cell grows throughout interphase. The G_1 and G_2 phases are periods of protein synthesis and organelle production, while the S phase is when DNA replication occurs. (p. 214)

Mitosis

- Chromatin condensation continues into prophase. The spindle apparatus is assembled, and sister chromatids are linked to opposite poles of the cell by microtubules. The nuclear envelope breaks down. (p. 215)
- During metaphase, chromosomes align in the center of the cell along the metaphase plate. (p. 215)
- Anaphase begins when centromeres divide, freeing the two sister chromatids from each other. Sister chromatids are pulled to opposite poles as the attached microtubules shorten. (pp. 216–217)
- In telophase, the spindle apparatus disassembles, and the nuclear membrane begins to re-form. (p. 217)

Cytokinesis

- Cytokinesis is the phase of the cell cycle when the cell actually divides. Cytokinesis generally involves the cleavage of the cell into roughly equal halves, forming two daughter cells. (p. 218)

11.4 The cell cycle is carefully controlled.

General Strategies of Cell Cycle Control

- A cell uses three main checkpoints to both assess the internal state of the cell and integrate external signals. The G_1/S checkpoint is the primary point at which the cell decides to divide; the G_2/M checkpoint represents a commitment to mitosis; and the spindle checkpoint ensures that all chromosomes are attached to the spindle in preparation for anaphase. (p. 219)

Molecular Mechanisms of Cell Cycle Control

- Two groups of proteins, cyclins and Cdk's, interact and regulate the cell cycle. (p. 220)
- Cells also receive protein signals (growth factors) that affect cell division. (p. 222)

Cancer and the Control of Cell Proliferation

- Cancer is failure of cell division control. (p. 223)
- It is believed that a malfunction in the *p53* gene may allow cells to go through repeated cell division without being stopped at the appropriate checkpoints. (p. 223)
- Proto-oncogenes are normal cellular genes that become oncogenes when mutated. Proto-oncogenes can encode growth factors, protein relay switches, and kinase enzyme. (p. 224)
- Tumor-suppressor genes can also lead to cancer when they are mutated. (p. 224)

Self Test

1. Bacterial cells divide by
   a. mitosis.
   b. replication.
   c. cytokinesis.
   d. binary fission.
2. Most eukaryotic organisms have _____ chromosomes in their cells.
   a. 1–5
   b. 10–50
   c. 100–500
   d. over 1000
3. Replicate copies of each chromosome are called _____ and are joined at the _____.
   a. homologues/centromere
   b. sister chromatids/kinetochore
   c. sister chromatids/centromere
   d. homologues/kinetochore
4. During which phase of the cell cycle is DNA synthesized?
   a. G_1
   b. G_2
   c. S
   d. M
5. Chromosomes are visible under a light microscope
   a. during mitosis.
   b. during interphase.
   c. when they are attached to their sister chromatids.
   d. All of these are correct.
6. During mitosis, the sister chromatids are separated and pulled to opposite poles during which stage?
   a. interphase
   b. metaphase
   c. anaphase
   d. telophase
7. Cytokinesis is
   a. the same process in plant and animal cells.
   b. the separation of cytoplasm and the formation of two cells.
   c. the final stage of mitosis.
   d. the movement of kinetochores.
8. The eukaryotic cell cycle is controlled at several points; which of these statements is *not* true?
   a. Cell growth is assessed at the G_1/S checkpoint.
   b. DNA replication is assessed at the G_2/M checkpoint.
   c. Environmental conditions are assessed at the G_0 checkpoint.
   d. The chromosomes are assessed at the spindle checkpoint.
9. What proteins are used to control cell growth specifically in *multicellular* eukaryotic organisms?
   a. Cdk
   b. MPF
   c. cyclins
   d. growth factors
10. What causes cancer in cells?
    a. damage to genes
    b. chemical damage to cell membranes
    c. UV damage to transport proteins
    d. All of these cause cancer in cells.

Test Your Visual Understanding

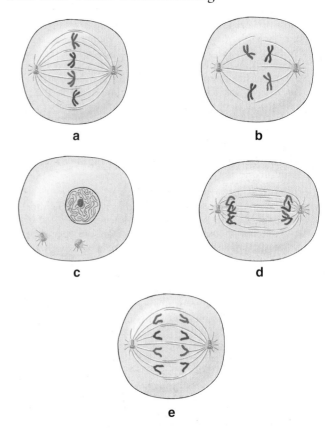

a   b   c   d   e

1. Match the mitotic and cell cycle phases with the appropriate figure.
   anaphase
   interphase
   metaphase
   prophase
   telophase

Apply Your Knowledge

1. An ancient plant called horsetail contains 216 chromosomes. How many homologous pairs of chromosomes does it contain? How many chromosomes are present in its cells during metaphase?
2. Colchicine is a poison that binds to tubulin and prevents its assembly into microtubules; cytochalasins are compounds that bind to the ends of actin filaments and prevent their elongation. What effects would these two substances have on cell division in animal cells?
3. If you could construct an artificial chromosome, what elements would you introduce into it, at a minimum, so that it could function normally in mitosis?

12

Sexual Reproduction and Meiosis

Concept Outline

12.1 Meiosis produces haploid cells from diploid cells.

Discovery of Reduction Division. Sexual reproduction does not increase chromosome number because gamete production by meiosis involves a decrease in chromosome number. Individuals produced from sexual reproduction inherit chromosomes from two parents.

12.2 Meiosis has unique features.

Meiosis. Three unique features of meiosis are synapsis, homologous recombination, and reduction division.

12.3 The sequence of events during meiosis involves two nuclear divisions.

Prophase I. Homologous chromosomes pair intimately and undergo crossing over that locks them together.
Metaphase I. Spindle microtubules align the chromosomes in the central plane of the cell.
Completing Meiosis. The second meiotic division is like a mitotic division, but has a very different outcome. Homologues separate at anaphase I, but sister chromatids do not. Sister chromatids separate at anaphase II, producing four haploid cells.

12.4 The evolutionary origin of sex is a puzzle.

Why Sex? Sex may have evolved as a mechanism to repair DNA, or perhaps as a means for contagious elements to spread. Sexual reproduction increases genetic variability by shuffling combinations of genes.

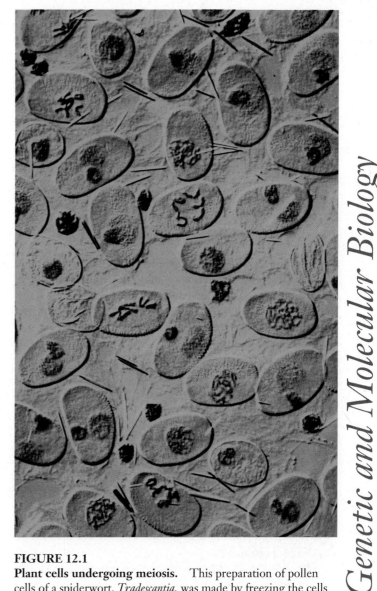

FIGURE 12.1
Plant cells undergoing meiosis. This preparation of pollen cells of a spiderwort, *Tradescantia*, was made by freezing the cells and then fracturing them. It shows several stages of meiosis (600×).

Most animals and plants reproduce sexually. Gametes of opposite sex unite to form a cell that, dividing repeatedly by mitosis, eventually gives rise to an adult body with some 100 trillion cells. The gametes that form the initial cell are the products of a special form of cell division called meiosis (figure 12.1), the subject of this chapter. Meiosis is far more intricate than mitosis, and the details behind it are not as well understood. The basic process, however, is clear. Also clear are the profound consequences of sexual reproduction: It plays a key role in generating the tremendous genetic diversity that is the raw mat̶ evolution.

Genetic and Molecular Biology

Part III

228

12.1 Meiosis produces haploid cells from diploid cells.

Discovery of Reduction Division

Only a few years after Walther Fleming's discovery of chromosomes in 1882, Belgian cytologist Pierre-Joseph van Beneden was surprised to find different numbers of chromosomes in different types of cells in the roundworm *Ascaris*. Specifically, he observed that the **gametes** (eggs and sperm) each contained two chromosomes, while the **somatic cells** (nonreproductive cells) of embryos and mature individuals each contained four.

Fertilization

From his observations, van Beneden proposed in 1887 that an egg and a sperm, each containing half the complement of chromosomes found in other cells, fuse to produce a single cell called a **zygote**. The zygote, like all of the somatic cells ultimately derived from it, contains two copies of each chromosome. The fusion of gametes to form a new cell is called **fertilization**, or **syngamy**.

Reduction Division

It was clear even to early investigators that gamete formation must involve some mechanism that reduces the number of chromosomes to half the number found in other cells. If it did not, the chromosome number would double with each fertilization, and after only a few generations, the number of chromosomes in each cell would become impossibly large. For example, in just 10 generations, the 46 chromosomes present in human cells would increase to over 47,000 (46×2^{10}).

The number of chromosomes does not explode in this way because of a special reduction division that occurs during gamete formation, producing cells with half the normal number of chromosomes. The subsequent fusion of two of these cells ensures a consistent chromosome number from one generation to the next. This reduction division process, known as **meiosis**, is the subject of this chapter.

The Sexual Life Cycle

Meiosis and fertilization together constitute a cycle of reproduction. Two sets of chromosomes are present in the somatic cells of adult individuals, making them **diploid** cells (Greek *diploos*, "double," + *eidos*, "form"), but only one set is present in the gametes, which are thus **haploid** (Greek *haploos*, "single," + *ploion*, "vessel"). Reproduction that involves this alternation of meiosis and fertilization is called **sexual reproduction**. Its outstanding characteristic is that offspring inherit chromosomes from *two* parents (figure 12.2). You, for example, inherited 23 chromosomes from your mother (maternal homologues), contributed by the egg fertilized at your conception, and 23 from your father (paternal homologues), contributed by the sperm that fertilized that egg.

The life cycles of all sexually reproducing organisms follow a pattern of alternation between diploid and haploid chromosome numbers, but there is some variation in the life cycles. Many types of algae spend the majority of their life cycle in a haploid state, the zygote undergoing meiosis to produce haploid cells that then undergo mitosis (figure 12.3*a*). In most animals, the diploid state dominates; the zygote first undergoes mitosis to produce diploid cells, and later in the life cycle, some of these diploid cells undergo meiosis to produce haploid gametes (figure 12.3*b*). Some plants and some algae alternate between a multicellular haploid phase and a multicellular diploid phase (figure 12.3*c*).

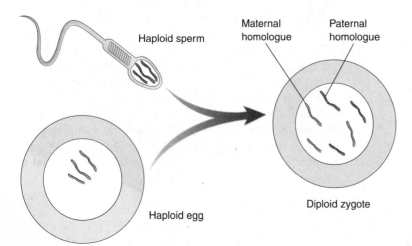

FIGURE 12.2
Diploid cells carry chromosomes from two parents.
A diploid cell contains two versions of each chromosome, a maternal homologue contributed by the haploid egg of the mother, and a paternal homologue contributed by the haploid sperm of the father.

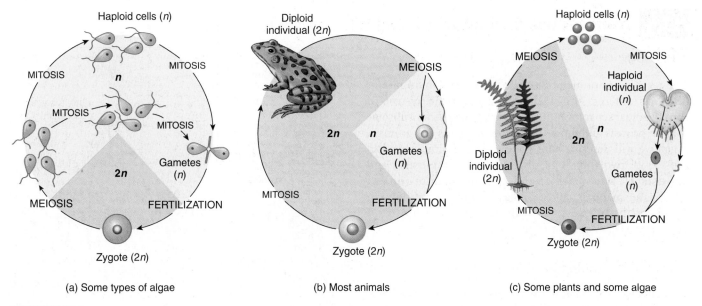

FIGURE 12.3
Three types of sexual life cycles. In sexual reproduction, haploid cells or organisms alternate with diploid cells or organisms.

(a) Some types of algae

(b) Most animals

(c) Some plants and some algae

Germ-Line Cells. In animals, the single diploid zygote undergoes mitosis to give rise to all of the cells in the adult body. The cells not destined to form gametes are called somatic cells, from the Latin word for "body." The cells that will eventually undergo meiosis to produce gametes are set aside from somatic cells early in the course of development. These cells are often referred to as germ-line cells. Both the somatic cells and the gamete-producing germ-line cells are diploid, but while somatic cells undergo mitosis to form genetically identical, diploid daughter cells, gamete-producing germ-line cells undergo meiosis to produce haploid gametes (figure 12.4).

> Meiosis is a process of cell division in which the number of chromosomes in certain cells is halved during gamete formation. In the sexual life cycle, alternation of diploid and haploid generations occurs.

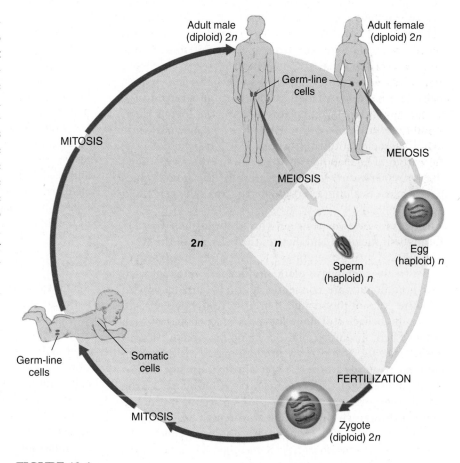

FIGURE 12.4
The sexual life cycle in animals. In animals, the zygote undergoes mitotic divisions and gives rise to all the cells of the adult body. Germ-line cells are set aside early in development and undergo meiosis to form the haploid gametes (eggs or sperm). The rest of the body cells are called somatic cells.

12.2 Meiosis has unique features.

Meiosis

The mechanism of meiotic cell division varies in important details in different organisms. This is particularly true of chromosomal separation mechanisms, which differ substantially in protists and fungi from those in plants and animals, which we will describe here. Meiosis in a diploid organism consists of two rounds of division, called **meiosis I** and **meiosis II**, with each round containing a prophase, metaphase, anaphase, and telophase. Before examining the details of this process, we will examine the features of meiosis that distinguish it from mitosis.

Pairing of Homologous Chromosomes

To understand the distinctions between mitosis and meiosis, you must understand the behavior of chromosomes during each process. The most important difference in the behavior of chromosomes occurs early in prophase I of meiosis when homologous chromosomes find each other and become closely associated, a process called pairing, or **synapsis** (figure 12.5a). Despite the long history of our knowledge of this process, its exact molecular details remain unclear. Evidence from a variety of sources, including electron micrographs, data from genetic crosses, and some biochemistry, all shed some light on synapsis, but thus far this knowledge has not been integrated into a complete picture. What is clear is that homologous chromosomes find their proper partners and become intimately associated during prophase I. This includes the formation of an elaborate structure in many species called the *synaptonemal complex*, consisting of the homologues paired closely along a lattice of proteins between them (see figure 12.7).

The association between the homologues persists throughout meiosis I and dictates the behavior of the chromosomes. During metaphase I, the paired homologues move to the metaphase plate and become oriented such that homologues are connected by their kinetochores to opposite poles of the spindle. This is in contrast to mitosis, where the behavior of homologues is independent of each other. Then, during anaphase I, homologues move to opposite poles for each pair of chromosomes. This is again in contrast to mitosis, where sister chromatids, not homologues, move to opposite poles. This is also why we call the first division the "reduction division"—it results in one homologue from each chromosome pair in each daughter cell. While the number of chromatids has not been reduced, the number of chromosomes has: Remember, we count chromosomes by counting centromeres.

During the period of intimate association during prophase I, another process unique to meiosis occurs: genetic recombination, or **crossing over.** This process

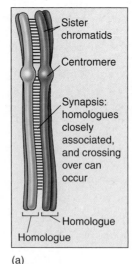
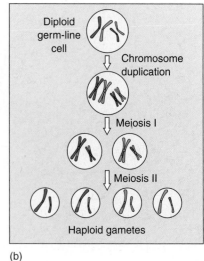

(a)   (b)

FIGURE 12.5
Unique features of meiosis. (*a*) Synapsis draws homologous chromosomes together, creating a situation where the two chromosomes can physically exchange parts, a process called crossing over. (*b*) A reductive division in meiosis I and the omission of chromosome duplication before meiosis II produce haploid gametes, thus ensuring that chromosome number remains stable during the reproduction cycle.

literally allows the homologues to exchange chromosomal material. The observation of this cytologically is called crossing over, while its detection genetically is called recombination. The sites of crossing over are called *chiasmata*, and these sites of contact are maintained until anaphase I. This continued association of homologues until anaphase I is critical to the accurate segregation of homologues, because it allows them to be oriented by connection of homologous centromeres to opposite poles.

Two Divisions, One Round of Replication

The most obvious distinction between meiosis and mitosis is the simple observation that meiosis involves two successive divisions with no replication of genetic material between them. The behavior of chromosomes during meiosis I puts the resulting cells in a position where a division that acts like mitosis with no replication will produce cells with half the original number of chromosomes (figure 12.5b). This is the last key to understanding meiosis: The second meiotic division is like mitosis except that it is not preceded by chromosome duplication. Figure 12.6 provides a comparison of meiosis and mitosis.

In meiosis, homologous chromosomes become intimately associated and do not replicate between the two nuclear divisions.

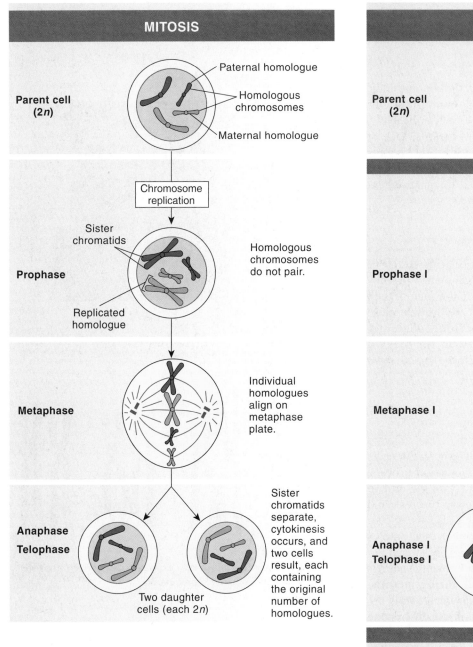

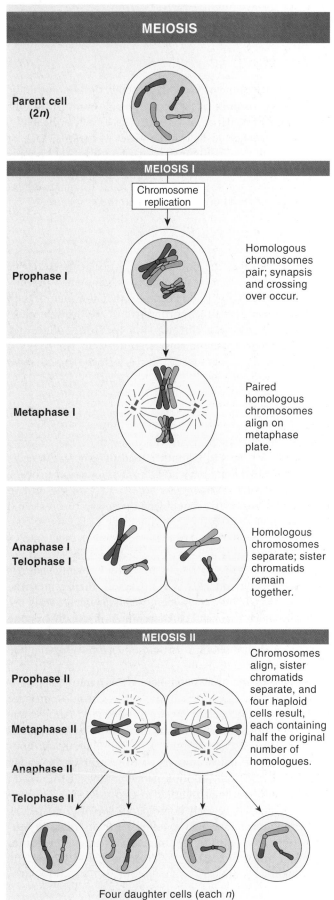

FIGURE 12.6

A comparison of meiosis and mitosis. Meiosis involves two nuclear divisions with no DNA replication between them. It thus produces four daughter cells, each with half the original number of chromosomes. Crossing over occurs in prophase I of meiosis. Mitosis involves a single nuclear division after DNA replication. It thus produces two daughter cells, each containing the original number of chromosomes.

12.3 The sequence of events during meiosis involves two nuclear divisions.

Prophase I

In prophase I of meiosis, the DNA coils tighter, and individual chromosomes first become visible under the light microscope as a matrix of fine threads. Because the DNA has already replicated before the onset of meiosis, each of these threads actually consists of two sister chromatids joined at their centromeres. In prophase I, homologous chromosomes become closely associated in synapsis, exchange segments by crossing over, and then separate.

Synapsis

During prophase, the ends of the chromatids attach to the nuclear envelope at specific sites. The sites the homologues attach to are adjacent, so that the members of each homologous pair of chromosomes are brought close together. They then line up side by side, apparently guided by heterochromatin sequences, in the process called synapsis. This association is referred to as *sister chromatid cohesion*.

Crossing Over

Along with the synaptonemal complex that forms during prophase I (figure 12.7), another kind of structure appears that correlates in timing with the recombination process. These are called *recombination nodules* and are thought to contain the enzymatic machinery necessary to break and rejoin homologous chromatids. The details of crossing over are not well understood, but the process involves a complex series of events in which DNA segments are exchanged between nonsister chromatids (figure 12.8). Crossing over between sister chromatids is suppressed during meiosis. Reciprocal crossovers between nonsister chromatids are controlled such that each chromosome arm usually has one or a few crossovers per meiosis, no matter what the size of the chromosome. Human chromosomes typically have two or three.

When crossing over is complete, the synaptonemal complex breaks down, and the homologous chromosomes are released from the nuclear envelope and begin to move away from each other. At this point, there are four chromatids for each type of chromosome (two homologous chromosomes, each of which consists of two sister chromatids). The four chromatids do not separate completely, however, because they are held together in two ways: (1) The two sister chromatids of each homologue, recently created by DNA replication, are held together by their common centromeres; and (2) the paired homologues are held together at the points where crossing over occurred within the synaptonemal complex.

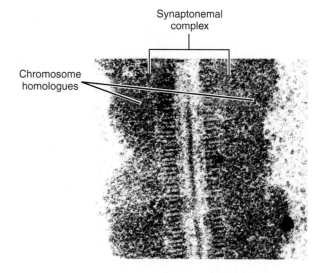

FIGURE 12.7
Structure of the synaptonemal complex. A portion of the synaptonemal complex of the ascomycete *Neotiella rutilans*, a cup fungus.

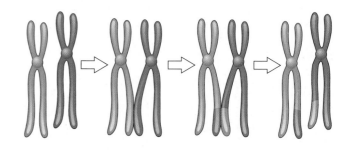

FIGURE 12.8
The results of crossing over. During crossing over, nonsister chromatids may exchange segments.

Chiasma Formation

Evidence of crossing over can often be seen under the light microscope as an X-shaped structure known as a **chiasma** (Greek, "cross"; plural, *chiasmata*). A chiasma is a crossover between nonsister chromatids, stabilized by sister chromatid cohesion. The presence of a chiasma indicates that two chromatids (one from each homologue) have exchanged parts. Like small rings moving down two strands of rope, the chiasmata move to the end of the chromosome arm before metaphase I.

Synapsis takes place early in prophase I of meiosis. Crossing over occurs between the paired DNA strands, creating the chromosomal configurations known as chiasmata. The two homologues are locked together by these exchanges, and they do not disengage readily.

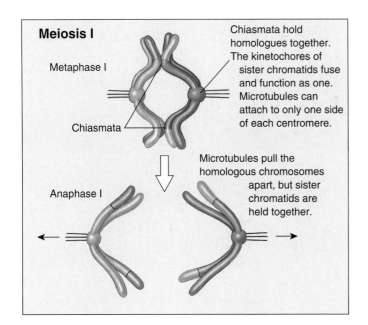

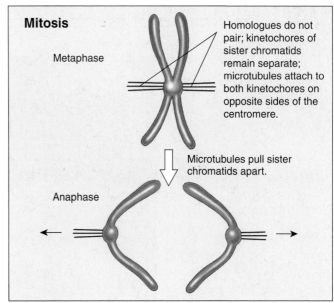

FIGURE 12.9

Chiasmata created by crossing over have a key impact on how chromosomes align in metaphase I. In metaphase I, the chiasmata hold homologous chromosomes together; consequently, the spindle microtubules bind to only one side of each centromere. By the end of meiosis I, mocrotubules shorten, breaking the chiasmata and pulling the homologous chromosomes apart, but sister chromatids are held together by their centromeres. In mitosis, microtubules attach to both sides of each centromere; when the microtubules shorten, the sister chromatids are split and drawn to opposite poles.

Metaphase I

By metaphase I, the second stage of meiosis I, the nuclear envelope has dispersed, and the microtubules form a spindle, just as in mitosis. During prophase I, the chiasmata move down the paired chromosomes from their original points of crossing over, eventually reaching the ends of the chromosomes. At this point, they are called *terminal chiasmata*. Terminal chiasmata hold the homologous chromosomes together in metaphase I, so that only one side of each centromere faces outward from the complex; the other side is turned inward toward the other homologue (figure 12.9). Consequently, spindle microtubules are able to attach to kinetochore proteins only on the outside of each centromere, and the centromeres of the two homologues attach to microtubules originating from opposite poles. This one-sided attachment is in marked contrast to the attachment in mitosis, when kinetochores on *both* sides of a centromere bind to microtubules.

Each joined pair of homologues then lines up on the metaphase plate. The orientation of each pair on the spindle axis is random; either the maternal or the paternal homologue may orient toward a given pole (figure 12.10; see also figure 12.11).

Chiasmata play an important role in aligning the chromosomes on the metaphase plate.

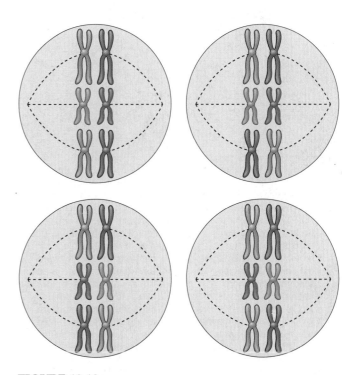

FIGURE 12.10

Random orientation of chromosomes on the metaphase plate. The number of possible chromosome orientations equals 2 raised to the power of the number of chromosome pairs. In this hypothetical cell with three chromosome pairs, eight (2^3) possible orientations exist, four of them illustrated here. Each orientation produces gametes with different combinations of parental chromosomes.

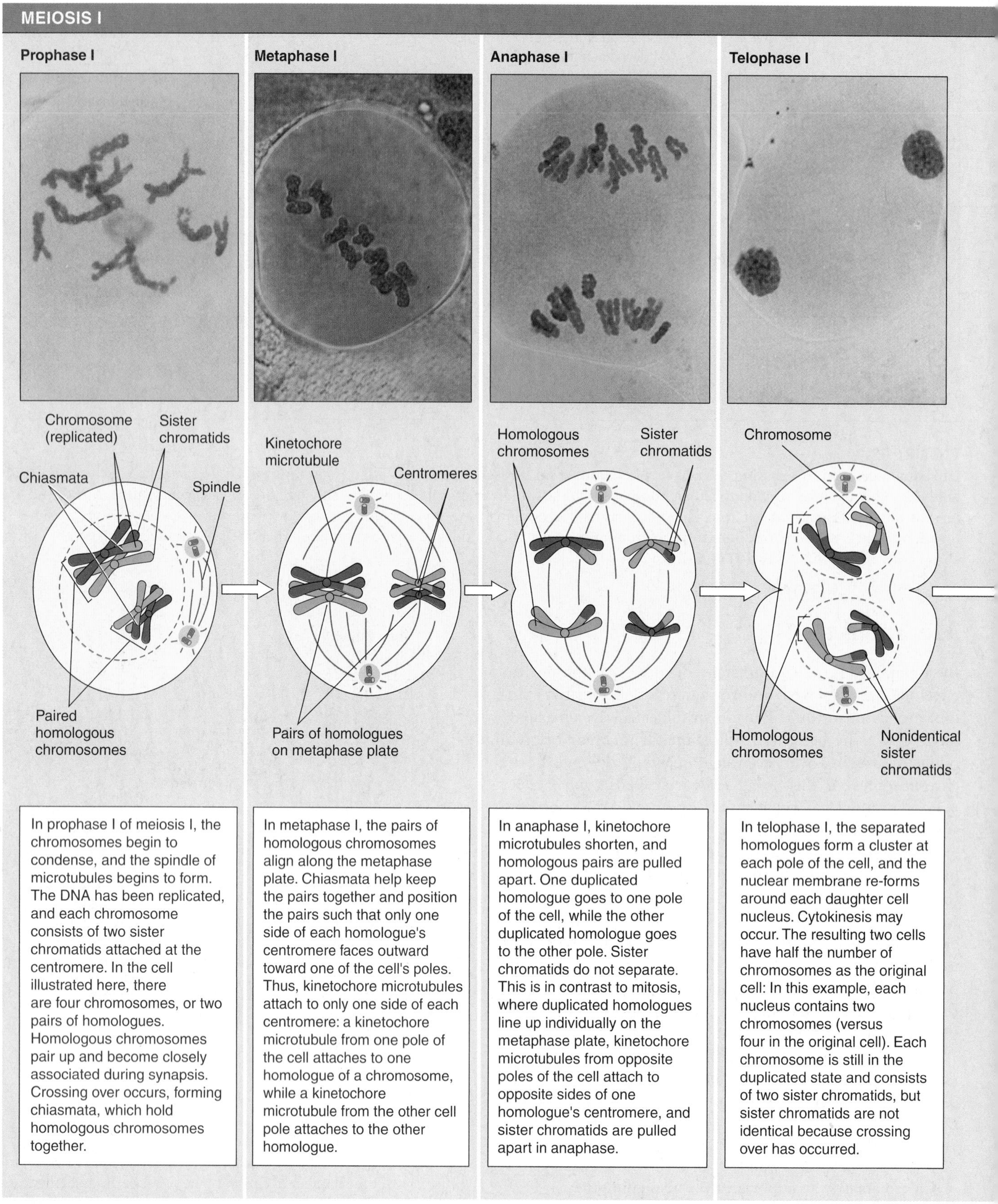

Prophase I

Chromosome (replicated)
Sister chromatids
Chiasmata
Spindle
Paired homologous chromosomes

In prophase I of meiosis I, the chromosomes begin to condense, and the spindle of microtubules begins to form. The DNA has been replicated, and each chromosome consists of two sister chromatids attached at the centromere. In the cell illustrated here, there are four chromosomes, or two pairs of homologues. Homologous chromosomes pair up and become closely associated during synapsis. Crossing over occurs, forming chiasmata, which hold homologous chromosomes together.

Metaphase I

Kinetochore microtubule
Centromeres
Pairs of homologues on metaphase plate

In metaphase I, the pairs of homologous chromosomes align along the metaphase plate. Chiasmata help keep the pairs together and position the pairs such that only one side of each homologue's centromere faces outward toward one of the cell's poles. Thus, kinetochore microtubules attach to only one side of each centromere: a kinetochore microtubule from one pole of the cell attaches to one homologue of a chromosome, while a kinetochore microtubule from the other cell pole attaches to the other homologue.

Anaphase I

Homologous chromosomes
Sister chromatids

In anaphase I, kinetochore microtubules shorten, and homologous pairs are pulled apart. One duplicated homologue goes to one pole of the cell, while the other duplicated homologue goes to the other pole. Sister chromatids do not separate. This is in contrast to mitosis, where duplicated homologues line up individually on the metaphase plate, kinetochore microtubules from opposite poles of the cell attach to opposite sides of one homologue's centromere, and sister chromatids are pulled apart in anaphase.

Telophase I

Chromosome
Homologous chromosomes
Nonidentical sister chromatids

In telophase I, the separated homologues form a cluster at each pole of the cell, and the nuclear membrane re-forms around each daughter cell nucleus. Cytokinesis may occur. The resulting two cells have half the number of chromosomes as the original cell: In this example, each nucleus contains two chromosomes (versus four in the original cell). Each chromosome is still in the duplicated state and consists of two sister chromatids, but sister chromatids are not identical because crossing over has occurred.

FIGURE 12.11
The stages of meiosis. Meiosis in plant cells (photos) and animal cells (drawings) is shown.

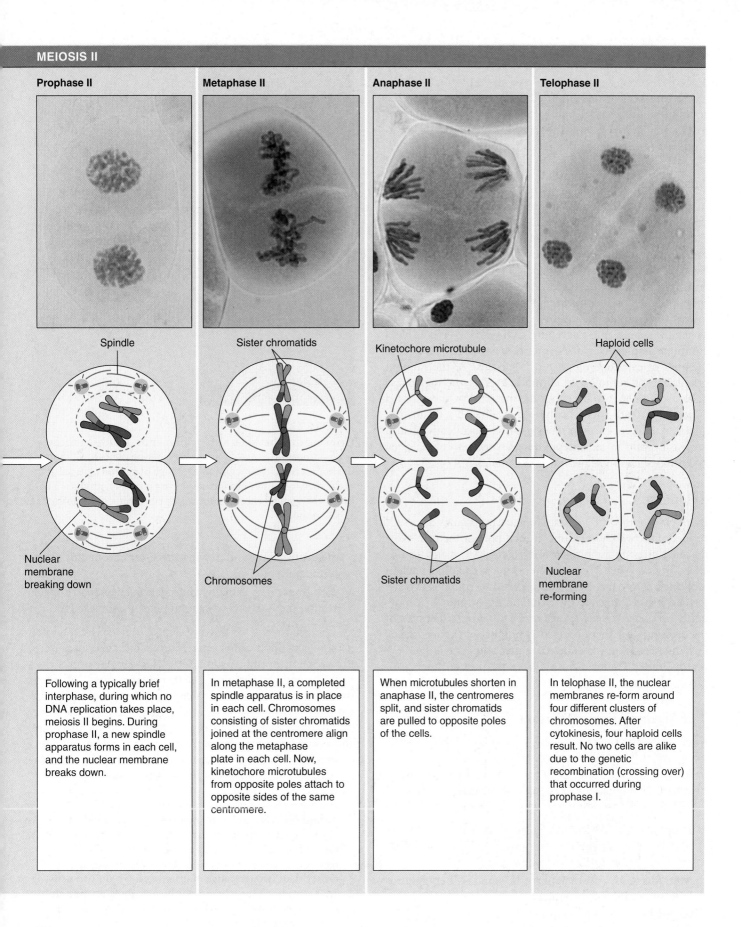

Prophase II

Spindle

Nuclear membrane breaking down

Following a typically brief interphase, during which no DNA replication takes place, meiosis II begins. During prophase II, a new spindle apparatus forms in each cell, and the nuclear membrane breaks down.

Metaphase II

Sister chromatids

Chromosomes

In metaphase II, a completed spindle apparatus is in place in each cell. Chromosomes consisting of sister chromatids joined at the centromere align along the metaphase plate in each cell. Now, kinetochore microtubules from opposite poles attach to opposite sides of the same centromere.

Anaphase II

Kinetochore microtubule

Sister chromatids

When microtubules shorten in anaphase II, the centromeres split, and sister chromatids are pulled to opposite poles of the cells.

Telophase II

Haploid cells

Nuclear membrane re-forming

In telophase II, the nuclear membranes re-form around four different clusters of chromosomes. After cytokinesis, four haploid cells result. No two cells are alike due to the genetic recombination (crossing over) that occurred during prophase I.

Completing Meiosis

Together, prophase and metaphase take up 90% or more of the duration of meiosis I. After that, meiosis I rapidly concludes. Anaphase I and telophase I proceed quickly, followed—without an intervening period of DNA synthesis—by the second meiotic division.

Anaphase I

In anaphase I, the microtubules of the spindle fibers begin to shorten. As they shorten, they break the chiasmata and pull the centromeres toward the poles, dragging the chromosomes along with them. Anaphase I comes about by the release of sister chromatid cohesion along the chromosome arms but not the centromeres. As a result of this release of cohesion, the homologues are pulled apart. Each centromere moves to one pole, taking both sister chromatids with it. When the spindle fibers have fully contracted, each pole has a complete haploid set of chromosomes consisting of one member of each homologous pair. Because of the random orientation of homologous chromosomes on the metaphase plate, a pole may receive either the maternal or the paternal homologue from each chromosome pair. As a result, the genes on different chromosomes assort independently; that is, meiosis I results in the **independent assortment** of maternal and paternal chromosomes into the gametes.

Telophase I

By the beginning of telophase I, the chromosomes have segregated into two clusters, one at each pole of the cell. Now the nuclear membrane re-forms around each daughter nucleus. Because each chromosome within a daughter nucleus replicated before meiosis I began, each now contains two sister chromatids attached by a common centromere. Importantly, *the sister chromatids are no longer identical* because of the crossing over that occurred in prophase I (figure 12.12). Cytokinesis may or may not occur after telophase I. The second meiotic division, meiosis II, occurs after an interval of variable length.

The Second Meiotic Division

After a typically brief interphase, in which no DNA synthesis occurs, the second meiotic division begins.

Meiosis II resembles a normal mitotic division. Prophase II, metaphase II, anaphase II, and telophase II follow in quick succession (see figure 12.11).

Prophase II.   At the two poles of the cell, the clusters of chromosomes enter a brief prophase II, each nuclear envelope breaking down as a new spindle forms.

Metaphase II.   In metaphase II, spindle fibers bind to both sides of the centromeres.

Anaphase II.   The spindle fibers contract, and centromeric sister chromatid cohesion is released, splitting

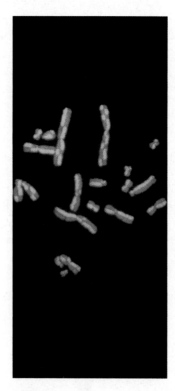
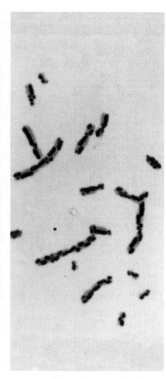

FIGURE 12.12
Crossing over between sister chromatids.   Exchange between sister chromatids is usually suppressed during meiosis. Without this suppression, the reciprocal exchange of genetic material occurs between sister chromatids during meiosis I, producing so-called "harlequin" chromosomes, each containing one fluorescent DNA strand.

the centromeres and moving the sister chromatids to opposite poles.

Telophase II.   Finally, the nuclear envelope re-forms around the four sets of daughter chromosomes.

The final result of this division is four cells containing haploid sets of chromosomes. No two are alike, because of the crossing over in prophase I. Nuclear envelopes then form around each haploid set of chromosomes. The cells that contain these haploid nuclei may develop directly into gametes, as they do in animals. Alternatively, they may themselves divide mitotically, as they do in plants, fungi, and many protists, eventually producing greater numbers of gametes or, as in some plants and insects, adult individuals with varying numbers of chromosome sets.

During meiosis I, homologous chromosomes move toward opposite poles in anaphase I, and individual chromosomes cluster at the two poles in telophase I. At the end of meiosis II, each of the four haploid cells contains one copy of every chromosome in the set, rather than two. Because of crossing over, no two cells are the same. These haploid cells may develop directly into gametes, as in animals, or they may divide by mitosis, as in plants, fungi, and many protists.

12.4 The evolutionary origin of sex is a puzzle.

Why Sex?

Not all reproduction is sexual. In **asexual reproduction,** an individual inherits all of its chromosomes from a single parent and is, therefore, genetically identical to its parent. Prokaryotic cells reproduce asexually, undergoing binary fission to produce two daughter cells containing the same genetic information (see chapter 11). Most protists reproduce asexually except under conditions of stress; then they switch to sexual reproduction. Among plants, asexual reproduction is common, and many other multicellular organisms are also capable of reproducing asexually. In animals, asexual reproduction often involves the budding off of a localized mass of cells, which grows by mitosis to form a new individual.

Even when meiosis and the production of gametes occur, reproduction may still take place without sex. The development of an adult from an unfertilized egg, called **parthenogenesis,** is a common form of reproduction in arthropods. Among bees, for example, fertilized eggs develop into diploid females, but unfertilized eggs develop into haploid males. Parthenogenesis even occurs among the vertebrates. Some lizards, fishes, and amphibians are capable of reproducing in this way; their unfertilized eggs undergo a mitotic nuclear division without cell cleavage to produce a diploid cell, which then develops into an adult.

Recombination Can Be Destructive

If reproduction can occur without sex, why does sex occur at all? This question has generated considerable discussion, particularly among evolutionary biologists. Sex is of great evolutionary advantage for populations or species, which benefit from the variability generated in meiosis by random orientation of chromosomes and by crossing over. However, evolution occurs because of changes at the level of *individual* survival and reproduction, rather than at the population level, and no obvious advantage accrues to the progeny of an individual that engages in sexual reproduction. In fact, recombination is a destructive as well as a constructive process in evolution. The segregation of chromosomes during meiosis tends to disrupt advantageous combinations of genes more often than it creates new, better-adapted combinations; as a result, some of the diverse progeny produced by sexual reproduction will not be as well adapted as their parents were. In fact, the more complex the adaptation of an individual organism, the less likely that recombination will improve it, and the more likely that recombination will disrupt it. It is, therefore, a puzzle to know what a well-adapted individual gains from participating in sexual reproduction, because *all* of its progeny could maintain its successful gene combinations if that individual simply reproduced asexually.

The Origin and Maintenance of Sex

There is no consensus among evolutionary biologists regarding the evolutionary origin or maintenance of sex. Conflicting hypotheses abound. Alternative hypotheses seem to be correct to varying degrees in different organisms.

The DNA Repair Hypothesis. If recombination is often detrimental to an individual's progeny, then what benefit promoted the evolution of sexual reproduction? Although the answer to this question is unknown, we can gain some insight by examining the protists. Meiotic recombination is often absent among the protists, which typically undergo sexual reproduction only occasionally. Often the fusion of two haploid cells occurs only under stress, creating a diploid zygote.

Why do some protists form a diploid cell in response to stress? Several geneticists have suggested that this occurs because only a diploid cell can effectively repair certain kinds of chromosome damage, particularly double-strand breaks in DNA. Both radiation and chemical events within cells can induce such breaks. As organisms became larger and lived longer, it must have become increasingly important for them to be able to repair such damage. The synaptonemal complex, which in early stages of meiosis precisely aligns pairs of homologous chromosomes, may well have evolved originally as a mechanism for repairing double-strand damage to DNA, using the undamaged homologous chromosome as a template to repair the damaged chromosome. A transient diploid phase would have provided an opportunity for such repair. In yeasts, mutations that inactivate the repair system for double-strand breaks of the chromosomes also prevent crossing over, suggesting a common mechanism for both synapsis and repair processes.

The Contagion Hypothesis. An unusual and interesting alternative hypothesis to explain the origin of sex is that it arose as a secondary consequence of the infection of eukaryotes by mobile genetic elements. Suppose a replicating transposable element were to infect a eukaryotic lineage. If it possessed genes promoting fusion with uninfected cells and synapsis, the transposable element could readily copy itself onto homologous chromosomes. It would rapidly spread by infection through the population, until all members contained it. The bizarre mating type "alleles" found in many fungi are very nicely explained by this hypothesis. Each of several mating types is in fact not an allele but an "idiomorph." Idiomorphs are genes occupying homologous positions on the chromosome but having such dissimilar sequences that they cannot be of homologous origin. These idiomorph genes may simply be the relics of several ancient infections by transposable elements.

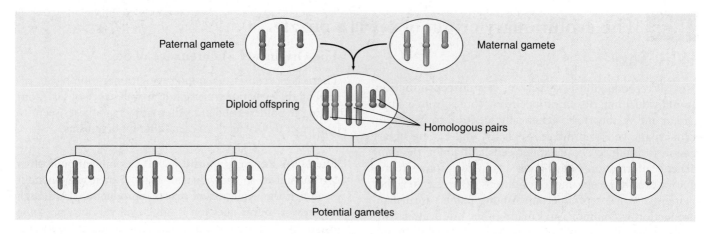

FIGURE 12.13

Independent assortment increases genetic variability. Independent assortment contributes new gene combinations to the next generation because the orientation of chromosomes on the metaphase plate is random. For example, in cells with three chromosome pairs, eight different gametes can result, each with different combinations of parental chromosomes.

The Red Queen Hypothesis. One evolutionary advantage of sex may be that it allows populations to "store" recessive alleles that are currently bad but have promise for reuse at some time in the future. Because populations are constrained by a changing physical and biological environment, selection is constantly acting against such alleles, but sexual species can never get rid of those sheltered in heterozygotes. Thus, the evolution of most sexual species, most of the time, manages to keep pace with ever-changing physical and biological constraints. This "treadmill evolution" is sometimes called the "Red Queen hypothesis," after the Queen of Hearts in Lewis Carroll's *Through the Looking Glass*, who tells Alice, "Now, here, you see, it takes all the running you can do, to keep in the same place."

Muller's Ratchet. The geneticist Herman Muller pointed out in 1965 that asexual populations incorporate a kind of mutational ratchet mechanism: Once harmful mutations arise, asexual populations have no way of eliminating them, and they accumulate over time, like turning a ratchet. Sexual populations, on the other hand, can employ recombination to generate individuals carrying fewer mutations, which selection can then favor. Sex may just be a way to keep the mutational load down.

The Evolutionary Consequences of Sex

While our knowledge of how sexual reproduction evolved is sketchy, it clearly has had an enormous impact on how species evolve, because of its ability to rapidly generate new genetic combinations. Independent assortment (figure 12.13), crossing over, and random fertilization each help generate genetic diversity.

Whatever the forces that led to sexual reproduction, its evolutionary consequences have been profound. No genetic process generates diversity more quickly. And, as you will see in later chapters, genetic diversity is the raw material of evolution—the fuel that drives it and determines its potential directions. In many cases, the pace of evolution appears to increase as the level of genetic diversity increases. For example, programs for selecting larger stature in domesticated animals such as cattle and sheep proceed rapidly at first, but then slow as the existing genetic combinations are exhausted; further progress must then await the generation of new gene combinations. Racehorse breeding provides another graphic example: Thoroughbred racehorses are all descendants of a small initial number of individuals, and selection for speed has accomplished all it can with this limited amount of genetic variability. As proof, the winning times in major races ceased to improve decades ago.

Paradoxically, the evolutionary process is both revolutionary and conservative. It is revolutionary in that the pace of evolutionary change is quickened by genetic recombination, much of which results from sexual reproduction. It is conservative in that evolutionary change is not always favored by selection, which may instead preserve existing combinations of genes. These conservative pressures appear greatest in certain asexually reproducing organisms that do not move around freely and that live in especially demanding habitats. In vertebrates, on the other hand, the evolutionary premium appears to have been on versatility, and sexual reproduction is the predominant mode of reproduction by an overwhelming margin.

The close association between homologous chromosomes that occurs during meiosis may have evolved as a mechanism to repair chromosomal damage, although several alternative mechanisms have also been proposed.

Concept Review

For interactive testing, visit the Online Learning Center with PowerWeb at www.mhhe.com/Raven7

12.1 Meiosis produces haploid cells from diploid cells.

Discovery of Reduction Division

- The fusion of gametes to form a new cell is referred to as fertilization, or syngamy. (p. 228)
- Gamete formation must involve some mechanism to reduce the number of chromosomes to half that found in other cells. (p. 228)
- Sexual reproduction involves an alternation of meiosis and fertilization, and follows a pattern of alternation between diploid and haploid chromosome numbers. (p. 228)

12.2 Meiosis has three unique features.

Meiosis

- Meiosis in diploid organisms consists of two rounds of division, meiosis I and meiosis II, but replication only occurs at the beginning of meiosis I. (p. 230)
- In prophase I of meiosis, homologues pair in synapsis, and crossing over occurs, allowing nonsister chromatids to exchange chromosomal material. (p. 230)

12.3 The sequence of events during meiosis involves two nuclear divisions.

Prophase I

- During prophase, the ends of the chromatids attach to the nuclear envelope at specific sites, and synapsis occurs. (p. 232)
- Crossing over occurs between nonsister chromatids, forming a chiasma. (p. 232)

Metaphase I

- By metaphase I, the nuclear envelope has dispersed, and microtubules form a spindle. (p. 233)
- Terminal chiasmata hold the homologous chromosomes together, and each joined pair of homologues lines up on the metaphase plate. (p. 233)

Completing Meiosis

- In anaphase I, the microtubules of the spindle fibers begin to shorten, pulling the centromeres toward the poles, and thus dragging the chromosomes toward the poles as well. (p. 236)
- Because of the random orientation of homologous chromosomes on the metaphase plate, meiosis I results in independent assortment of maternal and paternal chromosomes into the gametes. (p. 236)
- In telophase I, the chromosomes have segregated into clusters at each pole, and the nuclear membrane re-forms around each new daughter nucleus. (p. 236)
- Meiosis II resembles a normal mitotic division, except that at the end, each of the four haploid cells contains only one set of every chromosome instead of two sets. (p. 236)

12.4 The evolutionary origin of sex is a puzzle.

Why Sex?

- Several hypotheses exist as to the origin and maintenance of sex, including the DNA repair, contagion, Red Queen, and Muller's Ratchet hypotheses. (pp. 237–238)
- Paradoxically, the evolutionary process is both revolutionary and conservative. (p. 238)

Self Test

1. Fertilization results in
   a. a zygote.
   b. a diploid cell.
   c. a cell with a new genetic combination.
   d. All of these are correct.
2. The diploid number of chromosomes in humans is 46. The haploid number is
   a. 138.
   b. 92.
   c. 46.
   d. 23.
3. After chromosome replication and during synapsis,
   a. homologous chromosomes pair along their lengths.
   b. sister chromatids pair at the centromeres.
   c. homologous chromosomes pair at their ends.
   d. sister chromatids pair along their lengths.
4. During which stage of meiosis does crossing over occur?
   a. prophase I
   b. anaphase I
   c. prophase II
   d. telophase II
5. Synapsis is the process whereby
   a. homologous pairs of chromosomes separate and migrate toward a pole.
   b. homologous chromosomes exchange chromosomal material.
   c. homologous chromosomes become closely associated.
   d. the daughter cells contain half of the genetic material of the parent cell.
6. Terminal chiasmata are seen during which phase of meiosis?
   a. anaphase I
   b. prophase I
   c. metaphase I
   d. metaphase II
7. Which of the following occurs during anaphase I?
   a. Chromosomes cluster at the two poles of the cell.
   b. Crossing over occurs.
   c. Chromosomes align down the center of the cell.
   d. One version of each chromosome moves toward a pole.
8. Mitosis results in two _____ cells, while meiosis results in _____ haploid cells.
   a. haploid/four
   b. diploid/two
   c. diploid/four
   d. haploid/two
9. Genetic diversity is greatest in
   a. parthenogenesis.
   b. sexual reproduction.
   c. asexual reproduction.
   d. binary fission.
10. Which of the following is not a hypothesis about the evolution of sex?
    a. It evolved to repair damaged DNA.
    b. It evolved as a way to eliminate individuals.
    c. It evolved as a way to eliminate mutations.
    d. It evolved as a way to "store" recessive alleles that may prove beneficial in the future.

Test Your Visual Understanding

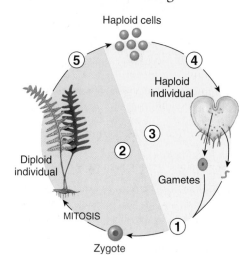

1. Match the following labels with the correct numbers on the figure. (Some labels may be used more than once.)
   a. diploid stage (2n)
   b. meiosis
   c. fertilization
   d. haploid stage (n)
   e. mitosis
2. Human cells spend most of their life cycles in a diploid stage, with only a selected number of cells undergoing meiosis to become haploid. Other organisms spend most of their life cycles in the haploid stage. Look ahead in this text to figures 28.14, 29.13, 29.15, and 30.10. Which of these organisms has a life cycle that is primarily haploid?

Apply Your Knowledge

1. An organism has 56 chromosomes in its diploid stage. Indicate how many chromosomes are present in the following, and explain your reasoning:
   a. somatic cells
   b. metaphase (mitosis)
   c. metaphase I (meiosis)
   d. metaphase II (meiosis)
   e. gametes
2. Humans have 23 pairs of chromosomes. Ignoring the effects of crossing over, what proportion of a woman's eggs contains only chromosomes she received from her mother?
3. Many sexually reproducing lizard species are able to generate local populations that reproduce by parthenogenesis. Would the sex of these local parthenogenetic populations be male, female, or neuter? Explain your reasoning.

13
Patterns of Inheritance

FIGURE 13.1
Human beings are extremely diverse in appearance. The differences among us are partly inherited and partly the result of environmental factors we encounter.

Concept Outline

13.1 Mendel solved the mystery of heredity.

Early Ideas About Heredity: The Road to Mendel. Before Mendel, the basis of inheritance was unknown.

Mendel and the Garden Pea. Mendel experimented with heredity in edible peas and counted his results.

What Mendel Found. Mendel discovered that alternative traits for a character segregated among second-generation progeny in the ratio 3:1.

How Mendel Interpreted His Results. Mendel found that one allele could mask the other in heterozygotes.

Mendelian Inheritance Is Not Always Easy to Analyze. A variety of factors can influence the Mendelian segregation of alleles.

13.2 Human genetics follows Mendelian principles.

Most Gene Disorders Are Rare. Tay-Sachs disease is due to a recessive allele.

Multiple Alleles: The ABO Blood Groups. The human ABO blood groups are determined by three *I* alleles.

Patterns of Inheritance Can Be Deduced from Pedigrees. Pedigrees track inheritance.

Gene Disorders Can Be Due to Simple Alterations of Proteins. Sickle cell anemia is caused by a single amino acid change.

Some Defects May Soon Be Curable: Gene Therapy. Gene therapy involves replacing mutant genes with healthy ones.

More Promising Vectors. New vectors may increase the success of gene therapy treatments.

13.3 Genes are on chromosomes.

Chromosomes: The Vehicles of Mendelian Inheritance. Mendelian segregation reflects the random assortment of chromosomes in meiosis.

Genetic Recombination. Crossover frequency reflects the physical distance between genes.

Human Chromosomes. Humans possess 23 pairs of chromosomes.

Human Abnormalities Due to Alterations in Chromosome Number. Loss or addition of chromosomes has serious consequences.

Genetic Counseling. Some gene defects can be detected early in pregnancy.

Every living creature is a product of the long evolutionary history of life on earth. While all organisms share this history, only humans wonder about the processes that led to their origin. We are still far from understanding everything about our origins, but we have learned a great deal. Like a partially completed jigsaw puzzle, the boundaries have fallen into place, and much of the internal structure is becoming apparent. In this chapter, we discuss one piece of the puzzle—the enigma of heredity. Why do groups of people from different parts of the world often differ in appearance (figure 13.1)? Why do the members of a family tend to resemble one another more than they resemble members of other families?

Mendel solved the mystery of heredity.

Early Ideas About Heredity: The Road to Mendel

As far back as written records go, patterns of resemblance among the members of particular families have been noted and commented on (figure 13.2). Some familial features are unusual, such as the protruding lower lip of the European royal family Hapsburg, evident in pictures and descriptions of family members from the thirteenth century onward. Other characteristics, such as the occurrence of redheaded children within families of redheaded parents, are more common (figure 13.3). Inherited features, the building blocks of evolution, are our concern in this chapter.

Classical Assumption 1: Constancy of Species

Before the twentieth century, two concepts provided the basis for most of the thinking about heredity. The first is that *heredity occurs within species.* For a very long time, people believed it was possible to obtain bizarre composite animals by breeding (crossing) widely different species. The minotaur of Cretan mythology, a creature with the body of a bull and the torso and head of a man, is one example. The giraffe was thought to be another; its scientific name, *Giraffa camelopardalis,* arose from the belief that it had resulted from a cross between a camel and a leopard. From the Middle Ages onward, however, people discovered that such extreme crosses were not possible and that variation and heredity occur mainly within the boundaries of a particular species. In addition, species were thought to have been maintained without significant change from the time of their creation.

Classical Assumption 2: Direct Transmission of Traits

The second early concept related to heredity is that *traits are transmitted directly.* When variation is inherited by off-spring from their parents, *what* is transmitted? The ancient Greeks suggested that elements from each of the parents' body parts were transmitted directly to their offspring. Hippocrates called this type of reproductive material *gonos,* meaning "seed." Hence, a characteristic such as a misshapen limb was the result of material that came from the misshapen limb of a parent. Information from each part of the body was supposedly passed along independently of the information from the other parts, and the child was formed after the hereditary material from all parts of the parents' bodies had come together.

This idea was predominant until fairly recently. For example, in 1868, Charles Darwin proposed that all cells

FIGURE 13.2
Heredity and family resemblance. Family resemblances are often strong—a visual manifestation of the mechanism of heredity. This is the Johnson family, the wife and daughters of one of the authors. While each daughter is different, all clearly resemble their mother.

FIGURE 13.3
An inherited trait. Many different traits are inherited in human families. This redhead is exhibiting one of these traits.

and tissues excrete microscopic granules, or "gemmules," that are passed to offspring, guiding the growth of the corresponding part in the developing embryo. Most similar theories of the direct transmission of hereditary material assumed that the male and female contributions *blend* in the offspring. Thus, parents with red and brown hair would produce children with reddish-brown hair, and tall and short parents would produce children of intermediate height.

Koelreuter Demonstrates Hybridization Between Species

Taken together, however, these two classical assumptions lead to a paradox. If no variation enters a species from outside and if the variation within each species blends in every generation, then all members of a species should soon have the same appearance. Obviously, this does not happen. Individuals within most species differ widely from each other, and they differ in characteristics that are transmitted from generation to generation.

How could this paradox be resolved? Actually, the resolution was provided long before Darwin, in the work of the German botanist Josef Koelreuter. In 1760, Koelreuter carried out successful **hybridizations** of plant species by crossing different strains of tobacco and obtaining fertile offspring. The hybrids differed in appearance from both parent strains. When individuals within the hybrid generation were crossed, their offspring were highly variable. Some of these offspring resembled plants of the hybrid generation (their parents), but a few resembled the original strains (their grandparents).

The Classical Assumptions Fail

Koelreuter's work represents the beginning of modern genetics, the first clues pointing to the modern theory of heredity. Koelreuter's experiments provided an important clue about how heredity works: The traits he was studying could be masked in one generation, only to reappear in the next. This pattern contradicts the theory of direct transmission. How could a trait that is transmitted directly disappear and then reappear? Nor were the traits of Koelreuter's plants blended. A contemporary account stated that the traits reappeared in the third generation "fully restored to all their original powers and properties."

It is worth repeating that the offspring in Koelreuter's crosses were not identical to one another. Some resembled

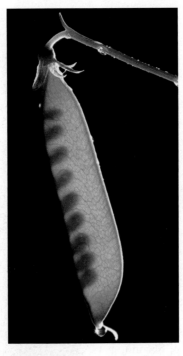

FIGURE 13.4
The garden pea, *Pisum sativum*. Easy to cultivate and able to produce many distinctive varieties, the garden pea was a popular experimental subject in investigations of heredity as long as a century before Gregor Mendel's experiments.

the hybrid generation, while others did not. The alternative forms of the characters Koelreuter was studying were distributed among the offspring. Referring to a heritable feature as a **character,** a modern geneticist would say the alternative forms of each character were **segregating** among the progeny of a mating, meaning that some offspring exhibited one alternative form of a character (for example, hairy leaves), while other offspring from the same mating exhibited a different alternative (smooth leaves). This segregation of alternative forms of a character, or **traits,** provided the clue that led Gregor Mendel to his understanding of the nature of heredity.

Knight Studies Heredity in Peas

Over the next hundred years, other investigators elaborated on Koelreuter's work. Prominent among them were English gentleman farmers trying to improve varieties of agricultural plants. In one such series of experiments, carried out in the 1790s, T. A. Knight crossed two true-breeding varieties (varieties that remain uniform from one generation to the next) of the garden pea, *Pisum sativum* (figure 13.4). One of these varieties had purple flowers, and the other had white flowers. All of the progeny of the cross had purple flowers. Among the offspring of these hybrids, however, were some plants with purple flowers and others, less common, with white flowers. Just as in Koelreuter's earlier studies, a trait from one of the parents had disappeared in one generation only to reappear in the next.

Within these deceptively simple results were the makings of a scientific revolution. Nevertheless, another century passed before the process of gene segregation was fully appreciated. Why did it take so long? One reason is that early workers did not quantify their results. A numerical record of results proved to be crucial to understanding the process. Knight and later experimenters who carried out other crosses with pea plants noted that some traits had a "stronger tendency" to appear than others, but they did not record the numbers of the different classes of progeny. Science was young then, and it was not obvious that the numbers were important.

Early geneticists demonstrated that some forms of an inherited character can: (1) disappear in one generation only to reappear unchanged in future generations; (2) segregate among the offspring of a cross; and (3) be more likely to be represented than their alternatives.

Mendel and the Garden Pea

The first quantitative studies of inheritance were carried out by Gregor Mendel, an Austrian monk (figure 13.5). Born in 1822 to peasant parents, Mendel was educated in a monastery and went on to study science and mathematics at the University of Vienna, where he failed his examinations for a teaching certificate. He returned to the monastery and spent the rest of his life there, eventually becoming abbot. In the garden of the monastery (figure 13.6), Mendel initiated a series of experiments on plant hybridization. The results of these experiments would ultimately change our views of heredity irrevocably.

Why Mendel Chose the Garden Pea

For his experiments, Mendel chose the garden pea, the same plant Knight and many others had studied. The choice was a good one for several reasons. First, many earlier investigators had produced hybrid peas by crossing different varieties. Mendel knew that he could expect to observe segregation of traits among the offspring. Second, a large number of true-breeding varieties of peas were available. Mendel initially examined 32. Then, for further study, he selected lines that differed with respect to seven easily distinguishable traits, such as round versus wrinkled seeds and purple versus white flowers, the latter a character that Knight had studied. Third, pea plants are small and easy to grow, and they have a relatively short generation time. Thus, a researcher can conduct experiments involving numerous plants, grow several generations in a single year, and obtain results relatively quickly.

A fourth advantage of studying peas is that both the male and female sexual organs are enclosed within the pea flower (figure 13.7), as are those of many flowering plants. Furthermore, the gametes produced by the male and female parts of the same flower, unlike those of many flowering plants, can fuse to form viable offspring. Fertilization takes place automatically within an individual flower if it is not disturbed, resulting in offspring that are the progeny from a single individual. Therefore, one can either let individual flowers engage in **self-fertilization,** or remove the flower's male parts before fertilization and introduce pollen from a strain with a different trait, thus performing *cross-pollination* that results in **cross-fertilization.**

FIGURE 13.5
Gregor Johann Mendel. Cultivating his plants in the garden of a monastery in Brunn, Austria (now Brno, Czech Republic), Mendel studied how differences among varieties of peas were inherited when the varieties were crossed. Similar experiments had been done before, but Mendel was the first to quantify the results and appreciate their significance.

FIGURE 13.6
Where Mendel worked. Gregor Mendel carried out his key plant-breeding experiments in this small garden in a monastery.

Mendel's Experimental Design

Mendel was careful to focus on only a few specific differences between the plants he was using and to ignore the countless other differences he must have seen. He also had the insight to realize that the differences he selected to analyze must be comparable. For example, he appreciated that trying to study the inheritance of round seeds versus tall height would be useless.

Mendel usually conducted his experiments in three stages:

1. First, he allowed pea plants of a given variety to produce progeny by self-fertilization for several generations. Mendel thus was able to assure himself that the traits he was studying were indeed **pure-breeding**, transmitted unchanged from generation to generation. Pea plants with white flowers, for example, when crossed with each other, produced only offspring with white flowers, regardless of the number of generations.

2. Mendel then performed crosses between varieties exhibiting alternative forms of characters. For example, he removed the male parts from the flower of a plant that produced white flowers and fertilized it with pollen from a purple-flowered plant. He also carried out the reciprocal cross, using pollen from a white-flowered individual to fertilize a flower on a pea plant that produced purple flowers (figure 13.8).

3. Finally, Mendel permitted the hybrid offspring produced by these crosses to self-fertilize for several generations. By doing so, he allowed the alternative forms of a character to segregate among the progeny. This was the same experimental design that Knight and others had used much earlier. But Mendel went an important step further: He counted the numbers of offspring exhibiting each trait in each succeeding generation. No one had ever done that before. The quantitative results Mendel obtained proved to be of supreme importance in revealing the process of heredity.

Mendel's experiments with the garden pea involved crosses between pure-breeding varieties, followed by a generation or more of self-fertilizing.

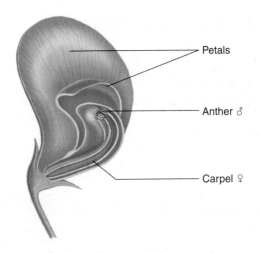

FIGURE 13.7
Structure of the pea flower (longitudinal section). In a pea plant flower, the petals enclose the male anther (containing pollen grains, which give rise to haploid sperm) and the female carpel (containing ovules, which give rise to haploid eggs). This ensures that self-fertilization will take place unless the flower is disturbed.

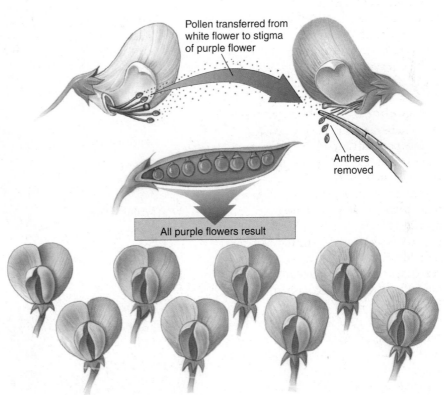

FIGURE 13.8
How Mendel conducted his experiments. Mendel pushed aside the petals of a white flower and collected pollen from the anthers. He then placed that pollen onto the stigma (part of the carpel) of a purple flower whose anthers had been removed, causing cross-fertilization to take place. All the seeds in the pod that resulted from this pollination were hybrids of the white-flowered male parent and the purple-flowered female parent. After planting these seeds, Mendel observed the pea plants they produced. All of the progeny of this cross had purple flowers.

What Mendel Found

The seven characters Mendel studied in his experiments possessed two variants that differed from one another in ways that were easy to recognize and score (table 13.1). We will examine in detail Mendel's crosses with flower color. His experiments with other characters were similar, and they produced similar results.

The F_1 Generation

When Mendel crossed two contrasting varieties of peas, such as white-flowered and purple-flowered plants, the hybrid offspring he obtained did not have flowers of intermediate color, as a hypothesis of blending inheritance would predict. Instead, in every case the flower color of the offspring resembled that of one of their parents. It is customary to refer to these offspring as the **first filial generation,**

Table 13.1 Seven Characters Mendel Studied and His Experimental Results

Character			F_2 Generation	
DOMINANT FORM	×	RECESSIVE FORM	DOMINANT:RECESSIVE	RATIO
Purple flowers	×	White flowers	705:224	3.15:1
Yellow seeds	×	Green seeds	6022:2001	3.01:1
Round seeds	×	Wrinkled seeds	5474:1850	2.96:1
Green pods	×	Yellow pods	428:152	2.82:1
Inflated pods	×	Constricted pods	882:299	2.95:1
Axial flowers	×	Terminal flowers	651:207	3.14:1
Tall plants	×	Dwarf plants	787:277	2.84:1

or F_1 (*filius* is Latin for "son"). Thus, in a cross of white-flowered with purple-flowered plants, the F_1 offspring all had purple flowers, just as Knight and others had reported earlier.

Mendel referred to the form or trait expressed in the F_1 plants as **dominant** and to the alternative form that was not expressed in the F_1 plants as **recessive.** For each of the seven pairs of contrasting traits that Mendel examined, one of the pair proved to be dominant and the other recessive.

The F_2 Generation

After allowing individual F_1 plants to mature and self-fertilize, Mendel collected and planted the seeds from each plant to see what the offspring in the **second filial generation,** or F_2, would look like. He found, just as Knight had earlier, that some F_2 plants exhibited white flowers, the recessive trait. After being hidden in the F_1 generation, the recessive form reappeared among some F_2 individuals.

Believing the proportions of the F_2 types would provide some clue about the mechanism of heredity, Mendel counted the numbers of each type among the F_2 progeny (figure 13.9). In the cross between the purple-flowered F_1 plants, he counted a total of 929 F_2 individuals (see table 13.1). Of these, 705 (75.9%) had purple flowers, and 224 (24.1%) had white flowers. Approximately $\frac{1}{4}$ of the F_2 individuals exhibited the recessive form of the character. Mendel obtained the same numerical result with the other six characters he examined: $\frac{3}{4}$ of the F_2 individuals exhibited the dominant trait, and $\frac{1}{4}$ displayed the recessive trait. In other words, the dominant:recessive ratio among the F_2 plants was always close to 3:1. Mendel carried out similar experiments with other traits, such as wrinkled versus round seeds (figure 13.10), and obtained the same result.

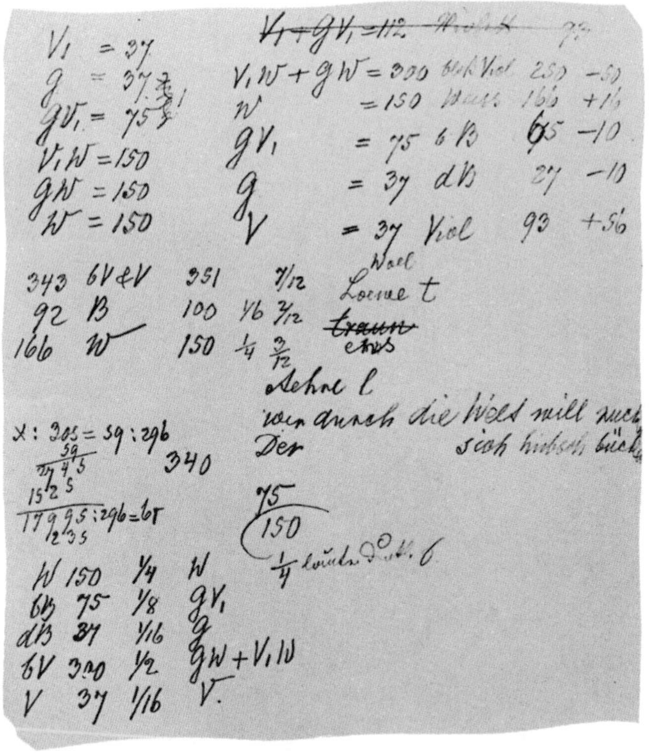

FIGURE 13.9
A page from Mendel's notebook.

FIGURE 13.10
Seed shape: a Mendelian character. One of the differences Mendel studied involved the shape of pea plant seeds. In some varieties, the seeds were round, while in others, they were wrinkled.

A Disguised 1:2:1 Ratio

Mendel went on to examine how the F$_2$ plants passed traits to subsequent generations. He found that the recessive $\frac{1}{4}$ were always pure-breeding. For example, in the cross of white-flowered with purple-flowered plants, the white-flowered F$_2$ individuals reliably produced white-flowered offspring when they were allowed to self-fertilize. By contrast, only $\frac{1}{3}$ of the dominant, purple-flowered F$_2$ individuals ($\frac{1}{4}$ of all F$_2$ offspring) proved pure-breeding, while $\frac{2}{3}$ were not. This last class of plants produced dominant and recessive individuals in the third filial (F$_3$) generation in a 3:1 ratio. This result suggested that, for the entire sample, the 3:1 ratio that Mendel observed in the F$_2$ generation was really a disguised 1:2:1 ratio: $\frac{1}{4}$ pure-breeding dominant individuals, $\frac{1}{2}$ not-pure-breeding dominant individuals, and $\frac{1}{4}$ pure-breeding recessive individuals (figure 13.11).

Mendel's Model of Heredity

From his experiments, Mendel was able to understand four things about the nature of heredity: (1) The plants he crossed did not produce progeny of intermediate appearance, as a hypothesis of blending inheritance would have predicted. Instead, different plants inherited each alternative intact, as a discrete characteristic that either was or was not visible in a particular generation. (2) For each pair of alternative forms of a character, one alternative was not expressed in the F$_1$ hybrids, although it reappeared in some F$_2$ individuals. *The trait that "disappeared" must therefore be latent (present but not expressed) in the F$_1$ individuals.* (3) The pairs of alternative traits examined segregated among the progeny of a particular cross, some individuals exhibiting one trait and some the other. (4) These alternative traits were expressed in the F$_2$ generation in the ratio of $\frac{3}{4}$ dominant to $\frac{1}{4}$ recessive. This characteristic 3:1 segregation is often referred to as the **Mendelian ratio.**

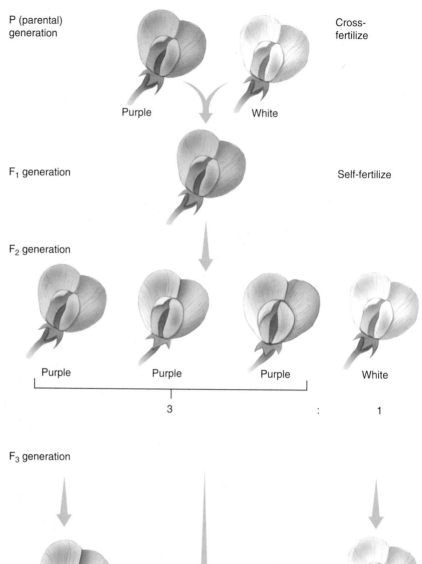

FIGURE 13.11
The F$_2$ generation is a disguised 1:2:1 ratio. By allowing the F$_2$ generation to self-fertilize, Mendel found from the offspring (F$_3$) that the ratio of F$_2$ plants was one pure-breeding dominant, two not-pure-breeding dominant, and one pure-breeding recessive.

Table 13.2 Some Dominant and Recessive Traits in Humans

Recessive Traits	Phenotypes	Dominant Traits	Phenotypes
Albinism	Lack of melanin pigmentation	Mid-digital hair	Presence of hair on middle segment of fingers
Alkaptonuria	Inability to metabolize homogentisic acid	Brachydactyly	Short fingers
Red-green color blindness	Inability to distinguish red or green wavelengths of light	Huntington disease	Degeneration of nervous system, starting in middle age
Cystic fibrosis	Abnormal gland secretion, leading to liver degeneration and lung failure	Phenylthiocarbamide (PTC) sensitivity	Ability to taste PTC as bitter
Duchenne muscular dystrophy	Wasting away of muscles during childhood	Camptodactyly	Inability to straighten the little finger
Hemophilia	Inability to form blood clots	Hypercholesterolemia (the most common human Mendelian disorder—1 in 500)	Elevated levels of blood cholesterol and risk of heart attack
Sickle cell anemia	Defective hemoglobin that causes red blood cells to curve and stick together	Polydactyly	Extra fingers and toes

To explain these results, Mendel proposed a simple model that has become one of the most famous in the history of science, containing simple assumptions and making clear predictions. The model has five elements:

1. Parents do not transmit physiological traits directly to their offspring. Rather, they transmit discrete information about the traits, what Mendel called "factors." These factors later act in the offspring to produce the trait. In modern terms, we would say that information about the alternative forms of characters that an individual expresses is *encoded* by the factors it receives from its parents.

2. Each individual receives two factors that may code for the same trait or for two alternative traits for a character. We now know that two factors for each character are present in each individual because these factors are carried on chromosomes, and each adult individual is *diploid*. When the individual forms gametes (eggs or sperm), the gametes contain only one of each kind of chromosome (see chapter 12); the gametes are *haploid*. Therefore, only one factor for each character of the adult organism is contained in the gamete. Which of the two factors ends up in a particular gamete is randomly determined.

3. Not all copies of a factor are identical. In modern terms, the alternative forms of a factor, leading to alternative forms of a character, are called **alleles**. When two haploid gametes containing exactly the same allele of a factor fuse during fertilization to form a zygote, the offspring that develops from that zygote is said to be **homozygous**; when the two haploid gametes contain different alleles, the individual offspring is **heterozygous**.

In modern terminology, Mendel's factors are called **genes**. We now know that each gene is composed of a particular DNA nucleotide sequence (see chap-

ter 3). The particular location of a gene on a chromosome is referred to as the gene's **locus** (plural, *loci*).

4. The two alleles, one contributed by the male gamete and one by the female, do not influence each other in any way. In the cells that develop within the new individual, these alleles remain discrete. They neither blend with nor alter each other. (Mendel referred to them as "uncontaminated.") Thus, when the individual matures and produces its own gametes, the alleles for each gene segregate randomly into these gametes.

5. The presence of a particular allele does not ensure that the trait encoded by it will be expressed in an individual carrying that allele. In heterozygous individuals, only one allele (the dominant one) is expressed, while the other (recessive) allele is present but unexpressed. To distinguish between the presence of an allele and its expression, modern geneticists refer to the totality of alleles that an individual contains as the individual's **genotype** and to the physical appearance of that individual as its **phenotype**. The phenotype of an individual is the observable outward manifestation of its genotype, the result of the functioning of the enzymes and proteins encoded by the genes it carries. In other words, the genotype is the blueprint, and the phenotype is the visible outcome.

These five elements, taken together, constitute Mendel's model of the hereditary process. Many traits in humans also exhibit dominant or recessive inheritance, similar to the traits Mendel studied in peas (table 13.2).

When Mendel crossed two contrasting varieties, he found that in the second generation 25% of the offspring were pure-breeding for the dominant trait, 50% were hybrid for the two traits and exhibited the dominant trait, and 25% were pure-breeding for the recessive trait.

How Mendel Interpreted His Results

Does Mendel's model predict the results he actually obtained? To test his model, Mendel first expressed it in terms of a simple set of symbols, and then used the symbols to interpret his results. It is very instructive to do the same. Consider again Mendel's cross of purple-flowered with white-flowered plants. We will assign the symbol P to the dominant allele, associated with the production of purple flowers, and the symbol p to the recessive allele, associated with the production of white flowers. By convention, genetic traits are usually assigned a letter symbol referring to their more common forms, in this case "P" for purple flower color. The dominant allele is written in uppercase, as P; the recessive allele (white flower color) is assigned the same symbol in lower case, p.

In this system, the genotype of an individual that is pure-breeding for the recessive white-flowered trait would be designated pp. In such an individual, both copies of the allele specify the white-flowered phenotype. Similarly, the genotype of a pure-breeding purple-flowered individual would be designated PP, and a heterozygote would be designated Pp (dominant allele first). Using these conventions and denoting a cross between two strains with ×, we can symbolize Mendel's original cross as $pp \times PP$.

The F₁ Generation

Employing these simple symbols, we can now go back and reexamine the crosses Mendel carried out. Because a white-flowered parent (pp) can produce only p gametes, and a pure purple-flowered (homozygous dominant) parent (PP) can produce only P gametes, the union of an egg and a sperm from these parents can produce only heterozygous Pp offspring in the F₁ generation. Because the P allele is dominant, all of these F₁ individuals are expected to have purple flowers. The p allele is present in these heterozygous individuals, but it is not phenotypically expressed. This is the basis for the latency Mendel saw in recessive traits.

The F₂ Generation

When F₁ individuals are allowed to self-fertilize, the P and p alleles segregate randomly during gamete formation. Their subsequent union at fertilization to form F₂ individuals is also random, not being influenced by which alternative alleles the individual gametes carry. What will the F₂ individuals look like? The possibilities may be visualized in a simple diagram called a **Punnett square**, named after its originator, the English geneticist Reginald Crundall Punnett (figure 13.12). Mendel's model, analyzed in terms of a Punnett square, clearly predicts that the F₂

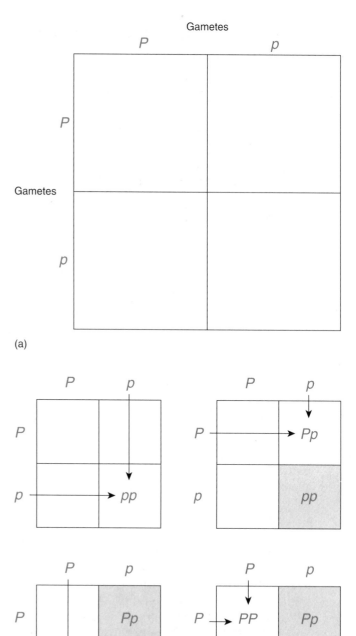

(a)

(b)

FIGURE 13.12

A Punnett square. (*a*) To make a Punnett square, place the different possible types of female gametes along the top of a square and the different possible types of male gametes along the side. (*b*) Each potential zygote can then be represented as the intersection of a vertical line and a horizontal line.

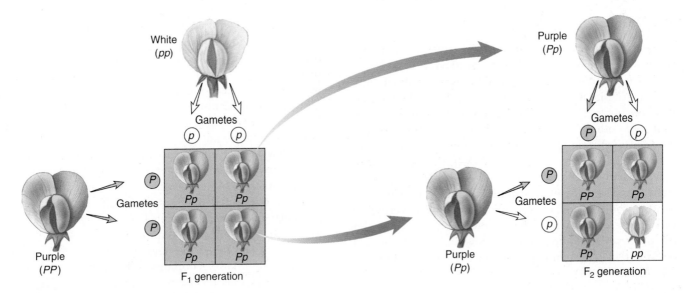

FIGURE 13.13
Mendel's cross of pea plants differing in flower color. All of the offspring of the first cross (the F_1 generation) are *Pp* heterozygotes with purple flowers. When two heterozygous F_1 individuals are crossed or self-fertilized, three kinds of F_2 offspring are possible: *PP* homozygotes (purple flowers); *Pp* heterozygotes (also purple flowers); and *pp* homozygotes (white flowers). Therefore, in the F_2 generation, the ratio of dominant to recessive phenotypes is 3:1. However, the ratio of genotypes is 1:2:1 (1 *PP*: 2 *Pp*: 1 *pp*).

generation should consist of $\frac{3}{4}$ purple-flowered plants and $\frac{1}{4}$ white-flowered plants, a phenotypic ratio of 3:1 (figure 13.13).

The Laws of Probability Can Predict Mendel's Results

A different way to express Mendel's result is to say that there are three chances in four ($\frac{3}{4}$) that any particular F_2 individual will exhibit the dominant trait, and one chance in four ($\frac{1}{4}$) that an F_2 individual will express the recessive trait. Stating the results in terms of probabilities allows simple predictions to be made about the outcomes of crosses. If both F_1 parents are *Pp* (heterozygotes), the probability that a particular F_2 individual will be *pp* (homozygous recessive) is the probability of receiving a *p* gamete from the male ($\frac{1}{2}$) times the probability of receiving a *p* gamete from the female ($\frac{1}{2}$), or $\frac{1}{4}$. This is the same operation we perform in the Punnett square illustrated in figure 13.12. The ways probability theory can be used to analyze Mendel's results are discussed in detail on page 253 (see Box: Probability and Allele Distribution).

As you can see in figure 13.13, there are really three kinds of F_2 individuals: $\frac{1}{4}$ are pure-breeding, white-flowered individuals (*pp*); $\frac{1}{2}$ are heterozygous, purple-flowered individuals (*Pp*); and $\frac{1}{4}$ are pure-breeding, purple-flowered individuals (*PP*). The 3:1 phenotypic ratio is really a disguised 1:2:1 genotypic ratio.

Mendel's First Law of Heredity: Segregation

Mendel's model thus accounts in a neat and satisfying way for the segregation ratios he observed. Its central assumption—that alternative alleles of a character segregate from each other in heterozygous individuals and remain distinct—has since been verified in many other organisms. It is commonly referred to as Mendel's first law of heredity, or the **Law of Segregation.** As we saw in chapter 12, the segregational behavior of alternative alleles has a simple physical basis, the random alignment of chromosomes on the metaphase plate during meiosis I, and the subsequent disjunction of the homologues in anaphase I of meiosis. It is a tribute to the intellect of Mendel's analysis that he arrived at the correct scheme with no knowledge of the cellular mechanisms of inheritance; neither chromosomes nor meiosis had yet been described.

The Testcross

To test his hypothesis further, Mendel devised a simple and powerful procedure called the **testcross**. Consider a purple-flowered plant. It is impossible to tell whether such a plant is homozygous or heterozygous simply by looking at its phenotype. To learn its genotype, you must cross it with some other plant. What kind of cross would provide the answer? If you cross it with a homozygous dominant individual, all of the progeny will show the dominant phenotype whether the test plant is homozygous or heterozygous. It is also difficult (but not impossible) to distinguish between the two possible test plant genotypes by crossing with a heterozygous individual. However, if you cross the test plant with a homozygous recessive individual, the two possible test plant genotypes will give totally different results (figure 13.14):

Alternative 1: Unknown individual homozygous dominant (PP)
$PP \times pp$: All offspring have purple flowers (Pp).

Alternative 2: Unknown individual heterozygous (Pp)
$Pp \times pp$: $\frac{1}{2}$ of offspring have white flowers (pp), and $\frac{1}{2}$ have purple flowers (Pp).

To perform his testcross, Mendel crossed heterozygous F_1 individuals back to the parent homozygous for the recessive trait. He predicted that the dominant and recessive traits would appear in a 1:1 ratio, and that is what he observed. For each pair of alleles he investigated, Mendel observed phenotypic F_2 ratios of 3:1 (see figure 13.13) and testcross ratios very close to 1:1, just as his model had predicted.

Testcrosses can also be used to determine the genotype of an individual when two genes are involved. Mendel carried out many two-gene crosses, some of which we will discuss. He often used testcrosses to verify the genotypes of particular dominant-appearing F_2 individuals. Thus, an F_2 individual showing both dominant traits ($A\_\ B\_$) might have any of the following genotypes: $AABB$, $AaBB$, $AABb$, or $AaBb$. By crossing dominant-appearing F_2 individuals with homozygous recessive individuals (that is, $A\_\ B\_ \times aabb$), Mendel was able to determine if either or both of the traits bred pure among the progeny, and so to determine the genotype of the F_2 parent:

$AABB$	Trait A breeds pure	Trait B breeds pure
$AaBB$	———	Trait B breeds pure
$AABb$	Trait A breeds pure	———
$AaBb$	———	———

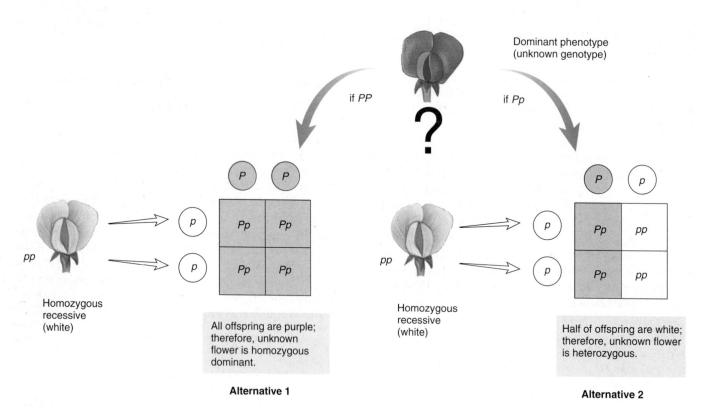

FIGURE 13.14

A testcross. To determine whether an individual exhibiting a dominant phenotype, such as purple flowers, is homozygous or heterozygous for the dominant allele, Mendel crossed the individual in question with a plant that he knew to be homozygous recessive—in this case, a plant with white flowers.

Probability and Allele Distribution

Many, although not all, alternative alleles produce discretely different phenotypes. Mendel's pea plants were tall or dwarf, had purple or white flowers, and produced round or wrinkled seeds. The eye color of a fruit fly may be red or white, and the skin color of a human may be pigmented or albino. When only two alternative alleles exist for a given character, the distribution of phenotypes among the offspring of a cross is referred to as a **binomial distribution.**

As an example, consider the distribution of sexes in humans. Imagine that a couple has chosen to have three children. How likely is it that two of the children will be boys and one will be a girl? The frequency of any particular possibility is referred to as its *probability* of occurrence. Let p symbolize the probability of having a boy at any given birth and q symbolize the probability of having a girl. Since any birth is equally likely to produce a girl or boy:

$$p = q = \tfrac{1}{2}$$

Table 13.A shows eight possible gender combinations among the three children. The sum of the probabilities of the eight possible combinations must equal one. Thus:

$$p^3 + 3p^2q + 3pq^2 + q^3 = 1$$

The probability that the three children will be two boys and one girl is:

$$3p^2q = 3 \times (\tfrac{1}{2})^2 \times (\tfrac{1}{2}) = \tfrac{3}{8}$$

To test your understanding, try to estimate the probability that two parents heterozygous for the recessive allele producing albinism (a) will have one albino child in a family of three. First, set up a Punnett square:

		Mother's Gametes	
		A	a
Father's	A	AA	Aa
Gametes	a	Aa	aa

You can see that one-fourth of the children are expected to be albino (aa). Thus, for any given birth, the probability of an albino child is $\tfrac{1}{4}$. This probability can be symbolized by q. The probability of a nonalbino child is $\tfrac{3}{4}$, symbolized by p. Therefore, the probability that there will be one albino child among the three children is:

$$3p^2q = 3 \times (\tfrac{3}{4})^2 \times (\tfrac{1}{4}) = \tfrac{27}{64}, \text{ or } 42\%$$

This means that the chance of having one albino child in the three is 42%.

Table 13.A Binomial Distribution of the Sexes of Children in Human Families

Composition of Family	Order of Birth	Calculation	Probability	
3 boys	bbb	$p \times p \times p$	p^3	
2 boys and 1 girl	bbg	$p \times p \times q$	p^2q	
	bgb	$p \times q \times p$	p^2q	$3p^2q$
	gbb	$q \times p \times p$	p^2q	
1 boy and 2 girls	ggb	$q \times q \times p$	pq^2	
	gbg	$q \times p \times q$	pq^2	$3pq^2$
	bgg	$p \times q \times q$	pq^2	
3 girls	ggg	$q \times q \times q$	q^3	

Mendel's Second Law of Heredity: Independent Assortment

After Mendel had demonstrated that different traits of a given character (alleles of a given gene) segregate independently of each other in crosses, he asked whether different genes also segregate independently. Mendel set out to answer this question in a straightforward way. He first established a series of pure-breeding lines of peas that differed in just two of the seven characters he had studied. He then crossed contrasting pairs of the pure-breeding lines to create heterozygotes. In a cross involving different seed shape alleles (round, R, and wrinkled, r) and different seed color alleles (yellow, Y, and green, y), all the F_1 individuals were identical, each one heterozygous for both seed shape (Rr) and seed color (Yy). The F_1 individuals of such a cross are **dihybrids,** individuals heterozygous for both genes.

The third step in Mendel's analysis was to allow the dihybrids to self-fertilize. If the alleles affecting seed shape and seed color were segregating independently, then the probability that a particular pair of seed shape alleles would occur together with a particular pair of seed color alleles would be simply the product of the individual probabilities that each pair would occur separately. Thus, the probability that an individual with wrinkled green seeds ($rryy$) would appear in the F_2 generation would be equal to the probability of observing an individual with wrinkled seeds ($\tfrac{1}{4}$) times the probability of observing one with green seeds ($\tfrac{1}{4}$), or $\tfrac{1}{16}$.

Because the gene controlling seed shape and the gene controlling seed color are each represented by a pair of alternative alleles in the dihybrid individuals, four types of gametes are expected: RY, Ry, rY, and ry. Therefore, in the F_2 generation, there are 16 possible combinations of alleles, each of them equally probable. Of these, 9 possess at least one dominant allele for each gene (signified $R\_\_Y\_\_$, where the underscore indicates the presence of either allele) and, thus, should have round, yellow seeds. Of the rest, 3 possess at least one dominant R allele but are homozygous

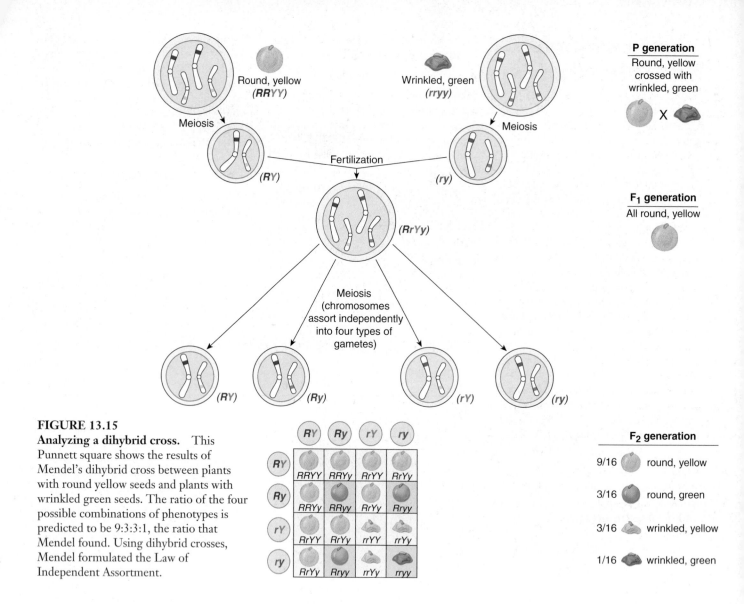

FIGURE 13.15
Analyzing a dihybrid cross. This Punnett square shows the results of Mendel's dihybrid cross between plants with round yellow seeds and plants with wrinkled green seeds. The ratio of the four possible combinations of phenotypes is predicted to be 9:3:3:1, the ratio that Mendel found. Using dihybrid crosses, Mendel formulated the Law of Independent Assortment.

recessive for color (*R__yy*); 3 others possess at least one dominant *Y* allele but are homozygous recessive for shape (*rrY__*); and 1 combination among the 16 is homozygous recessive for both genes (*rryy*) (figure 13.15). The hypothesis that color and shape genes assort independently thus predicts that the F$_2$ generation will display a 9:3:3:1 phenotypic ratio: nine individuals with round, yellow seeds, three with round, green seeds, three with wrinkled, yellow seeds, and one with wrinkled, green seeds.

What did Mendel actually observe? From a total of 556 seeds from dihybrid plants he had allowed to self-fertilize, he observed: 315 round yellow (*R__Y__*), 108 round green (*R__yy*), 101 wrinkled yellow (*rrY__*), and 32 wrinkled green (*rryy*). These results are very close to a 9:3:3:1 ratio (which would be 313:104:104:35). Consequently, the two genes appeared to assort completely independently of each other. Note that this independent assortment of different genes in no way alters the independent segregation of individual pairs of alleles. Round versus wrinkled seeds occur in a ratio of approximately 3:1 (423:133); so do yellow versus green seeds (416:140). Mendel obtained similar results for other pairs of traits.

Mendel's discovery is often referred to as Mendel's second law of heredity, or the **Law of Independent Assortment.** Like segregation, independent assortment can be understood by examining the behavior of chromosomes during meiosis (see figure 13.15). At metaphase I, homologues of each chromosome are oriented toward opposite poles, leading to segregation of alleles in a heterozygote. The independent behavior of each chromosome pair leads to the independent assortment of loci on different chromosomes.

Mendel summed up his discoveries about heredity in two laws. Mendel's first law of heredity states that alternative alleles of a trait segregate independently; his second law of heredity states that genes located on different chromosomes assort independently.

Mendelian Inheritance Is Not Always Easy to Analyze

Although Mendel's results did not receive much notice during his lifetime, three different investigators independently rediscovered his pioneering paper in 1900, 16 years after his death. They came across it while searching the literature in preparation for publishing their own findings, which closely resembled those Mendel had presented more than three decades earlier. In the decades following the rediscovery of Mendel's ideas, many investigators set out to test them. However, scientists attempting to confirm Mendel's theory often had trouble obtaining the same simple ratios he had reported. Often, the expression of the genotype is not straightforward. Most phenotypes reflect the action of many genes, and the phenotype can be affected by alleles that lack complete dominance as well as by the environment.

Continuous Variation

Few phenotypes result from the action of only one gene. Instead, most characters reflect multiple additive contributions to the phenotype by several genes. When multiple genes act jointly to influence a character such as height or weight, the character often shows a range of small differences. Because all of the genes that play a role in determining phenotypes such as height or weight segregate independently of one another, we see a gradation in the degree of difference when many individuals are examined (figure 13.16). We call this gradation **continuous variation,** and we call such traits *quantitative traits*. The greater the number of genes that influence a character, the more continuous the expected distribution of the versions of that character.

How can we describe the variation in a character such as the height of the individuals in figure 13.16? Humans range from quite short to very tall, with average heights more common than either extreme. Often, variations are grouped into categories—in this case, by measuring the heights of the individuals in inches and rounding fractions of an inch to the nearest whole number. Each height, in inches, is a separate phenotypic category. Plotting the numbers in each height category produces a histogram, such as that in figure 13.16. The histogram approximates an idealized bell-shaped curve, and the variation can be characterized by the mean and spread of that curve.

Pleiotropic Effects

Often, an individual allele will have more than one effect on the phenotype. Such an allele is said to be **pleiotropic.** When the pioneering French geneticist Lucien Cuenot studied yellow fur in mice, a dominant trait, he was unable to obtain a pure-breeding yellow strain by crossing individual yellow mice with each other. Individuals homozygous

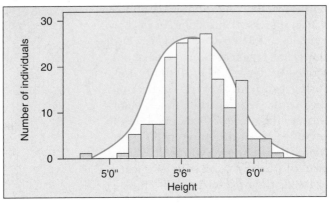

FIGURE 13.16
Height is a continuously varying trait. The photo shows variation in height among students of the 1914 class at the Connecticut Agricultural College. Because many genes contribute to height and tend to segregate independently of one another, the cumulative contribution of different combinations of alleles to height forms a *continuous* distribution of possible heights, in which the extremes are much rarer than the intermediate values.

for the yellow allele died, because the yellow allele was pleiotropic: One effect was yellow coat color, but another was a lethal developmental defect. A pleiotropic allele may be dominant with respect to one phenotypic consequence (yellow fur) and recessive with respect to another (lethal developmental defect). In pleiotropy, one gene affects many traits, in marked contrast to polygeny, where many genes affect one trait. Pleiotropic effects are difficult to predict, because the genes that affect a trait often perform other functions we may know nothing about.

Pleiotropic effects are characteristic of many inherited disorders, including cystic fibrosis and sickle cell anemia, both discussed in section 13.2. In these disorders, multiple symptoms can be traced back to a single gene defect. Cystic fibrosis patients exhibit clogged blood vessels, overly sticky mucus, salty sweat, liver and pancreas failure, and a battery of other symptoms. All are pleiotropic effects of a single defect, a mutation in a gene that encodes a chloride ion transmembrane channel. In sickle cell anemia patients, a defect in the oxygen-carrying hemoglobin molecule causes anemia, heart failure, increased susceptibility to pneumonia, kidney failure, enlargement of the spleen, and many other symptoms. It is usually difficult to deduce the nature of the primary defect from the range of a gene's pleiotropic effects.

Incomplete Dominance

Not all alternative alleles are fully dominant or fully recessive in heterozygotes. Instead, some pairs of alleles produce a heterozygous phenotype that is either intermediate between those of the parents (incomplete dominance) or representative of both parental phenotypes (codominance, discussed in section 13.2). For example, in the cross of red and white flowering Japanese four o'clocks described in figure 13.17, all the F₁ offspring had pink flowers—indicating that neither red nor white flower color was dominant. Does this example of incomplete dominance argue that Mendel was wrong? Not at all. When two of the F₁ pink flowers were crossed, they produced red-, pink-, and white-flowered plants in a 1:2:1 ratio. Heterozygotes are simply intermediate in color.

Environmental Effects

The degree to which an allele is expressed may depend on the environment. Some alleles are heat-sensitive, for example. Traits influenced by such alleles are more sensitive to temperature or light than are the products of other alleles. The arctic foxes in figure 13.18 make fur pigment only when the weather is warm. Similarly, the *ch* allele in Himalayan rabbits and Siamese cats encodes a heat-sensitive version of tyrosinase, one of the enzymes mediating the production of melanin, a dark pigment. The ch version of the enzyme is inactivated at temperatures above about 33°C. At the surface of the body and head, the temperature is above 33°C and the tyrosinase enzyme is inactive, while it is more active at body extremities such as the tips of the ears and tail, where the temperature is below 33°C. The dark melanin pigment this enzyme produces causes the ears, snout, feet, and tail of Himalayan rabbits and Siamese cats to be black.

FIGURE 13.17
Incomplete dominance. In a cross between a red-flowered Japanese four o'clock (genotype $C^R C^R$) and a white-flowered one ($C^W C^W$), neither allele is dominant. The heterozygous progeny have pink flowers and the genotype $C^R C^W$. If two of these heterozygotes are crossed, the phenotypes of their progeny occur in a ratio of 1:2:1 (red:pink:white).

(a)

(b)

FIGURE 13.18
Environmental effects on an allele. (*a*) An arctic fox in winter has a coat that is almost white, so it is difficult to see the fox against a snowy background. (*b*) In summer, the same fox's fur darkens to a reddish brown, so that it resembles the color of the surrounding tundra. Heat-sensitive alleles control this color change.

Epistasis

In the tests of Mendel's ideas that followed the rediscovery of his work, scientists had trouble obtaining Mendel's simple ratios, particularly with dihybrid crosses (when individuals heterozygous for two different genes mate). In a dihybrid cross, four different phenotypes are possible among the progeny: Offspring may display the dominant phenotype for both genes, for either one of the genes, or for neither gene. Sometimes, however, it is not possible for an investigator to identify successfully each of the four phenotypic classes, because two or more of the classes look alike. Such situations proved confusing to investigators following Mendel.

One example of such difficulty in identification is seen in the analysis of particular varieties of corn, *Zea mays*. Some commercial varieties exhibit a purple pigment called anthocyanin in their seed coats, while others do not. In 1918, geneticist R. A. Emerson crossed two pure-breeding corn varieties, neither exhibiting anthocyanin pigment. Surprisingly, all of the F_1 plants produced purple seeds.

When two of these pigment-producing F_1 plants were crossed to produce an F_2 generation, 56% were pigment producers and 44% were not. What was happening? Emerson correctly deduced that two genes were involved in producing pigment, and that the second cross had thus been a dihybrid cross. Mendel had predicted 16 equally possible ways gametes could combine. How many of these were in each of the two types Emerson obtained? He multiplied the fraction that were pigment producers (0.56) by 16 to obtain 9, and multiplied the fraction that were not (0.44) by 16 to obtain 7. Thus, Emerson had a **modified ratio** of 9:7 instead of the usual 9:3:3:1 ratio.

Why Was Emerson's Ratio Modified? When genes act sequentially, as in a biochemical pathway, an allele expressed as a defective enzyme early in the pathway blocks the flow of material through the rest of the pathway. This makes it impossible to judge whether the later steps of the pathway are functioning properly. Such gene interaction, in which one gene can interfere with the expression of another gene, is the basis of the phenomenon called **epistasis.**

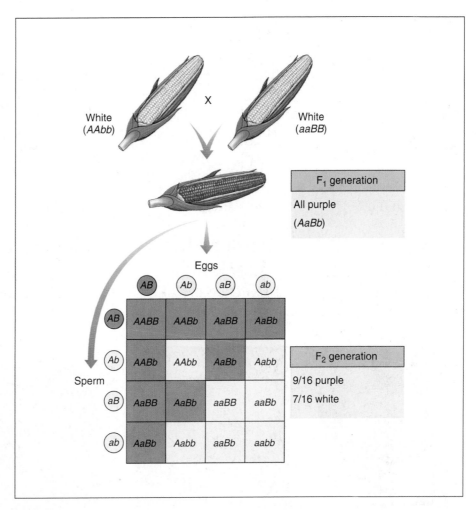

FIGURE 13.19

How epistasis affects grain color. The purple pigment found in some varieties of corn is the product of a two-step biochemical pathway. Unless both enzymes are active (the plant has a dominant allele for each of the two genes, *A__B__*), no pigment is expressed.

The pigment anthocyanin is the product of a two-step biochemical pathway:

$$\text{starting molecule} \xrightarrow{\text{enzyme 1}} \text{intermediate} \xrightarrow{\text{enzyme 2}} \textbf{anthocyanin}$$
$$\text{(colorless)} \qquad \text{(colorless)} \qquad \text{(Purple)}$$

To produce pigment, a plant must possess at least one functional copy of each enzyme gene (figure 13.19). The dominant alleles encode functional enzymes, but the recessive alleles encode nonfunctional enzymes. Of the 16 genotypes predicted by random assortment, 9 contain at least one dominant allele of both genes; they produce purple progeny. The remaining 7 genotypes lack dominant alleles at either or both loci (3 + 3 + 1 = 7) and so are phenotypically the same (nonpigmented), giving the phenotypic ratio of 9:7 that Emerson observed. The inability to see the effect of enzyme 2 when enzyme 1 is nonfunctional is an example of epistasis.

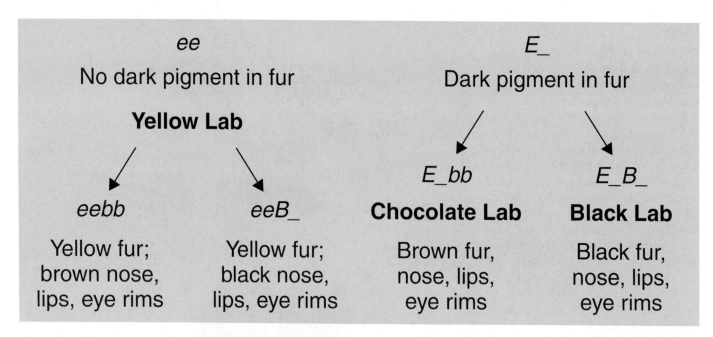

ee		*E_*	
No dark pigment in fur		**Dark pigment in fur**	
Yellow Lab			
eebb	*eeB_*	*E_bb*	*E_B_*
		Chocolate Lab	**Black Lab**
Yellow fur; brown nose, lips, eye rims	Yellow fur; black nose, lips, eye rims	Brown fur, nose, lips, eye rims	Black fur, nose, lips, eye rims

FIGURE 13.20
The effect of epistatic interactions on coat color in dogs. The coat color seen in Labrador retrievers is an example of the interaction of two genes, each with two alleles. The *E* gene determines if the pigment will be deposited in the fur, and the *B* gene determines how dark the pigment will be.

Other Examples of Epistasis

In many animals, coat color is the result of epistatic interactions among genes. Coat color in Labrador retrievers, a breed of dog, is due primarily to the interaction of two genes. The *E* gene determines whether dark pigment (eumelanin) will be deposited in the fur or not. If a dog has the genotype *ee*, no pigment will be deposited in the fur, and it will be yellow. If a dog has the genotype *EE* or *Ee* (*E_*), pigment will be deposited in the fur.

A second gene, the *B* gene, determines how dark the pigment will be. This gene controls the distribution of melanosomes in a hair. Dogs with the genotype *E_bb* will have brown fur and are called chocolate labs. Dogs with the genotype *E_B_* will have black fur. But, even in yellow dogs, the *B* gene does have some effect. Yellow dogs with

the genotype *eebb* will have brown pigment on their nose, lips, and eye rims, while yellow dogs with the genotype *eeB_* will have black pigment in these areas. The interaction among these alleles is illustrated in figure 13.20. The genes for coat color in this breed have been found, and a genetic test is available to determine the coat colors in a litter of puppies.

A variety of factors can disguise the Mendelian segregation of alleles. Among them are the continuous variation that results when many genes contribute to a trait, incomplete dominance that produces heterozygotes unlike either parent, environmental influences on the expression of phenotypes, and gene interactions that produce epistasis.

Human genetics follows Mendelian principles.

Random mutations are constantly occurring in the genes of any population. These gene changes rarely improve the functioning of the proteins encoded by genes, but they are the source of all new alleles. As we have learned more about the genomes of humans and other organisms, a key lesson has been the amount of variation that exists. The database of human mutations, the Human Gene Mutation Database, contains a catalog of 1163 mutant genes that result in clinical manifestations—some 3% of the genome. The vast majority of human mutations have not been studied because they do not produce clinical symptoms. Similar arrays of mutations occur in other organisms.

Most Gene Disorders Are Rare

Although the preceding description may make it seem that humans are largely at risk for genetic diseases of all kinds, the reality is that genetic diseases are actually rare. This is because even though many possible disease alleles may be present in a population, the actual frequency with which any one allele occurs is very low. To examine this, consider *Tay-Sachs disease*, which is an incurable genetic disorder affecting the nervous system. Affected children are normal at birth, first developing symptoms at about 8 months of age when mental deterioration appears. As degeneration progresses, blindness and death follow, usually by age 5. The disease is caused by a single Mendelian recessive allele. The Tay-Sachs allele encodes a mutant form of the enzyme hexosaminidase A that is involved in the breakdown of gangliosides, a class of lipids found in lysosomes in brain cells. The loss of the ability to break down this class of lipids leads to their buildup in lysosomes until the lysosomes burst, releasing their toxic contents into the cell and killing it.

Tay-Sachs is rare in most populations, occurring in only 1 in 300,000 live births in the United States. However, the disease has a high incidence among Ashkenazi Jews, those who are originally from eastern and central Europe; this group includes many of the Jewish people in the United States. In these populations, the incidence rises to as high as 1 in 3500, with an estimated 1 in 28 who actually carry the Tay-Sachs allele. This illustrates two important points about genetic diseases: Their frequencies can vary in different populations with different histories, and even lethal recessive alleles are not entirely removed from a population by natural selection acting against them. The Tay-Sachs allele remains in the population because of a threshold effect: The heterozygote has 50% enzyme activity remaining, compared to the almost complete lack of activity in the homozygous Tay-Sachs individual. This level of enzyme activity is enough to prevent the buildup of gangliosides and subsequent cell death in heterozygotes.

Not All Gene Defects Are Recessive

Most genetic disorders are recessive. A recent cataloging of human disease-causing genes according to biological function sheds some light on this tendency: Enzyme-encoding genes are the largest category by far (over 30% of some 900 genes analyzed). In the case of many enzymes, the amount of residual activity left from a single good allele in a heterozygote is enough to get the job done, as is seen with Tay-Sachs. These mutant alleles are thus recessive.

Despite this tendency, some genetic diseases are due to the action of dominant alleles. *Huntington disease* is a genetic disorder caused by a dominant allele that encodes a variant form of a protein (called huntingtin) found in nerve cells of the brain. In individuals who carry the Huntington allele, these neurons are gradually lost by a mechanism that is thought to be due to deposition of fragments of the altered protein. This leads to neural degeneration, and eventually to death. The search for the Huntington gene took many years of work by many researchers. The approach concentrated on a population in Venezuela with a high incidence of the disease allele and used genetic mapping techniques. It led to a surprise. When the mutant allele was located and analyzed, it proved to be the first case of a new kind of mutation—a set of three bases (CAG) that were duplicated many times over relative to the nonmutant allele. Now known as "triplet expansion," these mutations are discussed in chapter 20.

About 1 in 24,000 individuals in the United States develops Huntington disease. Because it is caused by a dominant allele that is lethal, this raises an obvious question: How can a dominant lethal allele be maintained in a population? In the case of Huntington, it is due to the age of onset of the disease. Individuals carrying the Huntington allele do not show any clinical symptoms until early middle age. Therefore, an individual who is carrying the dominant lethal allele has ample time to reproduce. Given the dominant nature of inheritance, any child of a Huntington sufferer has a 50% probability of receiving the Huntington allele from the affected parent. This can lead to an ethical dilemma for a carrier. In a famous case, Woody Guthrie died of Huntington, and thus his son Arlo, not knowing if he was a carrier, decided not to have children as a young man. We can now test for the presence of the most common Huntington allele, which raises an entirely different dilemma for a person faced with the possibility of carrying the allele: Do you want to know, and if you do, how will it affect your behavior? We will return to the idea of genetic testing later in this chapter.

Most gene defects are rare recessives, although some are dominant.

Multiple Alleles: The ABO Blood Groups

Most genes possess several different alleles in a population, and often no single allele is dominant; instead, each allele has its own effect, and the alleles are considered **codominant.** Codominance can be distinguished from incomplete dominance (discussed in section 13.1) because the effects of both alleles can be seen in codominance.

An example of a human gene with more than one codominant allele is the gene that determines ABO blood type. This gene encodes an enzyme that adds sugar molecules to lipids on the surface of red blood cells. These sugars act as recognition markers for the immune system. The gene that encodes the enzyme, designated I, has three common alleles: I^B, whose product adds galactose; I^A, whose product adds galactosamine; and i, which codes for a protein that does not add a sugar.

Different combinations of the three I gene alleles occur in different individuals because each person possesses two copies of the chromosome bearing the I gene and may be homozygous for any allele or heterozygous for any two. An individual heterozygous for the I^A and I^B alleles produces both forms of the enzyme and adds both galactose and galactosamine to the surfaces of red blood cells. Because both alleles are expressed simultaneously in heterozygotes, the I^A and I^B alleles are codominant. Both I^A and I^B are dominant over the i allele because both I^A and I^B alleles lead to sugar addition and the i allele does not. The different combinations of the three alleles produce four different phenotypes (figure 13.21):

1. Type A individuals add only galactosamine. They are either I^AI^A homozygotes or I^Ai heterozygotes.
2. Type B individuals add only galactose. They are either I^BI^B homozygotes or I^Bi heterozygotes.
3. Type AB individuals add both sugars and are I^AI^B heterozygotes.
4. Type O individuals add neither sugar and are ii homozygotes.

These four different cell surface phenotypes are called the **ABO blood groups.** A person's immune system can distinguish between these four phenotypes. If a type A individual receives a transfusion of type B blood, the recipient's immune system recognizes that the type B blood cells possess a "foreign" antigen (galactose) and attacks the donated blood cells, causing them to clump, or agglutinate. This also happens if the donated blood is type AB. However, if the donated blood is type O, no immune attack will occur, because there are no galactose antigens on the surfaces of blood cells produced by the type O donor. In general, any individual's immune system will tolerate a transfusion of type O blood. Because neither galactose nor galactosamine is foreign to type AB individuals (whose red blood cells have both sugars), those individuals may receive any type of blood.

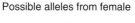

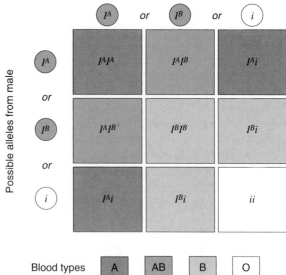

FIGURE 13.21

Multiple alleles control the ABO blood groups. Different combinations of the three I gene alleles result in four different blood type phenotypes: type A (either I^AI^A homozygotes or I^Ai heterozygotes), type B (either I^BI^B homozygotes or I^Bi heterozygotes), type AB (I^AI^B heterozygotes), and type O (ii homozygotes).

The Rh Blood Group

Another set of cell surface markers on human red blood cells are the **Rh blood group** antigens, named for the rhesus monkey in which they were first described. About 85% of adult humans have the Rh cell surface marker on their red blood cells, and are called Rh-positive. Rh-negative persons lack this cell surface marker because they are homozygous for the recessive gene encoding it.

If an Rh-negative person is exposed to Rh-positive blood, the Rh surface antigens of that blood are treated like foreign invaders by the Rh-negative person's immune system, which proceeds to make antibodies directed against the Rh antigens. This most commonly happens when an Rh-negative woman gives birth to an Rh-positive child (whose father is Rh-positive). At birth, some fetal red blood cells cross the placental barrier and enter the mother's bloodstream, where they induce the production of "anti-Rh" antibodies. In subsequent pregnancies with Rh-positive children, the mother's antibodies can cross the placenta to the new fetus and cause its red blood cells to clump, leading to a potentially fatal condition called erythroblastosis fetalis.

Many blood group genes possess multiple alleles that are codominant. Several of these alleles may be common in a population.

Patterns of Inheritance Can Be Deduced from Pedigrees

We cannot perform controlled crosses in humans the way Mendel did with pea plants, so how can we analyze patterns of human inheritance? The answer is that we have to use crosses that have been performed already; we have to look at families and their histories. The organized methodology we use is a **pedigree,** a consistent graphical representation of matings over multiple generations for a particular trait. Geneticists can use the information in the pedigree to deduce a model for the mode of inheritance of the trait.

An example is the pedigree showing the inheritance of hemophilia in the European royal families. **Hemophilia** is a disease that affects a single protein in a cascade of proteins involved in the formation of blood clots. Thus, in an untreated hemophiliac, even simple cuts will not stop bleeding. Genetically, hemophilia is an X-linked recessive allele. This means that women who are heterozygous will be asymptomatic carriers, and men who inherit the disease allele will show the trait. The most famous instance of hemophilia is the so-called Royal hemophilia that can be traced back to Queen Victoria of England (figure 13.22). In the five generations after Victoria, ten of her male descendants have had hemophilia as shown in the pedigree in figure 13.23. Ironically, this has not affected the British royal family, since Victoria's son, King Edward VII, did not receive the hemophilia allele, and all of the subsequent rulers of England are his descendants. However, the Russian house of Romanov was not so lucky. Czar Nicholas's son inherited the hemophilia allele through his mother, a granddaughter of Queen Victoria. The entire family was executed during the Russian revolution. Recently, a woman who had long claimed to be Anastasia, a surviving daughter of the Czar, was shown not to be a Romanov using modern genetic techniques to test her remains.

> **Family pedigrees can reveal the mode of inheritance of a hereditary trait.**

FIGURE 13.22
Queen Victoria of England in 1894, surrounded by some of her descendants. Of Victoria's four daughters who lived to bear children, two, Alice and Beatrice, were carriers of Royal hemophilia. Two of Alice's daughters are standing behind Victoria (wearing feathered boas): Princess Irene of Prussia (*right*) and Alexandra (*left*), who would soon become czarina of Russia. Both Irene and Alexandra were also carriers of hemophilia.

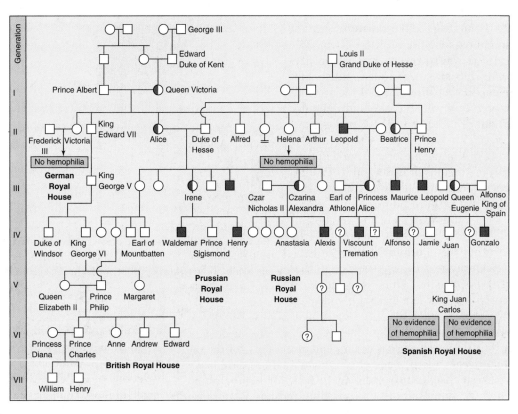

FIGURE 13.23
The Royal hemophilia pedigree. Queen Victoria's daughter Alice introduced hemophilia into the Russian and Austrian royal houses, and Victoria's daughter Beatrice introduced it into the Spanish royal house. Victoria's son Leopold, himself a victim, also transmitted the disorder in a third line of descent. Half-shaded symbols represent carriers with one normal allele and one defective allele; fully shaded symbols represent affected individuals.

Gene Disorders Can Be Due to Simple Alterations of Proteins

The change of a single amino acid in a protein can lead to a serious clinical phenotype. As we will see in chapter 14, this can be due to a change in a single base. The first disease in which such a change was shown to occur is **sickle cell anemia,** which is caused by a defect in the oxygen carrier hemoglobin that leads to impaired oxygen delivery to tissues. The defective hemoglobin molecules stick to one another, leading to stiff, rodlike structures that alter the shape of the red blood cells that carry them. These red blood cells take on a characteristic shape that led to the name "sickle cell" (figure 13.24). Individuals with sickle-shaped red blood cells exhibit intermittent illness and reduced life span. What is the molecular nature of this defect? A single amino acid in the 574-amino-acid protein β-globin is changed from glutamic acid to valine. This amino acid is not in the oxygen-binding region of the protein, but still has a catastrophic effect on its function. The replacement of a charged amino acid with a nonpolar amino acid on the surface of the protein makes it sticky. This stickiness is due to the tendency of nonpolar amino acids to aggregate in water, leading to the stiff, rodlike structures seen in sickled red blood cells.

Individuals heterozygous for the sickle cell allele are indistinguishable from normal individuals in a normal oxygen environment, although they do exhibit reduced ability to carry oxygen. The sickle cell allele is particularly prevalent in people of African descent. In some regions of Africa, up to 45% of the population is heterozygous for the trait, and 6% are homozygous. What has led to this extremely high level of an allele that is obviously deleterious? It turns out that a beneficial effect of heterozygosity is greater resistance to the malaria parasite. Thus, in regions of central Africa where malaria is endemic, the sickle cell allele also occurs at a high frequency (figure 13.25).

The sickle cell allele is not the end of the story for the β-globin gene; a large number of other alterations have been observed that lead to anemias. In fact, for hemoglobin, which is composed of two α-globins and two β-globins, over 700 structural variants have been cataloged. It is estimated that 7% of the human population worldwide are carriers for different inherited hemoglobin disorders.

The Human Gene Mutation Database, previously mentioned, has cataloged the nature of many disease alleles, including the sickle cell allele. The majority of alleles seem to

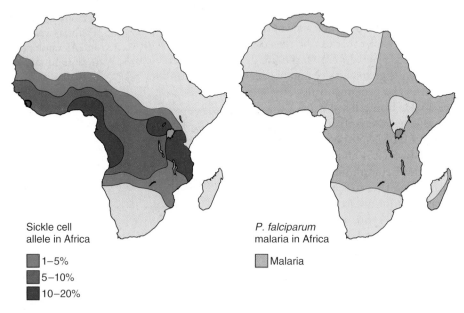

FIGURE 13.24
Sickle cell anemia. In individuals homozygous for the sickle cell trait, many of the red blood cells have sickled or irregular shapes, such as the cell on the far right.

Sickle cell allele in Africa

■ 1–5%
■ 5–10%
■ 10–20%

P. falciparum malaria in Africa

■ Malaria

FIGURE 13.25
The sickle cell allele increases resistance to malaria. The distribution of sickle cell anemia closely matches the occurrence of malaria in central Africa. This is not a coincidence. The sickle cell allele, when heterozygous, increases resistance to malaria, a very serious disease.

be simple changes. Almost 60% of the close to 28,000 alleles in the Human Gene Mutation Database are single base substitutions. Another 23% are due to small insertions or deletions of less than 20 bases. The rest of the alleles are made of more complex alterations. All of this indicates that simple changes have profound effects.

> Sickle cell anemia is caused by a single-nucleotide change in the gene for hemoglobin, which produces a protein with a nonpolar amino acid on its surface that tends to make the molecules clump together.

Some Defects May Soon Be Curable: Gene Therapy

The search for a way to introduce "healthy" genes into humans who lack them has gone on for nearly 40 years. In 1964, a trio of Nobel Prize winners (Ed Tatum, Joshua Lederberg, and Arthur Kornberg) suggested that it should be possible to cure often-fatal genetic disorders such as cystic fibrosis, muscular dystrophy, and multiple sclerosis by replacing the defective gene with a functional one.

Early Success

In 1990, two girls were cured of a rare blood disorder caused by a defective gene for the enzyme adenosine deaminase. In a procedure called **gene transfer therapy,** scientists isolated working copies of this gene and introduced them into bone marrow cells taken from each girl. The gene-modified bone marrow cells were allowed to proliferate, and then each girl's cells were injected back into her body. Both girls recovered and stayed healthy. For the first time, a genetic disorder had been cured by gene therapy.

The Rush to Cure Cystic Fibrosis

Like hounds to a hot scent, researchers set out to apply the new approach to one of the big killers, *cystic fibrosis*. The defective gene, labelled *cf*, had been isolated in 1989. Five years later, in 1994, researchers successfully transferred a healthy *cf* gene into a mouse with a defective one—in effect, they cured cystic fibrosis in the mouse. They achieved this remarkable result by adding the *cf* gene to a virus that infected the lungs of the mouse, carrying the gene with it "piggyback" into the lung cells. The virus chosen as the "vector" was adenovirus, which causes colds and is very infective of lung cells. To avoid any complications, the immune systems of the lab mice had been disabled.

Very encouraged by these well-publicized preliminary trials with mice, several labs set out early in 1995 to attempt to cure cystic fibrosis by transferring healthy copies of the *cf* gene into human patients. Confident of success, researchers added the human *cf* gene to adenovirus and then squirted the gene-bearing virus into the lungs of cystic fibrosis patients. For eight weeks, the gene therapy seemed successful, but then disaster struck. The gene-modified cells in the patients' lungs came under attack by the patients' own immune systems. The "healthy" *cf* genes were lost, and with them any chance of a cure.

Problems with the Vector

Other early attempts at gene therapy also met with failure. In retrospect, although it was not obvious then, the problem with these early attempts seems predictable. Adenovirus causes colds. Do you know anyone who has *never* had a cold? Due to previous colds, all of us have antibodies

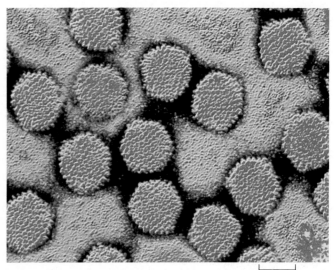

.17 μm

FIGURE 13.26
Adenovirus. This virus (200,000×), which causes the common cold, has been used to carry healthy genes in clinical trials of gene therapy. Its use as a vector is problematic, however, because it is usually attacked and destroyed by the immune system of the host. In addition, it can cause severe immune reactions and insert into the host's DNA at random places to cause mutations.

directed against adenovirus (figure 13.26). Thus, these researchers had introduced therapeutic genes through a vector their subjects' bodies were primed to destroy.

Later, in 1995, the newly appointed head of the National Institutes of Health (NIH), Nobel Prize winner Harold Varmus, held a comprehensive review of human gene therapy trials. Three problems became evident in the review: (1) The adenovirus vector being used in most trials elicits a strong immune response, leading to rejection of the added gene. (2) Adenovirus infection can, in rare instances, produce a very severe immune reaction, enough to kill. If many patients are treated, such instances can be expected to occur. (3) When the adenovirus infects a cell, it inserts its DNA into the human chromosome. Unfortunately, it does so at a random location. This means that the insertion events will cause mutations—by jumping into the middle of a gene, the virus inactivates that gene. Because the spot where adenovirus inserts is random, some of the mutations that result can be expected to cause cancer, certainly an unacceptable consequence.

Faced with these findings, Varmus called a halt to all further human clinical trials involving gene therapy. Go back to work in the laboratory, he told researchers, until you get a vector that works.

In principle, it should be possible to cure hereditary disorders such as cystic fibrosis by transferring a healthy gene into an affected individual. Early attempts using adenovirus vectors were not often successful.

Disorder	Symptom	Defect	Dominant/ Recessive	Frequency Among Human Births
Cystic fibrosis	Mucus clogs lungs, liver, and pancreas	Failure of chloride ion transport mechanism	Recessive	1/2500 (Caucasians)
Sickle cell anemia	Blood circulation is poor	Abnormal hemoglobin molecules	Recessive	1/625 (African Americans)
Tay-Sachs disease	Central nervous system deteriorates in infancy	Defective enzyme (hexosaminidase A)	Recessive	1/3500 (Ashkenazi Jews)
Phenylketonuria	Brain fails to develop in infancy	Defective enzyme (phenylalanine hydroxylase)	Recessive	1/12,000
Hemophilia	Blood fails to clot	Defective blood-clotting factor VIII	Sex-linked recessive	1/10,000 (Caucasian males)
Huntington disease	Brain tissue gradually deteriorates in middle age	Production of an inhibitor of brain cell metabolism	Dominant	1/24,000
Muscular dystrophy (Duchenne)	Muscles waste away	Degradation of myelin coating of nerves stimulating muscles	Sex-linked recessive	1/3700 (males)
Hypercholesterolemia	Excessive cholesterol levels in blood lead to heart disease	Abnormal form of cholesterol cell surface receptor	Dominant	1/500

Table 13.3 Some Important Genetic Disorders

More Promising Vectors

Within a few years, researchers had found a much more promising vector. This new gene carrier, a tiny parvovirus called *adeno-associated virus* (AAV), has only two genes and needs adenovirus to replicate. To create a vector for gene transfer, researchers remove both of the AAV genes. The shell that remains is still quite infective and can carry human genes into patients. Importantly, AAV enters the human DNA far less frequently than the adenovirus; thus, it is much less likely to produce cancer-causing mutations. In addition, AAV does not elicit a strong immune response—cells infected with AAV are not eliminated by a patient's immune system. Finally, since AAV never elicits a dangerously strong immune response, it can be safely administered to patients.

Success with the AAV Vector

In 1999, AAV successfully cured anemia in rhesus monkeys. In monkeys, as well as humans and other mammals, red blood cell production is stimulated by a protein called erythropoietin (EPO). People with anemia (that is, low red blood cell counts), such as dialysis patients, get regular injections of EPO. Using AAV to carry a souped-up EPO gene into the monkeys, scientists were able to greatly elevate their red blood cell counts, curing the monkeys of anemia—and keeping them cured.

A similar experiment using AAV cured dogs of a hereditary disorder leading to retinal degeneration and blindness. These dogs were blind due to a defective gene that produced a mutant form of a protein associated with the retina of the eye. Injection of AAV bearing the needed gene into the fluid-filled compartment behind the retina restored their sight (figure 13.27).

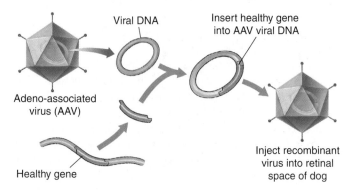

FIGURE 13.27
Using AAV to cure canine retinal degeneration. A healthy retinal gene was inserted into the AAV DNA, which was then injected into the retinal space of a dog with retinal degeneration. The healthy gene was expressed and the dog was cured.

Human clinical trials are now under way again. In 2000, scientists performed the first gene therapy experiment for muscular dystrophy, injecting genes into a 35-year-old South Dakota man. He is an early traveler on what is likely to become a well-used therapeutic highway. Table 13.3 lists other hereditary disorders that may someday be treated with gene therapy. Trials are currently being conducted for cystic fibrosis, rheumatoid arthritis, hemophilia, and a wide variety of cancers. The way seems open, the possibility of progress tantalizingly close.

New virus vectors such as AAV avoid the problems of vectors used earlier and offer promise of gene transfer therapy cures.

13.3 Genes are on chromosomes.

Chromosomes: The Vehicles of Mendelian Inheritance

Chromosomes are not the only structures that segregate regularly when eukaryotic cells divide. Centrioles also divide and segregate in a regular fashion, as do mitochondria and chloroplasts (when present). Thus, in the early twentieth century, it was by no means obvious that chromosomes were the vehicles of hereditary information.

The Chromosomal Theory of Inheritance

A central role for chromosomes in heredity was first suggested in 1900 by the German geneticist Karl Correns, in one of the papers announcing the rediscovery of Mendel's work. Soon after, observations that similar chromosomes paired with one another during meiosis led directly to the **chromosomal theory of inheritance**, first formulated by the American Walter Sutton in 1902.

Several pieces of evidence supported Sutton's theory. One was that reproduction involves the initial union of only two cells, egg and sperm. If Mendel's model were correct, then these two gametes must make equal hereditary contributions. Sperm, however, contain little cytoplasm, suggesting that the hereditary material must reside within the nuclei of the gametes. Furthermore, while diploid individuals have two copies of each pair of homologous chromosomes, gametes have only one. This observation was consistent with Mendel's model, in which diploid individuals have two copies of each heritable gene and gametes have one. Finally, chromosomes segregate during meiosis, and each pair of homologues orients on the metaphase plate independently of every other pair. Segregation and independent assortment were two characteristics of the genes in Mendel's model.

A Problem with the Chromosomal Theory

However, investigators soon pointed out one problem with this theory. If Mendelian characters are determined by genes located on the chromosomes, and if the independent assortment of Mendelian traits reflects the independent assortment of chromosomes in meiosis, why does the number of characters that assort independently in a given kind of organism often greatly exceed the number of chromosome pairs the organism possesses? This seemed a fatal objection, and it led many early researchers to have serious reservations about Sutton's theory.

Morgan's White-Eyed Fly

The essential correctness of the chromosomal theory of heredity was demonstrated long before this paradox was re-

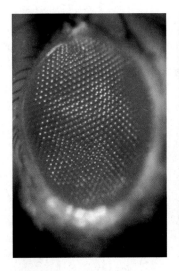

FIGURE 13.28
Red-eyed (normal) and white-eyed (mutant) *Drosophila.*
The white-eyed defect is hereditary, the result of a mutation in a gene located on the X chromosome. By studying this mutation, Morgan first demonstrated that genes are on chromosomes.

solved. A single small fly provided the proof. In 1910, Thomas Hunt Morgan, studying the fruit fly *Drosophila melanogaster*, detected a **mutant** male fly, one that differed strikingly from normal flies of the same species; its eyes were white instead of red (figure 13.28).

Morgan immediately set out to determine if this new trait would be inherited in a Mendelian fashion. He first crossed the mutant male to a normal female to see if red or white eyes were dominant. All of the F_1 progeny had red eyes, so Morgan concluded that red eye color was dominant over white. Following the experimental procedure that Mendel had established long ago, Morgan then crossed the red-eyed flies from the F_1 generation with each other. Of the 4252 F_2 progeny Morgan examined, 782 (18%) had white eyes. Although the ratio of red eyes to white eyes in the F_2 progeny was greater than 3:1, the results of the cross nevertheless provided clear evidence that eye color segregates. However, something about the outcome was strange and totally unpredicted by Mendel's theory—*all of the white-eyed F_2 flies were males!*

How could this be explained? Perhaps it is impossible for a white-eyed female fly to exist; such individuals might not be viable for some unknown reason. To test this idea, Morgan testcrossed the female F_1 progeny with the original white-eyed male. He obtained both white-eyed and red-eyed males and females in a 1:1:1:1 ratio, just as Mendel's theory had predicted. Hence, a female could have white eyes. Why, then, were there no white-eyed females among the progeny of the original cross?

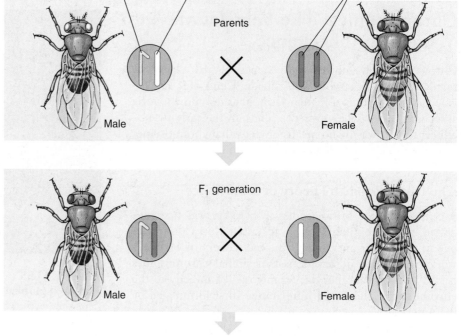

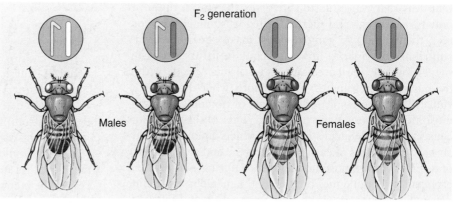

FIGURE 13.29
Morgan's experiment demonstrating the chromosomal basis of sex linkage in *Drosophila*. The white-eyed mutant male fly was crossed with a normal female. The F₁ generation flies all exhibited red eyes, as expected for flies heterozygous for a recessive white-eye allele. In the F₂ generation, all of the white-eyed flies were male.

Sex Linkage

The solution to this puzzle involved sex. In *Drosophila*, the sex of an individual is determined by the number of copies it has of a particular chromosome, the **X chromosome.** A fly with two X chromosomes is a female, and a fly with only one X chromosome is a male. In males, the single X chromosome pairs in meiosis with a dissimilar partner called the **Y chromosome.** The female thus produces only X gametes, while the male produces both X and Y gametes. When fertilization involves an X sperm, the result is an XX zygote, which develops into a female; when fertilization involves a Y sperm, the result is an XY zygote, which develops into a male.

The solution to Morgan's puzzle is that the gene causing the white-eye trait in *Drosophila* resides only on the X chromosome—it is absent from the Y chromosome. (We now know that the Y chromosome in flies carries almost no functional genes.) A trait determined by a gene on the X chromosome is said to be **sex-linked.** Knowing the white-

eye trait is recessive to the red-eye trait, we can now see that Morgan's result was a natural consequence of the Mendelian assortment of chromosomes (figure 13.29).

Morgan's experiment was one of the most important in the history of genetics because it presented the first clear evidence that the genes determining Mendelian traits do indeed reside on the chromosomes, as Sutton had proposed. The segregation of the white-eye trait has a one-to-one correspondence with the segregation of the X chromosome. In other words, Mendelian traits such as eye color in *Drosophila* assort independently because chromosomes do. When Mendel observed the segregation of alternative traits in pea plants, he was observing a reflection of the meiotic segregation of chromosomes.

Mendelian traits assort independently because they are determined by genes located on chromosomes that assort independently in meiosis.

Genetic Recombination

Morgan's experiments led to the general acceptance of Sutton's chromosomal theory of inheritance. Scientists then attempted to explain why there are many more independently assorting Mendelian genes than chromosomes. Obviously, each chromosome carries more than one gene, so how could two genes assort independently if they are located on the same chromosome? In 1903, the Dutch geneticist Hugo de Vries suggested that this paradox could be resolved only by assuming that homologous chromosomes exchange elements during meiosis. In 1909, the French cytologist F. A. Janssens provided evidence to support this suggestion. Investigating chiasmata produced during amphibian meiosis, Janssens noticed that, of the four chromatids involved in each chiasma, two crossed each other and two did not. He suggested that this crossing of chromatids reflected a switch in chromosomal arms between the paternal and maternal homologues, involving one chromatid in each homologue. His suggestion was not accepted widely, primarily because it was difficult to see how two chromatids could break and rejoin at exactly the same position.

Crossing Over

Later experiments clearly established that Janssens was indeed correct. One of these experiments, performed in 1931 by the American geneticist Curt Stern, is described in figure 13.30. Stern studied two sex-linked eye characters in *Drosophila* strains whose X chromosomes were visibly abnormal at both ends. He first examined many flies and identified those in which an exchange had occurred with respect to the two eye characters. He then studied the chromosomes of those flies to see if their X chromosomes had exchanged arms. Stern found that all of the individuals that had exchanged eye traits also possessed chromosomes that had exchanged abnormal ends. The conclusion was inescapable: Genetic exchanges of characters such as eye color involve the physical exchange of chromosome arms, a phenomenon called **crossing over.** Crossing over creates new combinations of genes, and is thus a form of **genetic recombination.**

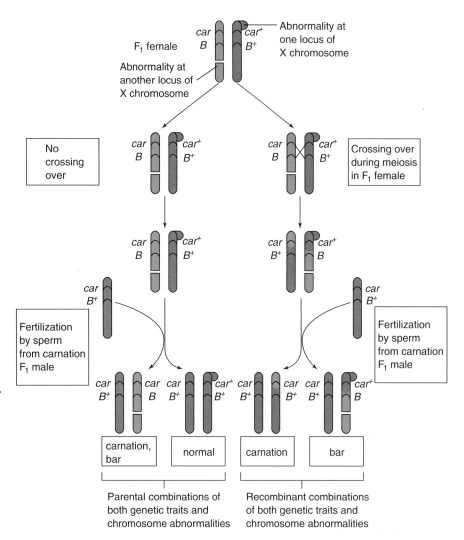

FIGURE 13.30

Stern's experiment demonstrating the physical exchange of chromosomal arms during crossing over. Stern monitored crossing over between two genes, the recessive carnation eye color (*car*) and the dominant bar-shaped eye (*B*), on chromosomes with physical peculiarities visible under a microscope. Whenever these genes recombined through crossing over, the chromosomes recombined as well. Therefore, the recombination of genes reflects a physical exchange of chromosome arms. The "+" notation on the alleles refers to the wild-type allele, the most common allele at a particular gene.

The chromosomal exchanges Stern demonstrated provide the solution to the paradox, because crossing over can occur between homologues anywhere along the length of the chromosome, in locations that seem to be randomly determined. Thus, if two different genes are located relatively far apart on a chromosome, crossing over is more likely to occur somewhere between them than if they are located close together. Two genes can be on the same chromosome and still show independent assortment if they are located so far apart on the chromosome that crossing over occurs regularly between them.

Using Recombination to Make Genetic Maps

Mendel observed independent assortment because he studied traits that appeared on different chromosomes or appeared far enough apart on the same chromosome that crossing over occurred (figure 13.31). Because crossing over is more frequent between two genes that are relatively far apart than between two that are close together, the frequency of crossing over can be used to map the relative positions of genes on chromosomes. In a cross, the proportion of progeny exhibiting an exchange between two genes is a measure of the frequency of crossover events between them, and thus indicates the relative distance separating them. The results of such crosses can be used to construct a **genetic map** that measures the distance between genes in terms of the frequency of recombination. One "map unit" is defined as the distance within which a crossover event is expected to occur in an average of 1% of gametes. A map unit is now called a **centimorgan,** after Thomas Hunt Morgan.

In addition to genetic maps, geneticists construct a variety of physical maps. These include the sites cut by DNA-cleaving restriction enzymes, sites amplified by the polymerase chain reaction (PCR), sequence-tagged sites (STS), and cytological maps (see chapter 16). The ultimate power of genetics is to integrate these physical maps with genetic maps based on recombination.

The Three-Point Cross. In constructing a genetic map, a researcher simultaneously monitors recombination among three or more genes located on the same chromosome. When genes are close enough together on a chromosome that they do not assort independently, they are said to be *linked* to one another. A cross involving three linked genes is called a **three-point cross.** Data obtained by Morgan on traits encoded by genes on the X chromosome of *Drosophila* were used by his student, A. H. Sturtevant, to draw the first genetic map (figure 13.32). By convention, the most common allele of a gene is often denoted with the symbol "+" and is designated as **wild type,** as shown in figure 13.30. All other alleles are denoted with just the specific letters.

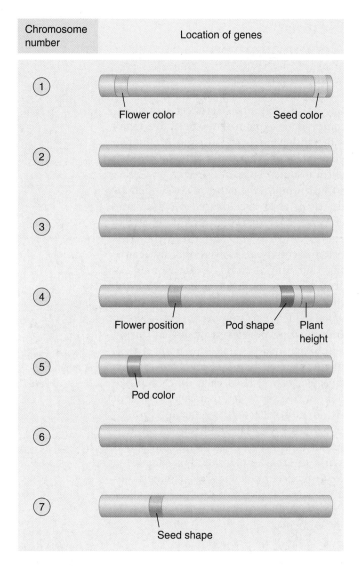

FIGURE 13.31
The chromosomal locations of the seven genes studied by Mendel in the garden pea. The genes for plant height and pod shape are very close to each other and rarely recombine. Plant height and pod shape were not among the characters Mendel examined in dihybrid crosses. One wonders what he would have made of the linkage he surely would have detected had he tested these characters together.

FIGURE 13.32
The first genetic map. This map of the X chromosome of *Drosophila* was prepared in 1913 by A. H. Sturtevant, a student of Morgan. On it he located the relative positions of five recessive traits that exhibited sex linkage by estimating their relative recombination frequencies in genetic crosses. The higher the recombination frequency, the farther apart the two genes.

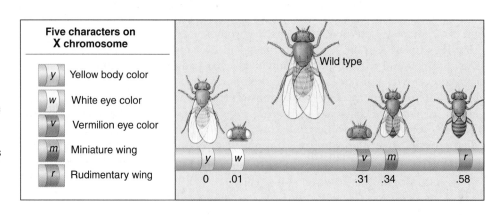

The Human Genetic Map

Given that we cannot perform controlled crosses in humans, can we map human genes? The answer is "yes," but we must use historical data from pedigrees. The principle is the same—genetic distance is still proportional to recombination frequency—but the analysis requires the use of complex statistics. The methodology is to look at a pair of genes, construct models of inheritance for linked versus unlinked cases, and then compare these predictions to the actual pedigree, using a statistic called LOD for *log of odds ratio*. LOD is a ratio of the probability that two genes are linked versus the probability that they are not. This method allows both the detection of linked genes, and the determination of map distances between them (figure 13.33). This process requires multiple repetitive calculations and has been automated with the use of computer programs.

To appreciate the enormity of the task of constructing a human genetic map, a little history is required. If we look at other animals with extensive genetic maps, the majority of genetic markers were found on alleles that caused morphological changes, such as variant eye color, body color, or wing morphology in flies. In humans, such alleles generally correspond to disease states. Thus, as late as the early 1980s, the number of markers for the human genome was in the hundreds. Given the size of the genome, this would never allow very dense coverage. Additionally, these are genes that we wish to map, but the disease alleles are at low frequencies in the population. Any one family would never have very many disease alleles segregating to allow for mapping.

This all changed with the development of anonymous markers, genetic markers that can be detected using molecular techniques, but do not cause a detectable phenotype. The nature of these markers has evolved with technology over the years and now has led to a standardized set of markers that are scattered throughout the genome, can be detected using techniques that are easy to automate, and have a relatively high density. Most of these markers are regions amplifiable by the polymerase chain reaction (PCR) (discussed in chapter 16) that include a region of two or three bases repeated in tandem arrays. The number of repeats varies among individuals, with no change in phenotype. Thus, we now have several thousand markers, instead of hundreds, and a genetic map that would have been unthinkable 25 years ago.

Investigators wishing to study a particular gene can use techniques described in chapter 16 to screen a library of gene fragments and rapidly determine which fragment carries the gene of interest. They will then be able to analyze that fragment in detail.

The next step has been to use the information developed from sequencing the human genome to identify and map single bases that differ between individuals. We call any differences between individuals in populations polymor-

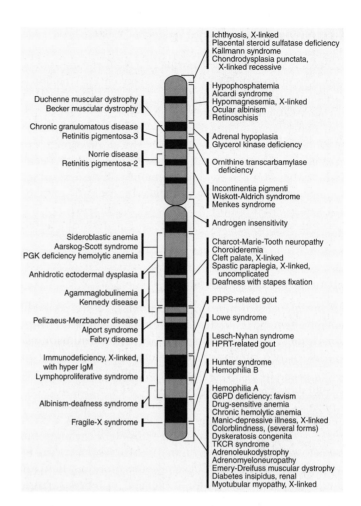

FIGURE 13.33

The human X chromosome gene map. More than 59 diseases have been traced to specific segments of the X chromosome. Many of these disorders are also influenced by genes on other chromosomes.

phisms, so these are now called single nucleotide polymorphisms, or SNPs. Over 2 million such differences have been identified and are being placed on both the genetic map and the human genome sequence. This confluence of techniques will allow the ultimate resolution of genetic analysis.

The recent progress in gene mapping applies to more than just the relatively small number of genes that show simple Mendelian inheritance. The development of a high resolution genetic map, and the characterization of millions of SNPs, opens up the possibility of beginning the genetic dissection of complex quantitative traits in humans as well.

Gene maps locate the relative positions of different genes on the chromosomes of an organism. Traditionally produced by analyzing the relative amounts of recombination in genetic crosses, gene maps are increasingly being made by using new techniques.

Human Chromosomes

Each human somatic cell normally has 46 chromosomes, which in meiosis form 23 pairs. By convention, the chromosomes are divided into seven groups (designated A through G), each characterized by a different size, shape, and appearance. The differences among the chromosomes are most clearly visible when the chromosomes are arranged in order in a karyotype (figure 13.34). Techniques that stain individual segments of chromosomes with different-colored dyes make the identification of chromosomes unambiguous. Like a fingerprint, each chromosome always exhibits the same pattern of colored bands.

Human Sex Chromosomes

Of the 23 pairs of human chromosomes, 22 are perfectly matched in both males and females and are called **autosomes.** The remaining pair, the **sex chromosomes,** consist of two similar chromosomes in females and two dissimilar chromosomes in males. In humans, females are designated XX and males XY. One of the sex chromosomes in the male (the Y chromosome) is highly condensed. Because few genes on the Y chromosome are expressed, recessive alleles on a male's single X chromosome have no *active* counterpart on the Y chromosome. Some of the active genes on the Y chromosome are responsible for the features associated with "maleness" in humans. Consequently, any individual with *at least* one Y chromosome is a male.

Sex Chromosomes in Other Organisms

The structure and number of sex chromosomes vary in different organisms (table 13.4). In the fruit fly, *Drosophila*, females are XX and males XY, as in humans and most other vertebrates. However, in birds, the male has two Z chromosomes, and the female has a Z and a W chromosome. Some insects, such as grasshoppers, have no Y chromosome—females are XX and males are characterized as XO (the O indicating the absence of a chromosome).

Sex Determination

In humans, a specific gene on the Y chromosome known as *SRY* plays a key role in the development of male sexual characteristics. This gene is expressed early in development, and acts to masculinize genitalia and secondary sexual organs that would otherwise be female. Lacking a Y chromosome, females fail to undergo these changes.

Among fishes and some species of reptiles, environmental factors can cause changes in the expression of this sex-determining gene and thus in the sex of the adult individual.

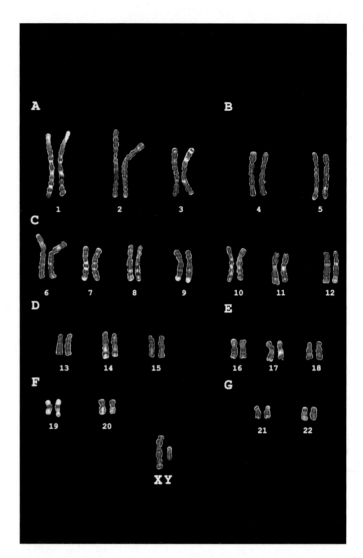

FIGURE 13.34
A human karyotype. This karyotype shows the colored banding patterns, arranged by class, A–G.

Table 13.4	Sex Determination in Some Organisms		Female	Male
Humans, *Drosophila*			XX	XY
Birds			ZW	ZZ
Grasshoppers			XX	XO
Honeybees			Diploid	Haploid

Barr Bodies

Although males have only one copy of the X chromosome and females have two, female cells do not produce twice as much of the proteins encoded by genes on the X chromosome. Instead, one of the X chromosomes in females is inactivated early in embryonic development, shortly after the embryo's sex is determined. Which X chromosome is inactivated varies randomly from cell to cell. If a woman is heterozygous for a sex-linked trait, some of her cells will express one allele and some the other. The inactivated and highly condensed X chromosome is visible as a darkly staining **Barr body** attached to the nuclear membrane.

Why Are X and Y Chromosomes so Different?

Like the other 22 pairs of human chromosomes, the X chromosome is packed with some 1000 genes. Biologists surmise that the reason females possess two copies of the X chromosome and of autosomes is to allow for the repair of the inevitable gene damage that occurs over time due to wear-and-tear, chemical damage, and mistakes in copying.

How can a cell detect and edit out a mutation involving only one or a few nucleotides in one strand of DNA? How does it know which of the two strands is the "correct" version and which the altered one? This neat trick is achieved by every cell having two nearly-identical copies of each chromosome. By comparing the two versions with each other, a cell can identify the "typos" and fix them.

Here is how it works: When a cell detects a chromosomal DNA duplex with a difference between its two DNA strands, that duplex is "repaired" by the rather Draconian expedient of chopping out the entire region, on both strands of the DNA molecule. No effort is made by the cell to determine which strand is correct—both are discarded. The gap that this creates is filled by copying off the sequence present at that region on the other chromosome, a process called **gene conversion.** All this editing happens when the two versions of the chromosome are paired closely together in the early stages of meiosis.

Most biologists think the need for DNA repair is the reason sex evolved. Asexual organisms, which do not undergo meiosis, accumulate mutational damage in a process of irreversible genetic decay that population geneticists call Muller's Ratchet, a progressive loss of genes that can lead to eventual extinction (see chapter 12).

So what are we to make of males? Males, in contrast to females, have only one copy of the X chromosome, not two. The other chromosome of the pair in males, the Y chromosome, is much smaller than the X. Biologists thought until very recently that the Y chromosome had only a few active genes. Because there is no other Y to serve as a pairing partner in meiosis, most of its genes had been thought to have decayed, the victims of Muller's ratchet, leaving the Y chromosome a genetic wasteland with only a very few active genes surviving on it.

We now know this view to have been way too simple. In June of 2003, researchers reported the full gene sequence of the human Y chromosome, and it was nothing like biologists had expected. The human Y chromosome contains not one or two active genes, but 78! One is concerned with male development, most of the others with sperm production and fertility. A few have no obvious role in sex. One of this latter group makes a component of ribosomes (complex tiny engines in the cell which assemble proteins), meaning that every ribosome in a man's body is slightly different from those in a woman's.

Taking all these genes into account, geneticists conclude that men and women differ by 1 to 2 percent of their genomes—which is the same as the difference between a man and a male chimpanzee (or a woman and a female chimpanzee). So we are going to have to reexamine the basis of the differences between the sexes. A lot more of it may be built into the genes than we had supposed.

The Y chromosome is much smaller than the X, and can only pair up with the X at the tips. Thus, there can be no close pairing between X and Y during meiosis, the sort of pairing that allows the proofreading and editing discussed above. We can now see that there is a very good reason evolution has acted to prevent the close pairing of X and Y—those 78 Y chromosome genes. As close pairing allows the exchange of large as well as small segments, any association of X and Y would lead to gene swapping, and the male-determining genes of the Y chromosome would sneak into the X, making everybody male.

One mystery remains. If the Y chromosome cannot pair with the X chromosome, how does it make do without copy-editing to prevent the accumulation of mutations? Why hasn't Muller's ratchet long ago driven males to extinction? The answer is right there for us to see in the Y chromosome sequence, and an elegant answer it is. Most of the 78 active genes on the Y chromosome lie within eight vast palindromes, regions of the DNA sequence that repeat the same sequence twice, running in opposite directions, like the sentence "Madam, I'm Adam," or Napoleon's quip "Able was I ere I saw Elba."

A palindrome has a very neat property: it can bend back on itself, forming a hairpin in which the two strands are aligned with nearly identical DNA sequences. This is the same sort of situation—alignment of nearly identical stretches of chromosomes—which permits the copy-edit of the X chromosome during meiosis. Thus, in the Y chromosome, mutations can be "corrected" by conversion to the undamaged sequence preserved on the other arm of the palindrome. Damage does not accumulate, Muller's ratchet is avoided, and males persist.

In humans, sex is determined by the presence of the Y chromosome, which contains at least 78 transcribed genes.

Human Abnormalities Due to Alterations in Chromosome Number

The failure of homologues or sister chromatids to separate properly during meiosis is called **nondisjunction**. This leads to gametes with the gain or loss of a chromosome, a condition called **aneuploidy**. The frequency of aneuploidy in humans is surprisingly high, being estimated to occur in 5% of conceptions.

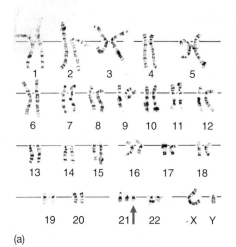

(a)

(b)

FIGURE 13.35

Down syndrome. (*a*) As shown in this male karyotype, Down syndrome is associated with trisomy of chromosome 21. (*b*) A child with Down syndrome sitting on his father's knee.

Nondisjunction Involving Autosomes

Almost all humans of the same sex have the same karyotype, for the same reason that all automobiles have engines, transmissions, and wheels: Other arrangements don't work well. Humans who have lost even one copy of an autosome are called **monosomics** and do not survive development. In all but a few cases, humans who have gained an extra autosome (called **trisomics**) also do not survive. Data from clinically recognized spontaneous abortions indicate levels of aneuploidy as high as 35%. However, five of the smallest autosomes—those numbered 13, 15, 18, 21, and 22—can be present in humans as three copies and still allow the individual to survive for a time. The presence of an extra chromosome 13, 15, or 18 causes severe developmental defects, and infants with such a genetic makeup die within a few months. In contrast, individuals who have an extra copy of chromosome 21 or, more rarely, chromosome 22, usually survive to adulthood. In these people, the maturation of the skeletal system is delayed, so they generally are short and have poor muscle tone. Their mental development is also affected, and children with trisomy 21 or trisomy 22 are always mentally retarded.

Down Syndrome. The developmental defect produced by trisomy 21 (figure 13.35*a*) was first described in 1866 by J. Langdon Down; for this reason, it is called **Down syndrome** (formerly "Down's syndrome"). About 1 in every 750 children exhibits Down syndrome, and the frequency is comparable in all racial groups. Similar conditions also occur in chimpanzees and other related primates. In humans, the defect occurs when a particular small portion of chromosome 21 is present in three copies instead of two. In 97% of the human cases examined, all of chromosome 21 is present in three copies. In the other 3%, a small portion of chromosome 21 containing the critical segment has been added to another chromosome by a process called *translocation* (see chapter 20); it exists along with the normal two copies of chromosome 21. This condition is known as *translocation Down syndrome.*

Not much is known about the developmental role of the genes whose extra copies produce Down syndrome, although clues are beginning to emerge from current research. Some researchers suspect that the gene or genes that produce Down syndrome are similar or identical to some of the genes associated with cancer and Alzheimer disease. The reason for this suspicion is that one of the human cancer-causing genes (to be described in chapter 20) and the gene causing Alzheimer disease are located on the segment of chromosome 21 associated with Down syndrome. Moreover, cancer is more common in children with Down syndrome. The incidence of leukemia, for example, is 11 times higher in children with Down syndrome than in unaffected children of the same age.

How does Down syndrome arise? In humans, it comes about almost exclusively as a result of primary nondisjunction of chromosome 21 during egg formation. The cause of these primary nondisjunctions is not known, but their incidence, like that of cancer, increases with maternal age (figure 13.36). In mothers younger than 20 years of age, the risk of giving birth to a child with Down syndrome is about 1 in 1700; in mothers 20 to 30 years old, the risk is only about 1 in 1400. However, in mothers 30 to 35 years old, the risk rises to 1 in 750, and by age 45, the risk is as high as 1 in 16!

Primary nondisjunctions are far more common in women than in men because all of the eggs a woman will ever produce have developed to the point of prophase in meiosis I by the time she is born. By the time she has children, her eggs are as old as she is. In contrast, men produce new sperm daily. Therefore, there is a much greater chance for problems of various kinds, including those that cause primary nondisjunction, to accumulate over time in the gametes of women than in those of men. For this reason, the age of the mother is more critical than that of the father for couples contemplating childbearing.

Nondisjunction Involving the Sex Chromosomes

Individuals who gain or lose a sex chromosome do not generally experience the severe developmental abnormalities caused by similar changes in autosomes. While such individuals have somewhat abnormal features, they often reach maturity.

The X Chromosome. When X chromosomes fail to separate during meiosis, some of the gametes produced possess both X chromosomes, and so are XX gametes; the other gametes that result from such an event have no sex chromosome and are designated "O" (figure 13.37).

If an XX gamete combines with an X gamete, the resulting XXX zygote develops into a female with one functional X chromosome and two Barr bodies. She is sterile but usually normal in other respects. If an XX gamete instead combines with a Y gamete, the effects are more serious. The resulting XXY zygote develops into a sterile male who has many female body characteristics and, in some cases, diminished mental capacity. This condition, called *Klinefelter syndrome*, occurs in about 1 out of every 500 male births.

If an O gamete fuses with a Y gamete, the resulting OY zygote is nonviable and fails to develop further because humans cannot survive when they lack the genes on the X chromosome. If, on the other hand, an O gamete fuses with an X gamete, the XO zygote develops into a sterile female of short stature, with a webbed neck and sex organs that never fully mature during puberty. The mental abilities of an XO individual are in the low-normal range. This condition, called *Turner syndrome*, occurs roughly once in every 5000 female births.

The Y Chromosome. The Y chromosome can also fail to separate in meiosis, leading to the formation of YY gametes. When these gametes combine with X gametes, the XYY zygotes develop into fertile males of normal appearance. The frequency of the XYY genotype (Jacob syndrome) is about 1 per 1000 newborn males, but it is approximately 20 times higher among males in penal and mental institutions. This observation has led to the highly controversial suggestion that XYY males are inherently antisocial, a suggestion supported by some studies but not by others. In any case, most XYY males do not develop patterns of antisocial behavior.

> **Gene dosage plays a crucial role in development, so humans do not tolerate the loss or addition of chromosomes well. Autosome loss is always lethal, and an extra autosome is, with few exceptions, lethal too. Additional sex chromosomes have less serious consequences, although they can lead to sterility.**

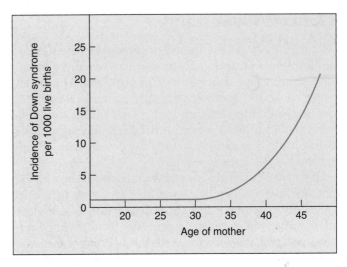

FIGURE 13.36
Correlation between maternal age and the incidence of Down syndrome. As women age, the chances they will bear a child with Down syndrome increase. After a woman reaches 35, the frequency of Down syndrome rises rapidly.
Over a five-year period between ages 20 and 25, the incidence of Down syndrome increases 0.1 per thousand; over a five-year period between ages 35 and 40, the incidence increases to 8.0 per thousand, 80 times as great. The period of time is the same in both instances. What has changed?

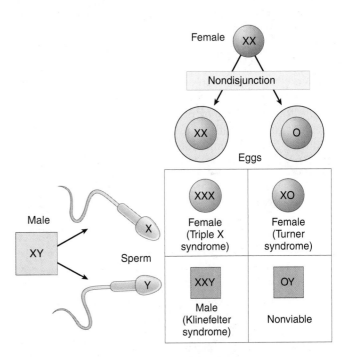

FIGURE 13.37
How nondisjunction can produce abnormalities in the number of sex chromosomes. When nondisjunction occurs in the production of female gametes, the gamete with two X chromosomes (XX) produces Klinefelter males (XXY) and triple-X females (XXX). The gamete with no X chromosome (O) produces Turner females (XO) and nonviable OY males lacking any X chromosome.

Genetic Counseling

Although most genetic disorders cannot yet be cured, we are learning a great deal about them, and progress toward successful therapy is being made in many cases. In the absence of a cure, however, the only recourse is to try to avoid producing children with these conditions. The process of identifying parents at risk for producing children with genetic defects and assessing the genetic state of early embryos is called *genetic counseling.*

If a genetic defect is caused by a recessive allele, how can potential parents determine the likelihood that they carry the allele? One way is through pedigree analysis, often employed as an aid in genetic counseling. By analyzing a person's pedigree, it is sometimes possible to estimate the likelihood that the person is a carrier for certain disorders. For example, if one of your relatives has been afflicted with a recessive genetic disorder such as cystic fibrosis, it is possible that you are a heterozygous carrier of the recessive allele for that disorder. When a couple is expecting a child, and pedigree analysis indicates that both of them have a significant probability of being heterozygous carriers of a recessive allele responsible for a serious genetic disorder, the pregnancy is said to be high-risk. In such cases, there is a significant probability that the child will exhibit the clinical disorder.

Another class of high-risk pregnancy is that in which the mothers are more than 35 years old. As we have seen, the frequency of infants with Down syndrome increases dramatically in the pregnancies of older women (see figure 13.36).

When a pregnancy is diagnosed as high-risk, many women elect to undergo **amniocentesis,** a procedure that permits the prenatal diagnosis of many genetic disorders. In the fourth month of pregnancy, a sterile hypodermic needle is inserted into the expanded uterus of the mother, removing a small sample of the amniotic fluid that bathes the fetus (figure 13.38). Within the fluid are free-floating cells derived from the fetus; once removed, these cells can be grown in cultures in the laboratory. During amniocentesis, the position of the needle and that of the fetus are usually observed by means of *ultrasound.* The sound waves used in ultrasound are not harmful to mother or fetus, and they permit the person withdrawing the amniotic fluid to do so without damaging the fetus. In addition, ultrasound can be used to examine the fetus for signs of major abnormalities.

In recent years, physicians have increasingly turned to a new, less invasive procedure for genetic screening called **chorionic villi sampling.** Using this method, the physician removes cells from the chorion, a membranous part of the placenta that nourishes the fetus. This procedure can be used earlier in pregnancy (by the eighth week) and yields results much more rapidly than does amniocentesis.

To test for certain genetic disorders, genetic counselors look for three characteristics in the cultures of cells obtained from amniocentesis or chorionic villi sampling. First, analysis of the karyotype can reveal aneuploidy (extra or missing chromosomes) and gross chromosomal alterations. Second, in many cases it is possible to test directly for the proper functioning of enzymes involved in genetic disorders. The lack of normal enzymatic activity signals the presence of the disorder. Thus, the lack of the enzyme responsible for breaking down phenylalanine signals PKU (phenylketonuria); the absence of the enzyme responsible for the breakdown of gangliosides indicates Tay-Sachs disease; and so forth.

With the changes in human genetics brought about by the Human Genome Project (see chapter 17), it is possible to design tests for many more diseases. Difficulties still exist in discerning the number and frequency of disease-causing alleles, but these are not insurmountable. At present, there are tests for at least 13 genes with alleles that lead to clinical syndromes. This number is bound to rise and include alleles that do not directly lead to disease states but predispose a person for a particular disease. Chapter 17 treats the many ethical questions that arise.

Many gene defects can be detected early in pregnancy, allowing for appropriate planning by the prospective parents.

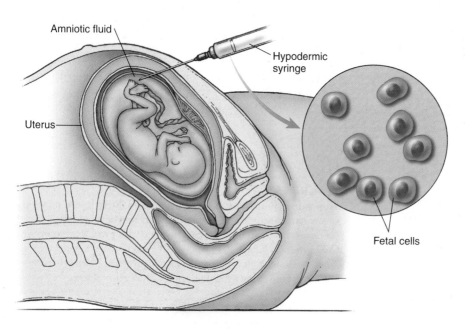

FIGURE 13.38
Amniocentesis. A needle is inserted into the amniotic cavity, and a sample of amniotic fluid, containing some free cells derived from the fetus, is withdrawn into a syringe. The fetal cells are then grown in culture, and their karyotype and many of their metabolic functions are examined.

For interactive testing, visit the Online Learning Center with PowerWeb at www.mhhe.com/Raven7

13.1 Mendel solved the mystery of heredity.

Early Ideas About Heredity: The Road to Mendel

- Two classical assumptions regarding heredity prior to Mendel were that species were constant and that traits were transmitted directly. (p. 242)
- Early researchers discovered that some inherited characters can disappear and appear again in later generations; some characters can segregate among different offspring; and some forms are more likely to appear than others. (p. 243)

Mendel and the Garden Pea

- Gregor Mendel, an Austrian monk, conducted hybridization studies on garden peas at his monastery. (p. 244)
- Mendel chose garden peas for several reasons: He knew he could produce hybrids by crossing varieties; many varieties were available; and pea plants are easy to grow and have short generation times. In addition, if left alone, the flowers can self-fertilize. (p. 244)
- Mendel conducted his experiment in three stages: (1) He allowed all plants to self-fertilize to produce pure-breeding individuals; (2) he performed crosses between varieties exhibiting alternate character forms; and (3) he allowed hybrid offspring to self-fertilize for several generations. (p. 245)

What Mendel Found

- From his experiments, Mendel derived four conclusions concerning heredity: (1) The plant crosses did not produce progeny of intermediate appearance. (2) For each pair of alternative forms of a character, one alternative was not expressed in the F_1 generation, but did reappear in some F_2 individuals. (3) Pairs of alternative traits segregated among progeny of a particular cross. (4) Alternative traits were expressed in a 3:1 ratio in the F_2 generation, which actually was a 1:2:1 ratio for pure-breeding dominant: non-pure-breeding dominant: pure-breeding recessive. (p. 248)
- Mendel's model of heredity contains five basic elements: (1) Parents do not transmit physiological traits directly to their offspring. (2) Each individual receives two factors that may code for the same trait or two alternative traits. (3) Not all copies of a factor are identical. (4) The alleles from each parent do not influence each other in any way. (5) The presence of a particular allele does not ensure the expression of that trait. (p. 249)

How Mendel Interpreted His Results

- By convention, dominant traits are assigned capital letters and recessive traits lowercase letters, which can easily be put into a Punnett square for analysis. (p. 250)
- Mendel proposed two laws of heredity: the Law of Segregation (alternative alleles segregate independently) and the Law of Independent Assortment (genes on different chromosomes assort independently). (pp. 251–253)

Mendelian Inheritance Is Not Always Easy to Analyze

- Most phenotypes reflect the action of many genes. (p. 255)
- Quantitative traits produce continuous variation. (p. 255)
- Pleiotropic alleles have more than one effect on the phenotype. (p. 255)
- Incomplete dominance occurs when alternative alleles are not fully dominant or recessive, and offspring express both parental phenotypes. (p. 256)
- Environmental effects can affect allele expression to produce varying phenotypes. (p. 256)
- Epistasis can occur when one gene interferes with the expression of another gene. (pp. 257–258)

13.2 Human genetics follows Mendelian principles.

Most Gene Disorders Are Rare

- Most genetic disorders are recessive (e.g., Tay-Sachs), although some are dominant (e.g., Huntington disease). (p. 259)

Multiple Alleles: The ABO Blood Groups

- Some genes, such as those determining ABO blood groups and Rh blood groups, exhibit codominance in which each allele has its own effect. (p. 260)

Patterns of Inheritance Can Be Deduced from Pedigrees

- Family pedigrees can be used to deduce the mode of inheritance of a trait through a family, such as hemophilia in European royal families. (p. 261)

Gene Disorders Can Be Due to Simple Alterations of Proteins

- Sickle cell anemia is caused by a single-nucleotide change in the gene coding for hemoglobin. (p. 262)

Some Defects May Soon Be Curable: Gene Therapy

- Theoretically, some heredity disorders should be able to be cured by inserting a healthy version of a damaged gene into an affected individual. Early attempts have met with varied results. (p. 263)

More Promising Vectors

- New vectors such as AAV may offer greater promise in gene therapy. (p. 264)

13.3 Genes are on chromosomes.

Chromosomes: The Vehicles of Mendelian Inheritance

- Mendelian traits assort independently because the genes determining them are located on chromosomes that assort independently. (p. 266)

Genetic Recombination

- Crossing over creates new genetic combinations. If two different genes are located relatively far apart on a chromosome, crossing over is more likely to occur somewhere between them than if they are located closer together. (p. 267)
- A genetic map can be constructed that measures the distance between genes in terms of the frequency of recombination. (p. 268)

Human Chromosomes

- The twenty-third pair of chromosomes in humans carries the genes that determine sex. The Y chromosome and appears to contain 78 genes. (pp. 270–271)

Human Abnormalities Due to Alterations in Chromosome Number

- Nondisjunction refers to the failure of homologues to separate properly during meiosis and can lead to aneuploidy (incorrect number of chromosomes), which can cause a number of genetic conditions, such as Down syndrome. (p. 272)

Genetic Counseling

- Many genetic abnormalities can be detected early in pregnancy using amniocentesis, chorionic villi sampling, and many newer techniques. (p. 274)

Self Test

1. In order to ensure that he had pure-breeding plants for his experiments, Mendel
   a. cross-fertilized each variety with each other.
   b. let each variety self-fertilize for several generations.
   c. removed the female parts of the plants.
   d. removed the male parts of the plants.
2. When two parents are crossed, the offspring are referred to as the
   a. recessives.
   b. testcross.
   c. F_1 generation.
   d. F_2 generation.
3. A cross between two individuals results in a ratio of 9:3:3:1 for four possible phenotypes. This is an example of a
   a. dihybrid cross.
   b. monohybrid cross.
   c. testcross.
   d. none of these.
4. Human height shows a continuous variation from the very short to the very tall. Height is most likely controlled by
   a. epistatic genes.
   b. environmental factors.
   c. sex-linked genes.
   d. multiple genes.
5. In the human ABO blood grouping, the four basic blood types are type A, type B, type AB, and type O. The blood proteins A and B are
   a. simple dominant and recessive traits.
   b. incomplete dominant traits.
   c. codominant traits.
   d. sex-linked traits.
6. Which of the following describes symptoms of sickle cell anemia?
   a. poor blood circulation due to abnormal hemoglobin molecules
   b. sterility in females
   c. failure of blood to clot
   d. failure of chloride ion transport mechanism
7. What finding finally determined that genes were carried on chromosomes?
   a. heat sensitivity of certain enzymes that determined coat color
   b. sex-linked eye color in fruit flies
   c. the finding of complete dominance
   d. establishing pedigrees
8. A genetic map can be used to determine
   a. the relative position of alleles on chromosomes.
   b. restriction sites on chromosomes.
   c. the frequency of recombination between two genes.
   d. all of these.
9. A Barr body is a(n)
   a. result of primary nondisjunction.
   b. inactivated Y chromosome.
   c. gene that plays a key role in male development.
   d. inactivated X chromosome.
10. Down syndrome in humans is due to
    a. three copies of chromosome 21.
    b. monosomy.
    c. two Y chromosomes.
    d. three X chromosomes.

Test Your Visual Understanding

F_2 generation

9/16   round, yellow

3/16   round, green

3/16   wrinkled, yellow

1/16   wrinkled, green

1. The figure shows the results of a dihybrid cross in which Mendel was examining the inheritance of two traits: seed shape (R and r) and seed color (Y and y). Consider the results if he had examined three traits. Using these two traits and plant height (T and t), predict:
   a. the genotypes and phenotypes of the parents.
   b. the genotypic and phenotypic ratios of the F_1 generation.
   c. the genotypes of the eggs and sperm.
   d. the phenotypic ratios of the F_2 generation.

Apply Your Knowledge

1. Phenylketonuria (PKU) is a genetic disorder caused by the mutation of an enzyme, phenylalanine hydroxylase, that breaks down phenylalanine in the cell. If identified at birth, dietary restrictions can control the disease. If not identified, buildup of phenylalanine interferes with brain development. The frequency of PKU in the population is 1 in 12,000 births. How many people with PKU would you expect to find in a population of 250,000?
2. How might Mendel's results and the model he formulated have been different if the traits he chose to study were governed by alleles exhibiting incomplete dominance or codominance?

Mendelian Genetics Problems

1. The following illustration describes Mendel's cross of *wrinkled* and *round* seed characters. What is wrong with this diagram? (*Hint:* Do you expect all the seeds in a pod to be the same?)

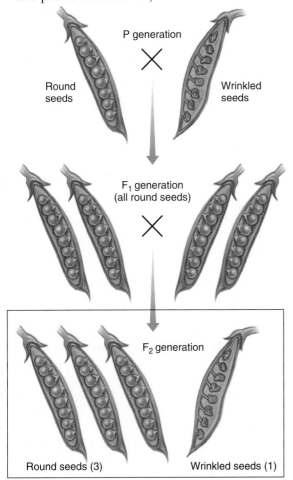

P generation
X

Round seeds

Wrinkled seeds

F₁ generation (all round seeds)
X

F₂ generation

Round seeds (3)   Wrinkled seeds (1)

2. The annual plant *Haplopappus gracilis* has two pairs of chromosomes (1 and 2). In this species, the probability that two characters, *a* and *b*, selected at random will be on the same chromosome is equal to the probability that they will both be on chromosome 1 ($\frac{1}{2} \times \frac{1}{2} = \frac{1}{4}$, or 0.25), plus the probability that they will both be on chromosome 2 (also $\frac{1}{2} \times \frac{1}{2} = \frac{1}{4}$, or 0.25), for an overall probability of $\frac{1}{2}$, or 0.5. In general, the probability that two randomly selected characters will be on the same chromosome is equal to $\frac{1}{n}$ where *n* is the number of chromosome pairs. Humans have 23 pairs of chromosomes. What is the probability that any two human characters selected at random will be on the same chromosome?

3. Among Hereford cattle, there is a dominant allele called *polled*; the individuals having this allele lack horns. Suppose you acquire a herd consisting entirely of polled cattle, and you carefully determine that no cow in the herd has horns. Some of the calves born that year, however, grow horns. You remove those calves from the herd and make certain that no horned adult has gotten into your pasture. Despite your efforts, more horned calves are born the next year. What is the reason for the appearance of the horned calves? If your goal is to maintain a herd consisting entirely of polled cattle, what should you do?

4. An inherited trait among humans in Norway causes affected individuals to have very wavy hair, not unlike that of a sheep. The trait, called *woolly*, is very evident when it occurs in families; no child possesses woolly hair unless at least one parent does. Imagine that you are a Norwegian judge, and a woolly-haired man comes before you to sue his normal-haired wife for divorce because their first child has woolly hair but their second child has normal hair. The husband claims this constitutes evidence of his wife's infidelity. Do you accept his claim? Justify your decision.

5. In human beings, Down syndrome, a serious developmental abnormality, results from the presence of three copies of chromosome 21 rather than the usual two copies. If a female exhibiting Down syndrome mates with a normal male, what proportion of her offspring would you expect to be affected?

6. Many animals and plants bear recessive alleles for *albinism*, a condition in which homozygous individuals lack certain pigments. An albino plant, for example, lacks chlorophyll and is white, and an albino human lacks melanin. If two normally pigmented persons heterozygous for the same albinism allele marry, what proportion of their children would you expect to be albino?

7. You inherit a racehorse and decide to put him out to stud. In looking over the studbook, however, you discover that the horse's grandfather exhibited a rare disorder that causes brittle bones. The disorder is hereditary and results from homozygosity for a recessive allele. If your horse is heterozygous for the allele, it will not be possible to use him for stud because the genetic defect may be passed on. How would you determine whether your horse carries this allele?

8. In the fruit fly *Drosophila*, the allele for dumpy wings (*d*) is recessive to the normal long-wing allele (*d⁺*), and the allele for white eye (*w*) is recessive to the normal red-eye allele (*w⁺*). In a cross of $d^+d^+w^+w \times d^+dww$, what proportion of the offspring are expected to be "normal" (long wings, red eyes)? What proportion are expected to have dumpy wings and white eyes?

9. Your instructor presents you with a *Drosophila* with red eyes, as well as a stock of white-eyed flies and another stock of flies homozygous for the red-eye allele. You know that the presence of white eyes in *Drosophila* is caused by homozygosity for a recessive allele. How would you determine whether the single red-eyed fly was heterozygous for the white-eye allele?

10. Some children are born with recessive traits (and, therefore, must be homozygous for the recessive allele specifying the trait), even though neither of the parents exhibits the trait. What can account for this?

11. You collect two individuals of *Drosophila*, one a young male and the other a young, unmated female. Both are normal in appearance, with the red eyes typical of *Drosophila*. You keep the two flies in the same bottle, where they mate. Two weeks later, the offspring they have produced all have red eyes. From among the offspring, you select 100 individuals, some male and some female. You cross each individually with a fly you know to be homozygous for the recessive allele *sepia*, which produces black eyes when homozygous. Examining the results of your 100 crosses, you observe that in about half of the crosses, only red-eyed flies were produced. In the other half, however, the progeny of each cross consists of about 50% red-eyed flies and 50% black-eyed flies. What were the genotypes of your original two flies?

12. Hemophilia is a recessive sex-linked human blood disease that leads to failure of blood to clot normally. One form of hemophilia has been traced to the royal family of England, from which it spread throughout the royal families of Europe. For the purposes of this problem, assume that it originated as a mutation either in Prince Albert or in his wife, Queen Victoria.

   a. Prince Albert did not have hemophilia. If the disease is a sex-linked recessive abnormality, how could it have originated in Prince Albert, a male, who would have been expected to exhibit sex-linked recessive traits?

   b. Alexis, the son of Czar Nicholas II of Russia and Empress Alexandra (a granddaughter of Victoria), had hemophilia, but their daughter Anastasia did not. Anastasia died, a victim of the Russian revolution, before she had any children. Can we assume that Anastasia would have been a carrier of the disease? Would your answer be different if the disease had been present in Nicholas II or in Alexandra?

13. In 1986, *National Geographic* magazine conducted a survey of its readers' abilities to detect odors. About 7% of Caucasians in the United States could not smell the odor of musk. If neither parent could smell musk, none of their children were able to smell it. On the other hand, if the two parents could smell musk, their children generally could smell it too, but a few of the children in those families were unable to smell it. Assuming that a single pair of alleles governs this trait, is the ability to smell musk best explained as an example of dominant or recessive inheritance?

14. A couple with a newborn baby is troubled that the child does not resemble either of them. Suspecting that a mixup occurred at the hospital, they check the blood type of the infant. It is type O. Because the father is type A and the mother type B, they conclude that a mixup must have occurred. Are they correct?

15. Mabel's sister died of cystic fibrosis as a child. Mabel does not have the disease, and neither do her parents. Mabel is pregnant with her first child. If you were a genetic counselor, what would you tell her about the probability that her child will have cystic fibrosis?

16. How many chromosomes would you expect to find in the karyotype of a person with Turner syndrome?

17. A woman is married for the second time. Her first husband has blood type A, and her child by that marriage has type O. Her new husband has type B blood, and when they have a child, its blood type is AB. What is the woman's blood genotype and blood type?

18. Two intensely freckled parents have five children. Three eventually become intensely freckled, and two do not. Assuming this trait is governed by a single pair of alleles, is the expression of intense freckles best explained as an example of dominant or recessive inheritance?

19. Total color blindness is a rare hereditary disorder among humans. Affected individuals can see no colors, only shades of gray. It occurs in individuals homozygous for a recessive allele, and it is not sex-linked. A man whose father is totally color blind intends to marry a woman whose mother is totally color blind. What are the chances they will produce offspring who are totally color blind?

20. A normally pigmented man marries an albino woman. They have three children, one of whom is an albino. What is the genotype of the father?

21. Four babies are born in a hospital, and each has a different blood type: A, B, AB, and O. The parents of these babies have the following pairs of blood groups: A and B, O and O, AB and O, and B and B. Which baby belongs to which parents?

22. A couple both work in an atomic energy plant, and both are exposed daily to low-level background radiation. After several years, they have a child who has Duchenne muscular dystrophy, a recessive genetic defect caused by a mutation on the X chromosome. Neither the parents nor the grandparents have the disease. The couple sues the plant, claiming that the abnormality in their child is the direct result of radiation-induced mutation of their gametes, and that the company should have protected them from this radiation. Before reaching a decision, the judge hearing the case insists on knowing the sex of the child. Which sex would be more likely to result in an award of damages, and why?

14

DNA: The Genetic Material

Concept Outline

14.1 What is the genetic material?

The Hammerling Experiment: Cells Store Hereditary Information in the Nucleus.

Transplantation Experiments: Each Cell Contains a Full Set of Genetic Instructions.

The Griffith Experiment: Hereditary Information Can Pass Between Organisms.

The Avery and Hershey–Chase Experiments: The Active Principle Is DNA.

14.2 What is the structure of DNA?

The Chemical Nature of Nucleic Acids. Nucleic acids are polymers containing four nucleotides.

The Three–Dimensional Structure of DNA. Watson and Crick concluded that the DNA molecule is a double helix, with two antiparallel strands held together by basepairing.

14.3 How does DNA replicate?

The Meselson–Stahl Experiment: DNA Replication Is Semiconservative.

The Replication Process. DNA is replicated by the enzyme DNA polymerase III, working in concert with many other proteins. DNA replicates by assembling a complementary copy of each strand semidiscontinuously.

Eukaryotic DNA Replication. Eukaryotic chromosomes consist of many origins of replication.

14.4 What is a gene?

The One-Gene/One-Polypeptide Hypothesis. A gene encodes all the information needed to express a functional protein.

How DNA Encodes Protein Structure. The nucleotide sequence of a gene dictates the amino acid sequence of the protein it encodes.

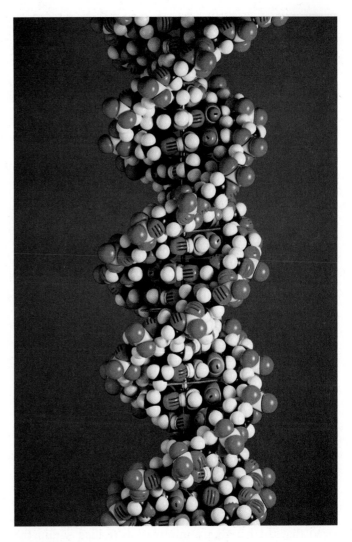

FIGURE 14.1
DNA. The hereditary blueprint in each cell of all living organisms is a very long, slender molecule called deoxyribonucleic acid (DNA).

The realization that patterns of heredity can be explained by the segregation of chromosomes in meiosis raised a question that occupied biologists for over 50 years: What is the exact nature of the connection between hereditary traits and chromosomes? This chapter describes the chain of experiments that have led to our current understanding of DNA and the molecular mechanisms of heredity (figure 14.1). These experiments are among the most elegant in science. Just as in a good detective story, each discovery has led to new questions. The intellectual path has not always been a straight one, the best questions not always obvious. But however erratic and lurching the course of the experimental journey, our picture of heredity has become progressively clearer, the image more sharply defined.

14.1 What is the genetic material?

The Hammerling Experiment: Cells Store Hereditary Information in the Nucleus

Perhaps the most basic question we can ask about hereditary information is where it is stored in the cell. To answer this question, Danish biologist Joachim Hammerling, working at the Max Planck Institute for Marine Biology in Berlin in the 1930s, cut cells into pieces and observed the pieces to see which were able to express hereditary information. For this experiment, Hammerling needed cells large enough to operate on conveniently and differentiated enough to distinguish the pieces. He chose the unicellular green alga *Acetabularia*, which grows up to 5 centimeters, as a model organism for his investigations. Just as Mendel used pea plants and Sturtevant used fruit flies, Hammerling picked a model organism that was suited to the specific experimental question he wanted to answer, assuming that what he learned could then be applied to other organisms.

Individuals of the genus *Acetabularia* have distinct foot, stalk, and cap regions; all are differentiated parts of a single cell. The nucleus is located in the foot. As a preliminary experiment, Hammerling amputated the caps of some cells and the feet of others. He found that when he amputated the cap, a new cap regenerated from the remaining portions of the cell (foot and stalk). When he amputated the foot, however, no new foot regenerated from the cap and stalk. Hammerling, therefore, hypothesized that the hereditary information resided within the foot of *Acetabularia*.

Surgery on Single Cells

To test his hypothesis, Hammerling selected individuals from two species of the genus *Acetabularia* in which the caps look very different from one another. *A. mediterranea* has a disk-shaped cap, and *A. crenulata* has a branched, flowerlike cap. Hammerling grafted a stalk from *A. crenulata* to a foot from *A. mediterranea* (figure 14.2). The cap that regenerated looked somewhat like the cap of *A. crenulata*, though not exactly the same.

Hammerling then cut off this regenerated cap and found that a disk-shaped cap exactly like that of *A. mediterranea* formed in the second regeneration and in every regeneration thereafter. This experiment supported Hammerling's hypothesis that the instructions specifying the kind of cap are stored in the foot of the cell, and that these instructions must pass from the foot through the stalk to the cap.

In this experiment, the initial flower-shaped cap was somewhat intermediate in shape, unlike the disk-shaped

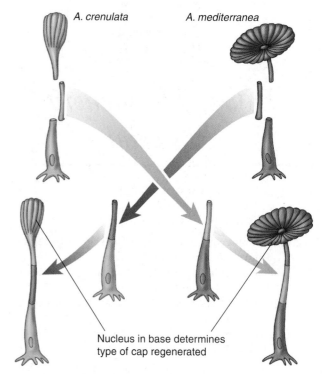

FIGURE 14.2
Hammerling's *Acetabularia* reciprocal graft experiment. Hammerling grafted a stalk of each species of *Acetabularia* onto the foot of the other species. In each case, the cap that eventually developed was dictated by the nucleus-containing foot rather than by the stalk.

caps of subsequent generations. Hammerling speculated that this initial cap, which resembled that of *A. crenulata*, was formed from instructions already present in the transplanted stalk when it was excised from the original *A. crenulata* cell. In contrast, all of the caps that subsequently regenerated used new information derived from the foot of the *A. mediterranea* cell the stalk had been grafted onto. In some unknown way, the original instructions that had been present in the stalk were eventually "used up." We now understand that genetic instructions (in the form of messenger RNA, discussed in chapter 15) pass from the nucleus in the foot upward *through the stalk* to the developing cap.

Hereditary information in *Acetabularia* is stored in the foot of the cell, where the nucleus resides.

Transplantation Experiments: Each Cell Contains a Full Set of Genetic Instructions

Because the nucleus is in the foot of *Acetabularia*, Hammerling's experiments suggested that the nucleus is the repository of hereditary information in a cell. A direct test of this hypothesis was carried out in 1952 by the American embryologists Robert Briggs and Thomas King. Using a glass pipette drawn to a fine tip and working with a microscope, Briggs and King removed the nucleus from a frog egg. Without the nucleus, the egg did not develop. However, when they replaced the nucleus with one removed from a more advanced frog embryo cell, the egg developed into an adult frog. Clearly, the nucleus was directing the egg's development (figure 14.3).

Successfully Transplanting Nuclei

Can every nucleus in an organism direct the development of an entire adult individual? The experiment of Briggs and King did not answer this question definitively, because the nuclei they transplanted from frog embryos into eggs often caused the eggs to develop abnormally. Two experiments performed soon afterward gave a clearer answer to the question. In the first, John Gurdon, working with another species of frog at Oxford and Yale, transplanted nuclei from tadpole cells into eggs from which the nuclei had been removed. The experiments were difficult—it was necessary to synchronize the division cycles of donor and host. However, in many experiments, the eggs went on to develop normally, indicating that the nuclei of cells in later stages of development retain the genetic information necessary to direct the development of all other cells in an individual.

Totipotency in Plants

In the second experiment, F. C. Steward at Cornell University in 1958 placed small fragments of fully developed carrot tissue (isolated from a part of the vascular system called the phloem) in a flask containing liquid growth medium. Steward observed that when individual cells broke away from the fragments, they often divided and developed into multicellular roots. When he immobilized the roots by placing them in a solid growth medium, they went on to develop normally into entire, mature plants. Steward's experiment makes it clear that, even in adult tissues, the nuclei of individual plant cells are "totipotent," meaning that each contains a full set of hereditary instructions and can generate an entire adult individual. As you will learn in chapter 19, animal cells, like plant cells, can be totipotent, and a single adult animal cell can generate an entire adult animal.

Hereditary information is stored in the nucleus of eukaryotic cells. Each nucleus in any eukaryotic cell contains a full set of genetic instructions.

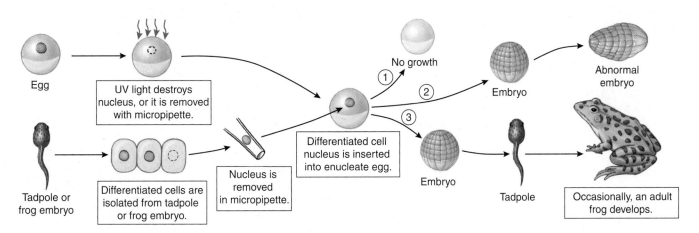

FIGURE 14.3
Nuclear transplant experiments. The nucleus was removed from a frog egg, either by sucking the egg nucleus into a micropipette or, more simply, by destroying it with ultraviolet light. A nucleus obtained from a differentiated cell from a frog embryo or tadpole was then injected into this enucleate egg. The hybrid egg was allowed to develop. One of three results was obtained: (1) No growth occurred, perhaps reflecting damage to the egg cell during the nuclear transplant operation; (2) normal growth and development occurred up to an early embryo stage, but subsequent development was not normal and the embryo did not survive; and (3) normal growth and development occurred, eventually leading to the development of an adult frog. That frog was of the strain that contributed the nucleus and not of the strain that contributed the egg. Only a few experiments gave this third result, but they clearly establish that the nucleus directs frog development.

The Griffith Experiment: Hereditary Information Can Pass Between Organisms

The identification of the nucleus as the repository of hereditary information focused attention on the chromosomes, which were already suspected to be the vehicles of Mendelian inheritance. Specifically, biologists wondered how the **genes,** the units of hereditary information studied by Mendel, were actually arranged in the chromosomes. They knew that chromosomes contained both protein and deoxyribonucleic acid (DNA). Which of these held the genes? Starting in the late 1920s and continuing for about 30 years, a series of investigations addressed this question.

In 1928, the British microbiologist Frederick Griffith made a series of unexpected observations while experimenting with pathogenic (disease-causing) bacteria. When he infected mice with a virulent strain of *Streptococcus pneumoniae* bacteria (then known as *Pneumococcus*), the mice died of blood poisoning. However, when he infected similar mice with a mutant strain of *S. pneumoniae* that lacked the virulent strain's polysaccharide coat, the mice showed no ill effects. The coat was apparently necessary for virulence. The normal pathogenic form of this bacterium is referred to as the S form because it forms smooth colonies on a culture dish. The mutant form, which lacks an enzyme needed to manufacture the polysaccharide coat, is called the R form because it forms rough colonies.

To determine whether the polysaccharide coat itself had a toxic effect, Griffith injected heat-killed bacteria of the virulent S strain into mice; the mice remained perfectly healthy. As a control, he injected mice with a mixture containing heat-killed S bacteria of the virulent strain and live coatless R bacteria, each of which by itself did not harm the mice (figure 14.4). Unexpectedly, the mice developed disease symptoms, and many of them died. The blood of the dead mice was found to contain high levels of live, virulent *Streptococcus* type S bacteria, which had surface proteins characteristic of the live (previously R) strain. Somehow, the information specifying the polysaccharide coat had passed from the dead, virulent S bacteria to the live, coatless R bacteria in the mixture, permanently transforming the coatless R bacteria into the virulent S variety. This transfer of genetic material from one cell to another, called **transformation,** can alter the genetic makeup of the recipient cell.

Hereditary information can pass from dead cells to living ones, transforming them.

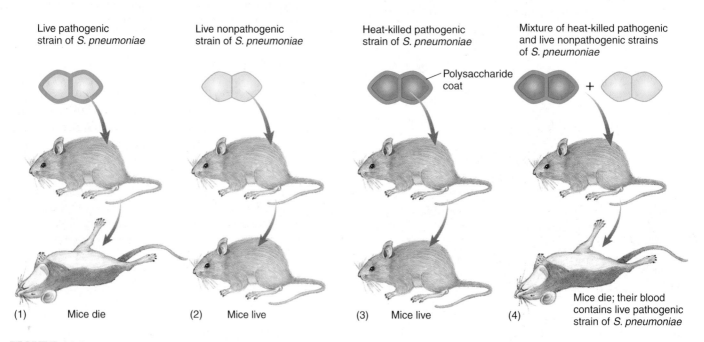

Live pathogenic strain of *S. pneumoniae*

Live nonpathogenic strain of *S. pneumoniae*

Heat-killed pathogenic strain of *S. pneumoniae*

Mixture of heat-killed pathogenic and live nonpathogenic strains of *S. pneumoniae*

Polysaccharide coat

(1) Mice die (2) Mice live (3) Mice live (4) Mice die; their blood contains live pathogenic strain of *S. pneumoniae*

FIGURE 14.4

Griffith's discovery of transformation. (*1*) The pathogenic strain of the bacterium *Streptococcus pneumoniae* kills many of the mice it is injected into. The bacterial cells are covered with a polysaccharide coat, which the bacteria themselves synthesize. (*2*) Interestingly, an injection of live, coatless bacteria produced no ill effects. However, the coat itself is not the agent of disease. (*3*) When Griffith injected mice with dead bacteria that possessed polysaccharide coats, the mice were unharmed. (*4*) But when Griffith injected a mixture of dead bacteria with polysaccharide coats and live bacteria without such coats, many of the mice died, and virulent bacteria with coats were recovered. Griffith concluded that the live cells had been "transformed" by the dead ones; that is, genetic information specifying the polysaccharide coat had passed from the dead cells to the living ones.

The Avery and Hershey–Chase Experiments: The Active Principle Is DNA

The Avery Experiments

The agent responsible for transforming *Streptococcus* went undiscovered until 1944. In a classic series of experiments, Oswald Avery and his co-workers Colin MacLeod and Maclyn McCarty characterized what they referred to as the "transforming principle" from Griffith's experiment. They first prepared the mixture of dead S *Streptococcus* and live R *Streptococcus* that Griffith had used. Then Avery and his colleagues removed as much of the protein as they could from their preparation, eventually achieving 99.98% purity. Despite the removal of nearly all protein, the transforming activity was not reduced. Moreover, the properties of the transforming principle resembled those of DNA in several ways:

1. When the purified principle was analyzed chemically, the array of elements agreed closely with DNA.
2. When spun at high speeds in an ultracentrifuge, the transforming principle migrated to the same level (density) as DNA.
3. Extracting the lipid and protein from the purified transforming principle did not reduce its activity.
4. Protein-digesting enzymes did not affect the principle's activity, nor did RNA-digesting enzymes.
5. The DNA-digesting enzyme DNase destroyed all transforming activity.

The evidence was overwhelming. The researchers concluded that "a nucleic acid of the deoxyribose type is the fundamental unit of the transforming principle of *Pneumococcus* Type III"—in essence, that DNA is the hereditary material for this bacterial species.

The Hershey–Chase Experiment

Avery's results were not widely accepted at first, because many biologists preferred to believe that proteins were the repository of hereditary information. Additional evidence supporting Avery's conclusion was provided in 1952 by Alfred Hershey and Martha Chase, who experimented with **bacteriophages,** viruses that attack bacteria. Viruses, described in more detail in chapter 33, consist of either DNA or RNA (ribonucleic acid) surrounded by a protein coat. When a *lytic* (potentially cell-rupturing) bacteriophage infects a bacterial cell, it first binds to the cell's outer surface and then injects its hereditary information into the cell. There, the hereditary information directs the production of thousands of new viruses within the bacterium. The bacterial cell eventually ruptures, or *lyses*, releasing the newly made viruses.

To identify the hereditary material injected into bacterial cells at the start of an infection, Hershey and Chase used the bacteriophage T2, which contains DNA rather than RNA. They labeled the two parts of the viruses—the DNA and the

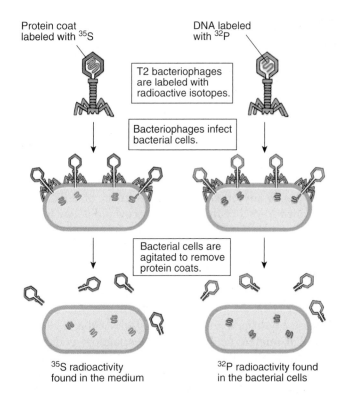

Protein coat labeled with ^{35}S

DNA labeled with ^{32}P

T2 bacteriophages are labeled with radioactive isotopes.

Bacteriophages infect bacterial cells.

Bacterial cells are agitated to remove protein coats.

^{35}S radioactivity found in the medium

^{32}P radioactivity found in the bacterial cells

FIGURE 14.5

The Hershey–Chase experiment. Hershey and Chase found that ^{35}S radioactivity did not enter infected bacterial cells and ^{32}P radioactivity did. They concluded that viral DNA, not protein, was responsible for directing the production of new viruses.

protein coat—with different radioactive isotopes that would serve as tracers. In some experiments, the viruses were grown on a medium containing an isotope of phosphorus, ^{32}P, and the isotope was incorporated into the phosphate groups of newly synthesized DNA molecules. Because phosphorus is not found in protein, this label is specific for DNA. In other experiments, the viruses were grown on a medium containing ^{35}S, an isotope of sulfur, which is incorporated into the amino acids of newly synthesized protein coats. Because sulfur is not found in DNA, this label is specific for protein. The ^{32}P and ^{35}S isotopes are easily distinguished from each other because they emit particles with different energies when they decay.

After the labeled viruses were permitted to infect bacteria, the bacterial cells were agitated violently to remove the protein coats of the infecting viruses from the surfaces of the bacteria. This procedure removed nearly all of the ^{35}S label (and thus nearly all of the viral protein) from the bacteria. However, the ^{32}P label (and thus the viral DNA) had transferred to the interior of the bacteria (figure 14.5) and was found in viruses subsequently released from the infected bacteria. Hence, the hereditary information injected into the bacteria that specified the new generation of viruses was DNA, not protein.

Avery's experiments demonstrate conclusively that DNA is Griffith's transforming material. The hereditary material of bacteriophages is DNA and not protein.

14.2 What is the structure of DNA?

The Chemical Nature of Nucleic Acids

A German chemist, Friedrich Miescher, discovered DNA in 1869, only four years after Mendel's work was published. Miescher extracted a white substance from the nuclei of human cells and fish sperm. The proportion of nitrogen and phosphorus in the substance was different from that in any other known constituent of cells, which convinced Miescher that he had discovered a new biological substance. He called this substance "nuclein," because it seemed to be specifically associated with the nucleus.

Levene's Analysis: DNA Is a Polymer

Because Miescher's nuclein was slightly acidic, it came to be called **nucleic acid.** For 50 years, biologists did little research on the substance because nothing was known of its function in cells. In the 1920s, the basic structure of nucleic acids was determined by the biochemist P. A. Levene, who found that a DNA molecule contains three main components (figure 14.6): (1) a five-carbon sugar; (2) a phosphate (PO_4) group; (3) a nitrogen-containing (nitrogenous) base. The base may be a **purine** (adenine, symbolized as A, or guanine, G) or a **pyrimidine** (thymine, T, or cytosine, C); RNA contains uracil, (U) instead of thymine. From the roughly equal proportions of these components, Levene concluded correctly that DNA and RNA molecules are made of repeating units of the three components. Each unit, consisting of a sugar attached to a phosphate group and a base, is called a **nucleotide.** The identity of the base distinguishes one nucleotide from another.

To identify the various chemical groups in DNA and RNA, it is customary to number the carbon atoms of the base and the sugar and then refer to any chemical group attached to a carbon atom by that number. In the sugar, four of the carbon atoms together with an oxygen atom form a five-membered ring. As illustrated in figure 14.7, the carbon atoms are numbered 1′ to 5′, proceeding clockwise from the oxygen atom; the prime symbol (′) indicates that the number refers to a carbon in a sugar rather than a base. Under this numbering scheme, the phosphate group is attached to the 5′ carbon atom of the sugar, and the base is attached to the 1′ carbon atom. In addition, a free hydroxyl (—OH) group is attached to the 3′ carbon atom.

The 5′ phosphate and 3′ hydroxyl groups allow DNA and RNA to form long chains of nucleotides, because these two groups can react chemically with each other. The reaction between the phosphate group of one nucleotide and the hydroxyl group of another is a dehydration synthesis, eliminating a water molecule and forming a covalent bond

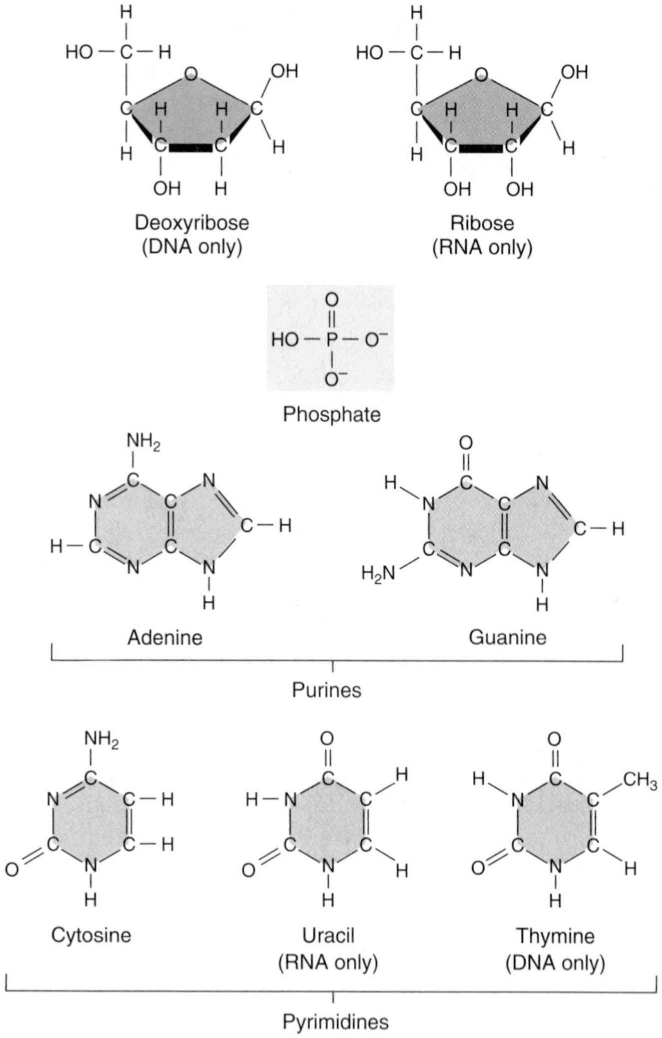

FIGURE 14.6
Nucleotide subunits of DNA and RNA. The nucleotide subunits of DNA and RNA are composed of three elements: (*top*) a five-carbon sugar (deoxyribose in DNA and ribose in RNA); (*middle*) a phosphate group; and (*bottom*) a nitrogenous base (either a purine or a pyrimidine).

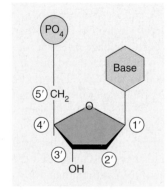

FIGURE 14.7
Numbering the carbon atoms in a nucleotide. The carbon atoms in the sugar of the nucleotide are numbered 1′ to 5′, proceeding clockwise from the oxygen atom. The "prime" symbol (′) indicates that the carbon belongs to the sugar rather than to the base.

Table 14.1 Chargaff's Analysis of DNA Nucleotide Base Compositions

Organism	Base Composition (mole percent)			
	A	T	G	C
Escherichia coli (K12)	26.0	23.9	24.9	25.2
Mycobacterium tuberculosis	15.1	14.6	34.9	35.4
Yeast	31.3	32.9	18.7	17.1
Herring	27.8	27.5	22.2	22.6
Rat	28.6	28.4	21.4	21.5
Human	30.9	29.4	19.9	19.8

Source: Data from E. Chargaff and J. Davidson (editors), *The Nucleic Acids*, 1955, Academic Press, New York, NY.

that links the two groups. The linkage is called a **phosphodiester bond** because the phosphate group is now linked to the two sugars by means of a pair of ester (P—O—C) bonds (figure 14.8). The two-unit polymer resulting from this reaction still has a free 5′ phosphate group at one end and a free 3′ hydroxyl group at the other, so it can link to other nucleotides. In this way, many thousands of nucleotides can join together in long chains.

Linear strands of DNA or RNA, no matter how long, will almost always have a free 5′ phosphate group at one end and a free 3′ hydroxyl group at the other. Therefore, every DNA and RNA molecule has an intrinsic directionality, and we can refer unambiguously to each end of the molecule. By convention, the sequence of bases is usually expressed in the 5′ to 3′ direction. Thus, the base sequence GTCCAT refers to the sequence:

5′ pGpTpCpCpApT—OH 3′

where the phosphates are indicated by "p." Note that this is not the same molecule as that represented by the reverse sequence:

5′ pTpApCpCpTpG—OH 3′

Levene's early studies indicated that all four types of DNA nucleotides were present in roughly equal amounts. This result, which later proved to be erroneous, led to the mistaken idea that DNA was a simple polymer in which the four nucleotides merely repeated (for instance, GCAT. . . . GCAT. . . . GCAT. . . . GCAT. . . .). If the sequence never varied, it was difficult to see how DNA might contain the hereditary information; this is why Avery's conclusion that DNA is the transforming principle was not readily accepted at first. It seemed more plausible that DNA was simply a structural element of the chromosomes, with proteins playing the central genetic role.

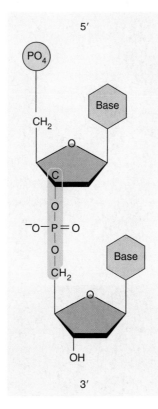

FIGURE 14.8
A phosphodiester bond.

Chargaff's Analysis: DNA Is Not a Simple Repeating Polymer

When Levene's chemical analysis of DNA was repeated using more sensitive techniques that became available after World War II, quite a different result was obtained. The four nucleotides were *not* present in equal proportions in DNA molecules after all. A careful study carried out by Erwin Chargaff showed that the nucleotide composition of DNA molecules varied in complex ways, depending on the source of the DNA (table 14.1). This strongly suggested that DNA was not a simple repeating polymer and might have the information-encoding properties genetic material requires. Despite DNA's complexity, however, Chargaff observed an important underlying regularity in double-stranded DNA: *The amount of adenine present in DNA always equals the amount of thymine, and the amount of guanine always equals the amount of cytosine.* These findings are commonly referred to as **Chargaff's rules:**

1. The proportion of A always equals that of T, and the proportion of G always equals that of C:

   A = T, and G = C

2. It follows that there is always an equal proportion of purines (A and G) and pyrimidines (C and T).

A single strand of DNA or RNA consists of a series of nucleotides joined together in a long chain. In all natural double-stranded DNA molecules, the proportion of A equals that of T, and the proportion of G equals that of C.

The Three-Dimensional Structure of DNA

As it became clear that DNA was the molecule that stored the hereditary information, investigators began to puzzle over how such a seemingly simple molecule could carry out such a complex function.

Franklin: X-ray Diffraction Patterns of DNA

The significance of the regularities pointed out by Chargaff were not immediately obvious, but they became clear when the British chemist Rosalind Franklin (figure 14.9a) carried out an X-ray diffraction analysis of DNA. In X-ray diffraction, a molecule is bombarded with a beam of X rays. When individual rays encounter atoms, their path is bent or diffracted, and the diffraction pattern is recorded on photographic film. The patterns resemble the ripples created by tossing a rock into a smooth lake (figure 14.9b). When carefully analyzed, they yield information about the three-dimensional structure of a molecule.

X-ray diffraction works best on substances that can be prepared as perfectly regular crystalline arrays. However, it was impossible to obtain true crystals of natural DNA at the time Franklin conducted her analysis, so she had to use DNA in the form of fibers. Franklin worked in the laboratory of the British biochemist Maurice Wilkins, who was able to prepare more uniformly oriented DNA fibers than anyone had previously. Using these fibers, Franklin succeeded in obtaining crude diffraction information on natural DNA. The diffraction patterns she obtained suggested that the DNA molecule had the shape of a helix, or corkscrew, with a diameter of about 2 nanometers and a complete helical turn every 3.4 nanometers (figure 14.9c).

Watson and Crick: A Model of the Double Helix

Learning informally of Franklin's results before they were published in 1953, James Watson and Francis Crick, two young investigators at Cambridge University, quickly worked out a likely structure for the DNA molecule (figure 14.10a), which we

(a)

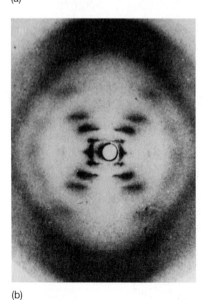
(b)

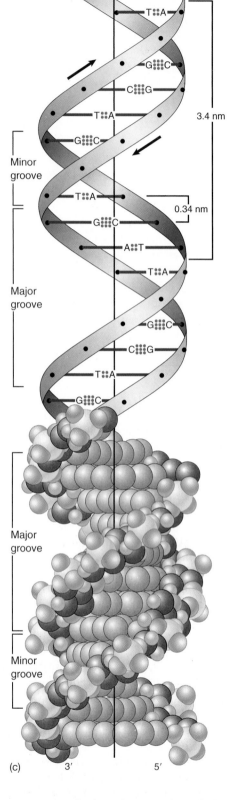
(c)

FIGURE 14.9
Rosalind Franklin's X-ray diffraction patterns. (*a*) Rosalind Franklin. (*b*) This telltale X-ray diffraction photograph of DNA fibers was made in 1953 by Rosalind Franklin in the laboratory of Maurice Wilkins. (*c*) The X-ray diffraction studies of Rosalind Franklin suggested the dimensions of the DNA double helix.

now know was substantially correct. The key to their model was Watson and Crick's understanding that each DNA molecule is actually made up of *two* chains of nucleotides that are intertwined—the double helix.

Backbone. As described in chapter 3, the two strands of the double helix are made up of long polymers of nucleotides. Each strand is made up of repeating sugar and phosphate units joined by phosphodiester bonds. These two strands are then wrapped about a common axis forming a double helix (see figure 14.9c). The helix is often compared to a spiral staircase, where the two strands of the double helix are the handrails on the staircase.

Complementarity. What holds the two strands together? Watson and Crick proposed that the bases can form particular sets of hydrogen bonds that lead to specific "base pairs." Thus, adenine (A) can form two hydrogen bonds with thymine (T) to form an A–T base-pair, and guanine (G) can form three hydrogen bonds with cytosine (C) to form a G–C base-pair (figure 14.10b). Note that this configuration also pairs a two-ringed purine with a single-ringed pyrimidine in each case, so that the diameter of each base-pair is the same. We refer to these combinations as *complementary*—for example, a strand that is ATGC would be complementary to a strand that is TACG. This seemingly simple concept has some profound implications. If we know the sequence of one strand, we automatically know the sequence of the other strand. So wherever there is an A in one strand, there must be a T in the other, and wherever there is a G in one strand, there must be a C in the other. This will become critical later when we look at how DNA is replicated and expressed.

The Watson–Crick model explained Chargaff's results: In a double helix, adenine forms two hydrogen bonds with thymine, but it will not form hydrogen bonds properly with cytosine. Similarly, guanine forms three hydrogen bonds with cytosine, but it will not form hydrogen bonds properly with thymine. Because of this base-pairing, adenine and thymine will always occur in the same proportions in any DNA molecule, as will guanine and cytosine.

Antiparallel Configuration. A single phosphodiester strand has an inherent polarity, meaning that one end terminates in a 3′ OH and the other end terminates in a 5′ PO_4. We thus refer to strands as having either a 5′ to 3′ or 3′ to 5′ polarity. There are obviously two ways we could put together two strands, with the polarity the same in each (parallel) or with the polarity opposite (antiparallel). Native double-stranded DNA always has the antiparallel configuration, with one strand running 5′ to 3′ and the other running 3′ to 5′ (figure 14.10b). This proves to have important implications for DNA replication.

The Watson–Crick DNA Molecule. To put together the entire structure of a DNA molecule as suggested by Watson and Crick, each DNA molecule is composed of two complementary phosphodiester strands that each form a helix with a common axis. These strands are antiparallel, with the bases extending into the interior of the helix. The bases from opposite strands form base-pairs with each other to join the two complementary strands (figures 14.9c and 14.10b). Although each individual base-pair is of low energy, the sum of many base-pairs has enough energy that the molecule is very stable. To return to our spiral staircase analogy, where the backbone is the handrails, the base-pairs are the stairs.

The DNA molecule is a double helix, the strands held together by base-pairing.

FIGURE 14.10
The DNA double helix.
(*a*) James Watson (*left*) and Francis Crick (*right*) deduced the structure of DNA in 1953 from Chargaff's rules and Franklin's diffraction studies. (*b*) In a DNA duplex molecule, only two base-pairs are possible: Adenine (A) can pair with thymine (T), and guanine (G) can pair with cytosine (C). An A–T base-pair has two hydrogen bonds, while a G–C base-pair has three.

(a)

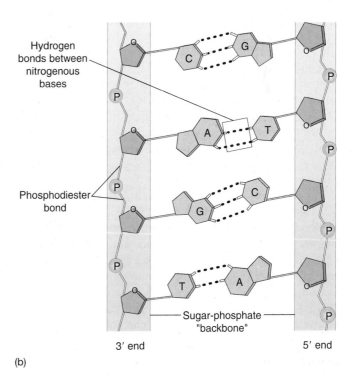

(b)

14.3 How does DNA replicate?

The Meselson–Stahl Experiment: DNA Replication Is Semiconservative

The Watson–Crick model immediately suggested that the basis for copying the genetic information is **complementarity.** One chain of the DNA molecule may have any conceivable base sequence, but this sequence completely determines the sequence of its partner in the duplex. For example, if the sequence of one chain is 5′-ATTGCAT-3′, the sequence of its partner *must* be 3′-TAACGTA-5′. Thus, each chain in the duplex is a complement of the other.

The complementarity of the DNA duplex provides a ready means of accurately duplicating the molecule. If we were to "unzip" the molecule, we would need only to assemble the appropriate complementary nucleotides on the exposed single strands to form two daughter duplexes with the same sequence. This form of DNA replication is called **semiconservative** because, although the sequence of the original duplex is conserved after one round of replication, the duplex itself is not. Instead, each strand of the duplex becomes part of another duplex.

Besides the semiconservative model, two other hypotheses of gene replication were also proposed. The *conservative* model stated that the parental double helix would remain intact and generate DNA copies consisting of entirely new molecules. The *dispersive* model predicted that parental DNA would become dispersed throughout the new copy so that each strand of all the daughter molecules would be a mixture of old and new DNA.

The three hypotheses of DNA replication were evaluated in 1958 by Matthew Meselson and Franklin Stahl of the California Institute of Technology. These two scientists grew bacteria in a medium containing the heavy isotope of nitrogen, ^{15}N, which became incorporated into the bases of the bacterial DNA. After several generations, the DNA of these bacteria was denser than that of bacteria grown in a medium containing the lighter isotope of nitrogen, ^{14}N. Meselson and Stahl then transferred the bacteria from the ^{15}N medium to the ^{14}N medium and collected the DNA at various intervals.

By dissolving the DNA they had collected in a heavy salt called cesium chloride and then spinning the solution at very high speeds in an ultracentrifuge, Meselson and Stahl were able to separate DNA strands of different densities. The enormous centrifugal forces generated by the ultracentrifuge caused the cesium ions to migrate toward the bottom of the centrifuge tube, creating a gradient of cesium concentration, and thus of density. Each DNA strand floats or sinks in the gradient until it reaches the position where its density exactly matches the density of

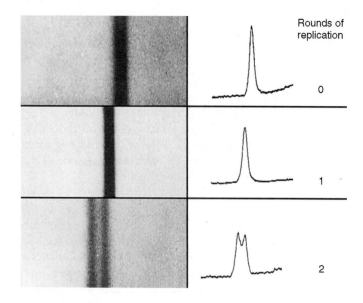

FIGURE 14.11
The key result of the Meselson–Stahl experiment. These bands of DNA, photographed on the left and scanned on the right, are from the density-gradient centrifugation experiment of Meselson and Stahl. At 0 generation, all DNA is heavy; after one replication, all DNA has a hybrid density; after two replications, all DNA is hybrid or light.

the cesium there. Because ^{15}N strands are denser than ^{14}N strands, they migrate farther down the tube to a denser region of the cesium gradient.

The DNA collected immediately after the transfer was all dense. However, after the bacteria completed their first round of DNA replication in the ^{14}N medium, the density of their DNA had decreased to a value intermediate between ^{14}N-DNA and ^{15}N-DNA. After the second round of replication, two density classes of DNA were observed, one intermediate and one equal to that of ^{14}N-DNA (figure 14.11).

Meselson and Stahl interpreted their results as follows: After the first round of replication, each daughter DNA duplex was a hybrid possessing one of the heavy strands of the parent molecule and one light strand; when this hybrid duplex replicated, it contributed one heavy strand to form another hybrid duplex and one light strand to form a light duplex (figure 14.12). Thus, this experiment clearly confirmed the prediction of the Watson–Crick model that DNA replicates in a semiconservative manner.

The basis for the great accuracy of DNA replication is complementarity. A DNA molecule is a duplex, containing two strands that are complementary mirror images of each other, so either one can be used as a template to reconstruct the other.

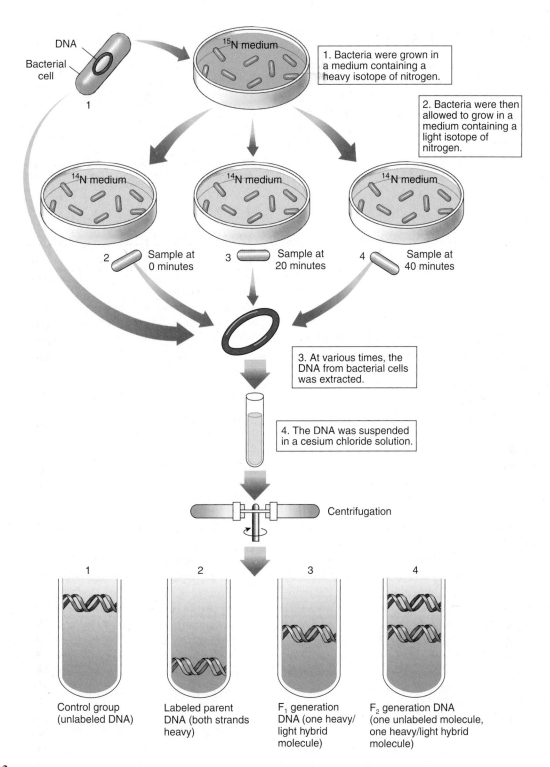

FIGURE 14.12
The Meselson–Stahl experiment: evidence demonstrating semiconservative replication. Bacterial cells were grown for several generations in a medium containing a heavy isotope of nitrogen (^{15}N) and then transferred to a new medium containing the normal lighter isotope (^{14}N). At various times thereafter, samples of the bacteria were collected, and their DNA was dissolved in a solution of cesium chloride, which was spun rapidly in a centrifuge. Because the cesium ion is so massive, it tends to settle toward the bottom of the spinning tube, establishing a gradient of cesium density. DNA molecules sink in the gradient until they reach a place where their density equals that of the cesium; they then "float" at that position. DNA containing ^{15}N is denser than that containing ^{14}N, so it sinks to a lower position in the cesium gradient. After one generation in ^{14}N medium, the bacteria yielded a single band of DNA with a density between that of ^{14}N-DNA and ^{15}N-DNA, indicating that only one strand of each duplex contained ^{15}N. After two generations in ^{14}N medium, two bands were obtained: one of intermediate density (in which one of the strands contained ^{15}N) and one of low density (in which neither strand contained ^{15}N). Meselson and Stahl concluded that replication of the DNA duplex involves building new molecules by separating strands and assembling new partners on each of these templates.

The Replication Process

To be effective, DNA replication must be fast and accurate. The process of DNA replication has been analyzed in great detail in *E. coli* and its viruses over the last 40 years. This has involved work by many researchers using all of the tools available to modern biologists: biochemistry, genetics, microscopy, and molecular biology. These researchers have built a detailed picture of the entire process, which we will present here in an overview.

Origin of Replication

Replication in *E. coli* begins at a specific site, the origin (called *OriC*), and ends at a specific site, the terminus. If you think about coordinating replication with cell division, it makes a lot of sense to have replication begin at a specific site to control the number of initiations and thus "rounds" of replication. The sequence of *OriC* consists of repeated nucleotides that bind a protein involved in initiation of replication, and an AT-rich sequence that can be opened easily during initiation of replication. (Remember, A–T base-pairs have only two hydrogen bonds compared with the three hydrogen bonds in G–C base-pairs.) After initiation, replication proceeds bidirectionally from this unique origin to the unique terminus (figure 14.13). The complete chromosome plus the origin can be thought of as a functional unit called a *replicon*.

Polymerases

The first polymerase, an enzyme that synthesizes nucleic acids, isolated in *E. coli* was called DNA polymerase I (pol I). It was assumed at first that this was the polymerase responsible for the bulk synthesis of DNA during replication. However, a mutant was isolated with no pol I activity that still showed replicative polymerase activity. Two additional polymerases were isolated from this strain of *E. coli* called DNA polymerase II (pol II) and DNA polymerase III (pol III). All three of these enzymes share a number of features, including the ability to synthesize polynucleotide strands. This activity itself has some features that were not expected. First, all known DNA polymerases require a **primer,** a short stretch of DNA or RNA nucleotides hydrogen-bonded to its complementary strand. Thus, DNA polymerases cannot begin synthesis of a new strand of DNA, but need a primer, an existing segment of nucleotides, on which to build. Second, these enzymes can only synthesize in one direction; they extend strands in a 5′ to 3′ direction by copying a template that is 3′ to 5′ (remember complementarity) (figure 14.14). The action of a DNA polymerase is to add new nucleotides to the 3′ OH of a primer. Interestingly, these enzymes also have the ability to remove nucleotides, and so are called "nucleases." Nucleases are classified as either **endonucleases** (cut DNA internally) or **exonucleases** (chew away at an end of DNA).

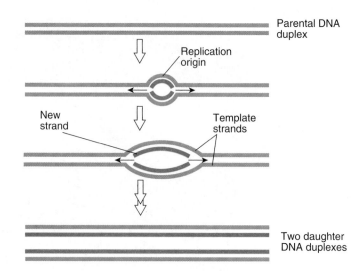

FIGURE 14.13
Origins of replication. At a site called the replication origin, the DNA duplex opens to create two separate strands, each of which can be used as a template for a new strand.

Each of these enzymes has a 3′ to 5′ exonuclease activity that is called a *proofreading function.* Mispaired bases can be removed by this activity, increasing the accuracy of the DNA replication process. The pol I enzyme also has a 5′ to 3′ exonuclease activity that is important in DNA replication, as we shall see later.

For many years, these three polymerases were thought to be the only DNA polymerases in *E. coli,* but recently several new polymerases have been identified. There are now five known polymerases, although all are not active in DNA replication. We will return to the roles of the different polymerases after considering another feature of replication.

Given the directionality of the polymerase enzymes and the antiparallel structure of DNA, it is obvious that replication cannot be the same on each strand. As the DNA double helix opens up, the strands will be antiparallel, so *both* cannot be replicated in a 5′ to 3′ direction as single continuous strands.

Leading and Lagging Strands

One strand, the one oriented in the 3′ to 5′ direction, can have a primer added complementary to its 3′ end, and then be extended 5′ to 3′ as one continuous strand. We call this the **leading strand.** The other strand must be replicated in short 5′ to 3′ stretches, filling in the open helix back to the end. As the helix is opened further, it can be replicated further, but the entire strand can only be replicated in this discontinuous fashion. We call this the **lagging strand.** The short stretches of newly synthesized DNA on the lagging strand have been named **Okazaki fragments** in honor of

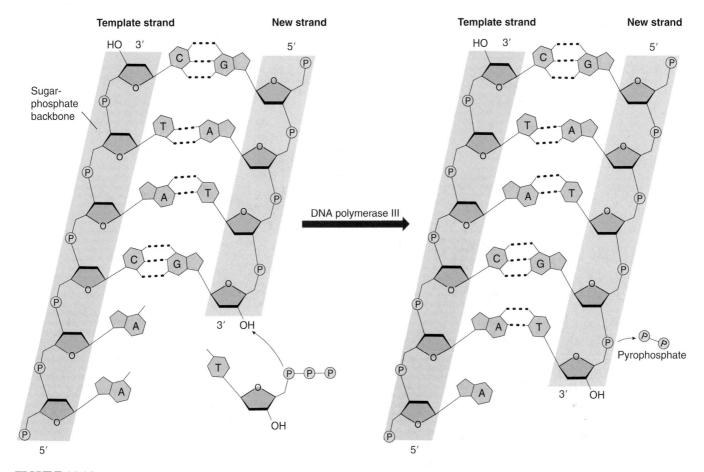

FIGURE 14.14
How nucleotides are added in DNA replication. DNA polymerase III, along with other enzymes, catalyzes the addition of nucleotides to the growing complementary strand of DNA. When a nucleotide is added, two of its phosphates are lost as pyrophosphate.

the man who first was able to detect them experimentally (figure 14.15).

Other Enzymatic Activities

A number of other enzymatic activities are needed to replicate DNA in addition to DNA polymerase. First, an enzyme is required to synthesize a primer to begin replication on the leading strand. There is also a need for continuous priming on the lagging strand. The enzyme responsible for these activities is *DNA primase*, which creates a short RNA primer complementary to a DNA template. It makes sense that the primer be RNA because RNA polymerases (enzymes that synthesize RNA molecules) do not require a primer to begin synthesis, so DNA primase is actually an RNA polymerase. Second, we need an enzyme to open up the helix in front of the polymerase. The enzyme responsible for this is *DNA helicase*, which can employ the energy from the hydrolysis of ATP to unwind DNA strands. Third, we need an enzyme that can remove the torsional strain introduced by opening the double helix. Think of opening the

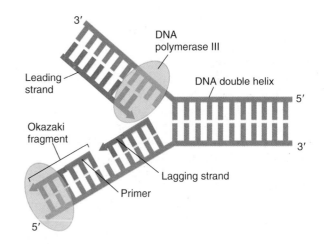

FIGURE 14.15
Synthesis of the leading and lagging strands.
DNA polymerase III synthesizes the leading strand as a continuous strand, but the lagging strand is synthesized in segments, called Okazaki fragments, each beginning with a primer.

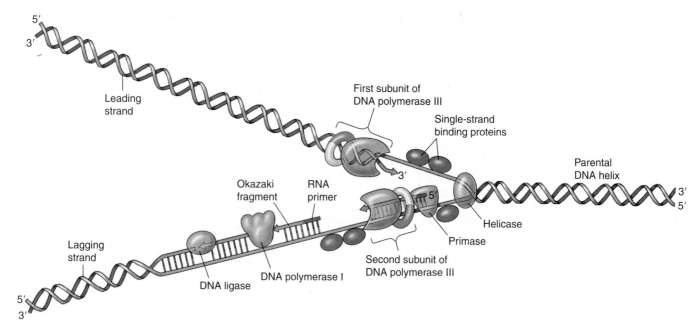

FIGURE 14.16

A DNA replication fork. Helicase enzymes separate the strands of the double helix, and single-stranded binding proteins stabilize the single-stranded regions. Replication occurs by two mechanisms. (1) *Continuous synthesis:* After primase adds a short RNA primer, DNA polymerase III adds nucleotides to the 3′ end of the leading strand. DNA polymerase I then replaces the RNA primer with DNA nucleotides. (2) *Discontinuous synthesis:* Primase adds a short RNA primer (*green*) ahead of the 5′ end of the lagging strand. DNA polymerase III then adds nucleotides to the primer until the gap is filled in. DNA polymerase I replaces the primer with DNA nucleotides, and DNA ligase attaches the short segment of nucleotides to the lagging strand.

double helix as unraveling a rope. As you pull apart the two strands of a rope, torsional strain causes the rope to coil up or form a knot. This strain must be relieved in order to open the DNA helix. The enzyme responsible for this is *DNA gyrase*, a form of topoisomerase, which is a group of enzymes that can alter the topological state of DNA. Fourth, the unwound single strands are not as stable in the aqueous environment of the cell as the double-stranded molecule, because the bases, which are highly hydrophobic, are exposed to the water. To solve this problem, as the DNA is unwound, the exposed single strands are covered by *single-strand binding protein* (ssb). Fifth, we need to remove the primers that are present at the beginning of the leading strand and in each Okazaki fragment on the lagging strand, and we need to join the fragments along the lagging strand. These two jobs are accomplished by DNA pol I and *DNA ligase*, respectively. DNA pol I removes primers with its 3′ to 5′ exonuclease activity and then replaces this with DNA, while DNA ligase creates a phosphodiester bond between adjacent Okazaki fragments. The various enzymes involved in replication are summarized in table 14.2.

The Replication Fork

How can we make sense of all these different activities for what had initially appeared to be a simple process? Instead of considering these many enzymes in isolation, we usually

think of the **replication fork,** the site of opening of the DNA strands where active replication occurs. All of the activities we have just discussed are found and are necessary for the replication fork. Figure 14.16 is meant to show schematically all the proteins in terms of function and how they are now thought to associate with each other. It is also important that all of the proteins involved act in a coordinated manner so that the whole process can proceed smoothly, an idea that will be developed further in the subsequent discussion.

Roles of the Polymerases

The three standard DNA polymerases are not all necessary for DNA replication during cell division. Pol III and pol I both have important roles to play, but pol II is thought to function primarily during DNA repair and not during replication. DNA pol I is made up of a single protein, and although it was the first identified polymerase, it is not the main polymerase at the replication fork. Rather, its role is to remove the repeated DNA primers that are necessary for DNA synthesis. The polymerase found at the replication fork is pol III, a large, multisubunit enzyme. The main catalytic subunit is called the α subunit, which along with two other proteins forms the catalytic core of the enzyme. While the core enzyme can synthesize DNA, it does not remain attached to its template for long stretches, a feature called *processivity*. The addition of the β subunit greatly increases the processivity of the core enzyme. The structure of the β subunit has been worked out, and it is a fascinating structure that has been called a sliding clamp. It literally forms a ring around the template strand, holding the catalytic core to the template (figure 14.17). The complete pol III enzyme is a large multiprotein complex made up of at least seven proteins.

The Replisome

Thus, DNA replication in *E. coli* is accomplished by a large, macromolecular assembly that is really a replication organelle, just as the ribosome is the protein synthetic organelle. We call this assembly the **replisome.** The replisome is a macromolecular protein machine that accomplishes the fast and accurate replication of DNA during cell division. The replisome has two main subcomponents, the *primosome* and a complex of two DNA pol III enzymes, one pol III for each strand. The primosome is composed of primase and helicase and a number of accessory proteins (see figure 14.16). The two pol III complexes include two synthetic core subunits, each with its own β subunit and a number of other proteins that hold the entire

Table 14.2 DNA Replication Enzymes of *E. coli*

Protein	Role	Size (kd)	Molecules per Cell
Helicase	Unwinds the double helix	300	20
Primase	Synthesizes RNA primers	60	50
Single-strand binding protein	Stabilizes single-stranded regions	74	300
DNA gyrase	Relieves torque	400	250
DNA polymerase III	Synthesizes DNA	≈900	20
DNA polymerase I	Erases primer and fills gaps	103	300
DNA ligase	Joins the ends of DNA segments; DNA repair	74	300

FIGURE 14.17
The DNA polymerase III complex. (*a*) The complex contains 10 kinds of protein chains. The protein is a dimer because both strands of the DNA duplex must be replicated simultaneously. The catalytic (α) subunits, the proofreading (ε) subunits, and the "sliding clamp" (β₂) subunits (*yellow and blue*) are labeled. (*b*) The "sliding clamp" units encircle the DNA template and (*c*) move it through the catalytic subunit like a rope drawn through a ring.

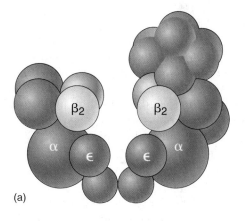

(a)

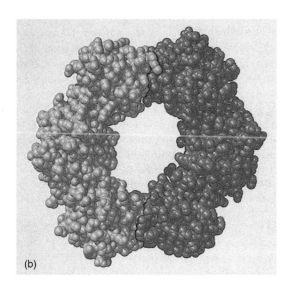

(b)

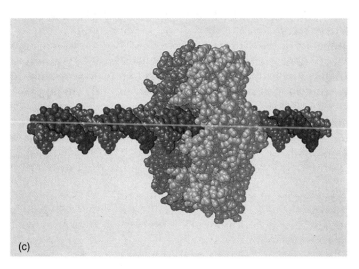

(c)

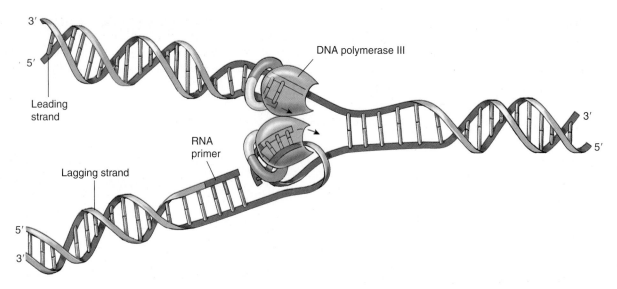

3′
5′

Leading
strand

DNA polymerase III

RNA
primer

Lagging strand

5′
3′

3′
5′

FIGURE 14.18
How DNA polymerase III works. This diagram presents a current view of how DNA polymerase III works. Note that the DNA on the lagging strand is folded to allow the dimeric DNA polymerase III molecule to replicate both strands of the parental DNA duplex simultaneously. This brings the 3′ end of each completed Okazaki fragment close to the start site for the next fragment.

complex together. The pol III form found in the replisome and at the replication fork is made up of over 14 individual proteins (see figure 14.17*a*). Even given the difficulties with lagging-strand synthesis, the two pol III enzymes in the replisome are active on both leading and lagging strands simultaneously. The need for constant priming on the lagging strand explains the need for the primosome complex as part of the entire replisome at the replication fork. How can the two strands be synthesized in the same direction? The model first proposed, still with us in some form, indicates that a loop forms in the lagging strand so that the polymerases can move in the same direction (figure 14.18).

Stages of Replication

Although a continuous process, replication can be logically broken down into stages, much as we earlier broke the continuous process of mitosis into stages.

Initiation.   Initiation always occurs at the same site, *OriC*. It is a complex process that involves first the recognition of specific sites within *OriC* by an initiator protein. This initiator protein is then joined by other proteins that are necessary only for initiation to open the helix at the AT-rich region within *OriC*. The primosome is then assembled onto the opened DNA strands, followed by the assembly of the entire replisome complex to complete the replication fork. Because replication is bidirectional from *OriC*, this is happening in both directions such that two replication forks proceed in opposite directions around the chromosome.

Elongation.   The majority of the time during replication is spent in elongation, during which the pol III molecules add successive new nucleotides based on complementarity to the template strand. This process is simple on the leading strand, because a single primer is all that is absolutely necessary (although multiple primings may occur). On the lagging strand, the process is considerably more complex. As the helix is opened, a new primer is synthesized, and pol III must release the template it has completed and begin with the "new" template. As the complex proceeds, primers are removed by pol I with its 5′ to 3′ exonuclease activity, and the strands are sealed by DNA ligase. Synthesis proceeds bidirectionally around the circular *E. coli* chromosome, creating two new DNA molecules.

Termination.   The exact details of the termination process are not clear, but it is known that a specific termination site is located roughly opposite *OriC* on the circular chromosome. There is also a need to keep the two new molecules from becoming intertwined by the replication process. This requires a topoisomerase enzyme such as DNA gyrase.

> **DNA replication in *E. coli* involves many different proteins that open and unwind the DNA double helix, stabilize the single strands, synthesize RNA primers, assemble new complementary strands on each exposed parental strand (one of them discontinuously), remove the RNA primer, and join new discontinuous segments on the lagging strand.**

Eukaryotic DNA Replication

The biggest difference between eukaryotic and prokaryotic replication is the sheer amount of DNA and how it is packaged (figure 14.19). Eukaryotes usually have multiple chromosomes that are each larger than the *E. coli* chromosome. Thus, the machinery could in principle be the same; however, if there were only a single unique origin for each chromosome, the length of the time necessary for replication would be prohibitive. This problem is solved by the use of multiple origins of replication for each chromosome, resulting in multiple *replicons*, which are sections of DNA replicated from individual origins (figure 14.20). The origins are not as sequence-specific as *OriC*, and seem to depend on chromatin structure as well as on sequence. The number of origins that "fire" can also be adjusted during the course of development so that early, when divisions need to be rapid, more origins are activated.

Although there are some differences in details, such as the names of the enzymes, the replication fork in eukaryotes appears substantially similar to the picture researchers have developed for *E. coli*. The main replication polymerase in eukaryotes is called DNA polymerase α; the sliding clamp subunit also exists in eukaryotes, but it is called PCNA (for Proliferating Cell Nuclear Antigen). This name reflects the fact that PCNA was first identified as an antibody-inducing protein in proliferating cells. The makeup of the primosome in eukaryotes is different in detail from that of the *E. coli* primosome, but its role is the same. Similarly, all the rest of the actors at the replication fork that we have seen in *E. coli* are also present in eukaryotes.

Eukaryotic chromosomes have multiple origins of replication.

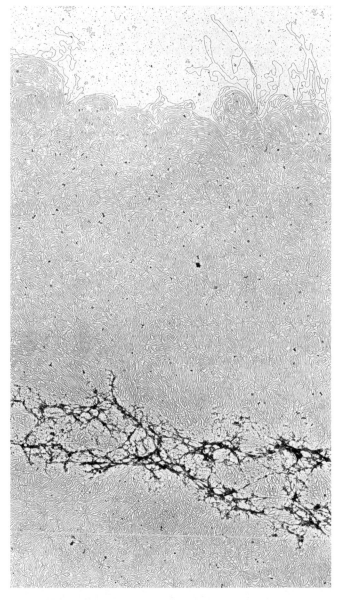

FIGURE 14.19
DNA of a single human chromosome. This chromosome has been "exploded," or relieved of most of its packaging proteins. The residual protein scaffolding appears as the dark material in the lower part of the micrograph.

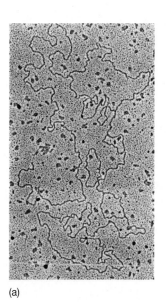
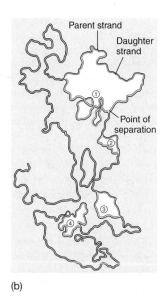

(a)   (b)

FIGURE 14.20
Eukaryotic chromosomes possess numerous replication forks spaced along their length. Four replication units (each with two replication forks) are producing daughter strands in (*a*) the electron micrograph. This is indicated in red in (*b*) the corresponding drawing.

14.4 What is a gene?

The One-Gene/One-Polypeptide Hypothesis

As the structure of DNA was being investigated, other biologists continued to puzzle over how the genes described by Mendel were related to DNA. What is the relationship between genotype and phenotype?

Garrod: Inherited Disorders Can Involve Specific Enzymes

In 1902, the British physician Archibald Garrod was working with one of the early Mendelian geneticists, his countryman William Bateson, when he noted that certain diseases among his patients seemed to be more prevalent in particular families. By examining several generations of these families, he found that some of the diseases behaved as if they were the product of simple recessive alleles. Garrod concluded that these disorders were Mendelian traits and that they had resulted from changes in the hereditary information in an ancestor of the affected families.

Garrod investigated several of these disorders in detail. In alkaptonuria, the patients produced urine that contained homogentisic acid (alkapton). This substance oxidized rapidly when exposed to air, turning the urine black. In normal individuals, homogentisic acid is broken down into simpler substances. With considerable insight, Garrod concluded that patients suffering from alkaptonuria lack the enzyme necessary to catalyze this breakdown. He speculated that many other inherited diseases might also reflect enzyme deficiencies.

Beadle and Tatum: Genes Specify Enzymes

From Garrod's finding, it took but a short leap of intuition to surmise that the information encoded within the DNA of chromosomes acts to specify particular enzymes. This point was not actually established, however, until 1941, when a series of experiments by Stanford University geneticists George Beadle and Edward Tatum provided definitive evidence. Beadle and Tatum deliberately set out to create Mendelian mutations in chromosomes and then studied the effects of these mutations on the organism (figure 14.21).

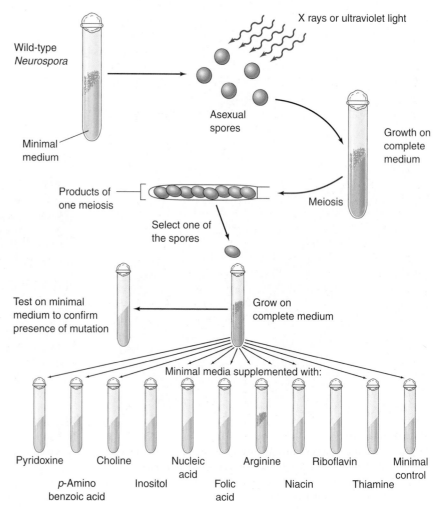

FIGURE 14.21
Beadle and Tatum's procedure for isolating nutritional mutants in *Neurospora*. This fungus grows easily on an artificial medium in test tubes. In this experiment, spores were irradiated to increase the frequency of mutation; they were then placed on a "complete" medium that contained all of the nutrients necessary for growth. Once the fungal colonies were established on the complete medium, individual spores were transferred to a "minimal" medium that lacked various substances the fungus could normally manufacture. Any spore that would not grow on the minimal medium but would grow on the complete medium contained one or more mutations in genes needed to produce the missing nutrients. To determine which gene had mutated, the minimal medium was supplemented with particular substances. The mutation illustrated here produced an arginine mutant, a collection of cells that had lost the ability to manufacture arginine. These cells will not grow on minimal medium but will grow on minimal medium with only arginine added.

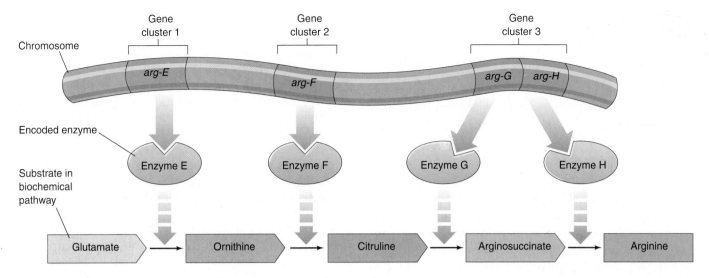

FIGURE 14.22
Evidence for the "one-gene/one-polypeptide" hypothesis. The chromosomal locations of the many arginine mutants isolated by Beadle and Tatum cluster in three places. These clusters correspond to the locations of the genes encoding the enzymes that carry out arginine biosynthesis.

A Defined System. One of the reasons Beadle and Tatum's experiments produced clear-cut results is that the researchers made an excellent choice of experimental organism. They chose the bread mold *Neurospora*, a fungus that can be grown readily in the laboratory on a defined medium (a medium that contains only known substances such as glucose and sodium chloride, rather than some uncharacterized mixture of substances, such as ground-up yeasts). Beadle and Tatum exposed *Neurospora* spores to X rays, expecting that the DNA in some of the spores would experience damage in regions encoding the ability to make compounds needed for normal growth (see figure 14.21).

Isolating Growth-Deficient Mutants. To determine whether any of the progeny of the irradiated spores had mutations causing metabolic deficiencies, Beadle and Tatum placed subcultures of individual fungal cells on a "minimal" medium that contained only sugar, ammonia, salts, a few vitamins, and water. Cells that had lost the ability to make other compounds necessary for growth would not survive on such a medium. Using this approach, Beadle and Tatum succeeded in identifying and isolating many growth-deficient mutants.

Identifying the Deficiencies. Next, the researchers added various chemicals to the minimal medium in an attempt to find one that would enable a given mutant strain to grow. This procedure allowed them to pinpoint the nature of the strain's biochemical deficiency. The addition of arginine, for example, permitted several mutant strains, dubbed *arg* mutants, to grow. When their chromosomal positions were located, the *arg* mutations were found to cluster in three areas.

One-Gene/One-Polypeptide

For each enzyme in the arginine biosynthetic pathway, Beadle and Tatum were able to isolate a mutant strain with a defective form of that enzyme, and the mutation was always located at one of a few specific chromosomal sites. Most importantly, they found there was a different site for each enzyme. Thus, each of the mutants they examined had a defect in a single enzyme, caused by a mutation at a single site on one chromosome. Beadle and Tatum concluded that genes produce their effects by specifying the structure of enzymes and that each gene encodes the structure of one enzyme. They called this relationship the **one-gene/one-enzyme hypothesis** (figure 14.22). Because many enzymes contain multiple protein or polypeptide subunits, each encoded by a separate gene, the relationship is today more commonly referred to as **one-gene/one-polypeptide.** This is a clear statement of the molecular relationship between genotype and phenotype.

Enzymes are responsible for catalyzing the synthesis of all the parts of an organism. They mediate the assembly of nucleic acids, proteins, carbohydrates, and lipids. Therefore, by encoding the structure of enzymes and other proteins, DNA specifies the structure of the organism itself.

As we learn more about genomes and gene expression, this clear relationship may not hold in all cases. In chapter 15, we will see that mRNA can be spliced differently to produce different products from the same gene, and that mRNAs can be edited. The basic principle that remains is: Genes encode proteins, and any single mRNA species encodes a single protein product.

> Genetic traits are expressed largely as a result of the activities of enzymes. Organisms store hereditary information by encoding the structures of enzymes and other proteins in their DNA.

How DNA Encodes Protein Structure

What kind of information must a gene encode to specify a protein? For some time, the answer to that question was not clear, because protein structure seemed impossibly complex.

Sanger: Proteins Consist of Defined Sequences of Amino Acids

The picture changed in 1953, the same year Watson and Crick unraveled the structure of DNA. The English biochemist Frederick Sanger, after many years of work, announced the complete sequence of amino acids in the protein insulin, a small protein hormone. This was the first protein for which the amino acid sequence was determined. Sanger's achievement was extremely significant because it demonstrated for the first time that proteins consist of definable sequences of amino acids—that is, for any given form of insulin, every molecule has the same amino acid sequence. Sanger's work soon led to the sequencing of many other proteins, and it became clear that all enzymes and other proteins are strings of amino acids arranged in a certain definite order. Therefore, the information needed to specify a protein such as an enzyme is an ordered list of amino acids.

Ingram: Single Amino Acid Changes in a Protein Can Have Profound Effects

Following Sanger's pioneering work, Vernon Ingram in 1956 discovered the molecular basis of sickle cell anemia, a protein defect inherited as a Mendelian disorder. By analyzing the structures of normal and sickle cell hemoglobin, Ingram, working at Cambridge University, showed that sickle cell anemia is caused by a change from glutamic acid to valine at a single position in the protein (figure 14.23). The alleles of the gene encoding hemoglobin differ only in their specification of this one amino acid in the hemoglobin amino acid chain.

These experiments and other related ones have finally brought us a clear understanding of the unit of heredity. The characteristics of sickle cell anemia and most other hereditary traits are defined by changes in protein structure brought about by an alteration in the sequence of amino acids that make up the protein. This sequence in turn is dictated by the order of nucleotides in a particular region of the chromosome. For example, the critical change leading to sickle cell disease is a mutation that replaces a single thymine with an adenine at the position that codes for glutamic acid, converting the position to valine. The sequence of nucleotides that determines the amino acid sequence of a protein is called a **gene**. Although most genes encode proteins or subunits of proteins, some genes are devoted to producing special forms of RNA, many of which play important roles in protein synthesis themselves.

A half-century of experimentation has made it clear that DNA is the molecule responsible for the inheritance of traits, and that this molecule is divided into functional units called genes.

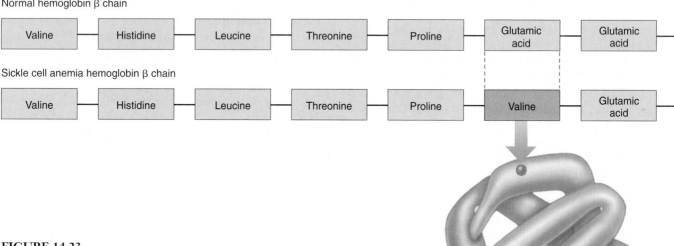

FIGURE 14.23
The molecular basis of a hereditary disease. Sickle cell anemia is produced by a recessive allele of the gene that encodes the hemoglobin β chains. It represents a change in a single amino acid, from glutamic acid to valine at the sixth position in the chains, which consequently alters the tertiary structure of the hemoglobin molecule, reducing its ability to carry oxygen.

For interactive testing, visit the Online Learning Center with PowerWeb at www.mhhe.com/Raven7

14.1 What is the genetic material?

The Hammerling Experiment: Cells Store Hereditary Information in the Nucleus

- Hammerling conducted a series of experiments and discovered that hereditary information in *Acetabularia* resided in the foot, which is also the location of the nucleus. (p. 280)

Transplantation Experiments: Each Cell Contains a Full Set of Genetic Instructions

- Later experiments in the mid-1950s showed that the nucleus of eukaryotic cells includes a full set of genetic information. (p. 281)

The Griffith Experiment: Hereditary Information Can Pass Between Organisms

- Griffith found that transformation occurs when genetic material is transferred from one cell to another, and that live cells can be transformed by dead cells. (p. 282)

The Avery and Hershey–Chase Experiments: The Active Principle Is DNA

- Avery provided conclusive evidence that DNA is the heredity material for the bacterial specimens under investigation. (p. 283)
- Hershey and Chase provided further evidence that heredity material in bacteriophages was found in DNA, not in proteins. (p. 283)

14.2 What is the structure of DNA?

The Chemical Nature of Nucleic Acids

- Both DNA and RNA are formed of nucleotides joined together in series. Each nucleotide is composed of a five-carbon sugar, a phosphate group, and a nitrogen-containing base. (p. 284)
- Chargaff's Rule states that in reference to the nitrogen-containing bases, adenine always pairs with thymine, and guanine always pairs with cytosine. Thus, there are always equal proportions of purines and pyrimidines. (p. 285)

The Three-Dimensional Structure of DNA

- Franklin was able to obtain the first glimpse of DNA using X-ray diffraction in 1953, while Watson and Crick theorized that DNA exists in a double-helical, antiparallel configuration. (pp. 286–287)
- Using a spiral staircase analogy, the handrails of the staircase represent the sugar-phosphate backbone of the DNA double helix, and the steps represent the hydrogen-bonded base pairs. (p. 287)

14.3 How does DNA replicate?

The Meselson–Stahl Experiment: DNA Replication Is Semiconservative

- Meselson and Stahl demonstrated that DNA replication is semiconservative because each strand of the original duplex becomes one of the two strands in each new duplex. (p. 288)

The Replication Process

- Replication of *E. coli* begins at a specific origin, proceeds bidirectionally, and ends at a specific terminus. (p. 290)
- Many enzymes function in DNA replication, including DNA primase, which creates a short RNA primer complementary to a DNA template; DNA helicase, which unwinds the helix in front of DNA polymerase, which then synthesizes new DNA by adding nucleotides to the growing strands; and DNA ligase, which creates phosphodiester bonds between adjacent Okazaki fragments. (pp. 292–293)
- Replication can be divided into three stages: initiation, elongation, and termination. (p. 294)

Eukaryotic DNA Replication

- The major difference between prokaryotic and eukaryotic replication is that eukaryotic chromosomes have multiple replication origins, whereas prokaryotic chromosomes have a single point of origin. (p. 295)

14.4 What is a gene?

The One-Gene/One-Polypeptide Hypothesis

- Beadle and Tatum concluded that genes produce their effects by specifying the structure of enzymes, and that each gene encodes the structure of one enzyme. Today, this is commonly referred to as the one-gene/one-polypeptide relationship. (p. 297)

How DNA Encodes Protein Structure

- Over 50 years of research has yielded clear evidence that DNA is the molecule responsible for the inheritance of traits from one generation to the next, and that DNA is divided into functional subunits, or genes, located on chromosomes. (p. 298)

Self Test

1. Which of the following experiments suggested that the nucleus is the repository for genetic information?
   a. Hammerling's experiment using *Acetabularia*
   b. Griffith's experiment using *S. pneumoniae* and mice
   c. the Hershey and Chase experiment using bacteriophages
   d. Franklin's X-ray diffraction
2. When Hershey and Chase differentially tagged the DNA and proteins of bacteriophages and allowed them to infect bacteria, what did the viruses transfer to the bacteria?
   a. radioactive phosphorus and sulfur
   b. radioactive sulfur
   c. DNA
   d. Both b and c are correct.
3. If one strand of a DNA molecule has the base sequence ATTGCAT, its complementary strand will have the sequence
   a. ATTGCAT
   b. TAACGTA
   c. GCCATGC
   d. CGGTACG
4. DNA is made up of building blocks called
   a. proteins.
   b. bases.
   c. nucleotides.
   d. deoxyribose.
5. X-ray diffraction experiments conducted by _____ led to the determination of the structure of DNA.
   a. Francis Crick
   b. James Watson
   c. Erwin Chargaff
   d. Rosalind Franklin
6. Meselson and Stahl proved that
   a. DNA is the genetic material.
   b. DNA is made from nucleotides.
   c. DNA replicates in a semiconservative manner.
   d. DNA is a double helix held together with base-pairing.
7. DNA polymerase III can only add nucleotides to an existing chain, so _____ is required.
   a. an RNA primer
   b. DNA polymerase I
   c. helicase
   d. a DNA primer
8. Okazaki fragments are
   a. synthesized in the 3′ to 5′ direction.
   b. found on the lagging strand.
   c. found on the leading strand.
   d. assembled as continuous replication.
9. Beadle and Tatum's experiment showed that each enzyme is specified by a single
   a. chromosome.
   b. gene.
   c. nucleotide.
   d. mutation.
10. What was significant about Ingram's work on sickle cell anemia?
    a. The gene for hemoglobin was missing.
    b. Proteins consisted of sequences of amino acids.
    c. A change in one amino acid can affect a protein's structure.
    d. One gene encodes one protein.

Test Your Visual Understanding

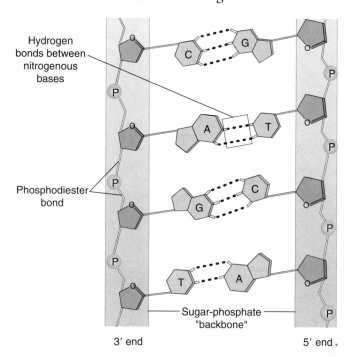

1. Based on the information in this figure, can you explain why pyrimidines don't base-pair with each other, and why purines don't base-pair with each other?
2. Can you also explain why adenine does not base-pair with cytosine, and why thymine does not base-pair with guanine?

Apply Your Knowledge

1. The human genome contains approximately 3 billion (3×10^9) nucleotide base-pairs, and each nucleotide in a strand of DNA takes up about 0.34 nanometers (0.34×10^{-9} meters). How long would the human genome be if it were fully extended?
2. From an extract of human cells growing in tissue culture, you obtain a white, fibrous substance. How would you distinguish whether it was DNA, RNA, or protein?
3. Cells were obtained from a patient with a viral infection. The DNA extracted from these cells consisted of two forms: double-stranded human DNA and single-stranded viral DNA. The base compositions of these two forms of DNA were as follows:

	A	C	G	T
Form 1	22.1%	27.9%	27.9%	22.1%
Form 2	31.3%	31.3%	18.7%	18.7%

Which form was the viral DNA, and which form was the human DNA? Explain your reasoning.

15

Genes and How They Work

Concept Outline

15.1 The Central Dogma traces the flow of gene-encoded information.

Cells Use RNA to Make Protein. The information in genes is expressed in two steps. First it is transcribed into RNA, and then the RNA is translated into protein.

15.2 Genes encode information in three-nucleotide code words.

The Genetic Code. The sequence of amino acids in a protein is encoded in the sequence of nucleotides in DNA such that three nucleotides encode each amino acid.

15.3 Genes are first transcribed, then translated.

Transcription in Prokaryotes. The enzyme RNA polymerase unwinds the DNA helix and synthesizes an RNA copy of one strand.

Transcription in Eukaryotes. Transcription in eukaryotes is more complex than that occurring in prokaryotes, requiring different types of enzymes and posttranscriptional modifications to the RNA.

Translation. mRNA is translated by activating enzymes that select tRNAs to match amino acids. Proteins are synthesized on ribosomes, which provide a framework for the interaction of tRNA and mRNA.

15.4 Eukaryotic gene transcripts are spliced.

The Discovery of Introns. Eukaryotic genes contain extensive material that is not translated.

Differences Between Prokaryotic and Eukaryotic Gene Expression. Gene expression is broadly similar in prokaryotes and eukaryotes, but it differs in some respects.

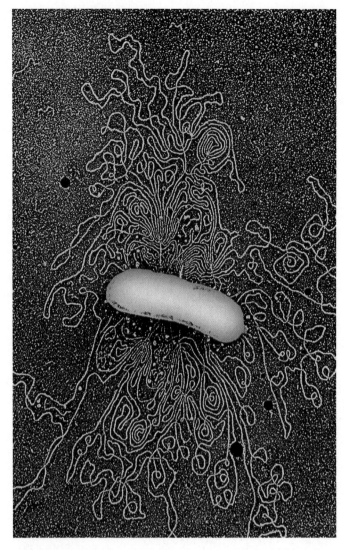

FIGURE 15.1
The unraveled chromosome of an *E. coli* bacterium. This complex tangle of DNA represents the full set of assembly instructions for the living organism *E. coli*.

Every cell in your body contains the hereditary instructions specifying that you will have arms rather than fins, hair rather than feathers, and two eyes rather than one. The color of your eyes, the texture of your fingernails, and all of the other traits you receive from your parents are recorded in the cells of your body. As we have seen, this information is contained in long molecules of DNA (figure 15.1). The essence of heredity is the ability of cells to use the information in their DNA to produce particular proteins, thereby affecting what the cells will be like. In that sense, proteins are the tools of heredity. In this chapter, we will examine how proteins are synthesized from the information in DNA, using both prokaryotes and eukaryotes as models.

15.1 The Central Dogma traces the flow of gene-encoded information.

Cells Use RNA to Make Protein

To find out how a eukaryotic cell uses its DNA to direct the production of particular proteins, you must first ask where in the cell the proteins are made. We can answer this question by placing cells in a medium containing radioactively labeled amino acids for a short time. The cells will take up the labeled amino acids and incorporate them into proteins. If we then look to see where in the cells radioactive proteins first appear, we will find them not in the nucleus, where the DNA is, but rather in the cytoplasm, on large RNA-protein aggregates called **ribosomes** (figure 15.2). These polypeptide-making factories are very complex, composed of two subunits, a small subunit (called 30S in prokaryotes and 40S in eukaryotes) and a large subunit (called 40S in prokaryotes and 60S in eukaryotes). Each subunit is composed of RNA and protein. The small subunit has over 20 proteins and one strand of RNA, and the large subunit has over 30 proteins and two strands of RNA (figure 15.3). Recently, the three-dimensional structure of both prokaryotic subunits and the complete prokaryotic ribosome was solved. These structures support the idea that the RNAs are the main catalytic units and the ribosomal proteins have a more structural role. Protein synthesis involves three different sites on the ribosome surface, called the P, A, and E sites, discussed later in this chapter.

Kinds of RNA

The class of RNA found in ribosomes is called **ribosomal RNA (rRNA)**. During polypeptide synthesis, rRNA provides the site where polypeptides are assembled. In addition to rRNA, there are two other major classes of RNA in cells. **Transfer RNA (tRNA)** molecules both transport the amino acids to the ribosome for use in building the polypeptides and position each amino acid at the correct place on the elongating polypeptide chain (figure 15.4). Human cells contain about 45 different kinds of tRNA molecules. **Messenger RNA (mRNA)** molecules are long strands of RNA that are transcribed from DNA and travel to the ribosomes to direct precisely *which* amino acids are assembled into polypeptides.

These RNA molecules, together with ribosomal proteins and certain enzymes, constitute a system that reads the genetic messages encoded by nucleotide sequences in the DNA and produces the polypeptides that those sequences specify. As we will see in chapter 18, biologists have recently discovered that RNA also plays many important roles in controlling how and when genes are used. Having learned to read the many messages of RNA, biologists are gaining a clearer picture of what genes are, how they are able to dictate what a protein will be like, and when the protein will be made.

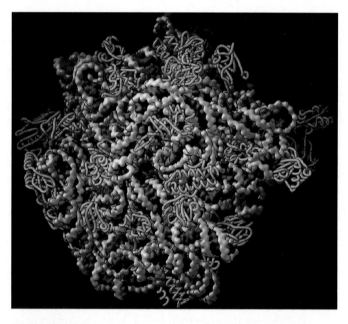

FIGURE 15.3
Ribosomes are very complex machines. The complete atomic structure of a prokaryotic large ribosomal subunit has been determined at 2.4 Å resolution. The RNA of the subunit is shown in gray and the proteins in gold. The subunit's RNA is twisted into irregular shapes that fit together like a three-dimensional jigsaw puzzle. The chemical reactions that form the peptide bond in protein synthesis are carried out deep in the interior by RNA. The ribosome is thus a ribozyme. Proteins are absent from the active site but abundant everywhere on the surface. The proteins stabilize the structure by interacting with adjacent RNA strands.

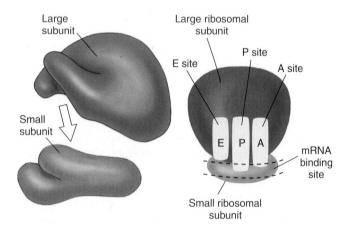

FIGURE 15.2
A ribosome is composed of two subunits. The smaller subunit fits into a depression on the surface of the larger one. The A, P, and E sites on the ribosome, discussed later in this chapter, play key roles in protein synthesis.

The Central Dogma

All organisms, from the simplest bacteria to humans, use the same basic mechanism of reading and expressing genes, a mechanism so fundamental to life as we know it that it is often referred to as the **Central Dogma:** Information passes from the genes (DNA) to an RNA copy of the gene, and the RNA copy directs the sequential assembly of a chain of amino acids (figure 15.5). Stated briefly,

$$DNA \rightarrow RNA \rightarrow protein$$

Transcription: An Overview

The first step of the Central Dogma is the transfer of information from DNA to RNA, which occurs when an mRNA copy of the gene is produced. Because the DNA sequence in the gene is transcribed into an RNA sequence, this stage is called **transcription.** Transcription is initiated when the enzyme **RNA polymerase** binds to a **promoter** binding site located at the beginning of a gene. Starting there, the RNA polymerase moves along the strand into the gene. As it encounters each DNA nucleotide, it adds the corresponding complementary RNA nucleotide to a growing mRNA strand. Thus, guanine (G), cytosine (C), thymine (T), and adenine (A) in the DNA would signal the addition of C, G, A, and uracil (U), respectively, to the mRNA.

When the RNA polymerase arrives at a transcriptional "stop" signal at the opposite end of the gene, it disengages from the DNA and releases the newly assembled RNA chain. This chain is a complementary transcript of the gene from which it was copied.

Translation: An Overview

The second step of the Central Dogma is the transfer of information from RNA to protein, which occurs when the information contained in the mRNA transcript is used to direct the sequence of amino acids during the synthesis of polypeptides by ribosomes. This process is called **translation** because the nucleotide sequence of the mRNA transcript is translated into an amino acid sequence in the polypeptide. Translation begins when an rRNA molecule within the ribosome recognizes and binds to a "start" sequence on the mRNA. The ribosome then moves along the mRNA molecule, three nucleotides at a time. Each group of three nucleotides is a code word that specifies which amino acid will be added to the growing polypeptide chain. The ribosome continues in this fashion until it encounters a translational "stop" signal; then it disengages from the mRNA and releases the completed polypeptide.

The two steps of the Central Dogma, taken together, are a concise summary of the events involved in the expression of an active gene. Biologists refer to this process as **gene expression.**

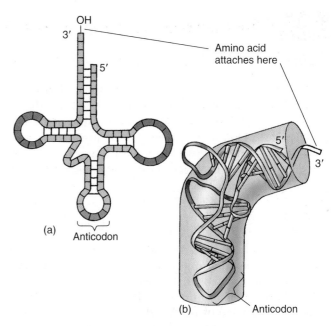

FIGURE 15.4
The structure of tRNA. (*a*) In this two-dimensional schematic, the three loops of tRNA are unfolded. Two of the loops bind to the ribosome during polypeptide synthesis, and the third loop contains the anticodon sequence, which is complementary to a three-base sequence on messenger RNA. Amino acids attach to the free, single-stranded —OH end. (*b*) In this three-dimensional structure, the loops of tRNA are folded.

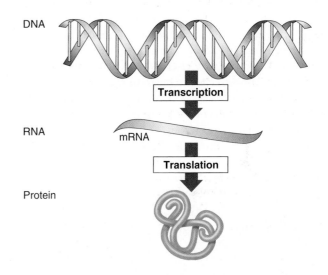

FIGURE 15.5
The Central Dogma of gene expression. DNA is transcribed to make mRNA, which is translated to make a protein.

The information encoded in genes is expressed in two phases: transcription, which produces an mRNA molecule whose sequence is complementary to the DNA sequence of the gene; and translation, which assembles a polypeptide whose amino acid sequence is specified by the nucleotide sequence in the mRNA.

15.2 Genes encode information in three-nucleotide code words.

The Genetic Code

The essential question of gene expression is, "How does the *order* of nucleotides in a DNA molecule encode the information that specifies the order of amino acids in a polypeptide?" The answer came in 1961, through an experiment led by Francis Crick. That experiment was so elegant and the result so critical to understanding the genetic code that we will describe it in detail.

Proving Code Words Have Only Three Letters

Crick and his colleagues reasoned that the genetic code most likely consisted of a series of blocks of information called **codons**, each corresponding to an amino acid in the encoded protein. They further hypothesized that the information within one codon was probably a sequence of three nucleotides specifying a particular amino acid. They arrived at the number three because a two-nucleotide codon would not yield enough combinations to code for the 20 different amino acids that commonly occur in proteins. With four DNA nucleotides (G, C, T, and A), only 4^2, or 16, different pairs of nucleotides could be formed. However, these same nucleotides can be arranged in 4^3, or 64, different combinations of three, more than enough to code for the 20 amino acids.

In theory, the codons in a gene could lie immediately adjacent to each other, forming a continuous sequence of transcribed nucleotides. Alternatively, the sequence could be punctuated with untranscribed nucleotides between the codons, like the spaces that separate the words in this sentence. It was important to determine which method cells employ because these two ways of transcribing DNA imply different translating processes.

To choose between these alternative mechanisms, Crick and his colleagues used a chemical to delete one, two, or three nucleotides from a viral DNA molecule and then asked whether a gene downstream of the deletions was transcribed correctly. When they made a single deletion or two deletions near each other, the **reading frame** of the genetic message shifted, and the downstream gene was transcribed as nonsense. However, when they made three deletions, the correct reading frame was restored, and the sequences downstream were transcribed correctly. They obtained the same results when they made additions to the DNA consisting of one, two, or three nucleotides. As shown in figure 15.6, these results could not have been obtained if the codons were punctuated by untranscribed nucleotides. Thus, Crick and his colleagues concluded that the genetic code is read in increments consisting of three nucleotides (in other words, it is a **triplet code**) and that reading occurs continuously without punctuation between the three-nucleotide units.

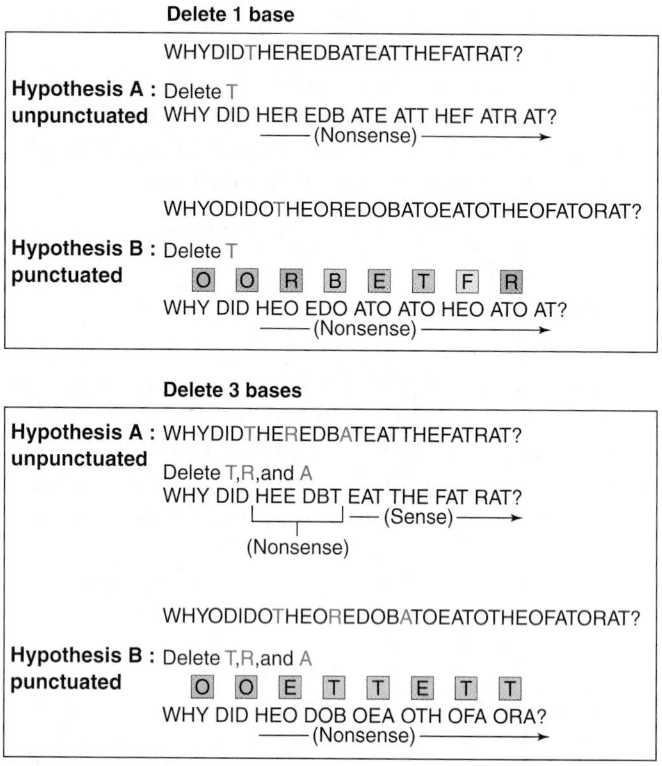

FIGURE 15.6
Using frameshift alterations of DNA to determine if the genetic code is punctuated. The hypothetical genetic message presented here is, "Why did the red bat eat the fat rat?" Under hypothesis B, which proposes that the message is punctuated, the three-letter words are separated by nucleotides that are not read (indicated by the letter "O").

Breaking the Genetic Code

Within a year of Crick's experiment, other researchers succeeded in determining the amino acids specified by particular three-nucleotide units. Marshall Nirenberg discovered in 1961 that adding the synthetic mRNA molecule polyU (an RNA molecule consisting of a string of uracil nucleotides) to cell-free systems resulted in the production of the polypeptide polyphenylalanine (a string of phenylalanine amino acids). Therefore, one of the three-nucleotide sequences specifying phenylalanine is UUU. In 1964, Nirenberg and Philip Leder developed a powerful **triplet binding assay** in which a specific triplet was tested to see which radioactive amino acid (complexed to tRNA) it would bind. Some 47 of the 64 possible triplets gave unambiguous results. Har Gobind Khorana decoded the remaining 17 triplets by constructing artificial mRNA molecules of defined sequence and examining what polypeptides they directed. In these ways, all 64 possible three-nucleotide sequences were tested, and the full genetic code was determined (table 15.1).

Table 15.1 The Genetic Code

First Letter	Second Letter								Third Letter
	U		**C**		**A**		**G**		
U	UUU	Phenylalanine	UCU	Serine	UAU	Tyrosine	UGU	Cysteine	U
	UUC		UCC		UAC		UGC		C
	UUA	Leucine	UCA		UAA	Stop	UGA	Stop	A
	UUG		UCG		UAG	Stop	UGG	Tryptophan	G
C	CUU	Leucine	CCU	Proline	CAU	Histidine	CGU	Arginine	U
	CUC		CCC		CAC		CGC		C
	CUA		CCA		CAA	Glutamine	CGA		A
	CUG		CCG		CAG		CGG		G
A	AUU	Isoleucine	ACU	Threonine	AAU	Asparagine	AGU	Serine	U
	AUC		ACC		AAC		AGC		C
	AUA		ACA		AAA	Lysine	AGA	Arginine	A
	AUG	Methionine; Start	ACG		AAG		AGG		G
G	GUU	Valine	GCU	Alanine	GAU	Aspartate	GGU	Glycine	U
	GUC		GCC		GAC		GGC		C
	GUA		GCA		GAA	Glutamate	GGA		A
	GUG		GCG		GAG		GGG		G

A codon consists of three nucleotides read in the sequence shown. For example, ACU codes for threonine. The first letter, A, is in the First Letter column; the second letter, C, is in the Second Letter column; and the third letter, U, is in the Third Letter column. Each of the mRNA codons is recognized by a corresponding anticodon sequence on a tRNA molecule. Some tRNA molecules recognize more than one codon in mRNA, but they always code for the same amino acid. In fact, most amino acids are specified by more than one codon. For example, threonine is specified by four codons, which differ only in the third nucleotide (ACU, ACC, ACA, and ACG).

The Code Is Practically Universal . . .

The genetic code is the same in almost all organisms. For example, the codon AGA specifies the amino acid arginine in bacteria, humans, and all other organisms whose genetic code has been studied. The universality of the genetic code is among the strongest evidence that all living things share a common evolutionary heritage. Because the code is universal, genes transcribed from one organism can be translated in another; the mRNA is fully able to dictate a functionally active protein. Similarly, genes can be transferred from one organism to another and be successfully transcribed and translated in their new host. This universality of gene expression is central to many of the advances of genetic engineering. Many commercial products, such as the insulin used to treat diabetes, are now manufactured by placing human genes into bacteria, which then serve as tiny factories to turn out prodigious quantities of insulin.

. . . But Not Quite

In 1979, investigators began to determine the complete nucleotide sequences of the mitochondrial genomes in humans, cattle, and mice. It came as something of a shock when these investigators learned that the genetic code used by these mammalian mitochondria was not quite the same as the "universal code" that has become so familiar to biologists. In the mitochondrial genomes, what should have been a "stop" codon, UGA, was instead read as the amino acid tryptophan; AUA was read as methionine rather than isoleucine; and AGA and AGG were read as "stop" rather than arginine. Furthermore, minor differences from the universal code have also been found in the genomes of chloroplasts and ciliates (certain types of protists).

Thus, it appears that the genetic code is not quite universal. Some time ago, presumably after they began their endosymbiotic existence, mitochondria and chloroplasts began to read the code differently, particularly the portion of the code associated with "stop" signals.

Within genes that encode proteins, the nucleotide sequence of DNA is read in blocks of three consecutive nucleotides, without punctuation between the blocks. Each block, or codon, codes for one amino acid.

15.3 Genes are first transcribed, then translated.

Transcription in Prokaryotes

The first step in gene expression is the production of an RNA copy of the DNA sequence encoding the gene, a process called **transcription.** To understand the mechanism behind transcription, it is useful to focus first on prokaryotes, where the process is well understood (figure 15.7).

RNA Polymerase

Prokaryotic RNA polymerase is very large and complex, consisting of five subunits: Two α subunits bind regulatory proteins, a β′ subunit binds the DNA template, a β subunit binds RNA nucleoside subunits, and a σ subunit recognizes the promoter and initiates synthesis. Only one of the two strands of DNA, called the **template strand,** is transcribed (that is, copied during transcription). The RNA transcript's sequence is complementary to the template strand. The strand of DNA that is not transcribed is called the **coding strand.** It has the same sequence as the RNA transcript, except T takes the place of U. The coding strand is also known as the *sense* (+) *strand*, and the template strand as the *antisense* (−) *strand*.

In prokaryotes, the polymerase adds ribonucleotides to the growing 3′ end of an RNA chain. No primer is needed, and synthesis proceeds in the 5′ to 3′ direction as in DNA replication.

Promoter

Transcription starts at RNA polymerase binding sites called **promoters** on the DNA template strand. A promoter is a short sequence that is not itself transcribed by the polymerase that binds to it. Striking similarities are evident in the sequences of different promoters. For example, two six-base sequences are common to many bacterial promoters: a TTGACA sequence called the **−35 sequence,** located 35 nucleotides upstream of the position where transcription actually starts, and a TATAAT sequence called the **−10 sequence,** located 10 nucleotides upstream of the "start" site.

Bacterial promoters differ widely in efficiency. Strong promoters cause frequent initiations of transcription—as often as every 2 seconds in some bacteria. Weak promoters may transcribe only once every 10 minutes. Most strong promoters have unaltered –35 and –10 sequences, while weak promoters often have substitutions within these sites.

Initiation

The binding of RNA polymerase to the promoter is the first step in gene transcription. In prokaryotes, a subunit of RNA polymerase called σ (sigma) recognizes the –10 sequence in

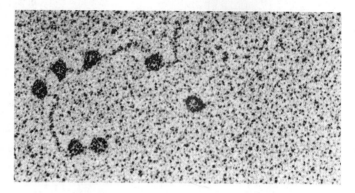

FIGURE 15.7
RNA polymerase. In this electron micrograph, the dark circles are RNA polymerase molecules bound to several promoter sites on bacterial virus DNA.

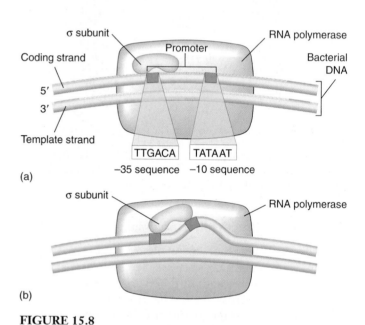

FIGURE 15.8
Initiation of transcription by the σ subunit. (*a*) RNA polymerase with its σ subunit slides along the DNA until it recognizes the promoter and binds to it. (*b*) Upon binding, the σ subunit begins unwinding the DNA double helix at the −10 sequence, exposing the template strand to be transcribed.

the promoter and binds RNA polymerase there. Importantly, this subunit can detect the –10 sequence without unwinding the DNA double helix.

Once bound to the promoter, the RNA polymerase begins to unwind the DNA helix (figure 15.8). Measurements indicate that bacterial RNA polymerase unwinds a segment approximately 17 base-pairs long, nearly two turns of the DNA double helix. This sets the stage for the assembly of the RNA chain.

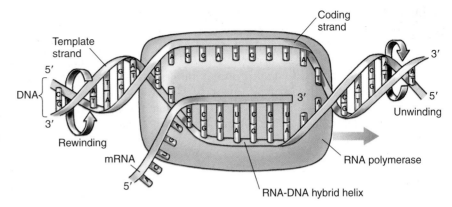

FIGURE 15.9
Model of a transcription bubble. The DNA duplex unwinds as it enters the RNA polymerase complex and rewinds as it leaves. One of the strands of DNA functions as a template, and nucleotide building blocks are assembled into RNA from this template.

Elongation

In prokaryotes, the transcription of the RNA chain usually starts with ATP or GTP. One of these forms the 5′ end of the chain, which grows in the 5′ to 3′ direction as ribonucleotides are added. Unlike DNA synthesis, a primer is not required. The region containing the RNA polymerase, DNA, and growing RNA transcript is called the **transcription bubble** because it contains a locally unwound "bubble" of DNA (figure 15.9). Within the bubble, the first 12 bases of the newly synthesized RNA strand temporarily form a helix with the template DNA strand. Corresponding to not quite one turn of the helix, this stabilizes the positioning of the 3′ end of the RNA so it can interact with an incoming ribonucleotide. The RNA-DNA hybrid helix rotates each time a nucleotide is added so that the 3′ end of the RNA stays at the catalytic site.

The transcription bubble moves down the bacterial DNA at a constant rate, about 50 nucleotides per second, leaving the growing RNA strand protruding from the bubble. After the transcription bubble passes, the now transcribed DNA is rewound as it leaves the bubble.

Unlike DNA polymerase, RNA polymerase has no proofreading capability. Transcription thus produces many more copying errors than replication. These mistakes, however, are not transmitted to progeny. Most genes are transcribed many times, so a few faulty copies are not harmful.

Termination

At the end of a bacterial gene are "stop" sequences that cause the formation of phosphodiester bonds to cease, the RNA-DNA hybrid within the transcription bubble to dissociate, the RNA polymerase to release the DNA, and the DNA within the transcription bubble to rewind. The simplest "stop" signal is a series of G–C base-pairs followed by a series of A–T base-pairs. The RNA transcript of this stop region forms a GC hairpin (figure 15.10),

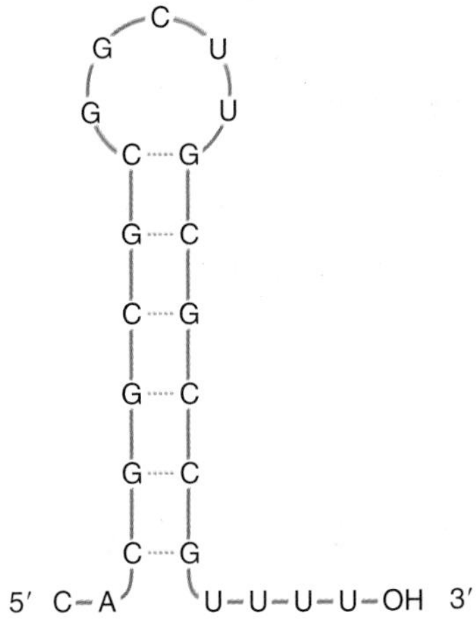

FIGURE 15.10
A GC hairpin. This structure stops gene transcription.

followed by four or more U ribonucleotides. How does this structure terminate transcription? The hairpin causes the RNA polymerase to pause immediately after the polymerase has synthesized it, placing the polymerase directly over the run of four uracils. The pairing of U with DNA's A is the weakest of the four hybrid base-pairs and is not strong enough to hold the hybrid strands together during the long pause. Instead, the RNA strand dissociates from the DNA within the transcription bubble, and transcription stops. A variety of protein factors aid hairpin loops in terminating transcription of particular genes.

> Transcription in prokaryotes is carried out by the enzyme RNA polymerase, which unwinds and transcribes the DNA.

Transcription in Eukaryotes

The basic mechanism of transcription by RNA polymerase is the same in eukaryotes as in prokaryotes. However, a number of details about the two processes differ enough that it is necessary to consider them separately. Here we will concentrate only on how eukaryotic systems differ from prokaryotic systems, such as the bacterial system just discussed. All other features may be assumed to be the same.

Multiple RNA Polymerases

Unlike prokaryotes, which have a single RNA polymerase enzyme, eukaryotes have three different RNA polymerases that are distinguished by both structure and function. The enzyme RNA polymerase I is specialized to transcribe rRNA and recognizes promoters that are unique to this enzyme. RNA polymerase II is specialized to transcribe mRNA and some small nuclear RNAs. We will concentrate more on this enzyme later, due to its importance in transcribing all mRNAs in the cell. RNA polymerase III transcribes tRNA and other small RNAs. It too has a specific promoter structure. Together, these three enzymes accomplish all transcription in the nucleus of eukaryotic cells.

Promoters

One way the activities of the three different RNA polymerases are controlled is by the differences in the structure of their promoters. RNA pol I promoters at first puzzled biologists, because comparisons of rRNA genes between species showed no similarities outside the coding region. The current view is that these promoters are specific for each species, and for this reason, cross-species comparisons do not yield similarities.

The RNA pol II promoters are the most complex of the three eukaryotic promoters, probably a reflection of the huge diversity of genes that are transcribed by this polymerase. A "core promoter" contains a so-called **TATA box** that resembles the −10 sequence from bacteria. Other conserved elements are found in many but not all RNA pol II transcribed genes, such as a CAAT box that makes up a second core promoter found in most genes. General transcription factors bind in this region to assemble the initiation complex. Additional upstream elements unique to individual genes are regulatory elements, binding sites for transcription factors that control tissue and developmentally specific expression.

Promoters for RNA pol III also were a source of surprise for biologists examining the control of gene expression in the early days of molecular biology. A common technique for analyzing regulatory regions was to make successive deletions from the 5′ end of genes to see when

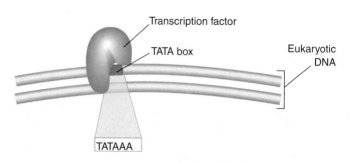

1. A transcription factor recognizes and binds to the TATA box sequence which is part of the core promoter.

2. Other transcription factors are then recruited, and the initiation complex begins to build.

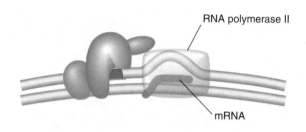

3. Ultimately, RNA polymerase II associates with the transcription factors and the DNA, forming the initiation complex, and transcription begins.

FIGURE 15.11
Eukaryotic initiation complex. Unlike transcription in prokaryotic cells, where the RNA polymerase recognizes and binds to the promoter, eukaryotic transcription requires the binding of transcription factors to the promoter before RNA polymerase II binds to the DNA. The association of transcription factors and RNA polymerase II at the promoter is called the initiation complex.

these deletions abolished specific transcription. The logic was conditioned from experiences with prokaryotes in which the regulatory regions had been found at the 5′ end of genes. In the case of tRNA genes, the 5′ deletions had no effect on expression! The promoters were found to actually be internal to the gene itself.

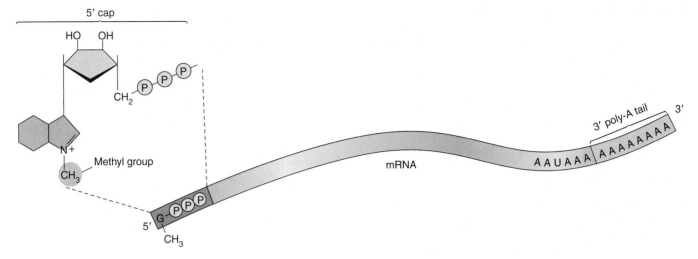

FIGURE 15.12
Posttranscriptional modifications. Eukaryotic mRNA molecules are modified in the nucleus with the addition of a methylated GTP to the 5′ end of the transcript, called the 5′ cap, and a long chain of adenine residues to the 3′ end of the transcript, called the poly-A tail.

Initiation

The initiation of transcription at RNA pol II promoters is analogous to prokaryotic initiation but much more complex. Instead of the polymerase simply recognizing and binding to the promoter, a host of general transcription factors interact at the promoter and RNA pol II to form an *initiation complex* (figure 15.11). We will explore this in detail in chapter 18 when we discuss the control of gene expression.

Posttranscriptional Modifications

There are a number of differences between the transcription of mRNAs in prokaryotes and eukaryotes. Some of these involve the packaging of eukaryotic DNA in the nucleus of the cell, but others were unexpected when first discovered. Between transcription in the nucleus and export of a mature mRNA to the cytoplasm, a number of modifications are performed in the initial transcripts made by RNA pol II.

5′ Cap. Eukaryotic transcripts have an unusual structure that is added to the 5′ end of mRNAs. The first base in the transcript is usually an adenine (A) or a guanine (G), and this is further modified by the addition of GTP to the 5′ PO₄ group, called a **5′ cap.** (figure 15.12). This

unusual 5′ to 5′ linkage is found only in this "cap" structure. The G in the GTP is also modified by the addition of a methyl group, so it is often called a methyl-G cap. The cap is added while transcription is still in progress. This cap protects the 5′ end of the mRNA from degradation and is also involved in translation initiation.

3′ Poly-A Tail. A major difference between transcription in prokaryotes and eukaryotes is that in eukaryotes, the end of the transcript is not the end of the mRNA. The eukaryotic transcript is cleaved downstream of a specific site (AAUAAA) that is not the termination site for trancription. A series of adenine (A) residues, called the **3′ poly-A tail,** are added after this cleavage by an enzyme other than RNA polymerase II such that the end of the mRNA is not the end of the transcript and is not created by RNA pol II (figure 15.12). The enzyme that adds the A's is called, appropriately, poly-A polymerase. The poly-A tail appears to play a role in the stability of mRNAs by protecting them from degradation. We will discuss mRNA stability in chapter 18 as well.

Transcription in eukaryotes differs from that in prokaryotes in that there are three RNA polymerase enzymes, an initiation complex forms at the promoter, and the RNAs are modified after transcription.

Translation

In prokaryotes, translation begins when the initial portion of an mRNA molecule binds to an rRNA molecule in a ribosome. The mRNA lies on the ribosome in such a way that only one of its codons is exposed at the polypeptide-making site at any time. A tRNA molecule possessing the complementary three-nucleotide sequence, or anticodon, binds to the exposed codon on the mRNA.

Because this tRNA molecule carries a particular amino acid, that amino acid and no other is added to the polypeptide in that position. As the mRNA molecule moves through the ribosome, successive codons on the mRNA are exposed, and a series of tRNA molecules bind one after another to the exposed codons. Each of these tRNA molecules carries an attached amino acid, which it adds to the end of the growing polypeptide chain (figure 15.13).

There are about 45 different kinds of tRNA molecules. Why are there 45 and not 64 tRNAs (one for each codon)? Because some tRNAs recognize more than one codon due to the fact that the third base-pair of a tRNA anticodon allows some "wobble."

How do particular amino acids become associated with particular tRNA molecules? The key translation step, which pairs the three-nucleotide sequences with appropriate amino acids, is carried out by a remarkable set of enzymes called *activating enzymes.*

Activating Enzymes

Particular tRNA molecules become attached to specific amino acids through the action of activating enzymes called **aminoacyl-tRNA synthetases,** one of which exists for each of the 20 common amino acids (figure 15.14). Therefore, these enzymes must correspond to specific anticodon sequences on a tRNA molecule as well as particular amino acids. Some activating enzymes correspond to only one anticodon and thus to only one tRNA molecule. Others recognize two, three, four, or six different tRNA molecules, each with a different anticodon but coding for the same amino acid (see table 15.1). If we consider the nucleotide sequence of mRNA a coded message, then the 20 activating enzymes are responsible for decoding that message. Or, to put it another way, the attachment of specific amino acids to the appropriate tRNA is the actual translation step.

"Start" and "Stop" Signals

For three of the 64 codons (UAA, UAG, and UGA), there exists no tRNA with a complementary anticodon. These codons, called **nonsense codons,** serve as "stop" signals in the mRNA message, marking the end of a polypeptide. The "start" signal that marks the beginning of a polypeptide within an mRNA message is the codon AUG, which also encodes the amino acid methionine. The ribosome will usually use the first AUG it encounters in the mRNA to signal the start of translation.

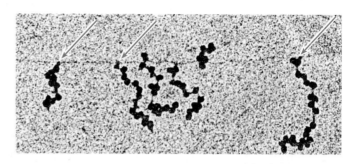

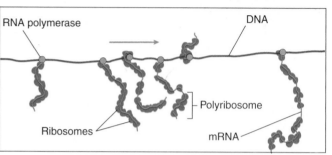

FIGURE 15.13
Translation in action.   Bacteria have no nucleus and hence no membrane barrier between the DNA and the cytoplasm. In this electron micrograph of genes being transcribed in the bacterium *Escherichia coli,* you can see every stage of the process. The arrows point to RNA polymerase enzymes. From each mRNA molecule dangling from the DNA, a series of ribosomes is assembling polypeptides. These clumps of ribosomes are sometimes called polyribosomes.

Initiation

In prokaryotes, polypeptide synthesis begins with the formation of an **initiation complex.** First, a tRNA molecule carrying a chemically modified methionine called *N*-formylmethionine (tRNA^{fMet}) binds to the small ribosomal subunit. Proteins called **initiation factors** position the tRNA^{fMet} on the ribosomal surface at the *P site* (for peptidyl), where peptide bonds will form. Nearby, two other sites will form: the *A site* (for aminoacyl), where successive amino-acid-bearing tRNAs will bind, and the *E site* (for exit), where empty tRNAs will exit the ribosome (figure 15.15). This initiation complex, guided by another initiation factor, then binds to the anticodon AUG on the mRNA. Proper positioning of the mRNA is critical because it determines the reading frame—that is, which groups of three nucleotides will be read as codons. Moreover, the complex must bind to the beginning of the mRNA molecule, so that all of the transcribed gene will be translated. In prokaryotes, the beginning of each mRNA molecule is marked by a *leader sequence* complementary to one of the rRNA molecules on the ribosome. This complementarity ensures that the mRNA is read from the beginning. Prokaryotes often include several genes within a single mRNA transcript (polycistronic mRNA), while each eukaryotic gene is transcribed on a separate mRNA (monocistronic mRNA).

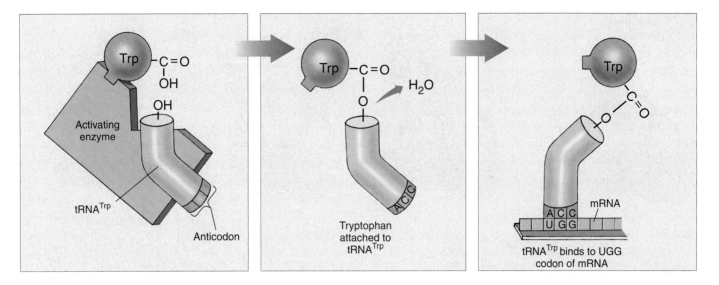

FIGURE 15.14
Activating enzymes "read" the genetic code. Each kind of activating enzyme recognizes and binds to a specific amino acid, such as tryptophan; it also recognizes and binds to the tRNA molecules with anticodons specifying that amino acid, such as ACC for tryptophan. In this way, activating enzymes link the tRNA molecules to specific amino acids.

Initiation in eukaryotes is similar, although it differs in two important ways. First, in eukaryotes, the initiating amino acid is methionine rather than N-formylmethionine. Second, the initiation complex is far more complicated than in prokaryotes, containing nine or more protein factors, many consisting of several subunits. Eukaryotic initiation complexes are discussed in detail in chapter 18.

Elongation

After the initiation complex has formed, the large ribosomal subunit binds, exposing the mRNA codon adjacent to the initiating AUG codon, and so positioning it for interaction with another amino-acid-bearing tRNA molecule. When a tRNA molecule with the appropriate anticodon appears, proteins called *elongation factors* assist in binding it to the exposed mRNA codon at the A site. When the second tRNA binds to the ribosome, it places its amino acid directly adjacent to the initial methionine, which is still attached to its tRNA molecule, which in turn is still bound to the ribosome. The two amino acids undergo a chemical reaction, catalyzed by the large ribosomal subunit, which releases the initial methionine from its tRNA and attaches it instead by a peptide bond to the second amino acid.

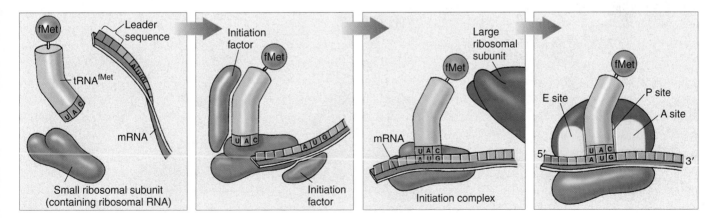

FIGURE 15.15
Formation of the initiation complex. In prokaryotes, proteins called initiation factors play key roles in positioning the small ribosomal subunit and the N-formylmethionine tRNA, or tRNAfMet, molecule at the beginning of the mRNA. When the tRNAfMet is positioned over the first AUG codon of the mRNA, the large ribosomal subunit binds, forming the P, A, and E sites where successive tRNA molecules bind to the ribosomes, and polypeptide synthesis begins.

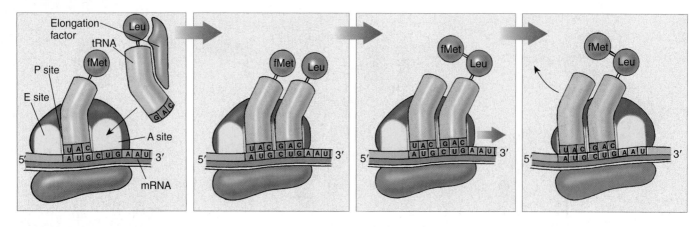

FIGURE 15.16

Translocation. The initiating tRNA^fMet in prokaryotes (tRNA^Met in eukaryotes) occupies the P site, and a tRNA molecule with an anticodon complementary to the exposed mRNA codon binds at the A site. fMet is transferred to the incoming amino acid (Leu in this case), as the ribosome moves three nucleotides to the right along the mRNA. The empty tRNA^fMet moves to the E site to exit the ribosome, the growing polypeptide chain moves to the P site, and the A site is again exposed and ready to bind the next amino-acid-laden tRNA.

Translocation

In a process called **translocation** (figure 15.16), the ribosome now moves (translocates) three more nucleotides along the mRNA molecule in the 5′ to 3′ direction, guided by other elongation factors. This movement relocates the initial tRNA to the E site and ejects it from the ribosome, repositions the growing polypeptide chain (at this point containing two amino acids) to the P site, and exposes the next codon on the mRNA at the A site. When a tRNA molecule recognizing that codon appears, it binds to the codon at the A site, placing its amino acid adjacent to the growing chain. The chain then transfers to the new amino acid, and the entire process is repeated.

Termination

Elongation continues in this fashion until a chain-terminating nonsense codon is exposed (for example, UAA in figure 15.17). Nonsense codons do not bind to tRNA, but they are recognized by **release factors,** proteins that release the newly made polypeptide from the ribosome.

> The first step in protein synthesis is the formation of an initiation complex. Each step of the ribosome's progress exposes a codon to which a tRNA molecule with the complementary anticodon binds. The amino acid carried by each tRNA molecule is added to the end of the growing polypeptide chain.

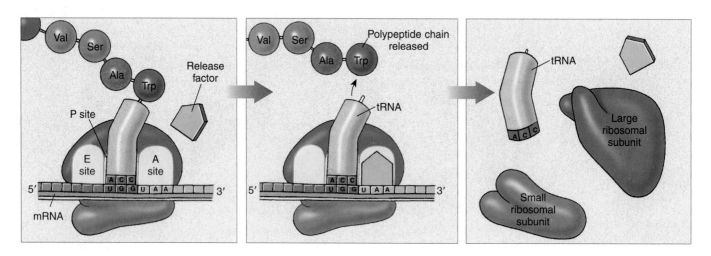

FIGURE 15.17

Termination of protein synthesis. There is no tRNA with an anticodon complementary to any of the three termination signal codons, such as the UAA nonsense codon illustrated here. When a ribosome encounters a termination codon, it therefore stops translocating. A specific release factor facilitates the release of the polypeptide chain by breaking the covalent bond that links the polypeptide to the P-site tRNA.

15.4 Eukaryotic gene transcripts are spliced.

The Discovery of Introns

The first genes isolated were prokaryotic genes found in *E. coli* and its viruses. A clear picture of the nature and some of the control of gene expression had been built up in these systems before the advent of the molecular cloning of eukaryotic genes. It was expected that while details would differ, the outline of gene expression would be similar. The world of biology was in for a shock when the first genes were isolated from eukaryotic organisms. Many of these genes appeared to contain sequences that were not represented in the mRNA. It is hard to exaggerate how unexpected this finding was. A basic tenant of molecular biology that had grown out of work on *E. coli* was that a gene was colinear with its protein product. This means that the sequence of bases in the gene corresponds to the bases in the mRNA, which in turn corresponds to the sequence of amino acids in the protein. In the case of eukaryotic genes, the genes are interrupted by sequences that are not represented in the mRNA and the protein. The term "split genes" was used at the time, but the nomenclature that has stuck describes the unexpected nature of sequences within genes. We call the noncoding DNA that interrupts the sequence of the gene "intervening sequences," or **introns,** and we call the coding sequences **exons** (figure 15.18).

Thus, the basic structure of eukaryotic genes is radically different from that of prokaryotic genes. It is still true that translated eukaryotic mRNA is colinear with its protein product, but the gene may not be. Imagine looking at an interstate highway from a satellite. Scattered randomly along the thread of concrete would be cars, some moving in clusters, others individually; most of the road would be bare. That is what a eukaryotic gene is like—scattered exons embedded within much longer sequences of introns. In humans, only 1 to 1.5% of the genome is devoted to the exons that encode proteins, while 24% is devoted to the noncoding introns within which these exons are embedded.

RNA Splicing

This raises the obvious question of how eukaryotic cells deal with the noncoding introns. The answer is that the product of transcription, called the primary transcript, in addition to the capping and poly-A tail reactions, is cut and put back together to produce the mature mRNA. We call this latter process *RNA splicing*, and it occurs in the nucleus prior to the export of the mRNA to the cytoplasm. The intron/exon junctions are recognized by small nuclear ribonuclearproteins called snRNPs (pronounced "snurps"). These snurps then cluster together forming a

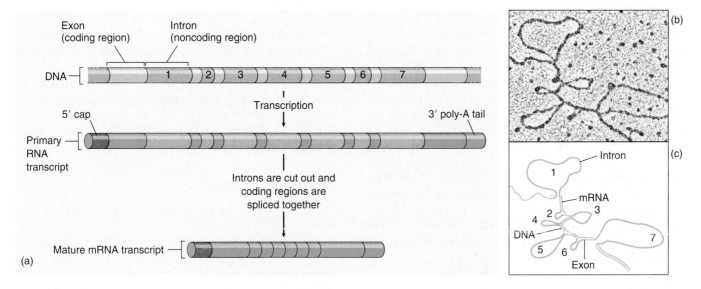

FIGURE 15.18
The eukaryotic ovalbumin gene is fragmented. (*a*) The ovalbumin gene and its primary RNA transcript contain seven segments not present in the mRNA that the ribosomes use to direct protein synthesis. Enzymes cut these segments (introns) out and splice together the remaining segments (exons). (*b*) By hybridizing the processed transcript to the DNA, introns within the DNA sequence can be visualized directly. The seven loops in the electron micrograph are the seven introns represented in (*c*) the schematic drawing.

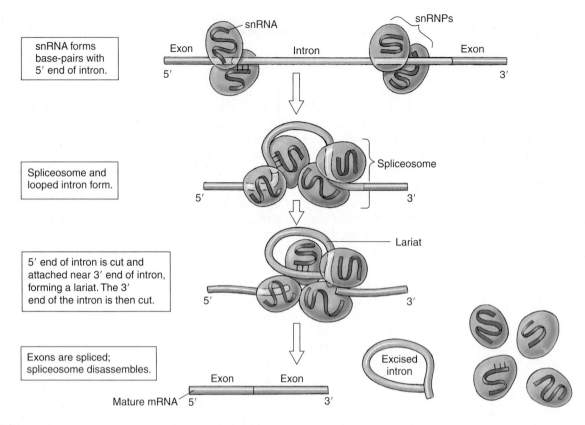

snRNA forms base-pairs with 5′ end of intron.

Exon   Intron   Exon
snRNA   snRNPs
5′   3′

Spliceosome and looped intron form.

Spliceosome
5′   3′

5′ end of intron is cut and attached near 3′ end of intron, forming a lariat. The 3′ end of the intron is then cut.

Lariat
5′   3′

Exons are spliced; spliceosome disassembles.

Excised intron
Exon   Exon
Mature mRNA   5′   3′

FIGURE 15.19

How spliceosomes process RNA. Particles called snRNPs contain snRNA that interacts with the 5′ end of an intron. Several snRNPs come together and form a spliceosome. As the intron forms a loop, the 5′ end is cut and linked to a site near the 3′ end of the intron. The intron forms a lariat that is excised, and the exons are spliced together. The spliceosome then disassembles and releases the mature mRNA.

larger complex called the **spliceosome,** which is responsible for the splicing, or removal, of the introns. The mechanism of splicing includes cleavage of the 5′ end of the intron such that this 5′ end becomes attached to the 2′ OH of an internal A nucleotide, forming a looped structure called a *lariat* (figure 15.19). The 3′ end of the first exon is then used to displace the 3′ end of the intron, joining the two exons together and releasing the intron.

There are no rules for the numbers of introns per gene or the sizes of introns and exons. Some genes have no introns; others may have 50 introns. The sizes of exons range from a few nucleotides to 7500 nucleotides, and the sizes of introns are equally variable. The presence of introns partly explains why so little of the genome is actually composed of "coding sequences" (see chapter 17 for results from the Human Genome Project).

The "how" of splicing is a lot easier to answer than the "why" of introns. In fact, "why are there introns?" may not even be a valid question. There is no consensus as to whether introns arrived early or late in the evolution of eukaryotic lineages; it may not be an all-or-none issue. Researchers have proposed a number of explanations for how

and why introns evolved. One of the more provocative explanations suggests that exons represent functional domains of proteins, and that the intron-exon arrangements found in genes represent the shuffling of these functional units over long periods of evolutionary time. This hypothesis, called *exon shuffling*, has support based on the structure of some genes, but is not supported by the structure of other genes. The degree to which shuffling has been an evolutionary mechanism remains unclear.

Alternative Splicing

One consequence of the splicing process is greater complexity in gene expression in eukaryotes. A single primary transcript can be spliced into different mRNAs by the inclusion of different sets of exons, a process called **alternative splicing.** Although many cases of alternative splicing have been documented, the recent completion of the draft sequence of the human genome, along with other large data sets of expressed sequences, now allow large-scale comparisons between sequences found in mRNAs and the genome. Three different computer-based

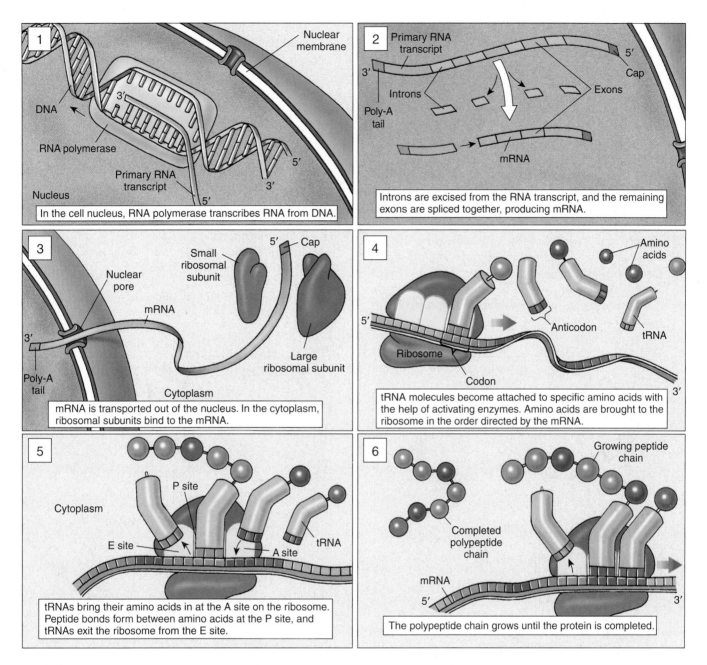

1 In the cell nucleus, RNA polymerase transcribes RNA from DNA.

Nuclear membrane

DNA

RNA polymerase

Primary RNA transcript

Nucleus

3′

5′

3′

5′

2 Introns are excised from the RNA transcript, and the remaining exons are spliced together, producing mRNA.

Primary RNA transcript

5′

Cap

3′

Introns

Poly-A tail

Exons

mRNA

3 mRNA is transported out of the nucleus. In the cytoplasm, ribosomal subunits bind to the mRNA.

Nuclear pore

Small ribosomal subunit

5′

Cap

mRNA

3′

Poly-A tail

Large ribosomal subunit

Cytoplasm

4 tRNA molecules become attached to specific amino acids with the help of activating enzymes. Amino acids are brought to the ribosome in the order directed by the mRNA.

Amino acids

5′

Anticodon

tRNA

Ribosome

Codon

3′

5 tRNAs bring their amino acids in at the A site on the ribosome. Peptide bonds form between amino acids at the P site, and tRNAs exit the ribosome from the E site.

Cytoplasm

P site

E site

A site

tRNA

6 The polypeptide chain grows until the protein is completed.

Growing peptide chain

Completed polypeptide chain

mRNA

5′

3′

FIGURE 15.20
An overview of gene expression in eukaryotes.

analyses have been performed, producing results that are in rough agreement. These initial genomic assessments indicate a range of 35 to 59% for human genes that exhibit some form of alternative splicing. If we pick the middle ground of around 40%, this still vastly increases the number of potential proteins encoded by the 30,000 genes in the human genome. It is important to point out that these analyses are entirely computer-based and have not been experimentally verified, nor have the functions of the possible spliced products been investigated. However, this does explain how the 30,000 genes

found in the human genome are able to encode the 120,000 different translated mRNAs reported to exist in human cells. The emerging field of proteomics will address the number and functioning of proteins encoded by the human genome. Figure 15.20 summarizes eukaryotic transcription and translation.

> Much of the eukaryotic gene is not translated. Noncoding segments called introns, scattered throughout the gene, are removed from the primary transcript before the mRNA is translated.

Differences Between Prokaryotic and Eukaryotic Gene Expression

1. Most eukaryotic genes possess introns. With the exception of a few genes in the Archaebacteria, prokaryotic genes lack introns (figure 15.21).

2. Individual prokaryotic mRNA molecules often contain transcripts of several genes. By placing genes with related functions on the same mRNA, prokaryotes coordinate the regulation of those functions. Eukaryotic mRNA molecules rarely contain transcripts of more than one gene. Regulation of eukaryotic gene expression is achieved in other ways.

3. Because eukaryotes possess a nucleus, their mRNA molecules must be completely formed and must pass across the nuclear membrane before they are translated. Prokaryotes, which lack nuclei, often begin translation of an mRNA molecule before its transcription is completed.

4. In prokaryotes, translation begins at an AUG codon preceded by a special nucleotide sequence. In eukaryotic cells, mRNA molecules are modified at the 5' leading end after transcription, adding a 5' cap, a methylated guanosine triphosphate. The cap initiates translation by binding the mRNA, usually at the first AUG, to the small ribosomal subunit.

5. Eukaryotic mRNA molecules are modified before they are translated: Introns are cut out, and the remaining exons are spliced together; a 5' cap is added; and a 3' poly-A tail consisting of some 200 adenine (A) nucleotides is added. These modifications can delay the destruction of the mRNA by cellular enzymes.

6. The ribosomes of eukaryotes are a little larger than those of prokaryotes.

Gene expression is broadly similar in prokaryotes and eukaryotes, although it differs in some details.

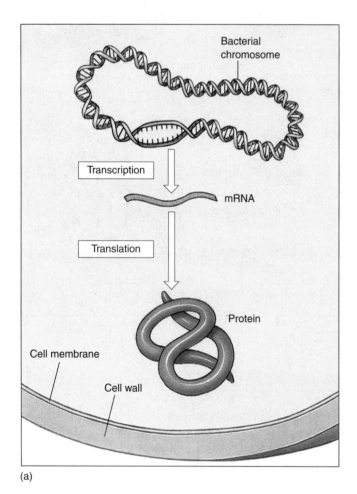

(a)

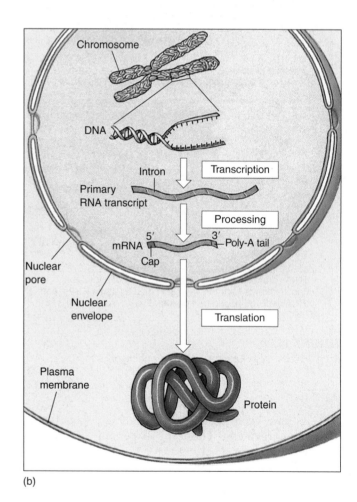

(b)

FIGURE 15.21
Gene information is processed differently in prokaryotes and eukaryotes. (*a*) Prokaryotic genes are transcribed into mRNA, which is translated immediately. Hence, the sequence of DNA nucleotides corresponds exactly to the sequence of amino acids in the encoded polypeptide. (*b*) Eukaryotic genes are typically different, containing long stretches of nucleotides called introns that do not correspond to amino acids within the encoded polypeptide. Introns are removed from the primary RNA transcript of the gene, and a 5' cap and 3' poly-A tail are added before the mRNA directs the synthesis of the polypeptide.

15.1 The Central Dogma traces the flow of gene-encoded information.

Cells Use RNA to Make Protein

- Ribosomes are composed of a large and a small subunit. (p. 302)
- Ribosomal RNA (rRNA) functions in polypeptide synthesis; transfer RNA (tRNA) transports amino acids to the ribosome for use in building polypeptides; and messenger RNA (mRNA) is transcribed from DNA and travels to ribosomes to direct polypeptide assembly. (p. 302)
- The Central Dogma is that information passes from DNA to mRNA (transcription), and then mRNA directs the sequential assembly of amino acids into proteins (translation). (p. 303)

15.2 Genes encode information in three-nucleotide code words.

The Genetic Code

- The genetic code is nearly universal. It is read in codons or triplets, increments of three nucleotides on mRNA, with each triplet coding for an amino acid. (pp. 304–305)
- There are 64 possible codons that code for 20 amino acids. (p. 304)

15.3 Genes are first transcribed, then translated.

Transcription in Prokaryotes

- Only the template strand of DNA is transcribed, while the coding strand is not. (p. 306)
- Transcription begins at the promoter site, and the transcription bubble moves along the DNA segment until the termination sequence is reached; the RNA strand then dissociates, and the DNA rewinds. (pp. 306–307)

Transcription in Eukaryotes

- Eukaryotic transcription differs from prokaryotic transcription because there are three different RNA polymerase enzymes, an initiation complex forms at the promoter, and posttranscriptional modification of the RNA occurs. (pp. 308–309)

Translation

- Translation begins when the initial portion of an mRNA molecule binds to an rRNA molecule in a ribosome. (p. 310)
- Each step of the ribosome's progress exposes a codon to which a tRNA molecule with the complementary anticodon binds. The amino acid carried by each tRNA molecule is added to the end of the growing polypeptide chain. (pp. 311–312)

15.4 Eukaryotic gene transcripts are spliced.

The Discovery of Introns

- Eukaryotic genes are interrupted by sequences not represented in the mature mRNA and the protein. Introns are noncoding sequences that interrupt coding sequences, or exons. (p. 313)
- Spliceosomes are responsible for splicing and removing the introns prior to mRNA translation. (p. 314)

Differences Between Prokaryotic and Eukaryotic Gene Expression

- Gene expression is similar in both prokaryotes and eukaryotes, although some differences do exist, such as ribosomal size and the possession of introns. (p. 316)

Self Test

1. The bases of RNA are the same as those of DNA with the exception that RNA contains
   a. cysteine instead of cytosine.
   b. uracil instead of thymine.
   c. cytosine instead of guanine.
   d. uracil instead of adenine.

2. Which one of the following is not a type of RNA?
   a. nRNA (nuclear RNA)
   b. mRNA (messenger RNA)
   c. rRNA (ribosomal RNA)
   d. tRNA (transfer RNA)

3. Each amino acid in a protein is specified by
   a. several genes.
   b. a promoter.
   c. an mRNA molecule.
   d. a codon.

4. The three-nucleotide codon system can be arranged into _____ combinations.
   a. 16
   b. 20
   c. 64
   d. 128

5. The TATA box in eukaryotes is a
   a. core promoter.
   b. −35 sequence.
   c. −10 sequence.
   d. 5′ cap.

6. The site where RNA polymerase attaches to the DNA molecule to start the formation of RNA is called a(n)
   a. promoter.
   b. exon.
   c. intron.
   d. GC hairpin.

7. When mRNA leaves the cell's nucleus, it next becomes associated with
   a. proteins.
   b. a ribosome.
   c. tRNA.
   d. RNA polymerase.

8. If an mRNA codon reads UAC, its complementary anticodon will be
   a. TUC.
   b. ATG.
   c. AUG.
   d. CAG.

9. The nucleotide sequences on DNA that actually have information encoding a sequence of amino acids are
   a. introns.
   b. exons.
   c. UAA.
   d. UGA.

10. Which of the following statements is correct about prokaryotic gene expression?
    a. Prokaryotic mRNAs must have introns spliced out.
    b. Prokaryotic mRNAs are often translated before transcription is complete.
    c. Prokaryotic mRNAs contain the transcript of only one gene.
    d. All of these statements are correct.

Test Your Visual Understanding

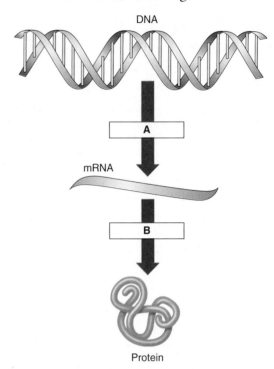

DNA

A

mRNA

B

Protein

1. Match the correct labels from the following list with the lettered boxes on the figure (not all labels will be used).
   replication
   RNA processing
   transcription
   translation
   a. What is this figure illustrating?
   b. Where does process A occur in the eukaryotic cell?
   c. Where does process B occur in the eukaryotic cell?

Apply Your Knowledge

1. Assume you had four copies per cell of mature mRNA for insulin, a protein made of two polypeptide chains with a total of 51 amino acids. Assume one ribosome can attach onto an mRNA every 20 nucleotides. One amino acid can be translated in 60 milliseconds. How many copies of insulin could be made in three minutes?

2. The nucleotide sequence of a hypothetical eukaryotic gene is:
   TACATACTAGTTAC G TCGCCCGGAAATATC

   If a mutation in this gene were to change the fifteenth nucleotide (blue boxed nucleotide) from guanine to thymine, what effect might it have on the expression of this gene?

16

Gene Technology

Concept Outline

16.1 Molecular biologists can manipulate DNA to clone genes.

    The DNA Manipulator's Toolbox. Enzymes that cleave DNA at specific sites allow DNA segments from different sources to be spliced together.

    Host/Vector Systems. Fragments produced by cleaving DNA with restriction endonucleases can be spliced into plasmids, phages, or other vectors, which can be used to insert the DNA into host cells.

    Using Vectors to Transfer Genes. Plasmids are used to transfer smaller fragments of DNA into host cells. Other vectors are used to transfer larger segments of DNA into hosts.

    DNA Libraries. A DNA library is a collection of DNA fragments that comprise an organism's genome.

16.2 Genetic engineering involves easily understood procedures.

    The Four Stages of a Genetic Engineering Experiment. Gene engineers cut DNA into fragments, which they then splice into vectors that carry the fragments into cells. The cells are then screened for the gene of interest.

    Working with Gene Clones. Gene technology is used in a variety of procedures involving DNA manipulation.

16.3 Biotechnology is producing a scientific revolution.

    Medical Applications. Many drugs and vaccines are now produced using gene technology, and gene technology is also being used in the treatment of genetic diseases.

    Agricultural Applications. Genetic engineers have developed crops resistant to pesticides and pests, as well as commercially superior animals.

    Risk and Regulation. Genetic engineering raises important questions about danger and privacy.

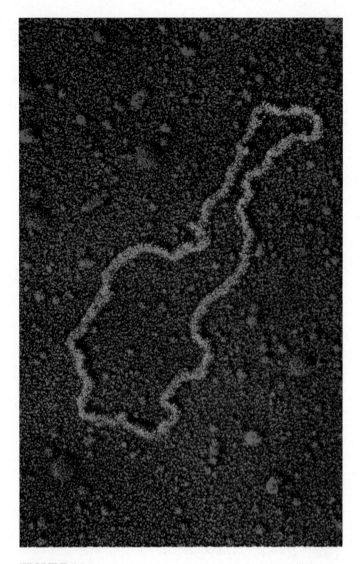

FIGURE 16.1
A famous plasmid. The circular molecule in this electron micrograph is pSC101, the first plasmid used successfully to clone a vertebrate gene. Its name comes from the fact that it was the 101st plasmid isolated by Stanley Cohen.

Over the past decades, the development of new and powerful techniques for studying and manipulating DNA has revolutionized genetics. The cloning plasmid shown in figure 16.1 is one example of the tools that are now available. These techniques have allowed biologists to intervene directly in the genetic fate of organisms for the first time. In this chapter, we explore these technologies and consider how they apply to specific problems of great practical importance. Few areas of biology will have as great an impact on our future lives.

16.1 Molecular biologists can manipulate DNA to clone genes.

The ability to directly isolate and manipulate genetic material was one of the most profound changes in the field of biology in the late twentieth century. Now, in the twenty-first century, we are already looking at entire genomes. So how did we get from cloning individual genes to determining the sequence of the human genome? We will begin to answer this question by describing the technology necessary to clone genes and how this technology is being applied.

The DNA Manipulator's Toolbox

Development of the technology behind cloning and manipulating DNA sequences began in 1975 with the construction of the first recombinant DNA molecules. From these first tentative steps, a technology evolved that has altered all of biology. To understand this technology, we need to first consider the tools required to work directly with DNA. From this, we will step through a simple genetic engineering experiment, and then look at how these techniques are being applied in medicine and agriculture.

One of the most basic requirements for manipulating DNA is a set of enzymes. The toolbox of the molecular biologist is filled with such enzymes. This toolbox has become extremely sophisticated over the years as more and more enzymes that alter DNA have been isolated. Notice that all of these enzymes already exist in nature, so that all of our manipulations are at some level merely imitations of what cells do naturally.

Restriction Endonucleases

The enzymes that catalyzed the revolution in molecular biology were those able to cleave DNA at specific sites: **restriction endonucleases.** Nucleases are enzymes that degrade DNA, and many were known prior to the isolation of the first restriction enzyme. But restriction endonucleases are different because they are able to fragment DNA at specific sites. This kind of activity, long sought by molecular biologists, eventually came out of basic research on why bacterial viruses can infect some cells and not others. The molecular basis for this "host restriction" proved to be enzymes that could cleave DNA at specific sequences. The host cells are able to avoid being cleaved by modifying their own DNA at the same sites by methylation. Since the initial discovery of these restriction endonucleases, hundreds more have been isolated that recognize and cleave different **restriction sites.**

The ability to cut DNA at specific sites is significant in two ways: First, it allows a form of physical mapping that was previously not possible, and second, it allows the creation of recombinant molecules. Physical maps can be constructed based on the positioning of cleavage sites for restriction enzymes. Such restriction maps provide crucial data for identifying and working with DNA molecules. The ability to construct recombinant molecules is even more important, because many steps in the process of cloning and manipulating DNA require the ability to put molecules from different sources together.

How do restriction enzymes allow the creation of recombinant molecules? The answer lies in the nature of the enzymes themselves. There are two types of restriction enzymes: type I and type II. Type I enzymes make simple cuts across both DNA strands, and are not often used in cloning and manipulating DNA. It is type II enzymes that allow creation of recombinant molecules. Type II enzymes recognize a specific DNA sequence, ranging from 4 bases to 12 bases, and cleave the DNA at a particular base within this sequence. Type II sites show what is called *dyad symmetry*, or twofold rotational symmetry. This can be seen in the sequence itself: It reads the same from 5' to 3' in one direction as it does on the other strand in the opposite direction 5' to 3'. Given this kind of sequence, cutting the DNA at the same base on either strand can lead to staggered cuts that will produce "sticky ends." This short, unpaired sequence will be the same for any DNA that is cut by the enzyme. Thus, these sticky ends allow DNAs from different sources to be easily joined together (figure 16.2).

Ligase

The two ends of a DNA molecule cut by a type II restriction enzyme have complementary sequences as we have seen, and so can pair to form a duplex. But to form a stable DNA molecule from the two fragments, we still need an enzyme to join the molecules. The enzyme **DNA ligase** catalyzes the formation of a phosphodiester bond between adjacent phosphate and hydroxyl groups of DNA nucleotides. For a proper link to form, there can be no gaps in either strand, just a "nick" in each strand. A nick is a break in one strand with no gap. The action of ligase is to seal nicks in one or both strands. This is the same enzyme that joins Okazaki fragments on the lagging strand during DNA replication (see chapter 14). In the toolbox of the molecular biologist, the action of ligase is necessary to create stable recombinant molecules from the fragments that restriction enzymes make possible.

The tools of the molecular biologist are enzymes that cut DNA into fragments and other enzymes that join the DNA fragments into complete molecules.

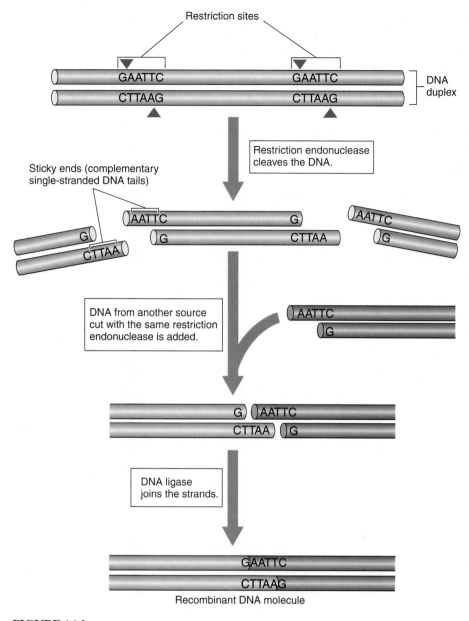

Restriction sites

GAATTC   GAATTC   DNA
CTTAAG   CTTAAG   duplex

Restriction endonuclease
cleaves the DNA.

Sticky ends (complementary
single-stranded DNA tails)

AATTC   G
G   CTTAA

AATTC
G

G
CTTAA

DNA from another source
cut with the same restriction
endonuclease is added.

AATTC
G

G   AATTC
CTTAA   G

DNA ligase
joins the strands.

GAATTC
CTTAAG

Recombinant DNA molecule

FIGURE 16.2
Many restriction endonucleases produce DNA fragments with "sticky ends." The restriction endonuclease *Eco*RI always cleaves the sequence GAATTC between G and A. Because the same sequence occurs on both strands, both are cut. However, the two sequences run in opposite directions on the two strands. As a result, single-stranded tails are produced that are complementary to each other, or "sticky."

Host/Vector Systems

Although we can synthesize short sequences of DNA in vitro, to clone large unknown sequences, it is important to be able to propagate recombinant DNA molecules in a cell. Thus, along with a toolbox of enzymes, molecular biologists use cells as factories to produce large amounts of recombinant DNA. The ability to propagate DNA in a host cell requires a **vector** (something to carry the recombinant DNA molecule) that can enter and replicate in the host.

Such host/vector systems are crucial to molecular biology.

The most flexible and common host used for routine cloning is the bacterium *E. coli*. This is not the only host, of course. We now routinely clone eukaryotic DNA, using mammalian tissue culture cells, yeast cells, and insect cells as host systems. Each kind of host/vector system allows particular uses of the cloned DNA.

The two most commonly used vectors are plasmids and phages. *Plasmids* are small, extrachromosomal DNAs that are dispensable to the cell. *Phages* are viruses that infect bacterial cells. The key to a vector is that it is not required by the cell, but it can be directly selected for by including a marker such as antibiotic resistance.

Plasmids

Plasmid vectors (small, circular chromosomes) are typically used to clone relatively small pieces of DNA, up to a maximum of about 10 kilobases (kb). A plasmid vector must have (1) an origin of replication to allow it to be replicated in *E. coli* independently of the chromosome, and (2) a selectable marker, usually antibiotic resistance. The selectable marker allows the presence of the plasmid to be easily identified through selection; cells that contain the marker will live when plated on antibiotic-containing growth media, and cells that lack the plasmid will not live (they are killed by the antibiotic). A fragment of DNA is inserted by the techniques described in figure 16.2 in a region of the plasmid called the **multiple cloning site (MCS)**. This region contains a number of unique restriction sites such that when the plasmid is cut with any of these enzymes, it will produce a linear plasmid. The DNA of interest can then be ligated into this site, after which plasmid DNA is introduced into cells.

Next, the presence of inserted DNA must be confirmed. This is typically done by screening for inactivation of another gene. For example, in one method, the presence of inserted DNA results in inactivation of the *LacZ'* gene. This event can be detected by plating cells on medium containing X-gal, a substrate for the enzyme encoded by *LacZ'*. Cells with an active version of the *LacZ'* gene will produce the enzyme and will be blue on this substrate, while cells that have an inactive *LacZ'* will lack the enzyme and will be white (figure 16.3*a*). Note that this is a "screen," not a "selection," because all cells survive. The use of selection or screening depends on the situation; the two approaches differ in their relative power. Selection allows the identification of extremely rare events because only the desired phenotype survives. Genetic screening preserves all the cells in an experiment, but is limited by the ability of the investigator to distinguish between two phenotypes—that is, all cells have to be checked, as opposed to looking for a survivor.

Phages

Phage vectors are larger than plasmid vectors, and can take inserts up to 40 kb. Most phage vectors are based on the well-studied virus phage lambda. Lambda-based vectors are used today primarily for cDNA libraries (that is, collections of DNA fragments complementary to the DNA being investigated). Lambda's use as a vector, while useful for cloning large fragments, has two requirements not shared with plasmid vectors: First, lambda requires live cells to replicate, so it is necessary to infect cells to recover lambda. Second, the lambda genome is linear, so instead of "opening up" a circle, the middle part of the lambda genome is removed and replaced with the inserted DNA. This means that after the inserted DNA is ligated to the two lambda "arms," it must be packaged into a phage head in vitro, and then used to infect *E. coli*. The two "arms" without any inserted DNA are not packaged into phage heads. This is a kind of selection for recombinant phages, because the arms alone will not be propagated (figure 16.3*b*).

Producing large quantities of recombinant DNA is accomplished by using vectors such as plasmids and phages. A gene of interest can be inserted into vector DNA, which infects and is replicated in a cell.

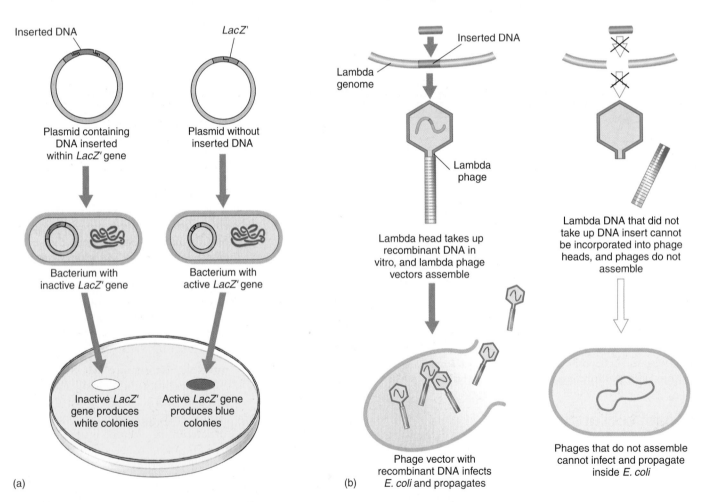

(a) (b)

FIGURE 16.3

Using plasmid and phage vectors. (*a*) Host bacteria can be screened for the presence of recombinant plasmids by observing the presence of a phenotype in the host, such as the lack of production of blue color when the bacteria are plated in a medium containing X-gal. (*b*) Phage vectors can be selected for the presence of recombinant DNA by the ability of the phage to assemble in vitro, infect a host, and propagate inside its host.

Using Vectors to Transfer Genes

A chimera is a mythical creature with the head of a lion, the body of a goat, and the tail of a serpent. Although no such creatures ever existed in nature, biologists have made chimeras of a more modest kind through genetic engineering.

Constructing pSC101

One of the first genetically engineered **chimeras** was manufactured from a bacterial plasmid called a resistance transfer factor by American geneticists Stanley Cohen and Herbert Boyer in 1973. Cohen and Boyer used a restriction endonuclease called *Eco*RI, which is obtained from *Escherichia coli*, to cut the plasmid into fragments. One fragment, 9000 nucleotides in length, contained both the origin of replication necessary for replicating the plasmid and a gene that conferred resistance to the antibiotic tetracycline (*tet*R). Because both ends of this fragment were cut by the same restriction endonuclease, they could be ligated to form a circle, a smaller plasmid Cohen dubbed pSC101.

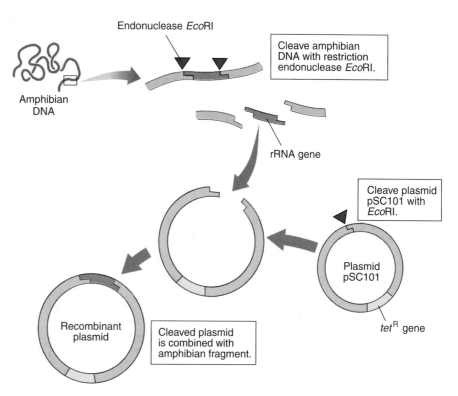

Endonuclease *Eco*RI

Cleave amphibian DNA with restriction endonuclease *Eco*RI.

Amphibian DNA

rRNA gene

Cleave plasmid pSC101 with *Eco*RI.

Plasmid pSC101

*tet*R gene

Recombinant plasmid

Cleaved plasmid is combined with amphibian fragment.

FIGURE 16.4
One of the first genetic engineering experiments. This diagram illustrates how Cohen and Boyer inserted an amphibian gene encoding rRNA into pSC101. The plasmid contains a single site cleaved by the restriction endonuclease *Eco*RI; it also contains *tet*R, a gene that confers resistance to the antibiotic tetracycline. The rRNA-encoding gene was inserted into pSC101 by cleaving the amphibian DNA and the plasmid with *Eco*RI and allowing the complementary sequences to pair.

Using pSC101 to Make Recombinant DNA

Cohen and Boyer also used *Eco*RI to cleave DNA that coded for rRNA that they had isolated from an adult amphibian, the African clawed frog, *Xenopus laevis*. They then mixed the fragments of *Xenopus* DNA with pSC101 plasmids that had been "reopened" by *Eco*RI, and allowed bacterial cells to take up DNA from the mixture (figure 16.4). Some of the bacterial cells immediately became resistant to tetracycline, indicating that they had incorporated the pSC101 plasmid with its antibiotic-resistance gene. Furthermore, some of these pSC101-containing bacteria also began to produce frog ribosomal RNA! Cohen and Boyer concluded that the frog rRNA gene must have been inserted into the pSC101 plasmids in those bacteria. In other words, the two ends of the pSC101 plasmid, produced by cleavage with *Eco*RI, had joined to the two ends of a frog DNA fragment that contained the rRNA gene, also cleaved with *Eco*RI.

The pSC101 plasmid containing the frog rRNA gene is a true chimera, an entirely new genome that never existed in nature and never would have evolved by natural means. It is a form of **recombinant DNA**—that is, DNA created in the laboratory by joining together pieces of different genomes to form a novel combination.

Other Vectors

The introduction of foreign DNA fragments into host cells has become common in molecular genetics. As mentioned previously, the genome that carries the foreign DNA into the host cell is called a vector. Plasmids, with names such as pUC18, can be induced to make hundreds of copies of themselves and thus of the foreign genes they contain. Much larger pieces of DNA can be introduced using a YAC (yeast artificial chromosome) as a vector instead of a plasmid. Not all vectors have bacterial targets. Animal viruses are serving as vectors to carry genes into monkey and human cells, and animal genes have even been introduced into plant cells.

One of the first recombinant genomes produced by genetic engineering was a bacterial plasmid into which an amphibian ribosomal RNA gene was inserted. Viruses or artificial chromosomes can also be used as vectors to insert foreign DNA into host cells and create recombinant genomes.

DNA Libraries

In order to place a particular gene or DNA sequence into a vector, you must first have a DNA source that contains that sequence. Typically, the source is a collection of DNA fragments representing all of the DNA from an organism. We call this a **DNA library.** A DNA library is a collection of DNAs from a specific source in a form that can be propagated in a host. In the case of a plasmid, this would be a collection of cells all harboring plasmids with different inserted DNAs such that the whole genome is represented. In the case of a phage, this would be a collection of phages, each carrying a different insert such that the entire genome is represented (figure 16.5).

Genomic Libraries

The simplest kind of DNA library that can be made is a **genomic library,** a representation of the entire genome in a vector. This entire genome is randomly fragmented, each fragment having been inserted into a vector and introduced into the host. Creating a genomic library is harder than it sounds, because it is difficult to truly randomly fragment DNA. The best methods use hydrodynamic shear forces, created by passing the DNA through a syringe. The experimenter then adds small synthetic linker sequences to the ends of the random fragments. These linkers contain a restriction site that can be used to connect the inserts to the vector. Genomic libraries are usually made in phage lambda due to the larger insert sizes. A genomic library would be a set of phage with random inserts that cover the entire genome.

cDNA Libraries

In addition to genomic libraries, investigators often wish to limit an experiment to all *expressed* genes. This is a much smaller amount of DNA than the entire genome. Such a gene-expression library is made possible by the use of another enzyme: **reverse transcriptase.** Reverse transcriptase was isolated from a class of viruses called retroviruses. The life cycle of a retrovirus requires making a DNA copy from the RNA genome of the virus. We can take advantage of the activity of the retrovirus enzyme to make DNA copies from mRNA. Such DNA copies of mRNA are called cDNA (figure 16.6). A cDNA library is made by first isolating mRNA from genes being expressed, and then using mRNA to make cDNA with the reverse transcriptase enzyme. The cDNA is then used to make a library, usually in phage lambda. These cDNA libraries are extremely useful and are often made for the genes expressed in specific tissues or cells.

DNA libraries are produced using plasmids or phages as vectors. The DNA library consists of all host cells that have taken up the recombinant DNA, representing all the DNA from the source DNA.

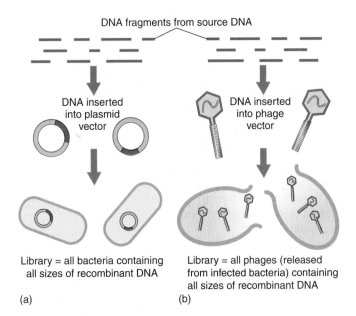

FIGURE 16.5
Creating DNA libraries. DNA libraries can be produced using (*a*) plasmid vectors or (*b*) phage vectors.

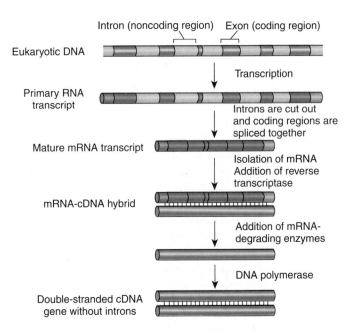

FIGURE 16.6
The formation of cDNA. A mature mRNA transcript is isolated from the cytoplasm of a cell. The enzyme reverse transcriptase is then used to make a DNA strand complementary to the processed mRNA. That newly made strand of DNA is the template for the enzyme DNA polymerase, which assembles a complementary DNA strand along it, producing cDNA, a double-stranded DNA version of the intron-free mRNA.

Genetic engineering involves easily understood procedures.

The Four Stages of a Genetic Engineering Experiment

Like the experiment of Cohen and Boyer, most genetic engineering experiments consist of four stages: DNA cleavage, production of recombinant DNA, cloning, and screening.

Stage 1: DNA Cleavage

A restriction endonuclease is used to cleave the source DNA into fragments. Because the endonuclease's recognition sequence is likely to occur many times within the source DNA, cleavage will produce a large number of different fragments. A different set of fragments will be obtained by employing endonucleases that recognize different sequences. The fragments can be separated from one another according to their size by gel electrophoresis, a procedure illustrated in figure 16.7.

Stage 2: Production of Recombinant DNA

The fragments of DNA are inserted into plasmids or viral vectors, which have been cleaved with the same restriction endonuclease as the source DNA.

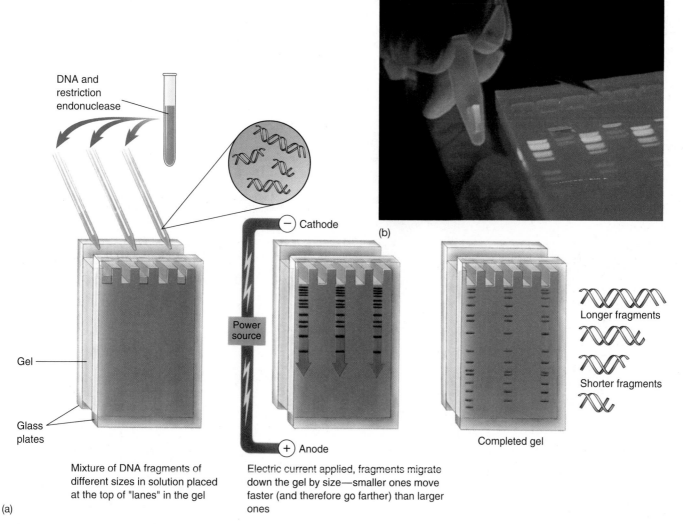

FIGURE 16.7
Gel electrophoresis. (*a*) After restriction endonucleases have cleaved the DNA, the fragments are loaded on a gel, and an electric current is applied. The DNA fragments migrate through the gel, with bigger ones moving more slowly. The fragments can be visualized easily because the migrating bands fluoresce in UV light when stained with ethidium bromide. (*b*) In the photograph, one band of DNA has been excised from the gel for further analysis and can be seen glowing in the tube the technician holds.

Stage 3: Cloning

The plasmids or viruses serve as vectors that can introduce the DNA fragments into cells—usually, but not always, bacteria (figure 16.8). As each cell reproduces, it forms a clone of cells that all contain the fragment-bearing vector. Each clone is maintained separately, and all of them together constitute a clone library of the original source DNA as described in section 16.1.

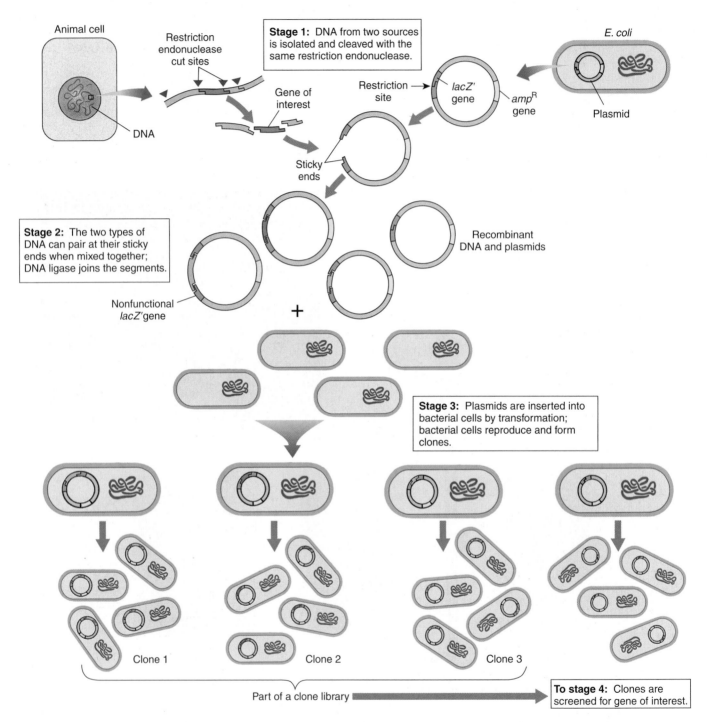

FIGURE 16.8

Stages in a genetic engineering experiment.   In stage 1, DNA containing the gene of interest (in this case, from an animal cell) and DNA from a plasmid are cleaved with the same restriction endonuclease. The genes *amp*^R and *lacZ'* are contained within the plasmid and used for screening a clone (stage 4). In stage 2, the two cleaved sources of DNA are mixed together and pair at their sticky ends. In stage 3, the recombinant DNA is inserted into a bacterial cell, which reproduces and forms clones. In stage 4, the bacterial clones will be screened for the gene of interest.

Stage 4: Screening

The clones containing a specific DNA fragment of interest, often a fragment that includes a particular gene, are identified from the clone library. Let's examine this stage in more detail, because it is generally the most challenging in any genetic engineering experiment.

4–I: Preliminary Screening of Clones. Investigators initially try to eliminate from the library any clones that do not contain vectors, as well as clones whose vectors do not contain fragments of the source DNA. The first category of clones can be eliminated by employing a vector with a gene that confers resistance to a specific antibiotic, such as tetracycline, penicillin, or ampicillin. In figure 16.9a, the gene *amp*^{R} is incorporated into the plasmid and confers resistance to the antibiotic ampicillin. When the clones are exposed to a medium containing that antibiotic, only clones that contain the vector will be resistant to the antibiotic and able to grow. This is an example of genetic selection for plasmids

containing cells. It is always desirable to design experiments able to provide at least one round of selection.

One way to eliminate clones with vectors lacking an inserted DNA fragment is to use a vector that, in addition to containing antibiotic-resistance genes, contains the *lacZ'* gene, which is required to produce β-galactosidase, an enzyme that enables the cells to metabolize the sugar X-gal. Metabolism of X-gal results in the formation of a blue reaction product, so any cells whose vectors contain a functional version of this gene will turn blue in the presence of X-gal (figure 16.9b). However, if we use a restriction endonuclease whose recognition sequence lies within the *lacZ'* gene, the gene will be interrupted when recombinants are formed, and the cell will be unable to metabolize X-gal. Therefore, cells with vectors that contain a fragment of source DNA should remain colorless in the presence of X-gal. This is an example of a genetic screen. Because all cells, both with and without an insert, survive, it is not selection. We often screen based on presence or absence of a particular phenotype (blue color on X-gal in this case).

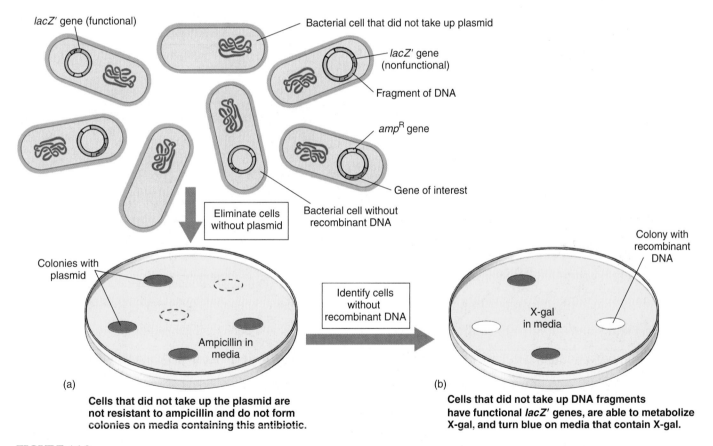

FIGURE 16.9

Stage 4-I: Identifying recombinant clones. Bacteria are transformed with recombinant plasmids that contain the gene for ampicillin resistance (*amp*^{R}). This allows selection on medium containing ampicillin. The plasmid also contains the *lacZ'* gene that encodes β-galactosidase, an enzyme involved in the metabolism of the sugar lactose. An artificial substrate called X-gal can be cleaved by the enzyme to produce a blue color. (*a*) Selection for *amp*^{R}: Only those bacteria that have been transformed and contain a plasmid will be able to grow on ampicillin. (*b*) Screening for β-galactosidase activity: The plasmid has been constructed such that inserted DNA will interrupt the *lacZ'* gene, making the β-galactosidase enzyme nonfunctional. Thus, when plated on medium containing X-gal, bacteria harboring plasmids with no insert will be blue, while cells that harbor plasmids that have inserts will not (they will be white).

Any cells that are able to grow in a medium containing the antibiotic but don't turn blue in the medium with X-gal must have incorporated a vector with a fragment of source DNA. Identifying cells that have a *specific* fragment of the source DNA is the next step in screening clones.

4–II: Finding the Gene of Interest.

A clone library may contain anywhere from a few dozen to many thousand individual fragments of source DNA. Many of those fragments will be identical, so to assemble a complete library of the entire source genome, several hundred thousand clones could be required. A complete *Drosophila* (fruit fly) library, for example, contains more than 40,000 different clones; a complete human library consisting of fragments 20 kilobases long would require close to a million clones. To search such an immense library for a clone that contains a fragment corresponding to a particular gene requires ingenuity, but many different approaches have been successful.

The most general procedure for screening clone libraries to find a particular gene is **hybridization** (figure 16.10). In this method, the cloned genes form base-pairs with complementary sequences on another nucleic acid. The complementary nucleic acid is called a **probe** because it is used to probe for the presence of the gene of interest.

At least part of the nucleotide sequence of the gene of interest must be known to be able to construct the probe.

In this method of screening, bacterial colonies containing an inserted gene are grown on agar. Some cells are transferred to a filter pressed onto the colonies, forming a replica of the plate. The filter is then treated with a solution that denatures the bacterial DNA and contains a radioactively labeled probe. The probe hybridizes with complementary single-stranded sequences on the bacterial DNA.

When the filter is laid over photographic film, areas that contain radioactivity will expose the film (autoradiography). Only colonies that contain the gene of interest hybridize with the radioactive probe and emit radioactivity onto the film. The pattern on the film is then compared to the original master plate, and the gene-containing colonies may be identified.

Genetic engineering generally involves four stages: cleaving the source DNA; making recombinants; cloning copies of the recombinants; and screening the cloned copies for the desired gene. Screening can be achieved by making the desired clones resistant to certain antibiotics and giving them other properties that make them readily identifiable.

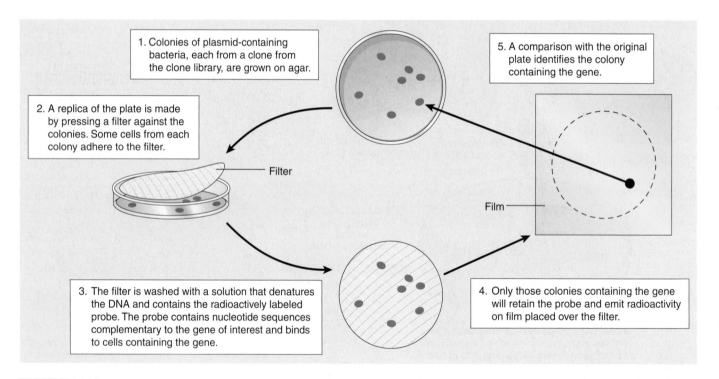

1. Colonies of plasmid-containing bacteria, each from a clone from the clone library, are grown on agar.

2. A replica of the plate is made by pressing a filter against the colonies. Some cells from each colony adhere to the filter.

Filter

3. The filter is washed with a solution that denatures the DNA and contains the radioactively labeled probe. The probe contains nucleotide sequences complementary to the gene of interest and binds to cells containing the gene.

4. Only those colonies containing the gene will retain the probe and emit radioactivity on film placed over the filter.

Film

5. A comparison with the original plate identifies the colony containing the gene.

FIGURE 16.10

Stage 4-II: Using hybridization to identify the gene of interest.   (*1*) Each of the colonies on these bacterial culture plates represents millions of clones descended from a single cell. To test whether a certain gene is present in any particular clone, it is necessary to identify colonies whose cells contain DNA that hybridizes with a probe containing DNA sequences complementary to the gene. (*2*) Pressing a filter against the master plate causes some cells from each colony to adhere to the filter. (*3*) The filter is then washed with a solution that denatures the DNA and contains the radioactively labeled probe. (*4*) Only those colonies that contain DNA that hybridizes with the probe, and thus contain the gene of interest, will expose film in autoradiography. (*5*) The film is then compared to the master plate to identify the gene-containing colony.

Working with Gene Clones

Once a gene has been successfully cloned, a variety of procedures are available to characterize it.

Getting Enough DNA to Work With: The Polymerase Chain Reaction

Once a particular gene is identified within the library of DNA fragments, the final requirement is to make multiple copies of it. One way to do this is to insert the identified fragment into a bacterium; after repeated cell divisions, millions of cells will contain copies of the fragment. A far more direct approach, however, is to use DNA polymerase to copy the gene sequence of interest through the **polymerase chain reaction** (**PCR;** figure 16.11). Kary Mullis developed PCR in 1983 while he was a staff chemist at the Cetus Corporation; in 1993, his discovery won him the Nobel Prize in chemistry. PCR can amplify specific sequences or add sequences (such as endonuclease recognition sequences) as primers to cloned DNA. There are three steps in PCR:

Step 1: Denaturation. First, an excess of primer (typically, a synthetic sequence of 20 to 30 nucleotides) is mixed with the DNA fragment to be amplified. This mixture of primer and fragment is heated to about 98°C. At this temperature, the double-stranded DNA fragment dissociates into single strands.

Step 2: Annealing of Primers. Next, the solution is allowed to cool to about 60°C. As it cools, the single strands of DNA reassociate into double strands. However, because of the large excess of primer, each strand of the fragment base-pairs with a complementary primer flanking the region to be amplified, leaving the rest of the fragment single-stranded.

Step 3: Primer Extension. Now a very heat-stable type of DNA polymerase, called Taq polymerase (after the thermophilic bacterium *Thermus aquaticus*, from which Taq is extracted) is added, along with a supply of all four nucleotides. Using the primer, the polymerase copies the rest of the fragment as if it were replicating DNA. When it is done, the primer has been lengthened into a complementary copy of the entire single-stranded fragment. Because *both* DNA strands are replicated, there are now two copies of the original fragment.

Steps 1 to 3 are now repeated, and the two copies become four. It is not necessary to add any more polymerase, because the heating step does not harm this particular enzyme. Each heating and cooling cycle, which can be as short as 1 or 2 minutes, doubles the number of DNA molecules. After 20 cycles, a single fragment produces more than one million (2^{20}) copies! In a few hours, 100 billion copies of the fragment can be manufactured.

PCR, now fully automated, has revolutionized many aspects of science and medicine because it allows the investigation of minute samples of DNA. In criminal investiga-

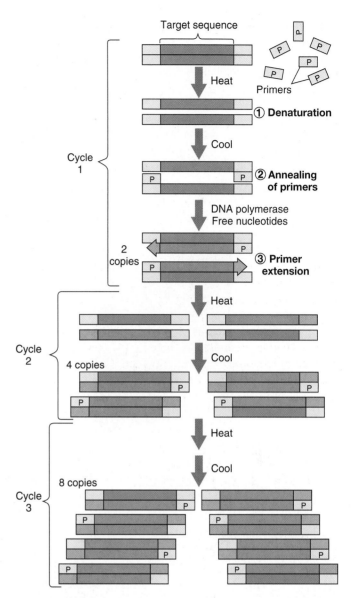

FIGURE 16.11

The polymerase chain reaction. (*1*) *Denaturation:* A solution containing primers and the DNA fragment to be amplified is heated so that the DNA dissociates into single strands. (*2*) *Annealing of primers:* The solution is cooled, and the primers bind to complementary sequences on the DNA flanking the region to be amplified. (*3*) *Primer extension:* DNA polymerase then copies the remainder of each strand, beginning at the primer. Steps 1–3 are then repeated with the replicated strands. This process is repeated many times, each time doubling the number of copies, until enough copies of the DNA fragment exist for analysis.

tions, *DNA fingerprints* are prepared from the cells in a tiny speck of dried blood or at the base of a single human hair. In pediatric medicine, physicians can detect genetic defects in very early embryos by collecting a few sloughed-off cells and amplifying their DNA. PCR could also be used to examine the DNA of historical figures, such as Abraham Lincoln, or that of now-extinct species, as long as even a minuscule amount of their DNA remains intact.

Identifying DNA: Southern Blotting

Once a gene has been cloned, it may be used as a probe to identify the same or a similar gene in another sample (figure 16.12). In this procedure, called a **Southern blot,** DNA from the sample is cleaved into restriction fragments with a restriction endonuclease, and the fragments are spread apart by gel electrophoresis. The double-stranded helix of each DNA fragment is then denatured into single strands by making the pH of the gel basic, and the gel is "blotted" with a sheet of nitrocellulose, transferring some of the DNA strands to the sheet. Next, a probe consisting of purified, single-stranded DNA corresponding to a specific gene (or mRNA transcribed from that gene) is poured over the sheet. Any fragment that has a nucleotide sequence complementary to the probe's sequence will hybridize (base-pair) with the probe (figure 16.13). If the probe has been labeled with ^{32}P, it will be radioactive, and the sheet will show a band of radioactivity where the probe hybridized with the complementary fragment.

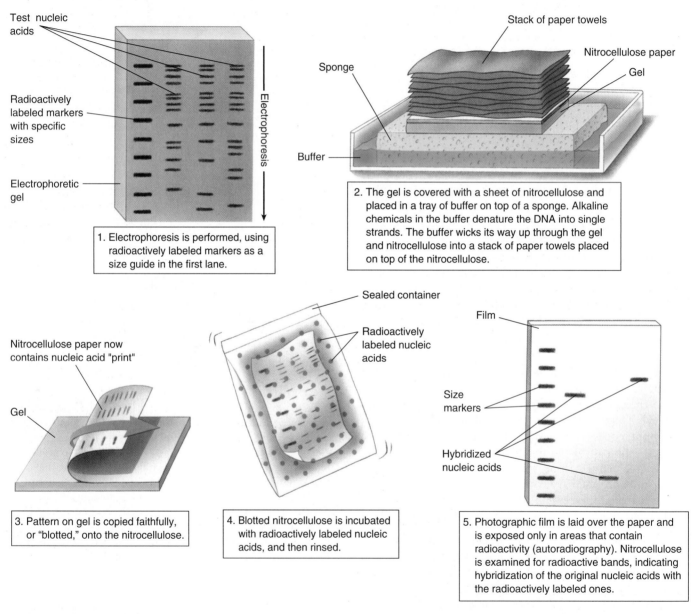

1. Electrophoresis is performed, using radioactively labeled markers as a size guide in the first lane.

2. The gel is covered with a sheet of nitrocellulose and placed in a tray of buffer on top of a sponge. Alkaline chemicals in the buffer denature the DNA into single strands. The buffer wicks its way up through the gel and nitrocellulose into a stack of paper towels placed on top of the nitrocellulose.

3. Pattern on gel is copied faithfully, or "blotted," onto the nitrocellulose.

4. Blotted nitrocellulose is incubated with radioactively labeled nucleic acids, and then rinsed.

5. Photographic film is laid over the paper and is exposed only in areas that contain radioactivity (autoradiography). Nitrocellulose is examined for radioactive bands, indicating hybridization of the original nucleic acids with the radioactively labeled ones.

FIGURE 16.12
The Southern blot procedure. E. M. Southern developed this procedure in 1975 to enable DNA fragments of interest to be visualized in a complex sample containing many other fragments of similar size. The DNA is separated on a gel, and then transferred ("blotted") onto a solid support medium such as nitrocellulose paper or a nylon membrane. Next, it is incubated with a radioactive single-stranded copy of the gene of interest, which hybridizes to the blot at the location(s) where there is a fragment with a complementary sequence. The positions of radioactive bands on the blot identify the fragments of interest.

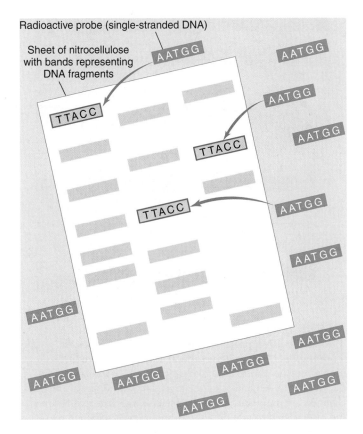

FIGURE 16.13
The hybridization step in Southern blot procedure. A probe (usually several hundred nucleotides in length) of single-stranded DNA (or an mRNA complementary to the gene of interest) is poured over the nitrocellulose sheet containing the DNA fragments. All DNA fragments that contain nucleotide sequences complementary to the probe will bind the probe.

Distinguishing Differences in DNA: RFLP Analysis

Often a researcher wishes not to find a specific gene, but rather to identify a particular *individual* using a specific gene as a marker. One powerful way to do this is by analyzing **restriction fragment length polymorphisms,** or **RFLPs** (figure 16.14). Point mutations, sequence repetitions, and transposons (see chapter 20) that occur within or between the restriction endonuclease recognition sites will alter the length of the DNA fragments (restriction fragments) the restriction endonucleases produce. DNA from different individuals rarely has exactly the same array of restriction sites and distances between sites, so the population is said to be polymorphic (having many forms) for their restriction fragment patterns. By cutting a DNA sample with a particular restriction endonuclease, separating the fragments according to length on an electrophoretic gel, and then using a radioactive probe to identify the fragments on the gel, we can obtain a pattern of bands often unique for each region of DNA analyzed. These are the **DNA fingerprints** mentioned previously as being used in forensic analysis during criminal investigations. RFLPs are also useful as markers to identify particular groups of people at risk for certain genetic disorders. This technique requires a large amount of DNA relative to PCR-based techniques, but is very reliable. The use of PCR has facilitated comparisons with smaller samples.

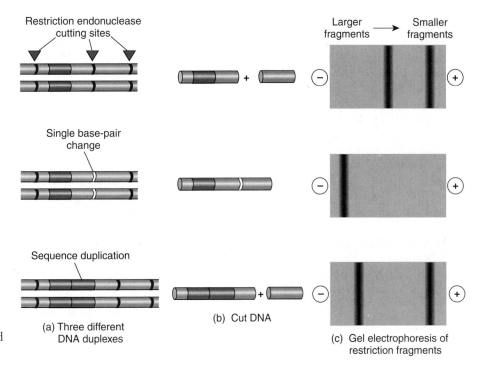

FIGURE 16.14
Restriction fragment length polymorphism (RFLP) analysis.
(*a*) Three samples of DNA differ in their restriction sites due to a single base-pair substitution in one case and a sequence duplication in another case. (*b*) When the samples are cut with a restriction endonuclease, different numbers and sizes of fragments are produced. (*c*) Gel electrophoresis separates the fragments, and different banding patterns result.

DNA Fingerprinting

As stated previously, two individuals rarely produce identical RFLP analyses; therefore, these DNA fingerprints can be used in criminal investigations. Figure 16.15 shows the DNA fingerprints a prosecuting attorney presented in a rape trial in 1987. They consist of autoradiographs, parallel bars on X-ray film resembling the line patterns of the universal price code found on groceries. Each bar represents the position of a DNA restriction endonuclease fragment produced by techniques similar to those described in figures 16.7 and 16.14. The lane with many bars represents a standardized control. Two different probes were used to identify the restriction fragments. A vaginal swab had been taken from the victim within hours of her attack; from it, semen was collected and its DNA analyzed for restriction endonuclease patterns.

Compare the restriction endonuclease patterns of the semen to that of the suspect, Andrews. You can see that the suspect's two patterns match that of the rapist (and are not at all like those of the victim). Clearly, the semen collected from the rape victim and the blood sample from the suspect came from the same person. The suspect was Tommie Lee Andrews, and on November 6, 1987, the jury returned a verdict of guilty. Andrews became the first person in the United States to be convicted of a crime based on DNA evidence.

Since the Andrews verdict, DNA fingerprinting has been admitted as evidence in more than 2000 court cases (figure 16.16). While some probes highlight profiles shared by many people, others are quite rare. Using several probes, identity can be clearly established or ruled out.

Just as fingerprinting revolutionized forensic evidence in the early 1900s, so DNA fingerprinting is revolutionizing it today. A hair, a minute speck of blood, or a drop of semen can all serve as sources of DNA to damn or clear a suspect. As the man who analyzed Andrews's DNA says, "It's like leaving your name, address, and social security number at the scene of the crime. It's that precise." Of course, laboratory analyses of DNA samples must be carried out properly—sloppy procedures could lead to a wrongful conviction. After widely publicized instances of questionable lab procedures, national standards are being developed.

> Techniques such as Southern blotting and PCR enable investigators to identify specific genes and produce them in large quantities, while RFLP analysis and DNA fingerprinting identify individuals and unknown gene sequences.

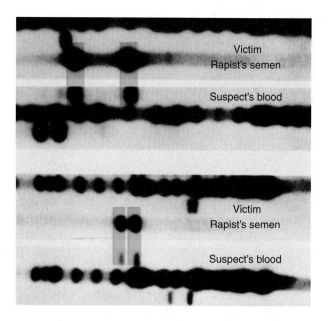

FIGURE 16.15
Two of the DNA profiles that led to the conviction of Tommie Lee Andrews for rape in 1987. The two DNA probes seen here were used to characterize DNA isolated from the victim, the semen left by the rapist, and the suspect. The dark channels are multiband controls. There is a clear match between the suspect's DNA and the DNA of the rapist's semen in these two profiles.

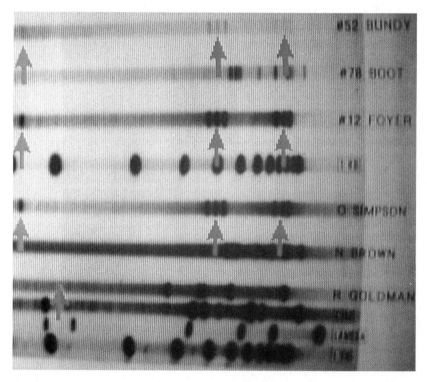

FIGURE 16.16
DNA profiles of O. J. Simpson and blood samples from the murder scene of his former wife. These patterns were used as evidence in the highly publicized and controversial 1995 murder trial.

16.3 Biotechnology is producing a scientific revolution.

Medical Applications

Pharmaceuticals

The first and perhaps most obvious commercial application of genetic engineering was the introduction of genes that encode clinically important proteins into bacteria. Because bacterial cells can be grown cheaply in bulk (fermented in giant vats, as is done with the yeasts that make beer), bacteria that incorporate recombinant genes can synthesize large amounts of the proteins those genes specify. This method has been used to produce several forms of human insulin and interferon, as well as other commercially valuable proteins, such as human growth hormone (figure 16.17) and erythropoietin, which stimulates red blood cell production.

Among the medically important proteins now manufactured by these approaches are **atrial peptides,** small proteins that may provide a new way to treat high blood pressure and kidney failure. Another is **tissue plasminogen activator,** a human protein synthesized in minute amounts that causes blood clots to dissolve and may be effective in preventing and treating heart attacks and strokes.

A problem with this general approach has been the difficulty of separating the desired protein from the others the bacteria make. The purification of proteins from such complex mixtures is both time-consuming and expensive, but it is still easier than isolating the proteins from the tissues of animals (for example, insulin from hog pancreases), which is how such proteins were formerly obtained. Recently, however, researchers have succeeded in producing RNA transcripts of cloned genes; they can then use the transcripts to produce only those proteins in a test tube containing the transcribed RNA, ribosomes, cofactors, amino acids, tRNA, and ATP.

Gene Therapy

In 1990, researchers first attempted to combat genetic defects by the transfer of human genes. When a hereditary disorder is the result of a single defective gene, an obvious way to cure the disorder is to add a working copy of the gene. This approach is being used in an attempt to combat cystic fibrosis (see chapter 13), and it offers potential for treating muscular dystrophy and a variety of other disorders (table 16.1). One of the first successful attempts was the transfer of a gene encoding the enzyme adenosine deaminase into the bone marrow of two girls suffering from a rare blood disorder caused by the lack of this enzyme. However, while many clinical trials are under way, no others have yet proven successful. This extremely promising approach will require a lot of additional effort. A more detailed account of gene therapy is presented in chapter 13.

FIGURE 16.17
Genetically engineered human growth hormone. These two mice are genetically identical, but the large one has one extra gene: the gene encoding human growth hormone. The gene was added to the mouse's genome by genetic engineers and is now a stable part of the mouse's genetic endowment.

Table 16.1 Diseases Being Treated in Clinical Trials of Gene Therapy
Disease
Cancer (melanoma, renal cell, ovarian, neuroblastoma, brain, head and neck, lung, liver, breast, colon, prostate, mesothelioma, leukemia, lymphoma, multiple myeloma)
SCID (severe combined immunodeficiency)
Cystic fibrosis
Gaucher disease
Familial hypercholesterolemia
Hemophilia
Purine nucleoside phosphorylase deficiency
Alpha-1 antitrypsin deficiency
Fanconi anemia
Hunter syndrome
Chronic granulomatous disease
Rheumatoid arthritis
Peripheral vascular disease
AIDS

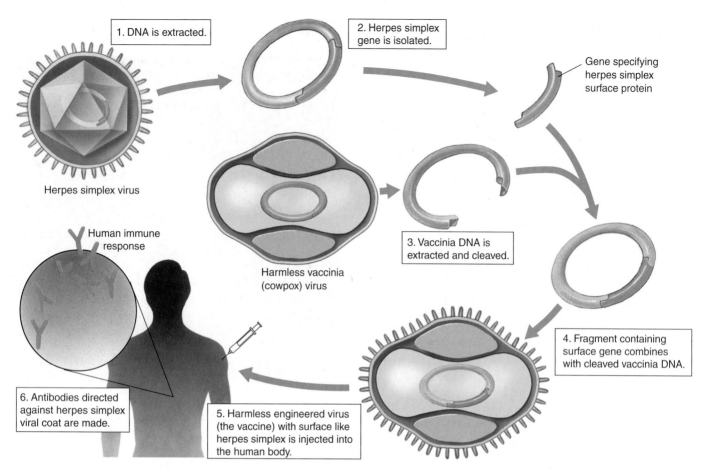

FIGURE 16.18
Strategy for constructing a subunit vaccine for herpes simplex.

1. DNA is extracted.

2. Herpes simplex gene is isolated.

Gene specifying herpes simplex surface protein

Herpes simplex virus

Human immune response

Harmless vaccinia (cowpox) virus

3. Vaccinia DNA is extracted and cleaved.

4. Fragment containing surface gene combines with cleaved vaccinia DNA.

5. Harmless engineered virus (the vaccine) with surface like herpes simplex is injected into the human body.

6. Antibodies directed against herpes simplex viral coat are made.

Piggyback Vaccines

Another area of potential significance involves the use of genetic engineering to produce **subunit vaccines** against viruses such as those that cause herpes and hepatitis. Genes encoding part of the protein-polysaccharide coat of the herpes simplex virus or hepatitis B virus are spliced into a fragment of the vaccinia (cowpox) virus genome (figure 16.18). The vaccinia virus, which British physician Edward Jenner used almost 200 years ago in his pioneering vaccinations against smallpox, is now used as a vector to carry the herpes or hepatitis viral coat gene into cultured mammalian cells. These cells produce many copies of the recombinant virus, which has the outside coat of a herpes or hepatitis virus. When this recombinant virus is injected into a mouse or rabbit, the immune system of the infected animal produces antibodies directed against the coat of the recombinant virus. It therefore develops an immunity to herpes or hepatitis virus. Vaccines produced in this way are harmless because the vaccinia virus is benign, and only a small fragment of the DNA from the disease-causing virus is introduced via the recombinant virus.

The great attraction of this approach is that it does not depend upon the nature of the viral disease. In the future,

similar recombinant viruses may be injected into humans to confer resistance to a wide variety of viral diseases.

In 1995, the first clinical trials began to test a novel new kind of **DNA vaccine,** one that depends not on antibodies but rather on the second arm of the body's immune defense, the so-called cellular immune response, in which blood cells known as killer T cells attack infected cells. The infected cells are attacked and destroyed when they stick fragments of foreign proteins onto their outer surfaces, which the T cells then detect. (The discovery by Peter Doherty and Rolf Zinkernagel that infected cells do this led to their receiving the Nobel Prize in Physiology or Medicine in 1996.) The first DNA vaccines spliced an influenza virus gene encoding an internal nucleoprotein into a plasmid, which was then injected into mice. The mice developed strong cellular immune responses to influenza. Although new and controversial, the approach offers great promise.

Genetic engineering has produced commercially valuable proteins and gene therapies, and will possibly give rise to new and powerful vaccines as well.

Agricultural Applications

Another major area of genetic engineering is manipulation of the genes of key crop plants. In plants, the primary experimental difficulty has been identifying a suitable vector for introducing recombinant DNA. Plant cells do not possess the many plasmids that bacteria have, so the choice of potential vectors is limited. The most successful results thus far have been obtained with the **Ti** (tumor-inducing) **plasmid** of the plant bacterium *Agrobacterium tumefaciens*, which infects broadleaf plants such as tomato, tobacco, and soybean. Part of the Ti plasmid integrates into the plant DNA, and researchers have succeeded in attaching other genes to this portion of the plasmid (figure 16.19). The characteristics of a number of plants have been altered using this technique, which should be valuable in improving crops and forests. Among the features scientists would like to affect are resistance to disease, frost, and other forms of stress; nutritional balance and protein content; and herbicide resistance. Unfortunately, *Agrobacterium* generally does not infect cereals such as corn, rice, and wheat, but alternative methods can be used to introduce new genes into them.

A recent advance in genetically manipulated fruit is Calgene's "Flavr Savr" tomato, which has been approved for sale by the United States Department of Agriculture (USDA).

The tomato has been engineered to inhibit genes that cause cells to produce ethylene. In tomatoes and other plants, ethylene acts as a hormone to speed fruit ripening. In Flavr Savr tomatoes, inhibition of ethylene production delays ripening. The result is a tomato that can stay on the vine longer and resists overripening and rotting during transport to market. A variety used for tomato paste has been engineered for antisense expression of the gene encoding polygalactaronidase to increase yield and reduce spoilage.

Nitrogen Fixation

A long-range goal of agricultural genetic engineering is to introduce the genes that allow soybeans and other legume plants to "fix" nitrogen into key crop plants. These so-called *nif* genes are found in certain symbiotic root-colonizing bacteria. Living in the root nodules of legumes, these bacteria break the powerful triple bond of atmospheric nitrogen gas, converting N_2 into NH_3 (ammonia). The plants then use the ammonia to make amino acids and other nitrogen-containing molecules. Other plants lack these bacteria and cannot fix nitrogen, so they must obtain their nitrogen from the soil. Farmland where these crops are grown soon becomes depleted of nitrogen, unless nitrogenous fertilizers are applied. Worldwide, farmers applied over 60 million metric tons of such

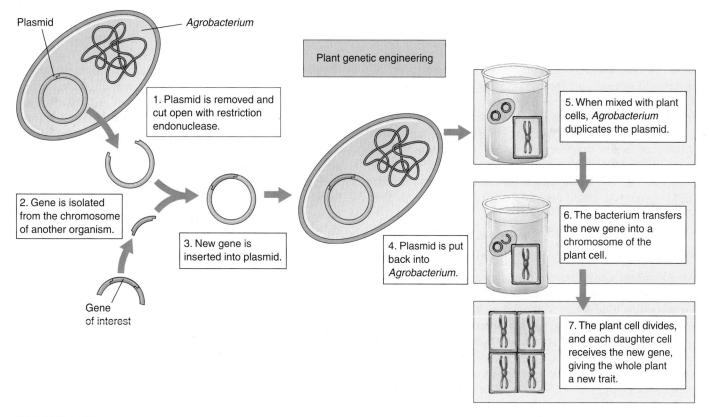

FIGURE 16.19
The Ti plasmid. This *Agrobacterium tumefaciens* plasmid is used in plant genetic engineering.

fertilizers in 1987, an expensive undertaking. Farming costs would be much lower if major crops such as wheat and corn could be engineered to carry out biological nitrogen fixation. However, introducing the nitrogen-fixing genes from bacteria into plants has proved difficult because these genes do not seem to function properly in eukaryotic cells, due to problems protecting nitrogenase from O_2. Researchers are actively experimenting with other species of nitrogen-fixing bacteria whose genes might function better in plant cells.

Herbicide Resistance

Recently, broadleaf plants have been genetically engineered to be resistant to **glyphosate,** the active ingredient in Roundup, a powerful, biodegradable herbicide that kills most actively growing plants (figure 16.20). Glyphosate works by inhibiting an enzyme called EPSP synthetase, which plants require to produce aromatic amino acids. Humans do not make aromatic amino acids; they get them from their diet, so they are unaffected by glyphosate. To make glyphosate-resistant plants, agricultural scientists used a Ti plasmid to insert extra copies of the EPSP synthetase genes into plants. These engineered plants produce 20 times the normal level of EPSP synthetase, enabling them to synthesize proteins and grow despite glyphosate's suppression of the enzyme. In later experiments, a bacterial form of the EPSP synthetase gene that differs from the plant form by a single nucleotide was introduced into plants via Ti plasmids; the bacterial enzyme in these plants is not inhibited by glyphosate.

These advances are of great interest to farmers because a crop resistant to Roundup would never have to be weeded if the field were simply treated with the herbicide. Because Roundup is a broad-spectrum herbicide, farmers would no longer need to employ a variety of different herbicides, most of which kill only a few kinds of weeds. Furthermore, glyphosate breaks down readily in the environment, unlike many other herbicides commonly used in agriculture. A plasmid is actively being sought for the introduction of the EPSP synthetase gene into cereal plants, making them also glyphosate-resistant.

Insect Resistance

Many commercially important plants are attacked by insects, and the traditional defense against such attacks is to apply insecticides. Over 40% of the chemical insecticides used today are targeted against boll weevils, bollworms, and other insects that eat cotton plants. Genetic engineers are now attempting to produce plants that are resistant to insect pests and thus remove the need to use many externally applied insecticides.

The approach is to insert into crop plants genes encoding proteins that are harmful to the insects that feed on the plants but harmless to other organisms. One such insecticidal

FIGURE 16.20
Genetically engineered herbicide resistance. All four of these petunia plants were exposed to equal doses of the herbicide Roundup. The two on top were genetically engineered to be resistant to glyphosate, the active ingredient in Roundup, while the two on the bottom were not.

protein has been identified in *Bacillus thuringiensis*, a soil bacterium. When the tomato hornworm caterpillar ingests this protein, enzymes in the caterpillar's stomach convert it into an insect-specific toxin, causing paralysis and death. Because these enzymes are not found in other animals, the protein is harmless to them. Using the Ti plasmid, scientists have transferred the gene encoding this protein into tomato and tobacco plants. They have found that these **transgenic** plants are indeed protected from attack by the insects that would normally feed on them. In 1995, the Environmental Protection Agency (EPA) approved altered forms of potato, cotton, and corn. The genetically altered potato can kill the Colorado potato beetle, a common pest. The altered cotton is resistant to cotton bollworm, budworm, and pink bollworm. The corn has been altered to resist the European corn borer and other mothlike insects.

Monsanto scientists screening natural compounds extracted from plant and soil samples have recently isolated a new insect-killing compound from a fungus, the enzyme cholesterol oxidase. Apparently, the enzyme disrupts membranes in the insect gut. The fungus gene, called the Bollgard gene after its discoverer, has been successfully inserted into a variety of crops. It kills a wide range of insects, including the cotton boll weevil and the Colorado potato beetle, both serious agricultural pests. Field tests began in 1996.

Some insect pests attack plant roots, and *B. thuringiensis* is being employed to counter that threat as well. This bacterium does not normally colonize plant roots, so biologists have introduced the *B. thuringiensis* insecticidal protein gene into root-colonizing bacteria, especially strains of *Pseudomonas*. Field testing of this promising procedure has been approved by the EPA.

The Real Promise of Plant Genetic Engineering

In the last decade, the cultivation of genetically modified crops of corn, cotton, and soybeans has become commonplace in the United States. In 1999, over half of the 72 million acres planted with soybeans in the United States were planted with seeds genetically modified to be herbicide resistant, with the result that less tillage has been needed and soil erosion has been greatly lessened. These improved crop varieties have mainly benefited farmers, enabling them to cultivate their crops more cheaply and efficiently. The food the public gets is the same; it just costs less to get it to the table.

Like the first act of a play, these developments have served mainly to set the stage for the real action, which is only now beginning to happen. The promise of plant genetic engineering is to produce genetically modified plants with desirable traits that directly benefit the consumer.

One recent advance, nutritionally improved rice, gives us a hint of what is to come. In developing countries, large numbers of people live on simple diets that are poor sources of vitamins and minerals (what botanists called "micronutrients"). Worldwide, the two major micronutrient deficiencies are iron, affecting 1.4 billion women (24% of the world population), and vitamin A, affecting 40 million children (7% of the world population). The deficiencies are especially severe in developing countries where the major staple food is rice. In recent research, Swiss bioengineer Ingo Potrykus and his team at the Institute of Plant Sciences, Zurich, have gone a long way toward solving this problem. Supported by the Rockefeller Foundation and with results to be made free to developing countries, their development of "golden rice" is a model of what plant genetic engineering can achieve.

To solve the problem of dietary iron deficiency among rice eaters, Potrykus first asked why rice is such a poor source of dietary iron. The problem, and the answer, proved to have three parts:

1. **Too little iron.** The proteins of rice endosperm contain unusually low amounts of iron. To solve this problem, a ferritin gene was transferred into rice from beans (figure 16.21). Ferritin is a protein with an extraordinarily high iron content, so its addition greatly increased the iron content of the rice.
2. **Inhibition of iron absorption by the intestine.** Rice contains an unusually high concentration of a chemical called phytate, which inhibits iron reabsorption in the intestine, meaning that it stops your body from taking up the iron in the rice. To solve this problem, a gene encoding an enzyme that destroys phytate was transferred into rice from a fungus.
3. **Too little sulfur for efficient iron absorption.** Sulfur is required for iron uptake, and rice has very little of it. To solve this problem, a gene encoding a particularly sulfur-rich metallothionin protein was transferred into rice from wild rice.

To solve the problem of vitamin A deficiency, the same approach was taken. First, the problem was identified. It turns out that rice only goes partway toward making β-carotene (provitamin A); there are no enzymes in rice to catalyze the last four steps. Thus, genes encoding these four enzymes were added to rice from a familiar flower, the daffodil.

Potrykus's development of transgenic rice to combat dietary deficiencies involved no subtle tricks, just straightforward bioengineering and the will to get the job done. The transgenic "golden" rice he has developed will directly improve the lives of millions of people. His work is

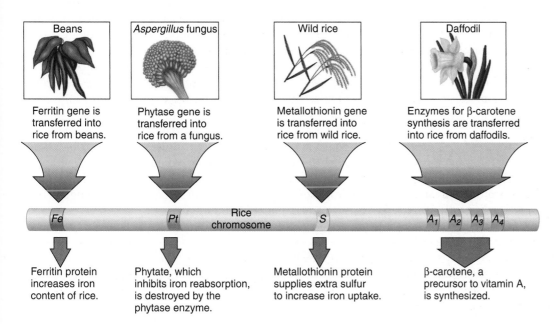

FIGURE 16.21
Transgenic rice.
Developed by Swiss bioengineer Ingo Potrykus, transgenic rice offers the promise of improving the diets of people in rice-consuming developing countries, where iron and vitamin A deficiencies are serious problems.

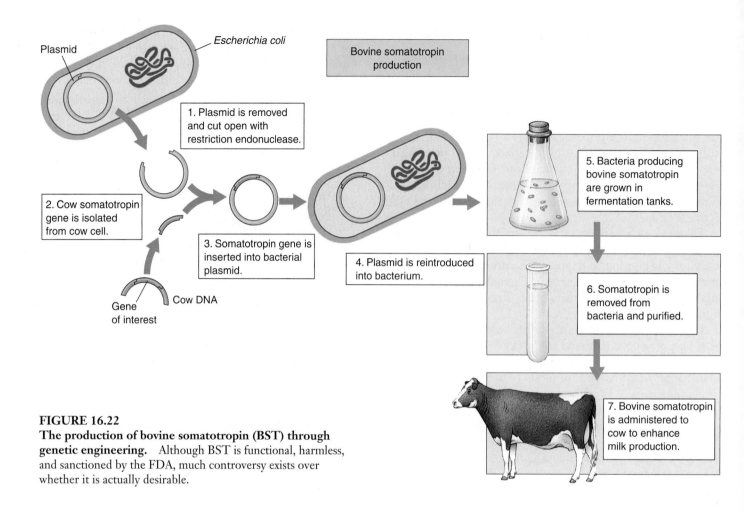

Plasmid

Escherichia coli

Bovine somatotropin production

1. Plasmid is removed and cut open with restriction endonuclease.

2. Cow somatotropin gene is isolated from cow cell.

3. Somatotropin gene is inserted into bacterial plasmid.

Gene of interest

Cow DNA

4. Plasmid is reintroduced into bacterium.

5. Bacteria producing bovine somatotropin are grown in fermentation tanks.

6. Somatotropin is removed from bacteria and purified.

7. Bovine somatotropin is administered to cow to enhance milk production.

FIGURE 16.22
The production of bovine somatotropin (BST) through genetic engineering. Although BST is functional, harmless, and sanctioned by the FDA, much controversy exists over whether it is actually desirable.

representative of the very real promise of genetic engineering to help meet the challenges of the new millennium.

The list of gene modifications that directly aid consumers will only grow. In Holland, Dutch bioengineers have announced that they are genetically engineering plants to act as vaccine-producing factories! To petunias they have added a gene for a vaccine against dog parvovirus, hiding the gene within the petunia genes that direct nectar production. The drug is produced in the nectar, collected by bees, and extracted from the honey. It is hard to believe this isn't science fiction. Clearly, the real promise of plant genetic engineering lies ahead, and not very far.

Farm Animals

The gene encoding the growth hormone somatotropin was one of the first to be cloned successfully. In 1994, Monsanto received federal approval to make its recombinant bovine somatotropin (BST) commercially available, and dairy farmers worldwide began to add the hormone as a supplement to their cows' diets, increasing the animals' milk production (figure 16.22). Genetically engineered somatotropin is also being tested to see if it increases the

muscle weight of cattle and pigs, and as a treatment for human disorders, such as dwarfism, in which the pituitary gland fails to make adequate levels of somatotropin. BST ingested in milk or meat has no effect on humans, because it is a protein and is digested in the stomach. Nevertheless, BST has met with some public resistance, due primarily to generalized fears about gene technology. Some people mistrust milk produced through genetic engineering, even though the milk itself is identical to other milk. Problems concerning public perception are not uncommon as gene technology makes an even greater impact on our lives.

Transgenic animals engineered to have specific desirable genes are becoming increasingly available to breeders. Now, instead of selectively breeding for several generations to produce a racehorse or a stud bull with desirable qualities, the process can be shortened by simply engineering such an animal right at the start.

Gene technology is revolutionizing agriculture, increasing yields and resistance to pests, improving nutritional value, and producing animals with desirable traits.

Risk and Regulation

The advantages afforded by genetic engineering are revolutionizing our lives. But what are the disadvantages—the potential costs and dangers of genetic engineering? Many people, including influential activists and members of the scientific community, have expressed concern that genetic engineers are "playing God" by tampering with genetic material. For instance, what would happen if we fragmented the DNA of a cancer cell, and then incorporated the fragments at random into vectors that were propagated within bacterial cells? Might there not be a danger that some of the resulting bacteria would transmit an infective form of cancer? Could genetically engineered products administered to plants or animals turn out to be dangerous for consumers after several generations? What kind of unforeseen impact on the ecosystem might "improved" crops have? Is it ethical to create "genetically superior" organisms, including humans?

How Do We Measure the Potential Risks of Genetically Modified Crops?

While the promise of genetic engineering is very much in evidence, this same technology has been the cause of outright war between researchers and protesters. In June 1999, British protesters attacked an experimental plot of genetically modified (GM) sugar beets; the following August, they destroyed a test field of GM canola (used for cooking oil and animal feed). The contrast could not be more marked between American acceptance of genetically modified crops on the one hand, and European distrust of genetically modified foods on the other. The intense feelings generated by this dispute point to the need to understand how we measure the risks associated with the genetic engineering of plants.

Two sets of risks need to be considered. The first stems from eating genetically modified foods, and the other concerns potential ecological effects.

Is Eating Genetically Modified Food Dangerous?
Protesters worry that genetically modified food may have been rendered somehow dangerous. To sort this out, it is useful to bear in mind that bioengineers modify crops in two quite different ways. One class of gene modification makes the crop easier to grow; a second class of modification is intended to improve the food itself.

The introduction of Roundup-resistant soybeans to Europe is an example of the first class of modification. This modification has been very popular with farmers in the United States, who planted half their crop with these soybeans in 1999. They like GM soybeans because the beans can be raised without intense cultivation (weeds are killed with Roundup herbicide instead), which both saves money and lessens soil erosion. But is the soybean that results nutritionally different? No. The gene that confers Roundup resistance in soybeans does so by protecting the plant's ability to manufacture so-called "aromatic" amino acids. In unprotected weeds, by contrast, Roundup blocks this manufacturing process, killing the weed. Because humans don't make any aromatic amino acids anyway (we get them in our diets), Roundup doesn't hurt us. The GM soybean we eat is nutritionally the same as an "organic" one, just cheaper to produce.

In the second class of modification, where a gene is added to improve the nutritional character of some food, the food will be nutritionally different. In each of these instances, it is necessary to examine the possibility that consumers may prove allergic to the product of the introduced gene. In one instance, for example, addition of a methionine-enhancing gene from the Brazil nut into soybeans (which are deficient in this amino acid) was discontinued when six of eight individuals allergic to Brazil nuts produced antibodies to the GM soybeans, suggesting the possibility of a reverse reaction. Instead, methionine levels in GM crops are being increased with genes from sunflowers. Screening for allergy problems is now routine.

On both scores, then, the risk of bioengineering to the food supply seems very slight. GM foods to date seem completely safe.

Are GM Crops Harmful to the Environment? What are we to make of the much-publicized report that Monarch butterflies might be killed by eating pollen blowing out of fields planted with GM corn? First, it should come as no surprise. The GM corn (so-called Bt corn) was engineered to contain an insect-killing toxin (harmless to people) in order to combat corn borer pests. Of course it will kill any butterflies or other insects in the immediate vicinity of the field. However, focus on the fact that the GM cornfields do not need to be sprayed with pesticide to control the corn borer. An estimated $9 billion in damage is caused annually by the application of pesticides in the United States, and billions of insects and other animals, including an estimated 67 million birds, are killed each year. This pesticide-induced murder of wildlife is far more damaging ecologically than any possible effects of GM crops on butterflies.

Will pests become resistant to the GM toxin? Not nearly as fast as they now become resistant to the far higher levels of chemical pesticide we spray on crops.

How about the possibility that introduced genes will pass from GM crops to their wild or weedy relatives? This sort of gene flow happens naturally all the time, and so this is a legitimate question. But if genes for resistance to Roundup herbicide spread from cultivated sugar beets to wild populations of sugar beets in Europe, why would that be a problem? Besides, there is almost never a potential relative around to receive the modified gene from the GM crop; no wild relatives of soybeans exist in Europe, for example. Thus, no genes can escape from GM soybeans in Europe, any more than genes can flow from you to other kinds of animals.

Calvin and Hobbes
by Bill Watterson

On either score, then, the risk of bioengineering to the environment seems very slight. Indeed, in some cases, it lessens the serious environmental damage produced by cultivation and agricultural pesticides.

Should We Label Genetically Modified Foods?

While there seems little tangible risk in the genetic modification of crops, it is important to assure the public that these risks are being carefully assessed. Few issues manage to raise the temperature of discussions about plant genetic engineering more than the labeling of genetically modified (GM) crops. Agricultural producers have argued that there are no demonstrable risks, so a GM label can only have the function of scaring off wary consumers. Consumer advocates respond that consumers have every right to make that decision and to have access to the information necessary to make it.

In considering this matter, it is important to separate two quite different issues, the *need* for a label and the *right* of the public to have one. Every serious scientific investigation of the risks of GM foods has concluded that they are safe—indeed, in the case of soybeans and many other crops modified to improve cultivation, the foods themselves are not altered in any detectable way, and no nutritional test could distinguish them from "organic" varieties. So there seems to be little if any health need for a GM label for genetically engineered foods.

The right of people to know what they are eating is a very different issue. Europeans fear genetic manipulation because it is unfamiliar. People there don't trust their regulatory agencies as we do in the United States, because their agencies have a poor track record of protecting them. When they look at genetically modified foods, they are haunted by past experiences of regulatory ineptitude. In England, people remember British regulators' failure to protect consumers from meat infected with mad cow disease.

It does no good whatsoever to tell a fearful European that there is no evidence to warrant fear, no trace of data supporting danger from GM crops. A European consumer will simply respond that the harm is not yet evident—that we don't know enough to see the danger lurking around the corner. "Slow down," the European consumers say. "Give research a chance to look around all the corners. Let's be sure." No one can argue against caution, but it is difficult to imagine what else researchers can look into, because safety has already been explored very thoroughly. The fear remains, though, for the simple reason that no amount of information can remove it. Like a child scared of a monster under the bed, looking under the bed again doesn't help—the monster still might be there next time. And that means we are going to have to have GM labels, for people have every right to be informed about something they fear.

What should these labels be like? A label that only says "GM FOOD" simply acts as a brand. Like a POISON label, it shouts a warning to the public of lurking danger. Why not instead have a GM label that provides information to the consumer—that tells what regulators know about that product? Here are some examples:

For Bt corn: The production of this food was made more efficient by the addition of genes that made plants resistant to pests so that fewer pesticides were required to grow the crop.

For Roundup-ready soybeans: Genes have been added to this crop to render it resistant to herbicides. This reduces soil erosion by lessening the need for weed-removing cultivation.

For high β-carotene rice: Genes have been added to this food to enhance its β-carotene content and thus combat vitamin A deficiency.

GM food labels that in each instance actually tell consumers what has been done to the gene-modified crop would go a long way toward hastening public acceptance of gene technology in the kitchen.

Genetic engineering affords great opportunities for progress in medicine and food production, although many people are concerned about possible risks. On balance, the risks appear slight, and the potential benefits substantial.

Concept Review

For interactive testing, visit the Online Learning Center with PowerWeb at www.mhhe.com/Raven7

16.1 Molecular biologists can manipulate DNA to clone genes.

The DNA Manipulator's Toolbox
- Restriction endonucleases are able to cleave DNA at specific restriction sites. (p. 320)
- DNA ligase joins DNA fragments by catalyzing the formation of phosphodiester bonds between DNA nucleotides. (p. 320)

Host/Vector Systems
- The ability to propagate DNA in a host cell requires a vector that can enter and replicate in the host. The two most commonly used vectors are plasmids and phages. (p. 321, p. 324)
- Using these vectors, the gene or genes of interest are introduced into the target organisms, where they infect cells and replicate the altered vector DNA. (p. 322)

Using Vectors to Transfer Genes
- Viruses and artificial chromosomes can also be used as vectors to insert foreign DNA into host cells, creating recombinant DNA and recombinant genomes. (p. 323)

DNA Libraries
- A DNA library is a collection of DNA fragments from a specific source in a form that can be propagated in a host. (p. 324)
- A genomic library is a representation of the entire genome in a vector. (p. 324)

16.2 Genetic engineering involves easily understood procedures.

The Four Stages of a Genetic Engineering Experiment
- The four stages of a genetic engineering experiment are (1) cleaving the source DNA, (2) producing recombinant DNA, (3) cloning copies of the recombinants, and (4) screening the cloned copies for the desired genes. (pp. 325–328)

Working with Gene Clones
- One approach to making enough DNA to work with is to use DNA polymerase to copy the gene of interest through the polymerase chain reaction (PCR). (p. 329)
- The three steps in PCR are denaturation, primer annealing, and primer extension. (p. 329)
- After 20 cycles, a single fragment produces more than one million copies of the target sequence. (p. 329)
- In the Southern Blotting process, DNA is cleaved into restriction fragments, and the fragments are then separated by gel electrophoresis. Fragments that contain the gene of interest can then be identified when a radioactively labeled probe is washed over paper that contains a copy of the gel's pattern (p. 330)
- Restriction fragment length polymorphisms (RFLP) analysis and DNA fingerprinting are both used to identify individuals and specific unknown gene sequences. (pp. 331–332)

16.3 Biotechnology is producing a scientific revolution.

Medical Applications
- Pharmaceuticals, gene therapy, and piggyback vaccines are all medical applications that have been developed from biotechnology advances. (pp. 333–334)

Agricultural Applications
- Increased nitrogen fixation, herbicide resistance, insect resistance, and dietary deficiencies are all agricultural problems that are currently being addressed via gene technology and transgenic species. (pp. 335–338)

Risk and Regulation
- Genetic engineering has the potential to create great medical and agricultural advances, although many people are concerned with possible risks of genetic engineering, especially those associated with transgenic species. (p. 340)
- In short, the potential benefits of genetic engineering appear to outweigh the potential risks. (p. 340)

Self Test

1. Cutting certain genes out of molecules of DNA requires the use of special
   a. degrading nucleases.
   b. restriction endonucleases.
   c. eukaryotic enzymes.
   d. viral enzymes.
2. Which of the following cannot be used as a vector?
   a. phage
   b. plasmid
   c. bacterium
   d. All can be used as vectors.
3. In Cohen and Boyer's recombinant DNA experiments, restriction endonucleases were used to
   a. isolate fragments of cloned bacterial plasmids.
   b. isolate fragments of frog DNA that contained an rRNA gene.
   c. cleave the bacterial plasmid.
   d. All of these are correct.
4. A DNA library is
   a. a general collection of all genes sequenced thus far.
   b. a collection of DNA fragments that make up the entire genome of a particular organism.
   c. a DNA fragment inserted into a vector.
   d. all DNA fragments identified with a probe.
5. A probe is used in which stage of genetic engineering?
   a. cleaving DNA
   b. recombining DNA
   c. cloning
   d. screening
6. The enzyme used in the polymerase chain reaction is
   a. restriction endonuclease.
   b. reverse transcriptase.
   c. DNA polymerase.
   d. RNA polymerase.
7. A method used to distinguish DNA of one individual from another is
   a. polymerase chain reaction.
   b. cDNA.
   c. reverse transcriptase.
   d. restriction fragment length polymorphism.
8. Inserting a gene encoding a pathogenic microbe's surface protein into a harmless virus produces a:
   a. piggyback vaccine
   b. virulent virus
   c. active disease-causing pathogen
   d. pharmaceutical human protein
9. Although the Ti plasmid has revolutionized plant genetic engineering, one limitation of its use is that it
   a. cannot infect broadleaf plants.
   b. cannot be used on fruit-bearing plants.
   c. cannot transmit prokaryotic genes.
   d. does not infect cereal plants such as corn and rice.
10. Which of the following is *not* an application of genetic engineering in plants?
    a. nitrogen fixation
    b. DNA vaccines
    c. resistance to glyphosate
    d. production of insecticidal proteins in plants

Test Your Visual Understanding

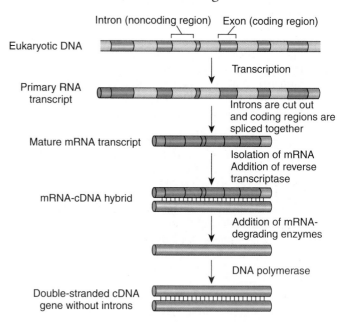

1. What process is illustrated in this figure? Where in the cell would you look to find the primary RNA transcript, and where would you look to find the mature mRNA transcript?
2. When might researchers use this process—that is, what would they be trying to accomplish?

Apply Your Knowledge

1. The human genome has about 3 billion base-pairs. Assume that you want to clone the entire human genome using various vectors, but there is a limit to the size of a DNA fragment that can be inserted in a vector. Following is a list of vectors along with their size limit of DNA fragment. Calculate how many vectors of each type would be needed to generate a library of the human genome.
   a. bacterial plasmid—18 kilo base-pairs
   b. phages—25 kilo base-pairs
   c. YACs—250 kilo base-pairs
2. A major focus of genetic engineering has been on attempting to produce large quantities of scarce human proteins by placing the appropriate genes into bacteria and thus turning the bacteria into protein production machines. Human insulin and many other proteins are produced this way. However, this approach does not work for producing human hemoglobin. Even if the proper clone is identified, the fragments containing the hemoglobin genes are successfully incorporated into bacterial plasmids, and the bacteria are infected with the plasmids, no hemoglobin is produced by the bacteria. Why doesn't this experiment work?

17

Genomes

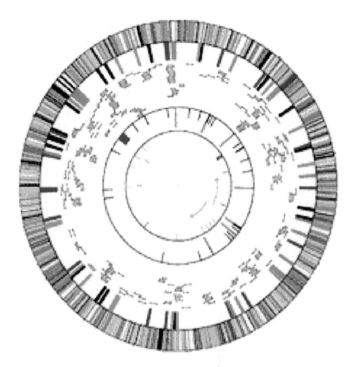

FIGURE 17.1
Haemophilus influenzae **genome.** The 1.8 million base-pairs of *H. influenzae* were sequenced in 1995. This bacterium causes respiratory infections and meningitis in children. It was the first free-living organism to be sequenced; its genome was found to be 10 times larger than any of the viral genomes that had previously been sequenced. Just five years later, the human genome was sequenced. Each color represents genes with similar functions.

Reprinted with permission from Fleishmann, et al., "Whole-Genome Random Sequencing LOOK UP," Science, 269: 496–512. Copyright © 1995 American Association for the Advancement of Science.

Concept Outline

17.1 Genomes can be mapped both genetically and physically.

    Genome Maps. The first genetic maps revealed the positional relationship among genes on chromosomes, but did not link them to specific DNA sequences. In physical maps, physical pieces of DNA are mapped to specific regions on chromosomes.

17.2 Genome sequencing produces the ultimate physical map.

    Sequencing. Automated sequencers and computer programs can be used to determine the entire sequence of a genome using either a clone-by-clone or shotgun method. Sequencers can only sequence relatively short segments of the genome at a time. Assembling these segments into a whole genome is like putting a jigsaw puzzle together.

17.3 Being more complex does not necessarily require more genes.

    Human Genome Project. The draft sequence for the human genome was released in February 2001. Perhaps the greatest surprise was how few genes comprise our genome.

    Genome Geography. In addition to functional genes, genomes contain a variety of repetitive DNA patterns that may have originally come from other organisms.

    Comparative Genomics. Synteny, conservation of large DNA sequences among related organisms, makes it possible to extend information about sequenced genomes to other genomes.

17.4 Genomics is opening a new window on life.

    Functional Genomics. Using DNA microarrays and inserting genes into organisms allow biologists to study the function of newly identified genes.

    Proteomics. The genome produces the proteome, all the proteins in an organism. Proteomics is the study of these proteins and how they function.

    Using Genomic Information. Intra- and interspecific genome comparisons can yield insight into evolution, crop improvement, and genetic disease.

Maps help you find specific locations. Individual countries can be located on a map of the world, and if there is enough detail, a city or perhaps even a famous landmark can be pinpointed. Genome maps work in a similar way, but are linear. Your genome is all of the DNA in the nucleus of one of your cells. Within this DNA world are chromosomes containing individual genes with specific base-pair sequences. For almost a century, geneticists have identified and located genes one by one. Creating a more comprehensive and detailed map required a breakthrough in DNA sequencing technology. The genomes of almost 100 organisms and many more viruses have been sequenced (figure 17.1). Now, more landmarks are needed to navigate genome maps and understand how the genome is used to build a functioning organism.

17.1 Genomes can be mapped both genetically and physically.

Genome Maps

News headlines celebrate the mapping of many genomes, including the human and rice genomes. Mapping genes is not new, but mapping an entire genome is a recent and astounding accomplishment (see figure 17.1). To understand how genome maps are made, we must first consider the different possible types of genome maps. Each type provides different information. **Genetic maps** are linkage maps (see chapter 13). They show the relative location of genes on a chromosome as determined by recombination frequencies. **Physical maps** are diagrams showing the relative positions of **landmarks** within specific DNA sequences. Landmarks include specific sequences of DNA and sites where restriction enzymes cut DNA. These landmarks help you find your way through the genome and identify parts of the genome that interest you.

Genetic Maps

The first genetic (linkage) map was made in 1911 when Sturtevant mapped five genes in *Drosophila* (see chapter 13, figure 13.32). Distances on a genetic map are measured in centimorgans (cM) in honor of the geneticist Morgan. One cM represents a 0.01% recombination frequency. Today, 13,744 genes have been mapped on the *Drosophila* genome. Linkage mapping can be done without knowing the DNA sequence of a gene. Computer programs make it possible to create a linkage map for 1000 genes at a time. There are a few limitations to genetic maps. Distances between genes determined by recombination frequencies between traits do not directly correspond with physical distance on a chromosome. The conformation of DNA between genes varies and can affect the frequency of recombination. Also, not all genes have obvious phenotypes that can be followed in segregating crosses.

Physical Maps

Distances between landmarks on a physical map are measured in base-pairs (1000 base-pairs equal 1 kilobase, kb). It is not necessary to know the DNA sequence of a segment of DNA in order to create a physical map or to know if that DNA codes for a specific gene. The first physical maps were created by cutting genomic DNA with different restriction enzymes (figure 17.2) and then putting the pieces back together, based on size and overlap, into a contiguous segment of the genome called a **contig**. Each restriction enzyme recognizes a very specific sequence of DNA that it cleaves, as described in chapter 16. The very first restriction enzymes to be isolated came from *Haemophilus*, which was also the first free-living genome to be sequenced (see figure 17.1).

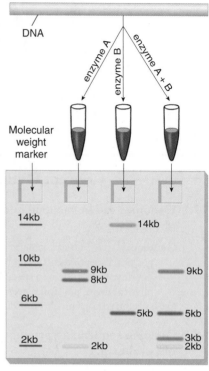

1. Multiple copies of a segment of DNA are cut with restriction enzymes.

2. The fragments produced by enzyme A only, by enzyme B only, and by enzymes A and B simultaneously are run out side-by-side on a gel, which separates them according to size, smaller fragments running faster.

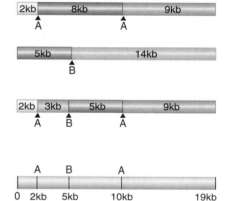

3. The fragments are arranged so that the smaller ones produced by the simultaneous cut can be grouped to generate the larger ones produced by the individual enzymes.

4. A physical map is constructed.

FIGURE 17.2
Restriction enzymes can be used to create a physical map. Comparing the sizes of DNA fragments produced by two restriction enzymes individually with the sizes of those produced by the two enzymes acting jointly allows you to deduce the locations where the enzymes cleave the DNA.

Homologous regions of DNA can vary slightly in base-pair composition among individuals in a population. When a homologous stretch of DNA is cut with restriction enzymes in different individuals, fragments of different lengths may be produced. These fragments are called restriction fragment length polymorphisms (RFLPs, see figure 16.14). When a RFLP segregates with a genetic trait, it becomes possible to link the genetic map (represented by the heritable trait) and the physical map (repre-

344   **Part III**   Genetic and Molecular Biology

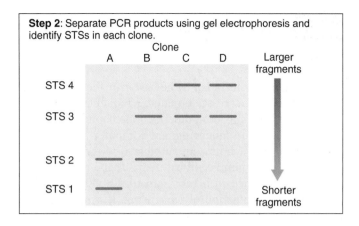

Step 1: Amplify parts of a cloned DNA segment with PCR, using primers that recognize particular STSs. Two primers per STS are used so that extension occurs in both directions on each double-stranded clone.

STS 1   STS 2   STS 3   STS 4

Clone A

Clone B

Clone C

Clone D

Primer =

Step 2: Separate PCR products using gel electrophoresis and identify STSs in each clone.

Clone
A   B   C   D   Larger fragments

STS 4
STS 3
STS 2
STS 1

Shorter fragments

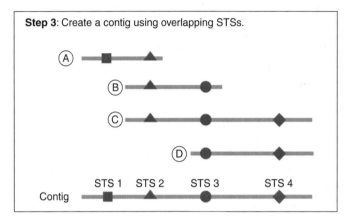

Step 3: Create a contig using overlapping STSs.

A
B
C
D

STS 1   STS 2   STS 3   STS 4
Contig

FIGURE 17.3
Creating a physical map with sequence-tagged sites. The presence of landmarks called sequence-tagged sites, or STSs, in the human genome made it possible to begin creating a physical map large enough in scale to provide a foundation for sequencing the entire genome. (*1*) Primers (shown as arrows) that recognize unique STSs are added to a cloned segment of DNA, followed by DNA replication via PCR. (*2*) The PCR products from each reaction are separated based on size on a DNA gel, and the STSs contained in each clone are identified. (*3*) The cloned DNA segments are then aligned based on overlapping STSs to create a contig.

sented by the specific RFLP). RFLPs are one of many types of physical landmarks in genomes.

Constructing a physical map for large genomes requires the efforts of many research labs. Suppose that two groups are each working on a segment of DNA that has been cloned (inserted into the genome of a virus or bacterium that can replicate), but has not been sequenced. How can the two groups tell if their clones contain overlapping DNA sequences? First, it would be helpful if everyone used the same type of landmark. Second, it would be helpful if these landmarks or tags could be completely described in a shared database so that researchers did not have to mail DNA clones back and forth. The solution for groups working on cloned DNA from the same species is to use **sequenced-tagged sites (STSs)**. An STS is a 100 to 500 base-pair sequence of a clone. The clone itself is much larger than the length of the tag, and it would take a much longer time to sequence this complete DNA segment. The clone needs to be from single-copy DNA so that the STS is unique within the genome. PCR (polymerase chain reaction, see figure 16.11) primers can be designed based on the STSs. Once you know the sequence of part of the land-

mark, you can use polymerase chain reactions (PCR) to determine if a cloned piece of DNA contains this STS. Primers are designed for PCR amplification of each known STS. Using these primers, sequences containing this STS can be amplified. The fragments produced by the extension of primers are separated using gel electrophoresis and stained with ethidium bromide to identify fragments based on size. The fragments can be pieced together to make a physical map by identifying the STSs in each fragment and overlapping them. Because of the high density of STSs in the human genome and the relative ease of identifying an STS in a DNA clone, it became possible to develop physical maps on the large scale of the 3.2-gigabase genome in the mid-1990s (figure 17.3). STSs make a scaffold for genome sequences.

Genetic maps provide information on the relative distances between genes. Distances are measured in centimorgans. Physical maps locate landmarks in the genome. Distance is measured in base-pairs. Linked physical and genetic maps provide a scaffold for whole genome sequencing.

17.2 Genome sequencing produces the ultimate physical map.

Sequencing

The ultimate physical map is the base-pair sequence of the entire genome. Large-scale genome sequencing uses the basic sequencing approach described in chapter 16, but relies on automated sequencing and computer analysis (figure 17.4). In a few hours, an automated sequencer can sequence the same number of base-pairs that a person could manually sequence in a year—up to 50,000 base-pairs. Without the automation of sequencing, it would not have been possible to sequence large, eukaryotic genomes like that of humans.

Automated Sequencers

While it would be ideal to isolate DNA from an organism, add it to a sequencer, and then come back in a week or two to pick up a computer-generated printout of the genome sequence for that organism, life is not quite that simple. Sequencers provide accurate sequences for DNA segments up to 500 base-pairs (bp) long. Even then, errors are possible. So, five to ten copies of a genome are sequenced to reduce errors.

DNA is prepared for sequencing by making many copies of it with fluorescently labeled adenines (A), cytosines (C), thymines (T), and guanines (G), mixed in with unlabeled nucleotides (you might want to review DNA replication in chapter 14). The fluorescent label for each nucleotide is distinct in color. These labeled nucleotides also lack an OH group in the 3′ position. Without the 3′ OH group, no more nucleotides can be added. As the DNA is replicated, the labeled nucleotides are incorporated randomly, resulting in sequences of DNA of different lengths. This labeled DNA is then added to one of 96 wells of a gel that has been

FIGURE 17.4
Automated sequencing. This sequence facility simultaneously runs multiple automated sequencers, each processing 96 samples at a time.

poured between two vertical glass plates spaced less than 0.5 millimeters apart. Each well is filled with different batches of labeled DNA. (Newer machines use 96 capillary tubes the diameter of a human hair.) The DNA then separates according to size, with the shorter sequences passing a laser beam detector first. The detector sends color information to a computer that translates the data into a sequence. If the smallest band terminates with a labeled A, it passes the detector first and A is recorded as the first base. If a labeled G is found in the largest band, that is recorded as the last base. Work through figure 17.5 to be sure you understand how a sequence is determined.

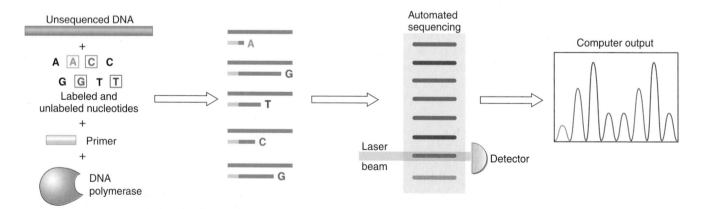

FIGURE 17.5
Sequencing DNA. DNA is replicated, and fluorescent nucleotides that halt replication are randomly inserted into different sequences. An automated sequencer determines the base-pair sequence.
Why is it important to know whether or not your primer hybridizes to the template strand in the replication step?

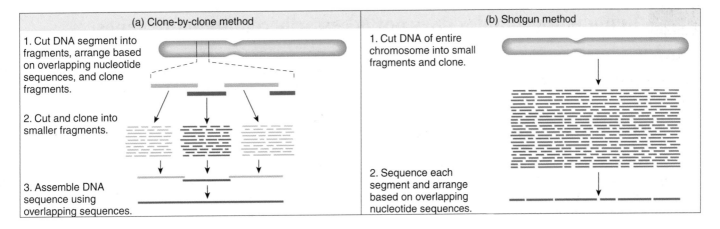

FIGURE 17.6
Comparison of sequencing methods. (*a*) The clone-by-clone method involves many rounds of cloning and arranging the fragments of a segment of DNA before assembling the final DNA sequence. (*b*) The shotgun method involves fragmenting and cloning the entire chromosome and then using a computer to assemble the final DNA sequence based on overlapping nucleotide sequences.

Artificial Chromosomes

The development of a new type of vector, called artificial chromosomes, has allowed scientists to clone larger pieces of DNA. The first generation of these new vectors were **yeast artificial chromosomes,** called **YACs.** These were constructed by using a yeast origin of replication and a yeast centromere sequence, with foreign DNA added to this construct. The origin of replication allows the artificial chromosome to replicate independently of the rest of the genome, and the centromere sequences make them mitotically stable. YACs were useful for cloning larger pieces of DNA but had many drawbacks, including a tendency to rearrange, or to lose portions of DNA by deletion. Despite the difficulties, the YACs were used early on to construct physical maps by restriction enzyme digestion of the YAC DNA.

The artificial chromosomes most commonly used now, particularly for large-scale sequencing, are made in *E. coli.* These are **bacterial artificial chromosomes,** called **BACs.** They are a logical extension of the use of bacterial plasmids to include much larger pieces of DNA that behave like a second bacterial chromosome. These BAC vectors accept DNA inserts between 100 and 200kb.

Sequencing by Whole Genomes

Clone-by-Clone Sequencing. The cloning of large inserts in BACs facilitates the analysis of entire genomes. The strategy most commonly pursued is to construct a physical map first, and then use it to place the site of BAC clones for later sequencing. The alignment of large portions of a chromosome requires identifying regions that overlap between clones. This can be done either by constructing restriction maps of each BAC clone or by identifying STSs between clones. If two BAC clones have the same STS, this means they must overlap.

Once a number of BAC clones have been aligned, this forms a contig, or contiguous stretch of DNA. The individual BAC clones can then be sequenced 500 base-pairs at a time to produce the sequence of the entire contig (figure 17.6*a*). This strategy of physical mapping followed by sequencing is called clone-by-clone sequencing.

Shotgun Sequencing. The idea of shotgun sequencing is to simply sequence all cloned fragments and use a computer to put together the overlaps. This approach is much less labor-intensive than the clone-by-clone method, but it requires much greater computer power to assemble the final sequence and very efficient algorithms to find overlaps. Unlike the clone-by-clone approach, shotgun sequencing does not tie the sequence to any other information about the genome. Many investigators have used both clone-by-clone and shotgun sequencing techniques, and hybrid approaches that use both are becoming the norm. This has the strength of tying the sequence to a physical map while greatly reducing the labor involved. The two methods are shown graphically in figure 17.6.

Assembler programs compare multiple copies of sequenced regions in order to *assemble* a **consensus sequence.** Consensus sequences reduce error. Although computer assemblers are incredibly powerful, final human analysis is required after both clone-by-clone and shotgun cloning to determine when a genome sequence is sufficiently accurate to be useful to researchers.

Genomes are cut into small pieces of DNA that are sequenced in automated sequencers. Shotgun cloning starts with small pieces of sequenced DNA and utilizes extensive computational analysis to determine the whole sequence from overlapping pieces. The clone-by-clone strategy starts with a physical map with many landmarks. Pieces of DNA from known sites on the map are sequenced.

17.3 Being more complex does not necessarily require more genes.

Human Genome Project

One of the defining characteristics of genome science (**genomics**) is the scale of the projects. In an era when large numbers are commonplace (for example, computers with 120-gigabyte hard drives), it is easy to underestimate the enormity of the task of mapping and sequencing the 3.2-gigabase (3,200,000,000-nucleotide) human genome. Conceptually, the approach is not all that different from the manual sequencing of the 5375-nucleotide genome of the X174 virus in 1977. As you saw in section 17.2, the real challenge is being able to rapidly sequence and assemble sequences for very large genomes. The limiting factor for sequencing projects has been genome size. Of the 100 or so sequenced genomes of free-living organisms, most are bacterial. Automated sequencing, new computer algorithms, STS maps, and BACs that can hold large DNA inserts have all made it possible to sequence our gigantic genome and, more recently, the even larger genome of rice. Perhaps the most startling finding in the analysis of the human genome is that the complexity of an organism is not necessarily reflected in the number of genes it contains. Humans have about twice as many genes as *Drosophila* and fewer genes than rice (figure 17.7).

The vast scale of genomics has ushered in a new way of doing biological research, involving large teams. While a single individual can clone a small genome manually, a huge genome like ours requires the collaborative efforts of hundreds of researchers. One example of such teamwork is the Human Genome Project, the results of which have illustrated key genomic concepts and offered tantalizing possibilities for understanding the genetic basis of disease and unlocking the mysteries of evolution.

The Human Genome Project originated in 1990 when a group of American scientists formed the International Human Genome Sequencing Consortium. The goal of this publicly funded effort was to use a clone-by-clone approach to sequence the human genome. Both genetic and physical maps were enhanced and published in the 1990s, and used as scaffolding to sequence each chromosome. Then, in May 1998, Craig Venter, who had sequenced *Haemophilus influenzae*, announced that he had formed a private company to sequence the human genome. He proposed to shotgun-clone the 3.2-gigabase genome in only two years. The consortium rose to the challenge, and the race to sequence the human genome began. The upshot was a tie of sorts. On June 26, 2000, the two groups jointly announced success, and each published its findings simultaneously in 2001. The consortium's draft alone included 248 names on its partial list of authors.

The draft sequence of the human genome is more of a beginning than an ending. Gaps in the sequence are still being filled, and the map is still being finished. More significantly, research on the whole genome can move ahead. Now that the ultimate physical map is in place and being integrated with the genetic map, diseases such as diabetes that result from defects in more than one gene can be addressed. Comparisons with other genomes are already changing our understanding of genome evolution (see chapter 24).

Sequencing of large genomes, including our own via the Human Genome Project, is changing a long-held belief that organisms with complex body plans and behaviors have more genes than simpler organisms.

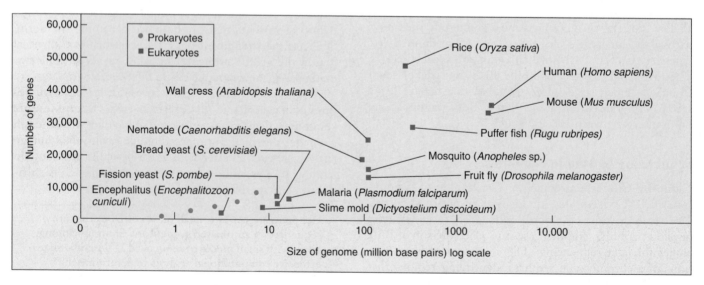

FIGURE 17.7
Size and complexity of genomes. In general, eukaryotic genomes are larger and have more genes than prokaryotic genomes, although the size of the organism is not the determining factor. The mouse genome is nearly as large as the human genome, and the rice genome contains more genes that the human genome.

Genome Geography

Finding Genes in the Genome

Once you have sequenced a genome, the next step is to determine which regions of the genome contain which genes and what those genes do. There is a lot of information to be mined from the sequence data. Using markers from physical maps and information from genetic maps, it is possible to find the sequence of the small percentage of genes that are identified by mutations with an observable effect. Information in the nucleotide sequence itself can also be used in the search for genes. A gene begins with a "start" codon such as ATG and contains no "stop" codons (UAA, UGA, or UAG) for a distance long enough to encode a protein. This coding region is referred to as an **open reading frame (ORF)**. While these sequences are likely to be genes, they may or may not actually be translated into a functional protein. Potential gene sequences need to be tested experimentally to determine whether they have a function. It is also possible to search genome databases for sequences that have homology to known genes in other species. Using computer programs to search for genes, to compare genomes, and to assemble genomes are only a few of the new genomics approaches falling under the heading of **bioinformatics.**

Another way to find genes that are actually expressed in an organism is to use a type of STS called an **expressed sequence tag (EST)**. ESTs are made by isolating mRNA, using reverse transcriptase to make cDNA copies of the mRNA, and then sequencing one or both ends of each cDNA. Just like STSs, PCR primers can be designed to make copies of genomic DNA or cDNA containing the EST. Unlike some STSs that serve only as physical landmarks on the genome, ESTs can identify functional genes that are transcribed to produce RNA. ESTs have been used to identify 87,000 cDNAs in different human tissues. About 80% of these cDNAs were previously unknown. This gene discovery approach allows the researcher to directly relate the new information to the map of the genome.

How is it possible that 87,000 cDNAs representing 87,000 different mRNAs could be identified in human tissue when there only appear to be 30,000 genes in the human genome? This was a puzzle for genome researchers who had predicted that the human genome would contain close to 100,000 genes. It turns out that the answer lies in the modularity of eukaryotic genes, which consist of exons interspersed with introns. Following transcription, the introns are removed, and exons are spliced together. In some cells, some of the splice sites are skipped, and one or more exons is removed along with the introns. This process, called **alternative splicing** (figure 17.8), yields different proteins that can have different functions. Thus, the added complexity of proteins in the human genome comes not from additional genes, but from new ways to put existing gene parts together.

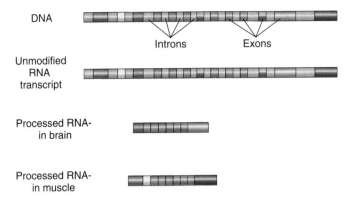

FIGURE 17.8
Alternative splicing can result in the transcription of different mRNAs from the same coding sequence. In some cells, exons can be excised along with neighboring introns, resulting in different proteins. Alternative splicing explains why 30,000 human genes can code for three to four times as many proteins.

Gene Organization in Eukaryotic Genomes

Four different classes of protein-encoding genes are found in the eukaryotic genomes, differing largely in gene copy number.

Single-copy genes. Many genes exist as single copies on a particular chromosome. Most mutations in these genes result in recessive Mendelian inheritance. Silent copies, inactivated by mutation, are called *pseudogenes.*

Segmental duplications. Sometimes whole blocks of genes are copied from one chromosome to another, resulting in *segmental duplication.* Blocks of similar genes in the same order are found throughout the human genome. Chromosome 19 seems to have been the biggest borrower, sharing blocks of genes with 16 other chromosomes.

Multigene families. As we have learned more about eukaryotic genomes, it has become apparent that many genes exist as parts of *multigene families,* groups of related but distinctly different genes that often occur together in clusters. These genes appear to have arisen from a single ancestral gene that duplicated during an uneven meiotic crossover in which genes were added to one chromosome and subtracted from the other.

Tandem clusters. Identical copies of genes can also be found in *tandem clusters.* These genes are all transcribed simultaneously, which increases the amount of mRNA available for protein production. For example, the genes encoding rRNA are typically present in clusters of several hundred copies.

Bioinformatic approaches can identify gene sequences in a sequenced genome. Genes are found as single copies as well as in arrays of closely related or exact copies in eukaryotic genomes.

Table 17.1   Classes of DNA Sequences Found in the Human Genome

Class	Frequency	Description
Protein-encoding genes	1%	Translated portions of the 30,000 genes scattered about the chromosomes
Introns	24%	Noncoding DNA that comprises the great majority of each human gene
Segmental duplications	5%	Regions of the genome that have been duplicated
Pseudogenes (inactive genes)	2%	Sequence that has characteristics of a gene, but is not a functional gene
Structural DNA	20%	Constitutive heterochromatin, localized near centromeres and telomeres
Simple sequence repeats	3%	Stuttering repeats of a few nucleotides such as CGG, repeated thousands of times
Transposable elements	45%	21%: Long interspersed elements (LINEs), which are active transposons
		13%: Short interspersed elements (SINEs), which are active transposons
		8%: Retrotransposons, which contain long terminal repeats (LTRs) at each end
		3%: DNA transposon fossils

Noncoding DNA in Eukaryotic Genomes

Sequencing of several eukaryotic genomes has now been completed, and one of the most notable characteristics is the amount of noncoding DNA they possess. The recent sequencing of the human genome reveals a particularly startling picture. Each of your cells has about 6 feet of DNA stuffed into it, but of that, less than 1 inch is devoted to genes! Nearly 99% of the DNA in your cells has little or nothing to do with the instructions that make you *you*.

True genes are scattered about the human genome in clumps among the much larger amount of noncoding DNA, like isolated hamlets in a desert. There are six major sorts of noncoding human DNA (table 17.1).

Noncoding DNA within genes. As discussed in chapter 15, a human gene is not simply a stretch of DNA, like the letters of a word. Instead, a human gene is made up of numerous fragments of protein-encoding information (exons) embedded within a much larger matrix of noncoding DNA (introns). Together, introns make up about 24% of the human genome, and exons, less than 1.5%.

Structural DNA. Some regions of the chromosomes remain highly condensed, tightly coiled, and untranscribed throughout the cell cycle. Called *constitutive heterochromatin*, these portions tend to be localized around the centromere or located near the ends of the chromosome, at the telomeres.

Simple sequence repeats. Scattered about chromosomes are *simple sequence repeats* (SSRs). An SSR is a one- to three-nucleotide sequence such as CA or CGG, repeated like a broken record thousands and thousands of times. SSRs can arise from DNA replication errors. SSRs make up about 3% of the human genome.

Segmental duplications. Blocks of genomic sequences composed of from 10,000 to 300,000 base-pairs have duplicated and moved either within a chromosome or to a nonhomologous chromosome.

Pseudogenes. These are inactive genes that may have lost function because of mutation.

Transposable elements. Fully 45% of the human genome consists of mobile bits of DNA called transposable elements. Discovered by Barbara McClintock in 1950 (she received the Nobel Prize for her discovery in 1983), transposable elements are bits of DNA that are able to jump from one location on a chromosome to another.

How do transposable elements accomplish this remarkable feat? In some cases, the transposon is duplicated, and the duplicated DNA moves to a new place in the genome, so the number of copies of the transposon increase. Other types of transposons are excised without duplication and insert themselves elsewhere in the genome.

Human chromosomes contain four sorts of transposable elements. Fully 21% of the genome consists of *long interspersed elements* (LINEs). An ancient and very successful element, LINEs are about 6000 base-pairs long, and contain all the equipment needed for transposition. LINEs encode a reverse transcriptase enzyme that can make a cDNA copy of the transcribed LINE RNA. The result is a double-stranded segment that can reinsert into the genome rather than undergo translation into a protein.

Short interspersed elements (SINEs) are similar to LINEs, but cannot transpose without using the transposition machinery of LINEs. Nested within the genome's LINEs are over half a million copies of a SINE element called ALu, 10% of the human genome. Like a flea on a dog, ALu moves with the LINE it resides within. Just as a flea sometimes jumps to a different dog, so ALu sometimes uses the enzymes of its LINE to move to a new chromosome location. Often ALu jumps right into genes, causing harmful mutations.

Two other sorts of transposable elements are also found in the human genome: 8% of the human genome is devoted to retrotransposons called *long terminal repeats* (LTRs). Although the transposition mechanism is a bit different from that of LINEs, LTRs also use reverse transcriptase

A Vocabulary of Genomics

alternative splicing Exons are combined in different ways so that one gene can produce multiple proteins.

annotate To determine what different regions of a sequenced genome, especially coding regions, do.

assembly Arranging sequenced pieces of DNA in the correct order in the genome.

BAC Bacterial artificial chromosome. Genomic DNA (about 150,000 bp) is inserted into bacterial DNA where many copies can be made.

clone-by-clone sequencing Hierarchical sequencing strategy that starts with physical and genetic maps and breaks DNA sequences from known locations into smaller pieces for sequencing and assembly.

contig Sequence of DNA assembled by identifying overlaps among smaller DNA segments. The term contig comes from "contiguous."

draft sequence An initial genome sequence that has some gaps but is useful to researchers.

EST Expressed sequence tag. Partially sequenced cDNAs used to identify genes in a genomic sequence.

finished sequence A genome sequence with minimal gaps, sequences in the right order, and no more than 1 error in 10,000 nucleotides.

genetic map Map of the relative position of genes based on linkage analysis—that is, frequency of recombination during meiosis.

genome All the genetic information (DNA) in a cell. In eukaryotes, the genome is equivalent to all of the chromosomes in a haploid cell. These cells also have a mitochondrial genome and, in the case of plants, a chloroplast genome.

PCR Polymerase chain reaction. Method for making many copies of a piece of DNA very rapidly.

proteome All the proteins that can be produced by a genome.

repetitive DNA Multiple copies of the same sequence of DNA. The length of the sequence varies. Large eukaryotic genomes consist mostly of repetitive DNA.

restriction enzyme An enzyme that recognizes a specific DNA sequence and cuts the DNA there.

RFLP Restriction fragment length polymorphism. Variation in a given DNA sequence in a population can result in fragments of different lengths when the DNA is cut with a restriction enzyme. RFLPs make good landmarks for physical maps.

shotgun sequencing All the DNA in a genome is broken into small pieces and directly sequenced. Computer programs assemble the sequences in order.

SNP Single nucleotide polymorphism. A single base-pair variation that can be used to differentiate between individuals of the same species.

STS Sequenced-tagged site. A unique sequence of nucleotides in the genome that is used for mapping. ESTs and SNPs are types of STSs.

synteny Blocks of DNA with genes in the same order that are conserved between species. Synteny makes it possible to use genome-mapped organisms to study related organisms.

---

to ensure that copies are double-stranded and can reintegrate into the genome. Some 3% is devoted to dead transposons, elements that have lost the signals for replication and can no longer jump.

Variation in the Human Genome

One fact that is becoming clear from analysis of the human genome is the huge amount of genetic variation that exists in our species. This information has practical use as we identify a specific kind of variation: **single nucleotide polymorphisms (SNPs)** in the human genome. SNPs are sites where individuals differ by only a single nucleotide. To be classified as a polymorphism, a SNP must be present in at least 1% of the population. At present, the International SNP Map Working Group has identified 50,000 SNPs in coding regions of the genome and an additional 1.4 million in noncoding DNA. It is estimated that this represents about 10% of the variation available.

These SNPs are being used to look for associations between genes. We expect that the genetic recombination occurring during meiosis randomizes all but the most tightly linked genes. We call the tendency for genes to not be randomized **linkage disequilibrium.** This kind of association can be used to map genes. The preliminary analysis of SNPs shows that many are in linkage disequilibrium. This unexpected result has led to the idea of genomic haplotypes, or regions of chromosomes that are not being exchanged by recombination. If these haplotypes stand up to further analysis, they could greatly aid in mapping the genetic basis of disease. The Human Genome Project is now working on a haplotype map of the genome.

Gene sequences in eukaryotes vary greatly in copy number, some occurring many thousands of times and others only once. Only about 1% of the human genome is devoted to protein-encoding genes. Much of the rest is composed of transposable elements.

Comparative Genomics

With the large number of sequenced genomes, it is now possible to make comparisons at both the gene and genome level (see figure 17.7). One of the striking lessons learned from the sequence of the human genome is how very like other organisms humans are. More than half of the genes of *Drosophila* have human counterparts. Among mammals, the differences are even fewer. Humans have only 300 genes that have no counterpart in the mouse genome.

The flood of information from different genomes has given rise to a new field: **genomics.** At this point, we have the complete sequences for close to 100 bacterial genomes. Among eukaryotes, we have the full genome sequences of both types of yeast used in genetics, *S. cerevisiae* and *S. pombe*, as well as the protist *Plasmodium*, the invertebrate animals *Drosophila* and *C. elegans*, and the vertebrates puffer fish, mouse, and human. In the plant kingdom, the genomes for *Arabidopsis* and rice have been completed. It is important to remember that most of these genomes are **draft sequences** that include many gaps in regions of highly repetitive DNA.

One active field of genomics is to determine the minimal genome that can support a cell. Studies have used a variety of approaches, including the analysis of parasitic bacteria with greatly reduced genomes, coupled with the analysis of cognate genes in known genomes. We have not yet been able to estimate the number of genes that is the minimal number necessary to support life, but we are closing in on that goal.

Another related ongoing project is to learn the true "language" of proteins. As more and more genomic sequences are available to analyze, it is becoming clear that only a limited number of basic kinds of proteins are encoded by genomes and that a limited number of domains have been used over and over by evolution.

The use of comparative genomics to ask evolutionary questions is also a field of great promise. For this reason, the next round of animal genomes being sequenced includes the chimp, our closest living relative. Comparison of human and chimp genomes will provide a clearer picture of our relationship. The comparison of the many prokaryotic genomes already indicates a greater degree of lateral gene transfer than was previously suspected.

Synteny

Similarities and differences between highly conserved genes in different species can be investigated on a gene-by-gene basis. Genome science allows for a much larger-scale approach to comparing genomes by taking advantage of **synteny.** Synteny refers to the conserved arrangements of segments of DNA in related genomes. Physical mapping can be used to look for conserved segments of DNA in genomes that have not been sequenced. Comparisons with the sequenced syntenous segment in another species can be very helpful. To illustrate this, consider rice, already sequenced, and its grain relatives maize, barley, and wheat, none of which have been fully sequenced. Even though these plants diverged more than 50 million years ago, the chromosomes of rice, corn, wheat, and other grass crops show extensive conserved arrangements of segments (synteny) (figure 17.9). In a genomic sense, "rice is wheat." Interestingly, the rice genome has more genes than the human genome. However, rice still has a much smaller genome than its grain relatives that represent a major food source for humans. By understanding the rice genome at the level of its DNA sequence, it should be much easier to identify and isolate genes from grains with larger genomes. DNA sequence analysis of cereal grains will be important for identifying genes associated with disease resistance, crop yield, nutritional quality, and growth capacity.

Organelle Genomes

Mitochondria and chloroplasts are bacterial relatives living in eukaryotes as a result of endosymbiosis. Their genomes have been sequenced in some species and are most like prokaryotic genomes. The chloroplast genome, having about 100 genes, is minute compared to the rice genome, with 32,000 to 55,000 genes.

The chloroplast is a plant organelle that functions in photosynthesis. It can independently replicate in the plant cell because it has its own genome. The DNA in the chloroplasts of all land plants have about the same number of genes, and they are present in about the same order. In contrast to the evolution of the DNA in the plant cell nucleus, chloroplast DNA has evolved at a more conservative pace, and therefore shows a more interpretable evolutionary pattern when scientists study DNA sequence similarities. Chloroplast DNA is also not subject to modification caused by transposable elements and mutations due to recombination.

Over time, there appears to have been some genetic exchange between the nuclear and chloroplast genomes. For example, the key enzyme (RUBISCO) in the Calvin cycle of photosynthesis consists of large and small subunits. The small subunit is encoded in the nuclear genome. The protein it encodes has a targeting sequence that allows it to enter the chloroplast and combine with large subunits, which are coded for and produced by the chloroplast. The evolutionary history of the localization of these genes is a puzzle. Comparative genomics and their evolutionary implications are explored in detail in chapter 24, after we have established the fundamentals of evolutionary theory.

Comparisons of whole genome maps reveal a surprising commonality of genes among organisms. Only a small percentage of the DNA of eukaryotic genomes codes for functional proteins.

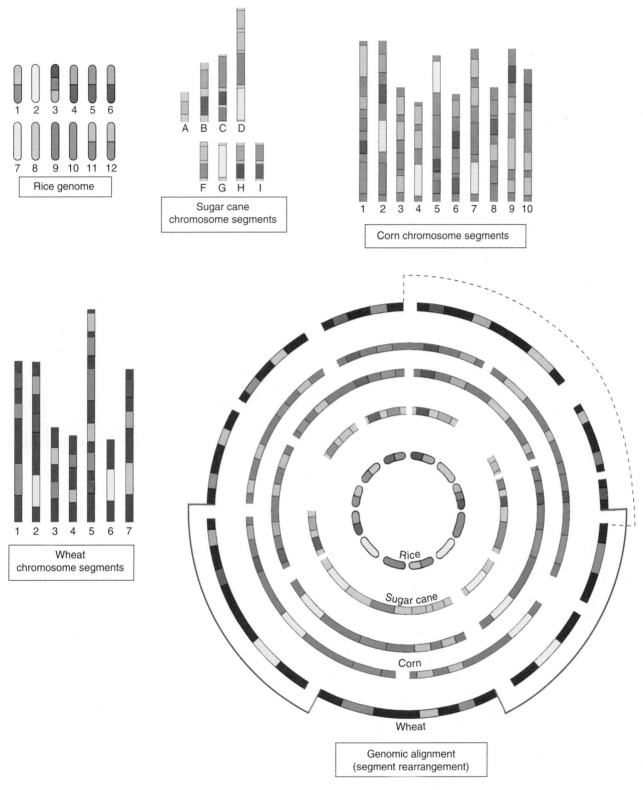

FIGURE 17.9

Grain genomes are rearrangements of similar chromosome segments. Shades of the same color represent pieces of DNA that are conserved among the different species but have been rearranged. By splitting the individual chromosomes of major grass species into segments and rearranging the segments, researchers have found that the genome components of rice, sugarcane, corn, and wheat are highly conserved. This implies that the order of the segments in the ancestral grass genome has been rearranged by recombination as the grasses have evolved.

17.4 Genomics is opening a new window on life.

Functional Genomics

Sequencing the human and rice genomes represents major technological accomplishment. The new field of bioinformatics takes advantage of high-end computer technology to analyze the growing gene databases, look for relationships among genomes, and hypothesize functions of genes based on sequence. Genomics is now shifting gears and moving back to hypothesis-driven science, to **functional genomics,** the study of the function of genes and their products. Bioinformatics approaches are being combined with newly developed technologies for analyzing genes and their functions. Like sequencing whole genomes, finding how these genomes work requires the efforts of a large team. For example, an international community of researchers have come together with a plan to assign function to all of the 20,000 to 25,000 *Arabidopsis* genes by 2010 (Project 2010). One of the first steps is to determine when and where these genes are expressed. Each step beyond that will require additional enabling technology.

DNA Microarrays

How can DNA sequences be made available to researchers, other than as databases of electronic information? DNA microarrays (figure 17.10) are a way to make DNA se-

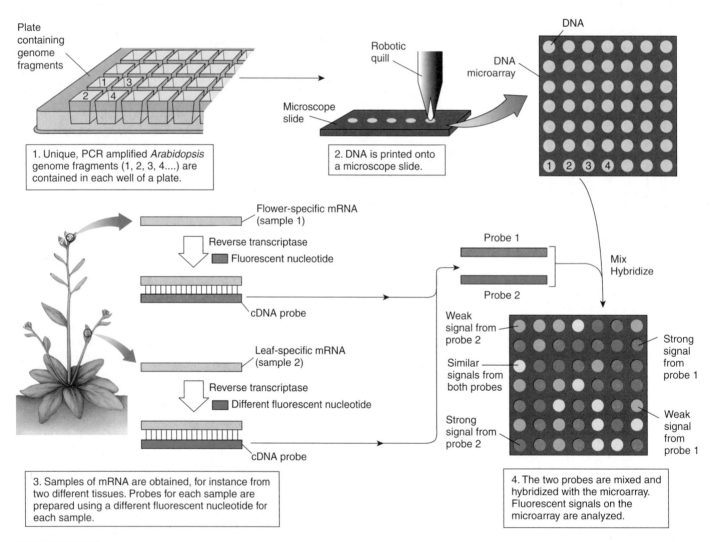

1. Unique, PCR amplified *Arabidopsis* genome fragments (1, 2, 3, 4....) are contained in each well of a plate.

2. DNA is printed onto a microscope slide.

3. Samples of mRNA are obtained, for instance from two different tissues. Probes for each sample are prepared using a different fluorescent nucleotide for each sample.

4. The two probes are mixed and hybridized with the microarray. Fluorescent signals on the microarray are analyzed.

FIGURE 17.10

Microarrays. Microarrays are created by robotically placing DNA onto a microscope slide. The microarray can then be probed with RNA from tissues of interest to identify expressed DNA. The microarray with hybridized probes is analyzed and often displayed as a false-color image. If a gene is frequently expressed in one of the samples, the fluorescent signal will be strong (*red or green*) where the gene is located on the microarray. If a gene is rarely expressed in one of the samples, the signal will be weak (*pink or light green*). A yellow color indicates genes that are expressed at similar levels in each sample.

FIGURE 17.11
Growth of a transgenic plant. DNA containing a gene for herbicide resistance was transferred into wheat (*Triticum aestivum*). The DNA also contains the *GUS* gene, which is used as a tag or label. The *GUS* gene produces an enzyme that catalyzes the conversion of a staining solution from clear to blue. (*a*) Embryonic tissue just prior to insertion of foreign DNA. (*b*) Following DNA transfer, callus cells containing the foreign DNA are indicated by color from the *GUS* gene *(blue spots)*. (*c*) Shoot formation in the transgenic plants growing on a selective medium. Here, the gene for herbicide resistance in the transgenic plants allows growth on the selective medium containing the herbicide. (*d*) Comparison of growth on the selection medium for transgenic plants bearing the herbicide resistance gene (*left*) and a nontransgenic plant (*right*).

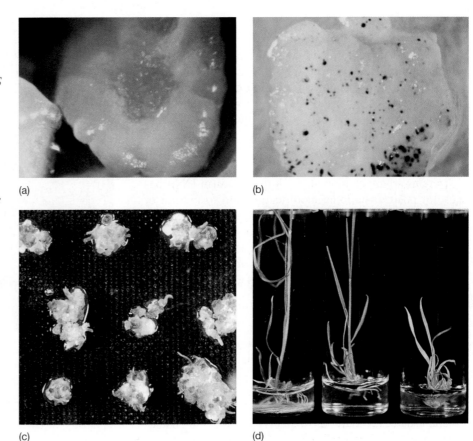

(a)　(b)　(c)　(d)

quences available to many researchers for the purpose of investigating gene function. To prepare a particular microarray, fragments of DNA are deposited on a microscope slide by a robot at indexed locations. Silicon chips instead of slides can also be arrayed. Researchers are currently using a chip with 24,000 *Arabidopsis* genes on it to identify genes that are expressed developmentally in certain tissues or in response to environmental factors. RNA from these tissues can be isolated and used as a probe for these microarrays. Only those sequences that are expressed in the tissues will be present and will hybridize to the microarray (see figure 17.10).

Microarrays are also being used to discern genetic differences among individuals of the same species. As we mentioned earlier, every human has a unique SNP profile that can be screened on arrays or screened using other emerging technology. A huge SNP database is being constructed for the human genome. About 1.42 million SNPs are distributed throughout our genome. On average, that is about one SNP in every 1200 base-pairs. SNPs can alter gene function or serve as markers embedded within repetitive DNA. About 85% of all human exons are less than 5 kilobases from a SNP. In addition to being fairly ubiquitous, SNPs have a very low rate of mutation. Medical applications of SNPs are discussed in chapter 20. SNPs are also valuable in following human migration and evolution.

Transgenics

How do you determine whether two genes that are from different species but have similar sequences have the same function? How can you be sure that a gene identified by an annotation program actually functions as a gene in the organism? One way to address these questions is to create transgenic organisms. The technology for creating transgenic organisms was discussed in chapter 16 and is illustrated for plants in figure 17.11. To test whether an *Arabidopsis* gene is homologous to a rice gene, it can be inserted into cells of rice, which are then regenerated into a rice plant. Different markers can be incorporated into the gene so that its protein product can be visualized or isolated in the transgenic plant. In some cases, the transgene (inserted foreign gene) may affect a visible phenotype. Of course, transgenics are but one of many ways to address questions about gene function.

Genomics has provided us with millions of new genes to investigate. Finding the function of these previously unknown genes depends on technologies, including microarrays, that can screen large numbers of genes very rapidly.

Proteomics

To fully understand how genes work, we need to characterize the proteins they produce. This information is essential in understanding cell biology, physiology, development, and evolution. How are similar genes used in different plants to create biochemically and morphologically distinct organisms? In many ways, we continue to ask the same questions that even Mendel asked, but at a much different level of organization.

Proteins are much more difficult to study than DNA because of posttranslational modification and formation of complexes of proteins. As we have seen, a single gene can code for multiple proteins using alternative splicing (see figure 17.8). While all the DNA in a genome can be isolated from a single cell, only a portion of the *proteome* (all the proteins coded for by the genome) is expressed in a single cell or tissue. There is an additional level of complexity to consider in **proteomics** (the study of the proteome). Just because a gene is being transcribed into RNA in a cell at a specific time, this does not mean that the transcript is necessarily being translated into a protein right then. Thus, an intermediate step in characterizing the proteome is studying the *transcriptome*, all the RNA present in a cell or tissue at a specific time.

The use of new methods to quickly identify and characterize large numbers of proteins is the distinguishing feature between traditional protein biochemistry and proteomics. As with genomics, the challenge is one of scale. Ideally, a researcher would like to be able to examine a nucleotide sequence and know what sort of functional protein the gene specifies. Databases of protein structures in different organisms can be searched to predict the structure and function of genes known only by sequence, as identified in genome projects. We are getting a clearer picture of how gene sequence relates to protein shape and function. Having a greater number of DNA sequences available allows for more extensive comparisons and identification of common structural patterns as groups of proteins continue to emerge.

Fortunately, while there may be as many as a million different proteins, most are just variations on a handful of themes. The same shared structural motifs—barrels, helices, molecular zippers—are found in the proteins of plants, insects, and humans (figure 17.12; also see chapter 3 for more information on protein motifs). The maximum number of distinct motifs has been estimated to be fewer than 5000. About 1000 of these motifs have already been cataloged. Both publicly and privately financed efforts are now under way to detail the shapes of all the common motifs.

Protein arrays, comparable to arrays of DNA, are being used to analyze large numbers of proteins simultaneously. Making a protein array starts with isolating the transcriptome of a cell or tissue. Then cDNAs are constructed and reproduced by cloning them into bacteria or viruses. Tran-

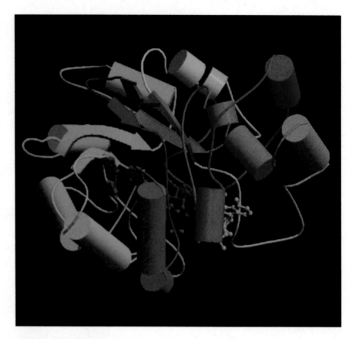

FIGURE 17.12
Computer-generated model of an enzyme. Searchable databases contain known protein structures, including human aldose reductase shown here.

scription and translation occur in the prokaryotic host, and micromolar quantities of protein are isolated and purified. These are then spotted onto glass slides. Protein arrays can be probed in at least three different ways. They can be screened with antibodies to specific proteins. Antibodies are labeled so that they can be detected, and the patterns on the protein array can be determined by computer analysis. An array of proteins can also be screened with another protein to detect binding or other protein interactions. Thousands of interactions can be tested simultaneously. For example, calmodulin (which mediates Ca^{++} function; see figure 7.9) was labeled and used to probe a yeast proteome array with 5800 proteins. The screen revealed 39 proteins that bound calmodulin. Of those 39, 33 were previously unknown! A third type of screen uses small molecules to assess whether or not they will bind to any of the proteins on the array. This approach shows promise for discovering new drugs that will inhibit proteins involved in disease.

Proteomics is a large-scale effort to identify and study the proteins coded for by the genome. Web-based databases allow researchers to compare new DNA sequences with known sequences and to predict the structure of the protein encoded in the DNA. New proteins are being identified by screening arrays of proteins.

Using Genomic Information

Swimming in a Sea of Genes

The genomics revolution has yielded millions of new genes to be investigated. The potential of genomics to improve human health through medical diagnostics and to improve nutrition through agriculture is enormous. Mutations in a single gene can explain some, but not most, heritable diseases. With entire genomes to search, the probability of unraveling human, animal, and plant diseases is greatly improved. While proteomics will lead to new pharmaceuticals, the immediate impact of genomics is being seen in diagnostics. Both improved technology and gene discovery are enhancing diagnosis of genetic abnormalities. Diagnostics are also being used to identify individuals. For example, SNPs were among the forensic diagnostic tools used to identify remains of victims of the terrorist attack on the World Trade Center in New York City.

The September 11 attacks were followed by an increased awareness and concern about biological weapons. When cases of anthrax began appearing in the fall of 2001, genome sequencing made it possible to explore possible sources of the deadly bacteria and determine whether or not they had been genetically engineered to increase their lethality. Substantial effort turned toward the use of genomic tools to distinguish between naturally occurring infections and intentional outbreaks of disease. The Centers for Disease Control and Prevention (CDC) prioritized bacteria and viruses that are likely targets for bioterrorism (table 17.2). Prokaryotes, with their small genomes, can be rapidly sequenced and readily altered to make them even more virulent. The focus on pathogens extends beyond bioterrorism. The spread of emerging viruses, including Ebola and Severe Acute Respiratory Syndrome (SARS), is increasingly problematic as global interactions among humans become more frequent.

Human evolutionary history and patterns of migration are subject to lively debate among scientists. The fossil record, mitochondrial DNA sequences, and now SNP comparisons are beginning to clarify a rather fuzzy picture. The worldwide pattern of SNPs to date indicates that Eurasian populations differ from African populations. However, the Eurasian SNP patterns in noncoding regions of DNA appear to be a subset of the African population. This is consistent with other work supporting an African origin for humans. The SNP data also raise an important caution about medical diagnostics based on SNP profiles; that is, the geographic origins of an individual must be considered.

Table 17.2 High-Priority Pathogens for Genomic Research

Pathogen	Disease	Genome*
Variola major	Smallpox	Complete
Bacillus anthracis	Anthrax	Complete
Yersinia pestis	Plague	In progress
Clostridium botulinum	Botulism	In progress
Francisella tularensis	Tularemia	Complete
Filoviruses	Ebola and Marburg hemorrhagic fever	Both are complete
Arenaviruses	Lassa fever and Argentine hemorrhagic fever	Both are complete

*There are multiple strains of these viruses and bacteria. "Complete" indicates that at least one has been sequenced. For example, the Florida strain of anthrax was the first to be sequenced.

FIGURE 17.13
Rice field. Most of the rice grown globally is directly consumed by humans and is the dietary mainstay of 2 billion people.

Feeding the World

Globally, nutrition is the greatest impediment to human health. Much of the excitement about the rice genome project is based on its potential for improving the yield and nutritional quality of rice and other cereals worldwide. The development of golden rice discussed in chapter 16 is a promising example of improved nutrition through genetic approaches. About one-third of the world population obtains half its calories from rice (figure 17.13). In some regions, individuals consume up to 1.5 kilograms of rice daily. More than 500 million tons of rice are produced each year, but this may not be enough food in the future. The current world population of 6 billion is expected to peak at 9 billion by 2070 and then decline to around 8.4 billion by 2100. Producing enough food to feed 9 billion people would be a major challenge. Food distribution can be problematic given the uneven distribution of good farmland throughout the world. One solution is to breed crops that are better adapted to regional environments with less than ideal soil and climate.

Due in large part to scientific advances in crop breeding and farming techniques, world food production has more than doubled, while world cropland has increased by only 9%. The world now farms an area the size of South America, but without the scientific advances of the past 40 years, the entire western hemisphere would need to be farmed to produce enough food for the world. Unfortunately, water usage for crops has tripled in that time period, and quality farmland is being lost to soil erosion. Many believe that conventional crop breeding programs may have reached their limit. The question is, How best to feed billions of additional people without destroying much of the planet in the process? Scientists are also concerned about the effects of global climate change on agriculture worldwide. Increasing the yield and quality of crops grown for both human and livestock consumption, especially on more marginal farmland, will depend on many factors. But genetic engineering built on the findings of genomics projects can contribute significantly to the solution.

Most crops grown in the United States produce less than half of their genetic potential because of environmental stresses (salt, water, and temperature), herbivores, and pathogens (figure 17.14). Identifying genes that can provide stress and pest protection is the focus of many current genomics research projects. Most likely, multiple genes will be involved. Having access to entire genomic sequences will enhance the probability of identifying critical genes.

Who Owns a Genome?

Genome science is also a source of ethical challenges and dilemmas. One example is the issue of gene patents. Actually, it is the use of a gene, not the gene itself, that is patentable. For a gene-related patent, the product and its function must be known. The public genome consortia, supported by federal funding, have been driven by the belief that the sequence of genomes should be freely available to all and should not be patented. Private companies patent gene functions, but often make sequence data available with certain restrictions. The physical sciences have negotiated the landscape of public and for-profit research for decades, but this is relatively new territory for biologists.

Another ethical issue involves privacy. How sequence data are used is the focus of thoughtful and ongoing discussions. The Universal Declaration on the Human Genome and Human Rights states, "The human genome underlies the fundamental unity of all members of the human family, as well as the recognition of their inherent dignity and diversity. In a symbolic sense, it is the heritage of humanity." While we talk about "the" human genome, each of us has subtly different genomes that can be used to identify us. We can already screen for genetic disorders such as cystic fibrosis and Huntington disease, but genomics will greatly increase the number of screenable traits. Behavioral genomics is an area that is also rich with possibilities and

FIGURE 17.14
Corn crop productivity well below its genetic potential due to drought stress. Corn production can be limited by water deficiencies due to the drought that occurs during the growing season in dry climates. Global climate change may increase drought stress in areas where corn is the major crop.
The corn genome has not been sequenced. How could you use information from the rice genome sequence to try to improve drought tolerance in corn?

dilemmas. Very few behavioral traits can be accounted for by single genes. Two genes have been associated with fragile-X mental retardation, and three with early-onset Alzheimer disease. Comparisons of multiple genomes will likely lead to the identification of multiple genes controlling a range of behaviors. Will this change the way we view acceptable behavior?

What if employers or insurance companies gain access to your personal SNP profile? Could you be discriminated against because you have a genetic tendency toward chemical addiction or heart disease? On a more positive note, the U.S. Armed Forces require DNA samples from members for possible casualty identification, and DNA-based identification brought peace of mind to some families of the World Trade Center victims.

In Iceland, the parliament has voted to have a private company create a database from pooled medical, genetic, and genealogical information about all Icelanders, a particularly fascinating population from a genetic perspective. Because minimal migration or immigration has occurred there over the last 800 years, the information that can be mined from the Icelandic database is phenomenal. Ultimately, the value of that information has to be weighed against any possible discrimination or stigmatization of individuals or groups.

Genomics offers tremendous benefits for human health, but also raises new questions about privacy and other issues.

17.1 Genomes can be mapped both genetically and physically.

Genome Maps

- Genetic maps show the relative location of genes on a chromosome as determined by recombination frequencies. Genetic maps measure distance in centimorgans. (p. 344)
- Physical maps show relative positions of landmarks within specific DNA sequences and measure distance in base-pairs. (p. 344)

17.2 Genome sequencing produces the ultimate physical map.

Sequencing

- Large-scale genome sequencing relies on an automated sequencer and computer analysis, but to reduce errors, genomes are cut into short segments for replication. (p. 346)
- Clone-by-clone sequencing starts by constructing a physical map with many landmarks and then sequencing known sites. (p. 347)
- Shotgun cloning sequences cloned fragments, and then uses computational analysis to determine the whole sequence from the overlapping sequences. (p. 347)

17.3 Being more complex does not necessarily require more genes.

Human Genome Project

- The Human Genome Project originated in 1990, and success was announced in 2001. (p. 348)
- Eukaryotic genomes appear to be larger than prokaryotic genomes, but the size of the organism does not appear to play a determining role. (p. 348)

Genome Geography

- Bioinformatic approaches can be used to identify gene sequences. (p. 349)
- Alternative splicing yields different proteins that can have different functions. (p. 349)
- Four different classes of protein-encoding genes found in eukaryotic genomes are (1) single-copy genes, (2) segmental duplications, (3) multigene families, and (4) tandem clusters. (p. 349)

- Only about 1% of the human genome is devoted to protein-encoding genes. (p. 350)
- Six major sorts of noncoding human DNA are (1) noncoding DNA within genes, (2) structural DNA, (3) simple sequence repeats, (4) segmental duplications, (5) pseudogenes, and (6) transposable elements. (p. 350)

Comparative Genomics

- Synteny refers to the conserved arrangements of segments of DNA in related genomes. (p. 352)
- Comparisons with the sequenced syntenous segment in another species can help in determining evolutionary relationships. (p. 352)

17.4 Genomics is opening a new window on life.

Functional Genomics

- Genomics is shifting toward functional genomics, the study of the function of genes and their products. (p. 354)
- DNA microarrays can screen large numbers of genes very rapidly. (p. 355)

Proteomics

- Proteins are much more difficult to study than DNA because of the posttranslational modification and formation of protein complexes. (p. 356)
- Proteonomics is an effort to identify and study proteins coded for by the genome, and is distinguished from traditional protein biochemistry by the use of new methods to quickly identify and characterize large numbers of proteins. (p. 356)

Using Genomic Information

- The potential of genomics to improve human health through medical diagnostics and to improve nutrition through agriculture is enormous. (p. 357)
- Genomic science is also a source of ethical challenges and dilemmas, such as patient rights and personal privacy issues. (p. 358)

Self Test

1. Researchers from many labs collaborated to determine the sequence of the human genome. How did labs avoid sequencing the same fragments multiple times?
   a. Each lab could isolate DNA from one particular chromosome to divide the sequencing projects.
   b. Using restriction fragment length polymorphisms, labs could ensure that they were not sequencing the same fragments.
   c. Using short sequences from their respective clones, sequenced-tagged sites (STSs), researchers could check to make sure their fragments were not already being sequenced by another group.
   d. By comparing sequences from collaborating labs, researchers could ensure that they were not sequencing the same fragments.

2. Some of your friends are trying to make sense of their genome. To help them out, you draw an analogy between our chromosomes and the interstate highway system. In this analogy, every interstate represents a single chromosome. How could you describe the relationship between chromosomes and genes to your friends using this analogy?
   a. Every mile marker would represent a gene.
   b. Every town would represent a gene.
   c. Every state would represent a gene.
   d. Genes would be defined by the twists and turns on the highway.

3. Imagine that you broke your mother's favorite vase and had to reconstruct it from the shattered pieces. To do this, you would have to look for pieces with similar ends to join and then progressively glue every piece together. What sequencing strategy does this most closely represent?
   a. shotgun sequencing
   b. contig sequencing
   c. clone-by-clone sequencing
   d. manual sequencing

4. Knowing the sequence of an entire genome
   a. completes our understanding of every gene's function in the organism.
   b. allows us to predict the genetic cause of every disease in the organism.
   c. provides a template for constructing an artificial life-form.
   d. provides the raw data that can then be used to identify specific genes.

5. If you were to look at the sequence of an entire chromosome, how could you identify which segments might contain a gene?
   a. You could identify large protein-coding regions (open reading frames).
   b. You could look for a match with an expressed sequence tag (EST).
   c. You could look for consensus regulatory sequences that could initiate transcription.
   d. All of these strategies could be used to identify possible genes.
   e. It is impossible to predict genes from sequence data alone.

6. You have been hired to characterize the genome of a novel organism, *Undergraduatus genomicus*. After fully sequencing the 10^6 base-pairs in the genome, you predict that this organism has approximately 10,000 genes. You have a collaborator on this project, however, who has identified 20,000 different expressed sequence tags from this organism. How can you resolve this conflict?
   a. You suggest that your collaborator is an idiot who counted every gene twice!
   b. You suggest that your collaborator may have identified multiple isoforms of the same gene that could arise by alternative splicing.
   c. You fear that you may have underestimated the number of genes, because you forgot that the organism is diploid and you did not count both copies of every gene in your total.
   d. You only identified genes with open reading frames, but most genes do not encode proteins, so your number will be low.

7. In addition to coding sequences, our genome contains
   a. noncoding DNA within genes (i.e., introns).
   b. structural DNA involved in telomeres and centromeres.
   c. simple repetitive DNA.
   d. DNA from transposable elements that have jumped around in the genome.
   e. All of these are present in genomic DNA.

8. Natural variation in the length of tandem repeat sequences (VNTRs) found in the genome can be used to identify individual people by their DNA fingerprint. Why is this possible?
   a. The statement is not true; such variability prevents this from being a useful identification tool.
   b. The changes in repeat length change the DNA synthesis pattern, so the cell cycles have different lengths, making the cells of different people different sizes.
   c. The changes in repeat length occur very infrequently, so there is only one pattern that everybody shares.
   d. The changes in repeat length occur very frequently, so everybody has a unique pattern of different lengths when several repeats are examined.

Test Your Visual Understanding

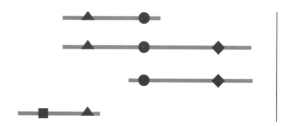

1. From the information given in the above diagram, construct a contig map of the region presented.

Apply Your Knowledge

1. Every cell in your body contains the same genomic DNA, yet the proteome of different tissues is unique. How can you explain this?

2. Chromosomes are much like interstate highways. Develop this analogy by assigning a chromosomal counterpart to the following:
   a. the beginning and end of a particular highway
   b. towns along the highway
   c. stretches of highway that pass through wilderness

3. If you are given a sample of DNA from an unknown organism, how could you determine the origin of the DNA sample?

18

Control of Gene Expression

Concept Outline

18.1 Gene expression is controlled by regulating transcription.

An Overview of Transcriptional Control. In prokaryotes, transcription is regulated by controlling access of RNA polymerase to the promoter in a flexible and reversible way; eukaryotes, by contrast, regulate many of their genes by turning them on and off in a more permanent fashion.

18.2 Regulatory proteins read DNA without unwinding it.

How to Read a Helix Without Unwinding It. Regulatory proteins slide special segments called DNA-binding motifs along the major groove of the DNA helix, reading the sides of the bases.

Four Important DNA-Binding Motifs. DNA-binding proteins contain structural motifs, such as the helix-turn-helix motif, which fit into the major groove of the DNA helix.

18.3 Prokaryotes regulate genes by controlling transcription initiation.

Prokaryotic Gene Regulation. Repressor proteins inhibit RNA polymerase's access to the promoter, while activators facilitate its binding.

18.4 Transcriptional control in eukaryotes operates at a distance.

Transcriptional Control in Eukaryotes. Eukaryotic genes use a complex collection of transcription factors and enhancers to aid the polymerase in transcription.

The Effect of Chromatin Structure on Gene Expression. The tight packaging of eukaryotic DNA into nucleosomes does not interfere with gene expression.

Posttranscriptional Control in Eukaryotes. Gene expression can be controlled at a variety of levels after transcription. Both proteins and small RNAs play major roles in regulating the stability and expression of gene transcripts.

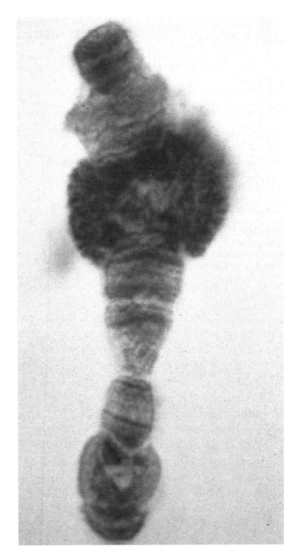

FIGURE 18.1
Chromosome puffs. In this chromosome of the fly *Drosophila melanogaster*, individual active genes can be visualized as "puffs" on the chromosomes. The RNA being transcribed from the DNA template has been radioactively labeled, and the dark specks indicate its position on the chromosome.

In an orchestra, all of the instruments do not play at the same time; if they did, all that would be produced is noise. Instead, a musical score determines which instruments in the orchestra play when. Similarly, all of the genes in an organism are not expressed at the same time, each gene producing the protein it encodes full tilt. Instead, different genes are expressed at different times, with a genetic score written in regulatory regions of the DNA determining which genes are active when (figure 18.1).

An Overview of Transcriptional Control

Control of gene expression is essential to all organisms. In prokaryotes, it allows the cell to take advantage of changing environmental conditions. In multicellular eukaryotes, it is critical for directing development and maintaining homeostasis.

Regulating Promoter Access

One way to control gene expression is to regulate the initiation of transcription. In order for a gene to be transcribed, RNA polymerase must have access to the DNA helix and must be capable of binding to the gene's **promoter,** a specific sequence of nucleotides at one end of the gene that tells the polymerase where to begin transcribing. How is the initiation of transcription regulated? Protein-binding nucleotide sequences on the DNA regulate the initiation of transcription by modulating the ability of RNA polymerase to bind to the promoter. These protein-binding sites are usually only 10 to 15 nucleotides in length (even a large regulatory protein has a "footprint," or binding area, of only about 20 nucleotides). Hundreds of these regulatory sequences have been characterized, and each provides a binding site for a specific protein able to recognize the sequence. Binding the protein to the regulatory sequence either *blocks* transcription by getting in the way of RNA polymerase, or *stimulates* transcription by facilitating the binding of RNA polymerase to the promoter.

Transcriptional Control in Prokaryotes

Control of gene expression is accomplished very differently in prokaryotes than in the cells of eukaryotes. Prokaryotic cells have been shaped by evolution to grow and divide as rapidly as possible, enabling them to exploit transient resources. Proteins in prokaryotes turn over rapidly. This allows them to respond quickly to changes in their external environment by changing patterns of gene expression. In prokaryotes, the primary function of gene control is to adjust the cell's activities to its immediate environment. Changes in gene expression alter which enzymes are present in the cell in response to the quantity and type of available nutrients and the amount of oxygen present. Almost all of these changes are fully reversible, allowing the cell to adjust its enzyme levels up or down as the environment changes.

Transcriptional Control in Eukaryotes

The cells of multicellular organisms, on the other hand, have been shaped by evolution to be protected from transient changes in their immediate environment. Most of them experience fairly constant conditions. Indeed, **homeostasis**—the maintenance of a constant internal environment—is considered by many to be the hallmark of multicellular organisms. Although cells in such organisms still respond to signals in their immediate environment (such as growth factors and hormones) by altering gene expression, in doing so they participate in regulating the body as a whole. In multicellular organisms with relatively constant internal environments, the primary function of gene control in a cell is not to respond to that cell's immediate environment, but rather to participate in regulating the body as a whole.

Some of these changes in gene expression compensate for changes in the physiological condition of the body. Others mediate the decisions that *produce* the body, ensuring that the right genes are expressed in the right cells at the right time during development. The growth and development of multicellular organisms entail a long series of biochemical reactions, each catalyzed by a specific enzyme. Once a particular developmental change has occurred, these enzymes cease to be active, lest they disrupt the events that must follow. To produce these enzymes, genes are transcribed in a carefully prescribed order, each for a specified period of time, following a fixed genetic program that may even lead to programmed cell death. The one-time expression of the genes that guide such a program is fundamentally different from the reversible metabolic adjustments prokaryotic cells make to the environment. In all multicellular organisms, changes in gene expression within particular cells serve the needs of the whole organism, rather than the survival of individual cells.

Posttranscriptional Control

Gene expression can be regulated at many levels. By far the most common form of regulation in both prokaryotes and eukaryotes is **transcriptional control**—that is, control of the transcription of particular genes by RNA polymerase. Other less common forms of control occur after transcription, influencing the mRNA that is produced from the genes or the activity of the proteins encoded by the mRNA. These controls, collectively referred to as **posttranscriptional controls,** will be discussed later.

Gene expression is controlled at the transcriptional and posttranscriptional levels. Transcriptional control, the more common type, is effected by the binding of proteins to regulatory sequences within the DNA.

18.2 Regulatory proteins read DNA without unwinding it.

How to Read a Helix Without Unwinding It

The ability of certain proteins to bind to *specific* DNA regulatory sequences provides the basic tool of gene regulation, the key ability that makes transcriptional control possible. To understand how cells control gene expression, it is first necessary to gain a clear picture of this molecular recognition process.

Looking into the Major Groove

Molecular biologists used to think that the DNA helix had to unwind before proteins could distinguish one DNA sequence from another; only in this way, they reasoned, could regulatory proteins gain access to the hydrogen bonds between base-pairs. We now know it is unnecessary for the helix to unwind, because proteins can bind to its outside surface, where the edges of the base-pairs are exposed. Careful inspection of a DNA molecule reveals two helical grooves winding around the molecule, one deeper than the other. Within the deeper groove, called the **major**

groove, the nucleotides' hydrophobic methyl groups, hydrogen atoms, and hydrogen bond donors and acceptors protrude. The pattern created by these chemical groups is unique for each of the four possible base-pair arrangements, providing a ready way for a protein nestled in the groove to read the sequence of bases (figure 18.2).

DNA-Binding Motifs

Protein-DNA recognition is an area of active research; so far, the structures of over 30 regulatory proteins have been analyzed. Although each protein is unique in its fine details, the part of the protein that actually binds to the DNA is much less variable. Almost all of these proteins employ one of a small set of **structural**, or **DNA-binding, motifs**, particular bends of the protein chain that permit it to interlock with the major groove of the DNA helix.

> Regulatory proteins identify specific sequences on the DNA double helix, without unwinding it, by inserting DNA-binding motifs into the major groove of the double helix where the edges of the bases protrude.

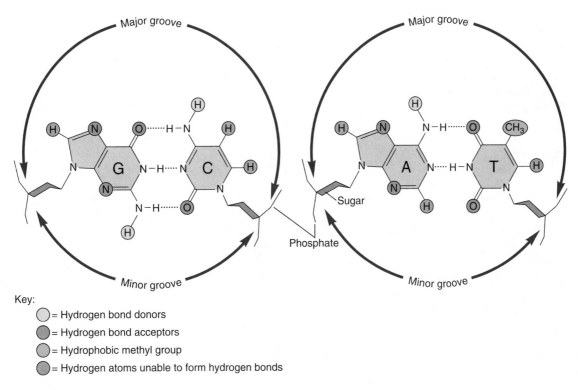

Key:
- ◯ = Hydrogen bond donors
- ● = Hydrogen bond acceptors
- ◯ = Hydrophobic methyl group
- ● = Hydrogen atoms unable to form hydrogen bonds

FIGURE 18.2
Reading the major groove of DNA. Looking down into the major groove of a DNA helix, we can see the edges of the bases protruding into the groove. Each of the four possible base-pair arrangements (two are shown here) extends a unique set of chemical groups into the groove, indicated in this diagram by differently colored balls. A regulatory protein can identify the base-pair arrangement by this characteristic signature.

Four Important DNA-Binding Motifs

The Helix-Turn-Helix Motif

The most common DNA-binding motif is the **helix-turn-helix,** constructed from two α-helical segments of the protein linked by a short, nonhelical segment, the "turn" (figure 18.3). As the first DNA-binding motif recognized, the helix-turn-helix motif has since been identified in hundreds of DNA-binding proteins.

A close look at the structure of a helix-turn-helix motif reveals how proteins containing such motifs are able to interact with the major groove of DNA. Interactions between the helical segments of the motif hold them at roughly right angles to each other. When this motif is pressed against DNA, one of the helical segments (called the recognition helix) fits snugly in the major groove of the DNA molecule, while the other butts up against the outside of the DNA molecule, helping to ensure the proper positioning of the recognition helix. Most DNA regulatory sequences recognized by helix-turn-helix motifs occur in symmetrical pairs. Such sequences are bound by proteins containing two helix-turn-helix motifs separated by 3.4 nanometers (nm), the distance required for one turn of the DNA helix (figure 18.4). Having *two* protein-DNA-binding sites doubles the zone of contact between protein and DNA, and so greatly strengthens the bond that forms between them.

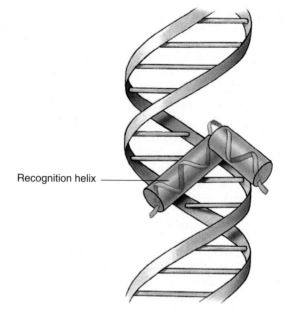

Recognition helix —

FIGURE 18.3
The helix-turn-helix motif. The recognition helix, one helical region of the motif, actually fits into the major groove of DNA. There it contacts the edges of base-pairs, enabling it to recognize specific sequences of DNA bases.

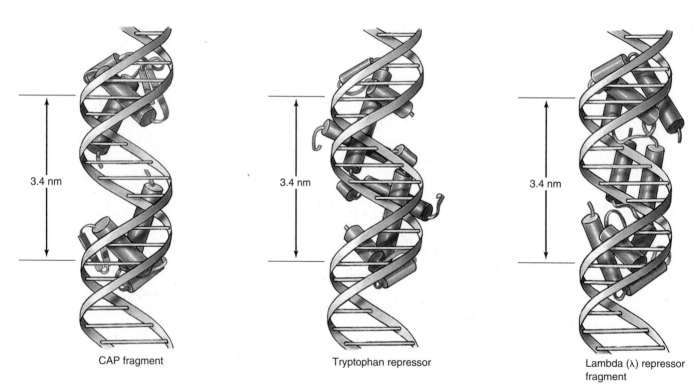

CAP fragment   Tryptophan repressor   Lambda (λ) repressor fragment

FIGURE 18.4
How the helix-turn-helix binding motif works. The three regulatory proteins illustrated here (*purple*) all bind to DNA using a pair of helix-turn-helix binding motifs. In each case, the two copies of the motif (*red*) are separated by 3.4 nanometers, precisely the spacing of one turn of the DNA helix. This allows the regulatory proteins to slip into two adjacent portions of the major groove in DNA, providing a strong attachment.

The Homeodomain Motif

A special class of helix-turn-helix motifs plays a critical role in development in a wide variety of eukaryotic organisms, including humans. These motifs were discovered when researchers began to characterize a set of homeotic mutations in *Drosophila* (mutations that alter how the parts of the body are assembled). They found that the mutant genes encoded regulatory proteins whose normal function was to initiate key stages of development by binding to developmental switch-point genes. More than 50 of these regulatory proteins have been analyzed, and they all contain a nearly identical sequence of 60 amino acids, the **homeodomain** (figure 18.5*b*). The center of the homeodomain is occupied by a helix-turn-helix motif that binds to the DNA. Surrounding this motif within the homeodomain is a region that always presents the motif to the DNA in the same way.

The Zinc Finger Motif

A different kind of DNA-binding motif uses one or more zinc atoms to coordinate its binding to DNA. Called **zinc fingers** (figure 18.5*c*), these motifs exist in several forms. In one form, a zinc atom links an α-helical segment to a β sheet segment so that the helical segment fits into the major groove of DNA. This sort of motif often occurs in clusters, the β sheets spacing the helical segments so that each helix contacts the major groove. The more zinc fingers in the cluster, the stronger the protein binds to the DNA. In other forms of the zinc finger motif, the β sheet's place is taken by another helical segment.

The Leucine Zipper Motif

In yet another DNA-binding motif, two different protein subunits cooperate to create a single DNA-binding site. This motif is created where a region on one of the subunits containing several hydrophobic amino acids (usually leucines) interacts with a similar region on the other subunit. This interaction holds the two subunits together at those regions, while the rest of the subunits are separated. Called a **leucine zipper,** this structure has the shape of a Y, with the two arms of the Y being helical regions that fit into the major groove of DNA (figure 18.5*d*). Because the two subunits can contribute quite different helical regions to the motif, leucine zippers allow for great flexibility in controlling gene expression.

> Regulatory proteins bind to the edges of base-pairs exposed in the major groove of DNA. Most contain structural motifs such as the helix-turn-helix, homeodomain, zinc finger, or leucine zipper.

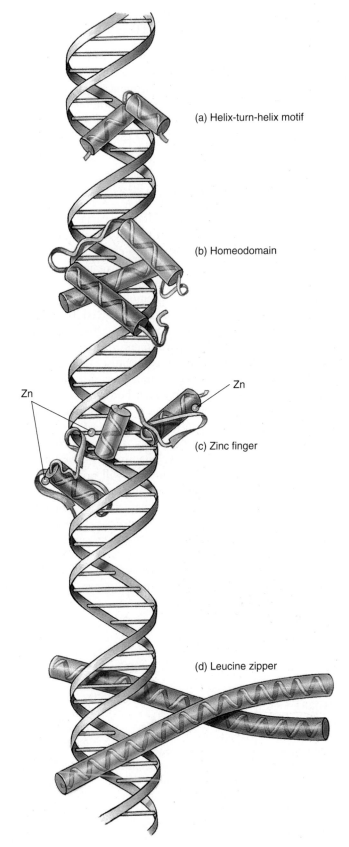

(a) Helix-turn-helix motif

(b) Homeodomain

Zn   Zn

(c) Zinc finger

(d) Leucine zipper

FIGURE 18.5
Major DNA-binding motifs.

18.3 Prokaryotes regulate genes by controlling transcription initiation.

Most organisms regulate gene expression by controlling the initiation of transcription. This is accomplished by the binding of proteins to regulatory sequences on the DNA. Prokaryotes and eukaryotes share some common themes, but they have some profound differences as well. We will focus first on prokaryotic systems and then look at how eukaryotic systems differ from this.

Prokaryotic Gene Regulation

Changes in the environments that bacteria and other prokaryotes encounter often result in changes in gene expression. How prokaryotes respond to changes depends in each instance on the nature of the proteins encoded by relevant genes. In general, genes encoding proteins involved in anabolic pathways (building up molecules) respond oppositely from genes encoding proteins involved in catabolic pathways (breaking down molecules). If a bacterium encounters a molecule in the environment that can be broken down for energy, such as the sugar lactose, then the bacterium will start to make the proteins necessary to utilize lactose. When lactose is not present, however, there is no need to make these proteins. Thus, we say that the synthesis of the proteins is *induced* by the presence of lactose. If a molecule available in the environment is the end product of a biosynthetic pathway, such as the amino acid tryptophan, then a bacterium in that environment will not synthesize the proteins necessary to make tryptophan. If tryptophan ceases to be available, then the bacterium will begin to make the proteins necessary for its synthesis. When bacteria *don't* make the proteins necessary for biosynthesis, we call this *repression*. In the case of both induction and repression, the bacterium is adjusting to produce the proteins that are optimal for its immediate environment.

Induction and repression are observed in a variety of prokaryotic gene systems. What is the nature of the genetic control circuits responsible for these adjustments? Knowing that control is probably at the level of initiation of transcription does not tell us the nature of the control. It might be either positive or negative. A regulatory molecule can increase the rate of initiation (positive) or decrease the rate of initiation (negative). On the surface, repression may appear to be negative and induction positive, but in fact both involve negative control.

For either mechanism to work, the molecule in the environment, such as lactose or tryptophan, must produce the proper effect on the gene being regulated. In the case of induction, it is necessary for the presence of lactose to prevent a negative regulator from binding to its regulatory sequence. In the case of repression, by contrast, the presence of tryptophan must make a negative regulator bind its regulatory sequence. These responses are opposite because the needs of the cell are opposite for anabolic versus catabolic pathways.

Operons

Prokaryotic genes are often organized into operons. **Operons** are multiple genes that are part of a single gene expression unit. The genes of an operon are all part of the same mRNA and thus are controlled by the same promoter. In bacteria, genes that are involved in the same metabolic pathway are often organized in this fashion. For example, the proteins necessary for the utilization of lactose are encoded by the *lac* operon (figure 18.6), and the proteins necessary for the synthesis of tryptophan are encoded by the *trp* operon. Both of these operons are regulated by systems of negative regulation.

FIGURE 18.6
The *lac* region of the *Escherichia coli* chromosome. The *lac* operon consists of a promoter, an operator, and three genes that code for proteins required for the metabolism of lactose. In addition, there is a binding site for the catabolite activator protein (CAP), which affects whether or not RNA polymerase will bind to the promoter. Gene *I* codes for a repressor protein, which will bind to the operator and block transcription of the *lac* genes. The genes *Z*, *Y*, and *A* encode the two enzymes and the permease involved in the metabolism of lactose.

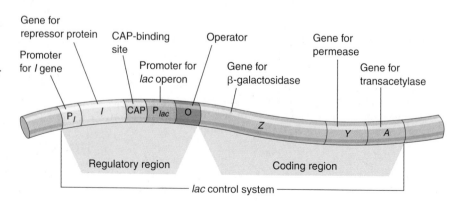

FIGURE 18.7

How the *trp* operon is controlled.
The tryptophan repressor cannot bind the operator (which is located *within* the promoter) unless tryptophan first binds to the repressor. Therefore, in the absence of tryptophan, the promoter is free to function, and RNA polymerase transcribes the operon. In the presence of tryptophan, the tryptophan-repressor complex binds tightly to the operator, preventing RNA polymerase from initiating transcription.

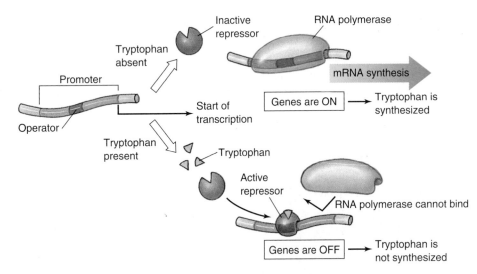

FIGURE 18.8

How the tryptophan repressor works. The binding of tryptophan to the repressor increases the distance between the two recognition helices in the repressor, allowing the repressor to fit snugly into two adjacent portions of the major groove in DNA.

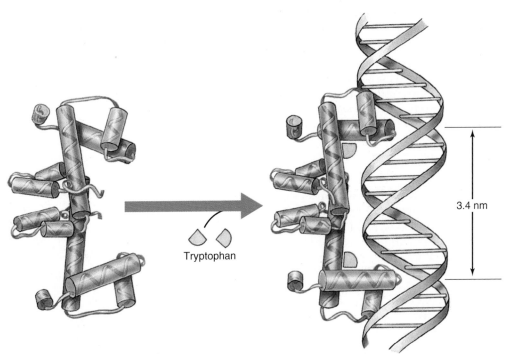

Repressors Are Off Switches

Repressors are proteins that bind to regulatory sites on DNA and prevent or decrease the initiation of transcription. Induction and repression are both mediated by a repressor molecule. These repressors do not act alone; each responds to specific effector molecules. Effector binding can alter the conformation of the repressor to either enhance or abolish its binding to DNA. These repressor proteins are allosteric proteins with an active site that binds DNA and a regulatory site that binds effectors.

trp Operon

In the *trp* operon, the operon is repressed in the presence of tryptophan and unrepressed in the absence of tryptophan. When tryptophan is present in the medium sur-

rounding the bacterium, the cell shuts off transcription of the *trp* genes by means of a tryptophan repressor, a helix-turn-helix regulatory protein that binds to the operator site located within the *trp* promoter (figure 18.7). Binding of the repressor to the operator prevents RNA polymerase from binding to the promoter. The key to the functioning of this control mechanism is that the tryptophan repressor cannot bind to DNA unless it has first bound to two molecules of tryptophan. The binding of tryptophan to the repressor alters the orientation of a pair of helix-turn-helix motifs in the repressor, causing their recognition helices to fit into adjacent major grooves of the DNA (figure 18.8).

Thus, the bacterial cell's synthesis of tryptophan depends upon the absence of tryptophan in the environment. When the environment lacks tryptophan, there is nothing to

Chapter 18 Control of Gene Expression **367**

RNA polymerase cannot transcribe *lac* genes

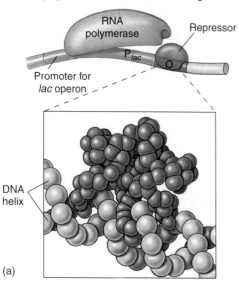

(a)

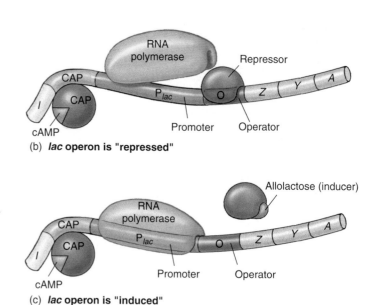

(b) ***lac* operon is "repressed"**

(c) ***lac* operon is "induced"**

FIGURE 18.9

How the *lac* repressor works. (*a*) The *lac* repressor. Because the repressor fills the major groove of the DNA helix, RNA polymerase cannot fully attach to the promoter, and transcription is blocked. (*b*) The *lac* operon is shut down (repressed) when the repressor protein is bound to the operator site. Because promoter and operator sites overlap, RNA polymerase and the repressor cannot functionally bind at the same time, any more than two people can sit in the same chair at once. (*c*) The *lac* operon is transcribed (induced) when CAP is bound and when allolactose binding to the repressor changes its shape so that it can no longer sit on the operator site and block RNA polymerase activity.

activate the repressor, so the repressor cannot prevent RNA polymerase from binding to the *trp* promoter. The *trp* genes are transcribed, and the cell proceeds to manufacture tryptophan from other molecules. On the other hand, when tryptophan is present in the environment, it binds to the repressor, which is then able to bind to the *trp* promoter. This blocks transcription of the *trp* genes, and the cell's synthesis of tryptophan halts.

lac Operon

The enzymes encoded by the *lac* operon are necessary only when lactose is available. The *lac* repressor binds DNA in the absence of lactose, but not in the presence of lactose. The DNA binding of the repressor is mediated by a metabolite of lactose, allolactose. In the absence of lactose, the operon is repressed, while in the presence of lactose, the synthesis of the proteins encoded by the operon is induced. As the level of lactose falls, the allolactose will no longer bind repressor, allowing the repressor to bind DNA again (figure 18.9).

Activators Are "On" Switches

In addition to repressors, there are proteins called **activators** that can bind DNA to stimulate the initiation of transcription. These allosteric proteins are the logical and physical opposites to repressors. All of the logic just ex-

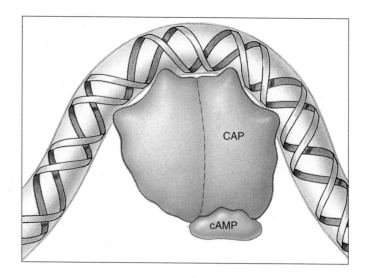

FIGURE 18.10

How CAP works. Binding of the catabolite activator protein (CAP) to DNA causes the DNA to bend around it. This increases the activity of RNA polymerase.

plained about repressors is reversed for activators: Effector molecules enhance activator binding to catabolic operons, increasing transcription.

Activators play a key role in the phenomenon of glucose repression, the preferential use of glucose in the presence of other sugars. Despite the name, glucose repression is

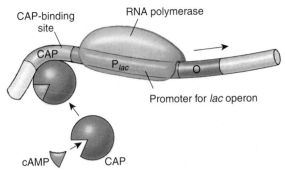

(a) **Glucose low, promoter activated**

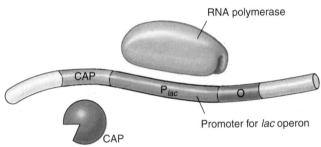

(b) **Glucose high, promoter not activated**

FIGURE 18.11
How the CAP site works. The CAP molecule can attach to the CAP-binding site only when the molecule is bound to cAMP. (*a*) When glucose levels are low, cAMP is abundant and binds to CAP. The cAMP-CAP complex binds to the CAP site, bends in the DNA, and gives RNA polymerase access to the promoter. (*b*) When glucose levels are high, cAMP is scarce, and CAP is unable to activate the promoter.

Glucose	Lactose	
		CAP-binding site — RNA-polymerase binding site (promoter) — Operator — *lacZ* gene
+	+	Operon OFF because CAP is not bound
+	−	Operon OFF both because *lac* repressor is bound and CAP is not
−	−	Operon OFF because *lac* repressor is bound
−	+	Operon ON because CAP is bound and *lac* repressor is not

FIGURE 18.12
Two regulatory proteins control the *lac* operon. Together, the *lac* repressor and CAP provide a very sensitive response to the cell's need to utilize lactose-metabolizing enzymes.

mediated by an activator protein, the **catabolite activator protein (CAP)**, which stimulates the transcription of all operons encoding proteins necessary for the catabolism of sugars. The CAP protein is another allosteric protein whose binding is controlled by an effector—in this case, cAMP (figure 18.10). Amounts of cAMP in the cell are inversely related to the level of glucose. Thus, when glucose levels are high, cAMP levels are low, and vice versa (figure 18.11). Since CAP must be bound to cAMP to bind to DNA, this prevents activation of other catabolic operons (glucose repression).

Combinations of Switches

From what we have seen, the *lac* operon is actually under both positive and negative control (figure 18.12). The negative control is due to the *lac* repressor protein, and the positive control is due to CAP. This combination of positive and negative control gives the cell fine control over the levels of the proteins needed to utilize lactose in a variety of environments. In the absence of lactose, the proteins are not made at all because the repressor will prevent initiation of transcription. In the presence of lactose, the repressor will no longer bind DNA, allowing transcription to initiate. However, if glucose is also present, then the level of cAMP will be low and CAP will be unable to bind the *lac* promoter. Transcription will occur but at a low level. In the absence of glucose, cAMP levels will be high and CAP will bind the *lac* promoter, resulting in maximal transcription.

Bacteria and other prokaryotes regulate gene expression transcriptionally through the use of repressor and activator "switches." The transcription of some clusters of genes is regulated by both repressors and activators.

18.4 Transcriptional control in eukaryotes operates at a distance.

Transcriptional Control in Eukaryotes

The control of transcription in eukaryotes is much more complex than in prokaryotes. The basic concepts of protein-DNA interactions are still valid, but the nature and number of interacting proteins is much greater due to some obvious differences in eukaryotes. First, eukaryotes have their DNA organized into chromatin, complicating protein-DNA interactions considerably. Second, eukaryotic transcription occurs in the nucleus, and translation occurs in the cytoplasm; this is in stark contrast to prokaryotes, where these processes are spatially and temporally coupled. As a consequence, the eukaryotic transcription apparatus is considerably more complex than the prokaryotic RNA polymerase, and the amount of DNA involved in regulating eukaryotic genes is much greater. The need for a fine degree of flexible control is important for eukaryotes with complex developmental programs and multiple tissue types. We will see, however, that general themes emerge from this complexity.

Eukaryotic Transcription Factors

Eukaryotic transcription requires a variety of proteins, or factors, that fall into two categories: basal transcription factors and specific transcription factors. The basal factors are necessary for the assembly of a transcription apparatus and recruitment of RNA pol II (see chapter 15) to a promoter. The specific factors increase the level of transcription in certain cell types or in response to specific signals.

Basal Transcription Factors. Transcription of RNA pol II templates (that is, genes that encode protein products) requires more than just RNA pol II to initiate transcription. A host of **basal transcription factors** are also necessary to establish productive initiation. These factors are required for transcription to occur, but do not increase the rate above this basal rate. They are given letter designations that follow the designator TFII, for transcription factor RNA pol II. The most important of these factors, TFIID, contains the TATA-binding protein that recognizes the TATA box sequence in the promoter (figure 18.13). Binding of TFIID is followed by binding of TFIIE, TFIIF, TFIIA, TFIIB, and TFIIH and a host of accessory factors called *transcription-associated factors*, TAFs. The *initiation complex* that results (figure 18.14) is clearly much more complex than a bacterial RNA polymerase, a single pol holoenzyme. And there is yet another level of complexity. The initiation complex, while capable of initiating synthesis at a basal level, will not achieve transcription at a high level without the participation of other specific factors.

Specific Transcription Factors. **Specific transcription factors** act in a tissue- or time-dependent manner to stimulate higher levels of transcription than the basal level. The number and diversity of these factors are overwhelming. Some sense can be made of this proliferation of factors by concentrating on the DNA-binding motif, as opposed to the specific factors. A key common theme that emerges from the study of these factors is that specific transcription factors, called **activators,** have a domain organization. Each factor consists of a DNA-binding domain

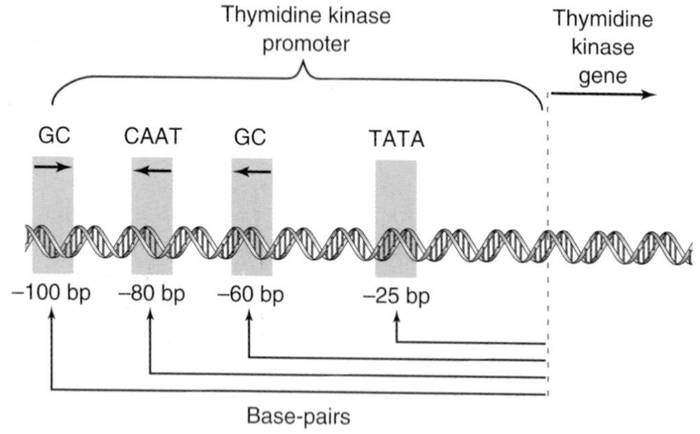

FIGURE 18.13

A eukaryotic promoter. This promoter for the gene encoding the enzyme thymidine kinase contains the TATA box that a transcription factor binds to, as well as three other DNA sequences that direct the binding of other elements of the transcription apparatus.

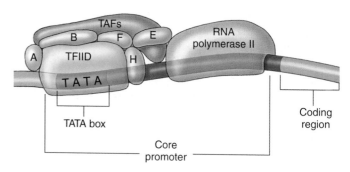

FIGURE 18.14

Formation of a eukaryotic initiation complex. The basal transcription factor, TFIID, binds to the TATA box and is joined by the other basal factors, TFIIE, TFIIF, TFIIA, TFIIB, and TFIIH. This complex is added to by a number of transcription-associated factors (TAFs) that together recruit the RNA pol II molecule to the core promoter.

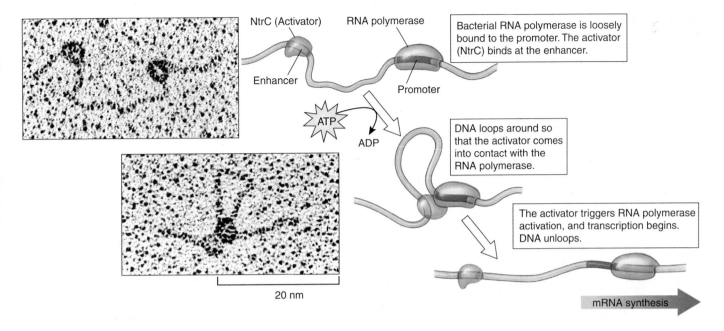

Bacterial RNA polymerase is loosely bound to the promoter. The activator (NtrC) binds at the enhancer.

NtrC (Activator)　RNA polymerase

Enhancer

Promoter

ATP

ADP

DNA loops around so that the activator comes into contact with the RNA polymerase.

The activator triggers RNA polymerase activation, and transcription begins. DNA unloops.

20 nm

mRNA synthesis

FIGURE 18.15

An enhancer in action. When the bacterial activator NtrC binds to an enhancer, it causes the DNA to loop over to a distant site where RNA polymerase is bound, activating transcription. While such enhancers are rare in prokaryotes, they are common in eukaryotes.

and a separate activating domain that interacts with the transcription apparatus. These domains are essentially independent in the protein, such that they can be "swapped" between different factors and retain their function. This has been demonstrated by domain-swapping experiments in which the activation domain of one factor is spliced to the DNA-binding domain of another factor. The resulting hybrid factor shows the DNA-binding specificity expected of the DNA-binding domain, while activating transcription to the level expected of the activating domain.

Enhancers. Enhancers were originally defined as DNA sequences necessary for high levels of transcription that can act in a position- and orientation-independent manner. At first, this seems counterintuitive, especially since we had been conditioned by prokaryotic systems to expect control regions to be immediately upstream of the coding region of genes. It turns out that enhancers are the binding site of the specific transcription factors. The ability of enhancers to act over large distances was at first puzzling to molecular biologists. We now think this is accomplished by DNA bending to form a loop, positioning the enhancer close to the promoter (figure 18.15). Distance separating two sites on the chromosome does not have to translate to great physical distance, because the flexibility of DNA allows bending and looping to bring the enhancer and the activator into contact with the transcription factors (figure 18.16).

Coactivators and Mediators. There are other factors that specifically mediate transcription factor action. These *coactivators* and *mediators* are also necessary for activation of transcription by the transcription factor. They act by bind-

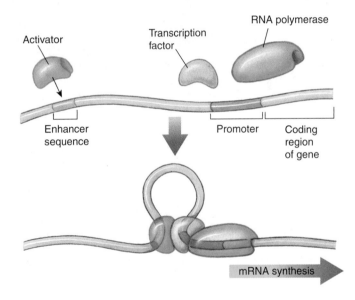

Activator　　Transcription factor　　RNA polymerase

Enhancer sequence　　Promoter　　Coding region of gene

mRNA synthesis

FIGURE 18.16

How enhancers work. The enhancer site is located far away from the gene being regulated. Binding of an activator (*red*) to the enhancer allows the activator to interact with the transcription factors (*green*) associated with RNA polymerase, activating transcription.

ing the transcription factor and then binding to another part of the transcription apparatus. These mediators are essential to the function of some transcription factors, but not all transcription factors require them. There is a much smaller number of coactivators than transcription factors because the same coactivator can be used with multiple transcription factors.

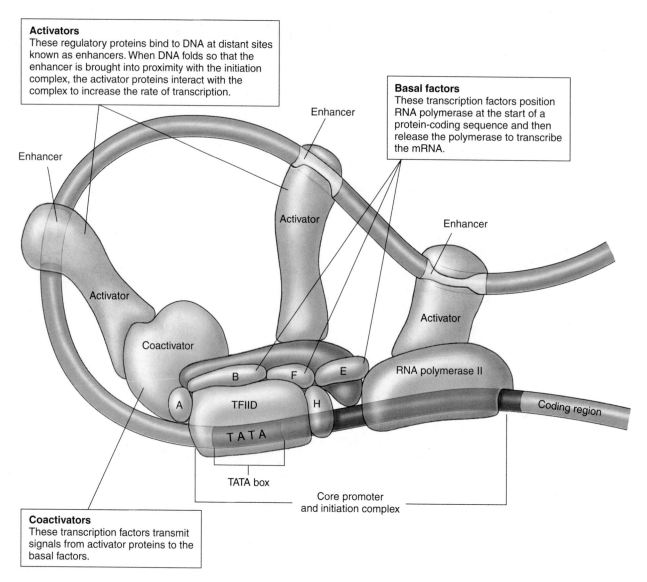

Activators
These regulatory proteins bind to DNA at distant sites known as enhancers. When DNA folds so that the enhancer is brought into proximity with the initiation complex, the activator proteins interact with the complex to increase the rate of transcription.

Basal factors
These transcription factors position RNA polymerase at the start of a protein-coding sequence and then release the polymerase to transcribe the mRNA.

Enhancer

Enhancer

Activator

Enhancer

Activator

Activator

Coactivator

B   F   E

A   TFIID   H   RNA polymerase II

Coding region

T A T A

TATA box

Core promoter
and initiation complex

Coactivators
These transcription factors transmit signals from activator proteins to the basal factors.

FIGURE 18.17
Interactions of various factors within the transcription complex. All specific transcription factors bind to enhancer sequences that may be distant from the promoter. These proteins can then interact with the initiation complex by DNA looping to bring the factors into proximity with the initiation complex. As detailed in the text, some transcription factors, called activators, can directly interact with the RNA polymerase II or the initiation complex, while others require additional coactivators.

Tying It All Together. How can we make sense of this extremely complicated situation? It is important to realize that although a few general principles apply to a broad range of situations, nearly every eukaryotic gene or group of coordinately regulated genes represents a unique case. Virtually all genes that are transcribed by RNA pol II need the same suite of basal factors to assemble an initiation complex, but the assembly of this complex and its ultimate level of transcription depend in each instance on the other specific factors involved that make up the *transcription complex* (figure 18.17). The makeup of eukaryotic promoters is thus either very simple, if you only consider that which is needed for the initiation complex, or very complex, if you consider all factors that may bind and affect transcription. This kind of combinatorial gene regulation leads to great flexibility in the control of gene expression because it provides great flexibility in responding to the many signals affecting transcription that a cell may receive, allowing integration of these signals.

Transcription factors and enhancers confer great flexibility on the control of gene expression in eukaryotes.

The Effect of Chromatin Structure on Gene Expression

As if the conditions described so far were not complex enough, eukaryotes have the additional hurdle of possessing DNA that is packaged into chromatin. The packaging of DNA first into nucleosomes (figure 18.18) and then into higher-order chromatin structures is now thought to be directly related to the control of gene expression. Chromatin structure at its lowest level is the organization of DNA and histone proteins into nucleosomes. These nucleosomes may block binding of transcription factors and RNA pol II at the promoter. The higher-order organization of chromatin, which is not completely understood, appears to depend on the state of the histones in nucleosomes. Histones can be modified to result in a greater condensation of chromatin-making promoters, even less accessible for protein-DNA interactions. There is also a chromatin remodeling complex that can make DNA more accessible.

DNA Methylation

Chemical **methylation** of the DNA was once thought to play a major role in gene regulation in vertebrate cells. The addition of a methyl group to cytosine creates 5-methylcytosine but has no effect on base-pairing with guanine (figure 18.19), just as the addition of a methyl group to uracil produces thymine without affecting base-pairing with adenine. Many inactive mammalian genes are methylated, and it was tempting to conclude that methylation caused the inactivation. However, methylation is now viewed as having a less direct role, blocking accidental transcription of "turned-off" genes. Vertebrate cells apparently possess a protein that binds to clusters of 5-methylcytosine, preventing transcriptional activators from gaining access to the DNA. DNA methylation in vertebrates thus ensures that once a gene is turned off, it stays off.

Chromatin Structure and Transcriptional Activators

As in prokaryotes, not all gene regulation involves repression of transcription. In at least some instances, the transcription of specific genes is activated. The activation of transcription by the activating domains of transcription factors has not been well characterized. Some activators seem to act by interacting directly with the initiation complex or with coactivators that interact with the initiation complex. Other cases are not so clear. The emerging consensus is that coactivators act by modifying the structure of chromatin by adding acetyl groups to amino acids, making DNA accessible to transcription factors. Thus, transcribed DNA is correlated with the presence of nucleosomes having histones that have been acetylated. Recently, some coactivators have been shown to be histone acetylases. In these cases, it appears that transcription is increased by removing higher-

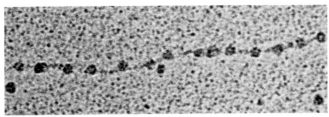

(a)

(b)

FIGURE 18.18
Nucleosomes. (*a*) In the electron micrograph, the individual nucleosomes have diameters of about 10 nanometers. (*b*) In the diagram of a nucleosome, the DNA double helix is wound around a core complex of eight histones; one additional histone binds to the outside of the nucleosome, exterior to the DNA.

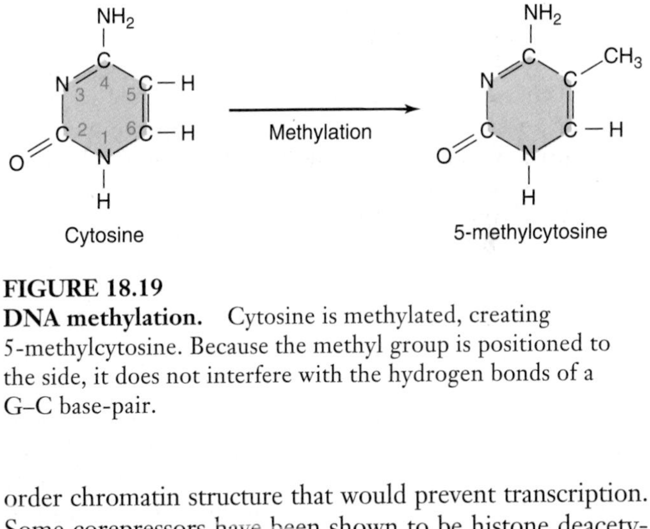

Cytosine   Methylation   5-methylcytosine

FIGURE 18.19
DNA methylation. Cytosine is methylated, creating 5-methylcytosine. Because the methyl group is positioned to the side, it does not interfere with the hydrogen bonds of a G–C base-pair.

order chromatin structure that would prevent transcription. Some corepressors have been shown to be histone deacetylases as well. This fits well with the view that histone modification affects chromatin structure, and that coactivators and corepressors then modulate this effect.

Transcriptional control of gene expression occurs in eukaryotes despite the tight packaging of DNA into nucleosomes.

Posttranscriptional Control in Eukaryotes

Thus far we have discussed gene regulation entirely in terms of transcription initiation—that is, when and how often RNA polymerase starts "reading" a particular gene. Most gene regulation appears to occur at this point. However, there are many other points after transcription where gene expression could be regulated in principle, and all of them serve as control points for at least some eukaryotic genes. In general, these posttranscriptional control processes involve the recognition of specific sequences on the primary RNA transcript by regulatory proteins and small RNA molecules.

Small RNAs

Recent experiments indicate that a class of RNA molecules loosely called *small RNAs* may play a major role in regulating gene expression by interacting directly with primary gene transcripts. Small RNAs are short segments of RNA ranging in length from 21 to 28 nucleotides. Researchers focusing on far larger messenger RNA (mRNA), transfer RNA (tRNA), and ribosomal RNA (rRNA) had not noticed these far smaller bits, tossing them out during experiments. The first hints of the existence of small RNAs emerged in 1993, when researchers reported in the nematode *Caenorhabditis elegans* the presence of tiny RNA molecules that don't encode any protein. These small RNAs appeared to regulate the activity of specific *C. elegans* genes.

Soon researchers found evidence of similar small RNAs in a wide range of other organisms. In the plant *Arabidopsis thaliana*, small RNAs seemed to be involved in the regulation of genes critical to early development, while in yeasts they were identified as the agents that silence genes in tightly packed regions of the genome. In the ciliated protozoan *Tetrahymena thermophila*, the loss of major blocks of DNA during development seems guided by small RNA molecules.

RNA Interference. What is going on here? How do small fragments of RNA act to regulate gene expression? The first clue emerged in 1998, when researchers injected small stretches of double-stranded RNA into *C. elegans*. Double-stranded RNA forms when a single strand folds back in a hairpin loop; this occurs because the two ends of the strand have a complementary nucleotide sequence, so that base-pairing holds the strands together much as it does the strands of a DNA duplex (figure 18.20). The result? The double-stranded RNA strongly inhibited the expression of the genes from which the double-stranded RNA had been generated. This kind of gene silencing, since seen in *Drosophila* and other organisms, is called **RNA interference.**

FIGURE 18.20
Small RNAs form double-stranded loops. These three RNA molecules fold back to form hairpin loops because the sequences of the left and right halves are complementary, and form base-pairs.

How Small RNAs Regulate Gene Expression. In 2001, researchers identified an enzyme, dubbed *dicer*, that appears to generate the small RNAs in the cell. Dicer chops double-stranded RNA molecules into little pieces. Two types of small RNA result: **microRNAs (miRNAs)** and **small interfering RNAs (siRNAs).**

miRNAs appear to act by binding directly to mRNAs and preventing their translation into protein. Researchers have identified over 100 different miRNAs, and are still trying to sort out how each functions and which miRNAs occur in which species.

siRNAs appear to be the main agents of RNA interference, acting to degrade particular messenger RNAs after they have been transcribed but before they can be translated by the ribosomes. The exact way they achieve this degradation of selected gene transcripts is not yet known. Current data suggest that dicer delivers siRNAs to an enzyme complex called RISC, which searches out and degrades any mRNA molecules with a complementary sequence (figure 18.21).

RNA interference appears to play a major role in *epigenetic change*, a change in gene expression that is passed from one generation to another but is not caused by changes in the DNA sequence of genes. Epigenetic regulation is in many cases the result of alterations in DNA packaging. As we saw in chapter 11, the DNA of eukaryotes is packaged in a highly compact form that enables it to fit into the cell nucleus. DNA is wrapped tightly around histone proteins to form nucleosomes (see figure 18.18), and then the strand of nucleosomes is twisted into higher-order filaments. By altering how the strands are twisted, siRNAs can alter which genes are accessible for gene expression. Just how siRNAs achieve this change in chromatin shape is not known.

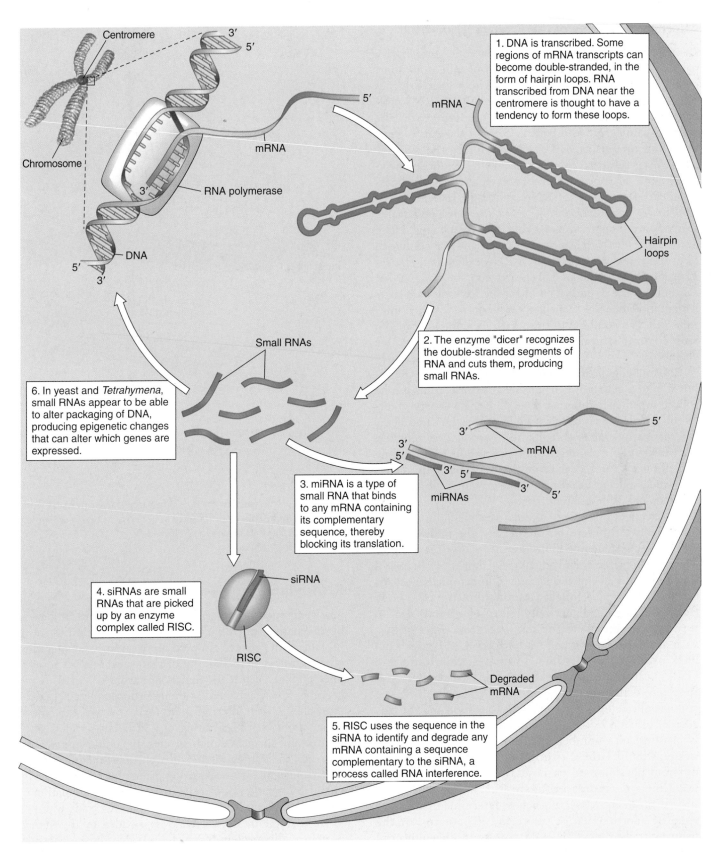

Centromere

3′
5′

mRNA

1. DNA is transcribed. Some regions of mRNA transcripts can become double-stranded, in the form of hairpin loops. RNA transcribed from DNA near the centromere is thought to have a tendency to form these loops.

Chromosome

5′

mRNA

RNA polymerase

Hairpin loops

5′
DNA
3′

Small RNAs

2. The enzyme "dicer" recognizes the double-stranded segments of RNA and cuts them, producing small RNAs.

6. In yeast and *Tetrahymena*, small RNAs appear to be able to alter packaging of DNA, producing epigenetic changes that can alter which genes are expressed.

5′

3′
3′
mRNA
5′
3′ 5′
3′
miRNAs
3′ 5′

3. miRNA is a type of small RNA that binds to any mRNA containing its complementary sequence, thereby blocking its translation.

siRNA

4. siRNAs are small RNAs that are picked up by an enzyme complex called RISC.

RISC

Degraded mRNA

5. RISC uses the sequence in the siRNA to identify and degrade any mRNA containing a sequence complementary to the siRNA, a process called RNA interference.

FIGURE 18.21
How small RNAs may act to regulate gene expression.   Small RNAs are produced when double-stranded hairpin loops of an RNA transcript are cut. Although the details are not well understood, two types of small RNA, miRNA and siRNA, are thought to block gene expression within the nucleus at the level of the mRNA gene transcript, a process called RNA interference. As the above diagram suggests, small RNAs are also thought to influence chromatin packaging in some organisms.

Alternative Splicing of the Primary Transcript

As we learned in chapter 15, most eukaryotic genes have a patchwork structure, being composed of numerous short coding sequences (exons) embedded within long stretches of noncoding sequences (introns). The initial mRNA molecule copied from a gene by RNA polymerase, the **primary transcript,** is a faithful copy of the entire gene, including introns as well as exons. Before the primary transcript is translated, the introns, which comprise on average 90% of the transcript, are removed in a process called *RNA processing,* or **RNA splicing.** Particles called *small nuclear ribonucleoproteins,* or *snRNPs* (more informally, **snurps**), are thought to play a role in RNA splicing. These particles reside in the nucleus of a cell and are composed of proteins and a special type of RNA called *small nuclear RNA,* or *snRNA.* One kind of snRNP contains snRNA that can bind to the 5′ end of an intron by forming base-pairs with complementary sequences on the intron. When multiple snRNPs combine to form a larger complex called a **spliceosome,** the intron loops out and is excised (see chapter 15).

RNA splicing provides a potential point at which the expression of a gene can be controlled, because exons can be spliced together in different ways, allowing a variety of different polypeptides to be assembled from the same gene. Alternative splicing is common in insects and vertebrates, with two or three different proteins produced from one gene. In many cases, gene expression is regulated by changing which splicing event occurs during different stages of development or in different tissues.

An excellent example of alternative splicing in action is found in two different human organs, the thyroid gland and the hypothalamus. The thyroid gland (see chapter 47) is responsible for producing hormones that control processes such as metabolic rate. The hypothalamus, located in the brain, collects information from the body (for example, salt balance) and releases hormones that in turn regulate the release of hormones from other glands, such as the pituitary gland (see chapter 47). The two organs produce two distinct hormones, calcitonin and CGRP (calcitonin gene-related peptide) as part of their function. Calcitonin is responsible for controlling the amount of calcium we take up from our food and the balance of calcium in tissues such as bone and teeth. CGRP is involved in a number of neural and endocrine functions. Although these two hormones are used for very different physiological purposes, the hormones are made using the same transcript (figure 18.22). The appearance of one product versus another is determined by tissue-specific factors that regulate the processing of the primary transcript. This ability offers another powerful way to control the expression of gene products, ranging from proteins with subtle differences to totally unrelated proteins.

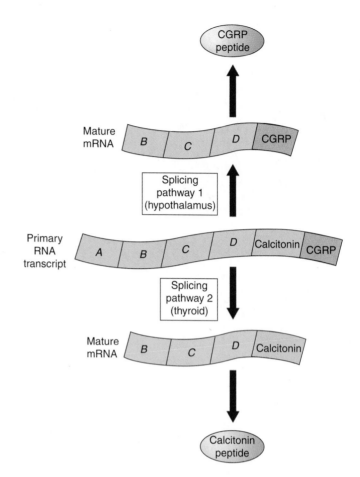

FIGURE 18.22
Alternative splicing products. The same transcript made from one gene can be spliced differently to give rise to two very distinct protein products, calcitonin and CGRP.

RNA Editing

The generation of multiple transcripts from the same gene by alternative splicing allows for the production of more proteins than "genes." An even more unexpected alteration of the message is the editing of mature transcripts to produce an altered mRNA that is not truly encoded in the genome. First discovered as the insertion of uracil residues into some RNA transcripts in protozoa, this was thought to be an anomaly. RNA editing of a different sort has since been found in mammalian species, including humans. In this case, the editing involves the chemical modification of a base to change its base-pairing properties. This occurs by either deamination of cytosine to uracil, or deamination of adenine to inosine (inosine pairs as G during translation).

An example of RNA editing is provided by the human protein apolipoprotein B, which is involved in the transport of cholesterol and triglycerides. The gene that encodes this protein (*apoB*) is large and complex, consisting of 39 introns scattered across almost 50 kilobases of DNA.

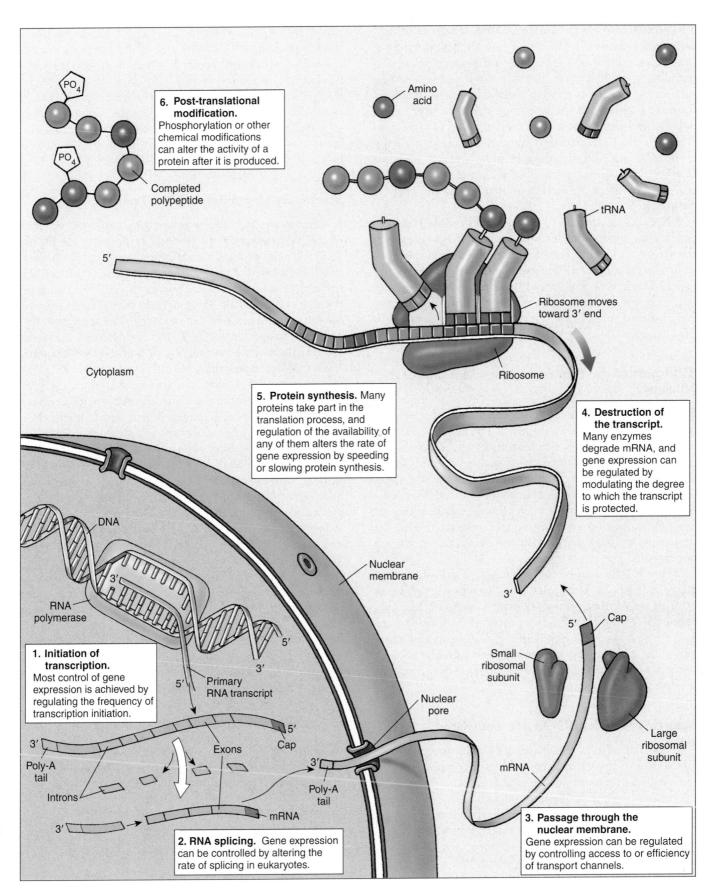

6. Post-translational modification. Phosphorylation or other chemical modifications can alter the activity of a protein after it is produced.

PO₄
PO₄
Completed polypeptide

Amino acid

tRNA

Ribosome moves toward 3′ end

Ribosome

5. Protein synthesis. Many proteins take part in the translation process, and regulation of the availability of any of them alters the rate of gene expression by speeding or slowing protein synthesis.

Cytoplasm

4. Destruction of the transcript. Many enzymes degrade mRNA, and gene expression can be regulated by modulating the degree to which the transcript is protected.

DNA

RNA polymerase

Nuclear membrane

Cap

Small ribosomal subunit

Large ribosomal subunit

1. Initiation of transcription. Most control of gene expression is achieved by regulating the frequency of transcription initiation.

Primary RNA transcript

Cap

Exons

Nuclear pore

mRNA

Poly-A tail

Introns

Poly-A tail

mRNA

3. Passage through the nuclear membrane. Gene expression can be regulated by controlling access to or efficiency of transport channels.

2. RNA splicing. Gene expression can be controlled by altering the rate of splicing in eukaryotes.

FIGURE 18.23
Six levels where gene expression can be controlled in eukaryotes.

The protein exists in two isoforms: a full-length APOB100 form, and a truncated APOB48 form. The truncated form is due to an alteration of the mRNA that changes a codon encoding glutamine to one that is a "stop" codon. Further, this editing occurs in a tissue-specific manner; the edited form only appears in the intestine, while the liver makes only the full-length form. The full-length APOB100 form is part of the LDL particle that carries cholesterol. High levels of serum LDL are thought to be a major predictor of atherosclerosis in humans. It does not appear that editing has any effect on the levels of the intestine-specific transcript.

RNA editing has also been observed in some brain receptors for opiates in humans. One of these receptors, the serotonin (5-HT) receptor, is edited at multiple sites to produce a total of 12 different isoforms of the protein. It is not clear how widespread these forms of editing are, but they are further evidence that the information encoded within genes is not the end of the story for protein production.

Transport of the Processed Transcript Out of the Nucleus

Processed mRNA transcripts exit the nucleus through the nuclear pores described in chapter 5. The passage of a transcript across the nuclear membrane is an active process that requires the transcript to be recognized by receptors lining the interior of the pores. Specific portions of the transcript, such as the poly-A tail, appear to play a role in this recognition. The transcript cannot move through a pore as long as any of the splicing enzymes remain associated with the transcript, ensuring that partially processed transcripts are not exported into the cytoplasm.

There is little hard evidence that gene expression is regulated at this point, although it could be. On average, about 10% of transcribed genes are exon sequences, but only about 5% of the total mRNA produced as primary transcript ever reaches the cytoplasm. This suggests that about half of the exon primary transcripts never leave the nucleus, but it is not clear whether the disappearance of this mRNA is selective.

Selecting Which mRNAs Are Translated

The translation of a processed mRNA transcript by the ribosomes in the cytoplasm involves a complex of proteins called *translation factors*. In at least some cases, gene expression is regulated by modification of one or more of these factors. In other instances, **translation repressor proteins** shut down translation by binding to the beginning of the transcript, so that it cannot attach to the ribosome. In humans, the production of ferritin (an iron-storing protein) is normally shut off by a translation repressor protein called aconitase. Aconitase binds to a 30-nucleotide sequence at the beginning of the ferritin mRNA, forming a stable loop to which ribosomes cannot bind. When the cell encounters iron, the binding of iron to aconitase causes the aconitase to dissociate from the ferritin mRNA, freeing the mRNA to be translated and increasing ferritin production 100-fold.

Selectively Degrading mRNA Transcripts

Another aspect that affects gene expression is the stability of mRNA transcripts in the cell cytoplasm (see figure 18.21). Unlike prokaryotic mRNA transcripts, which typically have a half-life of about 3 minutes, eukaryotic mRNA transcripts are very stable. For example, β-globin gene transcripts have a half-life of over 10 hours, an eternity in the fast-moving metabolic life of a cell. The transcripts encoding regulatory proteins and growth factors, however, are usually much less stable, with half-lives of less than 1 hour. What makes these particular transcripts so unstable? In many cases, they contain specific sequences near their 3′ ends that make them attractive targets for enzymes that degrade mRNA. A sequence of A and U nucleotides near the 3′ poly-A tail of a transcript promotes removal of the tail, which destabilizes the mRNA. Histone transcripts, for example, have a half-life of about 1 hour in cells that are actively synthesizing DNA; at other times during the cell cycle, the poly-A tail is lost, and the transcripts are degraded within minutes. Other mRNA transcripts contain sequences near their 3′ ends that are recognition sites for endonucleases, which causes these transcripts to be digested quickly. The short half-lives of the mRNA transcripts of many regulatory genes are critical to the function of those genes because they enable the levels of regulatory proteins in the cell to be altered rapidly. There are multiple pathways of decay, including deadenylation-dependent decay. *Cis*-acting elements can either stabilize or destabilize a transcript.

A review of various methods of posttranscriptional control of gene expression is provided in figure 18.23.

Posttranscriptional control of gene expression is exercised by both proteins and small RNAs. Proteins interacting with small nuclear RNA carry out alternative splicing of RNA transcripts. Chemical alterations of specific bases lead to RNA transcript editing. After the primary transcript is modified, further control of gene expression occurs by translation repression and selective degradation of mRNA transcripts.

18.1 Gene expression is controlled by regulating transcription.

An Overview of Transcriptional Control

- One way to control gene expression is to regulate the initiation of transcription by regulating promoter access. (p. 362)
- Gene expression is controlled at the transcriptional and posttranscriptional levels, with transcriptional control being the most common form of control (p. 362)

18.2 Regulatory proteins read DNA without unwinding it.

How to Read a Helix Without Unwinding It

- Regulatory proteins identify DNA sequences without unwinding the helix by inserting DNA-binding motifs into the major groove where the edges of the base-pairs are exposed. (p. 363)

Four Important DNA-Binding Motifs

- The four most important DNA-binding motifs are the helix-turn-helix motif, the homeodomain motif, the zinc finger motif, and the leucine zipper motif. (pp. 364–365)

18.3 Prokaryotes regulate genes by controlling transcription initiation.

Prokaryotic Gene Regulation

- How prokaryotes respond to changes depends on the nature of the proteins encoded by relevant genes. (p. 366)
- Prokaryotic genes are often organized into operons, multiple genes that are part of a single gene expression unit. (p. 366)
- Repressors are proteins that bind to regulatory sites on DNA and prevent or decrease initiation of transcription. (p. 367)
- Activators can bind to DNA to stimulate the initiation of transcription. (p. 368)

18.4 Transcriptional control in eukaryotes operates at a distance.

Transcriptional Control in Eukaryotes

- Eukaryotic transcription factors fall into two categories: basal transcriptional factors and specific transcriptional factors. Along with enhancers, coactivators, and mediators, these factors provide flexibility in controlling eukaryotic gene expression. (pp. 370–372)

The Effect of Chromatin Structure on Gene Expression

- Despite the additional complexity of DNA packaging in eukaryotes, transcriptional control of gene expression can still occur in eukaryotes. (p. 373)

Posttranscriptional Control in Eukaryotes

- Posttranscriptional control of gene expression is exercised by proteins and small RNAs. (p. 374)
- Small RNAs such as miRNAs and siRNAs are thought to regulate gene expression through RNA interference and, in some cases, through the alteration of DNA packaging. (pp. 374–375)
- Proteins interact with small nuclear RNA and carry out alternative splicing of RNA transcripts. (p. 376)
- RNA transcript editing can be caused by chemical alterations of specific bases. (pp. 376–378)
- Following modification of the primary transcript, further gene expression control occurs by translation repression and selective degradation of mRNA transcripts. (p. 378)

Self Test

1. Prokaryotes and eukaryotes use several methods to regulate gene expression, but the most common method is
   a. translational control.
   b. transcriptional control.
   c. posttranscriptional control.
   d. control of mRNA passage from the nucleus.
2. The two protein subunits of the leucine zipper are held together
   a. in the shape of a Y.
   b. by the interaction of leucine amino acids.
   c. by hydrophobic interactions.
   d. All of these are correct.
3. The helix-turn-helix motif contains two helical segments, and in order for the motif to bind DNA, the _____ fits into the major groove of the DNA.
   a. homeodomain
   b. recognition helix
   c. zinc finger
   d. leucine zipper
4. A(n) _____ is a piece of DNA with a group of genes that are transcribed together as a unit.
   a. promoter
   b. repressor
   c. operator
   d. operon
5. What effect would the addition of lactose have on a repressed *lac* operon?
   a. The operator site on the operon would move.
   b. It would reinforce the repression of that gene.
   c. The *lac* operon would be transcribed.
   d. It would have no effect whatsoever.
6. A type of DNA sequence that is located far from a gene but can promote its expression is a(n)
   a. promoter.
   b. activator.
   c. enhancer.
   d. TATA box.
7. Which of the following is *not* found in a eukaryotic transcription complex?
   a. activator
   b. RNA
   c. enhancer
   d. TATA-binding protein
8. DNA methylation of genes
   a. inhibits transcription by blocking the base-pairing between methylated cytosine and guanine.
   b. inhibits transcription by blocking the base-pairing between uracil and adenine.
   c. prevents transcription by blocking the TATA sequence.
   d. makes sure genes that are turned off remain turned off.
9. Which of the following are *not* matched correctly?
   a. RNA splicing—occurs in the nucleus
   b. snRNP—splicing out exons from the transcript
   c. poly-A tail—increased transcript stability
   d. All are matched correctly.
10. Which of the following is *not* a method of posttranscriptional control in eukaryotic cells?
    a. processing the transcript
    b. selecting the mRNA molecules that are translated
    c. digesting the DNA immediately after translation
    d. selectively degrading the mRNA transcripts

Test Your Visual Understanding

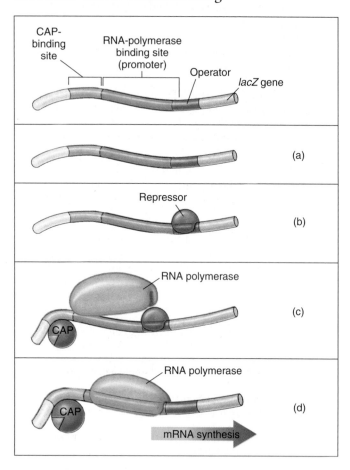

1. Match the following descriptions with the appropriate lettered panels in the figure, and explain what would be needed to activate the operon if it is not activated.
   i. Operon is OFF because *lac* repressor is bound.
   ii. Operon is OFF because CAP is not bound.
   iii. Operon is ON because CAP is bound and *lac* repressor is not.
   iv. Operon is OFF both because *lac* repressor is bound and because CAP is not.

Apply Your Knowledge

1. A method of posttranscriptional control is the selective degradation of mRNA transcripts. A transcript encoding a growth factor contains the terminal sequence AAGCUUGAAU and has a half-life of 40 minutes. Another transcript that encodes an immunoglobin has a terminal sequence of GGAUCGCCAGG and has a half-life of about 2 hours. The half-life is related to the rate of degradation by the equation: $t_{1/2} = 0.693/K$, where $t_{1/2}$ is the half-life, and K is the rate of degradation. Compare the degradation rates of the two transcripts.
2. All human beings have a rich growth of *E. coli* bacteria in their large intestine. Will the *lac* operon in the bacteria present in a lactose-intolerant individual who is careful never to consume anything containing lactose (milk sugar) be activated or repressed? Explain.

19

Cellular Mechanisms of Development

Concept Outline

19.1 Development is a regulated process.

Overview of Development. Studies of cellular mechanisms have focused on mice, fruit flies, flowering plants, and nematodes.

Vertebrate Development. Vertebrates develop in a highly orchestrated fashion.

Insect Development. Insect development is highly specialized, with many key events occurring in a fused mass of cells.

Plant Development. Unlike animal development, which is buffered from the environment, plant development is sensitive to environmental influences.

Nematode Development. The developmental history of every somatic cell of this animal is known.

19.2 Multicellular organisms employ the same basic mechanisms of development.

Cell Movement and Induction. Animal cells move by extending protein cables that they use to pull themselves past surrounding cells. Transcription within cells is influenced by signal molecules from other cells.

Determination. Cells become reversibly committed to particular developmental paths.

Pattern Formation. Diffusion of chemical inducers governs pattern formation in fly embryos.

Homeotic Genes. Master genes determine the form body segments will take.

Programmed Cell Death. Some genes, when activated, kill their cells.

19.3 Aging can be considered a developmental process.

Theories of Aging. While there are many ideas about why cells age, no single theory of aging is widely accepted.

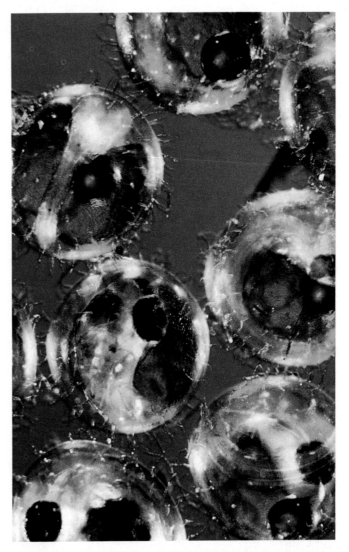

FIGURE 19.1
A collection of future fish undergo embryonic development. Inside a transparent fish egg, a single cell becomes millions of cells that form eyes, fins, gills, and other body parts.

In chapter 18, we explored gene expression from the perspective of an individual cell, examining the diverse mechanisms that a cell may employ to control the transcription of particular genes. Now we will broaden our perspective and look at the unique challenge posed by the development of a cell into a multicellular organism (figure 19.1). In the course of this developmental journey, a pattern of decisions about transcription are made that cause particular lines of cells to proceed along different paths, spinning an incredibly complex web of cause and effect. Yet, for all its complexity, this developmental program works with impressive precision. In this chapter, we explore the mechanisms multicellular organisms use to control their development and achieve this precision.

19.1 Development is a regulated process.

Overview of Development

Organisms in all three multicellular kingdoms—fungi, plants, and animals—realize cell specialization by orchestrating gene expression. That is, different cells express different genes at different times. To understand development, we need to focus on how cells determine which genes to activate, and when.

Among the fungi, the specialized cells are largely limited to reproductive cells. In basidiomycetes and ascomycetes (the so-called higher fungi), certain cells produce pheromones that influence other cells, but the basic design of all fungi is quite simple. For most of its life, a fungus has a two-dimensional body, consisting of long filaments of cells that are only imperfectly separated from each other. Fungal maturation is primarily a process of growth rather than specialization.

Development is far more complex in plants, in which the adult individuals contain a variety of specialized cells organized into tissues and organs. A hallmark of plant development is flexibility; as a plant develops, the precise array of tissues it achieves is greatly influenced by its environment.

In animals, development is complex and rigidly controlled, producing a bewildering array of specialized cell types through mechanisms that are much less sensitive to the environment. The subject of intensive study, animal development has in the last decades become relatively well understood.

Here we will focus on four developmental systems that researchers have studied intensively: (1) an animal with a very complexly arranged body, a vertebrate; (2) a less complex animal with an intricate developmental cycle, an insect; (3) a flowering plant; and (4) a very simple animal, a nematode (figure 19.2).

To begin our investigation, we will first examine the overall process of development in these four quite different organisms so that we can sort through differences in the gross process to uncover basic similarities in underlying mechanisms. We will start by describing the overall process in vertebrates, because it is the best understood among the animals. Then we will examine the very different developmental process carried out by insects, in which genetics has allowed us to gain detailed knowledge of many aspects. Finally, we will look at development in a third very different organism, a flowering plant, and then we will review development in a nematode worm.

Almost all multicellular organisms undergo development. The process has been well studied in plants and in animals, especially in vertebrates, insects, and nematodes.

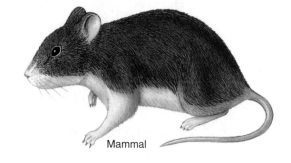

Mammal

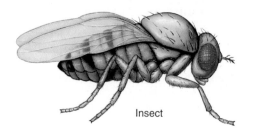

Insect

Flowering plant

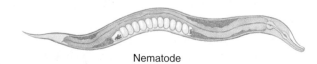

Nematode

FIGURE 19.2
Four developmental systems. Researchers studying the cellular mechanisms of development have focused on these four organisms.

Vertebrate Development

Vertebrate development is a dynamic process in which cells divide rapidly and move over each other as they first establish the basic geometry of the body (figure 19.3). At different sites, particular cells then proceed to form the body's organs, and then the body grows to a size and shape that will allow it to survive after birth. The entire process, described more fully in chapter 51, is traditionally divided into phases. As in mitosis, however, the boundaries between phases are somewhat artificial, and the phases, in fact, grade into one another.

Cleavage

Vertebrates begin development as a single fertilized egg, the zygote. Within an hour after fertilization, the zygote begins to divide rapidly into a larger and larger number of smaller and smaller cells called **blastomeres,** until a solid ball of cells is produced (figure 19.4). This initial period of cell division, termed **cleavage,** is not accompanied by any increase in the overall size of the embryo; rather, the contents of the zygote are simply partitioned into the daughter cells. The two ends of the zygote are traditionally referred to as the **animal pole** and the **vegetal pole.** In general, the blastomeres of the animal pole go on to form the external tissues of the body, while those of the vegetal pole form the internal tissues. The initial top-bottom (dorsal-ventral) orientation of the embryo is determined at fertilization by the location where the sperm nucleus enters the egg, a point that corresponds roughly to the future belly. After about 12 divisions, the burst of cleavage divisions slows, and transcription of key genes begins within the embryo cells.

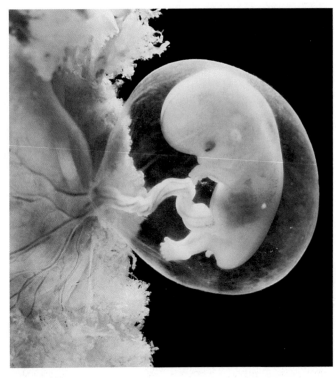

FIGURE 19.3
The miracle of development. This nine-week-old human fetus started out as a single cell: a fertilized egg, or zygote. The zygote's daughter cells have been repeatedly dividing and specializing to produce the distinguishable features of a fetus.

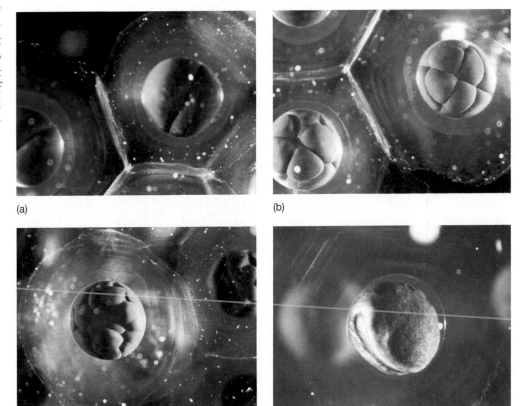

(a)

(b)

(c)

(d)

FIGURE 19.4
Cleavage divisions producing a frog embryo. (*a*) The initial divisions are, in this case, on the side of the embryo facing you, producing (*b*) a cluster of cells on this side of the embryo, which soon expands to become (*c*) a compact mass of cells. (*d*) This mass eventually invaginates into the interior of the embryo, forming a gastrula, and then a neurula.

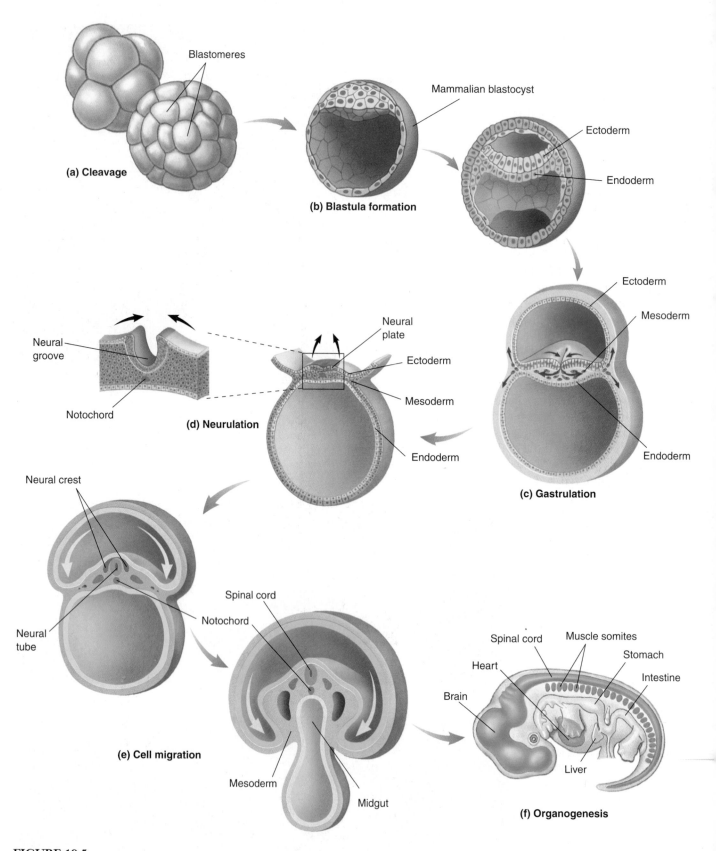

FIGURE 19.5

The path of vertebrate development. The major events in the development of *Mus musculus*, the house mouse, are (*a*) cleavage, (*b*) formation of blastula, (*c*) gastrulation, (*d*) neurulation, (*e*) cell migration, (*f*) organogenesis, and (*g*) growth.

Formation of the Blastula

The outermost blastomeres (figure 19.5*a*) in the ball of cells produced during cleavage are joined to one another by tight junctions, which, as you may recall from chapter 7, are belts of protein that encircle a cell and weld it firmly to its neighbors. These tight junctions create a seal that isolates the interior of the cell mass from the surrounding medium. At about the 16-cell stage, the cells in the interior of the mass begin to pump Na^+ from their cytoplasm into the spaces between cells. The resulting osmotic gradient causes water to be drawn into the center of the cell mass, enlarging the intercellular spaces. Eventually, the spaces coalesce to form a single large cavity within the cell mass. The resulting hollow ball of cells is called a **blastula,** or **blastocyst** in mammals (figure 19.5*b*).

Gastrulation

Some cells of the blastula then push inward, forming a **gastrula** that is invaginated. Cells move by using extensions called lamellipodia to crawl over neighboring cells, which respond by forming lamellipodia of their own. Soon a sheet of cells contracts on itself and shoves inward, starting the invagination. Called **gastrulation** (figure 19.5*c*), this process creates the main axis of the vertebrate body, converting the blastula into a bilaterally symmetrical embryo with a central gut. From this point on, the embryo has three **germ layers** whose organization foreshadows the future organization of the adult body. The cells that invaginate and form the tube of the primitive gut are endoderm; they give rise to the stomach, lungs, liver, and most of the other internal organs. The cells that remain on the exterior are ectoderm, and their derivatives include the skin on the outside of the body and the nervous system. The cells that break away from the invaginating cells and invade the space between the gut and the exterior wall are mesoderm; they eventually form the notochord, bones, blood vessels, connective tissues, and muscles.

Neurulation

Soon after gastrulation is complete, a broad zone of ectoderm begins to thicken on the dorsal surface of the embryo, an event triggered by the presence of the notochord beneath it. The thickening is produced by the elongation of certain ectodermal cells. Those cells then assume a wedge shape by contracting bundles of actin filaments at one end. This change in shape causes the neural tissue to roll up into a tube, which eventually pinches off from the rest of the ectoderm and gives rise to the brain and spinal cord. This tube is called the **neural tube,** and the process by which it forms is termed **neurulation** (figure 19.5*d*).

Cell Migration

During the next stage of vertebrate development, a variety of cells migrate to form distant tissues, following specific paths through the embryo to particular locations (figure 19.5*e*). These migrating cells include those of the **neural crest,** which pinch off from the neural tube and form a number of structures, including some of the body's sense organs; cells that migrate from central blocks of muscle tissue called **somites** and form the skeletal muscles of the body; and the precursors of blood cells and gametes. When a migrating cell reaches its destination, receptor proteins on its surface interact with proteins on the surfaces of cells in the destination tissue, triggering changes in the cytoskeleton of the migrating cell that cause it to cease moving.

Organogenesis and Growth

At the end of this wave of cell migration and colonization, the basic vertebrate body plan has been established, although the embryo is only a few millimeters long and has only about 10^5 cells. Over the course of subsequent development, tissues will develop into organs (figure 19.5*f*), and the embryo will grow to be a hundred times larger, with a million times as many cells (figure 19.5*g*).

Vertebrates develop in a highly orchestrated fashion. The zygote divides rapidly, forming a hollow ball of cells that then pushes inward, forming the main axis of an embryo that goes on to develop tissues and, after a process of cell migration, organs.

(g) Growth

Insect Development

Like all animals, insects develop through an orchestrated series of cell changes, but the path of development is quite different from that of a vertebrate. Many insects produce two different kinds of bodies during their development, the first a tubular eating machine called a **larva,** and the second a flying sex machine with legs and wings. The passage from one body form to the other, called **metamorphosis,** involves a radical shift in development. Here we describe development in the fruit fly *Drosophila* (figure 19.6), which is the subject of much genetic research.

FIGURE 19.6
The fruit fly, *Drosophila melanogaster.* A dorsal view of *Drosophila*, one of the most intensively studied animals in development.

Maternal Genes

The development of an insect like *Drosophila* begins before fertilization, with the construction of the egg. Specialized nurse cells that help the egg grow move some of their own mRNA into the end of the egg nearest them (figure 19.7*a*). As a result, mRNAs produced by maternal genes are positioned in particular locations in the egg, so that after repeated divisions subdivide the fertilized egg, different daughter cells will contain different maternal products. Thus, the action of maternal (rather than zygotic) genes determines the initial course of development.

Syncytial Blastoderm

After fertilization, 12 rounds of nuclear division without cytokinesis produce about 6000 nuclei, all within a single cytoplasm. All of the nuclei within this **syncytial blastoderm** (figure 19.7*b*) can freely communicate with one another, but nuclei located in different sectors of the egg experience different maternal products. The nuclei then space themselves evenly along the surface of the blastoderm, and membranes grow between them. Folding of the embryo and primary tissue development soon follow, in a process fundamentally similar to that seen in vertebrate development. The tubular body that results within a day of fertilization is a larva.

Larval Instars

The larva begins to feed immediately, and as it does so, it grows. However, its chitinous exoskeleton cannot stretch much, and within a day it sheds the exoskeleton. Before the new exoskeleton has had a chance to harden, the larva expands in size. A total of three larval stages, or **instars,** are produced over a period of four days (figure 19.7*c*).

Imaginal Discs

During embryonic growth, about a dozen groups of cells called **imaginal discs** are set aside in the body of the larva (figure 19.7*d*). Imaginal discs play no role in the life of the larva, but are committed to form key parts of the adult fly's body.

Metamorphosis

After the last larval stage, a hard outer shell forms, and the larva is transformed into a **pupa** (figure 19.7*e*). Within the pupa, the larval cells break down and release their nutrients, which are used in the growth and development of the various imaginal discs (eye discs, wing discs, leg discs, and so on). The imaginal discs then associate with one another, assembling themselves into the body of the adult fly (figure 19.7*f*). The metamorphosis of a *Drosophila* larva into a pupa and then into an adult fly takes about four days, after which the pupal shell splits and the fly emerges.

Drosophila development proceeds through two discrete phases, the first a larval phase that gathers food, and the second an adult phase that is capable of flight and reproduction.

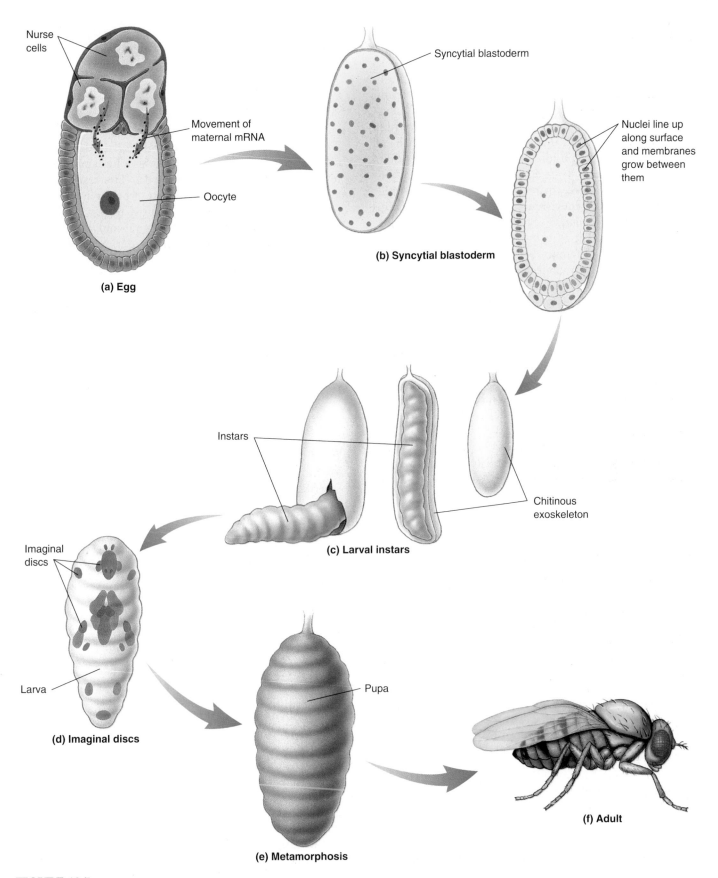

FIGURE 19.7
The path of insect development. The major events in the development of *Drosophila melanogaster* are (*a*) egg, (*b*) syncytial blastoderm, (*c*) larval instars, (*d*) imaginal discs, (*e*) metamorphosis, and (*f*) adult.

Plant Development

At the most basic level, the developmental paths of plants and animals share many key elements. However, the mechanisms used to achieve body form are quite different. While animal cells follow an orchestrated series of movements during development, plant cells are encased within stiff cellulose walls, and therefore cannot move. Each cell in a plant is fixed into position when it is created. Instead of using cell migration, plants develop by building their bodies outward, creating new parts from special groups of self-renewing cells called **meristems.** As meristem cells continually divide, they produce cells that can differentiate into the tissues of the plant.

Another major difference between animals and plants is that most animals are mobile and can move away from unfavorable circumstances, while plants are anchored in position and must simply endure whatever environment they experience. Plants compensate for this restriction by relaxing the rules of development to accommodate local circumstances. Instead of creating a body in which every part is specified to have a fixed size and location, a plant assembles its body from a few types of modules, such as leaves, roots, branch nodes, and flowers. Each module has a rigidly controlled structure and organization, but how the modules are utilized is quite flexible. As a plant develops, it simply adds more modules, with the environment having a major influence on the type, number, size, and location of what is added. In this way, the plant is able to adjust the path of its development to local circumstances.

Early Cell Division

The first division of the fertilized egg in a flowering plant is off-center, so that one of the daughter cells is small, with dense cytoplasm (figure 19.8*a*). That cell, the future embryo, begins to divide repeatedly, forming a ball of cells. The other daughter cell also divides repeatedly, forming an elongated structure called a **suspensor,** which links the embryo to the nutrient tissue of the seed. The suspensor also provides a route for nutrients to reach the developing embryo. Just as the animal embryo acquires its initial axis as a cell mass formed during cleavage divisions, so the plant embryo forms its root-shoot axis at this time. Cells near the suspensor are destined to form a root, while those at the other end of the axis ultimately become a shoot.

Tissue Formation

Three basic tissues differentiate while the plant embryo is still a ball of cells (figure 19.8*b*), analogous to the formation of the three germ layers in animal embryos, although in plants, no cell movements are involved. The outermost cells in a plant embryo become **epidermal cells.** The bulk of the embryonic interior consists of **ground tissue** cells that eventually function in food and water storage. Finally, cells at the core of the embryo are destined to form the future **vascular tissue.**

Seed Formation

Soon after the three basic tissues form, a flowering plant embryo develops one or two seed leaves called **cotyledons.** At this point, development is arrested, and the embryo is either surrounded by nutritive tissue or has amassed stored food in its cotyledons (figure 19.8*c*). The resulting package, known as a *seed,* is resistant to drought and other unfavorable conditions; in its dormant state, it is a vehicle for dispersing the embryo to distant sites and allows a plant embryo to survive in environments that might kill a mature plant.

Germination

A seed germinates in response to changes in its environment brought about by water, temperature, or other factors. The embryo within the seed resumes development and grows rapidly, its roots extending downward and its leaf-bearing shoots extending upward (figure 19.8*d*).

Meristematic Development

Plant development exhibits its great flexibility during the assembly of the modules that make up a plant body. Apical meristems at the root and shoot tips generate the large numbers of cells needed to form leaves, flowers, and all other components of the mature plant (figure 19.8*e*). At the same time, meristems ensheathing the stems and roots produce the wood and other tissues that allow growth in circumference. A variety of hormones produced by plant tissues influence meristem activity and, thus, the development of the plant body. Plant hormones (see chapter 41) are the tools that allow plant development to adjust to the environment.

Morphogenesis

The form of a plant body is largely determined by controlled changes in cell shape as cells expand osmotically after they form (see figure 19.8*e*). Plant growth-regulating hormones and other factors influence the orientation of bundles of microtubules on the interior of the plasma membrane. These microtubules seem to guide cellulose deposition as the cell wall forms around the outside of a new cell. The orientation of the cellulose fibers, in turn, determines how the cell will elongate as it increases in volume, and so determines the cell's final shape.

In a developing plant, leaves, flowers, and branches are added to the growing body in ways that are strongly influenced by the environment.

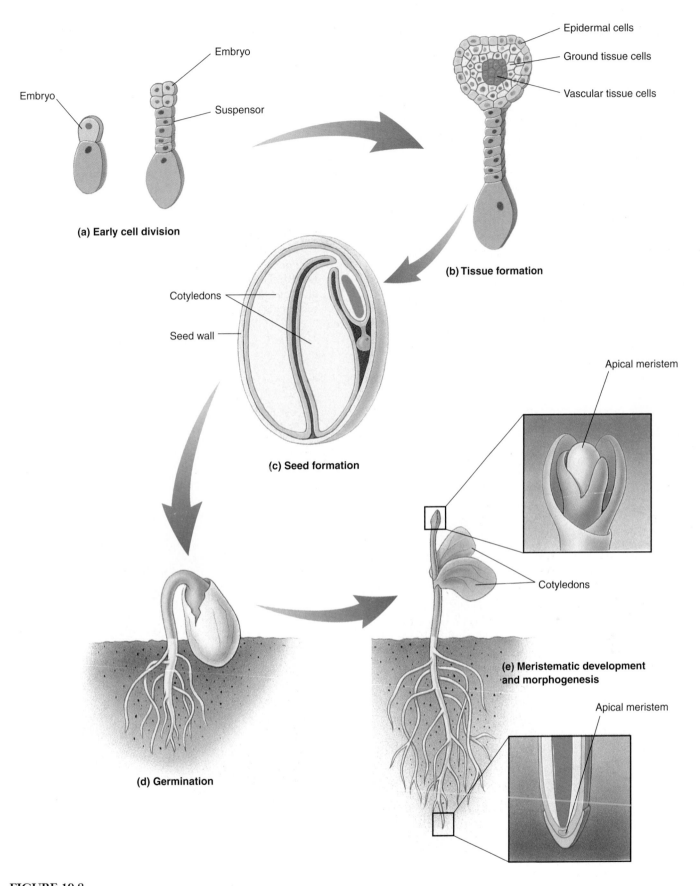

FIGURE 19.8
The path of plant development. The developmental stages of *Arabidopsis thaliana* are (*a*) early cell division, (*b*) tissue formation, (*c*) seed formation, (*d*) germination, and (*e*) meristematic development and morphogenesis.

Nematode Development

One of the most completely described models of development is the tiny nematode *Caenorhabditis elegans*. Only about 1 millimeter long, it consists of 959 somatic cells and has about the same amount of DNA as *Drosophila*. The entire genome has been mapped as a series of overlapping fragments, and a serious effort is under way to determine the complete DNA sequence of the genome.

Because *C. elegans* is transparent, individual cells can be followed as they divide. By observing them, researchers have learned how each of the cells that make up the adult worm is derived from the fertilized egg. As shown on the lineage map in figure 19.9*a*, the egg divides into two, and then its daughter cells continue to divide. Each horizontal line on the map represents one round of cell division. The length of each vertical line represents the time between cell divisions, and the end of each vertical line represents one fully differentiated cell. In figure 19.9*b*, the major organs of

the worm are color-coded to match the colors of the corresponding groups of cells on the lineage map.

Some of these differentiated cells, such as some of the cells that generate the worm's external cuticle, are "born" after only 8 rounds of cell division; other cuticle cells require as many as 14 rounds. The cells that make up the worm's pharynx, or feeding organ, are born after 9 to 11 rounds of division, while cells in the gonads require up to 17 divisions.

Exactly 302 nerve cells are destined for the worm's nervous system. Exactly 131 cells are programmed to die, mostly within minutes of their birth. The fate of each cell is the same in every *C. elegans* individual, except for the cells that will become eggs and sperm.

The nematode develops 959 somatic cells from a single fertilized egg in a carefully orchestrated series of cell divisions, which have been carefully mapped by researchers.

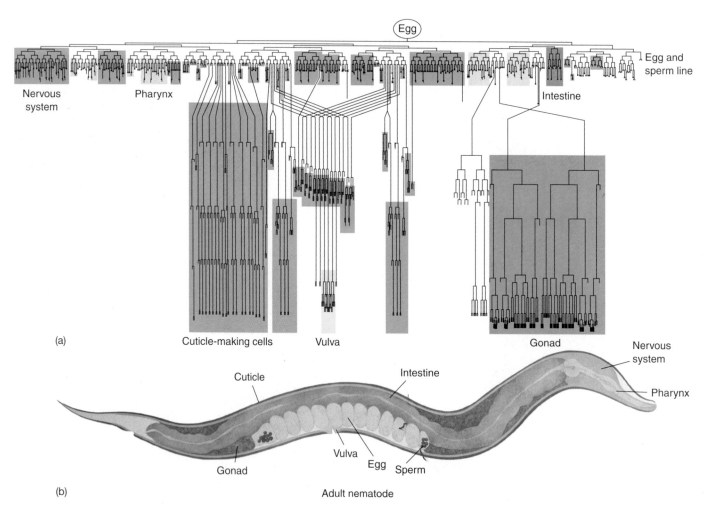

FIGURE 19.9
Studying development in the nematode. Development in *C. elegans* has been mapped out such that the fate of each cell from the single egg cell has been determined. (*a*) The lineage map shows the number of cell divisions from the egg, and the color coding links their placement in (*b*) the adult organism.

19.2 Multicellular organisms employ the same basic mechanisms of development.

Despite the many differences in the four developmental paths we have just discussed, it is becoming increasingly clear that most multicellular organisms develop according to molecular mechanisms that are fundamentally very similar. This observation suggests that these mechanisms evolved very early in the history of multicellular life. Here, we focus on six mechanisms that seem to be of particular importance in the development of a wide variety of organisms: cell movement, induction, determination, pattern formation, homeotic genes, and programmed cell death. We will consider them in roughly the order in which they first become important during development.

Cell Movement and Induction

Cell Movement

The migration of cells is important during many stages of development, from the early events of gastrulation to the formation of the nervous system during morphogenesis. The movement of cells involves both adhesion and loss of adhesion. Adhesion is necessary for cells to get "traction," but cells that are initially attached must lose this adhesion to be able to leave a site. Cell movement also involves both cell-to-cell interactions and cell-to-substrate interactions. Cell-to-cell interactions are often mediated through cadherins (introduced in chapter 7), while cell-to-substrate interactions often involve integrin-to-extracellular-matrix (ECM) interactions. Integrin was initially identified as the receptor for the ECM molecule fibronectin.

Cadherins have proved to be a large gene family, with over 80 members identified in humans. In the genomes of *Drosophila*, *C. elegans*, and humans, the cadherins can be sorted into four subfamilies that exist in all three genomes. These subfamilies can be further subdivided by both their function and species distribution. The cadherin family is not found in yeast or *Dictyostelium*, implying that they are a metazoan innovation. These proteins are all transmembrane proteins that share a common motif, the *cadherin domain*, a 110-amino-acid domain in the extracellular portion of the protein that mediates Ca^{++}-dependent binding between like cadherins.

Experiments in which cells are allowed to sort in vitro illustrate the function of cadherins. Cells with the same cadherins will adhere specifically while not adhering to other cells with different cadherins. If cell populations with different cadherins are disaggregated and then allowed to reaggregate, they will sort into two populations of cells based on the nature of the cadherins on their surface. This experiment can be extended by using cells with different amounts of the same cadherins: The cells will sort such that the cells with the most cadherins will be in the interior of the cell mass while the cells on the outside will have fewer cadherins.

An example of the action of cadherins can be seen in the development of the vertebrate nervous system. The neural-specific cadherin N-cadherin is expressed in the neuroectoderm during neurulation. The expression of N-cadherin is not essential for this process, but is involved in the sorting of future neural cells versus ectoderm. During later events in neurogenesis, a number of neural-specific cadherins are expressed in different subsets of neural tissue. These cadherins appear to be involved in the sorting of different subsets of the developing nervous system. This pattern of cadherins with spatially constrained expression is a common theme in development. It is not clear whether the pattern is causal or a correlation. The experiments being attempted to test this causality involve removing gene function. However, because "knocking out" a gene can cause multiple defects, the experimental results are as yet difficult to interpret.

In some tissues, such as connective tissue, much of the volume of the tissue is taken up by the spaces *between* cells. These spaces are not vacant, however. Rather, they are filled with a network of molecules secreted by surrounding cells, principally a matrix of long polysaccharide chains covalently linked to proteins (proteoglycans), within which are embedded strands of fibrous protein (collagen, elastin, and fibronectin). Migrating cells traverse this matrix by binding to it with cell surface proteins called integrins, which were also described in chapter 7. Integrins are attached to actin filaments of the cytoskeleton and protrude out from the cell surface in pairs, like two hands. The "hands" grasp a specific component of the matrix such as collagen or fibronectin, thus linking the cytoskeleton to the fibers of the matrix. In addition to providing an anchor, this binding can initiate changes within the cell, alter the growth of the cytoskeleton, and change the way the cell secretes materials into the matrix.

Thus, cell migration is largely a matter of changing patterns of cell adhesion. As a migrating cell travels, it continually extends projections that probe the nature of its environment. Tugged this way and that by different tentative attachments, the cell literally feels its way toward its ultimate target site.

Induction

In *Drosophila*, the initial cells created by cleavage divisions contain different developmental signals (called **determinants**) from the egg, setting individual cells off on different developmental paths. This pattern of development is called **mosaic development**. In mammals, by contrast, all of the blastomeres receive equivalent sets of determinants; body form is determined by cell–cell interactions, a pattern called **regulative development**.

We can demonstrate the importance of cell–cell interactions in development by separating the cells of an early blastula and allowing them to develop independently. Under these conditions, animal pole blastomeres develop features of ectoderm, and vegetal pole blastomeres develop features of endoderm, but none of the cells ever develop features characteristic of mesoderm. However, if animal pole and vegetal pole cells are placed next to each other, some of the animal pole cells will develop as mesoderm. The interaction between the two cell types triggers a switch in the developmental path of the cells! When a cell switches from one path to another as a result of interaction with an adjacent cell, **induction** has taken place (figure 19.10).

How do cells induce developmental changes in neighboring cells? Apparently, the inducing cells secrete proteins that act as intercellular signals. Signal molecules, which we discussed in detail in chapter 7, are capable of producing abrupt changes in the patterns of gene transcription.

In some cases, particular groups of cells called **organizers** produce diffusible signal molecules that convey positional information to other cells. Organizers can have a profound influence on the development of surrounding tissues (see chapter 51). Working as signal beacons, they inform surrounding cells of their distance from the organizer. The closer a particular cell is to an organizer, the higher the concentration of the signal molecule, or **morphogen,** it experiences (figure 19.11). Although only a few morphogens have been isolated, they are thought to be part of a widespread mechanism for determining relative position during development.

A single morphogen can have different effects, depending upon how far away from the organizer the affected cell is located. Thus, low levels of the morphogen activin will cause cells of the animal pole of an early *Xenopus* embryo to develop into epidermis, while slightly higher levels will induce the cells to develop into muscles, and levels a little higher than that will induce them to form notochord (figure 19.12).

Cells migrate by extending probes to neighboring cells, which they use to pull themselves along. Interactions between cells strongly influence the developmental paths they take. Signal molecules from an inducing cell alter patterns of transcription in the cells that come in contact with it.

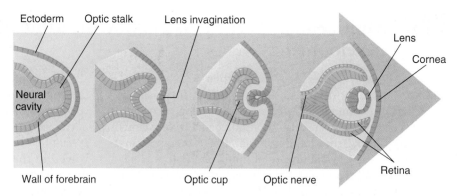

FIGURE 19.10
Development of the vertebrate eye by induction. The eye develops as an extension of the forebrain called the optic stalk that grows out until it contacts the ectoderm. This contact induces the formation of a lens from the ectoderm.

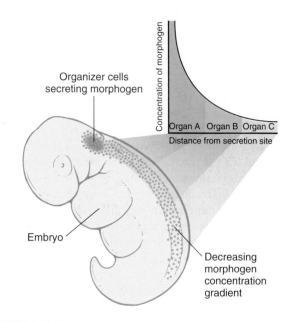

FIGURE 19.11
An organizer creates a morphogen gradient. As a morphogen diffuses from the organizer site, it becomes less concentrated. Different concentrations of the morphogen stimulate the development of different organs.

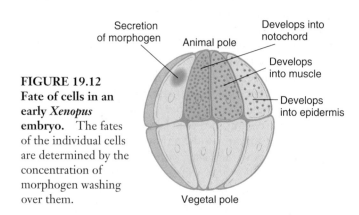

FIGURE 19.12
Fate of cells in an early *Xenopus* embryo. The fates of the individual cells are determined by the concentration of morphogen washing over them.

Determination

The vertebrate egg is symmetrical in its contents as well as its shape, so that all of the cells of an early blastoderm are equivalent up to the eight-cell stage. The cells are said to be **totipotent**, meaning that they are potentially capable of expressing all of the genes of their genome. If they are separated from one another, any one of them can produce a completely normal individual. Indeed, just this sort of procedure has been used to produce sets of four or eight identical offspring in the commercial breeding of particularly valuable lines of cattle. The reverse process works too; if cells from two different eight-cell-stage embryos are combined, a single normal individual results. Such an individual is called a **chimera**, because it contains cells from different genetic lines (figure 19.13).

Vertebrate cells start to become different after the eight-cell stage as a result of cell–cell interactions like those we just discussed. At this point, the pathway that will influence the future developmental fate of the cells is determined. The commitment of a particular cell to a specialized developmental path is called **determination**. A cell in the prospective brain region of an amphibian embryo at the early gastrula stage has not yet been determined; if transplanted elsewhere in the embryo, it will develop like its new neighbors (see chapter 51). By the late gastrula stage, however, determination has taken place, and the cell will develop as neural tissue no matter where it is transplanted. Determination must be carefully distinguished from **differentiation**, which is the cell specialization that occurs at the end of the developmental path. Cells may become *determined* to give rise to particular tissues long before they actually *differentiate* into those tissues. For example, the cells of a *Drosophila* eye imaginal disc are fully determined to produce an eye, but they remain totally undifferentiated during most of the course of larval development.

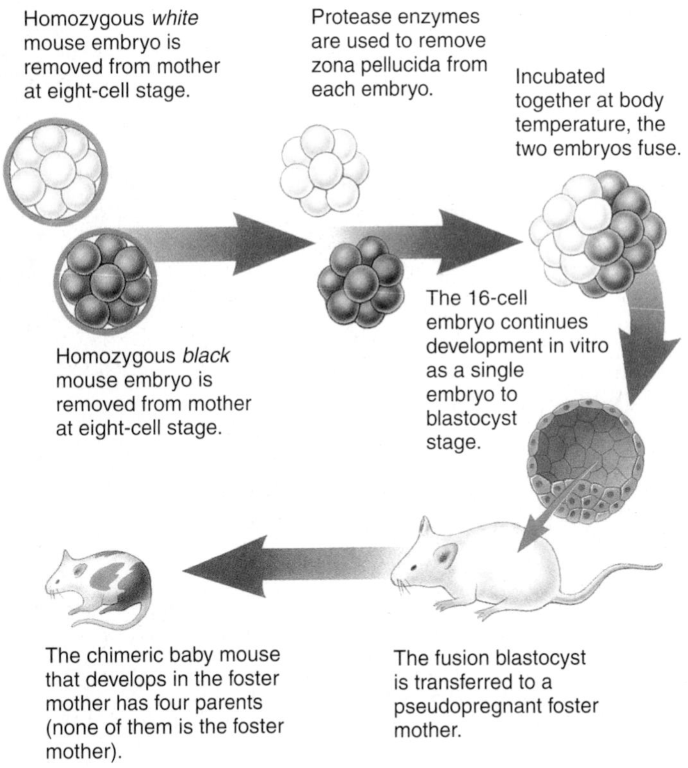

The Mechanism of Determination

What is the molecular mechanism of determination? The gene regulatory proteins discussed in detail in chapter 16 are the tools cells use to initiate developmental changes. When genes encoding these proteins are activated, one of their effects is to reinforce their own activation. This makes the developmental switch deterministic, initiating a chain of events that leads down a particular developmental pathway. Cells in which a set of regulatory genes have been activated may not actually undergo differentiation until some time later, when other factors interact with the regulatory protein and cause it to activate still other genes. Nevertheless, once the initial "switch" is thrown, the cell is fully committed to its future developmental path.

Often, before a cell becomes fully committed to a particular developmental path, it first becomes partially commit-

Homozygous *white* mouse embryo is removed from mother at eight-cell stage.

Homozygous *black* mouse embryo is removed from mother at eight-cell stage.

Protease enzymes are used to remove zona pellucida from each embryo.

Incubated together at body temperature, the two embryos fuse.

The 16-cell embryo continues development in vitro as a single embryo to blastocyst stage.

The fusion blastocyst is transferred to a pseudopregnant foster mother.

The chimeric baby mouse that develops in the foster mother has four parents (none of them is the foster mother).

FIGURE 19.13
Constructing a chimeric mouse. Cells from two eight-cell individuals fuse to form a single individual.

ted, acquiring **positional labels** that reflect its location in the embryo. These labels can have a great influence on how the pattern of the body subsequently develops. In a chicken embryo, if tissue at the base of the leg bud (which would normally give rise to the thigh) is transplanted to the tip of the identical-looking wing bud (which would normally give rise to the wing tip), that tissue will develop into a toe rather than a thigh! The tissue has already been determined as leg but is not yet committed to being a particular part of the leg. Therefore, it can be influenced by the positional signaling at the tip of the wing bud to form a tip (in this case, a tip of leg).

Is Determination Irreversible?

Until very recently, biologists thought determination was irreversible. Experiments carried out in the 1950s and 1960s by John Gurdon and others made what seemed a convincing case: Using very fine pipettes (hollow glass tubes) to suck the nucleus out of a frog or toad egg, these researchers replaced the egg nucleus with a nucleus sucked out of a body cell taken from another individual (see figure 14.3). If the transplanted nucleus was obtained from an advanced embryo, the egg went on to develop into a tadpole, but most died before becoming an adult.

Nuclear transplant experiments were attempted without success by many investigators, until finally, in 1984, Steen Willadsen, a Danish embryologist working in Texas, succeeded in cloning a sheep using the nucleus from a cell of an early embryo. The key to his success was in picking a cell very early in development. This exciting result was soon replicated by others in a host of other organisms, including pigs and monkeys.

Only early embryo cells seemed to work, however. Researchers became convinced, after many attempts to transfer older nuclei, that animal cells become irreversibly committed after the first few cell divisions of the developing embryo.

We now know this conclusion to have been unwarranted. The key advance unraveling this puzzle was made in Scotland by geneticists Keith Campbell and Ian Wilmut, who reasoned that perhaps the egg and the donated nucleus needed to be at the same stage in the cell cycle. They removed mammary cells from the udder of a six-year-old sheep. The cells were grown in tissue culture; then, in preparation for cloning, the researchers substantially reduced for five days the concentration of serum nutrients on which the sheep mammary cells were subsisting. Starving the cells caused them to pause at the beginning of the cell cycle. In parallel preparation, eggs obtained from a ewe were enucleated (figure 19.14).

Mammary cells and egg cells were surgically combined in January of 1996. A little over five months later, on July 5, 1996, one sheep gave birth to a lamb named Dolly, the first clone generated from a fully differentiated animal cell. Dolly established beyond all dispute that determination is reversible—that with the right techniques, the fate of a fully differentiated cell *can* be altered.

Since Dolly, clones have been created for the following organisms: mouse, cow, pig, goat, cat, and rabbit. All of these used some form of adult cell. The efficiency in all cases is quite low, on the order of 3 to 5% viable adults from transferred nuclei. As discussed in chapter 20, we have much to learn about the events necessary to reprogram an adult nucleus to act as an embryonic nucleus.

The commitment of particular cells to certain developmental fates is fully reversible.

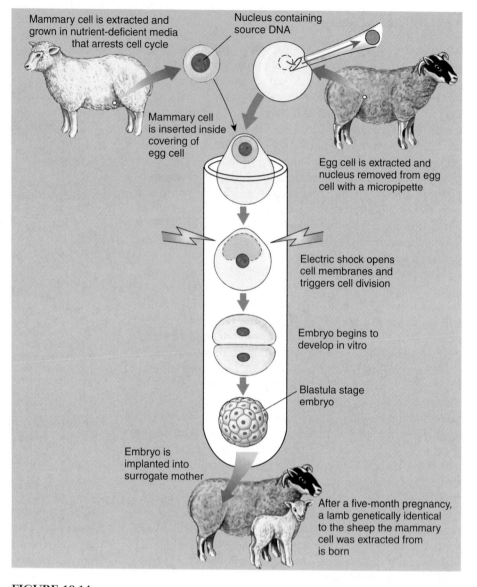

FIGURE 19.14
Proof that determination is reversible. This experiment by Campbell and Wilmut, the first successful cloning of an adult animal, shows that a differentiated adult cell can be used to drive all of development.

Pattern Formation

All animals seem to use positional information to determine the basic pattern of body compartments and, thus, the overall architecture of the adult body. How is positional information encoded in labels and read by cells? To answer this question, let us consider how positional labels are used in pattern formation in *Drosophila*. The Nobel Prize in Physiology or Medicine was awarded in 1995 for the unraveling of this puzzle (figure 19.15).

Pattern formation is an unfolding process during development. In the later stages, it may involve morphogenesis of organs, but during the earliest events of development, it lays down the basic body plan, the establishment of the

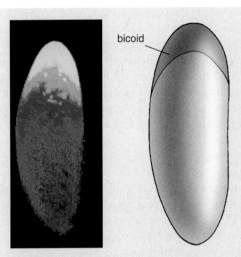

Establishing the polarity of the embryo: Fertilization of the egg triggers the production of bicoid protein from maternal RNA in the egg. The bicoid protein diffuses through the egg, forming a gradient. This gradient determines the polarity of the embryo, with the head and thorax developing in the zone of high concentration (*yellow* through *red* in photo).

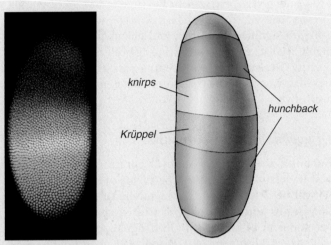

Setting the stage for segmentation: About 2½ hours after fertilization, bicoid protein turns on a series of brief signals from so-called gap genes. The gap proteins act to divide the embryo into large blocks. In this photo, fluorescent dyes in antibodies that bind to the gap proteins Krüppel *(red)* and hunchback *(green)* make the blocks visible; the region of overlap is yellow.

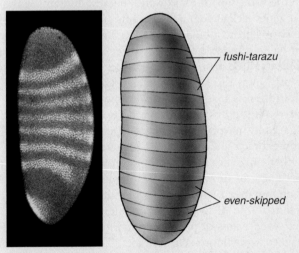

Laying down the fundamental regions: About ½ hour later, the gap genes switch on a so-called "pair-rule" gene called *hairy*. Hairy produces a series of boundaries within each block, dividing the embryo into seven fundamental regions.

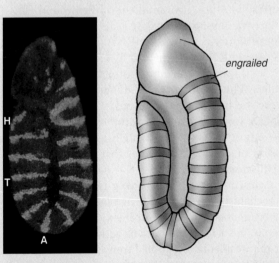

Forming the segments: The final stage of segmentation occurs when a "segment-polarity" gene called *engrailed* divides each of the seven regions into halves, producing 14 narrow compartments. Each compartment corresponds to one segment of the future body. There are three head segments (H, top left), three thoracic segments (T, lower left), and eight abdominal segments (A, from bottom left to upper right).

FIGURE 19.15
Body organization in an early *Drosophila* embryo. In these images by 1995 Nobel laureate Christiane Nüsslein-Volhard and Sean Carroll, we watch a *Drosophila* egg pass through the early stages of development, in which the basic segmentation pattern of the embryo is established. The drawings help illustrate what is occurring in the photos.

anterior/posterior (A/P) axis and the dorsal/ventral (D/V) axis. Thus, pattern formation can be considered the process of taking a radially symmetrical cell and imposing two perpendicular axes to define the basic body plan, which in this way becomes bilaterally symmetrical. The differences in the development of protostomes and deuterostomes initially led investigators to the idea that pattern formation must be fundamentally different processes in these divergent species. Even among vertebrates, there appeared to be significant differences between the amphibian *Xenopus*, which has an egg cell with significant polarity, and mammals, which exhibit highly regulative development and egg cells with no apparent polarity. Recently, some interesting parallels have been observed, although the mechanisms in different lineages remain distinctive.

Drosophila Pattern Formation

Forming the Axis. The invertebrate *Drosophila* is the best understood in terms of the events of early patterning. As will be described later, a hierarchy of gene expression that begins with maternally expressed genes controls the development of *Drosophila*. Two signaling pathways control the establishment of A/P and D/V polarity. The A/P axis is formed based on a gradient of the **bicoid** protein. This protein gradient is established by an interesting mechanism: Follicle cells provide maternally produced *bicoid* mRNA that becomes anchored at the anterior pole. Translation of this spatially localized message then produces an anterior-to-posterior gradient of the protein. The posterior determining protein **oskar** is also localized based on prior localization of the *oskar* mRNA. The localization of both of these mRNAs is achieved by attachment to microtubules, as shown by the action of agents that disrupt microtubules. The motor protein dynein appears to be necessary for the localization of bicoid, while oskar localization appears to be dependent on the motor protein kinesin. The action of the bicoid protein is then to control the expression of so-called gap genes that are expressed in the zygote (discussed later).

The dorsal/ventral axis in *Drosophila* is controlled by a different mechanism that does not involve localized cytoplasmic determinants like bicoid. Instead, an accumulation of message on one side of the nucleus activates a transcription factor that results in directional translation and the accumulation of protein product on that surface of the oocyte. The *gurken* mRNA accumulates on one surface of the nucleus such that, when translated, it accumulates on the membrane on one side of the cell. This will be the future dorsal side of the embryo; however, it is not directly specified by gurken—rather, *gurken* signals cause overlying follicle cells to adopt a dorsal fate. These cells then produce a signal that results in the selected transport of the transcription factor *dorsal* into the nucleus of ventral cells. This transcription factor then acts to produce a ventral to dorsal protein gradient similar to the signaling by bicoid.

While utilizing profoundly different mechanisms, the unifying factor controlling the establishment of both A/P and D/V polarity is that *gurken* and *bicoid* are both maternally expressed genes so that the polarity of the future embryo in both instances is laid down in the oocyte using information coming from the maternal genome. Both cases also involve an interaction between the future oocyte and the follicle cells. While the preceding is in some respects an oversimplification of the events that occur, the outline is clear: Polarity is established by an interaction between follicle cells and the oocyte, an interaction that results in the creation of morphogen gradients in the oocyte based on maternal information. These gradients then drive the expression of the zygotic genes that will actually pattern the embryo. This reliance on a hierarchy of regulatory genes, and on response to the interactions of cells, are unifying themes for all of development.

Producing the Body Plan. Bicoid protein exerts this profound effect on the organization of the embryo by activating genes that encode the first mRNAs to be transcribed after fertilization. Within the first two hours, before cellularization of the syncytial blastoderm, a group of six genes called the **gap genes** begins to be transcribed. These genes map out the coarsest subdivision of the embryo (see figure 19.15). One of them is a gene called *hunchback* (because an embryo without *hunchback* lacks a thorax and so, takes on a hunched shape). Although *hunchback* mRNA is distributed throughout the embryo, its translation is controlled by the protein product of another maternal mRNA called *nanos* (named after the Greek word for "dwarf," because mutants without *nanos* genes lack abdominal segments and, hence, are small). The **nanos** protein binds to *hunchback* mRNA, preventing it from being translated. The only place in the embryo where there is too little nanos protein to block translation of *hunchback* mRNA is the far anterior end. Consequently, hunchback protein is made primarily at the anterior end of the embryo. As it diffuses back toward the posterior end, it sets up a second morphogen gradient responsible for establishing the thoracic and abdominal segments.

Other gap genes act in more posterior regions of the embryo. They, in turn, activate 11 or more **pair-rule genes.** (When mutated, each of these genes alters every other body segment.) One of the pair-rule genes, named *hairy*, produces seven bands of protein, which look like stripes when visualized with fluorescent markers. These bands establish boundaries that divide the embryo into seven zones. Finally, a group of 16 or more **segment polarity genes** subdivide these zones. The *engrailed* gene, for example, divides each of the seven zones established by *hairy* into anterior and posterior compartments. The 14 compartments that result correspond to the three head segments, three thoracic segments, and eight abdominal segments of the embryo.

Thus, within three hours after fertilization, a highly orchestrated cascade of segmentation gene activity produces the fly embryo's basic body plan. The activation of these and other developmentally important genes (figure 19.16)

FIGURE 19.16

A gene controlling organ formation in *Drosophila*. Called *tinman*, this gene is responsible for the formation of gut musculature and the heart. The dye shows expression of *tinman* in (*a*) five-hour and (*b*) seventeen-hour *Drosophila* embryos. (*c*) The gut musculature then appears along the edges of normal embryos, but (*d*) is not present in embryos in which the gene has been mutated. (*e*) The heart tissue develops along the center of normal embryos but (*f*) is missing in *tinman* mutant embryos.

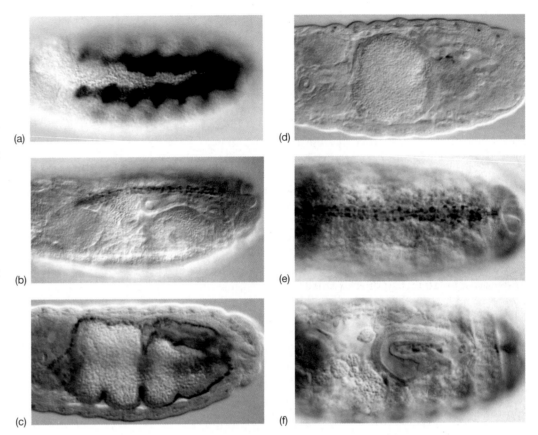

depends upon the free diffusion of morphogens that is possible within a syncytial blastoderm. In mammalian embryos with cell partitions, other mechanisms must operate.

Vertebrate Axis Formation

Development of at least one vertebrate, the frog *Xenopus*, now appears to have more in common with *Drosophila* than anticipated. The *Xenopus* oocyte has intrinsic polarity, an animal pole and a vegetal pole; these are not directly related to the future A/P or D/V axes but do imply the existence of cytoplasmic determinants. The D/V axis has always been the most interesting to embryologists because it contains the famous Spemann organizer. In experiments done in the 1920s, Spemann and Mangold showed that a bisected embryo that contained the dorsal half developed normally. This was refined to show that the dorsal lip of the blastopore alone could be transplanted to produce a second embryonic axis. The molecular nature of this organizer has been of interest ever since its discovery. The earliest events that lead to the determination of the D/V axis are related to the fertilization event. The side of the embryo opposite the sperm entry point will become the future dorsal side. The egg reorganizes based on a signal from the point of entry of the sperm. The egg rotates, and microtubules emanating from the point of sperm entry result in the movement of dorsal determinants from the vegetal pole to the side of the egg opposite the site of sperm entry. This

ultimately results in the formation of the organizer on the dorsal surface. The signaling events following this involve a transforming growth factor pathway, and the so-called *wnt* pathway. These pathways are also involved in *Drosophila* patterning downstream of the dorsal transcription factor.

The mammalian embryo has long been known to lack obvious cytoplasmic determinants. Mammals are an example of regulative development, in which development unfolds based on interactions between cells. Yet, even in this case, new evidence indicates that patterning may arise earlier than previously thought. The mouse embryo can be manipulated in a number of ways that do not perturb development: disruption of blastomeres, removal of polar cytoplasm, even removal of blastomeres up to the 8-cell stage. However, recent evidence indicates that the site of sperm entry is "remembered" by the developing embryo and influences later events. The site of sperm entry is later translated into the proximo-distal axis by signaling from the extraembryonic cells of the embryo during early development. All of this indicates that the differences in development we see represent evolution in different lineages, but that a limited number of actual mechanisms may have been used in many different ways. This will be amplified later in this chapter when we discuss homeobox-containing genes.

In *Drosophila*, diffusion of chemical inducers produces the embryo's basic body plan, a cascade of genes dividing it into 14 compartments.

Homeotic Genes

Role of Homeotic Genes

With the basic body plan laid down by the mechanisms described earlier, the next step is to give identity to the segments of the embryo. The outline of *Drosophila* pattern formation produces a segmented embryo with A/P and D/V polarity. The genes that then give rise to segment identity are the homeobox-containing genes. Mutations in homeotic genes lead to the appearance of perfectly normal body parts in unusual places. For example, mutations in *bithorax* (figure 19.17) cause a fly to grow an extra pair of wings, as if it had a double thoracic segment, and mutations in *Antennapedia* cause legs to grow out of the head in place of antennae! In the early 1950s, geneticist Edward Lewis discovered that several homeotic genes, including *bithorax*, map together on the third chromosome of *Drosophila* in a tight cluster called the **bithorax complex.** Mutations in these genes all affect body parts of the thoracic and abdominal segments, and Lewis concluded that the genes of the bithorax complex control the development of body parts in the rear half of the thorax and all of the abdomen. Most interestingly, the order of the genes in the bithorax complex mirrors the order of the body parts they control, as if the genes are activated serially! Genes at the beginning of the cluster switch on development of the thorax; those in the middle control the anterior part of the abdomen; and those at the end affect the tip of the abdomen. A second cluster of homeotic genes, the **Antennapedia complex,** was discovered in 1980 by Thomas Kaufmann. The Antennapedia complex governs the anterior end of the fly, and the order of genes in it also corresponds to the order of segments they control (figure 19.18).

FIGURE 19.17
Mutations in homeotic genes. Three separate mutations in the *bithorax* gene caused this fruit fly to develop an extra thoracic segment, with accompanying wings. Compare this photograph with that of the normal fruit fly in figure 19.6.

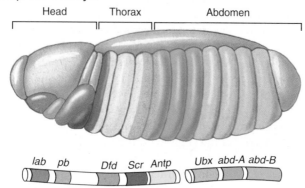

Drosophila embryo

Drosophila HOM genes

FIGURE 19.18
***Drosophila* homeotic genes.** Called the homeotic gene complex, or HOM complex, the genes are grouped into two clusters, the Antennapedia complex (anterior) and the bithorax complex (posterior).

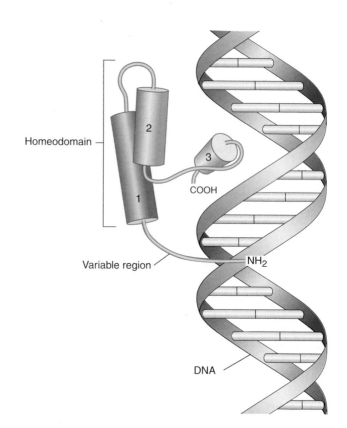

FIGURE 19.19
Homeodomain protein. This protein plays an important regulatory role when it binds to DNA and regulates expression of specific genes. The variable region of the protein determines the specific activity of the protein. The homeodomain, a 60-amino-acid sequence common to all proteins of this type, is coded for by the homeobox region of genes and is composed of three α-helices. One of the helices (number 3) recognizes and binds to a specific DNA sequence in target genes.

The Homeobox

Drosophila homeotic genes typically contain the **homeobox,** a sequence of 180 nucleotides that codes for a 60-amino-acid, DNA-binding peptide domain called the homeodomain (figure 19.19). As we saw in chapter 18, proteins that contain the homeodomain function as transcription factors, ensuring that developmentally related genes are transcribed at the appropriate time. Segmentation genes such as *bicoid* and *engrailed* also contain the homeobox sequence. Clearly, the homeobox distinguishes those portions of the genome devoted to pattern formation.

Evolution of Homeobox-Containing Genes

A large amount of research has been devoted to analyzing homeobox-containing genes, or *Hox* genes, in many different organisms. These investigations have led to a fairly coherent view of homeotic gene evolution. It is now clear that the *Drosophila* bithorax and Antennapedia complexes represent two parts of a single cluster of genes. There are four copies of *Hox* gene clusters in vertebrate species. As in *Drosophila*, the spatial domains of *Hox* gene expression correlate with the order of the genes on the chromosome (figure 19.20). The existence of four *Hox* clusters in vertebrates is viewed by many as evidence that two duplication events of the entire genome must have occurred in the vertebrate lineage. This raises the issue of when the original cluster arose. To answer this question, researchers have turned to more and more primitive organisms, first to the primitive chordate *Amphioxus* (now called *Branchiostoma*). The finding of one cluster of *Hox* genes in *Amphioxus* implies that indeed there have been two duplications of the vertebrate lineage, at least of the *Hox* cluster in the vertebrates. Given the single cluster in arthropods, this implies that the common ancestor to all bilaterians had a single *Hox* cluster as well. The next logical step is to look at the next available outgroup: cnidarians and ctenophores. Thus far, *Hox* genes have been found in a number of cnidarian species, but their organization is not yet clear. The analysis is confused by the presence of genes that contain homeoboxes but are not in a *Hox* cluster. Such genes exist in vertebrate genomes and *Drosophila* as well, but their relationship to the ancestral *Hox* cluster is not known.

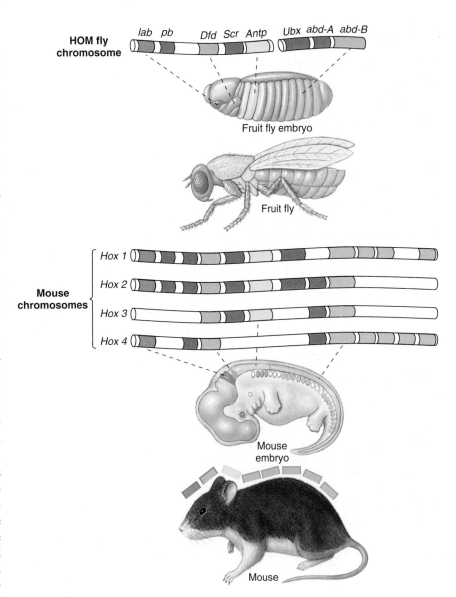

FIGURE 19.20

A comparison of homeotic gene clusters in the fruit fly *Drosophila melanogaster* and the mouse *Mus musculus.* Similar genes, the *Drosophila* HOM genes and the mouse *Hox* genes, control the development of front and back parts of the body. These genes are located on a single chromosome in the fly and on four separate chromosomes in mammals. On this illustration, the genes are color-coded to match the parts of the body in which they are expressed.

Homeotic genes encode transcription factors that activate blocks of genes specifying particular body parts.

Programmed Cell Death

Not every cell produced during development is destined to survive. For example, the cells between your fingers and toes die; if they did not, you would have paddles rather than digits. Vertebrate embryos produce a very large number of neurons, ensuring that there are enough neurons to make all of the necessary synaptic connections, but over half of these neurons never make connections and die in an orderly way as the nervous system develops. Unlike accidental cell deaths due to injury, these cell deaths are planned—and indeed required—for proper development. Cells that die due to injury typically swell and burst, releasing their contents into the extracellular fluid. This form of cell death is called **necrosis.** In contrast, cells programmed to die shrivel and shrink in a process called **apoptosis** (Greek, "shedding of leaves in autumn"), and their remains are taken up by surrounding cells.

Gene Control of Apoptosis

This sort of developmentally regulated cell suicide occurs when a "death program" is activated. All animal cells appear to possess such programs. In the nematode worm, for example, the same 131 cells always die during development in a predictable and reproducible pattern of apoptosis. Three genes govern this process. Two (*ced-3* and *ced-4*) constitute the death program itself; if either is mutant, those 131 cells do not die, and go on instead to form nervous and other tissue. The third gene (*ced-9*) represses the death program encoded by the other two (figure 19.21*a*). The same sorts of apoptosis programs occur in human cells: The *bax* gene encodes the cell death program, and another, an oncogene called *bcl-2*, represses it (figure 19.21*b*). The mechanism of apoptosis appears to have been highly conserved during the course of animal evolution. The protein made by *bcl-2* is 25% identical in amino acid sequence to that made by *ced-9*. If a copy of the human *bcl-2* gene is transferred into a nematode with a defective *ced-9* gene, *bcl-2* suppresses the cell death program of *ced-3* and *ced-4*!

How does *bax* kill a cell? The bax protein seems to induce apoptosis by binding to the permeability pores of the cell's mitochondria, increasing the mitochondrion's permeability and, in doing so, triggering cell death. How does *bcl-2* prevent cell death? One suggestion is that it prevents damage from *free radicals*, highly reactive fragments of atoms that can damage cells severely. Proteins or other molecules that destroy free radicals are called **antioxidants.** Antioxidants are almost as effective as *bcl-2* in blocking apoptosis.

> **Animal development involves programmed cell death (apoptosis), in which particular genes, when activated, kill their cells.**

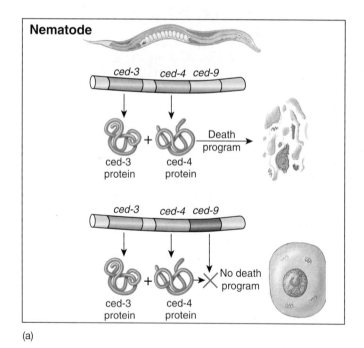

(a)

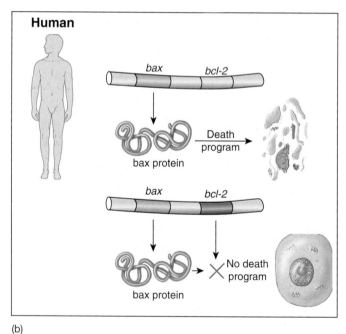

(b)

FIGURE 19.21
Programmed cell death. Apoptosis, or programmed cell death, is necessary for the normal development of all animals. (*a*) In the developing nematode, for example, two genes, *ced-3* and *ced-4*, code for proteins that cause the programmed cell death of 131 specific cells. In the other cells of the developing nematode, the product of a third gene, *ced-9*, represses the death program encoded by *ced-3* and *ced-4*. (*b*) In developing humans, the product of a gene called *bax* causes a cell death program in some cells and is blocked by the *bcl-2* gene in other cells.

19.3 Aging can be considered a developmental process.

Theories of Aging

All humans age, and eventually die. As you can see in figure 19.22, the "safest" age is around puberty: 10- to 15-year-olds have the lowest risk of dying. The death rate begins to increase rapidly after puberty as an exponential function of increasing age. Plotted on a log scale as in figure 19.22 (a so-called Gompertz plot), the mortality rate increases as a straight line from about 15 to 90 years, doubling about every eight years (the "Gompertz number"). By the time we reach 100, age has taken such a toll that the risk of dying reaches 50% per year.

A wide variety of theories have been advanced to explain why humans and other animals age. No single theory has gained general acceptance, but the following five are being intensively investigated.

Accumulated Mutation Hypothesis

The oldest general theory of aging is that cells accumulate mutations as they age, leading eventually to lethal damage. Careful studies have shown that somatic mutations do indeed accumulate during aging. As cells age, for example, they tend to accumulate the modified base 8-hydroxyguanine, in which an OH group is added to the base guanine. There is little direct evidence, however, that these mutations *cause* aging. No acceleration in aging occurred among survivors of Hiroshima and Nagasaki despite their enormous added mutation load, arguing against any general relationship between mutation and aging.

Telomere Depletion Hypothesis

In a seminal experiment carried out in 1961, Leonard Hayflick demonstrated that fibroblast cells growing in tissue culture will divide only a certain number of times (figure 19.23). After about 50 population doublings, cell division stops—the cell cycle is blocked just before DNA replication. If a cell sample is taken after 20 doublings and frozen, when thawed it resumes growth for 30 more doublings, and then stops.

An explanation of the "Hayflick limit" was suggested in 1986 when Howard Cooke first glimpsed an extra length of DNA at the ends of chromosomes. These **telomeres**, repeats of the sequence TTAGGG, were found to be substantially shorter in older somatic tissue, and Cooke speculated that a 100-base-pair portion of the telomere cap was lost by a chromosome during each cycle of DNA replication. Eventually, after some 50 replication cycles, the protective telomeric cap would be used up, and the cell line would then enter senescence, no longer able to proliferate. Cancer cells appear to avoid telomeric shortening.

Research reported in 1998 has confirmed Cooke's hypothesis, providing direct evidence for a causal relationship between telomeric shortening and cell senescence. Using genetic engineering, researchers transferred into human primary cell cultures a gene that leads to expression of telomerase, an enzyme that builds TTAGGG telomeric caps.

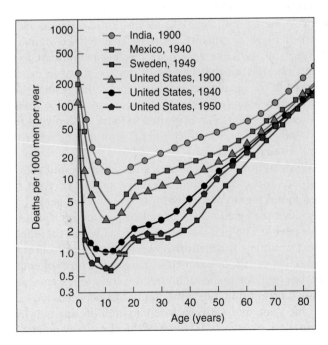

FIGURE 19.22
Gompertz curves. While human populations may differ 25-fold in their mortality rates before puberty, the slopes of their Gompertz curves are about the same in later years.
How do you explain this difference?

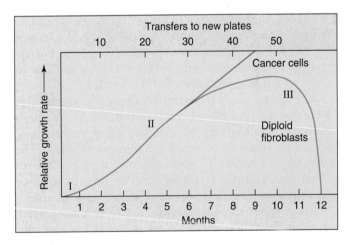

FIGURE 19.23
Hayflick's experiment. Fibroblast cells stop growing after about 50 doublings. Growth is rapid in phases I and II, but slows in phase III, as the culture becomes senescent, until the final doubling. Cancer cells, by contrast, do not "age."
How do cancer cells overcome this 50-division "Hayflick limit?"

The result was unequivocal. New telomeric caps were added to the chromosomes of the cells, and the cells with the artificially elongated telomeres did not senesce at the Hayflick limit, continuing to divide in a healthy and vigorous manner for more than 20 additional generations.

Wear-and-Tear Hypothesis

Numerous theories of aging focus in one way or another on the general idea that cells wear out over time, accumulating damage until they are no longer able to function. Loosely dubbed the "wear-and-tear" hypothesis, this idea implies that there is no inherent designed-in limit to aging, just a statistical one—that is, disruption, wear, and damage over time erode a cell's ability to function properly.

Considerable evidence indicates that aging cells do accumulate damage. Some of the most interesting evidence concerns free radicals, fragments of molecules or atoms that contain an unpaired electron. Free radicals are very reactive chemically and can be quite destructive in a cell. Free radicals are produced as natural by-products of oxidative metabolism, but most are mopped up by special enzymes that function to sweep the cell interior free of their destructive effects.

One of the most damaging free radical reactions that occurs in cells causes glucose to become linked to proteins, a nonenzymatic process called glycation. Two of the most commonly glycated proteins are collagen and elastin, key components of the connective tissues in our joints. Glycated proteins can cross-link to one another, reducing the flexibility of connective tissues in the joints and producing many of the other characteristic symptoms of aging.

Gene Clock Hypothesis

There is very little doubt that at least some aspects of aging are under the direct control of genes. Just as genes regulate the body's development, so they appear to regulate its rate of aging. Some genes appear to promote longevity. For example, people over 100 years old are five times as likely to carry a mutation in mitochondrial DNA called C150T.

Other genes produce premature aging. For example, in 1996, a gene was identified that is responsible for Werner's syndrome, which produces premature aging in the young and affects some 10 people per million worldwide. The syndrome is named after Otto Werner, who in 1904 reported on a German family affected by premature aging and believed a genetic component was at work. Werner's syndrome appears in adolescence, usually producing death before age 50 of heart attack or one of a variety of rare connective tissue cancers. Located on the short arm of chromosome 8, the gene seems to affect a helicase enzyme involved in the repair of DNA. The gene, which codes for a 1432-amino-acid protein, has been fully sequenced, and four mutant alleles identified. Helicase enzymes are needed to unwind the DNA double helix whenever DNA has to be replicated, repaired, or transcribed. The high incidence of certain cancers among Werner's syndrome patients leads investigators to speculate that the mutant helicase may fail to activate critical tumor-suppressor genes. The potential role of helicases in aging is the subject of heated research.

Research on aging in other animals strongly supports the hypothesis that genes regulate the rate of aging. A *Drosophila melanogaster* gene mutation called *Indy* ("I'm not dead yet") doubles the fruit fly life span from the usual 37 days to an average of 70 days. When researchers isolated the DNA of the *Indy* gene and compared its DNA sequence with the Human Genome Project sequences, they found that the *Indy* gene is 50% similar to a human gene called *dicarboxylate cotransporter*. In humans, dicarboxylate cotransporter proteins move preliminary products of food metabolism (dicarboxylic acids of the Krebs cycle) across membranes to where the food's processing takes place. In mutant *Indy* flies, poor dicarboxylic acid pumping means that less metabolic energy can be gleaned from the fly's food. In essence, the *Indy* mutation is the genetic equivalent of caloric restriction. Starving is known to prolong life in the nematode *Caenorhabditis elegans*, but *Indy*'s caloric restriction does not involve the unpleasantness of starving. The *Indy* mutation in effect puts flies on a severe diet, while the flies eat as much as normal and lead a normal vigorous life—for far longer.

Current Theories of Aging

An intriguing connection has been found recently that may shed light on the calorie restriction observations. In both *C. elegans* and *Drosophila*, there appears to be a connection between aging and signaling by insulin-like receptors. In both of these organisms, mutants that reduce signaling through insulin-like receptors lead to an increased life span. The relationship between calorie restriction experiments and the insulin-deficient mutants is that insulin is involved in a complex pathway that controls caloric intake. The effect in *Drosophila* is limited to females, implying that trade-offs between fecundity and life span may be related to this pathway.

Work in both of these systems has failed to find genetic evidence for a clocklike mechanism similar to what has been found for circadian rhythms. A number of genes have been identified that either extend or shorten life span, but these seem to be like the insulin case, in which there is no direct counting mechanism. The shortening of chromosomes due to lack of telomerase activity is a form of counting, but only for replications and not for a senescence program. Current thinking is that trade-offs between body size and life span, and between fecundity and life span (at least in females), may be at the heart of the genetics of aging.

Among the theories advanced to explain aging, many involve the progressive accumulation of damage to DNA. When genes affecting aging have been isolated, they affect DNA repair processes.

19.1 Development is a regulated process.

Overview of Development

- Organisms in all three multicellular kingdoms utilize gene expression (i.e., different cells express different genes at different times) to achieve cell specialization. (p. 382)

Vertebrate Development

- At different sites in the vertebrate embryo, particular cells proceed to form the body's tissues and organs, and then the body grows to a size and shape that will allow survival after birth. (p. 383)
- Vertebrate zygotes divide rapidly to form a blastula. Subsequent stages in vertebrate development include gastrulation, neurulation, cell migration, organogenesis, and growth. (pp. 383–385)

Insect Development

- Many insects go through metamorphosis from a larva to a fully functioning, reproductive adult. (p. 386)
- The development of *Drosophila melanogaster* proceeds from an egg to a syncytial blastoderm, to larval instars, which have imaginal discs, and finally through metamorphosis to an adult. (pp. 386–387)

Plant Development

- Plant cells are encased in cellulose walls. Instead of using cell migration, plants develop by building their bodies outward, creating new parts from meristems. (p. 388)
- Plants cope with environmental change by adjusting the type, number, size, and location of its structures to accommodate local circumstances. (p. 388)
- Three basic tissue types of plants are epidermal cells, ground tissue, and vascular tissue. (p. 388)
- Seeds germinate in response to environmental changes due to water, temperature, and other factors. (p. 388)

Nematode Development

- Nematodes develop a known number of somatic cells from a single fertilized egg in a complex series of cell divisions. (p. 390)

19.2 Multicellular organisms employ the same basic mechanisms of development.

Cell Movement and Induction

- Cell movement is largely a matter of changing patterns of cell adhesion. As a migrating cell travels, it continuously extends projections that probe the environment, and thus the cell feels its way toward a target. (p. 391)
- In some cases, particular groups of cells called organizers produce diffusible signal molecules that convey positional information to other cells. (p. 392)
- Induction occurs when a cell switches from one developmental path to another as a result of interactions with an adjacent cell. (p. 392)

Determination

- Totipotent cells are potentially capable of expressing all the genes of their genome. (p. 393)
- Determination is the commitment of a particular cell to a specialized developmental path. (p. 393)
- Recent experiments have shown that determination is reversible. (p. 394)

Pattern Formation

- All animals appear to use positional information to determine the basic pattern of body compartments and the overall body architecture. (p. 395)

Homeotic Genes

- Homeotic genes encode transcription factors that activate blocks of genes specifying particular body parts. (p. 399)

Programmed Cell Death

- Necrosis is cell death due to injury-caused swelling and bursting, while apoptosis refers to programmed cell death. (p. 400)
- The mechanism of apoptosis appears to have been highly conserved during the course of animal evolution. (p. 400)

19.3 Aging can be considered a developmental process.

Theories of Aging

- Several theories of aging that are being intensively investigated include the accumulated mutation hypothesis, the telomere depletion hypothesis, the wear-and-tear hypothesis, the gene clock hypothesis, and trade-offs between body size and life span, and between fecundity and life span. (p. 402)

Self Test

1. Which of the following series of events represents the path of vertebrate development?
   a. formation of blastula, cleavage, neurulation, cell migration, gastrulation, organogenesis, growth
   b. formation of blastula, cleavage, gastrulation, neurulation, cell migration, organogenesis, growth
   c. cleavage, formation of blastula, gastrulation, neurulation, cell migration, organogenesis, growth
   d. cleavage, gastrulation, formation of blastula, neurulation, cell migration, organogenesis, growth

2. Which of the following statements about *Drosophila* development is false?
   a. *Drosophila* go through four larval instar stages before undergoing metamorphosis.
   b. During the syncytial blastoderm stage, nuclei line up along the surface of the egg.
   c. Imaginal discs are groups of cells set aside that will give rise to key parts of the adult fly.
   d. Maternal, rather than zygotic, genes govern early *Drosophila* development.

3. If a plant embryo failed to form enough ground tissue, what function(s) would likely be directly affected in the corresponding mature plant?
   a. seed formation
   b. meristem development
   c. cotyledon formation
   d. food and water storage

4. *C. elegans* is a powerful developmental model because
   a. these nematodes are very small, so it is easy to maintain a large population in a laboratory.
   b. the fate of every cell has been mapped.
   c. the fates of cells that will become eggs and sperm are predetermined.
   d. these nematodes have the same amount of DNA as *Drosophila*.

5. Which of the following best describes a morphogen?
   a. a cell that secretes diffusible signaling molecules that play a role in specifying cell fate
   b. a diffusible signaling molecule that plays a role in specifying cell fate
   c. a protein that helps mediate direct cell–cell interaction
   d. a protein that enables cells to become totipotent

6. What would happen as a result of a transplantation experiment in a chick embryo in which cells determined to become a forelimb were replaced by cells determined to become a hindlimb?
   a. A hindlimb would form in the region where the forelimb should be.
   b. A forelimb would form in the region where the hindlimb should be.
   c. Nothing; the forelimb would form normally.
   d. Neither a forelimb nor a hindlimb would form because the cells were already determined.

7. Which group of genes, identified by Nusslein-Volhard and Caroll, is responsible for the final stages of segmentation in *Drosophila* embryos?
   a. morphogen gradient genes
   b. gap genes
   c. segment-polarity genes
   d. pair-rule genes

8. Suppose that during a mutagenesis screen to isolate mutations in *Drosophila*, you come across a fly with legs growing out of its head. What gene cluster is likely affected?
   a. *Bicoid*
   b. *Hunchback*
   c. *Bithorax*
   d. *Antennapedia*

9. What would be the likely result of a mutation of the *bcl-2* gene on the level of apoptosis?
   a. no change
   b. a decrease in apoptosis
   c. an increase in apoptosis
   d. First, it would increase, but later it would decrease.

10. The gene clock hypothesis is best described by which of the following explanations?
    a. Mutations accumulate partially through the addition of an −OH group to the base guanine.
    b. Specific genes exist to promote longevity.
    c. Free radicals can cause genetic mutations, particularly when we are sleeping.
    d. Calorie restriction leads to an increased life span.

Test Your Visual Understanding

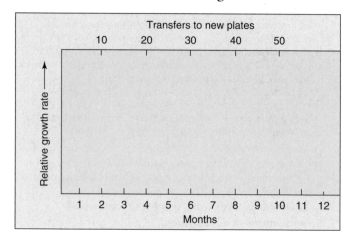

1. Hayflick's experiment revealed that noncancerous cells have a definitive life span, whereas cancerous cells do not have the same restrictions. Draw curves that represent the growth patterns of each cell type.

Apply Your Knowledge

1. You have generated a cell line that expresses an altered form of cadherin. This mutant cadherin has the 110-amino-acid extracellular domain that is required for interaction with other cadherins, but lacks a transmembrane domain. If you were to mix this cell population with other cells expressing a wild-type, or normal, form of cadherin, would you expect these two cell populations to aggregate with each other? Why or why not? Would the mutant cells be able to aggregate with other mutant cells?

20

Cancer Biology and Cell Technology

Concept Outline

20.1 Recombination alters gene location.

Gene Transfer. Plasmids can move between bacterial cells and carry bacterial genes.

Reciprocal Recombination. Reciprocal recombination can alter genes in several ways.

20.2 Mutations are changes in the genetic message.

Kinds of Mutation. Some mutations alter genes themselves; others alter the positions of genes.

DNA Repair. Cells repair damage to DNA effectively in a variety of ways.

20.3 Most cancer results from mutation of growth-regulating genes

What Is Cancer? Cancer is a growth disorder of cells.

Causes of Cancer. Many cancers are caused by chemicals that alter DNA.

Cancer and the Cell Cycle. Cancer results from mutation of genes regulating cell proliferation.

Smoking and Cancer. Smoking causes lung cancer.

Curing Cancer. New approaches offer promise of a cure.

20.4 Reproductive cloning of animals, once thought impossible, isn't.

The Challenge of Cloning. As recently as 1997, scientists thought it was impossible to clone an animal from an adult cell. They were wrong.

20.5 Therapeutic cloning is a promising but controversial possibility.

Stem Cells. Embryonic stem cells are capable of forming any tissue of the body. Tissue-specific adult stem cells may provide an alternative to embryonic stem cells.

Therapeutic Cloning. Somatic nuclear transfer of human embryonic stem cells seems to offer significant therapeutic promise.

Grappling with the Ethics of Stem Cell Research. Few issues in modern science raise as many difficult ethical issues as stem cell research.

FIGURE 20.1
Embryonic stem cells growing in tissue culture. Embryonic stem cells derived from early human embryos will grow indefinitely in tissue culture. When transplanted, they can sometimes be induced to form new cells of the adult tissue into which they have been placed. This suggests exciting therapeutic uses.

The last decade has seen remarkable advances in our understanding of the molecular events leading to cancer. In essence, cancer results from damage to grow-controlling genes. This damage can result from positional changes induced by recombination or, more commonly, from chemical changes induced by mutation. After reviewing these two processes briefly, this chapter explores in depth what we have learned about cancer. It then turns to another area in which landmark progress is being made—cell technology. While animal cloning, gene therapy, and stem cell research (figure 20.1) might have been treated within three different chapters, this chapter pulls them all together so that the broad sweep of what is occurring is most apparent. Advances in cell technology hold the promise of literally revolutionizing biology and medicine.

20.1 Recombination alters gene location.

Gene Transfer

Two very different sorts of recombination processes alter genes: gene transfer and reciprocal recombination. We will consider gene transfer first.

Genes are not fixed in their locations on chromosomes or on the circular DNA molecules of prokaryotes; they can move around. Some genes move because they are part of small, circular, extrachromosomal DNA segments called **plasmids.** Plasmids enter and leave the main genome at specific places where a nucleotide sequence matches one present on the plasmid. Plasmids occur primarily in prokaryotes, in which the main genomic DNA can interact readily with other DNA fragments. About 5% of the DNA that occurs in a bacterium is plasmid DNA. Some plasmids are very small, containing only one or a few genes, while others are quite complex and contain many genes. Other genes move within **transposons,** which jump from one genomic position to another at random in both prokaryotes and eukaryotes.

Gene transfer by plasmid movement was discovered by Joshua Lederberg and Edward Tatum in 1947. Three years later, transposons were discovered by Barbara McClintock. However, her work implied that the position of genes in a genome need not be constant. Researchers accustomed to viewing genes as fixed entities, like beads on a string, did not readily accept the idea of transposons. Therefore, while Lederberg and Tatum were awarded a Nobel Prize for their discovery in 1958, McClintock did not receive a Nobel Prize for hers until 1983.

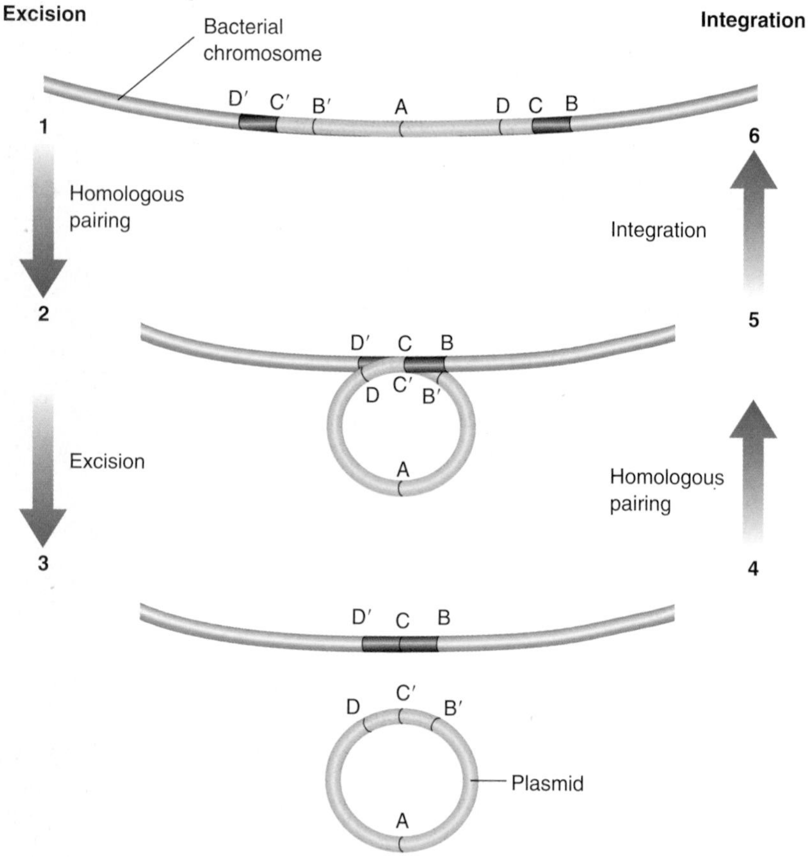

FIGURE 20.2
Excision and integration of a plasmid. Because the ends of the two sequences in the bacterial genome are the same (D′, C′, B′, and D, C, B), it is possible for the two ends to pair. Steps 1–3 show the sequence of events if the strands exchange during the pairing. The result is excision of the loop and a free circle of DNA—a plasmid. Steps 4–6 show the sequence when a plasmid integrates itself into a bacterial genome.

Plasmid Creation

To understand how plasmids arise, consider a hypothetical stretch of bacterial DNA that contains two copies of the same nucleotide sequence. It is possible for the two copies to base-pair with each other and create a transient "loop," or double duplex. All cells have recombination enzymes that can cause such double duplexes to undergo a **reciprocal exchange,** in which they exchange strands. As a result of the exchange, the loop is freed from the rest of the DNA molecule and becomes a plasmid (figure 20.2, steps 1–3). Any genes between the duplicated sequences (such as gene A in figure 20.2) are transferred to the plasmid.

Once a plasmid has been created by reciprocal exchange, DNA polymerase will replicate it if it contains a replication origin, often without the controls that restrict the main genome to one replication per cell division.

Integration

A plasmid created by recombination can reenter the main genome the same way it left. Sometimes the region of the plasmid DNA that was involved in the original exchange, called the **recognition site,** aligns with a matching sequence on the main genome. If a recombination event occurs anywhere in the region of alignment, the plasmid will integrate into the genome (figure 20.2, steps 4–6). Integration can occur wherever any shared sequences exist, so plasmids may be integrated into the main genome at positions other than the one from which they arose. If a plasmid is integrated at a new position, it transfers its genes to that new position.

Gene Transfer by Conjugation

One of the startling discoveries Lederberg and Tatum made was that plasmids can pass from one bacterium to another. The plasmid they studied was part of the genome of *Escherichia coli*. It was given the name F, for fertility factor, because only cells having that plasmid integrated into their DNA could act as plasmid donors. These cells are called Hfr cells (for "high-frequency recombination"). The F plasmid contains a DNA replication origin and several genes that promote its transfer to other cells. These genes encode protein subunits that assemble on the surface of the bacterial cell, forming a hollow tube called a **pilus.**

When the pilus of one cell (F$^+$) contacts the surface of another cell that lacks a pilus, and therefore does not contain an F plasmid (F$^-$), the pilus draws the two cells close together so that DNA can be exchanged (figure 20.3). First, the F plasmid binds to a site on the interior of the F$^+$ cell just beneath the pilus now called a *conjugation bridge*. Then, by a process called **rolling-circle replication,** the F plasmid begins to copy its DNA at the binding point. As it is replicated, the single-stranded copy of the plasmid passes into the other cell. There, a complementary strand is added, creating a new, stable F plasmid (figure 20.4). In this way, genes are passed from one bacterium to another. This transfer of genes between bacteria is called **conjugation.**

In an Hfr cell, with the F plasmid integrated into the main bacterial genome rather than free in the cytoplasm, the F plasmid can still organize the transfer of genes. In this case, the integrated F region binds beneath the pilus and initiates the *replication of the bacterial genome*, transferring the newly replicated portion to the recipient cell. Transfer proceeds as if the bacterial genome were simply a part of the F plasmid. By studying this phenomenon, researchers have been able to locate the positions of different genes in bacterial genomes (figure 20.5).

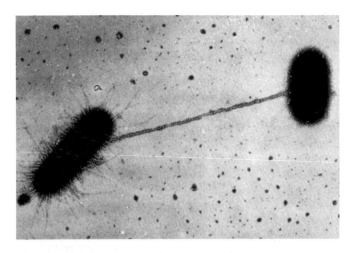

FIGURE 20.3
Contact by a pilus. The pilus of an F$^+$ cell connects to an F$^-$ cell and draws the two cells close together so that DNA transfer can occur.

Gene Transfer by Transposition

Transposons (figure 20.6) are small segments of DNA capable of moving from one location to another in the genome. After spending many generations in one position, a transposon may abruptly move to a new position in the genome, carrying various genes along with it. Transposons encode an enzyme called **transposase,** which inserts the transposon into the genome, a process known as **transposition** (figure 20.7). Because this enzyme usually does not recognize any particular sequence on the genome, transposons appear to move to random destinations.

The movement of any given transposon is relatively rare: It may occur perhaps once in 100,000 cell generations. Although low, this rate is still about 10 times as frequent as the rate at which random mutational changes

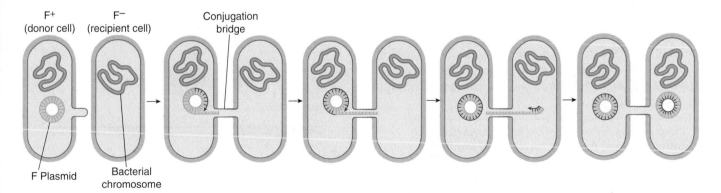

F$^+$
(donor cell)   F$^-$
(recipient cell)   Conjugation
bridge

F Plasmid   Bacterial
chromosome

FIGURE 20.4
Gene transfer between bacteria. Donor cells (F$^+$) contain an F plasmid that recipient cells (F$^-$) lack. The F plasmid replicates itself and transfers the copy across a conjugation bridge. The remaining strand of the plasmid serves as a template to build a replacement. When the single strand enters the recipient cell, it serves as a template to assemble a double-stranded plasmid. When the process is complete, both cells contain a complete copy of the plasmid.

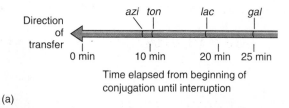

Direction
of
transfer

Time elapsed from beginning of
conjugation until interruption

(a)

FIGURE 20.5

A conjugation map of the *E. coli* chromosome. Scientists
have been able to break the *Escherichia coli* conjugation bridges by
agitating the cell suspension rapidly in a blender. By agitating at
different intervals after the start of conjugation, investigators can
locate the positions of various genes along the bacterial genome.
(*a*) The closer the genes are to the origin of replication, the
sooner one has to turn on the blender to block their transfer.
(*b*) Map of the *E. coli* genome developed using this method.

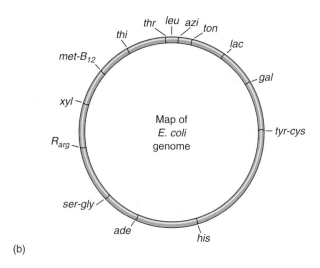

(b)

occur. Furthermore, most cells contain many transposons. Hence, over long periods of time, transposition
can have an enormous evolutionary impact.

One way this impact can be felt is through mutation.
The insertion of a transposon into a gene often destroys
the gene's function, resulting in what is termed **insertional
inactivation.** This phenomenon is thought to cause a significant number of the spontaneous mutations observed in
nature, and is thus a potential cause of cancer.

Transposition can also facilitate **gene mobilization,**
the bringing together in one place of genes that are usually located at different positions in the genome. In bacteria, for example, a number of genes encode enzymes that
make the bacteria resistant to antibiotics such as penicillin, and many of these genes are located on plasmids.
The simultaneous exposure of bacteria to multiple antibiotics, a common medical practice some years ago, favors
the persistence of plasmids that have managed to acquire
several resistance genes. Transposition can rapidly generate such composite plasmids, called **resistance transfer
factors (RTFs),** by moving antibiotic-resistance genes
from several plasmids to one. Bacteria possessing RTFs
are thus able to survive treatment with a wide variety of
antibiotics. RTFs are thought to be responsible for much
of the recent difficulty in treating hospital-engendered
Staphylococcus aureus infections and the new drug-resistant
strains of tuberculosis.

Plasmids transfer copies of bacterial genes (and even
entire genomes) from one bacterium to another.
Transposition is the one-way transfer of genes to a
randomly selected location in the genome. The genes
move because they are associated with mobile genetic
elements called transposons.

FIGURE 20.6

Transposon. Transposons form
characteristic stem-and-loop
structures called "lollipops" because
their two ends have the same
nucleotide sequence as inverted
repeats. These ends pair together to
form the stem of the lollipop.

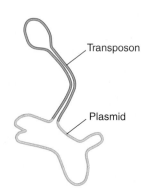

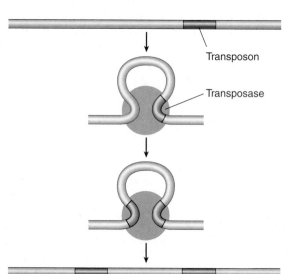

FIGURE 20.7

Transposition. Transposase does not recognize any particular
DNA sequence; rather, it selects one at random, moving the
transposon to a random location. Some transposons leave a copy
of themselves behind when they move.

Reciprocal Recombination

In the second major mechanism for producing genetic recombination, reciprocal recombination among eukaryotes, two homologous chromosomes exchange all or part of themselves during the process of meiosis.

Crossing Over

As we saw in chapter 12, crossing over occurs in the first prophase of meiosis, when two homologous chromosomes line up side by side within the synaptonemal complex. At this point, the homologues exchange DNA strands at one or more locations. This exchange of strands can produce chromosomes with new combinations of alleles.

Imagine, for example, that a giraffe has genes encoding neck length and leg length at two different loci on one of its chromosomes. Imagine further that a recessive mutation occurs at the neck length locus, leading after several rounds of independent assortment to some individuals that are homozygous for a variant "long-neck" allele. Similarly, a recessive mutation at the leg length locus leads to homozygous "long-leg" individuals.

It is very unlikely that these two mutations would arise at the same time in the same individual because the probability of two independent events occurring together is the product of their individual probabilities. If the spontaneous occurrence of both mutations in a single individual were the only way to produce a giraffe with both a long neck and long legs, it would be extremely unlikely that such an individual would ever occur. Because of recombination, however, a crossover in the interval between the two genes could in one meiosis produce a chromosome bearing both variant alleles. This ability to reshuffle gene combinations rapidly is what makes recombination so important to the production of natural variation.

Unequal Crossing Over

Reciprocal recombination can occur in any region along two homologous chromosomes with sequences similar enough to permit close pairing. Mistakes in pairing occasionally happen when several copies of a sequence exist in different locations on a chromosome. In such cases, one copy of a sequence may line up with one of the duplicate copies instead of with its homologous copy. Such misalignment causes slipped mispairing, which, as we will discuss later, can lead to small deletions and frameshift mutations. If a crossover occurs in the pairing region, it will result in unequal crossing over because the two homologues will exchange segments of unequal length.

In unequal crossing over, one chromosome gains extra copies of the multicopy sequences, while the other chro-

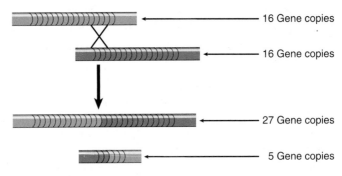

FIGURE 20.8
Unequal crossing over. When a repeated sequence pairs out of register, a crossover within the region will produce one chromosome with fewer gene copies and one with more. Much of the gene duplication that has occurred in eukaryotic evolution may well be the result of unequal crossing over.

mosome loses them (figure 20.8). This process can generate a chromosome with hundreds of copies of a particular gene, lined up side by side in tandem array.

Because the genomes of most eukaryotes possess multiple copies of transposons scattered throughout the chromosomes, unequal crossing over between copies of transposons located in different positions has had a profound influence on gene organization in eukaryotes. As we shall see later, most of the genes of eukaryotes appear to have been duplicated one or more times during their evolution.

Gene Conversion

Because the two homologues that pair within a synaptonemal complex are not identical, some nucleotides in one homologue are not complementary to their counterpart in the other homologue. These occasional nonmatching pairs of nucleotides are called **mismatch pairs.**

As you might expect, the cell's error-correcting machinery is able to detect mismatch pairs. If a mismatch is detected during meiosis, the enzymes that "proofread" new DNA strands during DNA replication correct it. The mismatched nucleotide in one of the homologues is excised and replaced with a nucleotide complementary to the one in the other homologue. Its base-pairing partner in the first homologue is then replaced, producing two chromosomes with the same sequence. This error correction causes one of the mismatched sequences to convert into the other, a process called **gene conversion.**

Unequal crossing over occurs between chromosomal regions that are similar in nucleotide sequence but not homologous. Gene conversion is the alteration of one homologue by the cell's error-detection and repair system to make it resemble the other homologue.

20.2 Mutations are changes in the genetic message.

Kinds of Mutation

Mutations are changes in the hereditary message of an organism. Mutations can have significant effects on the individual when they occur in somatic tissue, but are only inherited if they occur in germ-line tissue. Because mutations can occur randomly anywhere in a cell's DNA, mutations are often detrimental, just as making a random change in a computer program or a musical score usually worsens performance. The consequences of a detrimental mutation may be minor or catastrophic, depending on the function of the altered gene.

The effect of a mutation depends critically on the identity of the cell in which the mutation occurs. During the embryonic development of all multicellular organisms, there comes a point when cells destined to form gametes (germ-line cells) are segregated from those that will form the other cells of the body (somatic cells). Only when a mutation occurs within a germ-line cell is it passed to subsequent generations as part of the hereditary endowment of the gametes derived from that cell.

Mutations in germ-line tissue are of enormous biological importance because they provide the raw material from which natural selection produces evolutionary change. In a somatic cell, mutation may have drastic effects on the individual organism in which it occurs, because it is passed on to all the cells that are descended from the original mutant cell. Thus, if a mutant lung cell divides, all cells derived from it will carry the mutation. Somatic mutations of lung cells are, as we shall see, the principal cause of lung cancer in humans.

One category of mutational changes affects the message itself, producing alterations in the sequence of DNA nucleotides. A second category of mutation arises as the result of changes in gene location. Table 20.1 summarizes the sources and types of mutations.

Mutations Altering the Sequence of DNA

Mutational alterations of DNA arise in at least five ways:

1. **Base substitution.** Spontaneous pairing errors or polymerase mistakes may occur during DNA replication. These changes typically involve only one or a few base-pairs in the coding sequence, a so-called point mutation.
2. **Chemical modification.** A base may be chemically altered by a mutagenic chemical, another class of point mutation.
3. **DNA breaks.** Ionizing radiation can cause double-strand breaks in DNA, often resulting in the deletion (loss) of short segments.
4. **Slipped mispairing.** Deletions may also result from so-called slipped mispairing. When a sequence present in more than one copy on a chromosome pairs out of register—like a shirt buttoned wrong—the loop this mistake produces is sometimes excised by the cell's repair enzymes, producing a short deletion. Many of these deletions start or end in the middle of a codon, thereby shifting the reading frame by one or two bases. These so-called *frameshift mutations* cause the gene to be read in the wrong 3-base groupings, distorting the genetic message.
5. **Triplet expansion.** When a 3-base sequence is repeated several times in tandem within a gene, the number of repeats may become expanded. Thus, a CAG repeat, when expanded, produces a polyglutamine region within the resulting mutant protein. More than 15 human disease states are associated with triplet expansion within a gene.

Table 20.1 Types of Mutation

Mutation		Example Result
NO MUTATION		
A B C		Normal B protein is produced by the *B* gene.
SEQUENCE CHANGES		
Base substitution — A B C	Substitution of one or a few bases due to pairing error or chemical modification	B protein is inactive because changed amino acid disrupts function.
Triplet expansion — X 200 — CCGCCGCCGCCG	Additional copies of a repeated 3-base sequence	B protein is inactive because inserted material disrupts proper shape.
Deletion — A C	Loss of one or a few bases due to ionizing radiation or slipped mispairing	B protein is inactive because portion of protein is missing.
CHANGES IN GENE POSITION		
Chromosomal rearrangement — A C / B		*B* gene is inactivated or regulated differently in its new location on chromosome.
Insertional inactivation — A C	Addition of a transposon within a gene	B protein is inactive because inserted material disrupts gene translation or protein function.

Mutations Arising from Changes in Gene Position

Chromosome location is an important factor in determining whether genes are transcribed. Some genes cannot be transcribed if they are adjacent to a tightly coiled region of the chromosome, even though the same gene can be transcribed normally in any other location. Transcription of many chromosomal regions appears to be regulated in this manner; the binding of specific proteins regulates the degree of coiling in local regions of the chromosome, determining the accessibility RNA polymerase has to genes located within those regions.

Chromosomal Rearrangements. Chromosomes undergo several different kinds of gross physical alterations that have significant effects on the locations of their genes. There are two main kinds of alteration: translocations, in which a segment of one chromosome becomes part of another chromosome, and inversions, in which the orientation of a portion of a chromosome is reversed. Translocations often have significant effects on gene expression. Inversions, on the other hand, usually do not alter gene expression, but lead to serious problems in meiosis (figure 20.9): None of the gametes that contain chromatids produced following such a crossover event will have a complete set of genes.

Other chromosomal alterations change the number of gene copies an individual possesses. These include **duplication,** in which entire segments of the chromosome become duplicated; **deletions,** like the small deletions discussed earlier but involving much larger amounts of genetic material; **aneuploidy,** in which whole chromosomes are lost or gained; and **polyploidy,** in which entire sets of chromosomes are added. Even in diploid organisms, most deletions are harmful because they halve the number of gene copies within a diploid genome and thus seriously affect the level of transcription. Duplications cause gene imbalance and are also usually harmful.

Insertional Inactivation. As discussed in section 20.1, transposons are capable of moving from one location to another in the genome, using an enzyme to cut and paste themselves into new genetic neighborhoods. Transposons select their new locations at random, and are as likely to enter one segment of a chromosome as another. Inevitably, some transposons end up inserted into genes, and this almost always inactivates the gene. The encoded protein now has a large meaningless chunk within it, disrupting its structure. This form of mutation, called *insertional inactivation*, is common in nature. Other mutations result when insertion of a transposon at a new site affects the regulation of neighboring genes because it contains a strong promoter that turns on genes that would otherwise be silent.

As you might expect, a variety of human gene disorders are the result of transposition. The human transposon called *Alu*, for example, is responsible for an X-linked hemophilia, inserting into clotting factor IX and placing a premature "stop" codon there. It also causes inherited high levels of cholesterol (hypercholesterolemia) when *Alu* elements insert into the gene encoding the low-density lipoprotein (LDL) receptor. In one very interesting case, a *Drosophila* transposon called *Mariner* proves responsible for a rare human neurological disorder called Charcot-Marie-Tooth disease, in which the muscles and nerves of the legs and feet gradually wither away. The *Mariner* transposon is inserted into a key gene called *CMT* on chromosome 17, creating a weak site where the chromosome can break. No one knows how the *Drosophila* transposon got into the human genome.

Mutations result both from changes in the nucleotide sequence of genes and from the movement of genes to new locations on the chromosomes.

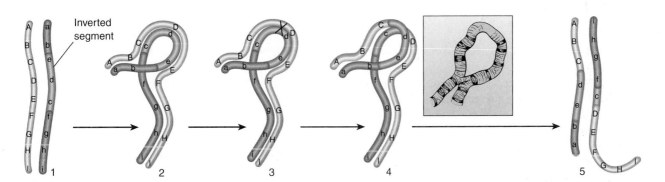

FIGURE 20.9
The consequence of inversion. (*1*) When a segment of a chromosome is inverted, (*2*) it can pair in meiosis only by forming an internal loop. (*3*) Any crossing over that occurs within the inverted segment during meiosis will result in nonviable gametes; some genes are lost from each chromosome, while others are duplicated (*4* and *5*). For clarity, only two strands are shown, although crossing over occurs in the four-strand stage. The pairing that occurs between inverted segments is sometimes visible under the microscope as a characteristic loop (*inset*).

DNA Repair

Given the possible number of errors that can occur during replication and the number of possible agents that can damage DNA, it is not surprising that systems have evolved to repair damaged DNA. These systems operate in a number of ways to safeguard the cell's hereditary information.

Mismatch Repair

First discovered in the bacterium *E. coli*, mismatch repair has now been found to be common in the cells of most organisms. These systems are designed to correct errors occurring during DNA replication. Even with their proofreading activities, the error rate of DNA polymerase is relatively high. Mismatch repair systems were first identified in strains of *E. coli* with unusually high mutation rates. A small number of genes proved to be responsible for the high mutation rates. These genes, labeled *Mut* for mutations, led to the identification of the activities involved in mismatch repair.

The **mismatch repair system** uses a clever strategy to identify sites of mutation. Specific enzymes in the cell recognize the sequence GATC on cellular DNA and add a methyl (CH_3) group to the adenine. Many mismatch repair mutants proved unable to methylate this adenine, and are dubbed *dam* mutations ("deficient in adenine methylation"). Methylation of GATC sequences allows the identification of the parental strand after DNA replication—it is the GATC-methylated strand. A little later, a maintenance methylase acts to methylate the adenines on the GATC sequences of the new daughter strand. However, in the short interval before it does so, the cell's mismatch repair enzymes act: (1) They locate replication errors as sites where the base mismatch between the two strands distorts the helix; (2) they identify the parental strand (the methylated one); and (3) they correct the opposite strand so that it is complementary to the methylated one.

Specific Repair Systems

Cells contain a number of repair systems that are specific for particular lesions in DNA. One such **UVR photorepair system** corrects damage caused by UV light. The energy in ultraviolet light is absorbed by thymine, causing the formation of cyclobutane links between adjacent thymines; because the resulting thymine dimers cannot be replicated, their creation leads to mutations. The damage is repaired by an enzyme that is able to absorb a photon of visible light and use the energy from this to cleave the cyclobutane bond.

Another specific repair system repairs any instances where uracil (an RNA nucleotide) is present in DNA by mistake, instead of thymine. An enzyme, uracil-N-glycosylase, cleaves the uracil side group from the nucleotide base without disrupting the DNA chain. This leaves an **apyrmidinic (AP) site** with a nucleotide that has no attached base. A specific nuclease then recognizes this AP site and removes the truncated nucleotide. The resulting one-base gap is then restored by the repair polymerase. Other chemically damaged bases are also repaired by similar two-step processes.

Excision Repair

DNA replication errors and chemical damage to DNA can also be repaired by **excision repair,** a nonspecific system that can repair a wide variety of lesions to DNA. In broad outline, excision repair involves two steps: (1) The damaged region of DNA is recognized and removed, (2) the repair polymerase fills in the resulting gap with newly synthesized DNA. In *E. coli*, this excision repair process is controlled by the UVR photorepair system. The genes that encode the relevant proteins were identified based on their unusual UV sensitivity, with the genes named *uvr* A, B, C, and D.

Because this nonspecific system responds only to distortions of the helix caused by damaged bases, excision repair can handle a wide variety of DNA damage and is generally "error-free"—that is, the DNA is restored to the wild-type condition after the damaged bases are removed.

Post-Replication Repair

Post-replication repair is the only repair system able to handle double-strand breaks in DNA. Repairing this sort of damage requires a second complete DNA duplex molecule with the same sequence as the damaged DNA. Because prokaryotes have only one copy of their DNA except during DNA replication, this form of repair is only possible during replication in a prokaryotic cell. In eukaryotes, this process is called **recombinational repair,** and occurs during prophase of meiosis I, when the two chromosomes are paired in close register. Many of the enzymes responsible for recombinational repair are also active in the recombination events of meiosis.

> **Cells are able to repair many potential mutations by recognizing the pairing mismatch that a base substitution creates or by recognizing damaged or inappropriate bases.**

20.3 Most cancer results from mutation of growth-regulating genes.

What Is Cancer?

Cancer is a growth disorder of cells. It starts when an apparently normal cell begins to grow in an uncontrolled and invasive way (figure 20.10). The result is a cluster of cells, called a **tumor,** that constantly expands in size. Cells that leave the tumor and spread throughout the body, forming new tumors at distant sites, are called **metastases** (figure 20.11). Cancer is perhaps the most pernicious disease. Of the children born in 1999, one-third will contract cancer at some time during their lives; one-fourth of the male children and one-third of the female children will someday die of cancer. Most of us have had family or friends affected by the disease. In 2002, an estimated 550,000 Americans died of cancer.

Not surprisingly, researchers are expending a great deal of effort to learn the cause of this disease. Scientists have made a great deal of progress in the last 20 years using molecular biological techniques, and the rough outlines of understanding are now emerging. We now know that cancer is a gene disorder of somatic tissue, in which damaged genes fail to properly control cell proliferation. The cell division cycle is regulated by a sophisticated group of proteins described in chapter 11. Cancer results from the mutation of the genes encoding these proteins.

Cancer can be caused by chemicals that mutate DNA or, in some instances, by viruses that circumvent the cell's normal proliferation controls. Whatever the immediate cause, however, all cancers are characterized by unrestrained growth and division. Cell division never stops in a cancerous line of cells. Cancer cells are virtually immortal—until the body in which they reside dies.

Cancer is unrestrained cell proliferation caused by damage to genes regulating the cell division cycle.

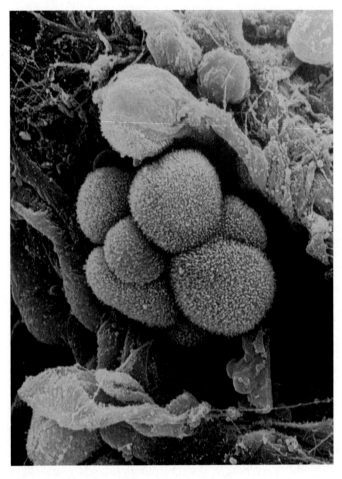

FIGURE 20.10
Lung cancer cells. These cells (530×) are from a tumor located in the alveolus (air sac) of a lung.

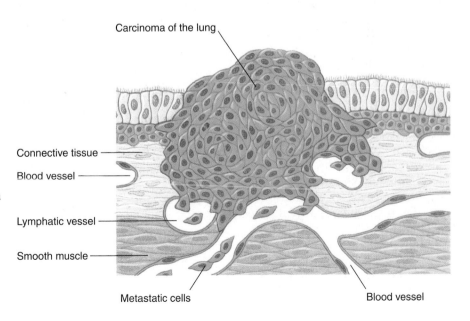

FIGURE 20.11
Portrait of a cancer. This ball of cells is a carcinoma (cancer tumor) developing from epithelial cells that line the interior surface of a human lung. As the mass of cells grows, it invades surrounding tissues, eventually penetrating lymphatic and blood vessels, both plentiful within the lung. These vessels carry metastatic cancer cells throughout the body, where they lodge and grow, forming new masses of cancerous tissue.

Carcinoma of the lung

Connective tissue

Blood vessel

Lymphatic vessel

Smooth muscle

Metastatic cells

Blood vessel

Causes of Cancer

Cancer can occur in almost any tissue, so a bewildering number of different cancers are possible. Tumors arising from cells in connective tissue, bone, or muscle are known as **sarcomas,** while those that originate in epithelial tissue such as skin are called **carcinomas.** In the United States, three of the deadliest human cancers are lung cancer, cancer of the colon and rectum, and breast cancer (table 20.2). Lung cancer, responsible for the most cancer deaths, is largely preventable; most cases result from smoking cigarettes. Colorectal cancers appear to be fostered by the high-meat diets so favored in the United States. The cause of breast cancer is still a mystery, although in 1994 and 1995 researchers isolated two genes responsible for hereditary susceptibility to breast cancer, *BRCA1* and *BRCA2* (breast cancer genes #1 and #2 located on human chromosomes 17 and 13); their discovery offers hope that researchers will soon be able to unravel the fundamental mechanism leading to hereditary breast cancer, which makes up about one-third of all breast cancers.

The association of particular chemicals with cancer, especially those that are potent mutagens, led researchers early on to suspect that cancer might be caused, at least in part, by chemicals, the so-called **chemical carcinogenesis theory.** Agents thought to cause cancer are called **carcinogens.** A simple and effective way to test whether a chemical is mutagenic is the Ames test (figure 20.12), named for its developer, Bruce Ames. The test uses a strain of *Salmonella* bacteria that has a defective histidine-synthesizing gene. Because these bacteria cannot make histidine, they cannot grow on media without it. Only a back-mutation that restores the ability to manufacture histidine will permit growth. Thus, the number of colonies of these bacteria that grow on histidine-free medium is a measure of the frequency of back-mutation. A majority of chemicals that cause back-mutations in this test are carcinogenic, and vice versa. To increase the sensitivity of the test, the strains of bacteria are altered to disable their DNA repair machinery. The search for the cause of cancer has focused in part on chemical carcinogens and other environmental factors, including ionizing radiation such as X rays.

Table 20.2 Incidence of Cancer in the United States in 2002*

Type of Cancer	New Cases	Deaths	% of Cancer Deaths
Lung	169,400	154,900	28
Colon and rectum	148,300	56,600	10
Leukemia/lymphoma	91,700	47,500	9
Breast	205,000	40,000	7
Prostate	189,000	30,200	5
Pancreas	30,300	29,700	5
Ovary	23,300	13,900	3
Liver	16,600	14,100	3
Nervous system/eye	19,200	13,300	2
Stomach	13,100	12,600	2
Bladder	56,500	12,600	2
Kidney	31,800	11,600	2
Cervix/uterus	52,300	10,700	2
Oral cavity	28,900	7,400	1
Malignant melanoma	53,600	7,400	1
Sarcoma (connective tissue)	10,400	5,800	1
All other cancers	145,500	87,200	16

*In the United States in 2002, there were an estimated 1,284,900 reported cases of new cancers and 555,500 cancer deaths, indicating that roughly half the people who develop cancer die from it.
Source: Data from the American Cancer Society, Inc., 2002.

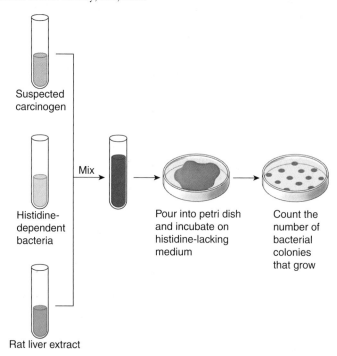

FIGURE 20.12
The Ames test. This test uses a strain of *Salmonella* bacteria that requires histidine in the growth medium due to a mutated gene. If a suspected carcinogen is mutagenic, it can reverse this mutation. Rat liver extract is added because it contains enzymes that can convert carcinogens into mutagens. The mutagenicity of the carcinogen can be quantified by counting the number of bacterial colonies that grow on a medium lacking histidine.

Some Tumors Are Caused by Chemicals

The chemical carcinogenesis theory was first advanced over 200 years ago in 1761 by Dr. John Hill, an English physician, who noted unusual tumors of the nose in heavy snuff users and suggested that tobacco had produced these cancers. In 1775, a London surgeon, Sir Percivall Pott, made a similar observation, noting that men who had been chimney sweeps exhibited frequent cancer of the scrotum, and suggesting that soot and tars might be responsible. British sweeps washed themselves infrequently and always seemed covered with soot. Chimney sweeps on the continent, who washed daily, had much less of this scrotal cancer. These and many other observations led to the hypothesis that cancer results from the action of chemicals on the body.

Demonstrating That Chemicals Can Cause Cancer

Over a century passed before this theory was directly tested. In 1915, the Japanese doctor Katsusaburo Yamagiwa applied extracts of coal tar to the skin of 137 rabbits every two or three days for three months. Then he waited to see what would happen. After a year, cancers appeared at the site of application in seven of the rabbits. Yamagiwa had induced cancer with the coal tar, the first direct demonstration of chemical carcinogenesis. In the decades that followed, this approach demonstrated that many chemicals were capable of causing cancer. Importantly, most of them were potent mutagens.

Because these were lab studies, many people did not accept that the results applied to real people. Do tars in fact induce cancer in humans? In 1949, the American physician Ernst Winder and the British epidemiologist Richard Doll independently reported that lung cancer showed a strong link to cigarette smoking, which introduces tars into the lungs. Winder interviewed 684 lung cancer patients and 600 normal controls, asking whether each had ever smoked. Cancer rates were 40 times higher in heavy smokers than in nonsmokers. Doll's study was even more convincing. He interviewed a large number of British physicians, noting which ones smoked, and then waited to see which would develop lung cancer. Many did. Overwhelmingly, those who did were smokers. From these studies, it seemed likely as long as 50 years ago that tars and other chemicals in cigarette smoke induce cancer in the lungs of persistent smokers. While this suggestion was (and is) resisted by the tobacco industry, the evidence that has accumulated since these pioneering studies makes a clear case, and there is no longer any real doubt. Chemicals in cigarette smoke cause cancer.

Carcinogens Are Common

In ongoing investigations over the last 50 years, many hundreds of synthetic chemicals have been shown capable of causing cancer in laboratory animals. Among them are

Table 20.3	Chemical Carcinogens in the Workplace	
Chemical	**Cancer**	**Workers at Risk for Exposure**
COMMON EXPOSURE		
Benzene	Myelogenous leukemia	Painters; dye users; furniture finishers
Diesel exhaust	Lung	Railroad and bus-garage workers; truckers; miners
Mineral oils	Skin	Metal machinists
Pesticides	Lung	Sprayers
Cigarette tar	Lung	Smokers
UNCOMMON EXPOSURE		
Asbestos	Mesothelioma, lung	Brake-lining, insulation workers
Synthetic mineral fibers	Lung	Wall and pipe insulation and duct wrapping users
Hair dyes	Bladder	Hairdressers and barbers
Paint	Lung	Painters
Polychlorinated biphenyls	Liver, skin	Users of hydraulic fluids and lubricants, inks, adhesives, insecticides
Soot	Skin	Chimney sweeps; bricklayers; firefighters; heating-unit service workers
RARE EXPOSURE		
Arsenic	Lung, skin	Insecticide/herbicide sprayers; tanners; oil refiners
Formaldehyde	Nose	Wood product, paper, textile, and metal product workers.

trichloroethylene, asbestos, benzene, vinyl chloride, arsenic, arylamide, and a host of complex petroleum products with chemical structures resembling chicken wire. People in the workplace encounter chemicals daily (table 20.3).

In addition to identifying potentially dangerous substances, what have the studies of potential carcinogens told us about the nature of cancer? What do these cancer-causing chemicals have in common? *They are all mutagens, each capable of inducing changes in DNA.*

Cancers occur in all tissues. Chemicals that produce mutations in DNA are often potent carcinogens. Tars in cigarette smoke, for example, are the direct cause of most lung cancers.

Cancer and the Cell Cycle

In chapter 11, we learned about the cell cycle and its control. Given the intricate interplay of regulatory elements involved in the cell cycle, it is not surprising that mutations that affect this regulatory system can lead to uncontrolled cell growth: cancer. Two general categories of genes are affected by such mutations: proto-oncogenes and tumor-suppressor genes. Before examining current attempts to counteract the effects of cancer-causing mutations to somatic DNA, we will briefly consider how mutations in cell cycle control genes can lead to the malignant transformation that is the basis of many forms of cancer in humans (table 20.4).

Oncogenes

Genes that when introduced into a normal cell cause it to become a cancer cell are called **oncogenes** (Greek *onco*, "tumor"). Oncogenes were originally discovered by the process of **transfection,** in which DNA isolated from one cell is introduced into another cell. In transfection experiments, DNA isolated from tumor-derived cells led to the transformation of normal cells into cancer cells. These same genes, in mutated form, were soon isolated from a variety of tumors. The normal, nonmutated forms of these genes came to be called **proto-oncogenes,** or genes that can be mutated to produce a tumor-forming oncogene. Subsequent study has shown that these genes all encode proteins that are necessary for the cell to interpret external signals for growth (figure 20.13). As discussed in chapter 11, these signals are necessary for a cell to leave G_0 and to pass the G_1 checkpoint. Mutations in proto-oncogenes accelerate the cell cycle by amplifying these signals.

If a growth factor receptor is mutated such that it is always "on" regardless of binding to the growth factor, the gene encoding the receptor will act as an oncogene. Similarly, all of the genes encoding the proteins necessary to carry this signal from the surface of the cell to the nucleus can be mutated to greatly amplify the signal, effectively converting the signal pathway genes to oncogenes.

As oncogenes produce their effect in the absence of the normal "divide" signal, they will be inherited as dominant alleles. That is, they will exert their cancer-causing effect even in cells heterozygous for the mutant oncogene.

Chromosomal abnormalities can lead to the activation of proto-oncogenes. A rearrangement that moves a strong enhancer near a proto-oncogene can lead either to over-expression or to expression in a tissue where the proto-oncogene normally is not expressed. This was first observed in chronic myelogenous leukemia where an abnormal chromosome, called the Philadelphia chromosome, is produced when chromosomes 9 and 22 exchange genetic information. The translocation of a portion of the chromosome creates a chimeric gene that includes the proto-oncogene *c-ABL*, causing *c-ABL* to be expressed and producing leukemia.

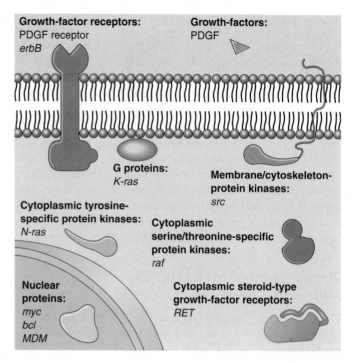

FIGURE 20.13
The main classes of oncogenes. Before they are altered by mutation to their cancer-causing condition, oncogenes are called proto-oncogenes (that is, genes able to become oncogenes). Illustrated here are the principal classes of proto-oncogenes, with some typical representatives indicated.

Among the most widely studied oncogenes are *myc* and *ras*. Expression of *myc* stimulates the production of cyclins and cyclin-dependent protein kinases (Cdk's), key elements in regulating the checkpoints of cell division.

The *ras* gene product is involved in the cellular response to a variety of growth factors, including EGF, an intercellular signal that normally initiates cell proliferation. When EGF binds to a specific receptor protein on the plasma membrane of epithelial cells, the portion of the receptor that protrudes into the cytoplasm stimulates the Ras protein to bind to GTP. The Ras protein/GTP complex in turn recruits and activates a protein called Raf to the inner surface of the plasma membrane, which in turn activates cytoplasmic kinases and so triggers an intracellular signaling system (see chapter 7). The final step is the activation of transcription factors that trigger cell proliferation. Cancer-causing mutations in *ras* greatly reduce the amount of EGF necessary to initiate cell proliferation.

Tumor-Suppressor Genes

If the first class of cancer-inducing mutations "steps on the accelerator" of cell division, the second class of cancer-inducing mutations "removes the brakes." Cell division is normally turned off in healthy cells by proteins that pre-

Table 20.4 Some Genes Implicated in Human Cancers

Gene	Product	Cancer
ONCOGENES		
Genes Encoding Growth Factors or Their Receptors		
erb-B	Receptor for epidermal growth factor	Glioblastoma (a brain cancer); breast cancer
erb-B2	A growth factor receptor (gene also called *neu*)	Breast cancer; ovarian cancer; salivary gland cancer
PDGF	Platelet-derived growth factor	Glioma (a brain cancer)
RET	A growth factor receptor	Thyroid cancer
Genes Encoding Cytoplasmic Relays in Intracellular Signaling Pathways		
K-ras	Protein kinase	Lung cancer; colon cancer; ovarian cancer; pancreatic cancer
N-ras	Protein kinase	Leukemias
Genes Encoding Transcription Factors That Activate Transcription of Growth-Promoting Genes		
c-myc	Transcription factor	Lung cancer; breast cancer; stomach cancer; leukemias
L-myc	Transcription factor	Lung cancer
N-myc	Transcription factor	Neuroblastoma (a nerve cell cancer)
Genes Encoding Other Kinds of Proteins		
bcl-2	Protein that blocks cell suicide	Follicular B cell lymphoma
bcl-1	Cyclin D1, which stimulates the cell cycle clock (gene also called *PRAD1*)	Breast cancer; head and neck cancers
MDM2	Protein antagonist of p53 tumor-suppressor protein	Wide variety of sarcomas (connective tissue cancers)
TUMOR-SUPPRESSOR GENES		
Genes Encoding Cytoplasmic Proteins		
APC	Step in a signaling pathway	Colon cancer; stomach cancer
DPC4	A relay in signaling pathway that inhibits cell division	Pancreatic cancer
NF-1	Inhibitor of Ras, a protein that stimulates cell division	Neurofibroma; myeloid leukemia
NF-2	Inhibitor of Ras	Meningioma (brain cancer); schwannoma (cancer of cells supporting peripheral nerves)
Genes Encoding Nuclear Proteins		
MTS1	p16 protein, which slows the cell cycle clock	A wide range of cancers
p53	p53 protein, which halts cell division at the G_1 checkpoint	A wide range of cancers
Rb	Rb protein, which acts as a master brake of the cell cycle	Retinoblastoma; breast cancer; bone cancer; bladder cancer
Genes Encoding Proteins of Unknown Cellular Locations		
BRCA1	?	Breast cancer; ovarian cancer
BRCA2	?	Breast cancer
VHL	?	Renal cell cancer

vent cyclins from binding to Cdk's. The genes that encode these proteins are called **tumor-suppressor genes.** Their mutant alleles are genetically recessive.

One of the first tumor-suppressor genes to be discovered was the *Rb* gene discussed in chapter 11. The role of *Rb* is to bind to the transcription factor E2F, preventing its activity. The E2F transcription factor is necessary for the expression of a number of cell-cycle-specific genes. In the normal course of the cell cycle, the action of cyclin-dependent kinase acts to phosphorylate Rb. Because phosphorylated Rb cannot bind E2F, this releases E2F from its inhibition and allows the cell cycle to progress.

A loss of normal Rb function leads to the loss of control over the cell cycle (figure 20.14). However, both normally functioning copies of *Rb* must be lost before control is removed. Thus, the cancer-inducing mutant alleles of *Rb* are genetically recessive.

Another well-characterized tumor-suppressor gene is *p53*, a gene found to be mutated in a large proportion of human cancers. The role of *p53* is to integrate signals that sense DNA damage during G_1 and G_2 (figure 20.15). When significant DNA damage is detected, *p53* induces apoptosis to remove the damaged cell. Loss of *p53* leads to accumulation of mutations that would have otherwise been removed, and thus failure to prevent transformation to cancerous growth. One of the reasons repeated smoking leads inexorably to lung cancer is that it induces *p53* mutations. Indeed, almost half of all cancers involve mutations of the *p53* gene.

Epigenetic Events and Cancer

Recent research on the relationship between chromatin structure and the regulation of gene expression also shed light on the process of malignant transformation. The role of methylation and histone deacetylation discussed in chapter 18 has a bearing on cancer as well. The majority of methylation of mammalian DNA is of cytosine at sites

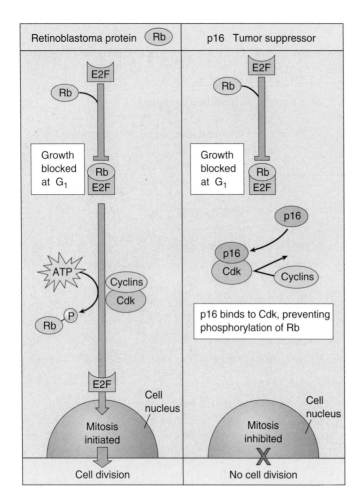

FIGURE 20.14

How the tumor-suppressor genes *Rb* and *p16* interact to block cell division. The retinoblastoma protein (Rb) binds to the transcription factor (E2F) that activates genes in the nucleus, preventing this factor from initiating mitosis. The G_1 checkpoint is passed when Cdk interacts with cyclins to phosphorylate Rb, releasing E2F. The p16 tumor-suppressor protein reinforces Rb's inhibitory action by binding to Cdk so that Cdk is not available to phosphorylate Rb.

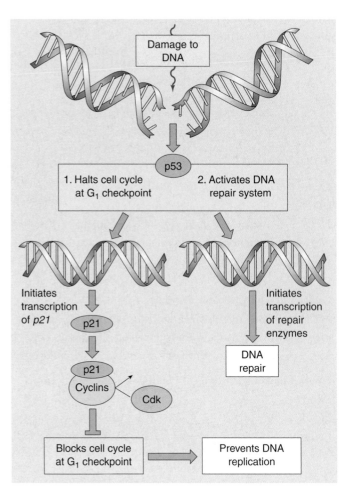

FIGURE 20.15

The role of tumor-suppressor *p53* in regulating the cell cycle. The p53 protein works at the G_1 checkpoint to check for DNA damage. If the DNA is damaged, p53 activates the DNA repair system and stops the cell cycle at the G_1 checkpoint (before DNA replication). This allows time for the damage to be repaired. p53 stops the cell cycle by inducing the transcription of *p21*. The p21 protein then binds to cyclins and prevents them from complexing with Cdk.

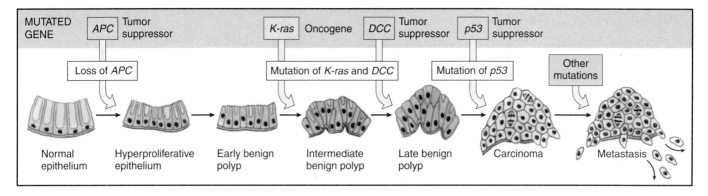

FIGURE 20.16

The progression of mutations that commonly lead to colorectal cancer. The fatal metastasis is the last of six serial changes that the epithelial cells lining the rectum undergo. One of these changes is brought about by mutation of a proto-oncogene, and three of them involve mutations that inactivate tumor-suppressor genes.

where cytosine is followed by guanine in the DNA sequence (called CpG repeat sites). Most of the CpG repeats found in humans are methylated. Those that are not, so-called "CpG islands," correlate with genes that are expressed. Recently, hypermethylation of CpG islands has been observed in a number of human cancers. Although it is not yet entirely clear that hypermethylation is a direct cause of cancer, rather than a result of it, the finding that key tumor-suppressor genes may exhibit this hypermethylation strengthens the case for this epigenetic process as a contributing cause. And even if the hypermethylation is not the cancer-initiating event, aberrant methylation patterns in cancer cells may shut off additional genes, accelerating the process. This provides another potential mechanism of inactivation of tumor-suppressor genes to add to our growing list: point mutation, structural alterations of chromosomes, and now epigenetic control of gene expression.

Cancer Is a Multistep Process

Cells control proliferation at several checkpoints, and all of these controls must be inactivated for cancer to be initiated. Therefore, the induction of most cancers involves the mutation of multiple genes; four to six is a typical number (figure 20.16). In many of the tissue culture cell lines used to study cancer, most of the controls are already inactivated, so that mutations in only one or a few genes transform the line into cancerous growth. The need to inactivate several regulatory genes almost certainly explains why most cancers occur in people over 40 years of age (figure 20.17); in older persons, individual cells have had more time to accumulate multiple mutations. It is now clear that mutations, including those in potentially cancer-causing genes, do accumulate over time. Using the polymerase chain reaction (PCR), researchers in 1994 searched for a certain cancer-associated gene mutation in the blood cells of 63 cancer-free people. They found that the mutation occurred 13 times more often in people over 60 years old than in people under 20.

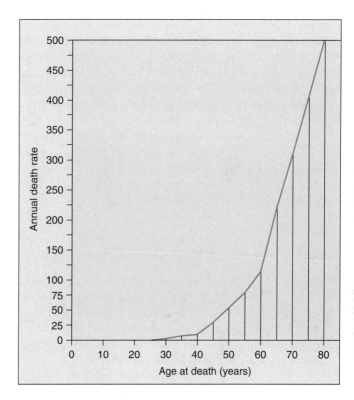

FIGURE 20.17

The annual death rate from cancer climbs with age. The rate of cancer deaths increases steeply after age 40 and even more steeply after age 60, suggesting that several independent mutations must accumulate to give rise to cancer.

Cancer is a disease in which the controls that normally restrict cell proliferation do not operate. In some cases, cancerous growth is initiated by the inappropriate activation of proteins that regulate the cell cycle; in other cases, it is initiated by the inactivation of proteins that normally suppress cell division.

Smoking and Cancer

How can we prevent cancer? The most obvious strategy is to minimize mutational insult. Anything that decreases exposure to mutagens can decrease the incidence of cancer because exposure has the potential to mutate a normal gene into an oncogene. It is no accident that the most reliable tests for the carcinogenicity of a substance are those that measure the substance's mutagenicity.

The Association between Smoking and Cancer

About one-third of all cases of cancer in the United States are directly attributable to cigarette smoking. The association between smoking and cancer is particularly striking for lung cancer (figure 20.18). Studies of male smokers show a highly positive correlation between the number of cigarettes smoked per day and the incidence of lung cancer (figure 20.19). For individuals who smoke two or more packs a day, the risk of contracting lung cancer is at least 40 times greater than it is for nonsmokers, whose risk level approaches zero. Clearly, an effective way to avoid lung cancer is not to smoke. Other studies have shown a clear relationship between cigarette smoking and reduced life expectancy (figure 20.20). Life insurance companies have calculated that smoking a single cigarette lowers one's life expectancy by 10.7 minutes—longer than it takes to smoke the cigarette! Every pack of 20 cigarettes bears an unwritten label:

"The price of smoking this pack of cigarettes is 3½ hours of your life."

Smoking Introduces Mutagens into the Lungs

Over half a million people died of cancer in the United States in 2002; about 28% of them died of lung cancer. In the 1980s, about 140,000 persons were diagnosed with lung cancer each year. Around 90% of them died within three years after diagnosis; 96% of them were cigarette smokers.

Smoking is a popular pastime. In 2001 in the United States, 23.5% of adults and 31% of teens smoked, and U.S. smokers consumed 420 billion cigarettes in 2000. The smoke emitted from these cigarettes contains some 3000 chemical components, including vinyl chloride, benzo[*a*]pyrenes, and nitroso-*nor*-nicotine, all potent mutagens. Smoking places these mutagens into direct contact with the tissues of the lungs.

Mutagens in the Lungs Cause Cancer

Introducing powerful mutagens into the lungs causes considerable damage to the genes of the epithelial cells that line the lungs and are directly exposed to the chemicals. Among the genes that are mutated as a result are some whose normal function is to regulate cell proliferation. When these genes are damaged, lung cancer results.

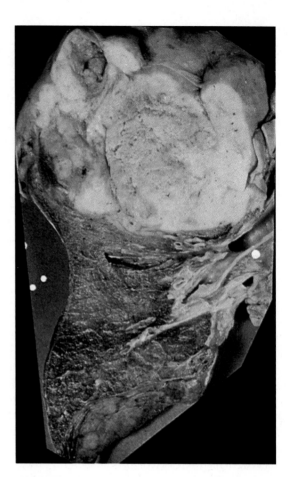

FIGURE 20.18
Photo of a cancerous human lung. The bottom half of the lung is normal, while a cancerous tumor has completely taken over the top half. The cancer cells will eventually break through into the lymph and blood vessels and spread through the body.

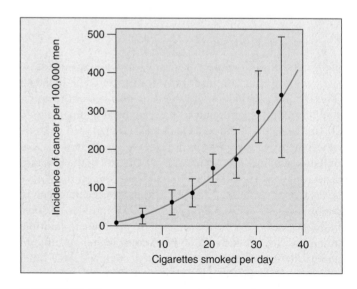

FIGURE 20.19
Smoking causes cancer. The annual incidence of lung cancer per 100,000 men clearly increases with the number of cigarettes smoked per day.

This process has been clearly demonstrated for benzo[*a*]pyrene (BP), one of the potent mutagens released into cigarette smoke from tars in the tobacco. The epithelial cells of the lung absorb BP from tobacco smoke and chemically alter it to a derivative form. This derivative form, benzo[*a*]pyrene-diolepoxide (BPDE), binds directly to the tumor-suppressor gene *p53* and mutates it to an inactive form. The protein encoded by *p53* oversees the G_1 cell cycle checkpoint described in chapter 11 and is one of the body's key mechanisms for preventing uncontrolled cell proliferation. The destruction of *p53* in lung epithelial cells greatly hastens the onset of lung cancer; *p53* is mutated to an inactive form in over 70% of lung cancers. When examined, the *p53* mutations in cancer cells almost all occur at one of three "hotspots." The key evidence linking smoking and cancer is that when the mutations of *p53* caused by BPDE from cigarettes are examined, they occur at the same three specific "hotspots!"

The Incidence of Cancer Reflects Smoking

Cigarette manufacturers argue that the causal connection between smoking and cancer has not been proved, and that somehow the relationship is coincidental. Look carefully at the data presented in figure 20.21, and see if you agree. The upper graph, compiled from data on American men, shows the incidence of smoking from 1900 to 1990 and the incidence of lung cancer over the same period. Note that as late as 1920, lung cancer was a rare disease. About 20 years after the incidence of smoking began to increase among men, lung cancer also started to become more common.

Now look at the lower graph, which presents data on American women. Because of social mores, significant numbers of American women did not smoke until after World War II, when many social conventions changed. As late as 1963, when lung cancer among males was near current levels, this disease was still rare in women. In the United States that year, only 6588 women died of lung cancer. But as more women smoked, more developed lung cancer, again with a lag of about 20 years. American women today have achieved equality with men in the numbers of cigarettes they smoke, and their lung cancer death rates are today approaching those for men. In 2002, more than 65,000 women died of lung cancer in the United States. The current annual rate of deaths from lung cancer in male and female smokers is 180 per 100,000, or about 2 out of every 1000 smokers *each year*.

The easiest way to avoid cancer is to avoid exposure to mutagens. The single greatest contribution one can make to a longer life is not to smoke.

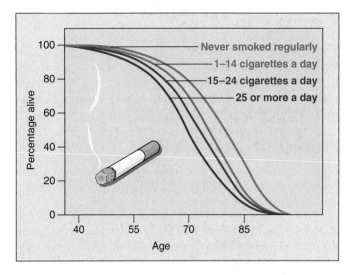

FIGURE 20.20
Tobacco reduces life expectancy. The world's longest-running survey of smoking, begun in 1951 in Britain, revealed that by 1994 the death rate for smokers had climbed to three times the rate for nonsmokers among men 35 to 69 years of age.
Source: Data from *New Scientist*, October 15, 1994.

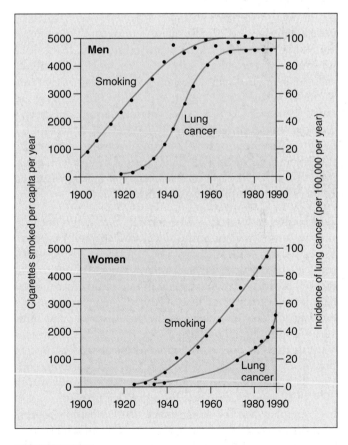

FIGURE 20.21
The incidence of lung cancer in men and women. What do these graphs indicate about the connection between smoking and lung cancer?

Curing Cancer

Potential cancer therapies are being developed on many fronts; eight of these targets are described here and shown in figure 20.22. Some therapies act to prevent the start of cancer within cells. Others act outside cancer cells, preventing tumors from growing and spreading.

Preventing the Start of Cancer

Many promising cancer therapies act within potential cancer cells, focusing on different stages of the cell's "Shall I divide?" decision-making process.

1. Receiving the Signal to Divide.   The first step in the decision process is the reception of a "divide" signal, usually a small protein called a growth factor released from a neighboring cell. The growth factor is received by a protein receptor on the cell surface. Mutations that increase the number of receptors on the cell surface amplify the division signal and so lead to cancer. For example, over 20% of breast cancer tumors have been proven to overproduce a protein called HER2 associated with the receptor for epidermal growth factor.

Therapies directed at this stage of the decision process utilize the human immune system to attack cancer cells. Special protein molecules called *monoclonal antibodies*, created by genetic engineering, are the therapeutic agents. Monoclonal antibodies are designed to seek out and stick to HER2. Like waving a red flag, the presence of the monoclonal antibody calls down attack by the immune system on the HER2 cell. Because breast cancer cells overproduce HER2, they are killed preferentially. The drug manufacturer Genentech's recently approved monoclonal antibody, called "herceptin," has given promising results in clinical tests. In other tests, the monoclonal antibody C225, directed against epidermal growth factor receptors, has succeeded in curing advanced colon cancer. Clinical trials of C225 have begun.

2. The Relay Switch.   The second step in the decision process is the passage of the signal into the cell's interior, the cytoplasm. This is carried out in normal cells by a protein called Ras that acts as a relay switch. When growth factor binds to a receptor such as EGF, the adjacent Ras protein acts like it has been "goosed," contorting into a new shape. This new shape is chemically active, and initiates a chain of reactions that passes the "divide" signal inward toward the nucleus. Mutated forms of the Ras protein behave like a relay switch stuck in the "on" position, continually instructing the cell to divide when it should not. In 30% of all cancers, a mutant form of Ras is present.

Therapies directed at this stage of the decision process take advantage of the fact that normal Ras proteins are inactive when made. Only after it has been modified by the special enzyme *farnesyl transferase* does Ras protein become able to function as a relay switch. In tests on animals, farnesyl transferase inhibitors induce the regression of tumors and prevent the formation of new ones.

3. Amplifying the Signal.   The third step in the decision process is the amplification of the signal within the cytoplasm. Just as a TV signal needs to be amplified in order to be received at a distance, so a "divide" signal must be amplified if it is to reach the nucleus at the interior of the cell, a very long journey at a molecular scale. Cells use an ingenious trick to amplify the signal. Ras, when "on," activates an enzyme, a protein kinase. This protein kinase activates other protein kinases that in their turn activate still others. The trick is that once a protein kinase enzyme is activated, it goes to work like a demon, activating hoards of others every second! And each and every one it activates behaves the same way, activating still more, in a cascade of ever-widening effect. At each stage of the relay, the signal is amplified 1000-fold. Mutations stimulating any of the protein kinases can dangerously increase the already amplified signal and lead to cancer. For example, 5% of all cancers have a mutant hyperactive form of the protein kinase Src.

Therapies directed at this stage of the decision process employ so-called "antisense RNA" directed specifically against Src or other cancer-inducing kinase mutations. The idea is that the *src* gene uses a complementary copy of itself to manufacture the Src protein (the "sense" RNA or messenger RNA), and a mirror image complementary copy of the sense RNA ("antisense" RNA) will stick to it, gumming it up so it can't be used to make Src protein. The approach appears promising. In tissue culture, antisense RNAs inhibit the growth of cancer cells, and some also appear to block the growth of human tumors implanted in laboratory animals. Human clinical trials are under way.

4. Releasing the Brake.   The fourth step in the decision process is the removal of the "brake" the cell uses to restrain cell division. In healthy cells, this brake, a tumor-suppressor protein called Rb, blocks the activity of a transcription factor protein called E2F. When free, E2F enables the cell to copy its DNA. Normal cell division is triggered to begin when Rb is inhibited, unleashing E2F. Mutations that destroy Rb release E2F from its control completely, leading to ceaseless cell division. A defective form of Rb is found in 40% of all cancers.

Therapies directed at this stage of the decision process are only now being attempted. They focus on drugs able to inhibit E2F, which should halt the growth of tumors arising from inactive Rb. Experiments in mice in which the E2F genes have been destroyed provide a model system to study such drugs, which are being actively investigated.

5. Checking That Everything Is Ready.   The final step in the decision process is the mechanism used by the cell to ensure that its DNA is undamaged and ready to divide. This job is carried out in healthy cells by the tumor-suppressor protein p53, which inspects the integrity of the DNA. When it detects damaged or foreign DNA, p53 stops cell division and activates the cell's DNA repair systems. If the damage doesn't get repaired in a reasonable time, p53 pulls the plug, triggering events that kill the cell. In this way, mutations such as those that cause cancer are either repaired, or the cells containing them eliminated. If p53 is itself destroyed by mutation, future damage accumulates unrepaired. Among this damage are mutations that lead to cancer. Approximately 50% of all cancers have a disabled p53. Fully 70 to 80% of lung cancers have a mutant inactive p53; the chemical benzo[*a*]pyrene in cigarette smoke is a potent mutagen of p53.

A promising new therapy using adenovirus (responsible for mild colds) is being targeted at cancers with a mutant p53. To grow in a host cell, adenovirus must use the product of its gene, *E1B*, to block the host cell's p53, thereby enabling replication of the adenovirus DNA. This means that while mutant adenovirus without *E1B* cannot grow in healthy cells, the mutants should be able to grow in, and destroy, cancer cells with defective p53. When human colon and lung cancer cells are introduced into mice lacking an immune system and allowed to produce substantial tumors, 60% of the tumors simply disappear when treated with E1B-deficient adenovirus, and do not reappear later. Initial clinical trials are very encouraging.

6. Stepping on the Gas.   Cell division starts with replication of the DNA. In healthy cells, another tumor suppressor "keeps the gas tank nearly empty" for the DNA replication process by inhibiting production of an enzyme called telomerase. Without this enzyme, a cell's chromosomes lose material from their tips, called telomeres. Every time a chromosome is copied, more tip material is lost. After some 30 divisions, so much is lost that copying is no longer possible. Cells in the tissues of an adult human have typically undergone 25 or more divisions. Cancer can't get very far with only the five remaining cell divisions, so inhibiting telomerase is a very effective natural brake on the cancer process. It is thought that almost all cancers involve a mutation that destroys the telomerase inhibitor, releasing this brake and making cancer possible. It should be possible to block cancer by reapplying this inhibition. Cancer therapies that inhibit telomerase are just beginning clinical trials.

Preventing the Spread of Cancer

7. Tumor Growth.   Once a cell begins cancerous growth, it forms an expanding tumor. As the tumor grows ever larger, it requires an increasing supply of food and nutrients, obtained from the body's blood. To facilitate this necessary grocery shopping, tumors leak substances into the surrounding tissues that encourage angiogenesis, the formation of small blood vessels. Chemicals that inhibit this process are called angiogenesis inhibitors. In mice, two such angiogenesis inhibitors, angiostatin and endostatin, caused tumors to regress to microscopic size. This very exciting result has proven controversial, but initial human trials seem promising.

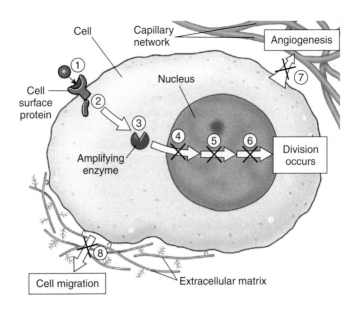

FIGURE 20.22
Potential targets for cancer therapies. New molecular therapies target eight different stages in the cancer process. (*1*) On the cell surface, a growth factor signals the cell to divide. (*2*) Just inside the cell, a protein relay switch passes on the divide signal. (*3*) In the cytoplasm, enzymes amplify the signal. In the nucleus, (*4*) a "brake" preventing DNA replication is released, (*5*) proteins check that the replicated DNA is not damaged, and (*6*) other proteins rebuild chromosome tips so DNA can replicate. (*7*) The new tumor promotes angiogenesis, the formation of growth-promoting blood vessels. (*8*) Some cancer cells break away from the extracellular matrix and invade other parts of the body.

8. Metastasis.   If cancerous tumors simply continued to grow where they form, many could be surgically removed, and far fewer would prove fatal. Unfortunately, many cancerous tumors eventually metastasize; that is, individual cancer cells break their moorings to the extracellular matrix and spread to other locations in the body where they initiate formation of secondary tumors. This process involves (1) metal-requiring protease enzymes that cleave the cell-matrix linkage, (2) components of the extracellular matrix such as fibronectin that also promote the migration of several noncancerous cell types, and (3) RhoC, a GTP-hydrolyzing enzyme that promotes cell migration by providing needed GTP. All of these components are necessary for metastasis to occur, and all offer promising targets for future anticancer therapy.

Therapies such as those described here are only part of a wave of potential treatments under development and clinical trial. The clinical trials will take years to complete, but in the coming decade we can expect cancer to become a curable disease.

Our understanding of how mutations produce cancer has progressed to the point where promising potential therapies can be tested.

20.4 Reproductive cloning of animals, once thought impossible, isn't.

The Challenge of Cloning

The promise of genetic engineering has had little impact on the one area of medicine where it might be expected to have had the greatest impact—reproduction. This is now changing, led by surprising advances in cloning commercial animals. The difficulty in using genetic engineering to improve livestock is in getting enough animals. Breeding genetically improved individuals produces offspring only slowly. Ideally, one would like to "Xerox" many exact genetic copies of the desirable strain, but adult animals can't be cloned—or can they? At one time, it was commonly accepted that they couldn't be, but we now know that conclusion was unwarranted.

Wilmut's Lamb

Although earlier attempts to clone animals using adult cells generally failed, the key advance was made in Scotland by Keith Campbell, a geneticist studying the cell cycle of agricultural animals. Campbell reasoned, "Maybe the egg and the donated nucleus need to be at the same stage in the cell cycle." This proved to be a key insight. In 1994, researcher Neil First, and in 1995 Campbell himself working with reproductive biologist Ian Wilmut, succeeded in cloning farm animals from advanced embryos by first starving the cells, so that they paused at the beginning of the cell cycle at the G_1 checkpoint. Two starved cells are thus synchronized at the same point in the cell cycle.

Wilmut then attempted the key breakthrough, the experiment that had eluded researchers since Spemann proposed it 59 years before: He set out to transfer the nucleus from an adult differentiated cell into an enucleated egg, and to allow the resulting embryo to grow and develop in a surrogate mother, hopefully producing a healthy animal.

Wilmut removed mammary cells from the udder of a six-year-old sheep (figure 20.23). (With tongue in cheek, the clone was later named "Dolly" after the country singer Dolly Parton.) The cells were grown in tissue culture; some were frozen so that in the future it would be possible with genetic fingerprinting to prove that a clone was indeed genetically identical to the six-year-old ewe.

In preparation for cloning, Wilmut's team reduced for five days the concentration of serum on which the sheep mammary cells were subsisting. In parallel preparation, eggs obtained from another ewe were enucleated, the nucleus of each egg carefully removed with a micropipette.

Mammary cells and egg cells were then surgically combined in January of 1996, a mammary cell being inserted inside the covering around each egg cell. Wilmut then applied a brief electrical shock. A neat trick, this causes the plasma membranes surrounding the two cells to become leaky, so that the contents of the mammary cell pass into the egg cell. The shock also jump-starts the cell cycle, initiating cell division.

After six days, in 30 of 277 tries, the dividing embryo reached the hollow-ball blastula stage, and 29 of these were transplanted into surrogate mother sheep. Approximately five months later, on July 5, 1997, one sheep gave birth to a lamb. This lamb, "Dolly," was the first successful clone generated from a differentiated animal cell.

Wilmut's successful cloning of fully differentiated sheep cells is a milestone event in gene technology. Even though his procedure proved inefficient (only one of 277 trials succeeded), it established the point beyond all doubt that adult animal cells *can* be cloned. In subsequent years, researchers

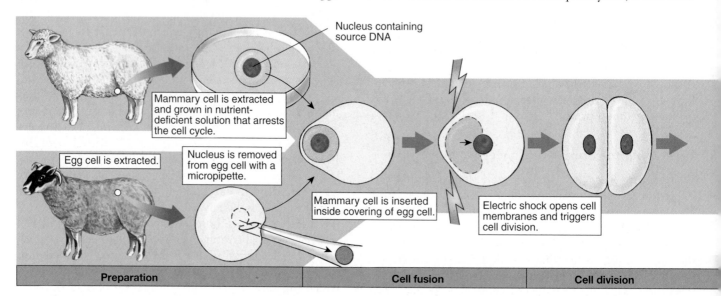

FIGURE 20.23

Wilmut's animal cloning experiment. Wilmut combined a nucleus from a mammary cell with an enucleated egg cell to successfully clone a sheep, named Dolly, that grew to be a normal adult and bore healthy offspring before dying of lung disease in 2003.

have succeeded in greatly improving the efficiency of cloning. Unfortunately, as we will see, problems often develop in the cloned animals.

Problems with Reproductive Cloning

Since Dolly's birth in 1997, scientists have successfully cloned sheep, mice, cattle, goats, and pigs. However, only a small percentage of the transplanted embryos survive to term; most die late in pregnancy. Those that survive to be born usually die soon thereafter. Many become oversized, a condition known as *large offspring syndrome.*

The few cloned offspring that reach childhood face an uncertain future, as their development into adults tends to go unexpectedly haywire.

The Importance of Genomic Imprinting

What is going wrong? It turns out that as human eggs and sperm mature, their DNA is conditioned by the parent female or male, a process called reprogramming. Chemical changes are made to the DNA that alter when particular genes are expressed without changing DNA sequences.

In the years since Dolly, scientists have learned a lot about reprogramming. It appears to occur by a process called *genomic imprinting.* While the details are complex, the basic mechanism of genomic imprinting is simple.

Like a book, a gene can have no impact unless it is read. Genomic imprinting works by blocking the cell's ability to read certain genes. A gene is locked in the "off" position by chemically altering some of the cytosine DNA units. Because this involves adding a $-CH_3$ group (a methyl group), the process is called *methylation.* After a gene has been methylated, the polymerase protein that is supposed to "read" the gene can no longer recognize it. The gene has been shut off.

Genomic imprinting can also lock genes in the "on" position, permanently activating them. This process also uses methylation; in this case, however, it is not the gene that is blocked. Rather, a DNA sequence that normally would have prevented the gene from being read is blocked.

Why Cloning Fails

Normal human development depends on precise genomic imprinting. This chemical reprogramming of the DNA, which occurs in adult reproductive tissue, takes months for sperm and years for eggs.

During cloning, by contrast, the reprogramming of the donor DNA must occur within a few minutes. After the donor nucleus is added to an egg whose nucleus has been removed, the reconstituted egg begins to divide within minutes, starting the process of making a new individual.

Cloning fails because there is simply not enough time in these few minutes to get the reprogramming job done properly. For example, Lorraine Young of the Roslin Institute in Scotland (Dolly's birthplace) reported in 2001 that in Large Offspring Syndrome sheep, many genes have failed to become properly methylated.

Human cloning will not be practical until scientists figure out how to reprogram a donor nucleus, as occurs in the DNA of sperm or eggs in our bodies. This reprogramming may be as simple as finding a way to postpone the onset of cell division after adding a donor nucleus to the enucleated egg, or may prove to be a much more complex process.

> **Recent experiments have demonstrated the possibility of cloning differentiated mammalian tissue. Reproductive cloning of animals from adult tissue usually fails for lack of proper gene conditioning.**

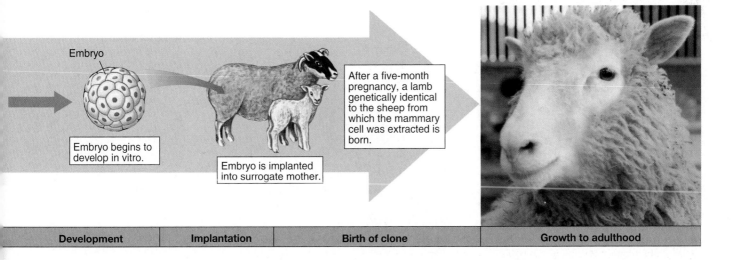

Embryo

Embryo begins to develop in vitro.

Embryo is implanted into surrogate mother.

After a five-month pregnancy, a lamb genetically identical to the sheep from which the mammary cell was extracted is born.

| Development | Implantation | Birth of clone | Growth to adulthood |

Therapeutic cloning is a promising but controversial possibility.

Stem Cells

Embryonic Stem Cells

In 1981, mouse stem cells were first discovered to be *pluripotent*—to have the ability to form any body tissue, and even an adult animal. This finding launched the era of stem cell research. After many years of failed attempts, human embryonic stem cells were isolated by James Thomson of the University of Wisconsin in 1998.

What is an embryonic stem cell? At the dawn of a human life, a sperm fertilizes an egg to create a single cell destined to become a child. As development commences, that cell begins to divide, producing after five or six days a small ball of a few hundred cells called a blastocyst. Described in chapter 51, a blastocyst consists of a protective outer layer destined to form the placenta, enclosing an inner cell mass of **embryonic stem cells** (figure 20.24). Each embryonic stem cell is capable by itself of developing into a healthy individual. In cattle breeding, for example, these cells are frequently separated by the breeder and used to produce multiple clones of valuable offspring.

Because they can develop into any tissue, these embryonic stem cells offer the exciting possibility of restoring damaged tissues, such as muscle or nerve tissue (figure 20.25). Experiments have already been tried successfully in mice. Heart muscle cells grown from mouse embryonic stem cells have been successfully integrated with the heart tissue of a living mouse. This suggests that the damaged heart muscle of heart attack victims might be repairable with stem cells. In other experiments with mice, damaged spinal neurons have been partially repaired, suggesting a

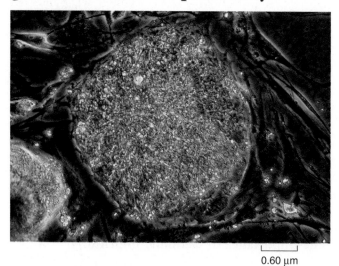

0.60 µm

FIGURE 20.24
Human embryonic stem cells (20×). Stem cells removed from a six-day blastocyst can be established in culture and then main-tained indefinitely. This mass is a colony of undifferentiated human embryonic stem cells surrounded by fibroblasts (elongated cells) that serve as a "feeder layer."

path to treating spinal injuries. Dopamine-producing neurons of mouse brains whose progressive loss is responsible for Parkinson disease have been successfully replaced with embryonic stem cells, as have the islet cells of the pancreas whose loss leads to juvenile diabetes.

These experiments in mice suggest that embryonic stem cell therapy may hold great promise in treating a wide variety of human illnesses involving damaged or lost tissues. This research, however, is quite controversial be-

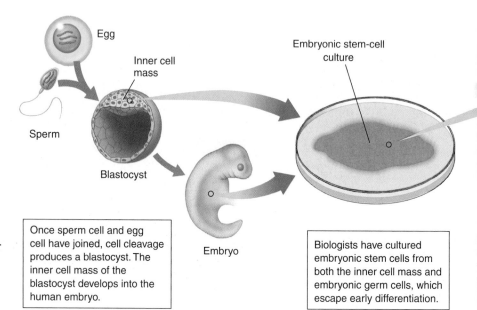

FIGURE 20.25 Using embryonic stem cells to restore damaged tissue.
Embryonic stem cells can develop into any body tissue. Methods are being developed for growing the tissue and using it to repair damaged tissue in adults, such as the brain cells of multiple sclerosis patients, heart muscle, and spinal nerves.

Egg

Sperm

Inner cell mass

Blastocyst

Embryo

Embryonic stem-cell culture

Once sperm cell and egg cell have joined, cell cleavage produces a blastocyst. The inner cell mass of the blastocyst develops into the human embryo.

Biologists have cultured embryonic stem cells from both the inner cell mass and embryonic germ cells, which escape early differentiation.

cause embryonic stem cells are typically isolated from embryos discarded by reproductive clinics, where in vitro fertilization procedures typically produce many more human embryos than can be successfully implanted in the prospective mother's womb. This raises serious ethical issues among those citizens who believe life starts at fertilization.

A second serious problem is associated with using embryonic stem cells to replace defective or lost tissues. All of the successful experiments just described were carried out in mice whose immune systems had been disabled. Had these mice possessed fully functional immune systems, they almost certainly would have rejected the implanted stem cells as foreign. For such stem cell therapy to work in humans, this problem needs to be addressed and solved.

Tissue-Specific Adult Stem Cells

New experimental results hint at a way around the ethical maze presented by the use of stem cells derived from embryos. Go back for a moment to our explanation of how a vertebrate develops. What happens to the embryonic stem cells? They start to take different developmental paths. Some become destined to form nerve tissue and, after this decision is taken, can never produce any other kind of cell. They are then called nerve stem cells. Others become specialized to produce blood, and still others muscle. Each major tissue is represented by its own kind of **tissue-specific stem cell.** Now here's the key point: As development proceeds, these tissue-specific stem cells persist—even in adults. So why not use these adult cells, rather than embryonic stem cells?

Transplanted Tissue-Specific Stem Cells Cure Multiple Sclerosis in Mice

In pathfinding 1999 laboratory experiments by Dr. Evan Snyder of Harvard Medical School, tissue-specific stem cells were able to restore lost brain tissue. He and his co-workers injected neural stem cells (immediate descendants of embryonic stem cells able to become any kind of neural cell) into the brains of newborn mice with a disease resembling multiple sclerosis (MS). These mice lacked the cells that maintain the layers of myelin insulation around signal-conducting nerves. The injected stem cells migrated all over the brain and were able to convert themselves into the missing type of cell. The new cells then proceeded to repair the ravages of the disease by replacing the lost insulation of signal-conducting nerve cells. Many of the treated mice fully recovered. In mice at least, tissue-specific stem cells offer a treatment for MS.

The approach seems very straightforward and should apply to humans. Indeed, blood stem cells are already routinely used in humans to replenish the bone marrow of cancer patients after marrow-destroying therapy. The problem with extending the approach to other kinds of tissue-specific stem cells is that it has not always been possible to find the kind of tissue-specific stem cell needed.

Human embryonic stem cells offer the possibility of replacing damaged or lost human tissues, although the procedures are controversial. Transplanted tissue-specific stem cells may allow us to replace damaged or lost tissue, offering cures for many disorders that cannot now be treated, while avoiding the ethical problems posed by the use of embryonic stem cells.

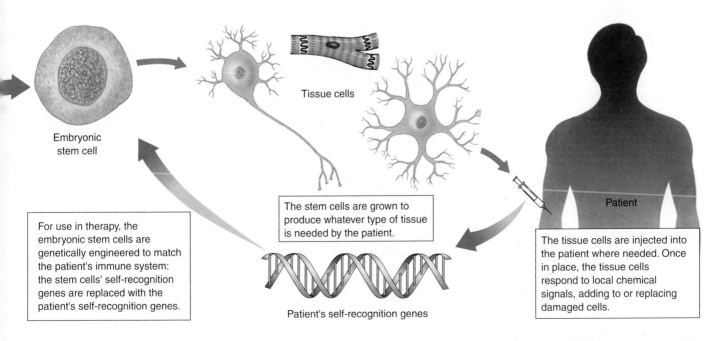

Embryonic stem cell

Tissue cells

Patient

For use in therapy, the embryonic stem cells are genetically engineered to match the patient's immune system: the stem cells' self-recognition genes are replaced with the patient's self-recognition genes.

The stem cells are grown to produce whatever type of tissue is needed by the patient.

Patient's self-recognition genes

The tissue cells are injected into the patient where needed. Once in place, the tissue cells respond to local chemical signals, adding to or replacing damaged cells.

Therapeutic Cloning

By surgically transplanting embryonic stem cells, scientists have performed the remarkable feat of repairing disabled body tissues in mice. The basic strategy for repairing damaged tissues is to surgically transfer embryonic stem cells to the damaged area, where the stem cells can form healthy replacement cells. Stem cells transferred into mouse heart muscle develop into heart muscle cells, replacing cells dead from heart attack. Stem cells transferred into a mouse brain form neurons, offering hope that we eventually will learn to use embryonic stem cells to repair spinal injury. Embryonic stem cells of mice have been induced to become insulin-secreting pancreas cells. The new cells produce only about 2% as much insulin as normal cells do, so there is still plenty to learn, but the take-home message is clear: Transplanted embryonic stem cells offer a path toward curing type-1 diabetes.

While exciting, these advances in stem cell research were all experiments carried out in strains of mice without functioning immune systems. This prevents the mice from rejecting transplanted stem cells as "foreign." A human with a normal immune system might well refuse to accept transplanted stem cells simply because they are from another individual.

Early in 2001, a research team at Rockefeller University reported a way around this potentially serious problem. Their solution? First, they isolate skin cells; then, using the same procedure that created Dolly, they create an embryo from those cells. After removing the nucleus from the skin cell, they insert it into an egg whose nucleus has already been removed. The egg with its skin cell nucleus is allowed to form a 120-cell embryo. The embryo is then destroyed, and its cells are used as embryonic stem cells for transfer to injured tissue.

Using this procedure, which they called **therapeutic cloning,** the researchers succeeded in making cells from the tail of a mouse convert into the dopamine-producing cells of the brain that are lost in Parkinson disease (figure 20.26).

Therapeutic cloning successfully addresses the key problem that must be solved before stem cells can be used to repair human tissues damaged by heart attack, nerve injury, diabetes, or Parkinson disease—the problem of immune acceptance. Since stem cells are cloned from the body's own tissues in therapeutic cloning, they pass the immune system's "self" identity check, and the body readily accepts them.

Two key problems remain unsolved, however. First, a means must be found to achieve proper genomic imprinting. Second, investigators must learn how to preserve the egg's mitotic machinery. When the first partially successful cloning of a human embryo from adult skin cells was reported in November 2001, the cloned cells did not survive long enough to make stem cells. In fact, no primate has ever been successfully cloned. From the very first step, the cells don't divide properly. Why? Chromosomes within the egg carry proteins that act as molecular motors during spindle formation. In humans and other primates, these proteins are so tightly bound to the chromosomes that cloning's first step of DNA removal pulls them out too, dooming hope of normal blastocyst formation.

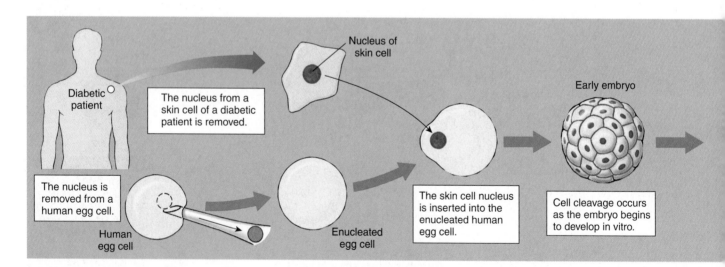

FIGURE 20.26

How human embryos might be used for therapeutic cloning. Therapeutic cloning differs from reproductive DNA cloning in that after the initial similar stages, the embryo is destroyed and its embryonic stem cells are extracted, grown in culture, and added to a tissue of the individual who provided the DNA. In reproductive cloning, by contrast, the embryo is preserved to be implanted and grown to term in a surrogate mother. This latter procedure was used in cloning Dolly the sheep. Human cells were first cloned in November 2001 in a failed attempt to obtain stem cells for therapeutic cloning procedures such as those outlined in this figure.

The difficult ethical issues raised by human therapeutic cloning could be largely avoided if the stem cells did not have to be harvested from an embryo. Imagine, for example, that it were possible to find pluripotent stem cells somewhere in the body of an *adult* human. In 2001, researchers claimed that they had found just such cells in the bone marrow of mice. They transplanted single stem cells from mouse bone marrow into the marrow of individuals whose marrow had been destroyed. After 11 months, the initial stem cell had given rise to descendant cells that had migrated throughout the body, forming new bone, blood, lung, esophagus, stomach, intestine, liver, and skin cells (figure 20.27). The bone marrow stem cells appear to have the properties of the long-sought pluripotent adult stem cells. Many labs are trying to repeat this preliminary result.

> **Therapeutic cloning involves initiating blastocyst development from a patient's tissue using nuclear transplant procedures, and then using these embryonic stem cells to replace the patient's damaged or lost tissue.**

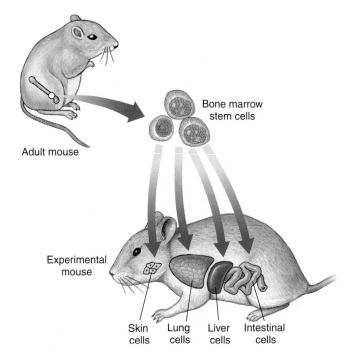

FIGURE 20.27
Pluripotent stem cells.   In May 2001, a single cell from the bone marrow of a mouse was claimed to have added functional cells to the lungs, liver, intestine, and skin of an experimental mouse.

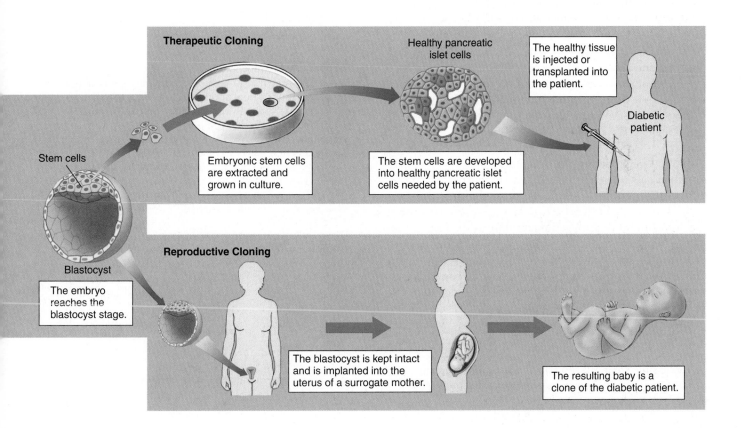

Grappling with the Ethics of Stem Cell Research

Human embryonic stem cells retain the potential to become any tissue in the body and thus have enormous promise for treating a wide range of diseases. Human embryonic stem cells are very difficult to isolate and establish in culture, but a few dozen lines have been successfully obtained from the inner cell mass of six-day blastocysts. It is important to isolate the cells at this early stage, before development begins the process of restricting what sorts of tissues the stem cells can become. The blastocysts are obtained from reproductive clinics, which routinely produce excess embryos in the process of helping infertile couples have children by in vitro fertilization.

However, obtaining embryonic stem cells destroys the early embryo in the process; for this reason, stem cell research raises profound ethical issues. The timeless question of when human life begins cannot be avoided when human embryos are being deliberately destroyed. What is the moral standing of a six-day human embryo? Resolving the tension between scientific knowledge and moral sensibilities brings into play religious, philosophical, and cultural issues. Table 20.5 illustrates the range of issues being discussed.

It will come as no surprise that government, which funds much of modern biomedical research, has become embroiled in the controversy. In Britain, reproductive cloning is banned, but stem cell research and therapeutic cloning to obtain clinically useful stem cells are both permitted. Because the research is funded by the government, there is careful ethical supervision of all research by a variety of governmental oversight committees. Britain's Human Fertilization and Embryology Authority (HFEA), for example, is a panel of scientists and ethicists accountable to Parliament, which oversees government-funded stem cell research. Similar arrangements are being established in Japan and France. Germany, by contrast, discourages all stem cell research.

In the United States, the situation is ambivalent. American stem cell research is chiefly carried out in private research labs using no government funds and thus subject to no ethical oversight. This leaves American scientists pretty much free to do what they want, as long as they use private money. Federal funds were made available in the summer of 2001 for research on the small number of existing human embryonic stem cell lines. In what has become a very political contest between those favoring increased stem cell research and those opposing all such research, it seems certain that federal government policies with regard to stem cell research will fluctuate for some time to come.

Embryonic stem cell research is quite controversial because it involves many ethical issues, not the least of which is when human life begins.

Table 20.5 The Ethics of Stem Cell Research

1. DESTRUCTION OF HUMAN EMBRYOS

Opponents: A human life begins at the moment egg and sperm are united, so destroying an embryo in order to harvest embryonic stem cells is simply murder, and morally wrong. Benefits to others, however great, cannot justify the destruction of a human life.

Proponents: While human embryos should be treated with respect, the potential for saving lives that embryonic stem cell research offers is also a strong moral imperative. The blastocysts being used to obtain stem cells were created to help infertile couples conceive, and would have been destroyed in any event. Besides, it is not clear that an individual life begins at fertilization. An early embryo can split, leading to the birth of identical twins, so it can be argued that individuality begins some days after fertilization.

2. POSSIBILITY OF FUTURE ABUSE

Opponents: Permitting embryonic stem cell research may open the door to further ethically objectionable research. Certainly the development of embryonic stem cell therapies will lead to a cry for therapeutic cloning, so that the therapies can actually be employed in clinical situations. This creation of embryos specifically for the production of embryonic stem cells is morally wrong. In addition, therapeutic cloning to obtain clinical embryonic stem cells is unnatural because it involves producing a viable human embryo without fertilization. If therapeutic use of the results of stem cell research is unacceptable, there is little use in carrying out the research in the first place. It only delays and excalibrates the morally difficult choice posed by therapeutic cloning. Even more disturbing, it opens the door to reproductive cloning—the production of human babies from cloned embryos. This moral nightmare will always be a threat if an absolute line is not drawn preventing all cloning.

Proponents: In practice, only a few hundred cell lines are likely to be required to carry out embryonic stem cell research. The derivation of human cell lines, which is difficult and expensive to do, would be limited to these few lines. Continuous destruction of embryos would be neither desirable nor likely. Therapeutic cloning presents a separate and more complex ethical issue. Because fertilization is not involved, the blastocyst might better be thought of as an "activated egg" rather than an embryo. Such a distinction has biological merit and avoids the ethical issues posed by human reproductive cloning, which should be banned.

3. ALTERNATIVE SOURCES OF STEM CELLS

Opponents: Why not use stem cells derived from adult tissues? These stem cells raise no difficult ethical issues, and can lead to the same medical benefits.

Proponents: Adult stem cells simply cannot do the job. By the time embryonic stem cells have developed into adult stem cells, they have lost much of their developmental versatility, and so lack the range of medical capabilities necessary for regenerative medicine. Also, adult stem cells are not very prolific, and have proven to be difficult to use in therapeutic procedures on experimental laboratory animals.

Concept Review

20.1 Recombination alters gene location.

Gene Transfer

- Some genes move because they are part of plasmids that transfer copies of bacterial genes from one bacterium to another, while others move within transposons that jump from one genomic position to another at random. (pp. 406–408)

Reciprocal Recombination

- Unequal crossing over occurs when homologues exchange segments of unequal length. One chromosome gains extra copies of the multicopy sequences, while the other chromosome loses them. (p. 409)
- Gene conversion occurs as error correction systems in a cell convert one of the mismatched sequences to resemble the other. (p. 409)

20.2 Mutations are changes in the genetic message.

Kinds of Mutation

- Mutations are changes in the hereditary message of an organism, and germ-line mutations provide raw material for evolutionary change. (p. 410)
- DNA mutational alterations arise in at least five ways: base substitution, chemical modification, DNA breaks, slipped mispairing, and triplet expansion. (p. 410)
- Some chromosomal alterations, such as duplications, deletions, aneuploidy, and polyploidy, can change the number of gene copies an individual possesses. (p. 411)

DNA Repair

- Cells can repair many potential genetic mutations by recognizing pairing mismatches created by a base substitution or by recognizing inappropriate bases. (p. 412)

20.3 Most cancer results from mutation of growth-regulating genes.

What Is Cancer?

- Cancer occurs when a cell begins to grow in an uncontrolled and invasive way. It can result in a cluster of cells (tumor) that might metastasize and form new tumors at distant sites. (p. 413).
- Cancer is a gene disorder of somatic tissue, in which damaged genes fail to properly control cell proliferation. (p. 413)

Causes of Cancer

- Agents thought to cause cancer are called carcinogens; for example, some chemicals are carcinogens. (p. 414)
- Chemicals in cigarette smoke, such as tar, have clearly been demonstrated to cause cancer. (p. 415)

Cancer and the Cell Cycle

- Oncogenes are genes that, when introduced, cause a normal cell to become cancerous. (p. 416)
- Proto-oncogenes are normal forms of genes that can be mutated to produce a tumor-forming oncogene. (p. 416)
- Chromosomal abnormalities can lead to the activation of proto-oncogenes. (p. 416)

- Cell division is normally turned off in healthy cells by tumor-suppressor genes. (pp. 416–417)
- Cells control proliferation at several checkpoints, and all of these controls must be inactivated for cancer to be initiated. Therefore, the induction of cancer typically involves the mutation of four to six genes. (p. 419)

Smoking and Cancer

- About one-third of all cancer cases in the United States are directly attributable to cigarette smoking. (p. 420)
- Introducing mutagens into the lungs causes damage to the genes of the epithelial cells that line the lungs and are directly exposed to the chemicals. (p. 420)
- The destruction of *p53* in lung epithelial cells hastens the onset of lung cancer; *p53* is mutated to an inactive form in over 70% of lung cancers. (p. 421)

Curing Cancer

- Cancer therapies are being developed on many fronts, which can be consolidated into two broad categories: preventing the start of cancer and preventing the spread of cancer. (pp. 422–423)

20.4 Reproductive cloning of animals, once thought impossible, isn't.

The Challenge of Cloning

- Cloning differentiated mammalian tissue is possible, but most attempts have failed due to improper gene reprogramming. (pp. 424–425)

20.5 Therapeutic cloning is a promising but controversial possibility.

Stem Cells

- Each embryonic stem cell is capable, by itself, of developing into a healthy individual. (p. 426)
- Because embryonic stem cells can develop into any tissue, they offer the possibility of replacing damaged or lost tissues. (p. 426)
- Stem cell research is very controversial because embryonic stem cells are typically isolated from embryos discarded by reproductive clinics. (p. 426)
- Tissue-specific stem cells may allow the replacement of damaged or lost tissues while avoiding the ethical considerations involved in using embryonic tissues. (p. 427)

Therapeutic Cloning

- Therapeutic cloning involves initiating blastocyst development from a patient's tissue, and then using these embryonic stem cells to replace the patient's damaged or lost tissue. (pp. 428–429)

Grappling with the Ethics of Stem Cell Research

- One of the most controversial issues involving embryonic stem cell research is deciding when human life begins. (p. 430)

For interactive testing, visit the Online Learning Center with PowerWeb at www.mhhe.com/Raven7

Self Test

1. Tumor-suppressor genes includes *p53* and *Rb*. How would a "gain-of-function" mutation likely affect the cell?
   a. The cell would divide constantly because of the loss of cell cycle repression.
   b. The cell would divide much less frequently because of the extra cell cycle repression.
   c. The cell would divide normally because these genes have no effect on cell cycle control.
   d. The cell would commit suicide by apoptosis.

2. In lab, you are studying cell cycle control in the fission yeast *S. pombe*. A student finds a new mutant that she wants to call "giant" because the cells are much larger than normal (suggesting that it is not dividing normally). What type of mutation do you think the student has isolated?
   a. a loss-of-function mutation in a tumor-suppressor gene
   b. a loss-of-function mutation in a cellular proto-oncogene
   c. a gain-of-function mutation in a tumor-suppressor gene
   d. a gain-of-function mutation in a cellular proto-oncogene
   e. Both *a* and *d* are possible.
   f. Both *b* and *c* are possible.

3. Which of the following would be an effective approach to a new cancer therapy?
   a. finding a way to stabilize *p53* specifically in tumor cells
   b. preventing nucleotide synthesis in tumor cells
   c. inactivating the HER2 receptor on tumor cells
   d. inhibiting growth of new blood vessels with endostatin
   e. All of these would help fight cancer.

4. How would the cell cycle be affected if you removed the phosphorylation sites in the Rb protein?
   a. The cell cycle would not be affected because pRb is not phosphorylated normally.
   b. The cell cycle would be blocked in G_1.
   c. The cell cycle would be blocked in G_2.
   d. The cell cycle would be shorter.

5. Embryonic stem (ES) cells are an attractive source of material for therapeutic cloning because
   a. they can be induced to assume any cell fate.
   b. ES cells are not targets for the host immune response, so tissue rejection is not an issue.
   c. there are no other sources of stem cells to use for therapeutic cloning, so ES cells are the only solution.
   d. ES cells will not work as a source of tissue for cloning.

6. How would growing cells in the presence of methyl-adenosine affect the mismatch repair system?
   a. The repair system would only repair half of the errors introduced by DNA polymerase.
   b. There would be no repair of mismatched DNA.
   c. Mismatch repair would be normal, but excision repair would fail.
   d. Methyladenosine would prevent DNA replication, so there would be no need for mismatch repair.

7. Too much time in a tanning booth probably causes DNA damage to epithelial cells. The most likely effect would be
   a. depurination.
   b. pyrimidine dimers.
   c. deamination.
   d. single-stranded nicks in the phosphodiester backbone.

8. Using a "car and driver" analogy, which of the following accurately describes the role of tumor-suppressor genes and proto-oncogenes in normal cells?
   a. Tumor-suppressor genes are the gas pedal, while proto-oncogenes are the brakes.
   b. Tumor-suppressor genes are the brakes, while proto-oncogenes are the gas.
   c. Both tumor-suppressor genes and proto-oncogenes are like the gas, but tumor-suppressors are like turbo and proto-oncogenes are like a regular carburetor.
   d. Tumor-suppressor genes are like the steering wheel, and proto-oncogenes are like the turn signals.

9. During the early years of cancer research, there were two schools of thought regarding the causes of cancer: 1) that cancer was caused entirely by environmental factors, and 2) that cancer was caused by genetic factors. Which was correct?
   a. #1 because we have identified many potential carcinogens
   b. #2 because we know of many proto-oncogenes
   c. #2 because we know of many tumor suppressor genes
   d. Both were correct; most chemical carcinogens function by altering genes.

10. If you found a specific chromosomal deletion in the genome from a tumor, what could be the cause of this specific cancer?
    a. The deletion likely affected a tumor-suppressor gene, leading to a loss of function in the tumor cells.
    b. The deletion likely affected a proto-oncogene, leading to a loss of function in the tumor cells.
    c. The deletion likely affected a tumor-suppressor gene, leading to a gain of function in the tumor cells.
    d. The deletion likely affected a proto-oncogene, leading to a gain of function in the tumor cells.

Test Your Visual Understanding

1. If you were to observe two bacterial cells as shown here, what would you suggest is happening?

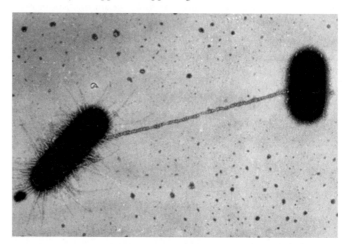

Apply Your Knowledge

1. The data in table 20.3 show the incidence of specific cancers following exposure to environmental carcinogens. Discuss how this type of chemical exposure leads to such a high proportion of skin and lung tumors. (*Hint:* Think about the points of contact with environmental chemicals.)

2. Pretend that you are preparing for a debate about the use of embryonic stem cells for therapeutic cloning, and list three pros and three cons of this technology.

Glossary

A

ABO blood group A set of four phenotypes produced by different combinations of three alleles at a single locus; blood types are A, B, AB, and O, depending on which alleles are expressed as antigens on the red blood cell surface.

abscission (L. *ab*, away, off, + *scissio*, dividing) In vascular plants, the dropping of leaves, flowers, fruits, or stems at the end of the growing season, as the result of the formation of a layer of specialized cells (the abscission zone) and the action of a hormone (ethylene).

acquired immune deficiency syndrome (AIDS) An infectious and usually fatal human disease caused by a retrovirus, HIV (human immunodeficiency virus), which attacks T cells. The affected individual is helpless in the face of microbial infections.

actin (Gr. *actis*, a ray) One of the two major proteins that make up vertebrate muscle; the other is myosin.

action potential A transient, all-or-none reversal of the electric potential across a membrane; in neurons, an action potential initiates transmission of a nerve impulse.

activation energy The energy that must be processed by a molecule in order for it to undergo a specific chemical reaction.

active site The region of an enzyme surface to which a specific set of substrates binds, lowering the activation energy required for a particular chemical reaction and so facilitating it.

active transport The pumping of individual ions or other molecules across a cellular membrane from a region of lower concentration to one of higher concentration (that is, against a concentration gradient); this transport process requires energy, which is typically supplied by the expenditure of ATP.

adaptation (L. *adaptare*, to fit) A peculiarity of structure, physiology, or behavior that promotes the likelihood of an organism's survival and reproduction in a particular environment.

adaptive radiation The evolution of several divergent forms from a primitive and unspecialized ancestor.

adenosine diphosphate (ADP) A nucleotide consisting of adenine, ribose sugar, and two phosphate groups; formed by the removal of one phosphate from an ATP molecule.

adenosine triphosphate (ATP) A nucleotide consisting of adenine, ribose sugar, and three phosphate groups; ATP is the energy currency of cellular metabolism in all organisms.

adenylyl cyclase An enzyme that produces large amounts of cAMP from ATP; the cAMP acts as a second messenger in a target cell.

adipose cells Fat cells, found in loose connective tissue, usually in large groups that form adipose tissue. Each adipose cell can store a droplet of fat (triacylglyceride).

adventitious (L. *adventicius*, not properly belonging to) Referring to a structure arising from an unusual place, such as stems from roots or roots from stems.

aerobic (Gr. *aer*, air, + *bios*, life) Requiring free oxygen; any biological process that can occur in the presence of gaseous oxygen (O_2).

alga, pl. algae A unicellular or simple multicellular photosynthetic organism lacking multicellular sex organs.

allantois (Gr. *allas*, sausage, + *eidos*, form) A membrane of the amniotic egg that functions in respiration and excretion in birds and reptiles and plays an important role in the development of the placenta in most mammals.

allele (Gr. *allelon*, of one another) One of two or more alternative states of a gene.

allometric growth (Gr. *allos*, other, + *ergon*, action) A pattern of growth in which different components grow at different rates.

allopatric speciation The differentiation of geographically isolated populations into distinct species.

allosteric site A part of an enzyme, away from its active site, that serves as an on/off switch for the function of the enzyme.

alternation of generations A reproductive cycle in which a haploid (*n*) phase, the gametophyte, gives rise to gametes, which, after fusion to form a zygote, germinate to produce a diploid (*2n*) phase, the sporophyte. Spores produced by meiotic division from the sporophyte give rise to new gametophytes, completing the cycle.

altruism Self-sacrifice for the benefit of others; in formal terms, the behavior that increases the fitness of the recipient while reducing the fitness of the altruistic individual.

alveolus, pl. alveoli (L., a small cavity) One of many small, thin-walled air sacs within the lungs in which the bronchioles terminate.

amino acid The subunit structure from which proteins are produced, consisting of a central carbon atom with a carboxyl group (—COOH), an amino group (—NH_2), a hydrogen, and a side group (*R* group); only the side group differs from one amino acid to another.

amniocentesis (Gr. *amnion*, membrane around the fetus, + *centes*, puncture) Examination of a fetus indirectly, by tests on cell cultures grown from fetal cells obtained from a sample of the amniotic fluid surrounding the developing embryo or tests on the fluid itself.

amnion (Gr., membrane around the fetus) The innermost of the extraembryonic membranes; the amnion forms a fluid-filled sac around the embryo in amniotic eggs.

amniotic egg An egg that is isolated and protected from the environment by a more or less impervious shell during the period of its development and that is completely self-sufficient, requiring only oxygen.

amyloplast (Gr. *amylon*, starch, + *plastos*, formed) A plant organelle called a plastid that specializes in storing starch.

anabolism (Gr. *ana*, up, + *bolein*, to throw) The biosynthetic or constructive part of metabolism; those chemical reactions involved in biosynthesis.

anaerobic (Gr. *an*, without, + *aer*, air, + *bios*, life) Any process that can occur without oxygen, such as anaerobic fermentation or H_2S photosynthesis.

analogous (Gr. *analogos*, proportionate) Structures that are similar in function but different in evolutionary origin, such as the wing of a bat and the wing of a butterfly.

anaphase In mitosis and meiosis II, the stage initiated by the separation of sister chromatids, during which the daughter chromosomes move to opposite poles of the cell; in meiosis I, marked by separation of replicated homologous chromosomes.

androecium (Gr. *andros*, man, + *oilos*, house) The floral whorl that comprises the stamens.

aneuploidy (Gr. *an*, without, + *eu*, good, + *ploid*, multiple of) The condition in an organism whose cells have lost or gained a chromosome; Down syndrome, which results from an extra copy of human chromosome 21, is an example of aneuploidy in humans.

angiosperms The flowering plants, one of five phyla of seed plants. In angiosperms, the ovules at the time of pollination are completely enclosed by tissues.

animal pole In fish and other aquatic vertebrates with asymmetrical yolk distribution in their eggs, the hemisphere of the blastula comprising cells relatively poor in yolk.

anion (Gr. *anienae*, to go up) A negatively charged ion.

anther (Gr. *anthos*, flower) In angiosperm flowers, the pollen-bearing portion of a stamen.

antheridium, pl. antheridia A sperm-producing organ.

antibody (Gr. *anti*, against) A protein called immunoglobulin that is produced by lymphocytes in response to a foreign substance (antigen) and released into the bloodstream.

anticodon The three-nucleotide sequence at the end of a transfer RNA molecule that is complementary to, and base-pairs with, an amino-acid–specifying codon in messenger RNA.

antigen (Gr. *anti*, against, + *genos*, origin) A foreign substance, usually a protein or polysaccharide, that stimulates an immune response.

anus The terminal opening of the gut; the solid residues of digestion are eliminated through the anus.

aorta (Gr. *aeirein*, to lift) The major artery of vertebrate systemic blood circulation; in mammals, carries oxygenated blood away from the heart to all regions of the body except the lungs.

apical meristem (L. *apex*, top, + Gr. *meristos*, divided) In vascular plants, the growing point at the tip of the root or stem.

apoptosis A process of programmed cell death, in which dying cells shrivel and shrink; used in all animal cell development to produce planned and orderly death of cells not destined to be present in the final tissue.

aposematic coloration An ecological strategy of some organisms that "advertise" their poisonous nature by the use of bright colors.

aquifers Permeable, saturated, underground layers of rock, sand, and gravel, which serve as reservoirs for groundwater.

archaebacteria A group of bacteria that are among the most primitive still in existence; characterized by the absence of peptidoglycan in their cell walls, a feature that distinguishes them from all other bacteria.

archegonium, pl. archegonia (Gr. *archegonos*, first of a race) The multicellular egg-producing organ in bryophytes and some vascular plants.

archenteron (Gr. *arche*, beginning, + *enteron*, gut) The principal cavity of a vertebrate embryo in the gastrula stage; lined with endoderm, it opens up to the outside and represents the future digestive cavity.

arteriole A smaller artery, leading from the arteries to the capillaries.

arteriosclerosis Hardening and thickening of the wall of an artery.

ascomycetes A large group comprising part of the "true fungi." They are characterized by separate hyphae, asexually produced conidiospores, and sexually produced ascospores within asci.

ascospore A fungal spore produced within an ascus.

ascus, pl. asci (Gr. *askos*, wineskin, bladder) A specialized cell, characteristic of the ascomycetes, in which two haploid nuclei fuse to produce a zygote that divides immediately by meiosis; at maturity, an ascus contains ascospores.

asexual reproduction The process by which an individual inherits all of its chromosomes from a single parent, thus being genetically identical to that parent; cell division is by mitosis only.

aster In animal cell mitosis, a radial array of microtubules extending from the centrioles toward the plasma membrane, possibly serving to brace the centrioles for retraction of the spindle.

atrioventricular (AV) node A slender connection of cardiac muscle cells that receives the heartbeat impulses from the sinoatrial node and conducts them by way of the bundle of His.

atrium (L., vestibule or courtyard) An antechamber; in the heart, a thin-walled chamber that receives venous blood and passes it on to the thick-walled ventricle; in the ear, the tympanic cavity.

autonomic nervous system (Gr. *autos*, self, + *nomos*, law) The involuntary neurons and ganglia of the peripheral nervous system of vertebrates; regulates the heart, glands, visceral organs, and smooth muscle.

autosome (Gr. *autos*, self, + *soma*, body) Any eukaryotic chromosome that is not a sex chromosome; autosomes are present in the same number and kind in both males and females of the species.

autotroph (Gr. *autos*, self, + *trophos*, feeder) An organism able to build all the complex organic molecules that it requires as its own food source, using only simple inorganic compounds.

auxin (Gr. *auxein*, to increase) A plant hormone that controls cell elongation, among other effects.

axon (Gr., axle) A process extending out from a neuron that conducts impulses away from the cell body.

B

bacteriophage (Gr. *bakterion*, little rod, + *phagein*, to eat) A virus that infects bacterial cells; also called a *phage*.

Barr body A deeply staining structure, seen in the interphase nucleus of a cell of an individual with more than one X chromosome, that is a condensed and inactivated X. Only one X remains active in each cell after early embryogenesis.

basal body A self-reproducing, cylindrical, cytoplasmic organelle composed of nine triplets of microtubules from which the flagella or cilia arise.

base-pair A complementary pair of nucleotide bases, consisting of a purine and a pyrimidine.

basidiospore A spore of the basidiomycetes, produced within and borne on a basidium after nuclear fusion and meiosis.

basidium, pl. basidia (L., a little pedestal) A specialized reproductive cell of the basidiomycetes, often club shaped, in which nuclear fusion and meiosis occur.

basophil A leukocyte containing granules that rupture and release chemicals that enhance the inflammatory response. Important in causing allergic responses.

Batesian mimicry A situation in which a palatable or nontoxic organism resembles another kind of organism that is distasteful or toxic. Both species exhibit warning coloration.

B cell A type of lymphocyte that, when confronted with a suitable antigen, is capable of secreting a specific antibody protein.

behavioral ecology The study of how natural selection shapes behavior.

biennial A plant that normally requires two growing seasons to complete its life cycle. Biennials flower in the second year of their lives.

bile salts A solution of organic salts that is secreted by the vertebrate liver and temporarily stored in the gallbladder; emulsifies fats in the small intestine.

binary fission (L. *binarius*, consisting of two things or parts, + *fissus*, split) Asexual reproduction by division of one cell or body into two equal or nearly equal parts.

binomial distribution The distribution of phenotypes seen among the progeny of a cross in which there are only two alternative alleles.

biodiversity The number of species and their range of behavioral, ecological, physiological, and other adaptations, in an area.

biomass (Gr. *bios*, life, + *maza*, lump or mass) The total mass of all the living organisms in a given population, area, or other unit being measured.

biome One of the major terrestrial ecosystems, characterized by climatic and soil conditions; the largest ecological unit.

bipedal Able to walk upright on two feet.

bipolar cell A specialized type of neuron connecting cone cells to ganglion cells in the visual system. Bipolar cells receive a hyperpolarized stimulus from the cone cell and then transmit a depolarization stimulus to the ganglion cell.

blade The broad, expanded part of a leaf; also called the lamina.

blastocoel (Gr. *blastos*, sprout, + *koilos*, hollow) The central cavity of the blastula stage of vertebrate embryos.

blastodisc (Gr. *blastos*, sprout, + *discos*, a round plate) In the development of birds, a disclike area on the surface of a large, yolky egg that undergoes cleavage and gives rise to the embryo.

blastomere (Gr. *blastos*, sprout, + *meros*, part) One of the cells of a blastula.

blastopore (Gr. *blastos*, sprout, + *poros*, a path or passage) In vertebrate development, the opening that connects the archenteron cavity of a gastrula stage embryo with the outside.

blastula (Gr., a little sprout) In vertebrates, an early embryonic stage consisting of a hollow, fluid-filled ball of cells one layer thick; a vertebrate embryo after cleavage and before gastrulation.

Bohr effect The release of oxygen by hemoglobin molecules in response to elevated ambient levels of CO_2.

Bowman's capsule In the vertebrate kidney, the bulbous unit of the nephron, which surrounds the glomerulus.

β-oxidation In the cellular respiration of fats, the process by which two-carbon acetyl groups are removed from a fatty acid and combined with coenzyme A to form acetyl-CoA, until the entire fatty acid has been broken down.

bronchus, pl. bronchi (Gr. *bronchos*, windpipe) One of a pair of respiratory tubes branching from the lower end of the trachea (windpipe) into either lung.

bryophytes Nonvascular plants, including mosses, hornworts, and liverworts.

bud An asexually produced outgrowth that develops into a new individual. In plants, an embryonic shoot, often protected by young leaves; buds may give rise to branch shoots.

C

C₃ photosynthesis The main cycle of the dark reactions of photosynthesis, in which CO_2 binds to ribulose 1,5-bisphosphate (RuBP) to form two three-carbon phosphoglycerate (PGA) molecules.

C₄ photosynthesis A process of CO_2 fixation in photosynthesis by which the first product is the four-carbon oxaloacetate molecule.

callus (L. *callos*, hard skin) Undifferentiated tissue; a term used in tissue culture, grafting, and wound healing.

Calvin cycle The dark reactions of C₃ photosynthesis; also called the Calvin-Benson cycle.

calyx (Gr. *kalyx*, a husk, cup) The sepals collectively; the outermost flower whorl.

capillary (L. *capillaris*, hairlike) The smallest of the blood vessels;

the very thin walls of capillaries are permeable to many molecules, and exchanges between blood and the tissues occur across them; the vessels that connect arteries with veins.

capsid The outermost protein covering of a virus.

carapace (Fr. from Sp. *carapacho*, shell) Shieldlike plate covering the cephalothorax of decapod crustaceans; the dorsal part of the shell of a turtle.

carbohydrate (L. *carbo*, charcoal, + *hydro*, water) An organic compound consisting of a chain or ring of carbon atoms to which hydrogen and oxygen atoms are attached in a ratio of approximately 2:1; having the generalized formula $(CH_2O)_n$; carbohydrates include sugars, starch, glycogen, and cellulose.

carbon fixation The conversion of CO_2 into organic compounds during photosynthesis; the first stage of the dark reactions of photosynthesis, in which carbon dioxide from the air is combined with ribulose 1,5-bisphosphate.

carpel (Gr. *karpos*, fruit) A leaflike organ in angiosperms that encloses one or more ovules.

carrying capacity The maximum population size that a habitat can support.

cartilage (L. *cartilago*, gristle) A connective tissue in skeletons of vertebrates. Cartilage forms much of the skeleton of embryos, very young vertebrates, and some adult vertebrates, such as sharks and their relatives.

catabolism (Gr. *ketabole*, throwing down) In a cell, those metabolic reactions that result in the breakdown of complex molecules into simpler compounds, often with the release of energy.

catalysis The process by which chemical subunits of larger organic molecules are held and positioned by enzymes that stress their chemical bonds, leading to the disassembly of the larger molecule into its subunits, often with the release of energy.

cation (Gr. *katienai*, to go down) A positively charged ion.

cecum (L. *caecus*, blind) In vertebrates, a blind pouch at the beginning of the large intestine.

cell cycle The repeating sequence of growth and division through which cells pass each generation.

cell plate The structure that forms at the equator of the spindle during early telophase in the dividing cells of plants and a few green algae.

cell surface receptor A cell surface protein that binds a signal molecule and converts the extracellular signal into an intracellular one.

cellular respiration The metabolic harvesting of energy by oxidation, ultimately dependent on molecular oxygen; carried out by the Krebs cycle and oxidative phosphorylation.

cellulose (L. *cellula*, a little cell) The chief constituent of the cell wall in all green plants, some algae, and a few other organisms; an insoluble complex carbohydrate formed of microfibrils of glucose molecules.

cell wall The rigid, outermost layer of the cells of plants, some protists, and most bacteria; the cell wall surrounds the plasma membrane.

central nervous system (CNS) That portion of the nervous system where most association occurs; in vertebrates, it is composed of the brain and spinal cord; in invertebrates, it usually consists of one or more cords of nervous tissue, together with their associated ganglia.

centriole (Gr. *kentron*, center, + L. *olus*, little one) A cytoplasmic organelle located outside the nuclear membrane, identical in structure to a basal body; found in animal cells and in the flagellated cells of other groups; divides and organizes spindle fibers during mitosis and meiosis.

centromere (Gr. *kentron*, center, + *meros*, a part) Condensed region on a eukaryotic chromosome where sister chromatids are attached to each other after replication. *See* kinetochore.

cerebellum (L., little brain) The hindbrain region of the vertebrate brain that lies above the medulla (brain stem) and behind the forebrain; it integrates information about body position and motion, coordinates muscular activities, and maintains equilibrium.

cerebral cortex The thin surface layer of neurons and glial cells covering the cerebrum; well developed only in mammals, and particularly prominent in humans. The cerebral cortex is the seat of conscious sensations and voluntary muscular activity.

cerebrum (L., brain) The portion of the vertebrate brain (the forebrain) that occupies the upper part of the skull, consisting of two cerebral hemispheres united by the corpus callosum. It is the primary association center of the brain. It coordinates and processes sensory input and coordinates motor responses.

chelicera, pl. chelicerae (Gr. *chele*, claw, + *keras*, horn) The first pair of appendages in horseshoe crabs, sea spiders, and arachnids—the chelicerates, a group of arthropods. Chelicerae usually take the form of pincers or fangs.

chemiosmosis The mechanism by which ATP is generated in mitochondria and chloroplasts; energetic electrons excited by light (in chloroplasts) or extracted by oxidation in the Krebs cycle (in mitochondria) are used to drive proton pumps, creating a proton concentration gradient; when protons subsequently flow back across the membrane, they pass through channels that couple their movement to the synthesis of ATP.

chiasma An X-shaped figure that can be seen in the light microscope during meiosis; evidence of crossing over, where two chromatids have exchanged parts; chiasmata move to the ends of the chromosome arms as the homologues separate.

chitin (Gr. *chiton*, tunic) A tough, resistant, nitrogen-containing polysaccharide that forms the cell walls of certain fungi, the exoskeleton of arthropods, and the epidermal cuticle of other surface structures of certain other surface structures of certain other surface structures of certain other animals.

chloroplast (Gr. *chloros*, green, + *plastos*, molded) A cell-like organelle present in algae and plants that contains chlorophyll (and usually other pigments) and carries out photosynthesis.

chorion (Gr., skin) The outer member of the double membrane that surrounds the embryo of reptiles, birds, and mammals; in placental mammals, it contributes to the structure of the placenta.

chromatid (Gr. *chroma*, color, + L. *-id*, daughters of) One of the two daughter strands of a duplicated chromosome that is joined by a single centromere.

chromatin (Gr. *chroma*, color) The complex of DNA and proteins of which eukaryotic chromosomes are composed; chromatin is highly uncoiled and diffuse in interphase nuclei, condensing to form the visible chromosomes in prophase.

chromosome (Gr. *chroma*, color, + *soma*, body) The vehicle by which hereditary information is physically transmitted from one generation to the next; in a bacterium, the chromosome consists of a single naked circle of DNA; in eukaryotes, each chromosome consists of a single linear DNA molecule and associated proteins.

cilium A short cellular projection from the surface of a eukaryotic cell, having the same internal structure of microtubules in a 9 + 2 arrangement as seen in a flagellum.

circadian rhythm An endogenous cyclical rhythm that oscillates on a daily (24-hour) basis.

cisterna A small collecting vessel that pinches off from the end of a Golgi body to form a transport vesicle that moves materials through the cytoplasm.

cladistics A taxonomic technique used for creating hierarchies of organisms that represent true phylogenetic relationship and descent.

class A taxonomic category between phyla and orders. A class contains one or more orders, and belongs to a particular phylum.

classical conditioning The repeated presentation of a stimulus in association with a response that causes the brain to form an association between the stimulus and the response, even if they have never been associated before.

clathrin A protein located just inside the plasma membrane in eukaryotic cells, in indentations called clathrin-coated pits.

cleavage In vertebrates, a rapid series of successive cell divisions of a fertilized egg, forming a hollow sphere of cells, the blastula.

climax vegetation Vegetation encountered in a self-perpetuating community of plants that has proceeded through all the stages of succession and stabilized.

cloaca (L., sewer) In some animals, the common exit chamber from the digestive, reproductive, and urinary system; in others, the cloaca may also serve as a respiratory duct.

cloning Producing a cell line or culture all of whose members contain identical copies of a particular nucleotide sequence; an essential element in genetic engineering.

coacervate (L. *coacervatus*, heaped up) A spherical aggregation of lipid molecules in water, held together by hydrophobic forces.

cochlea (Gr. *kochlios*, a snail) In terrestrial vertebrates, a tubular cavity of the inner ear containing the essential organs for hearing.

codon (L., code) The basic unit of the genetic code; a sequence of three adjacent nucleotides in DNA or mRNA that codes for one amino acid.

coenzyme (L. *co-*, together, + Gr. *en*, in, + *zyme*, leaven) A nonprotein organic molecule such as NAD^+ that plays an accessory role in enzyme-catalyzed processes, often by acting as a donor or acceptor of electrons.

coevolution (L. *co-*, together, + *e-*, out, + *volvere*, to fill) The simultaneous development of adaptations in two or more populations, species, or other categories that interact so closely that each is a strong selective force on the other.

cofactor One or more nonprotein components required by enzymes in order to function; many cofactors are metal ions, others are organic coenzymes.

collenchyma cell (Gr. *kolla*, glue, + *en*, in, + *chymein*, to pour) In plants, the cells that form a supporting tissue called collenchyma; often found in regions of primary growth in stems and in some leaves.

commensalism (L. *cum*, together with, + *mensa*, table) A relationship in which one individual lives close to or on another and benefits, and the host is unaffected; a kind of symbiosis.

community (L. *communitas*, community, fellowship) All of the species inhabiting a common environment and interacting with one another.

companion cell A specialized parenchyma cell that is associated with each sieve-tube member in the phloem of a plant.

competitive exclusion The hypothesis that two species with identical ecological requirements cannot exist in the same locality indefinitely, and that the more efficient of the two in utilizing the available scarce resources will exclude the other; also known as Gause's principle.

complementary Describes genetic information in which each nucleotide base has a complementary partner with which it forms a base-pair.

complement system The chemical defense of a vertebrate body that consists of a battery of proteins that become activated by the walls of bacteria and fungi.

concentration gradient The concentration difference of a substance across a distance; in a cell, a greater concentration of its molecules in one region than in another.

cone (1) In plants, the reproductive structure of a conifer. (2) In vertebrates, a type of light-sensitive neuron in the retina, concerned with the perception of color and with the most acute discrimination of detail.

conidia An asexually produced fungal spore.

conjugation (L. *conjugare*, to yolk together) Temporary union of two unicellular organisms, during which genetic material is transferred from one cell to the

other; occurs in bacteria, protists, and certain algae and fungi.

contractile vacuole In protists and some animals, a clear fluid-filled vacuole that takes up water from within the cell and then contracts, releasing it to the outside through a pore in a cyclical manner; functions primarily in osmoregulation and excretion.

conus arteriosus The anteriormost chamber of the embryonic heart in vertebrate animals.

convergent evolution The independent development of similar structures in organisms that are not directly related; often found in organisms living in similar environments.

cork cambium The lateral meristem that forms the periderm, producing cork (phellem) toward the surface (outside) of the plant and phelloderm toward the inside.

cornea (L. *corneus*, horny) The transparent outer layer of the vertebrate eye.

corolla (L., a little crown) The petals, collectively; usually the conspicuously colored flower whorl.

corpus callosum The band of nerve fibers that connects the two hemispheres of the cerebrum in humans and other primates.

corpus luteum (L., yellowish body) A structure that develops from a ruptured follicle in the ovary after ovulation.

cortex (L., bark) The outer layer of a structure; in animals, the outer, as opposed to the inner, part of an organ; in vascular plants, the primary ground tissue of a stem or root.

cotyledon (Gr. *kotyledon*, a cup-shaped hollow) A seed leaf that generally stores food in dicots or absorbs it in monocots, providing nourishment used during seed germination.

crassulacean acid metabolism (CAM) A mode of carbon dioxide fixation by which CO_2 enters open leaf stomata at night and is used in photosynthesis during the day, when stomata are closed to prevent water loss.

cross-current flow In bird lungs, the latticework of capillaries arranged across the air flow, at a 90° angle.

crossing over In meiosis, the exchange of corresponding chromatid segments between homologous chromosomes; responsible for genetic recombination between homologous chromosomes.

cuticle (L. *cuticula*, little skin) A waxy or fatty, noncellular layer (formed of a substance called cutin) on the outer wall of epidermal cells.

cyanobacteria A group of photosynthetic bacteria, sometimes called the "blue-green algae," that contain the chlorophyll pigments most abundant in plants and algae, as well as other pigments.

cyclic AMP (cAMP) A form of adenosine monophosphate (AMP) in which the atoms of the phosphate group form a ring; found in almost all organisms, cAMP functions as an intracellular second messenger that regulates a diverse array of metabolic activities.

cystic fibrosis An autosomal disorder that produces the most common fatal genetic disease in Caucasians, characterized by secretion of thick mucus that clogs passageways in the lungs, liver, and pancreas.

cytochrome (Gr. *kytos*, hollow vessel, + *chroma*, color) Any of several iron-containing protein pigments that serve as electron carriers in transport chains of photosynthesis and cellular respiration.

cytokinesis (Gr. *kytos*, hollow vessel, + *kinesis*, movement) Division of the cytoplasm of a cell after nuclear division.

cytoplasm (Gr. *kytos*, hollow vessel, + *plasma*, anything molded) The material within a cell, excluding the nucleus; the protoplasm.

cytoskeleton A network of protein microfilaments and microtubules within the cytoplasm of a eukaryotic cell that maintains the shape of the cell, anchors its organelles, and is involved in animal cell motility.

cytotoxic T cell (Gr. *kytos*, hollow vessel, + *toxin*) A special T cell activated during cell-mediated immune response that recognizes and destroys infected body cells.

D

dehydration synthesis (L. *co-*, together, + *densare*, to make dense) A type of chemical reaction in which two molecules join to form one larger molecule, simultaneously splitting out a molecule of water; one molecule is stripped of a hydrogen atom, and another is stripped of a hydroxyl group (—OH), resulting in the joining of the two molecules, while the H^+ and —OH released may combine to form a water molecule.

demography (Gr. *demos*, people, + *graphein*, to draw) The properties of the rate of growth and the age structure of populations.

denaturation The loss of the native configuration of a protein or nucleic acid as a result of excessive heat, extremes of pH, chemical modification, or changes in solvent ionic strength or polarity that disrupt hydrophobic interactions; usually accompanied by loss of biological activity.

dendrite (Gr. *dendron*, tree) A process extending from the cell body of a neuron, typically branched, that conducts impulses toward the cell body.

deoxyribonucleic acid (DNA) The genetic material of all organisms; composed of two complementary chains of nucleotides wound in a double helix.

depolarization The movement of ions across a plasma membrane that locally wipes out an electrical potential difference.

derived character A characteristic used in taxonomic analysis representing a departure from the primitive form.

desmosome A type of anchoring junction that links adjacent cells by connecting their cytoskeletons with cadherin proteins.

diaphragm (Gr. *diaphrassein*, to barricade) (1) In mammals, a sheet of muscle tissue that separates the abdominal and thoracic cavities and functions in breathing. (2) A contraceptive device used to block the entrance to the uterus temporarily and thus prevent sperm from entering during sexual intercourse.

diastolic pressure In the measurement of human blood pressure, the minimum pressure between heartbeats (repolarization of the ventricles). *Compare with* systolic pressure.

dicot Short for dicotyledon; a class of flowering plants generally characterized as having two cotyledons, net-veined leaves, and flower parts usually in fours or fives.

differentiation A developmental process by which a relatively unspecialized cell undergoes a progressive change to a more specialized form or function.

diffusion (L. *diffundere*, to pour out) The net movement of dissolved molecules or other particles from a region where they are more concentrated to a region where they are less concentrated.

digestion (L. *digestio*, separating out, dividing) The breakdown of complex, usually insoluble foods into molecules that can be absorbed into cells, where they are degraded to yield energy and the raw materials for synthetic processes.

dihybrid (Gr. *dis*, twice, + *hibridia*, mixed offspring) An individual heterozygous at two different loci; for example *A/a B/b*.

dikaryotic (Gr. *di*, two, + *karyon*, kernel) In fungi, having pairs of nuclei within each cell.

dioecious (Gr. *di*, two, + *oikos*, house) Having the male and female elements on different individuals.

diploid (Gr. *diploos*, double, + *eidos*, form) Having two sets of chromosomes (2*n*); in animals, twice the number characteristic of gametes; in plants, the chromosome number characteristic of the sporophyte generation; in contrast to haploid (*n*).

directional selection A form of selection in which selection acts to eliminate one extreme from an array of phenotypes.

disaccharide A carbohydrate formed of two simple sugar molecules bonded covalently.

disruptive selection A form of selection in which selection acts to eliminate rather than favor the intermediate type.

diurnal (L. *diurnalis*, day) Active during the day.

DNA ligase The enzyme that links together Okazaki fragments in DNA replication of the lagging strand; it also links other broken areas of the DNA backbone.

domain (1) A distinct modular region of a protein that serves a particular function in the action of the protein, such as a regulatory domain or a DNA-binding domain. (2) In taxonomy, the level higher than kingdom. The three domains currently recognized are Bacteria, Archaea, and Eukarya.

dominant An allele that is expressed when present in either the heterozygous or the homozygous condition.

double fertilization The fusion of the egg and sperm (resulting in a 2*n* fertilized egg, the zygote) and the simultaneous fusion of the second male gamete with the polar nuclei (resulting in a primary endosperm nucleus, which is often triploid, 3*n*); a unique characteristic of all angiosperms.

Down syndrome A congenital syndrome caused by the presence of an extra copy of chromosome 21.

duodenum (L. *duodeni*, 12 each—from its length, about 12 fingers' breadth) In vertebrates, the upper portion of the small intestine.

E

ecdysis (Gr. *ekdysis*, stripping off) Shedding of outer, cuticular layer; molting, as in insects or crustaceans.

ecdysone (Gr. *ekdysis*, stripping off) Molting hormone of arthropods, which triggers when ecdysis occurs.

ecology (Gr. *oikos*, house, + *logos*, word) The study of interactions of organisms with one another and with their physical environment.

ecosystem (Gr. *oikos*, house, + *systems*, that which is put together) A major interacting system that includes organisms and their nonliving environment.

ecotype (Gr. *oikos*, house, + L. *typus*, image) A locally adapted variant of an organism; differing genetically from other ecotypes.

ectoderm (Gr. *ecto*, outside, + *derma*, skin) One of the three embryonic germ layers of early vertebrate embryos; ectoderm gives rise to the outer epithelium of the body (skin, hair, nails) and to the nerve tissue, including the sense organs, brain, and spinal cord.

ectomycorrhizae Externally developing mycorrhizae that do not penetrate the cells they surround.

ectotherms Animals such as reptiles, fish, or amphibians, whose body temperature is regulated by their behavior or by their surroundings.

electronegativity A property of atomic nuclei that refers to the affinity of the nuclei for valence electrons; a nucleus that is more electronegative has a greater pull on electrons than one that is less electronegative.

electron transport chain The passage of energetic electrons through a series of membrane-associated electron-carrier molecules to proton pumps embedded within mitochondrial or chloroplast membranes. *See* chemiosmosis.

endergonic Describes a chemical reaction in which the products contain more energy than the reactants, so that free energy must be put into the reaction from an outside source to allow it to proceed.

endocrine gland (Gr. *endon*, within, + *krinein*, to separate) Ductless gland that secretes hormones into the extracellular spaces, from which they diffuse into the circulatory system.

endocytosis (Gr. *endon*, within, + *kytos*, hollow vessel) The uptake of material into cells by inclusion within an invagination of the plasma membrane; the uptake of solid material is phagocytosis, while that of dissolved material is pinocytosis.

endoderm (Gr. *endon*, within, + *derma*, skin) One of the three embryonic germ layers of early vertebrate embryos, destined to give rise to the epithelium that lines internal structures and most of the digestive and respiratory tracts.

endodermis (Gr. *endon*, within, + *derma*, skin) In vascular plants, a layer of cells forming the innermost layer of the cortex in roots and some stems.

endometrium (Gr. *endon*, within, + *metrios*, of the womb) The lining of the uterus in mammals; thickens in response to secretion of estrogens and progesterone and is sloughed off in menstruation.

endomycorrhizae Mycorrhizae that develop within cells.

endoplasmic reticulum (ER) An internal membrane system that forms a netlike array of channels and interconnections of organelles within the cytoplasm of eukaryotic cells.

endorphin One of a group of small neuropeptides produced by the vertebrate brain; like morphine, endorphins modulate pain perception.

endosperm (Gr. *endon*, within, + *sperma*, seed) A storage tissue characteristic of the seeds of angiosperms, which develops from the union of a male nucleus and the polar nuclei of the embryo sac. The endosperm is digested by the growing sporophyte either before maturation of the seed or during its germination.

endosymbiosis Theory that proposes that eukaryotic cells evolved from a symbiosis between different species of prokaryotes.

endotherm An animal capable of maintaining a constant body temperature. *See* homeotherm.

energy The capacity to do work.

enhancer A site of regulatory protein binding on the DNA molecule distant from the promoter and start site for a gene's transcription.

enthalpy In a chemical reaction, the energy contained in the chemical bonds of the molecule, symbolized as H; in a cellular reaction, the free energy is equal to the enthalpy of the reactant molecules in the reaction.

entropy (Gr. *en*, in, + *tropos*, change in manner) A measure of the randomness or disorder of a system; a measure of how much energy in a system has become so dispersed (usually as evenly distributed heat) that it is no longer available to do work.

enzyme (Gr. *enzymes*, leavened, from *en*, in, + *zyme*, leaven) A protein that is capable of speeding up specific chemical reactions by lowering the required activation energy.

epicotyl The region just above where the cotyledons are attached.

epidermis (Gr. *epi*, on or over, + *derma*, skin) The outermost layers of cells; in plants, the exterior primary tissue of leaves, young stems, and roots; in vertebrates, the nonvascular external layer of skin, of ectodermal origin; in invertebrates, a single layer of ectodermal epithelium.

epididymis (Gr. *epi*, on, + *didymos*, testicle) A sperm storage vessel; a coiled part of the sperm duct that lies near the testis.

epistasis (Gr. *epi*, on, + *stasis*, standing still) Interaction between two nonallelic genes in which one of them modifies the phenotypic expression of the other.

epithelium (Gr. *epi*, on, + *thele*, nipple) In animals, a type of tissue that covers an exposed surface or lines a tube or cavity.

equilibrium (L. *aequus*, equal, + *libra*, balance) A stable condition; the point at which a chemical reaction proceeds as rapidly in the reverse direction as it does in the forward direction, so that there is no further net change in the concentrations of products or reactants. In ecology, a stable condition that resists change and fairly quickly returns to its original state if disturbed by humans or natural events.

erythrocyte (Gr. *erythros*, red, + *kytos*, hollow vessel) Red blood cell, the carrier of hemoglobin.

erythropoiesis The manufacture of blood cells in the bone marrow.

estrous cycle The periodic cycle in which periods of estrus correspond to ovulation events.

estrus (L. *oestrus*, frenzy) The period of maximum female sexual receptivity, associated with ovulation of the egg.

ethology (Gr. *ethos*, habit or custom, + *logos*, discourse) The study of patterns of animal behavior in nature.

euchromatin (Gr. *eu*, good, + *chroma*, color) That portion of a eukaryotic chromosome that is transcribed into mRNA; contains active genes that are not tightly condensed during interphase.

eukaryote (Gr. *eu*, good, + *karyon*, kernel) A cell characterized by membrane-bounded organelles, most notably the nucleus, and one that possesses chromosomes whose DNA is associated with proteins; an organism composed of such cells.

eutrophic (Gr. *eutrophos*, thriving) Refers to a lake in which an abundant supply of minerals and organic matter exists.

evolution (L. *evolvere*, to unfold) Genetic change in a population of organisms; in general, evolution leads to progressive change from simple to complex.

exergonic Describes a chemical reaction in which the products contain less free energy than the reactants, so that free energy is released in the reaction.

exocrine gland (Gr. *ex*, out of, + *krinein*, to separate) A type of gland that releases its secretion through a duct, such as a digestive gland or a sweat gland.

exocytosis (Gr. *ex*, out of, + *kytos*, hollow vessel) A type of bulk transport out of cells in which a vacuole fuses with the plasma membrane, discharging the vacuole's contents to the outside.

exon (Gr. *exo*, outside) A segment of DNA that is both transcribed into RNA and translated into protein.

exoskeleton (Gr. *exo*, outside, + *skeletos*, hand) An external skeleton, as in arthropods.

exteroceptor (L. *exter*, outward, + *capere*, to take) A receptor that is excited by stimuli from the external world.

F

facilitated diffusion Carrier-assisted diffusion of molecules across a cellular membrane through specific channels from a region of higher concentration to one of lower concentration; the process is driven by the concentration gradient and does not require cellular energy from ATP.

fat A molecule composed of glycerol and three fatty acid molecules.

feedback inhibition Control mechanism whereby an increase in the concentration of some molecules inhibits the synthesis of that molecule.

fermentation (L. *fermentum*, ferment) The enzyme-catalyzed extraction of energy from organic compounds without the involvement of oxygen.

fertilization The fusion of two haploid gamete nuclei to form a diploid zygote nucleus.

fibroblast (L. *fibra*, fiber, + Gr. *blastos*, sprout) A flat, irregularly branching cell of connective tissue that secretes structurally strong proteins into the matrix between the cells.

First Law of Thermodynamics Energy cannot be created or destroyed, but can only undergo conversion from one form to another; thus, the amount of energy in the universe is unchangeable.

fitness The genetic contribution of an individual to succeeding generations. relative fitness refers to the fitness of an individual relative to other individuals in a population.

fixed action pattern A stereotyped animal behavior response, thought by ethologists to be based on programmed neural circuits.

foraging behavior A collective term for the many complex, evolved behaviors that influence what an animal eats and how the food is obtained.

founder effect The effect by which rare alleles and combinations of alleles may be enhanced in new populations.

fovea (L., a small pit) A small depression in the center of the retina with a high concentration of cones; the area of sharpest vision.

free energy Energy available to do work.

free radical An ionized atom with one or more unpaired electrons, resulting from electrons that have been energized by ionizing radiation being ejected from the atom; free radicals react violently with other molecules, such as DNA, causing damage by mutation.

fruit In angiosperms, a mature, ripened ovary (or group of ovaries), containing the seeds.

functional genomics Research into the function of specific genes in an organism.

fundamental niche Also referred to as the hypothetical niche, this is the entire niche an organism could fill if there were no other interacting factors (such as competition or predation).

G

gametangium, pl. **gametangia** (Gr. *gamein*, to marry, + L. *tangere*, to touch) A cell or organ in which gametes are formed.

gamete (Gr., wife) A haploid reproductive cell.

gametocytes Cells in the malarial sporozoite life cycle capable of giving rise to gametes when in the correct host.

gametophyte In plants, the haploid (*n*), gamete-producing generation, which alternates with the diploid (*2n*) sporophyte.

ganglion, pl. **ganglia** (Gr., a swelling) An aggregation of nerve cell bodies; in invertebrates, ganglia are the integrative centers; in vertebrates, the term is restricted to aggregations of nerve cell bodies located outside the central nervous system.

gap junction A junction between adjacent animal cells that allows the passage of materials between the cells.

gastrula (Gr., little stomach) In vertebrates, the embryonic stage in which the blastula with its single layer of cells turns into a three-layered embryo made up of ectoderm, mesoderm, and endoderm.

gene (Gr. *genos*, birth, race) The basic unit of heredity; a sequence of DNA nucleotides on a chromosome that encodes a protein, tRNA, or rRNA molecule, or regulates the transcription of such a sequence.

gene conversion Alteration of one homologous chromosome by the cell's error-detection and repair system to make it resemble the sequence on the other homologue.

genetic drift Random fluctuation in allele frequencies over time by chance.

genome The entire DNA sequence of an organism.

genomics The study of genomes as opposed to individual genes.

genotype (Gr. *genos*, birth, race, + *typos*, form) The genetic constitution underlying a single trait or set of traits.

genus, pl. **genera** (L., race) A taxonomic group that ranks below a family and above a species.

germination (L. *germinare*, to sprout) The resumption of growth and development by a spore or seed.

germ layers The three cell layers formed at gastrulation of the embryo that foreshadow the future organization of tissues; the layers, from the outside inward, are the ectoderm, the mesoderm, and the endoderm.

gill (1) In aquatic animals, a respiratory organ, usually a thin-walled projection from some part of the external body surface, endowed with a rich capillary bed and having a large surface area. (2) In basidiomycete fungi, the plates on the underside of the cap.

glomerular filtrate The fluid that passes out of the capillaries of each glomerulus.

glomerulus (L., a little ball) A cluster of capillaries enclosed by Bowman's capsule.

glucagon (Gr. *glykys*, sweet, + *ago*, to lead forward) A vertebrate hormone produced in the pancreas that acts to initiate the breakdown of glycogen to glucose subunits.

gluconeogenesis The synthesis of glucose from noncarbohydrates (such as proteins or fats).

glucose A common six-carbon sugar ($C_6H_{12}O_6$); the most common monosaccharide in most organisms.

glycocalyx A "sugar coating" on the surface of a cell resulting from the presence of polysaccharides on glycolipids and glycoproteins embedded in the outer layer of the plasma membrane.

glycogen (Gr. *glykys*, sweet, + *gen*, of a kind) Animal starch; a complex branched polysaccharide that serves as a food reserve in animals, bacteria, and fungi.

glycolipid Lipid molecule modified within the Golgi complex by having a short sugar chain (polysaccharide) attached.

glycolysis (Gr. *glykys*, sweet, + *lyein*, to loosen) The anaerobic breakdown of glucose; this enzyme-catalyzed process yields two molecules of pyruvate with a net of two molecules of ATP.

glycoprotein Protein molecule modified within the Golgi complex by having a short sugar chain (polysaccharide) attached.

glyoxysome A small cellular organelle or microbody containing enzymes necessary for conversion of fats into carbohydrates.

Golgi apparatus A collection of flattened stacks of membranes (each called a Golgi body) in the cytoplasm of eukaryotic cells; functions in collection, packaging, and distribution of molecules synthesized in the cell.

G protein A protein that binds guanosine triphosphate (GTP) and assists in the function of cell surface receptors. When the receptor binds its signal molecule, the G protein binds GTP and is activated to start a chain of events within the cell.

gravitropism (L. *gravis*, heavy, + *tropes*, turning) Growth response to gravity in plants; formerly called geotropism.

ground meristem (Gr. *meristos*, divisible) The primary meristem, or meristematic tissue, that gives rise to the plant body (except for the epidermis and vascular tissues).

guttation (L. *gutta*, a drop) The exudation of liquid water from leaves due to root pressure.

gymnosperm (Gr. *gymnos*, naked, + *sperma*, seed) A seed plant with seeds not enclosed in an ovary; conifers are gymnosperms.

gynoecium (Gr. *gyne*, woman, + *oikos*, house) The aggregate of carpels in the flower of a seed plant.

H

habitat (L. *hibitare*, to inhabit) The environment of an organism; the place where it is usually found.

habituation (L. *habitus*, condition) A form of learning; a diminishing response to a repeated stimulus.

haplodiploidy A phenomenon occurring in certain organisms such as wasps, wherein both haploid (male) and diploid (female) individuals are encountered.

haploid (Gr. *haploos*, single, + *ploion*, vessel) Having only one set of chromosomes (*n*), in contrast to diploid (2*n*).

Hardy-Weinberg equilibrium A mathematical description of the fact that allele and genotype frequencies remain constant in a random-mating population in the absence of inbreeding, selection, or other evolutionary forces; usually stated: if the frequency of allele *a* is *p* and the frequency of allele *b* is *q*, then the genotype frequencies after one generation of random mating will always be $p^2 + 2pq + q^2 = 1$.

Haversian canal After Clopton Havers, English anatomist. Narrow channels that run parallel to the length of a bone and contain blood vessels and nerve cells.

helper T cell A class of white blood cells that initiates both the cell-mediated immune response and the humoral immune response; helper T cells are the targets of the AIDS virus (HIV).

hemoglobin (Gr. *haima*, blood, + L. *globus*, a ball) A globular protein in vertebrate red blood cells and in the plasma of many invertebrates that carries oxygen and carbon dioxide.

hemophilia (Gr. *haima*, blood, + *philios*, friendly) A group of hereditary diseases characterized by failure of the blood to clot.

hemopoietic stem cell (Gr. *haima*, blood, + *poiesis*, a making) The cells in bone marrow where blood cells are formed.

hermaphroditism (Gr. *hermaphroditos*, containing both sexes) Condition in which an organism has both male and female functional reproductive organs.

heterochromatin (Gr. *heteros*, other, + *chroma*, color) The portion of a eukaryotic chromosome that is not transcribed into RNA; remains condensed in interphase and stains intensely in histological preparations.

heterokaryotic (Gr. *heteros*, other, + *karyon*, kernel) In fungi, having two or more genetically distinct types of nuclei within the same mycelium.

heterosporous (Gr. *heteros*, other, + *sporos*, seed) In vascular plants, having spores of two kinds, namely, microspores and megaspores.

heterotroph (Gr. *heteros*, other, + *trophos*, feeder) An organism that cannot derive energy from photosynthesis or inorganic chemicals, and so must feed on other plants and animals, obtaining chemical energy by degrading their organic molecules.

heterozygous Having two different alleles of the same gene; the term is usually applied to one or more specific loci, as in "heterozygous with respect to the *W* locus" (that is, the genotype is *W/w*).

histone (Gr. *histos*, tissue) One of a group of relatively small, very basic polypeptides, rich in arginine and lysine, forming the core of nucleosomes around which DNA is wrapped in the first stage of chromosome condensation.

holoblastic cleavage (Gr. *holos*, whole, + (*blastos*, germ) Process in vertebrate embryos in which the cleavage divisions all occur at the same rate, yielding a uniform cell size in the blastula.

homeobox A sequence of 180 nucleotides located in homeotic genes that produces a 60-amino-acid peptide sequence (the homeodomain) active in transcription factors.

homeostasis (Gr. *homos*, similar, + *stasis*, standing) The maintenance of a relatively stable internal physiological environment in an organism; usually involves some form of feedback self-regulation.

homeotherm (Gr. *homos*, same or similar, + *therme*, heat) An organism, such as a bird or mammal, capable of maintaining a stable body temperature independent of the environmental temperature. *See* endotherm.

homeotic gene One of a series of "master switch" genes that determine the form of segments developing in the embryo.

hominid (L. *homo*, man) Any primate in the human family, Hominidae. *Homo sapiens* is the only living representative.

hominoid (L. *homo*, man) Collectively, hominids and apes; the monkeys and hominoids constitute the anthropoid primates.

homokaryotic (Gr. *homos*, same, + *karyon*, kernel) In fungi, having nuclei with the same genetic makeup within a mycelium.

homologue One of a pair of chromosomes of the same kind located in a diploid cell; one copy of each pair of homologues comes from each gamete that formed the zygote.

homologous structure (Gr. *homologia*, agreement) A condition in which the similarity between two structures or functions is indicative of a common evolutionary origin.

homosporous (Gr. *homos*, same or similar, + *sporos*, seed) In some plants, production of only one type of spore rather than differentiated types. *Compare with* heterosporous.

homozygous Being a homozygote, having two identical alleles of the same gene; the term is usually applied to one or more specific loci, as in "homozygous with respect to the *W* locus" (that is, the genotype is *W/W* or *w/w*).

hormone (Gr. *hormaein*, to excite) A molecule, usually a peptide or steroid, that is produced in one part of an organism and triggers a specific cellular reaction in target tissues and organs some distance away.

human immunodeficiency virus (HIV) The virus responsible for AIDS, a deadly disease that destroys the human immune system. HIV is a retrovirus (its genetic material is RNA) that is thought to have been introduced to humans from chimpanzees.

hybridization The mating of unlike parents.

hydrostatic skeleton (Gr. *hydro*, water, + *skeletos*, hard) The skeleton of most soft-bodied invertebrates that have neither an internal nor an external skeleton. They use the relative incompressibility of the water within their bodies as a kind of skeleton.

hyperosmotic The condition in which a (hyperosmotic) solution has a higher osmotic concentration than that of a second solution. *Compare with* hypoosmotic.

hyperpolarization Above-normal negativity of a cell membrane during its resting potential.

hypersensitive response Plants respond to pathogens by selectively killing plant cells to block the spread of the pathogen.

hypha, pl. **hyphae** (Gr. *hyphe*, web) A filament of a fungus or oomycete; collectively, the hyphae comprise the mycelium.

hypocotyl The region immediately below where the cotyledons are attached.

hypoosmotic The condition in which a (hypoosmotic) solution has a lower osmotic concentration than that of a second solution. *Compare with* hyperosmotic.

hypothalamus (Gr. *hypo*, under, + *thalamos*, inner room) A region of the vertebrate brain just below the cerebral hemispheres, under the thalamus; a center of the autonomic nervous system, responsible for the integration and correlation of many neural and endocrine functions.

I

imaginal disk One of about a dozen groups of cells set aside in the abdomen of a larval insect and committed to forming key parts of the adult insect's body.

immune response In vertebrates, a defensive reaction of the body to invasion by a foreign substance or organism. *See* antibody and B cell.

immunoglobulin (L. *immunis*, free, + *globus*, globe) An antibody.

inbreeding The breeding of genetically related plants or animals; inbreeding tends to increase homozygosity.

inclusive fitness Describes the sum of the number of genes directly passed on in an individual's offspring and those genes passed on indirectly by kin (other than offspring) whose existence results from the benefit of the individual's altruism.

induction Process by which one group of embryonic cells affects an adjacent group, thereby inducing those cells to differentiate in a manner they otherwise would not have.

industrial melanism (Gr. *melas*, black) Phrase used to describe the evolutionary process in which initially light-colored organisms become dark as a result of natural selection.

inflammatory response (L. *inflammare*, to flame) A generalized nonspecific response to infection that acts to clear an infected area of infecting microbes and dead tissue cells so that tissue repair can begin.

initiation factor One of several proteins involved in the formation of an initiation complex in prokaryote polypeptide synthesis.

insertional inactivation Destruction of a gene's function by the insertion of a transposon.

instar A larval developmental stage in insects.

interferon In vertebrates, a protein produced in virus-infected cells that inhibits viral multiplication.

interneuron (association neuron) A nerve cell found only in the middle of the spinal cord that acts as a functional link between sensory neurons and motor neurons.

internode In plants, the region of a stem between two successive nodes.

interoceptor A receptor that senses information related to the body itself, its internal condition, and its position.

interphase The period between two mitotic or meiotic divisions in which a cell grows and its DNA replicates; includes G_1, S, and G_2 phases.

intron (L. *intra*, within) Portion of mRNA as transcribed from eukaryotic DNA that is removed by enzymes before the mature mRNA is translated into protein. *See* exon.

inversion (L. *invertere*, to turn upside down) A reversal in order of a segment of a chromosome; also, to turn inside out, as in embryogenesis of sponges or discharge of a nematocyst.

ion Any atom or molecule containing an unequal number of electrons and protons and therefore carrying a net positive or net negative charge.

ionizing radiation High-energy radiation that is highly mutagenic, producing free radicals that react with DNA; includes X rays and gamma rays.

isomer (Gr. *isos*, equal, + *meros*, part) One of a group of molecules identical in atomic composition but differing in structural arrangement; for example, glucose and fructose.

isosmotic The condition in which the osmotic concentrations of two solutions are equal, so that no net water movement occurs between them by osmosis.

K

karyotype (Gr. *karyon*, kernel, + *typos*, stamp or print) The morphology of the chromosomes of an organism as viewed with a light microscope.

keratin (Gr. *kera*, horn, + *in*, suffix used for proteins) A tough, fibrous protein formed in epidermal tissues and modified into skin, feathers, hair, and hard structures such as horns and nails.

kidney In vertebrates, the organ that filters the blood to remove nitrogenous wastes and regulates the balance of water and solutes in blood plasma.

kilocalorie Unit describing the amount of heat required to raise the temperature of a kilogram of water by one degree Celsius (°C); sometimes called a Calorie, equivalent to 1000 calories.

kinesis (Gr. *kinesis*, motion) Changes in activity level in an animal that are dependent on stimulus intensity. *See* kinetic energy.

kinetic energy The energy of motion.

kinetochore (Gr. *kinetikos*, putting in motion, + *choros*, chorus) Disc-shaped protein structure within the centromere to which the spindle fibers attach during mitosis or meiosis. *See* centromere.

kingdom The second highest commonly used taxonomic category.

kin selection Selection favoring relatives; an increase in the frequency of related individuals (kin) in a population, leading to an increase in the relative frequency in the population of those alleles shared by members of the kin group.

Krebs cycle Another name for the citric acid cycle; also called the tricarboxylic acid (TCA) cycle.

L

labrum (L., a lip) The upper lip of insects and crustaceans situated above or in front of the mandibles.

larynx The voice box; a cartilaginous organ that lies between the pharynx and trachea and is responsible for sound production in vertebrates.

lateral line system A sensory system encountered in fish, through which mechanoreceptors in a line down the side of the fish are sensitive to motion.

lateral meristems (L. *latus*, side, + Gr. *meristos*, divided) In vascular plants, the meristems that give rise to secondary tissue; the vascular cambium and cork cambium.

Law of Independent Assortment Mendel's second law of heredity, stating that genes located on nonhomologous chromosomes assort independently of one another.

Law of Segregation Mendel's first law of heredity, stating that alternative alleles for the same gene segregate from each other in production of gametes.

leaf primordium, pl. **primordia** (L. *primordium*, beginning) A lateral outgrowth from the apical meristem that will eventually become a leaf.

lenticels (L. *lenticella*, a small window) Spongy areas in the cork surfaces of stem, roots, and other plant parts that allow interchange of gases between internal tissues and the atmosphere through the periderm.

leucoplast (Gr. *leukos*, white, + *plasein*, to form) In plant cells, a colorless plastid in which starch grains are stored; usually found in cells not exposed to light.

leukocyte (Gr. *leukos*, white, + *kytos*, hollow vessel) A white blood cell; a diverse array of nonhemoglobin-containing blood cells, including phagocytic macrophages and antibody-producing lymphocytes.

lichen Symbiotic association between a fungus and a photosynthetic organism such as a green alga or cyanobacterium.

limbic system The hypothalamus, together with the network of neurons that link the hypothalamus to some areas of the cerebral cortex. Responsible for many of the most deep-seated drives and emotions of vertebrates, including pain, anger, sex, hunger, thirst, and pleasure.

lipase (Gr. *lipos*, fat, + *-ase*, enzyme suffix) An enzyme that catalyzes the hydrolysis of fats.

lipid (Gr. *lipos*, fat) A nonpolar hydrophobic organic molecule that is insoluble in water (which is polar) but dissolves readily in nonpolar organic solvents; includes fats, oils, waxes, steroids, phospholipids, and carotenoids.

lipid bilayer The structure of a cellular membrane, in which two layers of phospholipids spontaneously align so that the hydrophilic head groups are exposed to water, while the hydrophobic fatty acid tails are pointed toward the center of the membrane.

locus The position on a chromosome where a gene is located.

loop of Henle In the kidney of birds and mammals, a hairpin-shaped portion of the renal tubule in which water and salt are reabsorbed from the glomerular filtrate by diffusion.

lophophore A ring or U-shaped arrangement of tentacles, often ciliated, into which the coelom of the animal extends. The organ functions in feeding and gas exchange and is encountered in only a few phyla of small marine invertebrates.

luteal phase The second phase of the female reproductive cycle, during which the mature eggs are released into the fallopian tubes, a process called ovulation.

lymph (L. *lympha*, clear water) In animals, a colorless fluid derived from blood by filtration through capillary walls in the tissues.

lymphatic system In animals, an open vascular system that reclaims water that has entered interstitial regions from the bloodstream (lymph); includes the lymph nodes, spleen, thymus, and tonsils.

lymphocyte (L. *lympha*, water, + Gr. *kytos*, hollow vessel) A type of white blood cell. Lymphocytes are responsible for the immune response; there are two principal classes—B cells and T cells.

lymphokine A regulatory molecule that is secreted by lymphocytes. In the immune response, lymphokines secreted by helper T cells unleash the cell-mediated immune response.

lysis (Gr., a loosening) Disintegration of a cell by rupture of its plasma membrane.

M

macroevolution (Gr. *makros*, large, + L. *evolvere*, to unfold) The creation of new species and the extinction of old ones.

macromolecule (Gr. *makros*, large, + L. *moleculus*, a little mass) An extremely large biological molecule; refers specifically to proteins, nucleic acids, polysaccharides, lipids, and complexes of these.

macronutrients (Gr. *makros*, large, + L. *nutrire*, to nourish) Inorganic chemical elements required in large amounts for plant growth, such as nitrogen, potassium, calcium, phosphorus, magnesium, and sulfur.

macrophage A large phagocytic cell that is able to engulf and digest cellular debris and invading bacteria.

major histocompatibility complex (MHC) A set of protein cell-surface markers anchored in the plasma membrane, which the immune system uses to identify "self." All the cells of a given individual have the same "self" marker, called an MHC protein.

Malpighian tubules Blind tubules opening into the hindgut of terrestrial arthropods; they function as excretory organs.

mandibles (L. *mandibula*, jaw) In crustaceans, insects, and myriapods, the appendages immediately posterior to the antennae; used to seize, hold, bite, or chew food.

mantle The soft, outermost layer of the body wall in mollusks; the mantle secretes the shell.

marsupial (L. *marsupium*, pouch) A mammal in which the young are born early in their development, sometimes as soon as eight days after fertilization, and are retained in a pouch.

mass flow hypothesis The overall process by which materials move in the phloem of plants.

matrix (L. *mater*, mother) In mitochondria, the solution in the interior space surrounded by the cristae that contains the enzymes and other molecules involved in oxidative respiration; more generally, that part of a tissue within which an organ or process is embedded.

menstrual cycle (L. *mens*, month) In humans and higher primates, the cyclic changes in the ovaries and uterine endometrium; lasts about a month in humans.

menstruation   Periodic sloughing off of the blood-enriched lining of the uterus when pregnancy does not occur.

meristem   (Gr. *merizein*, to divide) Undifferentiated plant tissue from which new cells arise.

meroblastic cleavage   (Gr. *meros*, part, + *blastos*, sprout) A type of cleavage in the eggs of reptiles, birds, and some fish. Occurs only on the blastodisc.

mesoderm   (Gr. *mesos*, middle, + *derma*, skin) One of the three embryonic germ layers that form in the gastrula; gives rise to muscle, bone and other connective tissue, the peritoneum, the circulatory system, and most of the excretory and reproductive systems.

mesophyll   (Gr. *mesos*, middle, + *phyllon*, leaf) The photosynthetic parenchyma of a leaf, located within the epidermis.

messenger RNA (mRNA)   The RNA transcribed from structural genes; RNA molecules complementary to a portion of one strand of DNA, which are translated by the ribosomes to form protein.

metabolism   (Gr. *metabole*, change) The sum of all chemical processes occurring within a living cell or organism.

metamorphosis   (Gr. *meta*, after, + *morphe*, form, + *osis*, state of) Process in which a marked change in form takes place during postembryonic development as, for example, from tadpole to frog.

metaphase   (Gr. *meta*, after, + *phasis*, form) The stage of mitosis or meiosis during which microtubules become organized into a spindle and the chromosomes come to lie in the spindle's equatorial plane.

metastasis   The process by which cancer cells move from their point of origin to other locations in the body; also, a population of cancer cells in a secondary location, the result of movement from the primary tumor.

methanogens   Obligate, anaerobic archaebacteria that produce methane.

microarray   DNA sequences are placed on a microscope slide or chip with a robot. The microarray can then be probed with RNA from specific tissues to identify expressed DNA.

microbody   A cellular organelle bounded by a single membrane and containing a variety of enzymes; generally derived from endoplasmic reticulum; includes peroxisomes and glyoxysomes.

microevolution   (Gr. *mikros*, small, + L. *evolvere*, to unfold) Refers to the evolutionary process itself.

Evolution within a species. Also called adaptation.

micronutrient   (Gr. *mikros*, small, + L. *nutrire*, to nourish) A mineral required in only minute amounts for plant growth, such as iron, chlorine, copper, manganese, zinc, molybdenum, and boron.

micropyle   In the ovules of seed plants, an opening in the integuments through which the pollen tube usually enters.

microtubule   (Gr. *mikros*, small, + L. *tubulus*, little pipe) In eukaryotic cells, a long, hollow protein cylinder, composed of the protein tubulin; these influence cell shape, move the chromosomes in cell division, and provide the functional internal structure of cilia and flagella.

microvillus   (Gr. *mikros*, small, + L. *villus*, shaggy hair) Cytoplasmic projection from epithelial cells; microvilli greatly increase the surface area of the small intestine.

middle lamella   The layer of intercellular material, rich in pectic compounds, that cements together the primary walls of adjacent plant cells.

mimicry   (Gr. *mimos*, mime) The resemblance in form, color, or behavior of certain organisms (mimics) to other more powerful or more protected ones (models).

mitosis   (Gr. *mitos*, thread) Somatic cell division; nuclear division in which the duplicated chromosomes separate to form two genetically identical daughter nuclei.

monocot   Short for monocotyledon; flowering plant in which the embryos have only one cotyledon, the floral parts are generally in threes, and the leaves typically are parallel-veined.

monocyte   (Gr. *monos*, single, + *kytos*, hollow vessel) A type of leukocyte that becomes a phagocytic cell (macrophage) after moving into tissues.

monoecious   (Gr. *monos*, single, + *oecos*, house) A plant in which the staminate and pistillate flowers are separate, but borne on the same individual.

monophyletic   In phylogenetic classification, a group that includes the most recent common ancestor of the group and all its descendants. A clade is a monophyletic group.

monosaccharide   (Gr. *monos*, single, + L. *saccharum*, sugar) A simple sugar that cannot be decomposed into smaller sugar molecules.

monotreme   (Gr. *monos*, single, + *treme*, hole) An egg-laying mammal.

morphogen   A signal molecule produced by an embryonic

organizer region that informs surrounding cells of their distance from the organizer, thus determining relative positions of cells during development.

morphology   The form and structure of an organism.

morula   (L., a little mulberry) Solid ball of cells in the early stage of embryonic development.

mosaic development   A pattern of embryonic development in which initial cells produced by cleavage divisions contain different developmental signals (determinants) from the egg, setting the individual cells on different developmental paths.

motor (efferent) neuron   Neuron that transmits nerve impulses from the central nervous system to an effector, which is typically a muscle or gland.

Müllerian mimicry   After Fritz Müller, German biologist. A phenomenon in which two or more unrelated but protected species resemble one another, thus achieving a kind of group defense.

multigene family   A collection of related genes on a single chromosome or on different chromosomes.

muscle fiber   A long, cylindrical, multinucleated cell containing numerous myofibrils, which is capable of contraction when stimulated.

mutagen   (L. *mutare*, to change, + Gr. *genaio*, to produce) An agent that induces changes in DNA (mutations); includes physical agents that damage DNA and chemicals that alter DNA bases.

mutant   (L. *mutare*, to change) A mutated gene; alternatively, an organism carrying a gene that has undergone a mutation.

mutation   A permanent change in a cell's DNA; includes changes in nucleotide sequence, alteration of gene position, gene loss or duplication, and insertion of foreign sequences.

mutualism   (L. *mutuus*, lent, borrowed) A symbiotic association in which two (or more) organisms live together, and both members benefit.

mycelium, pl. **mycelia**   (Gr. *mykes*, fungus) In fungi, a mass of hyphae.

mycorrhiza, pl. **mycorrhizae**   (Gr. *mykes*, fungus, + *rhiza*, root) A symbiotic association between fungi and the roots of a plant.

myelin sheath   (Gr. *myelinos*, full of marrow) A fatty layer surrounding the long axons of motor neurons in the peripheral nervous system of vertebrates.

myofilament   (Gr. *myos*, muscle, + L. *filare*, to spin) A contractile

microfilament, composed largely of actin and myosin, within muscle.

myosin   (Gr. *myos*, muscle, + *in*, belonging to) One of the two protein components of microfilaments (the other is actin); a principal component of vertebrate muscle.

N

natural killer cell   A cell that does not kill invading microbes, but rather, the cells infected by them.

natural selection   The differential reproduction of genotypes; caused by factors in the environment; leads to evolutionary change.

negative feedback   A homeostatic control mechanism whereby an increase in some substance or activity inhibits the process leading to the increase; also known as feedback inhibition.

nephridium, pl. **nephridia**   (Gr. *nephros*, kidney) In invertebrates, a tubular excretory structure.

nephrid organ   A filtration system of many freshwater invertebrates in which water and waste pass from the body across the membrane into a collecting organ, from which they are expelled to the outside through a pore.

nephron   (Gr. *nephros*, kidney) Functional unit of the vertebrate kidney; one of numerous tubules involved in filtration and selective reabsorption of blood; each nephron consists of a Bowman's capsule, an enclosed glomerulus, and a long attached tubule; in humans, called a renal tubule.

nerve   A group or bundle of nerve fibers (axons) with accompanying neurological cells, held together by connective tissue; located in the peripheral nervous system.

neural crest   A special strip of cells that develops just before the neural groove closes over to form the neural tube in embryonic development.

neural groove   The long groove formed along the long axis of the embryo by a layer of ectodermal cells.

neural tube   The dorsal tube, formed from the neural plate, that differentiates into the brain and spinal cord.

neuroglia   (Gr. *neuron*, nerve, + *glia*, glue) Nonconducting nerve cells that are intimately associated with neurons and appear to provide nutritional support.

neuromuscular junction   The structure formed when the tips of axons contact (innervate) a muscle fiber.

neuron (Gr., nerve) A nerve cell specialized for signal transmission; includes cell body, dendrites, and axon.

neurotransmitter (Gr. *neuron*, nerve, + L. *trans*, across, + *mittere*, to send) A chemical released at the axon terminal of a neuron that travels across the synaptic cleft, binds a specific receptor on the far side, and depending on the nature of the receptor, depolarizes or hyperpolarizes a second neuron or a muscle or gland cell.

neurulation A process in early embryonic development by which a dorsal band of ectoderm thickens and rolls into the neural tube.

neutrophil An abundant type of granulocyte capable of engulfing microorganisms and other foreign particles; neutrophils comprise about 50 to 70% of the total number of white blood cells.

niche The role played by a particular species in its environment.

nicotinamide adenine dinucleotide (NAD⁺) A molecule that becomes reduced (to NADH) as it carries high-energy electrons from oxidized molecules and delivers them to ATP-producing pathways in the cell.

nociceptor A naked dendrite that acts as a receptor in response to a pain stimulus.

nocturnal (L. *nocturnus*, night) Active primarily at night.

node (L. *nodus*, knot) The part of a plant stem where one or more leaves are attached. *See* internode.

node of Ranvier After L.A. Ranvier, French histologist. A gap formed at the point where two Schwann cells meet and where the axon is in direct contact with the surrounding intercellular fluid.

nonassociative learning A learned behavior that does not require an animal to form an association between two stimuli, or between a stimulus and a response.

nonsense codon One of three codons (UAA, UAG, and UGA) that are not recognized by tRNAs, thus serving as "stop" signals in the mRNA message and terminating translation.

notochord (Gr. *noto*, back, + L. *chorda*, cord) In chordates, a dorsal rod of cartilage that runs the length of the body and forms the primitive axial skeleton in the embryos of all chordates.

nucellus (L. *nucella*, a small nut) Tissue composing the chief pair of young ovules, in which the embryo sac develops; equivalent to a megasporangium.

nucleic acid A nucleotide polymer; chief types are deoxyribonucleic acid (DNA), which is double-stranded, and ribonucleic acid (RNA), which is typically single-stranded.

nucleolus (L., a small nucleus) In eukaryotes, the site of rRNA synthesis; a spherical body composed chiefly of rRNA in the process of being transcribed from multiple copies of rRNA genes.

nucleosome (L. *nucleus*, kernel, + *soma*, body) The fundamental packaging unit of eukaryotic chromosomes; a complex of DNA and histone proteins in which the double-helical DNA winds around eight molecules of histone; chromatin is composed of long sequences of nucleosomes.

nucleotide A single unit of nucleic acid, composed of a phosphate, a five-carbon sugar (either ribose or deoxyribose), and a purine or a pyrimidine.

nucleus In atoms, the central core, containing positively charged protons and (in all but hydrogen) electrically neutral neutrons; in eukaryotic cells, the membranous organelle that houses the chromosomal DNA; in the central nervous system, a cluster of nerve cell bodies.

O

ocellus, pl. **ocelli** (L., little eye) A simple light receptor common among invertebrates.

Okazaki fragment A short segment of DNA produced by discontinuous replication elongating in the 5'→ 3' direction away from the replication.

olfaction (L. *olfactum*, smell) The process or function of smelling.

ommatidium, pl. **ommatidia** (Gr., little eye) The visual unit in the compound eye of arthropods; contains light-sensitive cells and a lens able to form an image.

oncogene (Gr. *oncos*, cancer, + *genos*, birth) A mutant form of a growth-regulating gene that is inappropriately "on," causing unrestrained cell growth and division.

oocyst The zygote in a sporozoan life cycle. It is surrounded by a tough cyst to prevent dehydration or other damage.

operant conditioning A learning mechanism in which the reward follows only after the correct behavioral response.

operator A site of gene regulation; a sequence of nucleotides overlapping the promoter site and recognized by a repressor protein; binding of the repressor prevents binding of the polymerase to the promoter site and so blocks transcription of the structural gene.

operculum A flat, bony, external protective covering over the gill chamber in fish.

operon (L. *operis*, work) A cluster of adjacent structural genes transcribed as a unit into a single mRNA molecule.

order A category of classification above the level of family and below that of class.

organ (Gr. *organon*, tool) A body structure composed of several different tissues grouped in a structural and functional unit.

organelle (Gr. *organella*, little tool) Specialized part of a cell; literally, a small cytoplasmic organ.

orthologs Genes that reflect the conservation of a single gene found in an ancestor.

osmoconformer An animal that maintains the osmotic concentration of its body fluids at about the same level as that of the medium in which it is living.

osmosis (Gr. *osmos*, act of pushing, thrust) The diffusion of water across a selectively permeable membrane (a membrane that permits the free passage of water but prevents or retards the passage of a solute); in the absence of differences in pressure or volume, the net movement of water is from the side containing a lower concentration of solute to the side containing a higher concentration.

osmotic pressure The potential pressure developed by a solution separated from pure water by a differentially permeable membrane. The higher the solute concentration, the greater the osmotic potential of the solution; also called *osmotic potential*.

osteoblast (Gr. *osteon*, bone, + *blastos*, bud) A bone-forming cell.

osteocyte (Gr. *osteon*, bone, + *kytos*, hollow vessel) A mature osteoblast.

outcrossing Breeding with individuals other than oneself or one's close relatives.

ovary (L. *ovum*, egg) (1) In animals, the organ in which eggs are produced. (2) In flowering plants, the enlarged basal portion of a carpel, which contains the ovule(s); the ovary matures to become the fruit.

oviduct (L. *ovum*, egg, + *ductus*, duct) In vertebrates, the passageway through which ova (eggs) travel from the ovary to the uterus.

oviparity (L. *ovum*, egg, + *parere*, to bring forth) Refers to a type of reproduction in which the eggs are developed after leaving the body of the mother, as in reptiles.

ovoviviparity Refers to a type of reproduction in which young hatch from eggs that are retained in the mother's uterus.

ovulation In animals, the release of an egg or eggs from the ovary.

ovum, pl. **ova** (L., egg) The egg cell; female gamete.

oxidation (Fr. *oxider*, to oxidize) Loss of an electron by an atom or molecule; in metabolism, often associated with a gain of oxygen or a loss of hydrogen.

oxidative respiration Process of cellular activity in which glucose or other molecules are broken down to water and carbon dioxide with the release of energy.

oxygen debt The amount of oxygen required to convert the lactic acid generated in the muscles during exercise back into glucose.

oxytocin (Gr. *oxys*, sharp, + *tokos*, birth) A hormone of the posterior pituitary gland that affects uterine contractions during childbirth and stimulates lactation.

ozone O₃, a stratospheric layer of the earth's atmosphere responsible for filtering out ultraviolet radiation supplied by the sun.

P

pacemaker A patch of excitatory tissue in the vertebrate heart that initiates the heartbeat.

palisade parenchyma (L. *palus*, stake) In plant leaves, the columnar, chloroplast-containing parenchyma cells of the mesophyll. Also called *palisade cells*.

papilla A small projection of tissue.

paracrine A type of chemical signaling between cells in which the effects are local and short-lived.

paralogs Two genes within an organism that arose from the duplication of one gene in an ancestor.

paraphyletic In phylogenetic classification, a group that includes the most recent common ancestor of the group, but not all its descendants.

parapodia (Gr. *para*, beside, + *pous*, foot) One of the paired lateral processes on each side of most segments in polychaete annelids.

parasexuality In certain fungi, the fusion and segregation of heterokaryotic haploid nuclei to produce recombinant nuclei.

parasitism (Gr. *para*, beside, + *sitos*, food) A living arrangement in which an organism lives on or in an organism of a different species and derives nutrients from it.

parenchyma cell The most common type of plant cell; characterized by large vacuoles, thin walls, and functional nuclei.

parthenogenesis (Gr. *parthenos*, virgin, + *genesis*, birth) The development of an egg without fertilization, as in aphids, bees, ants, and some lizards.

partial pressure The components of each individual gas—such as nitrogen, oxygen, and carbon dioxide—that together comprise the total air pressure.

pelagic Free-swimming, usually in open water.

pellicle A tough, flexible covering in ciliates and euglenoids.

peptide bond (Gr. *peptein*, to soften, digest) The type of bond that links amino acids together in proteins through a dehydration reaction.

perianth (Gr. *peri*, around, + *anthos*, flower) In flowering plants, the petals and sepals taken together.

pericycle (Gr. *peri*, around, + *kykos*, circle) In vascular plants, one or more cell layers surrounding the vascular tissues of the root, bounded externally by the endodermis and internally by the phloem.

periderm (Gr. *peri*, around, + *derma*, skin) Outer protective tissue in vascular plants that is produced by the cork cambium and functionally replaces epidermis when it is destroyed during secondary growth; the periderm includes the cork, cork cambium, and phelloderm.

peristalsis (Gr. *peristaltikos*, compressing around) In animals, a series of alternating contracting and relaxing muscle movements along the length of a tube such as the oviduct or alimentary canal that tend to force material such as an egg cell or food through the tube.

peroxisome A microbody that plays an important role in the breakdown of highly oxidative hydrogen peroxide by catalase.

petal A flower part, usually conspicuously colored; one of the units of the corolla.

petiole (L. *petiolus*, a little foot) The stalk of a leaf.

phagocyte (Gr. *phagein*, to eat, + *kytos*, hollow vessel) Any cell that engulfs and devours microorganisms or other particles.

phagocytosis (Gr., cell-eating) Endocytosis of a solid particle; the plasma membrane folds inward around the particle (which may be another cell) and engulfs it to form a vacuole.

pharynx (Gr., gullet) In vertebrates, a muscular tube that connects the mouth cavity and the esophagus; it serves as the gateway to the digestive tract and to the trachea.

phenotype (Gr. *phainein*, to show, + *typos*, stamp or print) The realized expression of the genotype; the physical appearance or functional expression of a trait.

pheromone (Gr. *pherein*, to carry, + *hormonos*, exciting, stirring up) Chemical substance released by one organism that influences the behavior or physiological processes of another organism of the same species. Some pheromones serve as sex attractants, as trail markers, and as alarm signals.

phloem (Gr. *phloos*, bark) In vascular plants, a food-conducting tissue basically composed of sieve elements, various kinds of parenchyma cells, fibers, and sclereids.

phosphodiester bond The type of bond that links nucleotides in a nucleic acid; formed when the phosphate group of one nucleotide binds to the 3′ hydroxyl group of the sugar of another.

phospholipid Similar in structure to a fat, but having only two fatty acids attached to the glycerol backbone, with the third space linked to a phosphorylated molecule; contains a polar hydrophilic "head" end (phosphate group) and a nonpolar hydrophobic "tail" end (fatty acids).

photoperiodism (Gr. *photos*, light, + *periodos*, a period) The tendency of biological reactions to respond to the duration and timing of day and night; a mechanism for measuring seasonal time.

photoreceptor (Gr. *photos*, light) A light-sensitive sensory cell.

photosystem An organized collection of chlorophyll and other pigment molecules embedded in the thylakoid of chloroplasts; traps photon energy and channels it as energetic electrons to the thylakoid membrane.

phototropism (Gr. *photos*, light, + *trope*, turning) In plants, a growth response to a light stimulus.

phycologist (Gr. *phykos*, seaweed) The scientist who studies algae.

phylogeny The evolutionary history of an organism, including which species are closely related and in what order related species evolved; often represented in the form of an evolutionary tree.

phylum, pl. phyla (Gr. *phylon*, race, tribe) A major category, between kingdom and class, of taxonomic classifications.

phytochrome (Gr. *phyton*, plant, + *chroma*, color) A plant pigment that is associated with the absorption of light; photoreceptor for red to far-red light.

phytoremediation The process of removing contamination from soil or water using plants.

pigment (L. *pigmentum*, paint) A molecule that absorbs light.

pilus, pl. pili Extensions of a bacterial cell enabling it to transfer genetic materials from one individual to another or to adhere to substrates.

pinocytosis (Gr. *pinein*, to drink, + *kytos*, hollow vessel, + *osis*, condition) The process of fluid uptake by endocytosis in a cell.

pistil (L. *pistillum*, pestle) Central organ of flowers, typically consisting of ovary, style, and stigma; a pistil may consist of one or more fused carpels and is more technically and better known as the gynoecium.

pith The ground tissue occupying the center of the stem or root within the vascular cylinder.

placenta, pl. placentae (L., a flat cake) (1) In flowering plants, the part of the ovary wall to which the ovules or seeds are attached. (2) In mammals, a tissue formed in part from the inner lining of the uterus and in part from other membranes, through which the embryo (later the fetus) is nourished while in the uterus and through which wastes are carried away.

plankton (Gr. *planktos*, wandering) Free-floating, mostly microscopic, aquatic organisms.

plasma (Gr., form) The fluid of vertebrate blood; contains dissolved salts, metabolic wastes, hormones, and a variety of proteins, including antibodies and albumin; blood minus the blood cells.

plasma cell An antibody-producing cell resulting from the multiplication and differentiation of a B lymphocyte that has interacted with an antigen.

plasma membrane The membrane surrounding the cytoplasm of a cell; consists of a single phospholipid bilayer with embedded proteins.

plasmid (Gr. *plasma*, a form or mold) A small fragment of extrachromosomal DNA, usually circular, that replicates independently of the main chromosome, although it may have been derived from it.

plasmodesmata In plants, cytoplasmic connections between adjacent cells.

plasmodium (Gr. *plasma*, a form or mold, + *eidos*, form) Stage in the life cycle of myxomycetes (plasmodial slime molds); a multinucleate mass of protoplasm surrounded by a membrane.

plastid (Gr. *plastos*, formed or molded) An organelle in the cells of photosynthetic eukaryotes that is the site of photosynthesis and, in plants and green algae, of starch storage.

platelet (Gr. dim. of *plattus*, flat) In mammals, a fragment of a white blood cell that circulates in the blood and functions in the formation of blood clots at sites of injury.

pleiotropy Condition in which an individual allele has more than one effect on production of the phenotype.

plumule The epicotyl of a plant with its two young leaves.

point mutation An alteration of one nucleotide in a chromosomal DNA molecule.

polar body Minute, nonfunctioning cell produced during the meiotic divisions leading to gamete formation in vertebrates.

pollen tube A tube formed after germination of the pollen grain; carries the male gametes into the ovule.

pollination The transfer of pollen from an anther to a stigma.

polyandry The condition in which a female mates with more than one male.

polyclonal antibody An antibody response in which an antigen elicits many different antibodies, each fitting a different portion of the antigen surface.

polygyny (Gr. *poly*, many, + *gyne*, woman, wife) A mating choice in which a male mates with more than one female.

polymer (Gr. *poly*, many, + *meris*, part) A molecule composed of many similar or identical molecular subunits; starch is a polymer of glucose.

polymerase chain reaction (PCR) A process by which DNA polymerase is used to copy a sequence of interest repeatedly, making millions of copies of the same DNA.

polymorphism (Gr. *poly*, many, + *morphe*, form) The presence in a population of more than one allele of a gene at a frequency greater than that of newly arising mutations.

polypeptide (Gr. *poly*, many, + *peptein*, to digest) A molecule consisting of many joined amino acids; not usually as complex as a protein.

polyphyletic In phylogenetic classification, a group that does not include the most recent common ancestor of all members of the group.

polyploidy Condition in which one or more entire sets of chromosomes is added to the diploid genome.

polysaccharide (Gr. *poly*, many, + *sakcharon*, sugar, from Latin *sakara*, gravel, sugar) A carbohydrate composed of many monosaccharide sugar subunits linked together in a long chain; examples are glycogen, starch, and cellulose.

polyunsaturated fat A fat molecule having at least two double bonds between adjacent carbons in one or more of the fatty acid chains.

population (L. *populus*, the people) Any group of individuals, usually of a single species, occupying a given area at the same time.

posttranscriptional control A mechanism of control over gene expression that operates after the transcription of mRNA is complete.

potential energy Energy that is not being used, but could be; energy in a potentially usable form; often called "energy of position."

precapillary sphincter A ring of muscle that guards each capillary loop and that, when closed, blocks flow through the capillary.

primary endosperm nucleus In flowering plants, the result of the fusion of a sperm nucleus and the (usually) two polar nuclei.

primary growth In vascular plants, growth originating in the apical meristems of shoots and roots; results in an increase in length.

primary immune response The first response of an immune system to a foreign antigen. If the system is challenged again with the same antigen, the memory cells created during the primary response will respond more quickly.

primary induction Inductions between the three primary tissue types—ectoderm, mesoderm, and endoderm.

primary nondisjunction Failure of chromosomes to separate properly at meiosis I.

primary phloem In plant phloem, the cells involved in food conduction.

primary productivity The amount of energy produced by photosynthetic organisms in a community.

primary structure The specific amino acid sequence of a protein.

primary tissues Tissues that comprise the primary plant body.

primary transcript The initial mRNA molecule copied from a gene by RNA polymerase, containing a faithful copy of the entire gene, including introns as well as exons.

primary wall In plants, the wall layer deposited during the period of cell expansion.

primate Monkeys and apes (including humans).

primitive streak (L. *primus*, first) In the early embryos of birds, reptiles, and mammals, a dorsal, longitudinal strip of ectoderm and mesoderm that is equivalent to the blastopore in other forms.

principle of parsimony Principle stating that scientists should favor the hypothesis that requires the fewest assumptions.

prions Infectious proteinaceous particles.

procambium (Gr. *pro*, before, + *cambiare*, to exchange) In vascular plants, a primary meristematic tissue that gives rise to primary vascular tissues.

prokaryote (Gr. *pro*, before, + *karyon*, kernel) A bacterium; a cell lacking a membrane-bounded nucleus or membrane-bounded organelles.

promoter A specific nucleotide sequence to which RNA polymerase attaches to initiate transcription of mRNA from a gene.

prophase (Gr. *pro*, before, + *phasis*, form) An early stage in nuclear division, characterized by the formation of a microtubule spindle along the future axis of division, the shortening and thickening of the chromosomes, and their movement toward the equator of the spindle (the "metaphase plate").

proprioceptor (L. *proprius*, one's own) In vertebrates, a sensory receptor that senses the body's position and movements.

prostaglandins (Gr. *prostas*, a porch or vestibule, + *glans*, acorn) A group of modified fatty acids that function as chemical messengers.

prostate gland (Gr. *prostas*, a porch or vestibule) In male mammals, a mass of glandular tissue at the base of the urethra that secretes an alkaline fluid that has a stimulating effect on the sperm as they are released.

protein (Gr. *proteios*, primary) A chain of amino acids joined by peptide bonds.

protein kinase An enzyme that adds phosphate groups to proteins, changing their activity.

proteomics The study of all proteins in an organism.

proton pump A protein channel in a membrane of the cell that expends energy to transport protons against a concentration gradient; involved in the chemiosmotic generation of ATP.

proto-oncogene A normal gene that promotes cell division, so called because mutations that cause these genes to become overexpressed convert them into oncogenes that produce excessive cellular proliferation (i.e., cancer).

protozoa (Gr. *protos*, first, + *zoon*, animal) The traditional name given to heterotrophic protists.

pseudogene (Gr. *pseudos*, false, + *genos*, birth) A copy of a gene that is not transcribed.

pseudopod (Gr. *pseudes*, false, + *pous*, foot) A nonpermanent cytoplasmic extension of the cell body. Also called a *pseudopodium*.

punctuated equilibrium A hypothesis about the mechanism of evolutionary change proposing that long periods of little or no change are punctuated by periods of rapid evolution.

pupa (L., girl, doll) A developmental stage of some insects in which the organism is nonfeeding, immotile, and sometimes encapsulated or in a cocoon; the pupal stage occurs between the larval and adult phases.

purine (Gr. *purinos*, fiery, sparkling) The larger of the two general kinds of nucleotide base found in DNA and RNA; a nitrogenous base with a double-ring structure, such as adenine or guanine.

pyrimidine (alt. of pyridine, from G. *pyr*, fire) The smaller of two general kinds of nucleotide base found in DNA and RNA; a nitrogenous base with a single-ring structure, such as cytosine, thymine, or uracil.

Q

quaternary structure The structural level of a protein composed of more than one polypeptide chain, each of which has its own tertiary structure; the individual chains are called subunits.

R

radicle (L. *radicula*, root) The part of the plant embryo that develops into the root.

radioactivity The emission of nuclear particles and rays by unstable atoms as they decay into more stable forms.

radula (L., scraper) Rasping tongue found in most mollusks.

realized niche The actual niche occupied by an organism when all biotic and abiotic interactions are taken into account.

receptor-mediated endocytosis Process by which specific macromolecules are transported into eukaryotic cells at clathrin-coated pits, after binding to specific cell-surface receptors.

receptor protein A highly specific cell-surface receptor embedded in a cell membrane that responds only to a specific messenger molecule.

recessive An allele that is only expressed when present in the homozygous condition, while being "hidden" by the expression of a dominant allele in the heterozygous condition.

reciprocal altruism Performance of an altruistic act with the expectation that the favor will be returned. A key and very controversial assumption of many theories dealing with the evolution of social behavior. *See* altruism.

reciprocal recombination A mechanism of genetic recombination that occurs only in eukaryotic organisms, in which two chromosomes trade segments; can occur between nonhomologous chromosomes as well as the more usual exchange between homologous chromosomes in meiosis.

recombinant DNA Fragments of DNA from two different species, such as a bacterium and a mammal, spliced together in the laboratory into a single molecule.

reduction (L. *reductio*, a bringing back; originally "bringing back" a metal from its oxide) The gain of an electron by an atom, often with an associated proton.

reflex (L. *reflectere*, to bend back) In the nervous system, a motor response subject to little associative modification; a reflex is among the simplest neural pathways, involving only a sensory neuron, sometimes (but not always) an interneuron, and one or more motor neurons.

reflex arc The nerve path in the body that leads from stimulus to reflex action.

refractory period The recovery period after membrane depolarization during which the membrane is unable to respond to additional stimulation.

replication fork The Y-shaped end of a growing replication bubble in a DNA molecule undergoing replication.

repolarization Return of the ions in a nerve to their resting potential distribution following depolarization.

repressor (L. *reprimere*, to press back, keep back) A protein that regulates DNA transcription by preventing RNA polymerase from attaching to the promoter and transcribing the structural gene. *See* operator.

residual volume The amount of air remaining in the lungs after the maximum amount of air has been exhaled.

resting membrane potential The charge difference (difference in electric potential) that exists across a neuron at rest (about 70 millivolts).

restriction endonuclease An enzyme that cleaves a DNA duplex molecule at a particular base sequence, usually within or near a palindromic sequence; also called a restriction enzyme.

restriction fragment length polymorphism (RFLP) Restriction enzymes recognize very specific DNA sequences. Alleles of the same gene or surrounding sequences may have base-pair differences, so that DNA near one allele is cut into a different length fragment than DNA near the other allele. These different fragments separate based on size on gels.

retina (L., a small net) The photosensitive layer of the vertebrate eye; contains several layers of neurons and light receptors (rods and cones); receives the image formed by the lens and transmits it to the brain via the optic nerve.

retrovirus (L. *retro*, turning back) An RNA virus. When a retrovirus enters a cell, a viral enzyme (reverse transcriptase) transcribes viral RNA into duplex DNA, which the cell's machinery then replicates and transcribes as if it were its own.

Rh blood group A set of cell surface markers (antigens) on the surface of red blood cells in humans and rhesus monkeys (for which it is named); although there are several alleles, they are grouped into two main types, called Rh-positive and Rh-negative.

rhizome (Gr. *rhizoma*, mass of roots) In vascular plants, a usually more or less horizontal underground stem; may be enlarged for storage or may function in vegetative reproduction.

ribonucleic acid (RNA) A class of nucleic acids characterized by the presence of the sugar ribose and the pyrimidine uracil; includes mRNA, tRNA, and rRNA.

ribosomal RNA (rRNA) A class of RNA molecules found, together with characteristic proteins, in ribosomes; transcribed from the DNA of the nucleolus.

ribosome The molecular machine that carries out protein synthesis; the most complicated aggregation of proteins in a cell, also containing three different rRNA molecules.

ribozyme An RNA molecule that can behave as an enzyme, sometimes catalyzing its own

assembly; rRNA also acts as a ribozyme in the polymerization of amino acids to form protein.

RNA polymerase An enzyme that catalyzes the assembly of an mRNA molecule, the sequence of which is complementary to a DNA molecule used as a template. *See* transcription.

RNA primer In DNA replication, a sequence of about 10 RNA nucleotides complementary to unwound DNA that attaches at a replication fork; the DNA polymerase uses the RNA primer as a starting point for addition of DNA nucleotides to form the new DNA strand; the RNA primer is later removed and replaced by DNA nucleotides.

RNA splicing A nuclear process by which intron sequences of a primary mRNA transcript are cut out and the exon sequences spliced together to give the correct linkages of genetic information that will be used in protein construction.

rod Light-sensitive nerve cell found in the vertebrate retina; sensitive to very dim light; responsible for "night vision."

root The usually descending axis of a plant, normally below ground, which anchors the plant and serves as the major point of entry for water and minerals.

rumen An "extra stomach" in cows and related mammals wherein digestion of cellulose occurs and from which partially digested material can be ejected back into the mouth.

S

saltatory conduction A very fast form of nerve impulse conduction in which the impulses leap from node to node over insulated portions.

saprobes Heterotrophic organisms that digest their food externally (such as most fungi).

sarcolemma The specialized cell membrane in a muscle cell.

sarcoma A cancerous tumor arising from cells of connective tissue, bone, or muscle.

sarcomere (Gr. *sarx*, flesh, + *meris*, part of) Fundamental unit of contraction in skeletal muscle; repeating bands of actin and myosin that appear between two Z lines.

sarcoplasmic reticulum (Gr. *sarx*, flesh, + *plassein*, to form, mold; L. *reticulum*, network) The endoplasmic reticulum of a muscle cell. A sleeve of membrane that wraps around each myofilament.

satellite DNA A nontranscribed region of the chromosome with a

distinctive base composition; a short nucleotide sequence repeated tandemly many thousands of times.

saturated fat A fat composed of fatty acids in which all the internal carbon atoms contain the maximum possible number of hydrogen atoms.

Schwann cells The supporting cells associated with projecting axons, along with all the other nerve cells that make up the peripheral nervous system.

sclereid (Gr. *skleros*, hard) In vascular plants, a sclerenchyma cell with a thick, lignified, secondary wall having many pits; not elongate like a fiber.

sclerenchyma cell (Gr. *skleros*, hard, + *en*, in, + *chymein*, to pour) A type of tissue made up of sclerenchyma cells.

scrotum (L., bag) The pouch that contains the testes in most mammals.

scuttellum The modified cotyledon in cereal grains.

secondary cell wall In plants, the innermost layer of the cell wall. Secondary walls have a highly organized microfibrillar structure and are often impregnated with lignin.

secondary growth In vascular plants, an increase in stem and root diameter made possible by cell division of the lateral meristems.

secondary immune response The swifter response of the body the second time it is invaded by the same pathogen because of the presence of memory cells, which quickly become antibody-producing plasma cells.

secondary induction An induction between tissues that have already differentiated.

secondary structure In a protein, hydrogen-bonding interactions between CO and NH groups of the primary structure.

Second Law of Thermodynamics A statement concerning the transformation of potential energy into heat; it says that disorder (entropy) is continually increasing in the universe as energy changes occur, so disorder is more likely than order.

second messenger A small molecule or ion that carries the message from a receptor on the target cell surface into the cytoplasm.

segregation The process by which alternative forms of traits are expressed in offspring rather than blending each trait of the parents in the offspring.

selection The process by which some organisms leave more offspring than competing ones,

and their genetic traits tend to appear in greater proportions among members of succeeding generations than the traits of those individuals that leave fewer offspring.

selectively permeable Condition in which a membrane is permeable to some substances but not to others.

self-fertilization The union of egg and sperm produced by a single hermaphroditic organism.

semen (L., seed) In reptiles and mammals, sperm-bearing fluid expelled from the penis during male orgasm.

semicircular canal Any of three fluid-filled canals in the inner ear that help to maintain balance.

semiconservative replication DNA replication in which each strand of the original duplex serves as the template for construction of a totally new complementary strand, so the original duplex is partially conserved in each of the two new DNA molecules.

senescent Aged, or in the process of aging.

sensory (afferent) neuron A neuron that transmits nerve impulses from a sensory receptor to the central nervous system or central ganglion.

sepal (L. *sepalum*, a covering) A member of the outermost floral whorl of a flowering plant.

septum, pl. septa (L., fence) A wall between two cavities.

seta, pl. setae (L., bristle) In an annelid, bristles of chitin that help anchor the worm during locomotion or when it is in its burrow.

sex-linked A trait determined by a gene carried on the X chromosome and absent on the Y chromosome.

sexual reproduction The process of producing offspring through an alternation of fertilization (producing diploid cells) and meiotic reduction in chromosome number (producing haploid cells).

sexual selection A type of differential reproduction that results from variable success in obtaining mates.

shoot In vascular plants, the aboveground portions, such as the stem and leaves.

sieve cell In the phloem of vascular plants, a long, slender sieve element with relatively unspecialized sieve areas and with tapering end walls that lack sieve plates.

sinoatrial (SA) node *See* pacemaker.

sinus (L., curve) A cavity or space in tissues or in bone.

sister chromatid   One of two identical copies of each chromosome, still linked at the centromere, produced as the chromosomes duplicate for mitotic division; similarly, one of two identical copies of each homologous chromosome present in a tetrad at meiosis.

sodium-potassium pump   Transmembrane channels engaged in the active (ATP-driven) transport of sodium ions, exchanging them for potassium ions, where both ions are being moved against their respective concentration gradients; maintains the resting membrane potential of neurons and other cells.

solute   A molecule dissolved in some solution; as a general rule, solutes dissolve only in solutions of similar polarity; for example, glucose (polar) dissolves in (forms hydrogen bonds with) water (also polar), but not in vegetable oil (nonpolar).

solvent   The medium in which one or more solutes is dissolved.

somatic cell   Any of the cells of a multicellular organism except those that are destined to form gametes (germ-line cells).

somatic mutation   A change in genetic information (mutation) occurring in one of the somatic cells of a multicellular organism, not passed from one generation to the next.

somatic nervous system   (Gr. *soma*, body) In vertebrates, the neurons of the peripheral nervous system that control skeletal muscle.

somite   (Gr. *soma*, body) One of the blocks, or segments, of tissue into which the mesoderm is divided during differentiation of the vertebrate embryo.

Southern blot   A procedure used for identifying a specific gene, in which DNA from the source being tested is cut into fragments with restriction enzymes and separated by gel electrophoresis, then blotted onto a sheet of nitrocellulose and probed with purified, labeled, single-stranded DNA corresponding to a specific gene; if the DNA matching the specific probe is present in the source DNA, it is visible as a band of radioactive label on the sheet.

species, pl. **species**   (L., kind, sort) A kind of organism; species are designated by binomial names written in italics.

spectrin   A scaffold of proteins that links plasma membrane proteins to actin filaments in the cytoplasm

of red blood cells, producing their characteristic biconcave shape.

sperm   (Gr. *sperma*, seed) A mature male gamete, usually motile and smaller than the female gamete.

spermatid   (Gr. *sperma*, seed) In animals, each of four haploid (*n*) cells that result from the meiotic divisions of a spermatocyte; each spermatid differentiates into a sperm cell.

spermatozoa   The male gamete, usually smaller than the female gamete, and usually motile.

sphincter   (Gr. *sphinkter*, band, from *sphingein*, to bind tightly) In vertebrate animals, a ring-shaped muscle capable of closing a tubular opening by constriction (such as between stomach and small intestine or between anus and exterior).

spindle apparatus   The assembly that carries out the separation of chromosomes during cell division; composed of microtubules (spindle fibers) and assembled during prophase at the equator of the dividing cell.

spiracle   (L. *spiraculum*, from *spirare*, to breathe) External opening of a trachea in arthropods.

spongy parenchyma   A leaf tissue composed of loosely arranged, chloroplast-bearing cells. *See* palisade parenchyma.

sporangium, pl. **sporangia**   (Gr. *spora*, seed, + *angeion*, a vessel) A structure in which spores are produced.

spore   A haploid reproductive cell, usually unicellular, capable of developing into an adult without fusion with another cell.

sporophyte   (Gr. *spora*, seed, + *phyton*, plant) The spore-producing, diploid (*2n*) phase in the life cycle of a plant having alternation of generations.

stabilizing selection   A form of selection in which selection acts to eliminate both extremes from a range of phenotypes.

stamen   (L., thread) The organ of a flower that produces the pollen; usually consists of anther and filament; collectively, the stamens make up the androecium.

starch   (Mid. Eng. *sterchen*, to stiffen) An insoluble polymer of glucose; the chief food storage substance of plants.

statocyst   (Gr. *statos*, standing, + *kystis*, sac) A sensory receptor sensitive to gravity and motion.

stele   The central vascular cylinder of stems and roots.

stem cell   A relatively undifferentiated cell in animal tissue that can divide to produce more differentiated tissue cells.

stereoscopic vision   (Gr. *stereos*, solid, + *opitkos*, pertaining to the eye) Ability to perceive a single, three-dimensional image from the simultaneous but slightly divergent two-dimensional images delivered to the brain by each eye.

stigma   (Gr., mark, tattoo mark) (1) In angiosperm flowers, the region of a carpel that serves as a receptive surface for pollen grains. (2) Light-sensitive eyespot of some algae.

stipules   Leaflike appendages that occur at the base of some flowering plant leaves or stems.

stolon   (L. *stolo*, shoot) A stem that grows horizontally along the ground surface and may form adventitious roots, such as runners of the strawberry plant.

stoma, pl. **stomata**   (Gr., mouth) In plants, a minute opening bordered by guard cells in the epidermis of leaves and stems; water passes out of a plant mainly through the stomata.

stratum corneum   The outer layer of the epidermis of the skin of the vertebrate body.

striated muscle   (L. *striare*, to groove) Skeletal voluntary muscle and cardiac muscle.

stromatolite   A fossilized mat of ancient bacteria formed as long as 2 billion years ago, in which the bacterial remains individually resemble some modern-day bacteria.

style   (Gr. *stylos*, column) In flowers, the slender column of tissue that arises from the top of the ovary and through which the pollen tube grows.

substrate   (L. *substratus*, strewn under) (1) The foundation to which an organism is attached. (2) A molecule upon which an enzyme acts.

succession   In ecology, the slow, orderly progression of changes in community composition that takes place through time.

summation   Repetitive activation of the motor neuron resulting in maximum sustained contraction of a muscle.

surface tension   A tautness of the surface of a liquid, caused by the cohesion of the molecules of liquid. Water has an extremely high surface tension.

surface-area-to-volume ratio   Relationship of the surface area of a structure, such as a cell, to the volume it contains.

swim bladder   An organ encountered only in the bony fish that helps the fish regulate its buoyancy by increasing or decreasing the amount of gas in

the bladder via the esophagus or a specialized network of capillaries.

symbiosis   (Gr. *syn*, together with, + *bios*, life) The condition in which two or more dissimilar organisms live together in close association; includes parasitism (harmful to one of the organisms), commensalism (beneficial to one, of no significance to the other), and mutualism (advantageous to both).

sympatric speciation   The differentiation of populations within a common geographic area into species.

synapomorphy   In systematics, a derived character that is shared by clade members.

synapse   (Gr. *synapsis*, a union) A junction between a neuron and another neuron or muscle cell; the two cells do not touch, the gap being bridged by neurotransmitter molecules.

synapsis   (Gr., contact, union) The point-by-point alignment (pairing) of homologous chromosomes that occurs before the first meiotic division; crossing over takes place during synapsis.

synaptic cleft   The space between two adjacent neurons.

synaptic vesicle   A vesicle of a neurotransmitter produced by the axon terminal of a nerve. The filled vesicle migrates to the presynaptic membrane, fuses with it, and releases the neurotransmitter into the synaptic cleft.

synaptonemal complex   A protein lattice that forms between two homologous chromosomes in prophase I of meiosis, holding the replicated chromosomes in precise register with each other so that base-pairs can form between nonsister chromatids for crossing over that is usually exact within a gene sequence.

syncytial blastoderm   A structure composed of a single large cytoplasm containing about 4000 nuclei in embryonic development of insects such as *Drosophila*.

syngamy   (Gr. *syn*, together with, + *gamos*, marriage) The process by which two haploid cells (gametes) fuse to form a diploid zygote; fertilization.

synteny   Extensive conserved regions of DNA among different species.

systematics   The reconstruction and study of evolutionary relationships.

systolic pressure   A measurement of how hard the heart is contracting. When measured during a blood pressure reading, ventricular systole (contraction) is what is being monitored.

T

tagma, pl. **tagmata** (Gr., arrangement, order, row) A compound body section of an arthropod resulting from embryonic fusion of two or more segments; for example, head, thorax, abdomen.

taxis, pl. **taxes** (Gr., arrangement) An orientation movement by a (usually) simple organism in response to an environmental stimulus.

taxonomy The science of classifying living things. By agreement among taxonomists, no two organisms can have the same name, and all names are expressed in Latin.

T cell A type of lymphocyte involved in cell-mediated immunity and interactions with B cells; the "T" refers to the fact that T cells are produced in the thymus.

telencephalon (Gr. *telos*, end, + *encephalon*, brain) The most anterior portion of the brain, including the cerebrum and associated structures.

telomere A specialized nontranscribed structure that caps each end of a chromosome.

tendon (Gr. *tendon*, stretch) A strap of cartilage that attaches muscle to bone.

tertiary structure The folded shape of a protein, produced by hydrophobic interactions with water, ionic and covalent bonding between side chains of different amino acids, and van der Waal's forces; may be changed by denaturation so that the protein becomes inactive.

testcross A mating between a phenotypically dominant individual of unknown genotype and a homozygous "tester," done to determine whether the phenotypically dominant individual is homozygous or heterozygous for the relevant gene.

testis, pl. **testes** (L., witness) In mammals, the sperm-producing organ.

tetanus Sustained forceful muscle contraction with no relaxation.

thalamus (Gr. *thalamos*, chamber) That part of the vertebrate forebrain just posterior to the cerebrum; governs the flow of information from all other parts of the nervous system to the cerebrum.

thermodynamics (Gr. *therme*, heat, + *dynamis*, power) The study of transformations of energy, using heat as the most convenient form of measurement of energy.

thigmotropism In plants, unequal growth in some structure that comes about as a result of physical contact with an object.

threshold The minimum amount of stimulus required for a nerve to fire (depolarize).

thylakoid (Gr. *thylakos*, sac, + *-oides*, like) A saclike membranous structure containing chlorophyll in cyanobacteria and the chloroplasts of eukaryotic organisms.

tight junction Region of actual fusion of plasma membranes between two adjacent animal cells that prevents materials from leaking through the tissue.

tissue (L. *texere*, to weave) A group of similar cells organized into a structural and functional unit.

totipotent A cell that possesses the full genetic potential of the organism.

trachea, pl. **tracheae** (L., windpipe) A tube for breathing; in terrestrial vertebrates, the windpipe that carries air between the larynx and bronchi (which leads to the lungs); in insects and some other terrestrial arthropods, a system of chitin-lined air ducts.

tracheids In plant xylem, dead cells that taper at the ends and overlap one another.

transcription (L. *trans*, across, + *scribere*, to write) The enzyme-catalyzed assembly of an RNA molecule complementary to a strand of DNA.

transcription factor One of a set of proteins required for RNA polymerase to bind to a eukaryotic promoter region, become stabilized, and begin the transcription process.

transfection The transformation of eukaryotic cells in culture.

transfer RNA (tRNA) (L. *trans*, across, + *ferre*, to bear or carry) A class of small RNAs (about 80 nucleotides) with two functional sites; at one site, an "activating enzyme" adds a specific amino acid, while the other site carries the nucleotide triplet (anticodon) specific for that amino acid.

translation (L. *trans*, across, + *latus*, that which is carried) The assembly of a protein on the ribosomes, using mRNA to specify the order of amino acids.

translocation (L. *trans*, across, + *locare*, to put or place) (1) In plants, the long-distance transport of soluble food molecules (mostly sucrose), which occurs primarily in the sieve tubes of phloem tissue. (2) In genetics, the interchange of chromosome segments between nonhomologous chromosomes.

transpiration (L. *trans*, across, + *spirare*, to breathe) The loss of water vapor by plant parts; most transpiration occurs through the stomata.

transposition Type of genetic recombination in which transposable elements (transposons) move from one site in the DNA sequence to another, apparently randomly.

transposon (L. *transponere*, to change the position of) A DNA sequence capable of transposition.

triglyceride (triacylglycerol) An individual fat molecule, composed of a glycerol and three fatty acids.

triploid Possessing three sets of chromosomes.

trochophore A free-swimming larval stage unique to the mollusks and annelids.

trophic level (Gr. *trophos*, feeder) A step in the movement of energy through an ecosystem.

trophoblast (Gr. *trephein*, to nourish, + *blastos*, germ) In vertebrate embryos, the outer ectodermal layer of the blastodermic vesicle; in mammals, it is part of the chorion and attaches to the uterine wall.

tropism (Gr. *trope*, a turning) A response to an external stimulus.

tropomyosin (Gr. *tropos*, turn, + *myos*, muscle) Low-molecular-weight protein surrounding the actin filaments of striated muscle.

troponin Complex of globular proteins positioned at intervals along the actin filament of skeletal muscle; thought to serve as a calcium-dependent "switch" in muscle contraction.

tubulin (L. *tubulus*, small tube, + *in*, belonging to) Globular protein subunit forming the hollow cylinder of microtubules.

tumor-suppressor gene A gene that normally functions to inhibit cell division; mutated forms can lead to the unrestrained cell division of cancer, but only when both copies of the gene are mutant.

turgor (L. *turgor*, a swelling) The pressure within a cell resulting from the movement of water into the cell; a cell with high turgor pressure is said to be turgid. *See* osmotic pressure.

U

unequal crossing over A process by which a crossover in a small region of misalignment at synapsis causes two homologous chromosomes to exchange segments of unequal length.

unsaturated fat A fat molecule in which one or more of the fatty acids contain fewer than the maximum number of hydrogens attached to their carbons.

urea (Gr. *ouron*, urine) An organic molecule formed in the vertebrate liver; the principal form of disposal of nitrogenous wastes by mammals.

urethra (Gr. from *ourein*, to urinate) The tube carrying urine from the bladder to the exterior of mammals.

uric acid Insoluble nitrogenous waste products produced largely by reptiles, birds, and insects.

urine (Gr. *ouron*, urine) The liquid waste filtered from the blood by the kidney and stored in the bladder pending elimination through the urethra.

uterus (L., womb) In mammals, a chamber in which the developing embryo is contained and nurtured during pregnancy.

V

vacuole A membrane-bounded sac in the cytoplasm of some cells, used for storage or digestion purposes in different kinds of cells; plant cells often contain a large central vacuole that stores water, proteins, and waste materials.

vascular cambium In vascular plants, a cylindrical sheath of meristematic cells, the division of which produces secondary phloem outwardly and secondary xylem inwardly; the activity of the vascular cambium increases stem or root diameter.

vascular tissue (L. *vasculum*, a small vessel) Containing or concerning vessels that conduct fluid.

vas deferens (L. *vas*, a vessel, + *ferre*, to carry down) In mammals, the tube carrying sperm from the testes to the urethra.

vasopressin A posterior pituitary hormone that regulates the kidney's retention of water.

vegetal pole The hemisphere of the zygote comprising cells rich in yolk.

vein (L. *vena*, a blood vessel) (1) In plants, a vascular bundle forming a part of the framework of the conducting and supporting tissue of a stem or leaf. (2) In animals, a blood vessel carrying blood from the tissues to the heart.

veliger The second larval stage of mollusks following the trochophore stage, during which the beginning of a foot, shell, and mantle can be seen.

ventricle A muscular chamber of the heart that receives blood from an atrium and pumps blood out to either the lungs or the body tissues.

vesicle (L. *vesicula*, a little bladder) A small intracellular, membrane-bounded sac in which various substances are transported or stored.

vessel element In vascular plants, a typically elongated cell, dead at maturity, which conducts water and solutes in the xylem.

vestibular apparatus The complicated sensory apparatus of the inner ear that provides for balance and orientation of the head in vertebrates.

villus, pl. villi (L., a tuft of hair) In vertebrates, one of the minute, fingerlike projections lining the small intestine that serve to increase the absorptive surface area of the intestine.

visceral mass (L., internal organs) Internal organs in the body cavity of an animal.

vitamin (L. *vita*, life, + *amine*, of chemical origin) An organic substance that cannot be synthesized by a particular organism but is required in small amounts for normal metabolic function.

viviparity (L. *vivus*, alive, + *parere*, to bring forth) Refers to reproduction in which eggs develop within the mother's body and young are born free-living.

voltage-gated ion channel A transmembrane pathway for an ion that is opened or closed by a change in the voltage, or charge difference, across the plasma membrane.

W

water potential The potential energy of water molecules. Regardless of the reason (e.g., gravity, pressure, concentration of solute particles) for the water potential, water moves from a region where water potential is greater to a region where water potential is lower.

wild type In genetics, the phenotype or genotype that is characteristic of the majority of individuals of a species in a natural environment.

X

xylem (Gr. *xylon*, wood) In vascular plants, a specialized tissue, composed primarily of elongate, thick-walled conducting cells, which transports water and solutes through the plant body.

Y

yolk plug A plug occurring in the blastopore of amphibians during formation of the archenteron in embryological development.

Z

zona pellucida An outer membrane that encases a mammalian egg.

zoospore A motile spore.

zooxanthellae—Symbiotic photosynthetic protists in the tissues of corals.

zygomycetes (Gr. *zygon*, yoke, + *mykes*, fungus) A type of fungus whose chief characteristic is the production of sexual structures called zygosporangia, which result from the fusion of two of its simple reproductive organs.

zygote (Gr. *zygotos*, paired together) The diploid (2*n*) cell resulting from the fusion of male and female gametes (fertilization).

Credits

Photo's

Bettmann; **13.24:** © Science Photo Library/Photo Researchers; **13.26:** © Alfred Paseika/Science Photo Library/Photo Researchers; **13.28 (both):** © Cabisco/Phototake; **13.34:** © Leonard Lessin/Peter Arnold; **13.36:** © Hans Reinhard/Okapia/Photo Researchers; **13.37 (left):** Courtesy of Loris McGavaran, Denver Children's Hospital; **13.37 (right):** © Richard Hutchings/Photo Researchers Inc.

Chapter 14
Figure 14.1: © PhotoDisc/Volume 29; **14.9 (both):** From "The Double Helix," by J.D. Watson, Atheneum Press, N.Y. 1968; **14.10a:** © Barrington Brown/Photo Researchers Inc.; **14.11:** From M. Meselson and F.W. Stahl/ *Proceedings of the Nat. Acad. of Sci.* 44 (1958):671; **14.17 (both):** From *Biochemistry* 4/e by Stryer © 1995 by Lubert Stryer. Used with permission of W.H. Freeman and Company; **14.19:** © Prof. Ulrich Laemmli/Photo Researchers Inc.; **14.20a:** Courtesy of Dr. David Wolstenholme

Chapter 15
Figure 15.1: © K.G. Murti/Visuals Unlimited; **15.3:** N. Ban, P. Nissen, J. Hansen, P.B. Moore & T.A. Steitz, "The Complete Atomic Structure of the Large Ribosomal subunit at 2.4A Resolution," Reprinted with permission from *Science*, v. 289 #5481, p917, © 2000 American Association for the Advancement of Science; **15.7:** From R.C. Williams; *Proc. Nat. Acad. of Sci.* 74 (1977):2313; **15.13:** Courtesy of Dr. Oscar L. Miller; **15.18b:** Courtesy of Dr. Bert O'Malley, Baylor College of Medicine

Chapter 16
Figure 16.1: © Stanley Cohen/Science Photo Library/Photo Researchers Inc.; **16.7b:** Courtesy of Bio-Rad Laboratories; **16.15:** Courtesy of Lifecodes Corp., Stamford, CT; **16.16:** AP/Wide World Photos; **16.17:** R.L. Brinster, U. of Pennsylvania Sch. of Vet. Med.; **16.20:** Courtesy of Monsanto

Chapter 17
Figure 17.1: Courtesy of Robert D. Fleischmann, The Institute for Genomic Research; **17.4:** Courtesy of Celera Genomics; **17.11 (all):** Reproduced with permission from Altpeter et al, *Plant Cell Reports* 16:12-17, 1996, photos

provided by Indra Vasil; **17.12:** Courtesy of Research Collaboratory for Structural Bioinformatics; **17.13:** © Corbis/R-F Website; **17.14:** © Grant Heilman/Grant Heilman Photography

Chapter 18
Figure 18.1: Courtesy of Dr. Claus Pelling; **18.15 (both):** Courtesy of Dr. Harrison Echols; **18.18a:** Courtesy of Dr. Victoria Foe

Chapter 19
Figure 19.1: © Cabisco/Visuals Unlimited; **19.3:** Photo Lennart Nilsson/Albert Bonniers Forlag AB, *A Child is Born*, Dell Publishing Company; **19.4 (all):** © Cabisco/Phototake; **19.6:** © Ed Lewis; **19.15 (top left):** Dr. Christiane Nusslein-Volhard/Max Planck Institute as published in From Egg to Adult, © 1992 HHMI; **19.15b-d:** James Langeland, Stephen Paddock & Sean Carroll as published in *From Egg to Adult*, © 1992 HHMI; **19.16 (all):** Courtesy of Manfred Frasch; **19.17:** Courtesy of E.B. Lewis

Chapter 20
Figure 20.1: © University of Wisconsin-Madison News & Public Affairs, Photo by Jeff Miller; **20.3:** Courtesy of Dr. Charles Brinton; **20.10:** © Custom Medical Stock Photo; **20.18:** Courtesy of American Cancer Society; **20.23:** AP/Wide World Photos; **20.25:** © University of Wisconsin-Madison News & Public Affairs

Chapter 21
Figure 21.1: © Corbis/R-F Website; **21.3:** Biological Photo Service; **21.19:** Courtesy of H. Rodd

Chapter 22
Figure 22.1: © PhotoDisc Website; **22.4 (both):** © Breck P. Kent/Animals Animals/Earth Scenes; **22.9 (both):** Courtesy of Lyudmilla N. Trut, Institute of Cytology & Genetics, Siberian Dept. of the Russian Academy of Sciences

Chapter 23
Figure 23.1: Jonathan Losos; **23.3:** © Porterfield/Chickering/Photo Researchers Inc.; **23.4:** © Barbara Gerlach/Visuals Unlimited; **23.6 (top left):** © John Shaw/Tom Stack & Associates; **23.6 (bottom left):** © Rob & Ann

Simpson/Visuals Unlimited; **23.6 (top right):** © Suzanne L. Collins & Joseph T. Collins/National Audubon Society Collection/Photo Researchers; **23.6 (bottom right):** © Phil A. Dotson/National Audubon Society Collection/Photo Researchers; **23.8 (left):** © Chas. McRae/Visuals Unlimited; **23.8 (right):** Jonathan Losos; **23.12 (left):** © Jeffrey Taylor; **23.12 (right):** © William P. Mull; **23.15:** © G.R. Roberts

Chapter 24
Figure 24.1: Courtesy of Richard P. Elinson; **p. 492a:** © Dr. R. Clark & M. Goff/Science Photo Library/Photo Researchers; **p. 492b:** © PhotoDisc Green/Getty Images; **p. 492c:** © BIOS(C. Ruoso)/Peter Arnold, Inc.; **p. 492d:** © Darwin Dale/Photo Researchers; **p. 492e:** Centers for Disease Control; **p. 492f:** © Paul G. Young/The Institute for Genomic Research; **p. 492g:** © Dr. Jeremy Burgess/Science Photo Library/Photo Researchers; **p. 492h:** © Science Photo Library/Photo Researchers; **p. 492I:** © PhotoDisc/Getty Images; **p. 492j:** © CNRI/Science Photo Library/Photo Researchers; **24.4:** © PhotoDisc/Getty Images; **24.10:** Courtesy of Anna Di Gregorio; **24.13 (top left):** © Image Bank/Getty Images; **24.13 (top right):** © Darwin Dale/Photo Researchers; **24.13 (bottom left):** © Aldo Brando/Peter Arnold, Inc.; **24.13 (bottom right):** © Tom E. Adams/Peter Arnold, Inc.; **24.14 (both):** Courtesy of Walter Gehring, reprinted with permission from Induction of Ectopic Eyes by Targeted Expression of the Eyeless Gene in Drosophila, G. Halder, P. Callaerts, Walter J. Gehring, *Science* Vol. 267, © 24 March 1995 American Association for the Advancement of Science; **24.15 (both):** Courtesy of Dr. William Jeffery

Chapter 25
Figure 25.1: © Corbis/Volume 8; **25.2 (top left):** © Corbis/Volume 102; **25.2 (top right):** © PhotoDisc/Volume 56; **25.2 (bottom left):** © PhotoDisc/Volume 1; **25.2 (bottom right):** © Corbis/Volume 8; **25.6:** Image # 5789, photo by D. Finnin/American Museum of Natural History

Chapter 26
Figure 26.1: © K.G. Murti/Visuals Unlimited; **26.3a:** © Dept. of Microbiology, Biozentrum/Science

Photo Library/Photo Researchers; **26.7:** Courtesy of Katherine Sutliff, from *Science* 300: 1377, May 30, 2003, p. 1377, © American Association for the Advancement of Science, Reprinted with permission; **26.8:** © Corbis/Volume 40

Chapter 27
Figure 27.1: © David M. Phillips/Visuals Unlimited; **p. 547 (above):** © Abraham & Beachey/BPS/Tom Stack & Associates; **p. 547 (below):** © F. Widell/Visuals Unlimited; **27.3a:** © Science Photo Library/Photo Researchers; **27.3b:** © University of Regensburg, Courtesy of Reinhard Rachel; **27.3c:** © Andrew Syred/Science Photo Library/Photo Researchers; **27.3d:** © Microfield Scientific Ltd./Science Photo Library/Photo Researchers; **27.3e:** © Alfred Paseika/Science Photo Library/Photo Researchers; **27.3g:** © S.W. Watson/ Visuals Unlimited; **27.3h:** © Dennis Kunkel Microscopy, Inc.; **27.3I:** Courtesy of Dr. Hans Reichenbach; **27.4:** © G.W. Willis; **27.5:** © Julius Adler/Visuals Unlimited; **27.6 (left):** © W. Watson/ Visuals Unlimited; **27.6 (right):** © Norma J. Lang/Biological Photo Service; **27.8:** © CNRI/SPL/Photo Researchers; **27.10:** © R. Calentine/ Visuals Unlimited; **27.11:** © Science/Visuals Unlimited

Chapter 28
Figure 28.1: © John D. Cunningham/Visuals Unlimited; **28.2:** Courtesy of Dr. Edward W. Daniels, Argonne National Lab, the University of Illinois College of Medicine at Chicago; **28.5:** © John D. Cunningham/ Visuals Unlimited; **28.6a:** © Michael Abby/ Visuals Unlimited; **28.7 (left):** © Manfred Kage/Peter Arnold Inc.; **28.7 (right):** © Edward S. Ross; **28.11a:** © Brian Parker/ Tom Stack & Associates; **28.12:** © Gregory Ochocki/Photo Researchers; **28.13:** © D.P. Wilson/Photo Researchers; **28.15:** © John D. Cunningham/ Visuals Unlimited; **28.16:** © Phil A. Harrington/Peter Arnold, Inc.; **28.18:** © Richard Rowan/Photo Researchers; **28.17:** © Manfred Kage/ Peter Arnold; **28.19:** © Edward S. Ross; **28.20a:** © John Shaw/Tom Stack & Associates; **28.20b:** © John Shaw/Tom Stack & Associates; **28.21:** © Genichiro Higuchi,

Higuchi Science Laboratory; **p.578:** © D.P. Wilson/Photo Researchers

Chapter 29

Figure 29.1: © Stephen J. Krasemann/DRK Photo; **29.4:** © Edward S. Ross; **29.6:** © Kirtley Perkins/Visuals Unlimited; **29.7:** © Kingsley R. Stern; **29.8:** Courtesy of Hans Steur, The Netherlands; **29.9:** Darrell Vodopich; **29.10:** © Kingsley R. Stern; **29.11, 29.12, 29.14:** © Edward S. Ross; **29.16 (left):** © Walter H. Hodge/Peter Arnold Inc.; **29.16 (center):** © Kjell Sandved/ Butterfly Alphabet; **29.16c:** © Runk/Schoenberger/Grant Heilman Photography; **29.17:** Courtesy of David Dilcher & Ge Sun; **29.18:** Courtesy of Sandra Floyd

Chapter 30

Figure 30.1: © Bill Keogh/Visuals Unlimited; **30.2a:** Courtesy of Dr. Peter Daszak; **30.2b:** © Cabisco/ Visuals Unlimited; **30.2c:** © Robert Simpson/Tom Stack & Associates; **30.2d:** © Michael & Patricia Fogden; **30.3:** Courtesy of E.C. Setliff & W. L. MacDonald; **30.4:** © Kjell Sandved/ Butterfly Alphabet; **30.5:** © Microfield Scientific Ltd/Photo Researchers; **30.6:** © L. West/Photo Researchers; **30.7:** Ralph Williams/USDA Forest Service; **30.9a:** © 1997 Regents of the University of Michigan; **30.10:** © Cabisco/Phototake; **30.11a:** Alexandra Lowry/The National Audubon Society Collection/ Photo Researchers; **30.12a:** © Ed Pembleton; **30.12b:** © Kjell Sandved/ Butterfly Alphabet; **30.13:** © David M. Phillips/ Visuals Unlimited; **30.14a:** © James Castner; **30.14b:** © Edward S. Ross; **30.15:** © Ed Reschke; **30.16b:** © D.H. Marx/ Visuals Unlimited; **30.17:** © Scott Camazine/Photo Researchers; **30.18:** Courtesy of Zoology Department/University of Canterbury, New Zealand; **30.19a:** Courtesy of Laura E. Sweets, University of Missouri; **30.19b:** © Manfred Kage/Peter Arnold, Inc.

Chapter 31

Figure 31.1: © Corbis/Volume 53; **p. 619a:** © Corbis/Volume 86; **p. 619b:** © Corbis/Volume 65; **p. 619c:** © David M. Phillips/Visuals Unlimited; **p. 619d:** © Royalty-Free/Corbis **p. 619e:** © Edward S. Ross; **p. 619f:** © Corbis; **p. 619g:** © Cleveland P. Hickman Jr.; **p. 619h:** © Cabisco/Phototake; **p. 619I:** © Ed Reschke; **31.2 (top left):** © Corbis/Volume 53; **31.2 (top right):** © Corbis; **31.2 (bottom left):** © Corbis; **31.2d:** © Corbis; **31.8:** Courtesy of Dr. Igor Eeckhaut

Chapter 32

Figure 32.1: © Denise Tackett/Tom Stack & Associates **32.4:** © Andrew J. Martinez/Photo Researchers; **32.9:** © Gwen Fidler/Tom Stack & Associates; **32.10:** © Kelvin Aitken/Peter Arnold Inc.; **32.11:** © Daniel Gotshall; **32.12:** © David Wrobel/Visuals Unlimited; **32.16:** © Kjell Sandved/Butterfly Alphabet; **32.17:** © Biology Media/R. Knauf/Photo Researchers; **32.19:** © T.E. Adams/Visuals Unlimited; **32.20:** Courtesy of Matthias Obst

Chapter 33

Figure 33.1: © James H. Robinson/Animals Animals/Earth Scenes; **33.2:** © Alex Kerstich/Visuals Unlimited; **33.4:** © A. Flowers & L. Newman/Photo Researchers; **33.7b:** © Kjell Sandved/Butterfly Alphabet; **33.8:** © Milton Rand/Tom Stack & Associates; **33.10:** © Fred Bavendam/Peter Arnold Inc.; **33.11:** © Visuals Unlimited; **33.13:** © Kjell Sandved/Butterfly Alphabet; **33.14:** © David Dennis/Tom Stack & Associates; **33.15:** © Cleveland P. Hickman; **33.17b:** © Robert Brons/Biological PhotoService; **33.18b:** © Fred Bavendam/Peter Arnold Inc.; **33.26:** © T.E. Adams/Visuals Unlimited; **33.27:** © Kjell Sandved/Butterfly Alphabet; **33.28a:** © Rod Planck/Tom Stack & Associates; **33.28b:** © Ann Moreton/Tom Stack & Associates; **33.29:** © Alex Kerstich/Visuals Unlimited; **33.30:** © Edward S. Ross; **33.31a:** © Cleveland P. Hickman; **33.31b:** © Kjell Sandved/Butterfly Alphabet; **33.31c:** © Norm Thomas/Photo Researchers; **33.31d:** © Valorie Hodgson/Visuals Unlimited; **33.31e:** © Corbis/Volume 6; **33.31f:** © Kjell Sandved/Butterfly Alphabet; **33.33:** © John Shaw/Tom Stack & Associates; **33.34:** © Kjell Sandved/Butterfly Alphabet; **33.35a:** © Alex Kerstitch/Visuals Unlimited; **33.35b:** © Randy Morse/Tom Stack & Associates; **33.35c:** © Daniel W. Gotshall/Visuals Unlimited; **33.38:** © Kjell Sandved/Butterfly Alphabet; **33.40:** © Daniel W. Gotshall/Visuals Unlimited; **33.41:** © Kjell Sandved/Visuals Unlimited

Chapter 34

Figure 34.1: © Corbis; **34.3:** © Eric N. Olson, PhD/The University of Texas MD Anderson Cancer Center; **34.5a:** © Rick Harbo Marine Images; **34.6:** © Heather Angel; **34.10:** © Corbis; **34.15:** © Corbis/Volume 53; **34.16:** © John D. Cunningham/Visuals Unlimited; **34.18:** © 1979 Peter Scoones/Contact Press Images/Woodfin Camp & Associates Inc.; **34.21:** © Natural History Museum/J. Sibbick; **34.22a:** © John Shaw/Tom Stack & Associates; **34.22b:** © Suzanne L. Collins & Joseph T. Collins/Photo Researchers; **34.22c:** © Jany Sauvanet/Photo Researchers; **34.24:** © Natural History Museum/J. Sibbick; **34.27:** Photo by Paul Sareno/University of Chicago; **34.31:** © William J. Weber/Visuals Unlimited; **34.32a:** © John Cancalosi/Tom Stack & Associates; **34.32b:** © Rod Planck/Tom Stack & Associates; **34.33:** © Corbis/Volume 6; **34.37:** © Corbis; **34.38:** © PhotoDisc/Volume 44; **34.41:** © Stephen Dalton/National Audubon Society Collection/Photo Researchers; **34.42:** © B.J. Alcock/Visuals Unlimited; **34.42b:** © Corbis/Volume 6; **34.42c:** © Stephen J. Krasemann/DRK Photo; **34.43:** © Alan Nelson/Animals Animals/Earth Scenes; **34.45 (all):** © David L. Brill; **34.47:** © 1985 David L. Brill/Fossil credit: National Museums of Kenya, Nairobi; **34.49:** AP/Wide World Photos

Chapter 35

Figure 35.1: Photo by Susan Singer; **35.2:** © Terry Ashley/Tom Stack & Associates; **35.4a:** © Hart-Davis/Science Photo Library/Photo Researchers; **35.4b:** © R. Calentine/Visuals Unlimited; **35.8 (both):** © Heidi Mullen; **35.9:** Courtesy of Liming Zhao & Fred Sack, Ohio State University; **35.10:** © Andrew Syred/Science Photo Library/Photo Researchers; **35.11 (both):** Courtesy of Alan Lloyd; **35.12a:** © Biophoto Associates/Photo Researchers Inc.; **35.12b:** © John D. Cunningham/Visuals Unlimited; **35.12c:** © Lawrence Mellinchamp/Visuals Unlimited; **35.13c:** Courtesy of Wilfred Cote, Suny College of Environmental Forestry; **35.14:** © Randy Moore/Visuals Unlimited; **35.15:** © E.J. Cable/Tom Stack & Associates; **35.16a:** Courtesy John Schiefelbein, from Myeong Min Lee & John Schiefelbein, "WEREWolf, MYB-related Protein in Arabidopsis," *Cell* V. 99:473-483, Nov. 24, 1999; **35.16b:** Courtesy of Dr. Philip Benfey, from Wysocka-Diller, J.W., Helariutta, Y., Fukaki, H., Malamy, J.E. and P.N. Benfey (2000) Molecular analysis of SCARECROW function reveals a radial patterning mechanism common to root and shoot Development, *Cell* 127, 595-603; **35.17:** Photomicrograph by G.S. Ellmore; **35.19a:** © Kingsley R. Stern; **35.19b:** Photomicrograph by G.S. Ellmore; **35.21a:** © Walter H. Hodge/Peter Arnold, Inc.; **35.21b:** Courtesy of Robert A. Schlising; **35.21c:** © Kingsley R. Stern; **35.22:** Courtesy of J.H. Troughton and L. Donaldson and Industrial Research Ltd.; **35.24:** © John D. cunningham/ Visuals Unlimited; **35.25 (both):** © Ed Reschke; **35.26:** © Ed Reschke/Peter Arnold Inc.; **35.27a:** © Ed Reschke; **35.27b:** © Jack M. Bostrack/ Visuals Unlimited; **35.29:** Richard Waites & Andrew Hudson, from Phantastica: a gene required for dorsoventrality of leaves in Antirrhinum majus, *Development* 121, 2143-2154 (1995) © The Company of Biologists Limited 1995; **35.30a:** © Kjell Sandved/Butterfly Alphabet; **35.30b:** © Pat Anderson/Visuals Unlimited; **35.31a:** © Edward S. Ross; **35.31b:** © Glenn M. Oliver/ Visuals Unlimited; **35.31c:** © Joel Arrington/Visuals Unlimited; **35.34:** © Ed Reschke; **35.35 (above):** © Michael P. Godomski/Photo Researchers; **35.35 (below):** © Ed Reschke/Peter Arnold, Inc.; **p.753:** © Kingsley R. Stern

Chapter 36

Figure 36.1: © Norm Thomas/The National Audubon Society Collection/Photo Researchers; **36.4:** Courtesy of E.C. Yeung & D.W. Meinke; **36.5:** Courtesy of Dr. Chun-Ming Liu; **36.6:** Courtesy of Kathy Barton & Jeff Lang; **36.7a(above):** © Runk/ Schoenberger/Grant Heilman Photography; **36.7b:** © Kevin & Betty Collins/Visuals Unlimited; **36.8:** © Jack M. Bostrack/Visuals Unlimited; **36.10a:** © Ed Reschke/Peter Arnold, Inc.;

36.10b: © David Sieren/Visuals Unlimited; 36.11 (top left): © Kingsley R. Stern; 36.11 (top center): © James Richardson/Visuals Unlimited; 36.11 (top right): © Kingsley R. Stern; 36.11 (middle left): © Kingsley R. Stern; 36.11 (center): © Kingsley R. Stern; 36.11 (middle right): © Barry L. Runk/Grant Heilman Photography; 36.11 (bottom left): Courtesy of Robert A. Schlising; 36.11 (bottom right): © Charles D.Winters/Photo Researchers; 36.12a: © Edward S. Ross; 36.12b, 36.13, 36.14: © James Castner; 36.16: Courtesy of Prof. Tuan-hua David Ho

Chapter 37

Figure 37.1: © Richard Rowan's Collection, Inc./Photo Researchers; 37.7: © John D. Cunningham/Visuals Unlimited; 37.8: © Terry Ashley/Tom Stack & Associates; 37.9: © Grant Heilman/Grant Heilman Photography; 37.10: © Bruce Iverson Photomicrography; 37.12a: © Andrew Syred/Science Photo Library/Photo Researchers; 37.12b: © Bruce Iverson/Science Photo Library/Photo Researchers

Chapter 38

Figure 38.1: © Scott T. Smith/Corbis 38.2 (all): Courtesy of Dr. Emmanuel Epstein; 38.4: © Michael P. Gadomski/Photo Researchers; 38.6 (both): Courtesy of Nicholas School of the Environment and Earth Sciences, Duke University; 38.7: Courtesy of Sharon Long; 38.9: © Kjell Sandved/Butterfly Alphabet; 38.10: © Runk/ Schoenberger/Grant Heilman Photography; 38.11: © Barry Rice; 38.12: © Don Albert; 38.14: U.S. EPA National Research Lab, Cincinnati; 38.15 (all): © AP/Wide World Photos; p.793: U.S. EPA National Research Lab, Cincinnati

Chapter 39

Figure 39.1: © Holt Studios International (Nigel Cattlin)/Photo Researchers; 39.2: © Jane Grushow/Grant Heilman Photography; 39.3: USDA/Agricultural Research Service; 39.4 (both): USDA/Agricultural Research Service; 39.6: © C. Allan Morgan/Peter Arnold, Inc.; 39.7: © Runk/Schoenberger/ Grant Heilman Photography

Chapter 40

Figure 40.1: © John D. Cunningham/Visuals Unlimited; 40.5: © Runk/Schoenberger/ Grant Heilman Photography; 40.6: © John D. Cunningham/Visuals Unlimited; 40.7: © S.J. Krasemann/Peter Arnold Inc.; 40.8: Courtesy of Frank B. Salisbury; 40.9a: © T. Walker; 40.9b: © R.J. Delorit, Agronomy Publications; 40.10a: © Jim Zipp/The National Audubon Society Collection/Photo Researchers; 40.10b: © Runk/Schoenberger/ Grant Heilman Photography; 40.11: © Prof. Malcolm B. Wilkins, Botany Dept., Glasgow University; 40.18 (all): © Prof. Malcolm B. Wilkins, Botany Dept., Glasgow University; 40.20: © Robert Calentine/Visuals Unlimited; 40.21: ©Runk/Schoenberger/ Grant Heilman Photography; 40.22: © Sylvan H. Wittwer/Visuals Unlimited; 40.25a: © John Solden/Visuals Unlimited; 40.25b: Courtesy of Donald R. McCarty, from "Molecular Analysis of viviparous-1: An Abscisic Acid-Insensitive Mutant of Maize," The Plant Cell, v.1, 523-532, © 1989 American Society of Plant Physiologists; 40.25c: © David M. Phillips/Visuals Unlimited;

Chapter 41

Figure 41.1: © Richard La Val/Animals Animals; 41.3a: © Stephen G. Maka/DRK Photo; 41.3b: © Heidi Mullen 41.4: Courtesy of Lingjing Chen & Renee Sung; 41.5 (both): Detlef Weigel & Ove Nilsson, The Salk Institute for Biological Studies; 41.7: © Jim Strawser/Grant Heilman Photography; 41.12 (all): Courtesy of John L. Bowman; 41.14: © John Bishop/Visuals Unlimited; 41.15: © Paul Gier/Visuals Unlimited; 41.16: Courtesy of Enrico Coen; 41.18a: Courtesy of William F. Chissoe, Noble Microscopy Lab, U. of Oklahoma; 41.18b: Courtesy of Dr. Joan Nowicke, Smithsonian Institution; 41.19: © Kingsley R. Stern; 41.20: © Edward S. Ross; 41.21: © Michael & Patricia Fogden; 41.22 (both): © Thomas Eisner; 41.23 © John D. Cunningham/Visuals Unlimited; 41.24: © Edward S. Ross; 41.25a: © David Sieren/Visuals Unlimited; 41.25b: © Barbara Gerlach/Visuals Unlimited; 41.28: © Jerome Wexler/Photo Researchers; 41.29 (all): Courtesy of Dr. Hans Ulrich Koop, from

Plant Cell Reports, 17:601-604; 41.30 (both): © Edward S. Ross

Chapter 42

Figure 42.1: Photo Lennart Nilsson/Albert Bonniers Forlag AB, Behold Man, Little Brown & Co.; p. 861 (top 3): © Ed Reschke; p. 861 (4th from top): © Fred Hossler/visuals Unlimited; p. 861 (bottom): © Ed Reschke; 42.6: © J. Gross/Science Photo Library/Photo Researchers; 42.7: © Biophoto Associates/ Photo Researchers; p. 863 (top): © Biophoto Associates/ Photo Researchers; p. 863 (2nd from top): © Cleveland P. Hickman; p. 863 (3rd from top): © Chuck Brown/Photo Researchers; p. 863 (4th from top): © Ed Reschke; p. 863 (bottom): © Ken Edward/Science Source/Photo Researchers; 42.8b: © Ed Reschke; 42.9: © Ed Reschke; 42.10: © David M. Phillips/Visuals Unlimited; p. 867 (all): © Ed Reschke; 42.13: © Anthony Bannister/ Animals Animals/Earth Scenes; 42.19, 42.20: © Dr. H.E. Huxley; 42.31: © Treat Davidson/Photo Researchers; p. 886: © Dr. H.E. Huxley

Chapter 43

Figure 43.1: © John Gerlach/Animals Animals/Earth Scenes; 43.20: From O.T. Avery, C.M. Macleod & M. McCarty, "Studies on the chemical nature of the substance inducing transformation of pneumococcal types," reproduced from the Journal of Experimental Medicine 79 (1944): 137-158, fig. 1 by copyright permission of the Rockefeller University Press, reproduced by permission. Photograph made by Mr. Joseph B. Haulenbeek

Chapter 44

Figure 44.1: © Professors P.M. Motta & S. Correr/Science Photo Library/Photo Researchers; 44.11: © Ed Reschke; 44.17 (all): Courtesy of Frank P. Sloop, Jr.; 44.18: © Frans Lanting/Minden Pictures

Chapter 45

Figure 45.1: Courtesy of David I. Vaney, University of Queensland, Australia; 45.4: © C.S. Raines/Visuals Unlimited; 45.14: © John Heuser, Washington University School of Medicine, St. Louis, MO; 45.16: © Ed Reschke; 45.18b: © E.R. Lewis, YY Zeevi, T.E. Everhart, U. of

California/Biological Photo Service; 45.27: Dr. Marcus E. Rachle, Washington University, McDonnell Center for High Brain Function; 45.28: Photo Lennart Nilsson/Albert Bonniers Forlag AB, Behold Man, Little Brown & Co.; 45.31: © E.R. Lewis/Biological Photo Service

Chapter 46

Figure 46.1: © Omikron/Photo Researchers; 46.6: © Ed Reschke; 46.23: © Leonard L. Rue, III

Chapter 47

Figure 47.1: © Francois Gohier/Science Source/Photo Researchers; 47.11: © Corbis/ Bettmann; 47.15: © John Paul Kay/Peter Arnold, Inc.; 47.20: © Robert & Linda Mitchell

Chapter 48

Figure 48.1: National Library of Medicine; 48.3: © Manfred Kage/Peter Arnold, Inc.; 48.7: © Visuals Unlimited; 48.9 (both): © Dr.Andrejs LiepinsScience Photo Library/Photo Researchers; 48.18: © Stuart Fox; 48.23: © CDC/Science Source/Photo Researchers

Chapter 49

Figure 49.1: © Belinda Wright/DRK Photo

Chapter 50

Figure 50.1: © Michael Fogden/DRK Photo; 50.2a: © Chuck Wise/Animals Animals/Earth Scenes; 50.2b: © Fred McConnaughey/ The National Audubon Society Collection/Photo Researchers; 50.4: © David Doubilet; 50.5: © Hans Pfletschinger/Peter Arnold Inc.; 50.7: © Cleveland P. Hickman Jr.; 50.8: Frans Lanting/Minden Pictures; 50.9a: © Jean Phillippe Varin/Jacana/Photo Researchers; 50.9b: © Tom McHugh/The National Audubon Society Collection/Photo Researchers; 50.9c: © Corbis/Volume 86; 50.12a: © David M. Phillips/Photo Researchers; 50.13: Photo Lennart Nilsson/Bonniers Forlag AB, A Child is Born, Dell Publishing Co.; 50.18: © Ed Reschke; 50.22 (all): © McGraw-Hill Higher Education/Bob Coyle, photographer

Chapter 51

Figure 51.1: Photo Lennart Nilsson/Bonniers Forlag AB, A

Child is Born, Dell Publishing Co.; **51.2b:** © P. Bagavandoss/Science Source/Photo Researchers; **51.2c:** © David M. Phillips/ Visuals Unlimited; **51.3b:** Courtesy of Dr. Everett Anderson; **51.6:** © David M. Phillips/Visuals Unlimited; **51.7:** © Cabisco/ Phototake; **51.8:** © David M. Phillips/Visuals Unlimited; **51.21:** Photo Lennart Nilsson/ Albert Bonniers Forlag AB, *A Child is Born*, Dell Publishing Company; **51.22 (all):** Photo Lennart Nilsson/Albert Bonniers Forlag AB, *A Child is Born*, Dell Publishing Company

Chapter 52

Figure 52.1: © K. Ammann/Bruce Coleman Inc.; **52.2 (left):** © UPI/ Corbis Bettmann; **52.2 (center):** © Corbis/ Bettmann; **52.2 (right):** © UPI/ Corbis-Bettmann; **52.3:** From J.L. Gould, Ehology, Norton 1982; **52.5:** © William C. Dilger, Cornell University; **52.6:** From J.R. Brown et al, "A defect in nurturing mice lacking . . . gene for fosB" *Cell* v. 86, 1996 pp 297-308, © Cell Press; **52.7 (all):** © Lee Boltin Picture Library; **52.8:** © William Grenfell/Visuals Unlimited; **52.9a:** Thomas McAvoy, Life Magazine/© Time, Inc.; **52.10 (both):** Grzimek's *Encyclopedia of Ethology* Van Nostrand-Reinhold Co.; **52.12:** © Roger Wilmshurst/ The National Audubon Society Collection/ Photo Researchers; **52.13a:** © Linda Koebner/Bruce Coleman; **52.13b:** © Jeff Foott/Tom Stack & Associates; **52.14 (all):** Superstock; **52.15:** Courtesy of Bernd Heinrich; **52.16b:** © Fred Bruenner/Peter Arnold Inc.; **52.16c:** © James L. Amos/Peter Arnold Inc.; **52.19:** © Dwight R. Kuhn/DRK Photo; **52.21** © Corbis/ Volume 53; **52.22:** © Sol Mednick; **52.23b:** © Dr. Mark Moffett/Minden Pictures; **52.24a:** © S. Osolinski /OSF/Animals Animals/Earth Scenes; **52.25:** Nina Leen, Life Magazine, © Time Inc.; **52.28:** © Bios(C.Thouvenin)/Peter Arnold, Inc.; **52.30b:** © B. Chudleigh/Vireo; **52.32:** © George D. Lepp/Corbis; **52.33a:** Courtesy of T.A. Burke, Reprinted by permission from *Nature*, "Parental care and mating behavior of polyandrous dunnocks," 338: 247-251, 1989; **52.35:** © Edward S. Ross; **52.37:** © Mark Moffett/ Minden Pictures; **52.38:** © Nigel Dennis/National Audubon Society Collection/ Photo Researchers

Chapter 53

Figure 53.1: © PhotoDisc/Volume 44; **53.2:** Courtesy of William J.

Hamilton III; **53.17:** Courtesy of Barry Sinervo; **53.23 (left):** © Jean Vie/Gamma; **53.23 (right):** © Gianni Tortole/National Audubon Society Collection/Photo Researchers

Chapter 54

Figure 54.1: © Corbis; **54.2:** © Tim Davis/Photo Researchers; **54.7 (all):** J.B. Losos; **54.11 (both):** © Edward S. Ross; **54.12 (both):** © Lincoln P. Brower; **54.13:** © Michael & Patricia Fogden/Corbis; **54.14:** © James L. Castner; **54.16:** © Merlin D. Tuttle/Bat Conservation International; **54.17:** © PhotoDisc/Volume 44; **54.18:** © Michael Fogden/DRK Photo; **54.19:** © Edward S. Ross; **54.21a:** © F. Stuart Westmorland/ Photo Researchers; **54.21b:** © Anne Wertheim/ Animals Animals/Earth Scenes; **54.24:** © David Hosking/National Audubon Society Collection/ Photo Researchers; **54.26 (all):** © Tom Bean; **54.27 (both):** Courtesy of Robert Whittaker; **54.28:** © Edward S. Ross

Chapter 55

Figure 55.1: © Corbis/Volume 46; **55.3:** © Martin Harvey, Gallo Images/Corbis; **55.7a:** U.S. Forest Service; **55.17a:** © Layne Kennedy/Corbis

Chapter 56

Figure 56.1: NASA; **56.10:** © Michael Graybill & Jan Hodder/Biological Photo Service; **56.11:** © IFA/Peter Arnold, Inc.; **56.15:** © Digital Vision/Picture Quest; **56.16a:** © Jim Church; **56.16b:** Courtesy of J. Frederick Grassel, Woods Hole Oceanographic Institution; **56.17:** © Edward S. Ross; **56.22:** © Gilbert S. Grant/ National Audubon Society Collection/Photo Researchers; **56.23a:** © Peter May/Peter Arnold Inc.; **56.23b:** © Frans Lanting/Minden Pictures; **56.24a:** NASA

Chapter 57

Figure 57.1: © Tom & Pat Leeson/Photo Researchers; **57.2:** Photo by Tom McHugh, © Natural History Museum of Los Angeles County/Photo Researchers; **57.6:** © Edward S. Ross; **57.9:** © Michael Fogden/ DRK Photo; **57.14:** © 1990 R.O. Bierregaard; **57.15:** Reprinted with permission from *Science*, Vol 282 Dec. © 1998 American Association for the Advancement of Science; **57.17:** © Jack Jeffrey; **57.18:** Mark Chandler; **57.20:** Merlin D.

Tuttle/Bat Conservation International; **57.21:** U.S. Fish & Wildlife Service; **57.23:** © Wm. J. Weber/Visuals Unlimited; **57.24 (both):** University of Wisconsin-Madison Arboretum

Text

Chapter 1

Box 1.1: From Howard Neverov, "The Consent" in *The Collected Works of Howard Nemerov*, 7th Edition, 1981. Reprinted by permission.

Chapter 5

Figure 5.6: Copyright © 2002 from *Molecular Biology of the Cell* by Bruce Alberts, et al. Reproduced by permission of Routledge/Taylor & Francis Books, Inc.

Chapter 6

Figure 6.10: Modified from Alberts, et al., *Molecular Biology of the Cell*, 3rd Edition, 1994 Garland Publishing, New York, NY.

Chapter 10

Figure 10.7: From Raven, et al., *Biology of Plants*, 5th edition. Reprinted by permission of Worth Publishers. **Figure 10.13:** From Lincoln Taiz and Eduardo Zeiger, *Plant Physiology*, 1991 Benjamin-Cummings Publishing. Reprinted with permission of the authors.

Chapter 11

Figure 11.4: Copyright © 2002 from *Molecular Biology of the Cell* by Bruce Alberts, et al. Reproduced by permission of Routledge/Taylor & Francis Books, Inc.

Chapter 14

Table 14.1: Data from E. Chargaff and J. Davidson (editors), *The Nucleic Acids*, 1955, Academic Press, New York, NY.

Chapter 16

TA 16.1: CALVIN AND HOBBES © 1995 Watterson. Reprinted with permission of Universal Press Syndicate. All rights reserved.

Chapter 17

Table 17.1: Reprinted with permission from *Nature*. Copyright Macmillan Magazines Limited. **Figure 17.9:** G. More, K.M. Devos, Z. Wang, and M.D. Gale: "Grasses, line up and form a circle," *Current Biology*, 1995, vol. 5, pp. 737-739. **Figure 17.11:** Modified from Keho Villiard and Sommerville, "DNA Microarrays for studies of

Photosynthetic Organisms," *Trends in Plant Science*, 1999.

Chapter 18

Figure 18.22: From an *Introduction of Genetic Analysis* 5/e by Anthony J.F. Griffiths, et al., Copyright © 1976, 1981, 1986, 1989, 1993 by W.H. Freeman and Company. Used with permission.

Chapter 19

Figure 19.5: Copyright © John Kochik for Howard Hughes Medical Institute. **Figure 19.9 (text):** From H. Robert Horvitz, *From Egg to Adult*, published by Howard Hughes Medical Institute. Copyright © 1992. Reprinted by permission. **Figure 19.9 (top):** M.E. Challinor illustration. From Howard Hughes Medical Institute © as published in *From Egg to Adult*, 1992. Reprinted by permission. **Figure 19.9 (bottom):** Illustration by: The studio of Wood Ronsaville Harlin, Inc. **Figure 19.20:** Modified from John Kochik for Howard Hughes Medical Institute.

Chapter 20

Table 20.2: Data from the American Cancer Society, Inc., 2002.

Chapter 21

Figure 21.8: Data from P.A. Powers, et al., "A Multidisciplinary Approach to the Selectionist/Neutralist Controversy." *Oxford Surveys in Evolutionary Biology.* Oxford University Press, 1993. **Figure 21.10:** From R.F. Preziosi and D.J. Fairbairn, "Sexual Size Dimorphism and Selection in the Wild in the Waterstrider *Aquarius remigis:* Lifetime Fecundity Selection on Female Total Length and Its Components," *Evolution, International Journal* of *Organic Evolution* 51:467-474, 1997. **Figure 21.11:** Data from M.R. MacNair in J.M. Bishops & L.M. Cook, *Genetic Consequences of Man-Made Change*, Academic Press, 1981, p. 177-207. **Figure 21.12:** Adapted from Clark, B. "Balanced Polymorphism and the Diversity of Sympatric Species." Syst. Assoc. Publ., Vol. 4, 1962.

Chapter 22

Figure 22.3A: Data from Grant, "Natural Selection and Darwin's Finches" in *Scientific American*, October 1991. **Figure 22.3b:** Data from Grant, "Natural Selection and Darwin's Finches" in *Scientific American*, October 1991. **Figure 22.5:** Data from Grant, et al., "Parallel Rise and Fall of Melanic Peppered Moths" in *Journal of*

Heredity, vol. 87, 1996, Oxford University Press. **Figure 22.6:** Data from G. Dayton and A. Roberson, *Journal of Genetics*, Vol. 55, p. 154, 1957.

Chapter 23

Figure 23.2: Data from R. Conant & J.T. Collins, *Reptiles & Amphibians of Eastern/Central North America*, 3rd edition, 1991. Houghton Mifflin Company. **Figure 23.10:** Data from B.M. Bechler, et al., *Birds of New Guinea*, 1986, Princeton University Press. **Figure 23.18:** Data from D. Futuyma, *Evolutionary Biology*, 1998, Sinauer.

Chapter 25

Figure 25.8: From Richard O. Prum and Alan H. Brush, "The Evolutionary Origin and Diversification of Feathers," *Quarterly Review of Biology* 77 (3): 261-295, September 2002. Reprinted with permission of The University of Chicago Press.

Chapter 27

Figure 27.9: Data from U.S. Centers for Disease Control and Prevention, Atlanta, GA.

Chapter 31

Figure 31.09: Courtesy of Joel Cracraft.

Chapter 36

Figure 36.3: From Ralph Quantrano, Washington University. **Figure 36.16:** From G.B. Fincher, "Molecular and Cellular Biology Associated with Endosperm Mobilization in Germinating Cereal Grains." Reprinted with permission from the *Annual Review of Plant Physiology and Plant Molecular Biology*, Volume 40. Copyright © 1989 by Annual Reviews www.annualreviews.org.

Chapter 40

Figure 40.10: Data from Hong et al., Arabidopsis *not* Mutants Define Multiple Functions Required for Acclimation to High Temperatures, *Plant Physiology*, Vol. 132, pp. 757–767, 2003.

Chapter 41

Figure 41.8: After McDaniel, 1996. **Figure 41.9:** After McDaniel, 1996. **Figure 41.10:** After McDaniel, 1996.

Chapter 42

Figure 42.32: From *New York Times*, December 15, 1998. Copyright © 1998 *New York Times*. Reprinted with permission.

Chapter 48

Figure 48.21: From Beck & Habicht, "Immunity and the Invertebrates" in *Scientific American*, November 1996. Reprinted by permission of Roberto Osti Illustrations. **Figure 48.25:** Data from U.S. Centers for Disease Control and Prevention, Atlanta, GA.

Chapter 49

Figure 49.4: Data from B. Heinrich, *Science*, American Association for the Advancement of Science.

Chapter 50

Table 50.2: Data from American College of Obstetricians and Gynecologists: Contraception, Patient Education Pamphlet No. AP005.ACOG, Washington, D.C., 1990.

Chapter 52

Figure 52.6: Data from J.R. Brown et al, "A Defect in Nurturing in Mice Lacking the Immediate Early Gene for fosB", Cell, 1996. **Figure 52.11:** Reprinted from *Animal Behaviour*, Vol. 51, M.D. Beecher, P.K. Stoddard, S.E. Campbell, and C.L. Horning, "Repertoire Matching Between Neighbouring Song Sparrows," pp. 917-923. Copyright © 1996, with permission from Elsevier. **Figure 52.20:** From John Alcock, *Animal Behavior*, 1989. **Figure 52.24b:** From John Alcock, *Journal of Animal Behavior*, 1988. Reprinted by permission of Academic Press, Ltd., London. **Figure 52.30C:** Data from M. Petrie, et al. "Peahens Prefer Peacocks with Elaborate Trains, *Animal Behavior*, 1991. **Figure 52.33B:** Data from H.L. Gibbs et al., Realized Reproductive Excess of Polygynous Red-Winged Blackbirds Revealed by NDA Markers," *Science*, 1990.

Chapter 53

Figure 53.24: From G.C. Varley, "Population Changes in German Forest Pests," *Journal of Animal Ecology*, Vol. 18, May 1949. Reprinted with permission of British Ecology Society. **Table 53.1:** Based on Table 14-11 in A.J. Vander, J.H. Sherman, and D.S. Luciano, *Human Physiology*, 5th ed. Copyright © 1997 McGraw-Hill Companies, Inc., Dubuque, Iowa. All Rights Reserved. **Figure 53.5:** After E.R. Pianka, *Evolutionary Ecology.*, 4th edition, New York, Harper & Row, 1987. **Figure 53.6:** Data from Brown & Lomolino, *Biogeorgraphy*, 3rd edition, 1998, Sinauer Associates, Inc. **Figure 53.7:** Data from Brown & Lomolino, *Biogeography*, 3rd edition, 1998, Sinauer Associates, Inc. After A.T. Smith *Ecology*, 1974. **Figure 53.8:** Data from Elizabeth Losos, Center for Tropical Forest Science, Smithsonian Tropical Research Institute. **Figure 53.10:** Data from *Patch Occupation and Population Size of the Glanville Fritillary in the Anland Islands*, Metapopulation Research Group, Helsinki, Finland. **Figure 53.11:** Data from Bonner, 1965. **Figure 53.2:** Modified from Ricklefs, 1997. **Figure 53.13:** Modified from Ricklefs, 1997. **Figure 53.16:** Data from C.M. Perrins, *Animal Ecology*, 1995. **Figure 53.20B:** Data from C.E. Goulden, L.L. Henry, and A.J. Tessier, *Ecology*, 1982. **Figure 53.22:** Data from Arcese & Smith J. *Animal Biology*, vol. 57, pp. 119-136, 1989 and Smith et al. 1991. **Table 53.3:** After E.R. Pianka, *Evolutionary Ecology*, 4th edition, 1987, New York, Harper & Row, 1987. **Figure 53.30:** Data from National Geographic, July 2001.

Chapter 54

Figure 54.3: From *The Economy of Nature* 4/e, by Robert E. Ricklefs. Copyright © 1973, 1979 by Chiron Press Inc.; Copyright © 1990, 2000 by W.H. Freeman and Company. Used with permission. **Figure 54.4:** From *The Economy of Nature* 4/e, by Robert E. Ricklefs. Copyright © 1973, 1979 by Chiron Press Inc.; Copyright © 1990, 2000 by W.H. Freeman and Company. Used with permission. **Figure 54.6:** Data from Begon et al., *Ecology*, 1996. After: W.B. Clapham, *Natural Ecosystems*, Clover, Macmillan. **Figure 54.8:** Data from E.J. Heske, et al., *Ecology*, 1994. **Figure 54.9:** Data from E.J. Heske, et al., *Ecology*, 1994. **Figure 54.22:** Data from D.W. Davidson et al., "Granivory in a Desert Ecosystem." *Ecology*, 1984.

Chapter 55

Figure 55.18: Data from F. Morrin, *Community Ecology*, Blackwell, 1999. **Figure 55.16:** Data from J.T. Wooten & M.E. Power, "Productivity, Consumers & the Structure of a River Food Chain," *Proceedings National Academic Sciences*, 1993. **Figure 55.13:** Data from Flecker, A.S. and Townsend, C.R., "Community-Wide Consequences of Trout Introduction in New Zealand Streams." In *Ecosystem Management: Selected Readings*, F.B. Samson and F.L. Knopf eds., Springer-Verlag, New York, 1996. **Figure 55.14:** Data from M. Power "Habitat Heterogenieity and the Functional Significance of Fish," *Ecology*, 1997. **Table 55.1:** After Whittaker, 1975.

Chapter 56

Figure 56.13: Map from "An Act of God," July 19, 1997, *The Economist.* Copyright © 1997 *The Economist* Newspaper Ltd. All rights reserved. Reprinted with permission. Further reproduction prohibited. www.economist.com. **Figure 56.25:** Data from Geophysical Monograph, American Geophysical Union, National Academy of Sciences, and National Center for Atmospheric Research.

Chapter 57

Figure 57.3: After Smith et al., 1993. **Figure 57.10:** IUCN 2000, AmphibiaWeb, Hero J.M. & L. Shoo, 2003. Chapter 7 in *Amphibian Conservation*, Smithsonian Press. Background biodiversity hotspots map from Myers et al, 2000. *Nature* 403:853-858 c/o Conservation International. Prepared by J.M. Hero, April 2002. **Figure 57.11:** After Green and Sussman, 1990. **Figure 57.12:** Data assembled by Pima, 1991. **Figure 57.15:** Data from Marra, Hobson, Holmes, "Linking Winter & Summer Events," in *Science*, Dec. 1998. **Figure 57.16:** Data from UNEP, *Environmental Data Report*, 1993, 1994. **Figure 57.22:** Data from H.L. Billington, "Effects of Population Size on a Genetic Variation in a Dioecious Conifer" in *Conservation Biology*, Blackwell Scientific Publication, Inc., 1991. **Figure 57.25:** Data from The Peregrine Fund. **Figure 57.27:** From "Quetzal and The Macaw," The Story of Costa Rica's National Parks, by David Rains Wallace, 1992 Sierra Club. Reprinted courtesy of David Rains Wallce. **Figure 57.4:** From *Nature's Place: Human Population and the Future of Biological Diversity*, by Richard P. Cincotta and Robert Engelman, 2000. Reprinted with permission of Population Action International. **Figure 57.5:** From *Nature's Place: Human Population and the Future of Biological Diversity*, by Richard P. Cincotta and Robert Engelman, 2000. Reprinted with permission of Population Action International. **Figure 57.26:** From *Nature and Resources, Vol. XXII*, 1986. Copyright © 1986 UNESCO Press. Reproduced by permission of UNESCO.

Index

B

Genetic counseling, 274
Genetic disease. *See* Gene disorder
Genetic drift, 437–39, 437t, 438–39f, 443, 443f, 479, 488, 1244
Genetic engineering, 325–33
  agricultural applications of, 335–38, 335–38f
  bacteria and, 558
  medical applications of, 333–34, 333–34f
  risk and regulation of, 339–40, 340f
  stages of
    cloning, 326, 326f
    DNA cleavage, 325, 326f
    finding gene of interest, 326f, 328, 328f
    preliminary screening of clones, 326–37f, 327
    production of recombinant DNA, 325, 326f
  working with gene clones, 329–32
Genetic map, **268**, **344**, 351
  of *Drosophila*, 344
  of garden pea, 268f
  of humans, 269, 269f
  using recombination to make maps, 268, 268f
Genetic marker, anonymous, 269
Genetic material. *See* Hereditary material
Genetic privacy, 358
Genetic recombination. *See* Recombination
Genetics
  behavioral, 1106, 1108–9, 1108–9f
  population, 433–34, **435**–50
  symbols used in, 250
Genetic screen, 327, 327f, 358
Genetic similarity, 466
Genetic system, 63
Genetic template, **1113**
Genetic testing, 259
Genetic variation, 13, 227, 237–38, 238f, 241f, 259, 351
  conditions for natural selection, 440
  in crop plants, 1232
  evolution and, 434–35, 434f, 450, 450f
  genes within populations, 433–50
  in human genome, 351
  loss of, 1234, 1244–45, 1244f
  measuring levels of, 435
  in nature, 435, 435f
  in prokaryotes, 552
Gene transfer, 406–8, 406–8f
  by conjugation, **407**, 407f
  lateral, **499**–500, 500f, 519
  by transposition, 407–8, 408f
  using vectors, 323, 323f
  vertical, **499**
Gene transfer therapy, **263**
Genistein, 799t, 800
Genital pore, 643f, 647f
Genome, **15**, 343–58. *See also specific organisms*
  of chloroplasts, 352
  eukaryotic, 348f
    gene organization in, 349, 350t
    noncoding DNA in, 350–51
  evolution of, 492–500
  finding genes in, 349, 349f
  human. *See* Human genome
  minimal size to support life, 352
  of mitochondria, 352
  origins of genomic differences, 497–501
  prokaryotic, 348f

  size and complexity of, 348, 348f, 494, 497
  of virus, 532–33
Genome map, 344–45, 344–45f
  genetic. *See* Genetic map
  physical. *See* Physical map
Genome sequencing, 346–47, 346–47f
  clone-by-clone method, 347, 347f, 351
  databases, 349
  DNA preparation for, 346
  draft sequence, 351–**52**, 494
  evolutionary relationships from, 519
  finished sequence, 351
  shotgun method, 347, 347f, 351
  using artificial chromosomes, 347
Genomic imprinting, 428
Genomic library, **324**
Genomics, **348**, **352**
  applications of, 357–58, 357–58f
  behavioral, 358
  comparative, 352, 353f, 491–96, 492–93t
  functional, **354**–58, 503
  vocabulary of, 351
Genotype, 249–50
  testcross to determine, 252, 252f
Genotype frequency, **436**–37
Genus, **510**–11
Geographic distribution, variation within species, 472–73, 472f
Geographic isolation, 473t, 488, 488f
Geography, of speciation, 480–81, 480–81f
Geological timescale, 75f, 694f
Geomagnetotropism, 811
Geometric progression, 11, 11f
Geranium, 744
Gerbil, 1050
German measles. *See* Rubella
Germ cells, 1068, 1069f
Germinal center, 914
Germinal epithelium, 1068
Germination, of seeds, 388, 389f, 596f, 756, 760, 763, **763**, 764f, 808, 814, 824, 827, 840
Germ layer, 384f, **385**, 624, 856, **1087**
  developmental fates of, 1087t, 1094f
  formation of, 1087–89, 1087–89f
Germ-line cells, 229, 229f, 410
GH. *See* Growth hormone
GHIH. *See* Growth hormone-inhibiting hormone
GHRH. *See* Growth hormone-releasing hormone
Giant clam (*Tridacna maxima*), 652, 653f
Giant ground sloth (*Megatherium*), 717t, 1228, 1228f
Giant ragweed, 851
Giant sequoia (*Sequoiadendron giganteum*), 851f, 1206f
Giant squid, 652
Giant tube worm, 623t
Giardia, 562
Gibberellic acid, 763, 764f
Gibberellin, 774, 816, 817t, **824**, 824f, 826, 833–34, 836, 838f
Gibbon (*Hylobates*), 720, 721f
Gibbs' free energy. *See* Free energy
Gigantism, 1002, 1002f
Gila monster, 708
Gill(s), 922
  of animals, 684
  of bivalves, 657
  of crustaceans, 668
  external, **924**
  of fish, 590, 915, 923f, 924–25, 925f
  of mollusks, 653f, **654**, 657f, 924
  of mushroom, 600f, 608, 608f

Gill arch, 693, 693f, **925**, 925f
Gill chamber, 1091
Gill cover, 697
Gill filament, 925, 925f
Gill raker, 925f
Gill slits, 464, 685f, 687, 687f, 890
Gilman, Alfred, 131
Gingiva, 891f
Gingko, 7, 7f, 585t, 590, 592, 592f, 594f
Gingko biloba. *See* Maidenhair tree
Ginkgophyta (phylum), 585t, 592, 592f
GIP. *See* Gastric inhibitory peptide
Giraffe (*Giraffa camelopardalis*), 62f, 242, 434, 434f, 464, 515f, 716, 719t, 1210
Girdling of tree, 738
Gizzard
  of annelids, 658
  of birds, 888, 890, 890f
  of earthworm, 660, 888, 888f
Glaciation, 485, 485f
Glacier, 1141, 1178, 1178–79f, 1224
Glacier Bay, Alaska, succession at, 1178–79f
Gladiolus, 747
Gland (animal)
  endocrine, **860**
  exocrine, **860**
Gland (plant), 748
Glanville fritillary butterfly, 1144, 1144f
Gleaning bird, 1173, 1173f
Gleason, H. A., 1162
Glenn, John, 810
Gliding bacteria, 546, 549f
Glioblastoma, 417t
Global climate change, 358, 488, 1163, 1186
  crop production and, 785–86
  prehistoric, 1223
Global warming, 785–86, **1222**–24, 1222–23f
  effect on humans, 1224
  effect on natural ecosystems, 1223
  effect on species, 1223, 1223f
  geographic variation in, 1222, 1222f
β-Globin gene, 262, 298, 298f, 378
Globular protein, 44f
Globulin, **910**
Glomales, 605, 607, 612
Glomeromycota (phylum), 605
Glomerular filtrate, 1048, **1053**, 1054–55
Glomerulus, 1048, **1052**, 1053–55f
Glomus, 606
Glottis, 891, 891f, 927, 927–28f
Glucagon, 895f, 902, 902f, 995t, 1008, 1008f
Glucocorticoid, 1007
Gluconate metabolism, 450
Gluconeogenesis, **902**, 1007
Glucose, 37t, 55
  alpha form of, **58**, 58f
  beta form of, **58**, 58f
  blood, 880, 993, 1008, 1054
    regulation of, 902, 902f, 1042–43, 1042f
  catabolism of, 160, 162–63
    regulation of, 177, 177f
  homeostasis, 1007
  metabolism of, 1007
  priming of, 164
  production in photosynthesis, 187, 187f, 189, 201f, 203
  reabsorption in kidney, 1054
  structure of, 55f
  urine, 1008, 1054
Glucose 1-phosphate, 202

Glucose 6-phosphate, 165f, 902
Glucose-6-phosphate dehydrogenase deficiency, 269f
Glucose repression, 368–69, 369f
Glucose transporter, 39t, 115, 120–21, 121f
Glutamate, 178, 178f, **950**
Glutamic acid, 41, 42f
Glutamine, 42f
Glycation, of proteins, 402
Glyceraldehyde, 55f
Glyceraldehyde 3-phosphate, 164, 165f, 166, 200–201f, 201–2
Glycerol, 37t, 52, 52f, 106, 106f, 179, 1138
Glycerol kinase deficiency, 269f
Glycerol phosphate, 36f
Glycine, 42f, **950**
Glycocalyx, 109
Glycogen, **54**, **57**
  breakdown of, 998, 998f
  liver, 902, 1008
  muscle, 880–81
  synthesis of, 903, 999, 1042
Glycogenolysis, 902
Glycolipid, 92, 109, 109t, 136
Glycolysis, **162**, 164–67, 164–65f, 167f
  ATP production in, 176, 176f
  evolution of, 166, 182
  regulation of, 177, 177f
Glycophorin, 109t
Glycoprotein, 92, **102**, 108f, 109, 109t, 536
Glycoprotein hormone, 993, 994t, 998
Glyoxysome, **94**
Glyphosate, **336**, 336f
Glyptodont, 10f
G_2/M checkpoint, 219–20, 219f, 221f
Gnetophyta (phylum), 585t, 590, 592, 592f, 594f
Gnetum, 585t, 592
GnRH. *See* Gonadotropin-releasing hormone
Goat, 716
Goblet cells, 860
Goiter, 1004, 1004f
Golden mean, 744
Golden plover, 1117f
Golden rice, 357
Goldenrod, 834f, 835
Golden toad (*Bufo periglenes*), 1061, 1061f, 1234–35, 1234f
Golgi, Camillo, 92
Golgi apparatus, 86–87f, 90t, 91–**92**, 92–94f, 99, 102t
Golgi body, **92**, 739
Golgi tendon organ, 974
Gompertz number, 401
Gompertz plot, 401, 401f
Gonad
  indifferent, 1063, 1063f
  of polychaetes, **660**
Gonadotropin, 1002
Gonadotropin-releasing hormone (GnRH), 1003, 1003f, 1071, 1071f
Gonorrhea, 555t, 556, 556f
Gonos, 242
Gonyaulax, 568f
Goose, 464, 711t, 882, 1066
  egg retrieval behavior in, 1106, 1107f
  imprinting in, 1112, 1112f
Gooseneck barnacle (*Lepas anatifera*), 669f
Gorgonian coral, 641
Gorilla (*Gorilla*), 14, 466, 495, 495f, 513, 720–21, 721f, 1121
Gould, James L., 1121
Gould, John, 454

Hydrogen, 26, 26f
  atomic structure of, 20f
  in plants, 782, 782t
  prebiotic chemistry, 67f
Hydrogenated oils, 54
Hydrogen bond, 28
  in DNA, 50, 50f, 287, 287f
  in proteins, 43–44f, 45
  in water, 28, 28f, 30, 30f
Hydrogen cyanide, prebiotic
    chemistry, 66, 67f
Hydrogen ion, 31–32
  excretion into urine, 1056, 1056f
Hydrogen peroxide, 94, 94f, 803
Hydrogen sulfide, 163, 182, 189, 196,
    546, 553, 1217
Hydroid, 638, 640, 640f
Hydrolysis, 37, 37f
Hydronium ion, 31
Hydrophilic molecule, 30, 106, 106f
Hydrophobic exclusion, 30, 45
Hydrophobic molecule, 30, 106, 106f
Hydrophyte, 1217
Hydroponic culture, 782–83, 783f
Hydrostatic pressure, 117, 117f
Hydrostatic skeleton, 646, 658,
    871, 871f
Hydrotropism, 811
Hydroxide ion, 31–32
Hydroxyapatite, 864
Hydroxyl group, 36, 36f
Hydrozoa (class), 640
Hyena, 180f
Hymenoptera (order), 675t
  social systems in, 1132–33, 1132–33f
Hyoseris longiloba, 843f
Hyperaccumulating plant, 791–92
Hypercholesterolemia, 119, 249t, 264t,
    333t, 411
Hyperosmotic solution, 116, 117f, 1044
Hyperpolarization, 944–45, 945f, 950
Hypersensitive response, in plants,
    803, 803–4f
Hypersensitivity, 1036
  delayed, 1036
  immediate, 1036, 1036f
Hypertension, 918, 921
Hypertonic solution, 1044
Hypertrophy, 881
Hyperventilation, 931–32
Hyphae, 600, 600f, 601–3, 797, 797f
Hypocotyl, 730, 756f, 764f
Hypolimnion, 1216–17, 1217f
Hypoosmotic solution, 116, 117f, 1044
Hypophosphatemia, 269f
Hypophyseal duct, 686f
Hypothalamohypophyseal portal
    system, 1003–4, 1003f
Hypothalamus, 955f, 955t, 956, 958,
    973, 993, 1000, 1000–1001f,
    1010, 1042, 1109
  control of anterior pituitary by,
    1003, 1003f
Hypothesis, 5–7, 6f
Hypothyroidism, 1005
Hypotonic solution, 1044
Hypoventilation, 931–32
Hyracotherium, 462–63, 462–63f

I

IAA. See Indoleacetic acid
IBA. See Indolebutyric acid
I band, 874–77, 874–77f
Ibis, 711t
Ice, 27f, 28f, 29, 30f
Ice Age, 1163, 1228
Ichthyosauria (order), 703t

Ichthyosis, 269f
Ichthyostega, 700, 700f
Icosahedron, 532
Identical twins, 1108
Idiomorph, 237
Ig. See Immunoglobulin
I gene (ABO blood group), 260, 260f
I gene (lac operon), 366f
Ig fold, 1032
Iguana, 708
  Galápagos, 471f
Iiwi, 1140f
Ileocecal valve, 897f
Ileum, 894
Illicium, 594f
Imaginal disc, 387f
Immediate hypersensitivity,
    1036, 1036f
Immigration, 437, 1150
Immovable joint, 872, 872f
Immune regulation, in vertebrates,
    1009–10
Immune response, 909, 1018, 1028f
  cell-mediated, 1018, 1031–32
  cells involved in, 1019, 1019t
  concepts of specific immunity, 1018
  discovery of, 1018
  humoral, 1018, 1023–24, 1023–24f,
    1031–32
  initiation of, 1020, 1020t
  primary, 1027, 1027f
  secondary, 1027, 1027f
Immune surveillance, 910f, 1022
Immune system, 857t, 858f, 909,
    1013–36, 1014
  cells of, 910f
  defeat of, 1033–36
  effect of HIV on, 536
  evolution of, 1031–32, 1032f
  of invertebrates, 1031, 1031f
  of vertebrates, 1032
Immunity
  active, 1018, 1026–27, 1027f
  passive, 1018
Immunization, passive, 1029
Immunochemistry, characterization of
    receptor proteins, 126
Immunocytochemistry, 83
Immunodeficiency, X-linked with
    hyper IgM, 269f
Immunoglobulin (Ig), 39t, 136,
    1023–24. See also Antibody
  classes of, 1024t
  structure of, 136f
Immunoglobulin A (IgA), 1024–25,
    1024t
Immunoglobulin D (IgD), 1024–25,
    1024t
Immunoglobulin E (IgE), 1024–25,
    1024t, 1036
Immunoglobulin G (IgG), 1024–25,
    1024f, 1024t
Immunoglobulin M (IgM), 1024–25,
    1024f, 1024t
Immunoglobulin superfamily, 1032
Immunological tolerance, 1026
Impala, 1126, 1173f
Implantation, 1074, 1074f, 1098
  prevention of, 1078
Imprinting (behavior), 1112, 1112f
  filial, 1112
  sexual, 1112
Imprinting, genomic, 428
Inchworm caterpillar (Necophora
    quernaria), 1170f
Incisor, 715, 715f, 890, 890–91f
Incomplete flower, 840
Incontinentia pigmenti, 269f
Incurrent siphon, 657f, 686f

Incus, 980, 981f
Independent assortment, 236, 238,
    238f, 267
  law of, 253–54, 254f
Indeterminate development, 626f, 627
Indian grass (Sorghastrum nutans), 835
Indian pipe (Hypopitys uniflora),
    789, 789f
Indifferent gonad, 1063, 1063f
Indigo bunting, 1116–17
Indirect effect, 1176
Individualistic concept, of
    communities, 1162
Indoleacetic acid (IAA), 820, 820f
Indolebutyric acid (IBA), 821
Induced fit, 150, 150f
Induced ovulator, 1067
Inducer T cells, 1019, 1019t, 1033
Induction (development), 391–92,
    392f, 1090, 1093, 1093f
  primary, 1093
  secondary, 1093
Induction of protein, 366–67
Inductive reasoning, 4
Industrialized countries. See Developed
    countries
Industrial melanism, 456
  agent of selection, 457
  in peppered moth, 456–57, 456–57f
  selection against melanism, 457, 457f
  selection for melanism, 456, 456f
Industrial pollution, 457, 1218
Industrial Revolution, 1155f
Indy gene, 402
Inert element, 24
Infant, growth of, 1102
Infection thread, 787f
Inferior vena cava, 917f, 918
Inflammatory response, 538, 910f,
    1017, 1017f, 1025
Influenza, 533t, 540
  pandemic of, 1013
Influenza virus, 524f, 531f, 533t
  antigen shifting in, 1035
  subtypes of, 540
  types of, 540
  vaccine for, 334
Information molecules, 48–51
Infrared radiation, sensing of,
    988, 988f
Ingenhousz, Jan, 188
Ingram, Vernon, 298
Inguinal canal, 1068, 1070
Inhalation. See Inspiration
"In heat," 1067
Inheritance
  of acquired characteristics, 434, 434f
  blending, 243, 248, 436
  chromosomal theory of, 265
  patterns of, 241–74
Inhibin, 1071, 1071f
Inhibitor, 152
  allosteric, 152, 153f
  competitive, 152, 153f
  noncompetitive, 152, 153f
Inhibitory postsynaptic potential
    (IPSP), 949f, 950–51, 951f, 963
Initiation complex, 308–9, 308f,
    310–11, 311f, 370, 370f,
    372f, 373
Initiation factor, 310, 311f
Injectable contraceptive, 1077t
Innate behavior, 1106–7, 1107f
Innate releasing mechanism, 1106
Inner cell mass, 1086, 1086f, 1089,
    1096–97
Inner ear, 978, 980, 981f
Inner membrane
  of chloroplasts, 96, 97f, 186f

of mitochondria, 96, 96f, 163f,
    176, 562
Inonotus tomentosus, 601f
Inositol triphosphate (IP₃), 132, 133f,
    135
Inositol triphosphate (IP₃)/calcium
    second-messenger system,
    998–99, 999f
Insect, 635, 651f, 664, 664–65f, 668t,
    672–74, 675t, 871
  chromosome number in, 210t
  coevolution of insects and plants, 844
  development in, 382f, 386, 386–87f
  digestive system of, 674
  diversity among, 672f
  evolution of, 527–28
  excretory organs in, 1046–47, 1047f
  external features of, 673, 673f
  eyes of, 450, 450f, 504–5, 504–5f,
    666–67, 666f, 673
  heart of, 908, 908f
  internal organization of, 674
  locomotion in, 883–84
  metamorphosis in, 674, 1009, 1009f
  molting in, 871, 1009, 1009f
  orders of, 675t
  pheromones of, 674
  pollination by, 595, 844, 1164
  respiration in, 923f, 926–27
  segmentation in, 665f
  selection for pesticide resistance in,
    441, 441f
  sense receptors of, 674
  sex chromosomes of, 270, 270t
  social, 1120, 1120f, 1132–33,
    1132–33f
  thermoregulation in, 1042, 1042f
  wings of, 503, 664, 665f, 673,
    673f, 884
Insecticide, 336
Insectivora (order), 719t
Insectivore, digestive system of, 899f
Insectivorous leaf, 751
Insect resistance, in transgenic
    plants, 336
Insertional inactivation, 408, 410t, 411
Insertion of muscle, 873
Inspiration, 929–32, 931f
Instar, 386, 387f
Instinct, 1106, 1111–13, 1113f
Insulin, 39t, 298, 402, 895f, 902, 902f,
    993, 995t, 999, 1008, 1008f,
    1042, 1042f
  genetically engineered, 333, 558, 1008
Insulin-like growth factor, 994, 1002
Insulin-like receptor, 402
Integrating center, 1040,
    1040–41f, 1057
Integrin, 102, 102f, 139, 391
Integument (flower), 589, 595–96,
    595f, 760, 838
Integumentary system, 857t, 859f
Intelligent design theory, against
    theory of evolution, 468
Interarterial pathway, 919
Intercalary meristem, 731, 756
Intercalated disc, 867t, 868
Intercostal muscles, 930–32,
    931f, 1101f
Interference competition, 1164
Interferon, 1022
  alpha-interferon, 1016
  beta-interferon, 1016
  gamma-interferon, 1016, 1022
Intergovernmental Panel on Climate
    Change, 785
Interleukin, 1021
Interleukin-1, 1017, 1020, 1021,
    1021f, 1023f, 1028f

Mosquito, 210t, 568, 569f, 673f, 884, 1224
Mosquito fish, 1064
Moss, 581–82, 582–83f, 1178, 1178–79f
Moss animal, 623t
Moth, 664f, 672f, 673, 675t, 844
"Mother figure," 1112
Mother-of-pearl, 652, 654
Motif, protein, 43, 44f, 45
Motility, 523t
Motion sensing, 971t
Motor cortex, primary, 958f
Motor effectors, 940
Motor neuron, 869t, 879, **940,** 940f, 961–62f, 963–66
    autonomic, **940**
    somatic, 879, **940**
Motor protein, 98–99, 99f, 214, 396, 428
Motor unit, **879,** 879f
Mountain ash (*Sorbus*), 749f
Mountain lion, 1195
Mountain zone, 1208, 1208f
Mouse (*Mus musculus*), 719t
    chimeric, 393f
    coat color in, 255
    development in, 382–85f, 383–85, 397, 684f
    genome of, 348f, 352, 492t, 494–95
    homeodomain proteins of, 16f
    homeotic genes in, 399f
    marsupial, 467f
    maternal care in, 1109, 1109f
    ob gene in, 903, 903f
    single gene effects on behavior, 1109
    transgenic, 333, 333f
Mouth, 888–91, 888f, 891f
Mouthparts
    of arthropods, 668t
    of insects, 665f, 673, 673f
Movement
    in animals, 870–73
    as characteristic of life, 62, 62f
MPF. *See* M-phase-promoting factor
M phase. *See* Mitosis
M-phase-promoting factor (MPF), 219–20
mRNA. *See* Messenger RNA
MSH. *See* Melanocyte-stimulating hormone
MTS1 gene, 417t
Mucilage, 751, 788
Mucilaginous lubricant, of roots, 739
Mucosa, of gastrointestinal tract, 889, 889f, 894f
Mucous cells, 893f
Mucus, 654, 860, 1014
Mud eel, 701
Mulberry, 761f, 846
Mule, 476
Mullein, 851
Müller, Fritz, 1171
Müllerian mimicry, **1171,** 1171f
Muller's ratchet, for origin of sex, 238
Mullis, Kary, 329
Multicellularity, **73**
    in animals, 618t
    evolution of, 73
    in sponges, 636
Multicellular organism, 522, 523t
    cell cycle control in, 221
    development in, 381–402
Multienzyme complex, 151, 151f, 168
Multigene family, 349
Multinucleate hypothesis, for origin of metazoans, **630**
Multiple alleles, 260, 260f
Multiple cloning site (MCS), **321**
Multiple fruit, 761f

Multiple sclerosis, 427
Multiregional Hypothesis, 725
Mummichog (*Fundulus heteroclitus*), 441, 441f
Muscle
    insertion of, **873**
    lactic acid accumulation in, 181, 181f
    length and tension of, 974
    metabolism during rest and exercise, 880–81
    origin of, **873**
Muscle cells, 81, 99
    cell cycle in, 222
Muscle contraction, 132, 868, 872–73, 949, 963
    control of, 878–79
    isometric, **873**
    isotonic, **873**
    sensing of, 971t, 974, 974f
    sliding filament model of, 874, 874–77f, **875–77**
Muscle fascicle, 874f
Muscle fatigue, **881**
Muscle fiber, 868, 868f, **874,** 874f, 879
    fast-twitch (type II), **880**–81, 881f
    slow-twitch (type I), **880,** 881f
    types of, 879–80
Muscle spindle, 961f, 962, 974, 974f
Muscle stretch reflex, 962
Muscle tissue, 856, 856f, 867–68, 867t, 868f
Muscular dystrophy
    Becker, 269f
    Duchenne, 249t, 264t, 269f
    Emery-Dreifuss, 269f
    gene therapy for, 264, 333
Muscularis, of gastrointestinal tract, 889, 889f, 894f
Muscular system, 857t, 859f
Mushroom, 599–601, 605t, 608
Musical ability, 960
Musk-oxen, 1211
Mussel, 652, 656–57, 1123, 1123f, 1215
    starfish predation on, 1176, 1176f
Mustard, 845, 1169
Mustard oil, 1169, 1169f
Mutagen, 410, 414f, 415, 420–21
Mutant, **265**
Mutation, **410,** 488, 497
    aging and, 401
    altering DNA sequence, 410, 410t
    arising from changes in gene position, 410t, 411
    cancer and, 224f, 413–23, 419f
    disorder-causing, 259
    evolution and, 437–38, 437t, 438f
    in germ-line tissue, 410
    interactions among evolutionary forces, 443, 443f
    kinds of, 410–11, 410t
    Muller's ratchet, 238
    in prokaryotes, 552, 552f
    somatic, 410, **1026**
Mutation rate, 438, 443, 494
Mut genes, 412
Mutualism, 557, 604, 611, **1172,** 1174, 1174f
    fungal-animal, 613, 613f
Myasthenia gravis, 1026
myb gene, 764f
Mycelium, **601,** 601f
    primary, 608, 608f
    secondary, 608, 608f
myc gene, 416, 417t
Mycobacterium leprae, 555t
Mycobacterium tuberculosis, 285t, 554, 554f, 555t
Mycologist, **600**

Mycoplasma, 555t
Mycorrhizae, **604, 612**–13, 612f, 789, 1172
    arbuscular, **612,** 612f
    ectomycorrhizae, **612**–13, 612f
Myc protein, 224
Myelin sheath, 869, 869f, **941,** 941f, 946–47, 946–47f, 947t, 963f
Myeloma cells, 1030, 1030f
Myocardium, 868
Myofibril, 868, 868f, **874,** 874–75f, 878f
Myofilament, 874, 874f
Myoglobin, 39, 39t, 43, **880**
Myosin, 39t, 40, 100, **876,** 876f. *See also* Thick myofilament
Myotubular myopathy, 269f
Myrica faya, 1241
Myxini (class), 689, 691f, 692, 693t
Myxobacteria, 549t
Myxomycota (phylum), 576f
Myzostoma mortenensi, 628f
Myzostomid, 628, 628f

N

NAA. *See* Naphthalene acetic acid
NAD⁺, 51, **153,** 904t
    as electron acceptor, 153
    in oxidation-reduction reactions, 145f
    regeneration of, 166–67, 167f, 173, 173f, 181, 181f
    structure of, 153, 153f, 173, 173f
NADH
    contributing electrons to electron transport chain, 162, 163f, 173–76, 174–76f
    from fatty acid catabolism, 179, 179f
    from glycolysis, 163–65f, 166
    inhibition of pyruvate dehydrogenase, 177, 177f
    from Krebs cycle, 163f, 169f, 170, 171f
    from pyruvate oxidation, 162, 163f, 168, 168f
    recycling into NAD⁺, 166–67, 167f, 173, 173f, 181, 181f
    structure of, 173, 173f
NADH dehydrogenase, **174,** 174f
NADP⁺, 904t
NADPH
    production in photosynthesis, 187, 187f, 189, 197–99, 197–99f
    use in Calvin cycle, 200–201, 200–201f
NADP reductase, 199
Naiad, 657
Naked mole rat, 1132, 1134
Naked-seeded plant. *See* Gymnosperm
Nanaloricus mysticus, 623t
Nannippus, 463f
Nanoarchaeum equitans, 519
Nanos protein, **396**
Naphthalene acetic acid (NAA), 821
Nasal cavity, 928f
Nasal passage, 976, 976f
Natural killer cells, 1015–**16,** 1016f, 1019t, 1022
Natural selection, 8, **12,** 433, **434,** 434f, **440,** 1123, 1130
    conditions needed for, 440
    ecological species concept and, 477
    evolution and, 454–59, 468, 486
    invention of theory of, 11–12, 11–12f
    maintenance of variation in populations, 444–45, 444–45f
    in speciation, 479
Nature-versus-nurture debate, 1106, 1108

Nauplius larva, **668,** 668f
Nautilus, 657
Navigation, 988, **1116**–17, 1117f
Neanderthals, 725–26, 725f
Near-shore habitat, Alaskan, 1242–43, 1242f
Nearsightedness, 984, 984f
Neck, of tapeworms, **645**
Necrosis, **400,** 796
Nectar, 595, 736, 844–45, 845f, 1120, 1124
Nectary, 595, 801, 844, 1174f
nef gene, 538, 539f
Negative feedback loop, 900, 1003, 1003f, 1040, 1040–41f, 1152
    blood glucose, 1042, 1042f
    thermoregulation, 1040, 1041f, 1042
Negative gravitropism, 810, 810f
Negative phototaxis, 1116
Negative pressure breathing, 927
Neisseria gonorrhoeae, 555t, 556, 556f
Nekton, 1215
Nematocyst, 622t, **638**–39, 639f, 656
Nematoda (phylum), 621f, 622t, 635, 646, 646–47f
Nematode, 634–35f, 646. *See also Caenorhabditis elegans*
    digestive tract of, 888, 888f
    disease-causing, 646f, 647
    eaten by fungi, 603, 603f
    plant parasites, 796, 797f
    root-knot, 796, 797f
Nemertea (phylum), 623t, 645, 645f
Neonate, 1101
Neotenic larva, 924
Neotiella rutilans, 232f
Neotyphodium, 613
Nephridia, 653f, **655,** 659, 659f, 1046f
Nephron, **1048,** 1048f, 1052–58
    organization of, 1048f
    structure and filtration, 1052–53
    transport processes in, 1055–56, 1055–56f
Nephrostome, **655,** 1046, 1046f
Nereis virens, 658f
Neritic zone, **1214,** 1214f
Nerve, 858f, 869, **941,** 963, 963f
    stimulation of muscle contraction, 879
Nerve cord, 647f, 954, 954f
    dorsal, **684,** 684–85f, 688, 689f, 1094f
Nerve gas, 950
Nerve growth factor (NGF), 222, 994
Nerve impulse, 869, 869f, 942–47
Nerve net, 954, 954f
Nerve ring, **678**
Nervous system, 523t, 857t, 858f, 939–66
    of annelids, 659f, 954, 954f
    of arthropods, 665–66f, 667, 954, 954f
    central. *See* Central nervous system
    of cephalopods, 657
    of cnidarians, 954, 954f
    of echinoderms, 954f
    of fish, 954–55, 955f
    of flatworms, 643, 954, 954f
    neural and endocrine interactions, 993
    neurons and supporting cells, 940–41, 940–41f
    peripheral. *See* Peripheral nervous system
    regulation of digestion, 900
Nervous tissue, 856, 856f, 869, 869f, 869t
Nest, of weaver bird, 1134
Nesting behavior, 1108, 1108f

O

Orchid, 594, 613, 743, 762, 841, 841f
Orchidaceae (family), 841f
Order, as characteristic of life, 2
Order (taxonomic), **511**
Oreaster occidentalis, 676f
ORF. *See* Open reading frame
Organ, 2, 3f, **857**, 857f
Organelle, **2**, 3f, 63f, **80**, 80f, **86**, 90t, 102t, 520t
Organism, 2, 3f
Organizer, 392, 392f, **1092**–93
Organochlorine pesticide, 1247
Organ of Corti, 971t, **980**, 981f
Organogenesis, 384f, **385**, 397, 397f, 1085t, 1098
Organ system, **2**, 3f, **857**, 857–59f, 857t
Orgasm, 1000
oriC site, 290, 294
Orientation, 988, **1116**
Origin of life, 19, 27, 61–62, 61f
    deep in earth's crust, 65
    at deep-sea vents, 65
    extraterrestrial, 64, 76, 76f
    Miller-Urey experiment, 66–67, 66–67f
    at ocean's edge, 62, 65, 68
    place where life began, 65
    scientific viewpoint, 64
    special creation, 8, 64
    spontaneous origin, 64
    within clay, 65
Origin of muscle, **873**
Origin of replication, 208, 208f, 290, 290f, 294–95, 321
Oriole, 1238
Ornithine transcarbamylase deficiency, 269f
Ornithischia (order), 703t
Orrorin tugenensis, 723f
Ortholog, **495**, 503
Orthoptera (order), 666f, 672f, 673, 675t
Oscillating selection, **444**, 486, 512
Osculum, 636, 636f
Oskar protein, **396**
Osmoconformer, **1044**
Osmolality, **1044**, 1045f
    plasma, 1000, 1000f
Osmoreceptor, 920, 1000f, 1057, 1057f
Osmoregulator, 1044
Osmoregulatory functions, of hormones, 1057–58
Osmoregulatory organs, 857t, 1046–47, 1046–47f
Osmosis, **116**–17, 116f, 122t, 768–69, 914
Osmotic balance, 117, 1044, 1045f
Osmotic concentration, **116**
Osmotic potential. *See* Solute potential
Osmotic pressure, **117**, 117f, **1044**
Osmotic protein, 39t
Osmotroph, **565**
Ossicle, 678, 980
Ossification, 696
Osteoblasts, 864–65, 865f
Osteoclasts, 865, 1006, 1006f
Osteocytes, 863t, 864–65, 865f
Ostracoderm, 692, 692f
Ostrich, 711t, 713, 1066
Otolith, 978f, 980
Otolith membrane, 978
Otter, 1239
Ottoia, 630f
Outcrossing, 846–47, 847f
Outer bark, 746
Outer ear, 980, 981f
Outer membrane
    of chloroplasts, 96, 97f, 186f
    of mitochondria, 96, 96f, 562

Outgroup, **512**
Outgroup comparison, **512**
Ovalbumin gene, 313f
Oval window, 980, 981f
Ovarian cancer, 414t, 417t
Ovarian cycle, 1010
Ovarian follicle, 1072
Ovary, 858–59f, 993f, 995t, 1072–73f
Ovary (plant), 589, **595**, 595–96f, **840**, 840f, 848f
Overexploitation, 1234t, 1239, 1239f
Overyielding, 1197
Oviduct. *See* Fallopian tube
Oviparity, 1064, 1066–67
Oviraptor, 516f
Ovoviviparity, 1064, 1066
Ovulation, 1067, 1074, 1074–75f
    prevention of, 1076–78
Ovulator
    cyclic, 1067
    induced, 1067
Ovule, 589, **590**, 591, **595**, 595f, 756, 838, **840**, 840f, 842f, 843, 848
Owl, 711t, 713, 982
Oxaloacetate, 169–70, 171f, 178, 204f
Oxidation, **23**, 23f, **144**–45, 145f, 156, 177
β-Oxidation, 178–79f, 179
Oxidation-reduction (redox) reaction, **144**–45, 145f, 153, 172
Oxpecker, 1173, 1173f
Oxygen
    in air, 924, 934
    atomic structure of, 20f
    diffusion from environment into cells, 923
    diffusion into tissues, 934
    as electron acceptor, 173–74, 174f
    exchange in capillaries, 913
    in freshwater ecosystems, 1217
    in marine ecosystems, 1214
    partial pressure in blood, 930, 932, 934
    from photosynthesis, 65, 71, 182, 187–89, 187f, 196, 198, 557
    in plants, 782, 782t
    transport in blood, 909–11, 910f, 922, 934–35, 934–35f
Oxygen free radicals, 1015
Oxyhemoglobin, **934**
Oxyhemoglobin dissociation curve, 934, 935f
Oxytocin, 39t, 994t, **1000**, 1001f, 1004, 1070t, 1101–2
Oyster, 622t, 652, 654–56, 1147
    shell shape in, 461, 461f
Oyster drill, 656
Oyster mushroom (*Pleurotus osteatus*), 603, 603f
Ozone, 826, **1221**
Ozone hole, 1221, 1221f
Ozone layer, 65, 71, 191, 1204

P

P$_{680}$, 196, 198
P$_{700}$, 196
P$_{870}$, 196
p16 gene, 418f
p21 gene, 418f
p53 gene, 223, 417t, 418, 418–19f, 421
p53 protein, 220, 223, 223–24f, 422–23
Paal, Arpad, 818
Pacemaker, cardiac, 917, 919, 919f
Pacific giant octopus (*Octopus defleini*), 657f
Pacific yew (*Taxus brevifolia*), 799t, 800
Pacinian corpuscle, 971t, **973**, 973f
Pain, perception of, 950, 971t, 973

Pair-bonding, 1076, 1102
Paired appendages, of fish, 690
Pair-rule genes, 395f, **396**
Pakicetus attocki, 461, 461f
Paleontologist, 461
Palisade mesophyll, **750**, 750f, 771f
Palm, 594
Palmately compound leaf, 749, 749f
Palm oil, 54
Palo Verde tree (*Cercidium floridum*), 814f
Pancreas, 858f, 888–89f, 894, 900t, 901–2f, 993f
    as endocrine organ, 995t, 1008, 1008f
    secretions of, 895, 895f
Pancreatic cancer, 414t, 417t
Pancreatic duct, 895, 895f, 1008
Pancreatic juice, 889, 895, 895f
Pangaea, 700
Panspermia, **64**
Pantothenic acid. *See* Vitamin B$_5$
Paper, 37t
Papermaking, 737
Paper wasp (*Polistes*), 651f
Papillae, of tongue, 975, 975f
Papillomavirus, human, 541
Papuan kingfisher (*Tanysiptera hydrocharis*), 480, 481f
Parabronchi, 929, 929f
Paracrine regulation, **992**, 992f, 994–95, 1093
Paracrine signaling, 127, 127f
Paradise whydah, 1126f
Parallax, 987
Paralog, **499**, 503
Paramecium, 63f, 82f, 117, 569–70, 569f, 1168f
    competitive exclusion among species of, 1165, 1165f
    killer strains of, 570
    life cycle of, 570f
Paramylon granule, 566f
Paramyxovirus, 533t
Paraphyletic group, **514**, 515f
Parapodia, 658f, **660**
Parasite, 604
    effect on competition, 1176
    external, 1175, 1175f
    internal, 1175
    manipulation of host behavior, 1175, 1175f
Parasitic plant, 789, 789f
Parasitic root, 743
Parasitism, 557, **1172**
    brood, 1113, 1113f
Parasitoid, **1175**
Parasitoid wasp, 801, 801f
Parastichopus parvimensis, 676f
Parasympathetic division, **940**, 940f, 964, 964f
Parathion, 950
Parathyroid gland, 993f, 994t, 1005–6, 1006f
Parathyroid hormone (PTH), 994t, 1006, 1006f
Paratyphoid fever, 554
Parazoa, **620**, 620–21f, 634, 634–37f, 636–37
Parenchyma, **732**, 736, 745–46f, 775f
Parenchyma cells, **736**, 736f, 737, 741, 768, 776
Parental care, in dinosaurs, 516, 516f
Parental investment, **1125**
Parent-offspring interactions, 1105f, 1112, 1112f
Parietal cells, 892, 893f, 901f
Parietal eye, 708
Parietal lobe, 957, 957f
Parietal pleural membrane, **930**

Park, Thomas, 1176
Parkinson disease, 426, 950, 958
Parmotrema gardneri, 611f
Parnassius imperator, 673f
Parrot, 711t, 713
Parrot feather (*Myriophyllum spicatum*), 791
Parsimony, principle of, **514**
Parsnip, 743
Parthenogenesis, **237**, **1062**, 1083
Partial pressure, **924**
Passenger pigeon, 1239
Passeriformes (order), 711t, 713
Passion vine, 798
Passive immunity, **1018**
Passive immunization, 1029
Pasteur, Louis, 1018
Patella, 961f
Patellar ligament, 961f
Patent, gene-related, 358
Paternity testing, 1128, 1129f
Pathogen, 604, 860, 1014, 1018
Pattern formation, 395–97, 395–97f
    in *Drosophila*, 395–98, 395f, 397f
Pavlov, Ivan, 1110
Pavlovian conditioning. *See* Classical conditioning
Pax6 gene, 501, 503–6, 505f
Payne, Robert, 1176, 1176f
PCB, 415t
PCNA. *See* Proliferating cell nuclear antigen
PCR. *See* Polymerase chain reaction
PDGF. *See* Platelet-derived growth factor
Pea, 593, 734f, 747, 759, 761f, 763, 787f, 836, 841
Peach, 761f
Peacock, 1126f, 1127
Peacock worm, 660
Peanut, 614
Peanut allergy, 1036
Pear, 736, 736f
Pearl, 652, 654
Peat, 1186
Peat moss (*Sphagnum*), 582
Peccary, 1195
Pectin, 57
Pectoral fin, 690, 695
Pectoralis major muscle, 859f, 1101
Pectoralis minor muscle, 1101
Pedicel, **594**, 595f, 663f
Pedigree analysis, **261**, 261f, 269, 274
Pedipalp, **670**
Peer review, 7
Peking man, 725f
Pelizaeus-Merzbacher disease, 269f
Pellagra, 904t
Pellicle, 566, 566f, 569, 569f
Pelomyxa palustris, 562, 562f
Pelvic bones, of baleen whale, 465, 465f
Pelvic fin, 690, 695
Pelvic girdle, 872f
Pelvic inflammatory disease (PID), 556
Pelvis, 721, 858f, 871f
Pelycosaur, 704, 704–6f
pen gene, 441, 441f
Penguin, 711t, 713, 882, 1066f
Penicillin, 84
Penicillin allergy, 1036
Penicillin resistance, 552
Penicillium, 603, 609
Penis, 859f, 1068, 1068f, 1070f
Pepper, 761f
Peppered moth (*Biston betularia*), industrial melanism and, 456–57, 456–57f